Advances in Karst Research: Theory, Fieldwork and Applications

Geological Society books refereeing procedures

The Society makes every effort to ensure that the scientific and production quality of its books matches that of its journals. Since 1997, all book proposals have been refereed by specialist reviewers as well as by the Society's Books Editorial Committee. If the referees identify weaknesses in the proposal, these must be addressed before the proposal is accepted.

Once the book is accepted, the Society Book Editors ensure that the volume editors follow strict guidelines on refereeing and quality control. We insist that individual papers can only be accepted after satisfactory review by two independent referees. The questions on the review forms are similar to those for *Journal of the Geological Society*. The referees' forms and comments must be available to the Society's Book Editors on request.

Although many of the books result from meetings, the editors are expected to commission papers that were not presented at the meeting to ensure that the book provides a balanced coverage of the subject. Being accepted for presentation at the meeting does not guarantee inclusion in the book.

More information about submitting a proposal and producing a book for the Society can be found on its website: www.geolsoc.org.uk.

It is recommended that reference to all or part of this book should be made in one of the following ways:

Parise, M., Gabrovsek, F., Kaufmann, G. & Ravbar, N. (eds) 2018. *Advances in Karst Research: Theory, Fieldwork and Applications*. Geological Society, London, Special Publications, **466**.

Naughton, O., McCormack, T., Gill, L. & Johnston, P. 2018. Groundwater flood hazards and mechanisms in lowland karst terrains. *In*: Parise, M., Gabrovsek, F., Kaufmann, G. & Ravbar, N. (eds) *Advances in Karst Research: Theory, Fieldwork and Applications*. Geological Society, London, Special Publications, **466**, 397–410. First published online December 11, 2017, https://doi.org/10.1144/SP466.9

GEOLOGICAL SOCIETY SPECIAL PUBLICATION NO. 466

Advances in Karst Research: Theory, Fieldwork and Applications

EDITED BY

M. PARISE
University Aldo Moro, Italy

F. GABROVSEK
Karst Research Institute ZRC SAZU, Slovenia

G. KAUFMANN
Freie Universität Berlin, Germany

and

N. RAVBAR
Karst Research Institute ZRC SAZU, Slovenia

2018
Published by
The Geological Society
London

THE GEOLOGICAL SOCIETY

The Geological Society of London (GSL) was founded in 1807. It is the oldest national geological society in the world and the largest in Europe. It was incorporated under Royal Charter in 1825 and is Registered Charity 210161.

The Society is the UK national learned and professional society for geology with a worldwide Fellowship (FGS) of over 10 000. The Society has the power to confer Chartered status on suitably qualified Fellows, and about 2000 of the Fellowship carry the title (CGeol). Chartered Geologists may also obtain the equivalent European title, European Geologist (EurGeol). One fifth of the Society's fellowship resides outside the UK. To find out more about the Society, log on to www.geolsoc.org.uk.

The Geological Society Publishing House (Bath, UK) produces the Society's international journals and books, and acts as European distributor for selected publications of the American Association of Petroleum Geologists (AAPG), the Indonesian Petroleum Association (IPA), the Geological Society of America (GSA), the Society for Sedimentary Geology (SEPM) and the Geologists' Association (GA). Joint marketing agreements ensure that GSL Fellows may purchase these societies' publications at a discount. The Society's online bookshop (accessible from www.geolsoc.org.uk) offers secure book purchasing with your credit or debit card.

To find out about joining the Society and benefiting from substantial discounts on publications of GSL and other societies worldwide, consult www.geolsoc.org.uk, or contact the Fellowship Department at: The Geological Society, Burlington House, Piccadilly, London W1J 0BG: Tel. +44 (0)20 7434 9944; Fax +44 (0)20 7439 8975; E-mail: enquiries@geolsoc.org.uk.

For information about the Society's meetings, consult *Events* on www.geolsoc.org.uk. To find out more about the Society's Corporate Affiliates Scheme, write to enquiries@geolsoc.org.uk.

Published by The Geological Society from:
The Geological Society Publishing House, Unit 7, Brassmill Enterprise Centre, Brassmill Lane, Bath BA1 3JN, UK

The Lyell Collection: www.lyellcollection.org
Online bookshop: www.geolsoc.org.uk/bookshop
Orders: Tel. +44 (0)1225 445046, Fax +44 (0)1225 442836

British Library Cataloguing in Publication Data

A catalogue record for this book is available from the British Library.
ISBN 978-1-78620-359-5
ISSN 0305-8719

Distributors
For details of international agents and distributors see:
www.geolsoc.org.uk/agentsdistributors

Typeset by Nova Techset Private Limited, Bengaluru & Chennai, India
Printed and bound by CPI Group (UK) Ltd, Croydon CR0 4YY

Contents

Karst modelling

Karst hazards and management

Dedication

Giovanni Badino passed away on 8 August 2017. He was still working and trying to complete several projects and articles, including the contribution that we have published in this volume. Giovanni was Associate Professor at the Department of Physics, Torino University, Italy. A specialist in cosmic ray physics, principally in underground supernova neutrino detection, in the last years his research focused on the thermodynamics of fluid transport underground, with specific studies focused on: (1) water transport and speleogenesis in glaciers; (2) the underground climate of alpine caves; (3) water transport in fractured rocks from a percolative viewpoint; and (4) the interaction between geothermal flux and caves.

Giovanni was a passionate and expert speleologist. His speleological activity began in the 1970s and, from that time, he soon became a reference point for European speleology and was one of the most famous speleologists, carrying out a number of explorations and discoveries in Italy and abroad. He participated in many speleological expeditions, from Nepal to Uzbekistan, Brazil, Argentina (Patagonia), Pakistan and many other countries. Furthermore, he worked as the scientific and technical co-ordinator in the Proyecto Naica, in Mexico. He also participated in ice cave research and exploration in Iceland and Antarctica (South Shetland) and was involved in Victoria Land with Progetto Nazionale di Ricerche in Antartide.

As a crucial part of his caving activities, Giovanni was a member of the Italian national alpine and speleological rescue service (CNSAS) for many years, and was involved in multiple rescue operations. In 1981 he received the Medaglia d'Argento al Valor Civile for his contribution to the rescue of three people trapped in a cave.

He was a member of the steering board and of the scientific board of Associazione La Venta, devoted to underground exploration all over the world and, from 1994 to 1999, was president of the Italian Speleological Society. From 2009 to 2017 he was a member of the Bureau of the International Union of Speleology.

Always ready to share his experience, Giovanni dedicated himself throughout his life to dissemination, through a variety of tools (such as notes on caving bulletins, scientific articles, books, short stories, interviews, lectures, videos).

An author of many books and publications on cave techniques, cave monitoring and cave exploration, Giovanni's work has undoubtedly left a profound mark on speleology and karst research, at the national as well as at the international level.

Giovanni monitoring the air flow in a gypsum cave in Sicily using a 'technologically advanced' tool (November 2016; photograph by C. Chiappino).

Recent advances in karst research: from theory to fieldwork and applications

MARIO PARISE[1,2]*, FRANCI GABROVSEK[3], GEORG KAUFMANN[4] & NATASA RAVBAR[3]

[1]*Department of Earth and Environmental Sciences, University Aldo Moro, Bari, Italy*

[2]*National Research Council, IRPI, Bari, Italy*

[3]*Karst Research Institute ZRC SAZU, Titov trg 2, Postojna, Slovenia*

[4]*Institute of Geological Sciences, Geophysics Section, Freie Universität Berlin, Germany*

**Correspondence: mario.parise@uniba.it*

Abstract: Karst landscapes and karst aquifers, which are composed of a variety of soluble rocks such as salt, gypsum, anhydrite, limestone, dolomite and quartzite, are fascinating areas of study. As karst rocks are abundant on the Earth's surface, the fast evolution of karst landscapes and the rapid flow of water through karst aquifers present challenges from a number of different perspectives. This collection of 25 papers deals with different aspects of these challenges, including karst geology, geomorphology and speleogenesis, karst hydrogeology, karst modelling, and karst hazards and management. Together these papers provide a state-of-the-art review of the current challenges and solutions in describing karst from a scientific perspective.

Karst describes the slow work of dissolution exerted by water, enriched with carbon dioxide, on soluble rocks such as carbonates, evaporites and halite (White 1988; Ford & Williams 2007; Palmer 2007). Morphological effects at the surface are represented by a variety of typical landforms, the most typical being dolines and sinkholes. Below the surface, conduits of different sizes are formed by the dissolution of the soluble rocks, leading to a network of voids, of which the parts large enough to be explored by humans are defined as caves.

The term karst stems from the German derivation of the local name of the region between Italy and Slovenia (*Carso* in Italian, *Kras* in Slovene). This region is the site where the discipline of karstology was born at the end of the seventeenth century. The work of scholars such as Cvijić (1893) greatly contributed to the introduction of the term karst in the international literature. The importance of the wider region, namely the Dinaric karst area, is recognized by the use of many local terms that have become accepted in international karst terminology – for example, doline, polje, uvala and ponor (Stevanović & Mijatovic 2005).

According to recent estimates (Chen *et al.* 2017), *c.* 14% of the Earth's land surface consists of the carbonate rock outcrops that form karst. In Europe, much of the karst occurs beneath some of the continent's most densely populated regions, where effective water resource management is especially crucial, such as England, northern and southern France, parts of Germany, central Italy, eastern Spain and large parts of the Mediterranean. On other continents, such as in the Middle East and Central Asia, karst terrains occupy *c.* 25% of the land surface. These karst terrains often host primary economic benefits, such as drinking water and building materials. Karst areas are valuable ecosystems and have high levels of biodiversity (Brancelj & Culver 2005; Pipan & Culver 2013). The numerous caves and other natural phenomena have played an important part in stimulating tourism, contributing to economic development in many countries (Ravbar & Šebela 2015).

Karst landscapes are of great importance because they host remarkable natural resources, including karst aquifers. Karst water resources have been important in the historical and economic development of many regions (Parise & Sammarco 2015) and karst groundwater is of crucial importance today for the sustainable development of the economy in many countries and regions. Karst aquifers currently supply *c.* 15% of the global population with drinking water and, in some regions, are the only water resource available.

Looking to the future, there are many concerns about karst regions. The predicted growth in population, which will mostly be concentrated in urban

From: Parise, M., Gabrovsek, F., Kaufmann, G. & Ravbar, N. (eds) 2018. *Advances in Karst Research: Theory, Fieldwork and Applications*. Geological Society, London, Special Publications, **466**, 1–24.
First published online May 3, 2018, https://doi.org/10.1144/SP466.26

areas, will increase pressures on karst aquifers in terms of the global supply of water, although such an increase might be partly mitigated by improvements in treatment technologies for tapping water from sources other than karst.

As a result of the peculiar intrinsic characteristics of karst and its specific processes, these resources are particularly susceptible to destruction and over-exploitation. The utilization of karst resources and populations living on karst landscapes pose specific scientific and practical challenges. There are many unresolved theoretical problems in karst processes, water flow, and mass and heat transport. The efficient and safe exploitation of geothermal resources from karst systems also needs novel practical solutions.

Following the 2007 Special Publication of the Geological Society, London, *Natural and Anthropogenic Hazards in Karst Areas: Recognition, Analysis and Mitigation* (Parise & Gunn 2007), this volume deals with the most recent advances in karst research and presents an updated state-of-the-art source of the main scientific activities currently taking place in this peculiar and vulnerable environment.

This volume presents the outcomes of several sessions dealing with karst held in the last few years on the occasion of the General Assembly of the European Geosciences Union. It is designed to provide readers with an overview of studies in the peculiar and highly fragile karst environment. The volume presents a collection of papers, subdivided into four sections (karst geology, geomorphology and speleogenesis; karst hydrogeology; karst modelling; and karst hazards and management), to provide a balanced overview of the diverse approaches available to elucidate the current hot topics in karstology. It includes both theoretical models and case studies and consists of 25 papers prepared by researchers from 14 countries, covering karst areas located in 16 countries. By integrating geomorphological, speleological, hydrogeological, geochemical, geophysical and engineering geological approaches, it provides readers with new insights into the subject of comprehensive karstology and reveals important links between surface and subsurface processes and between the karst environment and society. Geologists, engineers and geophysicists interested in karst from both academia and geological or engineering consulting companies will find papers of interest. The fourth section, partly dedicated to management of karst, will be of interest to land planners, developers, and managers of show caves, natural parks and reserves in karst terrains.

Karst geology, geomorphology and speleogenesis

Karst landscapes are known worldwide for their beauty and high aesthetic value, typically expressed by spectacular surface geomorphology and subterranean features visible in caves. Show caves are often among the most visited tourist sites in many parts of the world, which has led to karst landscapes being known by a wide range of people.

There has been much interest in karst research in recent years, from many different disciplines. Caves have been recognized as sites where remarkable remains and deposits are preserved (Sasowsky & Mylroie 2004), in contrast with the outside environment (the Earth's surface), where the same features are generally destroyed by a combination of natural (e.g. weathering, erosion or mass movements) and anthropogenic processes. The biodiversity of karst is progressively being recognized as of extreme importance and several studies involving experts in biology, medicine, astrobiology and planetary geology have started, with the aim of exploring the bacteria and other species living in difficult environments and under peculiar climatic and environmental conditions (Boston & Northup 2017). Many cave expeditions and explorations, in some of the most important cave systems in the world, are now carried out through the joint action of speleologists and karst and cave scientists. New technologies, such as LiDAR scanners, allow the high-resolution scanning of underground channels and their morphology. These areas of research have resulted in the acquisition of large amounts of data and many scientific publications in the last few decades and there has been a growing number of congresses dedicated to karst research and speleology at both national and international levels.

The cave environment has been recognized as a potential platform to create analogues for space exploration missions and caves are ideal sites for training astronauts (Fig. 1). The European Space Agency has developed a specific training programme and many astronauts from different space agencies around the world have been trained in caves over the last five years (Bessone *et al.* 2017).

The study of karst landscapes and their features is not just of interest in karst science, but has significant implications in many other fields. The multidisciplinary nature of karst studies is expressed by the number of workers from many different fields of science who are interested in carrying out research in this extraordinary environment.

Karst is relevant to oil research because carbonate rocks may host significant reservoirs of high economic importance. Many studies are being carried out to improve our knowledge of the economic aspects of karst. Karst-type reservoirs have been encountered in different parts of the world and consist of carbonate and evaporite formations with evidence of pervasive dissolution and cave formation, followed by infilling and collapse during subsequent burial (e.g. Wang *et al.* 2015; Uni Research 2016).

Fig. 1. Astronaut training at Su Benti cave (Sardinia, Italy) during the camp CAVES 2014 organized by the European Space Agency. Progression on ropes presents many operational and safety protocol analogies with extravehicular activities. Credits: European Space Agency/S. Sechi.

Palaeokarst reservoirs are difficult to characterize due to their extreme contrasts in porosity and permeability and their variability over short distances. Their analyses therefore require very detailed studies.

Linked to the issue of palaeokarst reservoirs, the first paper in this volume, by **Broughton (2017)**, reviews the origin and distribution of the regional-scale dissolution patterns that reconfigured the Middle Devonian Prairie Evaporite Formation salt basin in northeastern Alberta and southern Saskatchewan and interprets the impact of continental-scale tectonism on these dissolution processes. The dissolution trends in the formation strata accumulated across Western Canada resulted in the largest known hypogene halite karst collapse and subsidence structures. The structural configuration of the collapse subsidence features in the strata overlying the areas of salt removal resulted in traps that accumulated migrated oil in areas of northern Alberta and southern Saskatchewan (Holter 1969; Hein *et al.* 2001). These dissolution patterns exerted controls on the distribution of reservoirs accumulated as the Athabasca Oil Sands in northeastern Alberta. This is the largest, and commercially the most important, of the three Cretaceous bituminous sand deposits in northern Alberta, with 45 000 km^2 of deposits accumulated as a hydrocarbon resource of 2×10^{11} m^3 or of 10^{12} barrels. The highly detailed reconstruction of the removal of halite salt beds and the resulting collapse subsidence structures, emplaced against near-vertical normal fault planes between adjacent differentially subsided fault blocks, controlled the distribution, alignment and migration of trapped oil in this area.

In addition to the economic aspects of karst research, the analysis of karst features and landforms may offer insights into other sectors of the earth sciences. A good example is represented by the contribution of **Klimchouk (2017)**, dedicated to tafoni and honeycomb structures, which are some of the most enigmatic and puzzling geomorphological phenomena. Through the example of tafoni and honeycomb structures in the Crimean Piedmont, **Klimchouk (2017)** interpreted these features as related to hypogene karstification, with their formation being predetermined by the alteration of the host rocks along fractures and karst conduits by focused rising flow. The characteristic morphological expression occurs through the removal of the alterite, which may be the result of speleogenesis in subsurface conditions or exposure to atmospheric weathering. From these observations, **Klimchouk (2017)** proposes a new conceptual model for the formation of cavernous features, which could be applicable to other regions and to a wide range of lithologies. The proposed model also resolves the major issues inherent in previous interpretations of cavernous features, offering general applicability and important implications that go well beyond the reach of karst scholars and may also be of interest in geodynamic and palaeohydrogeological reconstructions.

The observation of karst landforms, both at the surface and underground, and the related surveys and data interpretation are often not a simple task because of problems with access and logistics and the presence of a dense cover of vegetation and/or rugged topography. The situation is even more difficult underground and specific expertise is required to survey the subterranean environment safely. A variety of logistical problems and difficulties in carrying and using delicate instruments and equipment in peculiar environments, characterized by high humidity and low temperatures, has always been at the origin of the inaccuracies in many past surveys. On more than one occasion this has resulted in wrong assumptions and conclusions based on incorrect cave surveys (Martimucci & Parise 2012). Nevertheless, both speleological documentation and more general surveys of karst areas are of extreme importance in providing the necessary data to better understand the landforms produced by karst processes, their evolution, and the interactions between surface and subsurface forms.

Recent advances in technology, with the availability of low-cost and light instruments, has enabled highly detailed surveys to be carried out, even in deep caves and difficult environments. It is now possible to check previous surveys in a short period of time, to verify their reliability and accuracy, and to correct previous biases. The paper by **Petrović *et al.* (2018)** is an example of this approach and reports the analysis of karst valleys, through-caves and natural bridges in the Carpatho-Balkan Mountains of eastern Serbia through the use of terrestrial laser scanning.

Analysis of the available data concerning the distribution and characteristics of caves is an indispensable tool in understanding the development and evolution of karst. The most common documents available for caves are maps, photographs and reports. Their reliability may not be high, however, especially for maps compiled decades ago, when the use of simple instruments to survey caves was common. Nevertheless, such documents represent the main tools through which cavers have been able to transfer information about the underground world to people who, for a variety of reasons, do not or cannot enter the cave. It is therefore interesting to critically analyse these documents, verify their validity and use them in the correct way. This latter point is essential when dealing with the use of cave maps by practitioners, planners and local managers.

The paper by **Oberender & Plan (2018)** presents a classification of caves based exclusively on genetic processes, which is made possible through the use of cave maps, photographs and reports. Three types of cave are distinguished in this classification: solution caves, mainly formed by the dissolving action of underground water; mechanical weathering and erosion caves; and deposition caves. Applying this classification to a dataset of >6000 caves in eastern Austria, **Oberender & Plan (2018)** found that mechanical weathering and erosion caves are almost as common as solution caves, an unexpected result because the vast majority of analysed caves are developed in carbonate rocks. Such a classification, based on available cave documents, can be applied to a large number of caves within a reasonable time. It can also be used by decision-makers non-specialized in cave genesis as an indicator of natural phenomena, such as erosion and mass movements, or of vulnerable karst areas, thus providing important insights to non-cavers.

Surveys play a crucial part in the production of documentation about karst. However, before starting surveying, any karst scholar, or any speleologist, has first to identify the specific features to explore, map and study. The peculiarity of karst landscapes, which may show an uninterrupted series of landforms such as dolines, swallow holes and blind valleys, may sometimes lead to an under-evaluation of some elements in the landscape, which may have provided remarkable insights for the understanding of complex karst systems. In this regard, recent work in southern Italy has shown the importance of mapping and taking into consideration even small-sized karst landforms. **Parise & Benedetto (2018)** describe a small swallet in Apulia (SE Italy), which at the ground surface looks like a simple site of infiltration of water during heavy rainstorms. After several weeks of excavation, with the permission of the landowner, access was found to a remarkable underground system, reaching the deep water table at −260 m depth and going down for several tens of metres underwater, as documented by speleo-diving exploration. This cave is the deepest in Apulia and, with a total depth of −324 m, represents an open window on the water table and is a perfect site at which to carry out hydrogeological investigations and to monitor groundwater quality.

A crucial element in karst is the direct connection between the surface and the underground, from which the importance of speleological explorations and research stems. Many international expeditions and explorations in recent years have combined their exploratory goals with research aims. A good example, which also involves the management of a show cave, is presented in the paper by **Badino *et al.* (2018)**.

Show caves have a great variety of different management plans, from high-impact activities dedicated to tourists in caves, to simple visits with minor changes to the natural environment. The latter situation relates to only a small number of show caves worldwide, but some are worth mentioning. For instance, the Puerto Princesa Underground River, one of the largest caves in the Philippine Islands, is the most visited show cave in the country, with *c.* 300 000 visitors each year. The cave, which is a UNESCO World Heritage Site, is a highly protected National Geological Monument and was declared as one of the New Seven Wonders of Nature in 2012. The cave has undergone no adaptation to tourism because it was appointed as a Natural Park and World Heritage Site before its first part was opened to large flows of tourists. As a consequence, safeguarding it has always been the main concern of local and national authorities and the cave has remained well preserved in its natural state. It is also of great scientific value, which primarily relies on the fact that it is one of the largest known underground estuaries in the world, with the effect of tides visible along >7 km of the cave length. As a consequence, the complex relationships between seawater and freshwater influence not only the hydrodynamics of the system and the speleogenetic processes presently active, but also its climate and ecosystem.

Badino *et al.* (2018) describe some of the exploration and research carried out in this coastal karst, focused on understanding the speleogenesis of the system, notably the initial phreatic solution followed by vadose erosion with periodic marine invasion, followed by saline–freshwater mixing processes during sea-level highstands. Research has also covered aspects such as speleothem analysis, mineralogy, the analysis of palaeontological remains and biospeleology, with two large populations of bats and swiftlets sustaining a highly complex subterranean ecosystem.

Coastal karst is of great interest for a variety of reasons, including, but not limited to, tourist pressure along karst coastlines, the mixing of freshwater and seawater (leading to the possibility of marine intrusion problems), coastal evolution related to the development of karst processes and interactions with marine processes, such as wave action.

In this setting, flank margin caves (Carew & Mylroie 1995; Mylroie & Mylroie 2007; Mylroie 2013) are excellent indicators of sea-level position at the time of their formation because the boundary between brackish and freshwater is intimately tied to sea-level position. They form in the distal margin of the freshwater lens within a carbonate island and remain as viable terrestrial signatures of old sea-level highstands, which may not be registered in other records such as fossil coral reefs.

Mylroie & Mylroie (2017), with the example of Guam, the largest island in Micronesia in the west Pacific, examine karst denudation on eogenetic carbonate coasts and how such denudation affects the interpretation of tectonic uplift and glacio-eustatic sea-level change. The speed at which flank margin caves form makes them reliable indicators of sea-level position, even when the sea-level is stable at a given elevation for only a few thousand years. These caves also persist through time, far longer than bio-erosion notches, sea caves and other surface features used to assess sea-level position. Flank margin caves are therefore a high-resolution and long-lasting indicator of sea-level position.

Moving to completely different climatic settings, the development of karstification in arid and semi-arid regions is generally constrained by water scarcity. In these peculiar climatic conditions, studying relict hypogene karst may be of great interest in understanding the past palaeohydrology and palaeoclimate. Ashalim Cave, a maze cave in the NW Negev Desert of Israel, is dealt with in two papers in this volume. **Frumkin & Langford (2017)** discuss the hypogene karst features of Ashalim Cave and of neighbouring caves. These features are shown to have developed as a result of the mixing of two types of groundwater flowing in opposite directions within two tiers of Cretaceous rock aquifers. Ashalim Cave can be considered as a window to understanding past groundwater flow in a desert zone, in which the groundwater flow was recharged a long distance away. Some similarities can be observed with the present situation because the general flow in the Ashalim region today seems similar, notwithstanding the drop in groundwater level.

Another field of importance for studying caves is palaeoclimatology. Speleothems provide one of the most valuable archives of palaeoclimatic data because of their ability to be dated with high precision and accuracy and the potential for high-resolution stable isotope records. **Vaks *et al.* (2017)** report the timing of humid periods from the speleothems record at Ashalim Cave, where the annual mean precipitation is 100–120 mm. No speleothem deposition is currently occurring, but the existence of old speleothems in the cave shows that water seeped into it in the past. The speleothems of Ashalim Cave in the central Negev Desert record humid periods during the last 3.1 myr, equivalent to an annual precipitation of >300 mm in an area that is currently a desert. The speleothem deposition record of Ashalim Cave, although more intermittent than that in caves further to the north, appears to be similar to other caves of the central Negev Desert and this has allowed the reconstruction of the palaeoclimate of the northern Saharan–Arabian desert margin.

Karst hydrogeology

Karst aquifers often provide abundant groundwater reserves, which are invaluable resources relevant to human health, food security and industry. The importance of aquifers with karst permeability, which already supply around a quarter of the global demand for drinking water, is increasing (Bakalowicz 2005; Ford & Williams 2007). In particular, groundwater in the Mediterranean basin is generally more abundant in karst than in other aquifers and has been extensively exploited. Five European capitals use water from large karst springs for drinking purposes: Rome, Sarajevo, Tirana, Skopje and Podgorica. Many of the major Alpine cities, such as Vienna, with a population of >1.5 million, are supplied by high-quality karst water derived from the Lower Austrian–Styrian Alps nearly 200 km away (Fig. 2). Vienna is supplied with 400 000 m^3 of fresh spring water daily, which, on its way to the city, flows through hydroelectric power stations, generating 65×10^6 kW h of power (Plan 2009).

In some countries, such as Slovenia, Croatia and Austria, karst water sources supply about half of the drinking water needs of the population. Many karst springs contribute to surface waters and play a major part in maintaining aquatic ecosystems and wetlands. Some of these springs discharge large amounts of water. The highest number of springs regularly discharging >2 $m^3\ s^{-1}$ are found in Bosnia-Herzegovina and Turkey (eight springs in each), followed by Montenegro (five springs; Fig. 3). These aquifers are increasingly threatened by numerous environmental problems such as pollution, over-exploitation and the effects of climate change (Stevanović 2015).

The management of karst aquifers and their water resources is confronted by many problems related to heterogeneity and the locally high permeability of karst. This results in difficulties in defining the boundaries of karst catchments (Gunn 2007; Parise 2016), uncertainties in determining their highly irregular discharge regimes (Price *et al.* 2000; Stevanović 2015) and a high vulnerability to pollution (Zwahlen 2004; Goldscheider & Drew 2007; Parise & Gunn 2007), making karst aquifers much more problematic than any other aquifer type. For example, the origin of the Danube is at the confluence of the two streams Brigach and Breg outflowing as karst springs. Near Immendingen and Fridingen, the water of the Danube sinks into the river bed in various places. The Danube loses about 6 $m^3\ s^{-1}$ of its flow into river bed sinks and this water rises from the Aach Spring, 12–18 km away and 135–175 m lower, at the head of a major tributary to the Rhine, causing bifurcation at the North Sea–Black Sea watershed (Käß *et al.* 1996; Fig. 4).

Similar flow bifurcations can be observed in many places of the Classical Karst area in Slovenia, located on the Adriatic–Black Sea watershed. The

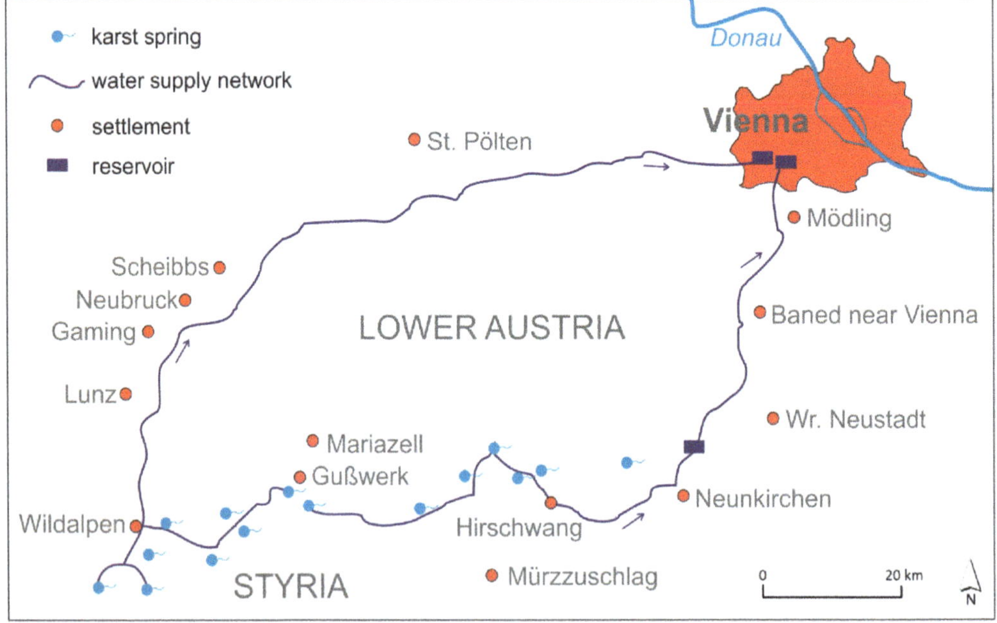

Fig. 2. Vienna is supplied with drinking water from karst springs nearly 200 km away via two long-distance mains. Modified from Vienna's Water Supply (2016).

Fig. 3. Sopot spring, located in Boka Kotorska Bay (Montenegro), has a discharge >50 $m^3 s^{-1}$ at high water. Photograph courtesy of Natasa Ravbar.

drainage of the Javorniki-Snežnik karst massive, which is recharged by rainwater and snowmelt, is also affected by great variabilities in flow characteristics and these are a function of the specific hydrological conditions (Fig. 5).

The geomorphological features acting as preferential pathways of intensive groundwater circulation, with turbulent flow in karst conduits and caves, and the extreme irregularity in the pattern of karst water resources, which are unequally distributed throughout the year, are at the origin of the many uncertainties existing in karst. Monitoring systems are necessary to fully understand the regime of karst aquifers and springs. This is a particular requirement in those areas supplied by karst water sources where a shortage of water may occur during the summer season, often as a result of increased demand from tourists. This might have consequences for the reserves of the spring, putting its minimum discharge in danger.

Deepening our knowledge about flow and storage mechanisms, and better protection and management of karst water resources, are currently the most important issues in karst hydrogeology. There is therefore a need to revise the European Water Framework Directive by 2019 (Voulvoulis *et al.* 2017) and to find the best solutions to the problems of water supply authorities.

Among major societal goals are the need to maintain the high quality of drinking water, to protect it

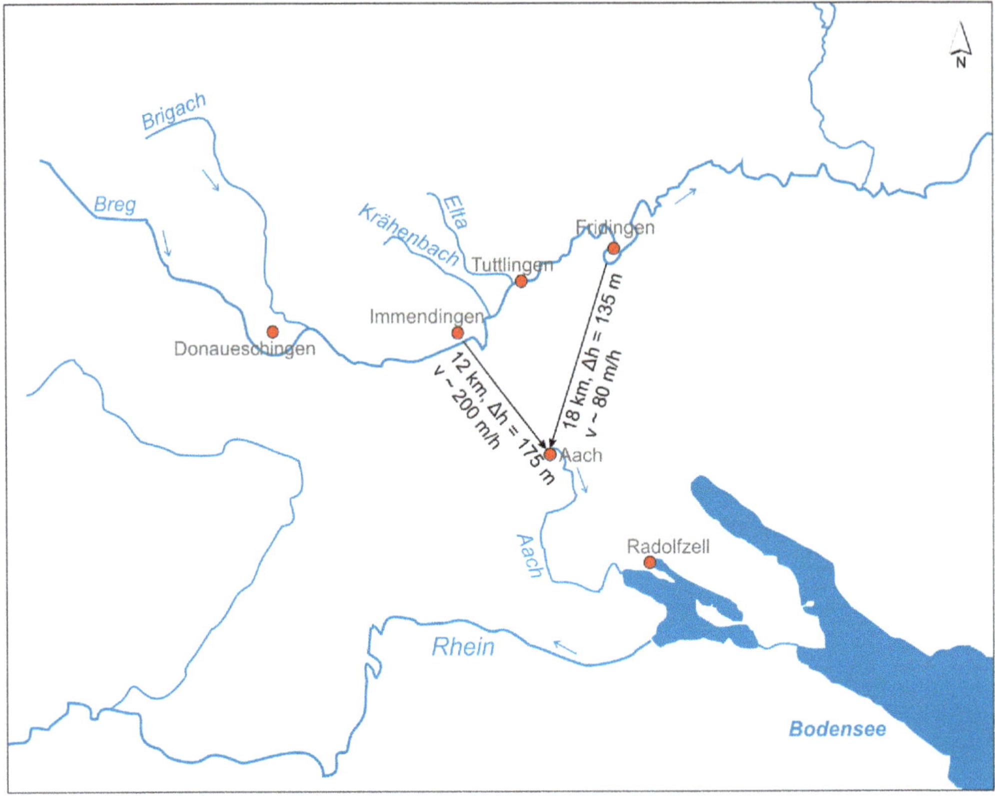

Fig. 4. The Danube River is associated with the Danube–Aach karst system of sinking and reappearing water.

from contamination and to ensure reliable access. This latter issue is affected by the great and diverse challenges of the characteristics of karst aquifers. **Stevanović (2018)** reports that an estimation and verification of groundwater reserves and their potential for replenishment can only be reached through a precise knowledge of the regional and local geological and hydrogeological properties. These include many different, not easily determined, parameters, such as the aquifer geometry, the groundwater table and the thickness of the saturated zone, the permeability, the storativity and the availability of water (reserves). Data from the systematic monitoring of groundwater quantity and quality and long-term pumping tests are also required and should be integrated with data on climate, hydrology and vegetation.

Most water in karst aquifers is transported through a network of solution conduits, which evolve as a result of the dissolution of the host rock along discontinuities in primary fissured aquifers (White 1988, 2002; Worthington 2009; Kaufmann *et al.* 2012; Fig. 6). Karst aquifers are characterized by fast diffuse and concentrated infiltration, high heterogeneity and anisotropy, mainly conditioned by the position of conduits, and fast transport from inputs to the springs (Jones & White 2012). Unfortunately, too many of these are contaminated by nitrates and pesticides (e.g. Liu *et al.* 2006; Huebsch *et al.* 2014; Musgrove *et al.* 2016) and even less knowledge is available on pollution by emerging chemicals such as fluorinated substances, medicines (e.g. antibiotics and endocrine disruptors) and micropollutants.

The link between the karst elements visible at the ground surface and those linked to the subterranean course of water highlights the complexity of the hydrogeology in karst. Interaction among shallow and deep karst systems often occurs in ways that are not fully understood. Analysis of infiltration points, such as sinkholes, is important and these need to be related to the recharge of the overall hydrogeological system. At the same time, dye tests need to be performed to obtain information about the deep flow paths by integration with recession curve analysis via discharge data from the springs. A more reasonable understanding of the whole system may be reached by combining the outcomes from these

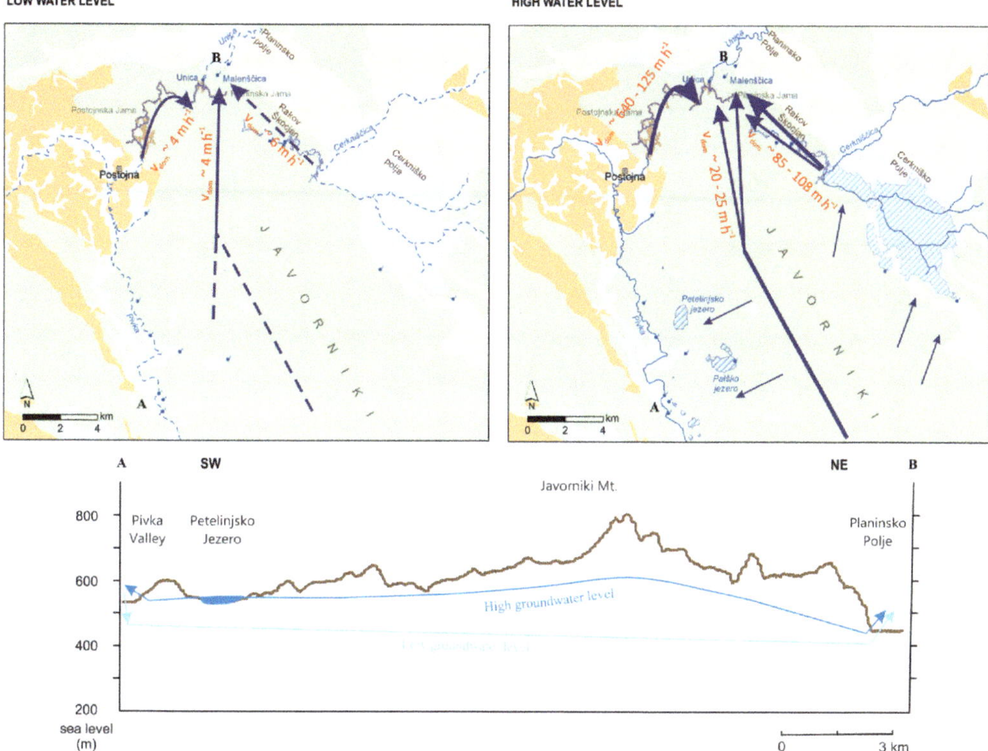

Fig. 5. Map of the water connections from the Slovene examples in Dinaric karst, characterized by significant temporal hydrological variability and consequent alterations of flow direction and groundwater divides, different types of surface–groundwater interactions, and changing flow velocities.

two different analyses. As an example of this type of study, looking at the northern rim of the Grand Canyon on the Kaibab Plateau in northern Arizona (southwestern USA), **Jones *et al.* (2017)** performed a careful study of the karst aquifer in this region, which supplies numerous large springs providing desert oasis habitats, drinking water and base flow to the Colorado River. The Grand Canyon is the second largest karst region in the US National Park System and contains >4000 km^2 of karst features. In this specific case study, two karst aquifers, with embedded non-karstic strata, are presented. The only way to understand the interconnection between the two aquifers is the interpretation of a dye tracer study and hydrograph analysis of discharge from the deep aquifer (Groves 2007; Fiorillo *et al.* 2012).

The work by **Jones *et al.* (2017)** once again confirms that the hydrogeology of karst aquifers strongly differs from that of aquifers developed in porous or fractured media (White 1988; Bakalowicz 1995, 2005; Ford & Williams 2007). The inherent heterogeneity and anisotropy of karst aquifers makes it necessary to develop studies through a wide range of approaches (Doctor *et al.* 2000; Goldscheider & Drew 2007), including: classical hydrological methods involving water balances and analysis of the spring hydrographs; the use of stable and radioactive isotopes, hydrochemical methods and tracer techniques with artificial and/or natural dyes; and specifically designed modelling experiments. These are all required to reconstruct the path followed by groundwater, which is often not accessible through speleological explorations. Analysis of the chemical composition of the water drained by springs adds to these important data dealing with the structure, storage capacity and dynamics of carbonate aquifers (Mudry 1987; Mudarra & Andreo 2011).

Combining and integrating the outcomes from such a variety of approaches is not a simple task, but has been proved to be the only possibility in attempts to fully characterize karst aquifers and move towards an understanding of the hydrogeological functioning of the systems, from the recognition and measurements of groundwater flow paths and velocities, to the identification of the recharge areas and groundwater residence times.

Fig. 6. Underground flow through a well-developed karst channel, Tkalca jama, Rakov Škocjan, SW Slovenia. Photograph courtesy of Matej Blatnik.

Groundwater vulnerability maps are a crucial tool in supporting decision-making and land planning, especially in environmentally protected areas, because they synthesize the main information available at the site from many different disciplines. Nevertheless, vulnerability maps may be of no help, or even provide questionable/wrong information, if they are not accompanied by the correct interpretation of the hydrogeological setting. This requires the validation of karst vulnerability maps, which may be pursued by applying several different techniques (Ravbar & Goldscheider 2009), such as tracer tests, isotope analyses and hydrograph studies.

Iván & Mádl-Szőnyi (2017) applied the Slovene approach to the Gömör-Torna Karst, a transboundary aquifer straddling the border of Hungary and Slovakia. The region is protected and lies within the Aggtelek National Park and the Slovak Karst National Park. Long-term hydrograph and recession curve analyses have been particularly useful in understanding the functioning of the main spring in the area (the Kis-Tohonya spring) and the complex karst system.

As a further example of the implementation of hydrogeological studies in karst areas, **Sánchez *et al.* (2017)** present a practical and critical overview of the most common hydrochemical and isotopic approaches used to analyse carbonate karst aquifers. An integration of techniques, from the hydrochemical and isotopic characterization of waters to mapping recharge areas, is applied to study the carbonate massif of Sierra Grazalema Natural Park in southern Spain. This is a strategic groundwater reserve located in a region that periodically suffers from droughts and water scarcity. The main outcomes from the study indicate that groundwater from the Sierra Grazalema aquifers shows marked differences in chemical composition, with low to intermediate mineralization and a chemical composition that is mainly controlled by the minerals forming the rocks through which the groundwater flows and by the residence time of water in the aquifer.

Water quality issues are closely related to the maintenance of healthy ecosystems (Bonacci *et al.* 2009). The complexity of karst water resources, which is related to the great variability in flow characteristics and is dependent on the hydrological conditions (Ravbar *et al.* 2011; Smith *et al.* 2015; McCormack *et al.* 2016), needs to be addressed. Fluctuations in karst groundwater can be several tens to hundreds of metres and, as a consequence, different types of surface–groundwater interaction can occur (Gabrovšek & Peric 2006). These are characteristic of areas such as the Dinaric karst, many lowland karst areas of Ireland, the karst in Estonia, the coastal karst of the Yucatan peninsula, the subtropical areas of the South China karst, and many others. Catchment boundaries and flow directions in karst aquifers may change due to the temporal hydrological conditions.

Given the dynamics of hydrological processes, karst aquifers are particularly vulnerable to the effects of environmental changes, but are also suitable for exploring these effects on a human timescale. Despite the important role of karst water resources, this topic is still poorly addressed. There have only been a few studies that have adequately evaluated the impact of predicted environmental changes on water quantity (e.g. Pinault *et al.* 2005; Hartmann *et al.* 2012; Finger *et al.* 2013; Jia *et al.* 2017).

The high vulnerability of karst aquifers, resulting from the spatial and temporal heterogeneity of water flow and storage, may increase further as a result of environmental changes, in particular those occurring in the hydrological cycle. Taking into account the predicted changes, such as an increase in the frequency of extreme drought periods, **Ravbar *et al.* (2017)** assessed the long-term climatological and hydrological trends and the short-term effects of such changes for the Mediterranean karst spring of Rižana in SW Slovenia. This is the most important water source for Slovene Istria, a region between Slovenia and Croatia containing a transboundary karst aquifer. The aquifer is important for both agriculture and tourism, and supplies water for 86 000–120 000 people.

The type of approach presented by **Ravbar *et al.* (2017)** shows, through detailed monitoring of the physical, chemical and microbiological parameters, that the flood pulses caused by precipitation events after a long dry period cause a significant deterioration in water quality, with a significant increase in the amount of coliform bacteria in the water. Therefore it is necessary to develop proper management strategies and actions for the continuous monitoring of karst water resources, aimed at their rational use and taking into account the anticipated climatological trends and the effects they could have on karst aquifers. Further work needs to be done to simulate site-specific hydraulic responses to different climate scenarios.

Different needs in karst areas (typically, protection issues v. tourist development) are often conflicting and it is not easy to find a balance between them. In this regard, **Bonacci (2017)** examines an interesting case study from the small Adriatic island of Cres, Croatia, where Vrana Lake is the only source of potable water for the whole archipelago. A dangerous drop in the water level of the lake started in 1983, driven by both global climate change and anthropogenic impacts as a consequence of over-exploitation. The lake is a complex hydrological–hydrogeological system with an average water volume of *c.* 220×10^6 m^3. The limited natural inflows and outflows of the lake are insufficient for self-cleaning once pollution has occurred (Ožanić & Rubinić 2003). This poses a serious risk of seawater intrusion, a widespread phenomenon in many sectors of the Mediterranean basin (Stevanović 2015) related to strong tourist pressure in specific areas during the long, hot, dry summer periods typical of the Mediterranean climate.

Interdisciplinary monitoring and co-operation between different specialists and professionals (e.g. hydrologists, hydrogeologists, ecologists and water resource managers) is clearly the only way to protect and manage these vulnerable and exceptionally valuable aquatic and terrestrial ecosystems.

The two contrasting needs – the tourist-dependent economy of Cres and the preservation of the valuable natural ecosystems, including the groundwater resources – require proper and balanced management of the water resources. This approach to management must address the need to protection and safeguard the vulnerable and valuable biological diversity of Cres and the surrounding islands, while recognizing that the landscape is one of the main factors attracting tourists to the natural beauty of the archipelago. Sustainable development is possible, but it must start with a strong emphasis on the natural resources based on a thorough knowledge of the environment through the establishment of a truly interdisciplinary approach involving different scientific disciplines (e.g. hydrology, climatology, hydrogeology, biology and ecology).

Further problems arise when karst aquifers expand over administrative borders that lack common political actions on an international basis, hampering the planning of possible encroachments or the international protection of groundwater (Turpaud *et al.* 2018). If we consider transboundary aquifer systems – those shared by two or more countries – ≥20% of internationally shared aquifers are found in karst (Stevanović *et al.* 2016). Another problem linked to water and its role in societal conflicts is that springs in many countries worldwide have been exposed to severe contamination during wars in the last few decades. Using water as a weapon is especially dangerous in the case of vulnerable karst aquifers (Shiva 2002). Another topical issue, at present not fully developed, but which is expected to be tackled in the long term, is the exploitation of geothermal resources from karst aquifers (Goldscheider *et al.* 2010).

Karst modelling

Karst offers great challenges in terms of groundwater and engineering problems, which typically involve many uncertainties that cannot easily be addressed by fieldwork. Large parts of karst systems are inaccessible to human exploration and sampling techniques only provide limited insights into the dynamics of a karst system, with either spot measurements (e.g. boreholes) or responses averaging over large areas (spring discharges). Quantitative approaches such as mathematical or numerical modelling provide a deeper understanding of both the evolution of karst landscapes and karst aquifer systems and the functioning of present day karst systems.

Understanding the evolution of solution conduits (i.e. speleogenesis) is one of the major challenges. By coupling equations of flow in single fractures, networks of fractures or general fractured–porous aquifers, it is possible to simulate conduit evolution in soluble rocks and search for the basic mechanisms guiding the evolution. The early work of Dreybrodt (1990), Palmer (1991) and Palmer *et al.* (1999) showed the feedback mechanisms between flow and dissolution rates in evolving fractures and introduced the breakthrough time as an important indicator of the intensity of conduit evolution. Assembling individual fractures into networks demonstrated how competition between different flow paths results in different geometries of conduit networks. Different complexities of the initial networks and approximations of the underlying processes were used. A comprehensive review of these models is given by Dreybrodt *et al.* (2005) and Kaufmann *et al.* (2012). Models were also extended to three-dimensional domains (Annable 2003; Kaufmann *et al.* 2010) and to open channel flow conditions (Perne *et al.* 2014). Some of the mechanisms of flow path competition in soluble rocks have been explored by Szymczak & Ladd (2009), who provided the mathematical background of wormhole formation in soluble porous and fractured medium.

To provide an example of karst modelling, a three-dimensional limestone evolution model is presented in Figure 7. A karst aquifer, 200 m long and 100 m wide, extends 20 m in the vertical direction. Structures in the karst aquifer are resolved to the metre scale, so that 1.15×10^6 fracture elements and 400 000 matrix elements are used. The aquifer is recharged by 2 m^3 day^{-1} from the left-hand side, with inflowing water reaching 90% saturation with respect to calcite. Along the right-hand side, a fixed-head boundary condition mimics a base level. Although the porous matrix has a low conductivity ($K_m = 10^{-7}$ m s^{-1}), the initial fracture width is assigned as a log-normal distribution. The resulting hydraulic flow field (top) indicates flow from left to right and the small-scale disturbance of the flow field results from the initial log-normal distribution of fracture widths.

The evolved karst aquifer is shown after 10 000 years for two different log-normal distributions. For a mean initial fracture width of 0.1 mm and a standard deviation of 0.0001 mm (lower left-hand panel), the enlarged fracture zones form clusters, mainly in the direction of the main pressure gradient, with no strong coupling between the different clusters. The central cluster of enlarged fractures becomes the most dominant and provides the first efficient connection between the input and output sides, with fractures enlarged to 10 cm in size.

When the standard deviation is increased to 0.05 mm, with all other parameters remaining the same, the evolution of the enlarged fractures is different (lower right-hand panel). As before, several clusters of enlarged fractures grow from left to right, but they now develop in a more sinusoidal way, leading to more interactions between clusters and a stronger exchange flow.

The evolution of karst landscapes and solutional rocky features has attracted less attention from modellers. Kaufmann & Braun (2001) presented a comprehensive model of landscape evolution by coupling surface creep and planar dissolution. Fleurant *et al.* (2008) modelled the evolution of cockpit karst and stressed the importance of spatial anisotropy in landscape evolution. Modern modelling approaches, such as computational fluid dynamics, have been used to model the evolution of scallops under turbulent flow conditions (Grm *et al.* 2017).

In most soluble rocks, the timescale of karst evolution is much longer than the timescale of processes observed in real time. This allows a separation between genetic models and present state models. Several approaches have been taken to model the functioning of karst aquifers. Lumped parameter (black box) models are often applied because the spatial distribution of the relevant parameters is not known (**Hartmann 2017**). Distributed models are used when the spatial distribution of parameters is known.

The majority of models study heat or mass transport in karst aquifers. A general theoretical framework of signal propagation through karst aquifers was proposed by Covington *et al.* (2012), who introduced the concept of process length scale to karst systems.

Numerical models are more often used with observational data from the surface (e.g. springs) and from cave systems through direct exploration by speleologists to estimate the cave geometry and conduit paths (Palmer *et al.* 1999; Gabrovšek & Peric 2006; Covington *et al.* 2013; Chen & Goldscheider 2014; Kaufmann *et al.* 2016). These numerical simulations can be extended with indirect measurements of the surface and subsurface structure by geophysical mapping techniques, such as gravity, electrical resistivity imaging and ground-penetrating radar surveys (e.g. Kaufmann *et al.* 2015; Kaufmann & Romanov 2016).

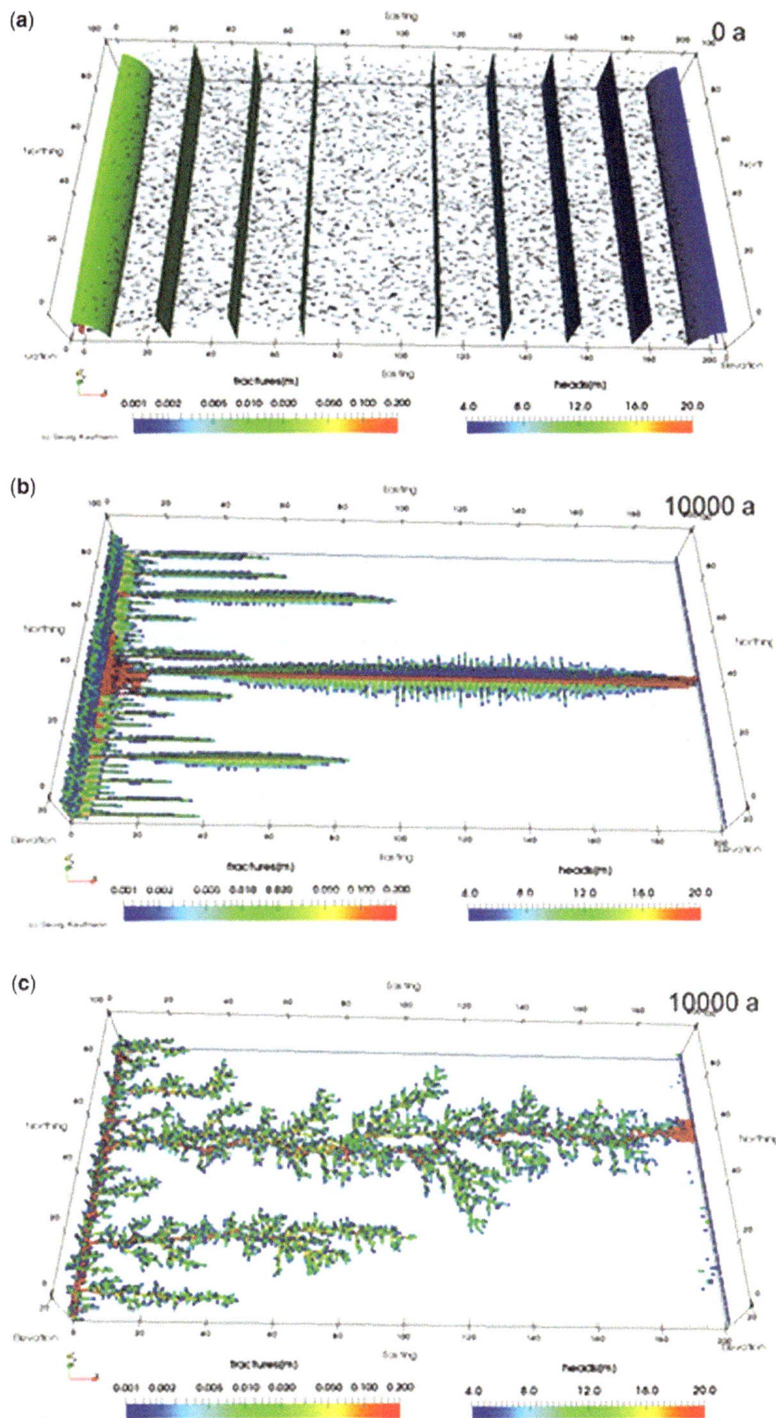

Fig. 7. (**a**) View towards the karst aquifer, with inflow on the left, base level on the right and hydraulic heads colour-coded. The small arrows indicate flow velocities. (**b**) Evolution of fractures after 10 000 years starting with an initial mean fracture width of 0.1 mm and a standard deviation of 0.0001 mm. (**c**) Evolution of fractures after 10 000 years starting with an initial mean fracture width of 0.1 mm and a standard deviation of 0.05 mm.

The main goals of modelling in karst are typically a better understanding of the hydraulic and hydrogeological flow of water and analysis of the microclimatic conditions characterizing cave systems.

Modelling of groundwater flow in karst aquifers is difficult because the position of solution conduits is largely unknown. When the geometry is known, assessed or assumed, different algorithms are used to account for the turbulent flows in conduits embedded into the fracture–matrix system. Several of these approaches are embedded into MODFLOW, a United States Geological Survey modelling environment, which is the model most widely used in groundwater modelling. An example of such an approach is given by **Kresic & Panday (2017)**, who present MODFLOW-USG (UnStructured Grid), which takes advantage of unstructured grids and finite volume numerical solutions. This new version of MODFLOW enables hydrogeologists to accurately translate even the most complex conceptual site models in karst into a numerical environment, thus eliminating the need for surrogate modelling solutions based on an equivalent porous medium approach.

As a significant percentage of high-quality, potable water comes from karst aquifers, the high vulnerability of karst waters must be taken into account, with particular emphasis on human-induced, often irreparable, pollution and over-exploitation. Sustainable management of these resources is not an easy task and should rely on the maximum possible amount of information from hydrogeological studies. These studies should represent the basis for developing hydrological models for karst, with the goal of both quantifying the water volumes available and assessing their sensitivity to changes in climate or land use.

Different modelling approaches are available to simulate karst hydrology, but guidance on the application of these approaches is scarce. **Hartmann (2017)** provides insights into the application, calibration and evaluation of lumped karst hydrological models by discussing model calibration and sensitivity analyses. He presents three case studies of karst model applications at different scales (plot, aquifer and continental scales) and with varying data availability, which apply model calibration and sensitivity analysis to obtain realistic simulations.

Aimed at identifying the likely hazards in karst, and at prevention measures to avoid catastrophic collapses, geophysical techniques may be successfully used to contribute to the mitigation of the sinkhole risk. Mechanically unstable structures close to the surface, of both natural and anthropogenic origin, are among the features highly prone to collapse. They therefore represent a clear hazard to the infrastructure above and may be located through geophysical investigations.

Kaufmann & Romanov (2017) compiled a large dataset of geophysical surveys in and above two caves in northern Germany, selected because they have a shallow overburden of 10–40 m and are characterized by both small passages and large rooms. In these settings, the known and surveyed cave passages can be traced with both gravity and electrical measurements, which reflect the lower density and the different electrical resistivities of the cave voids and host rock. Using these indirect geophysical observations, a structural model for both cave sites was obtained by numerical modelling and was successfully calibrated against the Bouguer gravity data.

Caves present a unique atmospheric environment, where several driving mechanisms (e.g. heat exchange between air, water and rock and changes in the external atmospheric pressure) force air exchange between the cave and the outside atmosphere. The microclimate of caves is receiving increasing attention from the scientific community because its understanding might help to reveal past climates from cave sediments and protect fragile recent ecosystems. Many important contributions on the physics of the underground environment have been published by Giovanni Badino (1953–2017). The present volume includes one of his last papers, in which he analyses the three major problems in cave micrometeorology: the concept of the temperature of a cave and its phenomenology; the internal energy flows and consequent local entropy production; and a non-hydrostatic physical model of the underground convective air circulation (**Badino 2018**). This study highlights how important it is to understand, and to find a correct way to measure and monitor, processes such as thermal stratification, seasonal variations and the effects of the external morphology on the cave environment. This fascinating topic, definitely worthy of more studies and analyses, will probably be one of the main issues in the physics of caves in the near future.

Karst hazards and management

The presence of karst landscapes in large parts of the world, and karst aquifers supplying important regions and many metropolitan areas with drinking water, is evidence of the great significance of karst and the need to preserve karst landscapes and their natural resources. At the same time, karst is recognized as an extremely fragile environment, susceptible to a variety of natural and anthropogenic hazards (Gutiérrez 2010; De Waele *et al.* 2011; Gutiérrez *et al.* 2014).

Sinkhole hazards are the most typical geohazard in karst. Even though several different genetic processes for sinkhole formation have been identified (Waltham *et al.* 2005; Yechieli *et al.* 2006; Gutiérrez *et al.* 2008, 2014), the most worrying, in terms of likely damage to human activities and infrastructures, are collapse and cover-collapse sinkholes, which generally occur in a catastrophic way with little or no warning. From studies of sinkholes in karst, related to natural karst caves and to the presence of underground conduits, the attention of many researchers has recently included anthropogenic sinkholes – that is, those related to the presence of artificial cavities (e.g. mines and quarries; see Parise *et al.* 2013), which are generally distributed below inhabited areas and pose greater risks to the population (Brinkmann & Parise 2012; Parise 2012, 2015). Efforts to conduct research to evaluate sinkhole hazards are important in civil defence and land planning, responding to a specific societal need and the need to mitigate the risk from geohazards. The occurrence of sinkholes in urban areas has become common in many built-up areas (Hermosilla 2012) as a consequence of the closure of these artificial voids and the loss of records of their presence.

Southern Italy, as a result of its geological features – including extensive areas of soluble rocks and of easily excavated volcanic and sedimentary rocks – is an interesting case study for sinkholes related to both natural caves and, because of the long history of the region, to artificial cavities, a high number of which were excavated in past epochs for a variety of purposes. **Fiore *et al.* (2018)**, by considering the instability problems and related sinkholes resulting from anthropogenic cavities in the karst of Apulia, the SE region forming the heel of the Italian boot, developed charts for use by practitioners and technicians through parametric analyses as an important step in an assessment of possible use by the public to classify anthropogenic cavities. This approach can give a preliminary assessment of the stability conditions and provide a basis on which further, site-specific, analyses can build, taking into account the local geometric, stratigraphic and geomechanical features. In addition to providing a preliminary evaluation of stability, this approach follows the recent Apulian regional law L.R. 33 2009 Safeguard and Promotion of the Geological and Speleological Heritage. This approach could be of practical use in the development of new opportunities for landowners to exploit these underground sites, which are considered as part of the cultural and historical heritage of the area, for tourism and to present favourable stability conditions.

The quantity and quality of karst aquifers makes them extremely important for society. Apart from their use as potable water, karst water resources are of interest for the generation of hydroelectric power and many dams have been built in karst areas to impound artificial lakes, facing severe problems in the construction phases to avoid losses through karst conduits and voids (Milanovic 2000, 2002). In general, the management of karst resources and the design of engineering works in karst are still difficult because of the often insufficient level of knowledge we have about the hydrogeological properties of aquifers and of the high vulnerability of karst aquifers to pollution (Milanovic 2000; Zhou & Beck 2011; Parise *et al.* 2015). As the water quality in karst aquifers deteriorates when pollutants are present (Calò & Parise 2009), the only way to manage such situations is to eliminate any possible sources of pollution or to establish sanitary protection zones (Doerfliger & Zwahlen 1997; Goldscheider 2005; Vìas *et al.* 2006; Ravbar & Goldscheider 2007).

Raising awareness in the local population living in karst regions about the importance of preserving the quality of water resources is crucial and should be pursued in every project located on karst lands. This is one of the main actions that actually protects karst and its resources.

Some human activities, such as land use modifications, the alteration of surface drainage and the opening or blocking of cave entrances, may have a great effect on karst hydrology or ecology. There is an urgent need to understand how human activities directly affect the recharge and flow of groundwater in karst because there are important consequences if karst lands are inundated during heavy and torrential rainstorms. This potentially leads to flash floods, which, in addition to sinkholes, represent the most common and dangerous hazard in karst (López-Chicano *et al.* 2002; Parise 2003; Bonacci *et al.* 2006; Jourde *et al.* 2007, 2014; Najib *et al.* 2008; Kovačič & Ravbar 2010). There is a great need to improve our understanding of flood dynamics in karst terrains to contribute to the assessment and management of flood risks (European Council 2007).

In such a delicate environment, the complex hydrological and hydrogeological behaviour of karst terrains poses unique challenges in flood risk management. Effective flood risk management requires an understanding of the recharge, storage and transport mechanisms governing water movement across the landscape during flood conditions. Lowland karst landscapes can be particularly susceptible to groundwater flooding due to a combination of low aquifer storage, high diffusivity and limited or absent surface drainage.

Naughton *et al.* (2017*b*) present a detailed example of the phenomenon of groundwater flooding in the lowland karst terrains of western Ireland following the dramatic floods that occurred during the winters of 2009 and 2015. These floods caused widespread damage and disruption to communities across the country, particularly in the extensive karstic limestone lowlands on the western seaboard (Naughton *et al.* 2017*a*). Backwater flooding of sinks, high water levels in ephemerally flooded basins (turloughs), the overtopping of depressions, and discharges from springs and resurgences have repeatedly been observed during these flood events. Field observations provide an understanding and description of the main hydrological and geomorphological characteristics influencing flooding in these complex lowland karst groundwater systems.

A future aspect of research in this region would involve the establishment of a permanent monitoring network to provide long-term quantitative data at flood-prone locations, together with the development of methodologies for improving groundwater flood hazard maps and real-time flood monitoring.

In addition to the direct and indirect effects of flood events, and because an increase in the frequency of extreme hydrological conditions affects various socioeconomic activities, it is necessary to evaluate the role of karst aquifers in flood attenuation and maintenance of the baseflow, which is especially important in terms of water supply.

The management of karst lands requires a deep knowledge of the natural processes acting in karst and a careful understanding of the likely effects of human activities on karst and its natural resources. To reach such a goal, a common effort must be pursued, involving decision-makers, legislators, land planners, karst scientists, speleologists and, last but not least, the population living in karst regions. The latter should be the first to appreciate the importance, but at the same time the fragility, of karst and their direct involvement is therefore of crucial importance in any action dedicated to sustainable development in karst. The main sites where karst information is transferred to the wider public is represented by show caves and by visits to reserves or protected areas and parks in karst.

Karst is known to the public essentially as a result of the beauty of caves, which can be visited and admired in show caves. Opening a show cave is seen by many scholars as 'killing a cave' because the operations necessary for making a cave available to the public change the environmental and climatic conditions of the site, inevitably altering the pristine conditions. Nevertheless, show caves represent the first and most direct visiting card of karst to people. When properly managed, show caves can also be used to transfer information from karst science to the public, such as the importance and fragility of the natural resources contained in karst (particularly groundwater), and therefore raise the environmental awareness of the local residents, who should be the first to act and live in a sustainable way in the surroundings of the cave.

The paper by **Debevec *et al.* (2017)** deals with one of the most important and historical show caves in the world: the Škocjan Caves in Slovenia. This is an important site for karst and for the history of karst research. At Škocjan Caves, the Reka River sinks underground and has created an impressive canyon in an area characterized by a number of large collapse sinkholes. This is a place of geological contact between flysch deposits (where the Reka River flows at the surface, from its source at the boundary between Croatia and Slovenia) and carbonate rocks (Jurkovsek *et al.* 2016). At Škocjan, the underground course of the river begins and extends for *c.* 40 km, before emerging at the Bocche del Timavo spring near Trieste in NE Italy. The river is accessible by many caves during its underground course and was explored by cavers between the end of the nineteenth and the first decades of the twentieth century (Guidi & Torelli 2017). A study by Gabrovšek *et al.* (2018) provides an in-depth analysis of groundwater dynamics along the flow path of the Reka–Timavo system.

The Škocjan Caves are included in UNESCO's World Heritage List and in the Ramsar Directory of Wetlands of International Importance. Together with their wider surface area, the site is also known as the UNESCO Karst Biosphere Reserve. The area of influence of the Škocjan Caves Regional Park encompasses the entire Reka River watershed and covers 450 km^2.

Debevec *et al.* (2017) describe the main steps of managing the reserve, aimed at protecting and preserving its outstanding universal value for future generations. Among these activities, monitoring of the water quality in the Reka River and meteorological surveys on the surface are of primary importance. In addition, monitoring of the cave microclimate focuses on measuring the effects of tourism and checking the levels of radon, with the aim of ensuring the safety of the park's employees and the status of the underground habitats and species laid down in the Natura 2000 management programme. The park staff carry out awareness-raising and educational activities, with an emphasis on understanding the importance of protecting the underground water and the unique and vulnerable karst land. At the same time, particular attention is paid to ensuring high-quality, safe visits to the caves.

Policy-based strategies for the management and protection of karst landscapes in developed countries are often inadequate and have, in many cases, only been implemented during the last few decades (Fleury 2009). This is particularly true with respect to the management of water resources and the natural environment.

The vulnerability of karst aquifers derives from their high permeability and rapid recharge, two factors which play a crucial part in making them susceptible to contamination. The unpredictability of groundwater movement in karst, and significant differences from water circulation in other settings, requires a specific approach to evaluate the possibility of pollution in karst settings. Although fairly straightforward, this simple concept has not been fully understood and in many countries groundwater vulnerability in karst is evaluated through methods and approaches that are not suitable in karst.

Aimed at presenting karst-specific approaches, the karst disturbance index (KDI) was proposed by van Beynen & Townsend (2005), and later modified by North *et al.* (2009), to assess the impact of human activities on the karst environment. The KDI covers five categories (geomorphology, atmosphere, hydrology, biota and cultural factors), each composed of several attributes, which are, in turn, subdivided into a number of indicators. Since its proposal, the KDI has been applied in a number of karst areas worldwide (Calò & Parise 2006; North *et al.* 2009; Day *et al.* 2011; Van Aken *et al.* 2014; Porter *et al.* 2016).

In this volume, **Kovarik & van Beynen (2017)** combine this approach with a groundwater vulnerability map to provide a tool for developing management strategies in a subcatchment of the Rio la Venta watershed in the state of Chiapas, southern Mexico. Their analysis shows that the opportunity exists to prevent major human impacts on vulnerable areas and the entire ecosystem of Reserva de la Biosfera Selva el Ocote, but only if local stakeholders are incorporated into the process of limiting development. Such a combination recognizes both human disturbance and how the physical nature of the karst will affect this impact. In practice, the composite model allows resource managers to quickly identify the areas of highest concern and the implementation of adaptive holistic management strategies to mitigate these problems. Providing a snapshot of the environmental state of the area will allow future studies to evaluate whether remediation efforts have been successful.

It is essential that local residents and natural resource managers work together to create a plan for mitigation by developing an awareness of the value and fragility of karst. Many case studies (e.g. Elke *et al.* 2007; Bocchino *et al.* 2014) have shown that environmental education of the locals greatly improves the possibility of success, basing the project on the relations between environmental issues and everyday life.

Auler *et al.* (2017) review the existing protocols for establishing the importance of caves and provide a statistical assessment of the criteria applied in Brazil to determine the level of importance. Despite the difficulties encountered in dealing with both quantitative and subjective criteria, their analysis is a useful example for decision-makers in clarifying the classification schemes and the selection of speleological sites worthy of environmental protection. A primary issue is represented by the selection of

Fig. 8. Cavers at work in the incredible setting of the gypsum giant crystals in the Naica Cave, Chihuahua, Mexico (see **Forti 2017**). Photograph courtesy of Paolo Patrignani/La Venta.

the parameters to be considered. The protocols vary significantly from country to country, but in many cases require working on a long, impractical lists of parameters, many of which actually have low significance or are extremely subjective. These drawbacks may lead to evaluations that differ in the function of the people involved and in the non-repeatability of the approach. To provide an example, 70 parameters are included in the analysis by **Auler *et al.* (2017)**. Of this large number, 30 deal with cave parameters, but on average less than five of these parameters occur together at each cave. This suggests that the relevant parameters should be selected carefully, avoiding those where high subjectivity is inherent and at the same time trying to balance biotic v. abiotic features.

An effort must be made by karst scientists to provide clear and repeatable classifications of caves because the need for these is increasing in many countries worldwide. **Auler *et al.* (2017)** state that, 'to achieve an objective and repeatable classification of the importance of caves and thus enabling the ranking of caves by conservation priority is a challenging proposition'. Karst scientists may contribute significantly to such a goal by sharing their experience to develop well-balanced, scientifically robust and non-debatable approaches.

An interesting example of the interaction between human activities and karst science is provided by **Forti (2017)**. The Naica caves in the state of Chihuahua, Mexico are probably one of the best examples of hypogenic speleogenesis (Klimchouk 2009) and have become famous as a result of the large gypsum crystals they host (Fig. 8). The caves were intersected by mine works in 2000 and, from 2006, became the object of a multidisciplinary research project with a variety of goals. Extensive research conducted over the past decade in the framework of the Naica Project, and briefly summarized by **Forti (2017)**, has made the thermal caves of Naica probably the most studied cavity systems in the world. Among the many outcomes of the project, it recorded the presence of ten new cave minerals and the fluid inclusions trapped in the gypsum crystals allowed the salinity and temperature of the feeding water at the time of their deposition to be determined, thus giving an insight into the palaeoenvironment and palaeoclimate of the Naica region (Garofalo *et al.* 2010). For the first time, well-preserved pollen grains were found inside euhedral crystals in the caves (Holden 2008). Such a wonder had, unfortunately, a short life, at least for scientific research. In 2015, mining works intersected a completely unexpected major thermal vein, which started flooding the lower levels of the mine. Because it was not possible to continue dewatering to stop the progressive rise of groundwater, procedures to close and abandon the mine were started at the end of 2015.

Concluding remarks

Karst terrains are, as a result of their particular nature, extremely fragile areas. At the same time, these landscapes represent a remarkable natural heritage and, due to the wide occurrence of carbonate rocks worldwide, have high economic importance, both nationally and regionally. Comprehensive economic and urban development in recent years has resulted in increased pressure on karst landscapes from the intensive and unsustainable spread of settlements, infrastructure and industry, the development of tourism and intensive agrarian land use. Once damaged, karst ecosystems (including the surface landforms and underground resources) take a long time to recover and the process is difficult. For this reason, karst must be holistically managed in an appropriate and careful manner.

There is therefore an urgent and continually increasing need to raise public awareness about karst. One of the main points on which the karst science community definitely agrees is that the population living in a karst area cannot be excluded from the process of learning to live sustainably in such a setting. Local residents must be the first to protect, with their daily habits, the natural resources contained in karst. This is a great challenge, which directly involves the scientific world in the effort to transfer our studies and research outcomes to people living in karst, in a language not intended for technicians only, but understandable to all, starting with the younger generation. Didactic actions should be performed in schools of different grades to show children what karst is, its importance and the need to preserve this treasure for generations to come.

References

Annable, W.K. 2003. *Numerical Analysis of Conduit Evolution in Karstic Aquifers*. University of Waterloo, Waterloo, ON.

Auler, A.S., Souza, T.A.R., Sé, D.C. & Soares, G.A. 2017. A review and statistical assessment of the criteria for determining cave significance. *In*: Parise, M., Gabrovsek, F., Kaufmann, G. & Ravbar, N. (eds) *Advances in Karst Research: Theory, Fieldwork and Applications*. Geological Society, London, Special Publications, **466**. First published online November 28, 2017, https://doi.org/10.1144/SP466.8

Badino, G. 2018. Models of temperature, entropy production and convective airflow in caves. *In*: Parise, M., Gabrovsek, F., Kaufmann, G. & Ravbar, N. (eds) *Advances in Karst Research: Theory, Fieldwork and Applications*. Geological Society, London, Special Publications, **466**. First published online January 31, 2018, https://doi.org/10.1144/SP466.24

Badino, G., De Vivo, A., Forti, P. & Piccini, L. 2018. The Puerto Princesa Underground River (Palawan, Philippines): some peculiar features of a tropical, high-energy coastal karst system. *In*: Parise, M., Gabrovsek, F.,

Kaufmann, G. & Ravbar, N. (eds) *Advances in Karst Research: Theory, Fieldwork and Applications*. Geological Society, London, Special Publications, **466**. First published online January 24, 2018, https://doi.org/10.1144/SP466.22

Bakalowicz, M. 1995. La zone d'infiltration des aquifers karstiques. Methodes d'etude. Structure et fonctionnement. *Hydrogeologie*, **4**, 3–21.

Bakalowicz, M. 2005. Karst groundwater: a challenge for new resources. *Hydrogeology Journal*, **13**, 148–160.

Bessone, L., De Waele, J. & Sauro, F. 2017. Speleology as an analogue to space exploration: five years of astronaut training, testing and operations in the ESA CAVES Program. *In*: Moore, K. & White, S. (eds) *Proceedings of the 17th International Congress of Speleology*, 23–29 July 2017, Sydney, Australia, Vol. 1. 374–379.

Bocchino, B., Del Vecchio, U. *et al.* 2014. Increasing people's awareness about the importance of karst landscapes and aquifers: an experience from southern Italy. *In*: Kukuric, N., Stevanovic, Z. & Kresic, N. (eds) *Proceedings of the International Conference and Field Seminar Karst Without Boundaries*, 11–15 June 2014, Trebinje–Dubrovnik, Bosnia-Herzegovina & Croatia, 398–405.

Bonacci, O. 2017. Preliminary analysis of the decrease in water level of Vrana Lake on the small carbonate island of Cres (Dinaric Karst, Croatia). *In*: Parise, M., Gabrovsek, F., Kaufmann, G. & Ravbar, N. (eds) *Advances in Karst Research: Theory, Fieldwork and Applications*. Geological Society, London, Special Publications, **466**. First published online November 6, 2017, https://doi.org/10.1144/SP466.6

Bonacci, O., Ljubenkov, I. & Roje-Bonacci, T. 2006. Karst flash floods: an example from the Dinaric karst, Croatia. *Natural Hazards and Earth System Sciences*, **6**, 195–203.

Bonacci, O., Pipan, T. & Culver, D.C. 2009. A framework for karst ecohydrology. *Environmental Geology*, **56**, 891–900.

Boston, P.J. & Northup, D. 2017. A theoretical approach to energy and materials flow and consequent biodiversity: predictions for caves on Earth and other planetary bodies. *In*: Moore, K. & White, S. (eds) *Proceedings of the 17th International Congress of Speleology*, 23–29 July 2017, Sydney, Australia, Vol. 1, 379.

Brancelj, A. & Culver, D.C. 2005. Epikarstic communities. *In*: Culver, D.C. & White, W.B. (eds) *Encyclopedia of Caves*. Academic Press, Waltham, MA, 223–229.

Brinkmann, R. & Parise, M. 2012. Karst environments: problems, management, human impacts, and sustainability. *Journal of Cave and Karst Studies*, **74**, 135–136.

Broughton, P.L. 2017. Orogeny and the collapse of the Devonian Prairie Evaporite karst in Western Canada: impact on the overlying Cretaceous Athabasca Oil Sands. *In*: Parise, M., Gabrovsek, F., Kaufmann, G. & Ravbar, N. (eds) *Advances in Karst Research: Theory, Fieldwork and Applications*. Geological Society, London, Special Publications, **466**. First published online November 8, 2017, https://doi.org/10.1144/SP466.7

Calò, F. & Parise, M. 2006. Evaluating the human disturbance to karst environments in southern Italy. *Acta Carsologica*, **35**, 47–56.

Calò, F. & Parise, M. 2009. Waste management and problems of groundwater pollution in karst environments in the context of a post-conflict scenario: the case of Mostar (Bosnia Herzegovina). *Habitat International*, **33**, 63–72.

Carew, J.L. & Mylroie, J.E. 1995. Quaternary tectonic stability of the Bahamian archipelago: evidence from fossil coral reefs and flank margin caves. *Quaternary Science Reviews*, **14**, 144–153.

Chen, Z. & Goldscheider, N. 2014. Modeling spatially and temporally varied hydraulic behavior of a folded karst system with dominant conduit drainage at catchment scale, Hochifen–Gottesacker, Alps. *Journal of Hydrology*, **514**, 41–52.

Chen, Z., Auler, A.S. *et al.* 2017. The World Karst Aquifer Mapping project: concept, mapping procedure and map of Europe. *Hydrogeology Journal*, **25**, 771–785.

Covington, M.D., Luhmann, A.J., Wicks, C.M. & Saar, M.O. 2012. Process length scales and longitudinal damping in karst conduits. *Journal of Geophysical Research*, **117**, 1–19.

Covington, M.D., Prelovsek, M. & Gabrovsek, F. 2013. Influence of × dynamics on the longitudinal variation of incision rates in soluble bedrock channels: feedback mechanisms. *Geomorphology*, **186**, 85–95.

Cvijić, J. 1893. *Das Karstphanomenon. Versuch einer morphologischen Monographie*. Vol. 5. Geographischen Abhandlung, Vienna.

Day, M., Halfen, A. & Chenoweth, S. 2011. The cockpit country, Jamaica: boundary issues in assessing disturbance and using a karst disturbance index in protected areas planning. *In*: van Beynen, P.E. (ed.) *Karst Management*. Springer, Dordrecht, 399–414.

De Waele, J., Gutiérrez, F., Parise, M. & Plan, L. 2011. Geomorphology and natural hazards in karst areas: a review. *Geomorphology*, **134**, 1–8.

Debevec, V., Peric, B., Šturm, S., Zorman, T. & Jovanovič, P. 2017. Škocjan Caves, Slovenia: an integrative approach to the management of a World Heritage Site. *In*: Parise, M., Gabrovsek, F., Kaufmann, G. & Ravbar, N. (eds) *Advances in Karst Research: Theory, Fieldwork and Applications*. Geological Society, London, Special Publications, **466**. First published online December 18, 2017, https://doi.org/10.1144/SP466.14

Doctor, D.H., Lojen, S. & Horvat, M. 2000. A stable isotope investigation of the Classical Karst aquifer: evaluating karst groundwater components for water quality preservation. *Acta Carsologica*, **29**, 79–82.

Doerfliger, N. & Zwahlen, F. 1997. EPIK, methode de cartographie de la vulnerabilite des aquiferes karstiques pour la delimitation des zones de protection. *In*: *Proceedings of the 12th International Congress of Speleology*, 10–17 August, La Chaux-de-Fonds, Switzerland, Vol. 2, 209–212.

Dreybrodt, W. 1990. The role of dissolution kinetics in the development of karst aquifers in limestone: a model simulation of karst evolution. *Journal of Geology*, **98**, 639–655.

Dreybrodt, W., Gabrovsek, F., Romanov, D., Bauer, S. & Birk, S. 2005. *Processes of Speleogenesis: a Modeling Approach (Carsologica)*. Založba, Ljubljana.

Elke, M., Glaser, D., Setty, K., Sussman, D. & Yocum, D. 2007. Design and Implementation of Sustainable

Water Resources Programs in San Cristobal de las Casas, Mexico, http://www2.bren.ucsb.edu/~keller/courses/GP_reports/FinalReport_Chiapas2.pdf

European Council 2007. *Directive 2007/60/EC of the European Parliament and of the Council on the Assessment and Management of Flood Risks*. European Environment Agency, Brussels, http://eur-lex.europa.eu/legal-content/EN/TXT/?uri=CELEX:32007L0060

Finger, D., Hugentobler, A. *et al.* 2013. Identification of glacial meltwater runoff in a karstic environment and its implication for present and future water availability. *Hydrology and Earth System Sciences*, **17**, 3261–3277.

Fiore, A., Fazio, N.L. *et al.* 2018. Evaluating the susceptibility to anthropogenic sinkholes in Apulian calcarenites, southern Italy. *In*: Parise, M., Gabrovsek, F., Kaufmann, G. & Ravbar, N. (eds) *Advances in Karst Research: Theory, Fieldwork and Applications*. Geological Society, London, Special Publications, **466**. First published online January 4, 2018, https://doi.org/10.1144/SP466.20

Fiorillo, F., Revellino, P. & Ventafridda, G. 2012. Karst aquifer draining during dry periods. *Journal of Cave and Karst Studies*, **74**, 148–156.

Fleurant, C., Tucker, G.E. & Viles, H.A. 2008. Modelling cockpit karst landforms. *In*: Gallagher, K., Jones, S.J. & Wainwright, J. (eds) *Landscape Evolution: Denudation, Climate and Tectonics over Different Time and Space Scales*. Geological Society, London, Special Publications, **296**, 47–62, https://doi.org/10.1144/SP296.4

Fleury, S. 2009. *Land Use Policy and Practice on Karst Terrains. Living on Limestone*. Springer, Dordrecht.

Ford, D.C. & Williams, P. 2007. *Karst Hydrogeology and Geomorphology*. Wiley, Chichester.

Forti, P. 2017. What will be the future of the giant gypsum crystals of Naica mine? *In*: Parise, M., Gabrovsek, F., Kaufmann, G. & Ravbar, N. (eds) *Advances in Karst Research: Theory, Fieldwork and Applications*. Geological Society, London, Special Publications, **466**. First published online November 6, 2017, https://doi.org/10.1144/SP466.4

Frumkin, A. & Langford, B. 2017. Arid hypogene karst in a multi-aquifer system: hydrogeology and speleogenesis of Ashalim Cave, Negev Desert, Israel. *In*: Parise, M., Gabrovsek, F., Kaufmann, G. & Ravbar, N. (eds) *Advances in Karst Research: Theory, Fieldwork and Applications*. Geological Society, London, Special Publications, **466**. First published online November 6, 2017, https://doi.org/10.1144/SP466.3

Gabrovšek, F. & Peric, B. 2006. Monitoring the flood pulses in the epiphreatic zone of karst aquifer: the case of Reka river system, karst plateau, SW Slovenia. *Acta Carsologica*, **35**, 35–45.

Gabrovšek, F., Peric, B. & Kaufmann, G. 2018. Hydraulics of epiphreatic flow of a karst aquifer. *Journal of Hydrology*, **560**, 56–74.

Garofalo, P.S., Fricker, M., Günther, D., Mercuri, A.M., Loreti, M., Forti, P. & Capaccioni, B. 2010. A climatic control on the formation of gigantic gypsum crystals within the hypogenic caves of Naica (Mexico). *Earth and Planetary Science Letters*, **289**, 560–569.

Goldscheider, N. 2005. Karst groundwater vulnerability mapping: application of a new method in the Swabian Alb, Germany. *Hydrogeology Journal*, **13**, 555–564.

Goldscheider, N. & Drew, D. (eds) 2007. *Methods in Karst Hydrogeology*. International Association of Hydrogeologists, Contributions to Hydrogeology, **26**. CRC Press, Boca Raton, FL.

Goldscheider, N., Mádl-Szőnyi, J., Erőss, A. & Schill, E. 2010. Thermal water resources in carbonate rock aquifers. *Hydrogeology Journal*, **18**, 1303–1318.

Grm, A., Šuštar, T., Rodič, T. & Gabrovšek, F. 2017. A numerical framework for wall dissolution modeling. *Mathematical Geosciences*, **49**, 657–675, https://doi.org/10.1007/s11004-016-9641-2

Groves, C. 2007. Hydrological methods. *In*: Goldscheider, N. & Drew, D. (eds) *Methods in Karst Hydrogeology*. International Association of Hydrogeologists, Contributions to Hydrogeology, **26**. CRC Press, Boca Raton, FL, 45–64.

Guidi, P. & Torelli, L. 2017. Lacus Timavi – le ricerche speleologiche. *Atti e Memorie Commissione Grotte 'E. Boegan'*, **47**, 109–134.

Gunn, J. 2007. Contributory area definition for groundwater source protection and hazard mitigation in carbonate aquifers. *In*: Parise, M. & Gunn, J. (eds) *Natural and Anthropogenic Hazards in Karst Areas: Recognition, Analysis and Mitigation*. Geological Society, London, **279**, 97–109, https://doi.org/10.1144/SP279.9

Gutiérrez, F. 2010. Hazards associated with karst. *In*: Alcantara, I. & Goudie, A. (eds) *Geomorphological Hazards and Disaster Prevention*. Cambridge University Press, Cambridge, 161–175.

Gutiérrez, F., Guerrero, J. & Lucha, P. 2008. A genetic classification of sinkholes illustrated from evaporite paleokarst exposures in Spain. *Environmental Geology*, **53**, 993–1006.

Gutiérrez, F., Parise, M., De Waele, J. & Jourde, H. 2014. A review on natural and human-induced geohazards and impacts in karst. *Earth-Science Reviews*, **138**, 61–88.

Hartmann, A. 2017. Experiences in calibrating and evaluating lumped karst hydrological models. *In*: Parise, M., Gabrovsek, F., Kaufmann, G. & Ravbar, N. (eds) *Advances in Karst Research: Theory, Fieldwork and Applications*. Geological Society, London, Special Publications, **466**. First published online December 6, 2017, https://doi.org/10.1144/SP466.18

Hartmann, A., Lange, J., Vivo Aguado, A., Mizyed, N., Smiatek, G. & Kunstmann, H. 2012. A multi-model approach for improved simulations of future water availability at a large Eastern Mediterranean karst spring. *Journal of Hydrology*, **468–469**, 130–138.

Hein, F., Langenberg, C., Kidston, C., Berhane, H. & Berezniuk, T. 2001. *A Comprehensive Field Guide for Facies Characterization of the Athabasca Oil Sands, Northeast Alberta*. Special Report, **13**. Alberta Geological Survey, Edmonton.

Hermosilla, R.G. 2012. The Guatemala City sinkhole collapses. *Carbonates and Evaporites*, **27**, 103–107.

Holden, C. 2008. Life in crystal. *Science*, **319**, 391.

Holter, M. 1969. *Middle Devonian Prairie Evaporite of Saskatchewan*. Saskatchewan Department of Mineral Resources, Reports, **123**.

Huebsch, M., Horan, B., Blum, P., Richards, K.G., Grant, J. & Fenton, O. 2014. Statistical analysis correlating changing agronomic practices with nitrate concentrations in a karst aquifer in Ireland. *WIT Transactions on Ecology and the Environment*, **182**, 99–109.

IVÁN, V. & MÁDL-SZŐNYI, J. 2017. Vulnerability assessment and its validation: the Gömör -Torna Karst, Hungary and Slovakia. *In*: PARISE, M., GABROVSEK, F., KAUFMANN, G. & RAVBAR, N. (eds) *Advances in Karst Research: Theory, Fieldwork and Applications*. Geological Society, London, Special Publications, **466**. First published online November 29, 2017, https://doi.org/10.1144/SP466.15

JIA, Z., ZANG, H., ZHENG, X. & XU, Y. 2017. Climate change and its influence on the karst groundwater recharge in the Jinci Spring Region, Northern China. *Water*, **9**, 267.

JONES, C.J.R., SPRINGER, A.E., TOBIN, B.W., ZAPPITELLO, S.J. & JONES, N.A. 2017. Characterization and hydraulic behaviour of the complex karst of the Kaibab Plateau and Grand Canyon National Park, USA. *In*: PARISE, M., GABROVSEK, F., KAUFMANN, G. & RAVBAR, N. (eds) *Advances in Karst Research: Theory, Fieldwork and Applications*. Geological Society, London, Special Publications, **466**. First published online November 6, 2017, https://doi.org/10.1144/SP466.5

JONES, W.K. & WHITE, W.B. 2012. Karst. *In*: WHITE, W.B. & CULVER, D.C. (eds) *Encyclopedia of Caves*. Academic Press, Waltham, MA, 430–438.

JOURDE, H., ROESCH, A., GUINOT, V. & BAILLY-COMTE, V. 2007. Dynamics and contribution of karst groundwater to surface flow during Mediterranean flood. *Environmental Geology*, **51**, 725–730.

JOURDE, H., LAFARE, A., MAZZILLI, N., BELAUD, G., NEPPEL, L., DOERFLIGER, N. & CERNESSON, F. 2014. Flash flood mitigation as a positive consequence of anthropogenic forcings on the groundwater resource in a karst catchment. *Environmental Earth Sciences*, **71**, 573–583.

JURKOVSEK, B., BIOLCHI, S. *ET AL*. 2016. Geology of the Classical Karst region (SW Slovenia – NE Italy). *Journal of Maps*, **12**, 352–362.

KÄß, W., LÖHNERT, E.P. & WERNER, A. 1996. Der jüngste Markierungsversuch im Karst von Paderborn (Nordrhein-Westfalen). *Grundwasser*, **1**, 83–89.

KAUFMANN, G. & BRAUN, J. 2001. Modelling karst denudation on a synthetic landscape. *Terra Nova*, **13**, 313–320.

KAUFMANN, G. & ROMANOV, D. 2016. Structure and evolution of collapse sinkholes: combined interpretation from physico-chemical modelling and geophysical field work. *Journal of Hydrology*, **540**, 688–698.

KAUFMANN, G. & ROMANOV, D. 2017. Geophysical observations and structural models of two shallow caves in gypsum/anhydrite-bearing rocks in Germany. *In*: PARISE, M., GABROVSEK, F., KAUFMANN, G. & RAVBAR, N. (eds) *Advances in Karst Research: Theory, Fieldwork and Applications*. Geological Society, London, Special Publications, **466**. First published online December 6, 2017, https://doi.org/10.1144/SP466.13

KAUFMANN, G., ROMANOV, D. & HILLER, T. 2010. Modeling three-dimensional karst aquifer evolution using different matrix-flow contributions. *Journal of Hydrology*, **388**, 241–250.

KAUFMANN, G., ROMANOV, D. & DREYBRODT, W. 2012. Modelling of caves. *In*: WHITE, W.B. & CULVER, D.C. (eds) *Encyclopedia of Caves*. 2nd edn. Academic Press, Waltham, MA, 508–511.

KAUFMANN, G., NIELBOCK, R. & ROMANOV, D. 2015. The unicorn cave, Southern Harz Mountains, Germany: from known passages to unknown extensions with the help of geophysical surveys. *Journal of Applied Geophysics*, **123**, 123–140.

KAUFMANN, G., GABROVSEK, F. & TURK, J. 2016. Modelling flow of subterranean Pivka river in Postojnska Jama, Slovenia. *Acta Carsologica*, **45**, 57–70.

KLIMCHOUK, A.B. 2009. Morphogenesis of hypogenic caves. *Geomorphology*, **106**, 100–117.

KLIMCHOUK, A. 2017. Tafoni and honeycomb structures as indicators of ascending fluid flow and hypogene karstification. *In*: PARISE, M., GABROVSEK, F., KAUFMANN, G. & RAVBAR, N. (eds) *Advances in Karst Research: Theory, Fieldwork and Applications*. Geological Society, London, Special Publications, **466**. First published online November 28, 2017, https://doi.org/10.1144/SP466.11

KOVAČIČ, G. & RAVBAR, N. 2010. Extreme hydrological events in karst areas of Slovenia, the case of the Unica River basin. *Geodinamica Acta*, **23**, 89–100.

KOVARIK, J.L. & VAN BEYNEN, P.E. 2017. Karst-specific composite model for informed resource management decisions on the Biosfera de la Reserva Selva el Ocote, Chiapas, Mexico. *In*: PARISE, M., GABROVSEK, F., KAUFMANN, G. & RAVBAR, N. (eds) *Advances in Karst Research: Theory, Fieldwork and Applications*. Geological Society, London, Special Publications, **466**. First published online November 6, 2017, https://doi.org/10.1144/SP466.1

KRESIC, N. & PANDAY, S. 2017. Numerical groundwater modelling in karst. *In*: PARISE, M., GABROVSEK, F., KAUFMANN, G. & RAVBAR, N. (eds) *Advances in Karst Research: Theory, Fieldwork and Applications*. Geological Society, London, Special Publications, **466**. First published online December 14, 2017, https://doi.org/10.1144/SP466.12

LIU, C.Q., LI, S.L., LANG, Y.C. & XIAO, H.Y. 2006. Using δ^{15}N-and δ^{18}O-values to identify nitrate sources in karst ground water, Guiyang, Southwest China. *Environmental Science & Technology*, **40**, 6928–6933.

LÓPEZ-CHICANO, M., CALVACHE, M.L., MARTÍN-ROSALES, W. & GISBERT, J. 2002. Conditioning factors in flooding of karstic poljes – the case of the Zafarraya polje (South Spain). *Catena*, **49**, 331–352.

MARTIMUCCI, V. & PARISE, M. 2012. Cave surveys, the representation of underground karst landforms, and their possible use and misuse. *Guide Book & Abstracts, 20th International Karstological School Karst Forms and Processes*, 18–21 June 2012, Postojna, Slovenia, 69–70.

MCCORMACK, T., NAUGHTON, O., JOHNSTON, P.M. & GILL, L.W. 2016. Quantifying the influence of surface water–groundwater interaction on nutrient flux in a lowland karst catchment. *Hydrology and Earth System Sciences*, **20**, 2119–2133, https://doi.org/10.5194/hess-20-2119-2016

MILANOVIC, P. 2000. *Geological Engineering in Karst*. Zebra Publishing, Belgrade.

MILANOVIC, P. 2002. The environmental impacts of human activities and engineering constructions in karst regions. *Episodes*, **25**, 13–21.

MUDARRA, M. & ANDREO, B. 2011. Relative importance of the saturated and the unsaturated zones in the hydrogeological functioning of karst aquifers: the case of

Alta Cadena (southern Spain). *Journal of Hydrology*, **397**, 263–280.

Mudry, J. 1987. *Apport du traça geophysico–chimique naturel à la connaissance hydrocinématique des aquifères carbonatés*. PhD thesis, University of Franche-Comté.

Musgrove, M., Opsahl, S.P., Mahler, B.J., Herrington, C., Sample, T.L. & Banta, J.R. 2016. Source, variability, and transformation of nitrate in a regional karst aquifer: Edwards aquifer, central Texas. *Science of the Total Environment*, **568**, 457–469.

Mylroie, J.E. 2013. Coastal karst development in carbonate rocks. *In*: Lace, M.J. & Mylroie, J.E. (eds) *Coastal Karst Landforms*. Coastal Research Library, **5**. Springer, Dordrecht, 77–109.

Mylroie, J.E. & Mylroie, J.R. 2007. Development of the carbonate island karst model. *Journal of Cave and Karst Studies*, **69**, 59–75.

Mylroie, J.E. & Mylroie, J.R. 2017. Role of karst denudation on the accurate assessment of glacio-eustasy and tectonic uplift on carbonate coasts. *In*: Parise, M., Gabrovsek, F., Kaufmann, G. & Ravbar, N. (eds) *Advances in Karst Research: Theory, Fieldwork and Applications*. Geological Society, London, Special Publications, **466**. First published online November 6, 2017, https://doi.org/10.1144/SP466.2

Najib, K., Jourde, H. & Pistre, S. 2008. A methodology for extreme groundwater surge predetermination in carbonate aquifers: groundwater flood frequency analysis. *Journal of Hydrology*, **352**, 1–15.

Naughton, O., Johnston, P.M., McCormack, T. & Gill, L.W. 2017*a*. Groundwater flood risk mapping and management: examples from a low land karst catchment in Ireland. *Journal of Flood Risk Management*, **10**, 53–64.

Naughton, O., McCormack, T., Gill, L. & Johnston, P. 2017*b*. Groundwater flood hazards and mechanisms in lowland karst terrains. *In*: Parise, M., Gabrovsek, F., Kaufmann, G. & Ravbar, N. (eds) *Advances in Karst Research: Theory, Fieldwork and Applications*. Geological Society, London, Special Publications, **466**. First published online December 11, 2017, https://doi.org/10.1144/SP466.9

North, L.A., van Beynen, P.E. & Parise, M. 2009. Interregional comparison of karst disturbance: west central Florida and southeast Italy. *Journal of Environmental Management*, **90**, 1770–1781.

Oberender, P. & Plan, L. 2018. A genetic classification of caves and its application in eastern Austria. *In*: Parise, M., Gabrovsek, F., Kaufmann, G. & Ravbar, N. (eds) *Advances in Karst Research: Theory, Fieldwork and Applications*. Geological Society, London, Special Publications, **466**. First published online January 29, 2018, https://doi.org/10.1144/SP466.21

Ožanić, N. & Rubinić, J. 2003. The regime of inflow and runoff from Vrana Lake and the risk of permanent water pollution. *RMZ-Materials and Geoenvironment*, **50**, 281–284.

Palmer, A.N. 1991. Origin and morphology of limestone caves. *Geological Society of America Bulletin*, **103**, 1–21.

Palmer, A.N. 2007. *Cave Geology*. Cave Books, Dayton, OH.

Palmer, A.N., Palmer, M.V. & Sasowsky, I.D. (eds) 1999. *Karst Modeling*. Karst Water Institute, Special Publications, **5**, Leesburg, VA, USA.

Parise, M. 2003. Flood history in the karst environment of Castellana-Grotte (Apulia, southern Italy). *Natural Hazards and Earth System Sciences*, **3**, 593–604.

Parise, M. 2012. A present risk from past activities: sinkhole occurrence above underground quarries. *Carbonates and Evaporites*, **27**, 109–118.

Parise, M. 2015. A procedure for evaluating the susceptibility to natural and anthropogenic sinkholes. *Georisk*, **9**, 272–285.

Parise, M. 2016. How confident are we about the definition of boundaries in karst? Difficulties in managing and planning in a typical transboundary environment. *In*: Stevanovic, Z., Kresic, N. & Kukuric, N. (eds) *Karst without Boundaries*. International Association of Hydrologist, Selected Papers on Hydrogeology, **23**. CRC Press, Boca Raton, FL, 27–38.

Parise, M. & Gunn, J. (eds) 2007. *Natural and Anthropogenic Hazards in Karst Areas: Recognition, Analysis and Mitigation*. Geological Society, London, Special Publications, **279**, http://sp.lyellcollection.org/content/279/1

Parise, M. & Sammarco, M. 2015. The historical use of water resources in karst. *Environmental Earth Sciences*, **74**, 143–152, https://doi.org/10.1007/s12665-014-3685-8

Parise, M. & Benedetto, L. 2018. Surface landforms and speleological investigation for a better understanding of karst hydrogeological processes: a history of research in southeastern Italy. *In*: Parise, M., Gabrovsek, F., Kaufmann, G. & Ravbar, N. (eds) *Advances in Karst Research: Theory, Fieldwork and Applications*. Geological Society, London, Special Publications, **466**. First published online January 25, 2018, https://doi.org/10.1144/SP466.25

Parise, M., Galeazzi, C., Bixio, R. & Dixon, M. 2013. Classification of artificial cavities: a first contribution by the UIS commission. *In*: Filippi, M. & Bosak, P. (eds) *Proceedings of the 16th International Congress of Speleology*. Vol. 2. Czech Speleological Society, Brno, 230–235.

Parise, M., Closson, D., Gutierrez, F. & Stevanovic, Z. 2015. Anticipating and managing engineering problems in the complex karst environment. *Environmental Earth Sciences*, **74**, 7823–7835.

Perne, M., Covington, M. & Gabrovšek, F. 2014. Evolution of karst conduit networks in transition from pressurized flow to free-surface flow. *Hydrology and Earth Systems Science*, **18**, 4617–4633, https://doi.org/10.5194/hess-18-4617-2014

Petrović, A.S., Ćalic, J., Spalević, A. & Pantić, M. 2018. Relations between surface and underground karst forms inferred from terrestrial laser scanning. *In*: Parise, M., Gabrovsek, F., Kaufmann, G. & Ravbar, N. (eds) *Advances in Karst Research: Theory, Fieldwork and Applications*. Geological Society, London, Special Publications, **466**. First published online January 31, 2018, https://doi.org/10.1144/SP466.23

Pinault, J.-L., Amraoui, N. & Golaz, C. 2005. Groundwater-induced flooding in macropore-dominated hydrological system in the context of climate changes. *Water Resources Research*, **41**, W05001.

Pipan, T. & Culver, D.C. 2013. Forty years of epikarst: what biology have we learned? *International Journal of Speleology*, **42**, 215–223.

Plan, L. 2009. Kläffer Spring – the major spring of the Vienna Water Supply (Austria). *In*: Kresic, N. & Stevanovic, Z. (eds) *Groundwater Hydrology of Springs: Engineering, Theory, Management and Sustainability*. Butterworth-Heinemann, Oxford, 411–427.

Porter, B.L., North, L.A. & Polk, J.S. 2016. Comparing and refining karst disturbance index methods through application in an island karst setting. *Environmental Management*, **58**, 1027–1045.

Price, M., Low, R.G. & Mc Cann, C. 2000. Mechanisms of water storage and flow in the unsaturated zone of the Chalk aquifer. *Journal of Hydrology*, **233**, 54–71.

Ravbar, N. & Goldscheider, N. 2007. Proposed methodology of vulnerability and contamination risk mapping for the protection of karst aquifers in Slovenia. *Acta Carsologica*, **36**, 397–411, https://doi.org/10.3986/ac.v36i3.174

Ravbar, N. & Goldscheider, N. 2009. Comparative application of four methods of groundwater vulnerability mapping in a Slovene karst catchment. *Hydrogeology Journal*, **17**, 725–733.

Ravbar, N. & Šebela, S. 2015. The effectiveness of protection policies and legislative framework with special regard to karst landscapes: insights from Slovenia. *Environmental Science Policy*, **51**, 106–116.

Ravbar, N., Engelhardt, I. & Goldscheider, N. 2011. Anomalous behaviour of specific electrical conductivity at a karst spring induced by variable catchment boundaries: the case of the Podstenjšek spring, Slovenia. *Hydrological Processes*, **25**, 2130–2140.

Ravbar, N., Kovačič, G., Petrič, M., Kogovšek, J., Brun, C. & Koželj, A. 2017. Climatological trends and anticipated karst spring quantity and quality: case study of the Slovene Istria. *In*: Parise, M., Gabrovsek, F., Kaufmann, G. & Ravbar, N. (eds) *Advances in Karst Research: Theory, Fieldwork and Applications*. Geological Society, London, Special Publications, **466**. First published online December 6, 2017, https://doi.org/10.1144/SP466.19

Sánchez, D., Barberá, J.A., Mudarra, M., Andreo, B. & Martín, J.F. 2017. Hydrochemical and isotopic characterization of carbonate aquifers under natural flow conditions, Sierra Grazalema Natural Park, southern Spain. *In*: Parise, M., Gabrovsek, F., Kaufmann, G. & Ravbar, N. (eds) *Advances in Karst Research: Theory, Fieldwork and Applications*. Geological Society, London, Special Publications, **466**. First published online November 28, 2017, https://doi.org/10.1144/SP466.16

Sasowsky, I.D. & Mylroie, J. (eds) 2004. *Studies of Cave Sediments. Physical and Chemical Recorders of Climate Change*. Kluwer Academic, New York.

Shiva, V. 2002. *Water Wars: Privatization, Pollution, and Profit*. South End Press, Cambridge, MA.

Smith, B.A., Hunt, B.B., Andrews, A.G., Watson, J.A., Gary, M.O., Wierman, D.A. & Broun, A.S. 2015. Surface water groundwater interactions along the Blanco River of central Texas, USA. *Environmental Earth Science*, **74**, 7633–7642, https://doi.org/10.1007/s12665-015-4630-1

Stevanović, Z. (ed.) 2015. *Karst Aquifers – Characterization and Engineering*. Professional Practice in Earth Science Series. Springer International, Basle.

Stevanović, Z. 2018. Global distribution and use of water from karst aquifers. *In*: Parise, M., Gabrovsek, F., Kaufmann, G. & Ravbar, N. (eds) *Advances in Karst Research: Theory, Fieldwork and Applications*. Geological Society, London, Special Publications, **466**. First published online January 4, 2018, https://doi.org/10.1144/SP466.17

Stevanović, Z. & Mijatovic, B. 2005. *Cvijic and Karst*. Board of Karst and Speleology, Belgrade.

Stevanović, Z., Kresic, N. & Kukuric, N. (eds) 2016. *Karst without Boundaries*. International Assiciation of Hydrologists, Selected Papers on Hydrogeology, **23**. CRC Press, Boca Raton, FL.

Szymczak, P. & Ladd, A.J.C. 2009. Wormhole formation in dissolving fractures. *Journal of Geophysical Research: Solid Earth*, **114**, https://doi.org/10.1029/2008jb006122

Turpaud, P., Zini, L., Ravbar, N., Cucchi, F., Petrič, M. & Urbanc, J. 2018. Development of a protocol for the karst water source protection zoning. Application to the Classical Karst Region (NE Italy and SW Slovenia). *Water Resources Management*, **32**, 1953–1968.

Uni Research 2016. Oil in ancient caves poses new challenges. *Science Daily*, 25 January 2016, www.sciencedaily.com/releases/2016/01/160125090756.htm

Vaks, A., Bar-Matthews, M., Ayalon, A., Matthews, A. & Frumkin, A. 2017. Pliocene–Pleistocene palaeoclimate reconstruction from Ashalin Cave speleothems, Negev Desert, Israel. *In*: Parise, M., Gabrovsek, F., Kaufmann, G. & Ravbar, N. (eds) *Advances in Karst Research: Theory, Fieldwork and Applications*. Geological Society, London, Special Publications, **466**. First published online December 15, 2017, https://doi.org/10.1144/SP466.10

Van Aken, M., Harley, G.L., Dickens, J.F., Polk, J.S. & North, L.A. 2014. A GIS-based modelling approach to predicting cave disturbance in karst landscapes: a case study from west-central Florida. *Physical Geography*, **35**, 123–133.

van Beynen, P.E. & Townsend, K. 2005. A disturbance index for karst environments. *Environmental Management*, **36**, 101–116.

Vías, J.M., Andreo, B., Perles, J.M., Carrasco, F. & Vadillo, I. 2006. Proposed method for groundwater vulnerability mapping in carbonate (karstic) aquifers: the COP method: application in two pilot sites in southern Spain. *Hydrogeology Journal*, **14**, 912–925.

Vienna's Water Supply 2016. https://www.wien.gv.at/ [last accessed 3 May 2016].

Voulvoulis, N., Arpon, K.D. & Giakoumis, T. 2017. The EU Water Framework Directive: from great expectations to problems with implementation. *Science of the Total Environment*, **575**, 358–366.

Waltham, T., Bell, F. & Culshaw, M. 2005. *Sinkholes and Subsidence*. Springer Science and Business Media, Dordrecht.

Wang, J., Zhao, L. *et al.* 2015. Buried hill karst reservoirs and their controls on productivity. *Petroleum Exploration and Development*, **42**, 852–860.

WHITE, W.B. 1988. *Geomorphology and Hydrology of Karst Terrains*. Oxford University Press, New York.

WHITE, W.B. 2002. Karst hydrology: recent developments and open questions. *Engineering Geology*, **65**, 85–105.

WORTHINGTON, S.R.H. 2009. Diagnostic hydrogeologic characteristics of a karst aquifer (Kentucky, USA). *Hydrogeology Journal*, **17**, 1665–1678, https://doi.org/10.1007/s10040-009-0489-0

YECHIELI, Y., ABELSON, M., BEIN, A., CROUVI, O. & SHTIVELMAN, V. 2006. Sinkholes 'swarms' along the Dead Sea coast: reflection of disturbance of lake and adjacent groundwater systems. *Geological Society of America Bulletin*, **118**, 1075–1087.

ZHOU, W. & BECK, B.F. 2011. Engineering issues on karst. *In*: VAN BEYNEN, P. (ed.) *Karst Management*. Springer, Dordrecht, 9–45.

ZWAHLEN, F. 2004. *Vulnerability and Risk Mapping for the Protection of Carbonate (Karstic) Aquifers*. Final Report COST Action, **620**. European Commission, Brussels.

Orogeny and the collapse of the Devonian Prairie Evaporite karst in Western Canada: impact on the overlying Cretaceous Athabasca Oil Sands

PAUL L. BROUGHTON

Broughton & Associates, PO Box 6976, Calgary, Alberta, Canada T2P 2G2
broughton@shawcable.com

Abstract: The Middle Devonian Prairie Evaporite Formation accumulated up to 200 m of halite-dominated beds across Western Canada in a salt basin extending from northern Alberta southeastwards into southern Saskatchewan and western North Dakota. A 1000 km long salt dissolution trend along the eastern basin margin resulted in the removal of 100–150 m of halite–anhydrite beds. A second salt dissolution trend removed up to 200 m of section across southern Saskatchewan. The removal of the halite-dominated beds and the collapse of the overlying strata responded to regional aquifer and vertical water flows driven by compaction up-structure to the NE towards the eastern margins of the foreland Alberta and intracratonic Williston constituent basins of the Western Canadian Sedimentary Basin. Flows resulting in dissolution trends in the deepening Alberta foreland basin were responses to Middle Jurassic–Early Cretaceous Columbian orogenic tectonism, in contrast with the multiple Paleozoic and Mesozoic dissolution stages across the southern Saskatchewan area of the northern Williston basin. Water flows along the Devonian Keg River strata migrated into the overlying Prairie Evaporite salt beds. The dissolution fronts advanced along multi-kilometre long NW- and NE-oriented sets of fault–fracture lineaments resulting from orogenic Precambrian block movements propagated into the overlying Devonian strata. The Athabasca Oil Sands accumulated above a 300 km long segment of the dissolution trend along the eastern margin of the evaporite basin. An unusually low 1:2 thickness ratio of removed salt beds to overlying strata resulted in significant structural controls on the deposition of the overlying Cretaceous McMurray Formation point bar complexes. Fluvial point bars up to 6 km long and tens of metres thick collinearly aligned along fairways tens of kilometres long, following the halite dissolution trends 200 m below. These sand reservoirs trapped Laramide oil migration into the area. Biodegradation followed, resulting in a trillion barrel resource. A final stage of salt removal occurred with the influx of subglacial meltwaters, rejuvenating 10 km^2 Cretaceous collapses across southern Saskatchewan.

The Prairie Evaporite Formation accumulated as a salt basin deposit of the Elk Point Group (Middle Devonian), extending across large areas of Western Canada and to the south into eastern Montana and western North Dakota (Holter 1969). The Presqu'ile barrier reef at the northern seaward end of the evaporite basin, located along the northern Alberta boundary with the Northwest Territories, allowed the sustained evaporation of this barred basin. Up to 200 m of halite-dominated beds accumulated across north-central Alberta and southern Saskatchewan (Fig. 1). These deposits consist mostly of halite (40–80%) variously mixed with anhydrite and calcareous mudstone. Potash deposits also accumulated in central Saskatchewan.

The halite-dominated salt beds have been affected by dissolution trends tens to hundreds of kilometres long, resulting in the largest known halite karst collapse subsidence structures. There were multiple regional and local post-burial water flow systems that variously affected the patterns of salt removal (Grobe 2000; Hein & Cotterill 2006). It is generally accepted that the dissolution stages resulted from subsurface water flows upwards into the Prairie Evaporite Formation salt beds (Broughton 2013, 2015; Hein *et al.* 2013). Uncertainty remains regarding the relative impact on the dissolution patterns of: (1) influxes of meteoric water flow into the subsurface along permeable beds coming into contact with shallowly buried salt beds; (2) regional aquifer water flow responding to tectonic compression and basin-deepening drives resulting in flows up-structure to the NE towards the basin margin; (3) compaction-driven formation water flows vertically up-section resulting from basin loading, mostly from Keg River Formation strata into the overlying Prairie Evaporite salt beds; and (4) topography-driven flow systems to the NE towards surface topography lows. It is uncertain whether aquifer flows up-structure to the NE were for tens to hundreds of kilometres, or as lesser incremental flows that did not flush the system. All of these

From: Parise, M., Gabrovsek, F., Kaufmann, G. & Ravbar, N. (eds) 2018. *Advances in Karst Research: Theory, Fieldwork and Applications*. Geological Society, London, Special Publications, **466**, 25–78.
First published online November 8, 2017, https://doi.org/10.1144/SP466.7

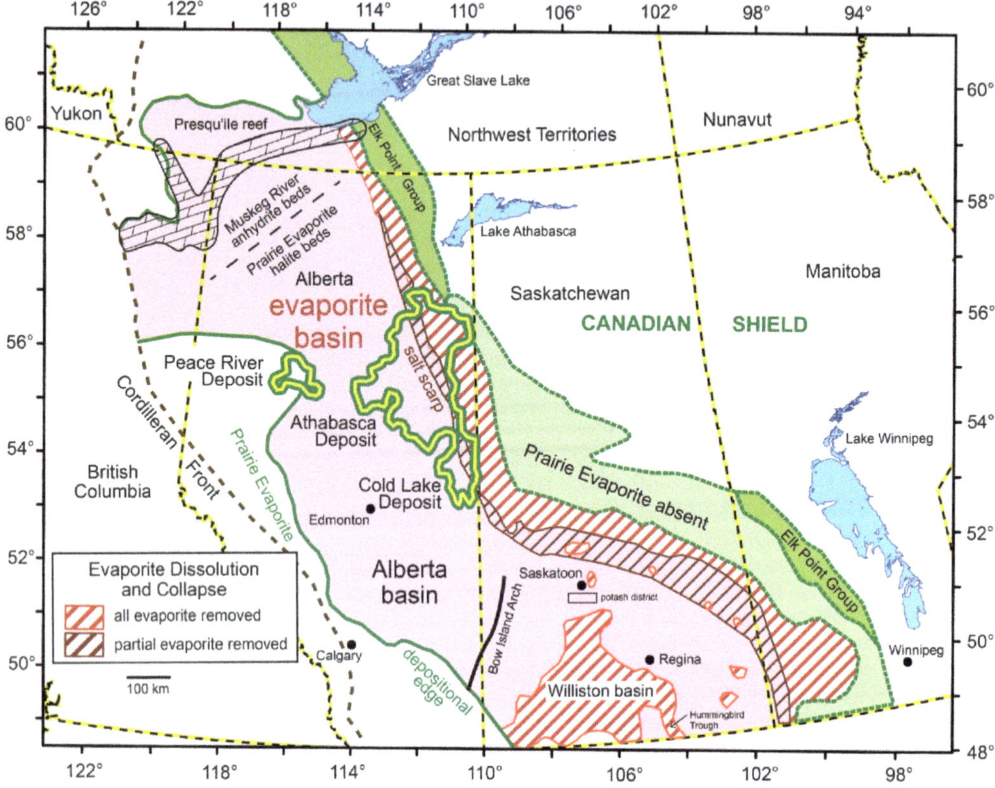

Fig. 1. The Prairie Evaporite Formation of the Elk Point Group (Middle Devonian) accumulated up to 200 m of halite-dominated beds across Western Canada. Post-burial structural trends controlled the salt dissolution patterns in the foreland Alberta and intracratonic Williston constituent basins of the Western Canada Sedimentary Basin. Paleozoic dissolution occurred in the southern Saskatchewan salt beds and along the Presqu'ile barrier carbonate reef during the Antler orogeny. In northeastern Alberta, 100–150 m of the salt beds were removed during the Middle Jurassic–Early Cretaceous Columbian orogeny and during the post-orogenic Aptian accumulation of the Athabasca deposit. The northern Athabasca and eastern Cold Lake oil sand deposits overlie a segment of the 1000 km long and 150 km wide dissolution trend extending along the eastern margin of the Western Canada Sedimentary Basin. Up to 200 m of halite were removed across the southern Saskatchewan area of the northern Williston basin during Paleozoic and Jurassic–Cretaceous orogenic events. Modified from Meijer Drees (1994) and Broughton (2013, 2015).

hypogene and epigene salt karst processes may have contributed to the widespread dissolution patterns in the halite salt basin, but the relative contribution by each process remains uncertain.

This paper reviews the origin and distribution of the regional-scale dissolution patterns that reconfigured the Prairie Evaporite Formation salt basin in northeastern Alberta and southern Saskatchewan and interprets the impact of continental-scale tectonism on these dissolution processes. Orogenic events exerted major regional controls on the development of the foreland Alberta and intracratonic Williston basins as constituents of the Western Canada Sedimentary Basin (WCSB). Tectonic compression and the hydraulic head, charged in the mountainous terrains to the west and SW, provided the drive for the basin water flows northeastwards concurrent with sediment loading (Garven & Freeze 1984; Garven 1989; Hein *et al.* 2013). Such flows variously mixed with influxes of surface-charged groundwater. Alternative models suggest that salt dissolution patterns resulted from mostly or entirely compaction-driven vertical water flows within the Devonian strata (Bachu & Underschultz 1993; Bachu *et al.* 1993; Bachu 1995, 1999).

The continental-scale tectonics are associated with the creation of the Pangaea supercontinent and subsequent fragmentation that resulted in North and Central America being separated from Europe and Western Africa. The tectonism broadly structured the Middle Devonian Prairie Evaporite Formation salt basin across Western Canada into regional sub-basins NE of the Rocky Mountain uplands: the foreland Alberta basin and the southern Saskatchewan area of the northern Williston basin, north of the Central Montana Uplift. The Sweetgrass Arch in northeastern Montana and its northern extension into Alberta, the Bow Island Arch, separate the basins. The arch partition was followed by karstic erosion during the Middle Jurassic to Early Cretaceous Columbian orogeny. As a result, the Alberta and Williston basins evolved separate, but similar, regional aquifer water flow systems linked by a multi-kilometre thick drape of Cretaceous strata onto these constituent sub-basins of the WCSB.

Burial and uplift cycles associated with the Paleozoic Antler orogeny affected the evaporite basin areas to the north and south of the incipient Alberta foreland. Influxes of meteoric charged groundwater resulted in the karstification of the Presqu'ile carbonate barrier reef at the northern end of the evaporite basin. Deeper burial of the barrier reef occurred during the Carboniferous towards the end of the Antler orogeny. There is little evidence for significant Paleozoic dissolution events across large areas of the foreland Alberta basin south of the Presqu'ile reef. This contrasts with salt removal patterns initiated as early as the Late Devonian in the southern Saskatchewan area of the northern Williston basin, where salt removal patterns occurred throughout the Paleozoic, with significant dissolution stages during the Carboniferous. The Paleozoic salt removal patterns across southern Saskatchewan become more widespread during Jurassic–Cretaceous Columbian–Laramide tectonism. By contrast, a regional-scale dissolution pattern occurred in the foreland Alberta basin during Columbian tectonism, but probably not prior to this, as the basin deepened and northeastern Alberta was uplifted, resulting in strata tilting to the SW. Tectonic compression and the hydraulic head were drivers for basin aquifer flows up-structure to the NE. The widespread erosion of most post-Devonian strata across the uplifted areas of the northeastern Alberta basin allowed deeper basin aquifer water flows to mix with shallow meteoric charged groundwater as they came into contact with the shallow buried Prairie Evaporite salt beds.

Multi-stage salt dissolution patterns in northeastern Alberta and southern Saskatchewan resulted from ascending water flows that were directed along cross-cutting fractures and fault lineaments dissecting the 100–200 m thick Middle Devonian evaporite beds. Collapse subsidence structures, tens to hundreds of kilometres long, reconfigured the overlying strata and controlled the concurrent deposition patterns at the surface. These dissolution and collapse patterns were linked to orogenic tectonism, resulting in NW- and NE-oriented fracture and fault lineament sets that emanated upwards from the deformed craton (Thomas 1974; Brown & Brown 1987; Anderson & Knapp 1993; Broughton 2013, 2015). Collapse subsidence structural configuration of the strata overlying the areas of salt removal resulted in traps that accumulated oil migrations into areas of northern Alberta and southern Saskatchewan (Holter 1969; Hein *et al.* 2013). These dissolution patterns exerted significant tectono-stratigraphic controls on the distribution of Lower Cretaceous fluvial–estuarine sandy point bar reservoirs accumulated as the Athabasca Oil Sands in northeastern Alberta, one of the largest known hydrocarbon resources (Broughton 2015).

Presqu'ile barrier and origin of the Middle Devonian evaporite basin

The Laurentia and Gondwana palaeocontinents were separated by the Rheic Ocean during the Devonian. The closure of the Rheic Ocean initiated during the Early Devonian was completed by the Mississippian Period. These Paleozoic movements resulted in compressional and extensional Antler orogenic tectonism across the proto-western North America as the two continental masses collided to form the Pangaea supercontinent. During the Middle Devonian, a southwards trending arm of the Panthalassic Ocean embayed the northern equatorial margin of Laurentia (Fig. 2). A shallow sea extended from the present day position of northern Alberta in Western Canada southwards into eastern Montana and western North Dakota. During the Givetian Stage, the 400 km long Presqu'ile barrier reef accumulated across the northern seaward end of this inland sea, resulting in the shallow Elk Point carbonate–evaporite barred basin (Fig. 3). This equatorial palaeogeography was favourable for evaporation cycles, resulting in the accumulation of up to 200 m of halite-dominated salt beds as the Prairie Evaporite Formation of the Elk Point Group. The Presqu'ile barrier reef separated the Muskeg River Formation anhydrite beds accumulated along the northern margin of the Prairie Evaporite salt basin from the seaward carbonate and calcareous mudstone deposits to the north, represented by strata of the Buffalo River and Windy Point formations along the northern margin of the reef complex (Fig. 4). The Horn River Formation (Givetian–Frasnian) shale accumulated to the west in northeastern British Columbia.

Early stage near-surface karstification of the Presqu'ile carbonate reef trend has been interpreted as a Devonian early Antler orogenic event (Rhodes *et al.* 1984). Deeper burial during the late Carboniferous phase of the Antler orogeny allowed the influx of metalliferous hydrothermal brines into the barrier reef (Rhodes *et al.* 1984). Alternative models suggest that near-surface epigenetic karstification during Paleozoic Antler tectonism was followed by

Fig. 2. Middle Devonian palaeogeography of the Laurentia palaeocontinent prior to the Late Paleozoic collision with Gondwana and closure of the Rheic Ocean, which resulted in the formation of the Pangaea supercontinent. A shallow inland sea across Laurentia extended from northern Alberta southwards into eastern Montana and western North Dakota. The Presqu'ile barrier reef at the northern seaward end of a shallow sea extending into the interior of the Laurentia palaeocontinent allowed the sustained evaporation of this barred basin. Up to 200 m of halite and anhydrite beds accumulated as the Prairie Evaporite Formation salt basin. Modified from a map by Ron Blakey, Global and Regional Paleogeography Project, Northern Arizona University.

hydrothermal hypogene karstification with the deeper burial associated with Cretaceous Laramide tectonism (Qing & Mountjoy 1992, 1994; Qing 1998). Several karstification stages controlled the dissolution and dolomitization patterns along the Presqu'ile barrier reef. Structural adjustments concurrent with carbonate deposition and during shallow burial controlled the karstic dissolution patterns, resulting in parallel near-linear dissolution trends along the length of the eastern reef complex. The early stage of finely crystalline dolomitization of the Pine Point Formation limestone beds was patchy, but the spar was widely disseminated in the overlying beds of the Sulphur Point Formation (Fig. 4). This finely crystalline dolomitization may have occurred as the saline basin water evaporated and the water level dropped, resulting in a hydraulic head that forced marine water to reflux through the barrier. Because seawater is under-saturated relative to gypsum and halite, this may account for the early stage near-surface dolomitization along the barrier reef (Bebout & Maiklem 1973). Second stage coarsely crystalline dolomitization occurred following deeper burial associated with Antler or, alternatively, Laramide tectonism, which resulted in the replacement of the Sulphur Point beds as the massive, coarsely crystalline Presqu'ile dolomite. The eastern end of the Presqu'ile dolomite overlies a segment of the McDonald basement fault system, which extends along the eastern Great Slave Lake shear zone and separates the Churchill and Slave cratons of the Canadian Shield (Fig. 3).

Deep-seated metalliferous hydrothermal brines flowed eastwards up-structure into the eastern end of the barrier reef complex at the base of the massive Presqu'ile dolomite, following previously formed multi-kilometre long karstic dissolution channels developed earlier within the carbonate reef (Rhodes *et al.* 1984). Geothermal sites along this basement fault array probably sourced Pb and Zn compounds

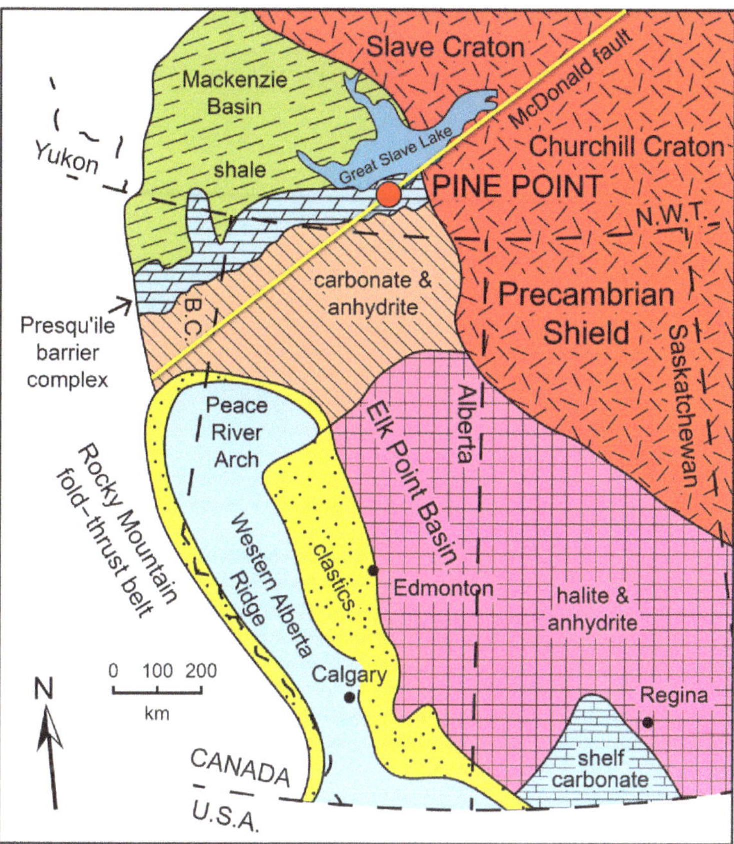

Fig. 3. Geological map of the northern Elk Point basin. The 400 km long Presqu'ile barrier reef barred the northern seaward end of the inland sea. Anhydrite and calcareous mudstone beds of the Muskeg River Formation (Middle Devonian) accumulated along the southern margin of the dolomitized carbonate reef. The anhydrite beds passed southwards into halite-dominated beds of the Prairie Evaporite Formation. Metalliferous hydrothermal brines flowed into the eastern end of the Presqu'ile barrier reef, resulting in strata-bound Mississippi Valley-type Pb–Zn sulphide deposits at Pine Point. The 100 sulphide ore bodies formed along parallel karstic dissolution trends at the eastern end of the reef where it was emplaced above a segment of the McDonald fault array and shear zone separating the Churchill and Slave cratons of the Canadian Shield.

into the hydrothermal brines that flowed eastwards into the Presqu'ile dolomite (Skall 1975; Krebs & Macqueen 1984). The hydrothermal brines flowed along structural and karstic permeability pathways at the eastern end of the reef trend resulting from aligned fault block patterns responding to incremental movement along the McDonald fault array. This resulted in a strata-bound Mississippi Valley-type Pb–Zn sulphide deposit at Pine Point, the largest known of this type in Canada. The metalliferous hydrothermal brine flows accumulated as 100 ore bodies distributed along two 30–50 km long and 2–3 km wide Pb Zn sulphide mineralized trends strata-bound along the base of the Presqu'ile dolomite. The metalliferous mineralization was distributed as tabular strata-bound ore pods, 100–250 m long and 10–20 m thick, and ovoid vertical bodies extending upwards for as much as 60 m cross-cutting the Presqu'ile dolomite and into the overlying Watt Mountain Formation beds.

Early researchers interpreted pervasive laminated textures of sulphide mineralization as evidence for hydrothermally altered stratified cave sediment (Rhodes *et al.* 1984). However, the nearly 100 ore bodies are not interconnected, although they have cave-like pod shapes and are distributed along parallel linear dissolution trends. Some mineralized zones have vertical orientations, often with collapse and crinkle breccia. Alternative interpretations suggest that the metalliferous thermal brines followed the earlier formed karstic dissolution trends, but largely replaced the surrounding dolomite host rock by

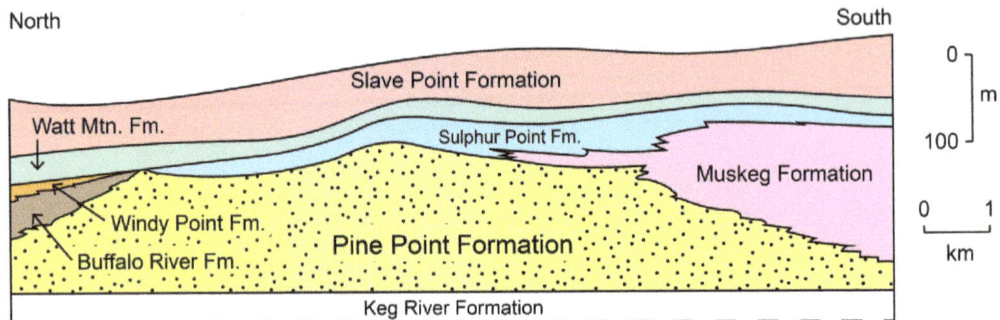

Fig. 4. Schematic cross-section of the Presqu'ile barrier reef at Pine Point illustrating the build-up of Pine Point Formation reef facies that barred the shallow saline sea to the south and allowed sustained evaporation. Burial of the reef trend resulted in massive, coarsely crystalline dolomitization of the overlying Sulphur Point Formation as the Presqu'ile dolomite. Metalliferous hydrothermal brines flowed along the base of the Presqu'ile dolomite at Pine Point, accumulating Mississippi Valley-type strata-bound ore bodies consisting of hydrothermal dolomite and Pb–Zn sulphides. Modified from Rhodes *et al.* (1984).

hydrothermal solution–replacement processes. This hydrothermal karstification resulted in detached stratiform ore pods. As such, the interbedded saddle dolomite and sheet-like layers of sulphide minerals do not represent the replacement of stratified cave-fill sediments, but resulted from solution–replacement back-filling precipitation as the hydrothermal alteration fronts advanced into the micro-fractured wall rock, beyond the earlier formed, but narrower, dissolution trends (Qing & Mountjoy 1992, 1994; Qing 1998).

Stratigraphic setting of the Prairie Evaporite Formation salt basin

Paleozoic stratigraphy

The Middle Devonian Prairie Evaporite basin was covered by a 1.5–2.5 km thick succession of Paleozoic and Mesozoic strata across southern Saskatchewan, in contrast with the preservation of only 200–300 m across some areas of northeastern Alberta (DeMille *et al.* 1964; Holter 1969; Christopher 1974; Broughton 1977; Irvine *et al.* 1978; Grobe 2000). In the foreland Alberta basin, the Cordilleran fold–thrust fault belt developed as the deformation front advanced eastwards. Northeast Alberta was uplifted by the Middle–Late Jurassic, responding to Columbian tectonism and concurrent buckling deformation of the craton underlying the foreland basin. As the basin trough to the west deepened, there was erosion or the non-deposition of most post-Devonian to pre-Cretaceous strata across uplifted northeastern Alberta. The thickness of the eroded post-Devonian strata is uncertain, but it may have been as much as 1–1.5 km. As a result, the pre-Cretaceous stratigraphic interval is only 200–400 m thick across northeastern Alberta and is mostly represented by Devonian strata. The pre-Cretaceous unconformity resulted in the sub-crop of Devonian strata across northeastern Alberta subsequently being covered by Lower Cretaceous strata of the McMurray Formation (Aptian), accumulated as the Athabasca Oil Sands deposit. By contrast, most of the Paleozoic stratigraphic column has been preserved across southern Saskatchewan (Fig. 5).

The upper Elk Point Group (Middle Devonian) accumulated as the Keg River and overlying Prairie Evaporite formations. The Prairie Evaporite Formation salt beds are only 200 m below the pre-Cretaceous unconformity surface in northeastern Alberta (Fig. 5). The Lower Cretaceous Athabasca sand deposits are mostly represented by the Aptian deposition of McMurray Formation strata accumulated on the Middle–Upper Devonian karstic limestone palaeotopography (Meijer Drees 1994; Grobe 2000). The Keg River Formation accumulated as a regional platform carbonate deposit overlain by dolomitized biohermal mounds and pinnacle reefs up to 80 m thick. The overlying Prairie Evaporite Formation salt beds accumulated along the Elk Point basin embayment into the Laurentia palaeocontinent (Fig. 2), extending across most of present day Western Canada (Holter 1969) and into eastern Montana and western North Dakota (LeFever & LeFever 2005). The anhydrite and calcareous mudstone beds of the Muskeg Formation accumulated along the southern margin of the Presqu'ile barrier reef and transitioned southwards into the halite-dominated beds of the Prairie Evaporite Formation. Low water levels and excessive evaporation in this barred basin resulted in the accumulation of halite and other evaporite minerals as sabkha, supratidal flat and coastal lagoon deposits (Maiklem 1971), or alternatively as

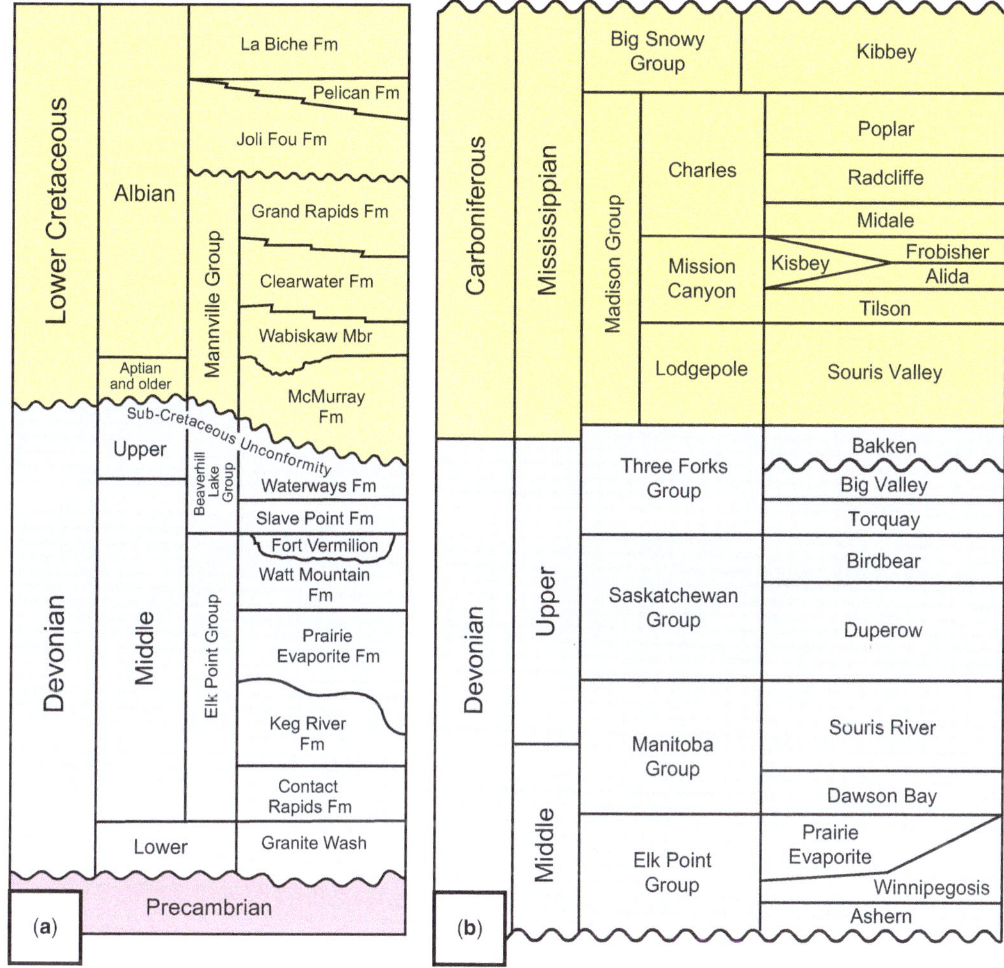

Fig. 5. Stratigraphy of the Western Canada Sedimentary Basin. (**a**) Devonian and Cretaceous formations of the Athabasca Oil Sands area, northeastern Alberta. (**b**) Devonian and Carboniferous formations of southern Saskatchewan.

hyperconcentrated subaqueous accumulations (Bebout & Maiklem 1973).

The Prairie Evaporite Formation in Alberta is divisible into distinctive members. The lowermost Aurora Member consists of dolostone, dololaminite and anhydrite beds. The Whitlow Member consists of coarsely crystalline halite beds up to 80 m thick, accumulated in trough areas between Keg River mounds. The overlying Shell Lake anhydrite deposits include dolostone beds. The uppermost Leofnard Member consists of halite beds with subordinate dolostone, anhydrite and mudstone. These upper interval beds pass into stratigraphically equivalent potash mineral deposits located in south-central Saskatchewan and western North Dakota (Meijer Drees 1986). Overall, these evaporite deposits are a regionally uniform succession up to 200 m thick across south-central Saskatchewan and central Alberta (Meijer Drees 1994; Grobe 2000). Middle–Upper Devonian strata overlying the Prairie Evaporite Formation are the Beaverhill Lake Group, mostly represented by up to 200 m of limestone and calcareous siltstone beds of the Waterways Formation (Schneider & Grobe 2013).

Mesozoic stratigraphy

The Middle–Upper Devonian palaeotopography was draped by Lower Cretaceous fluvial–estuarine deposits of the McMurray Formation (Aptian). The McMurray Formation sand deposits (up to 100 m thick) and the overlying relatively thin (less than

10 m) nearshore sand beds of the basal Wabiskaw Member of the Clearwater Formation accumulated as the Athabasca Oil Sands (Figs 1 & 6). These deposits were covered by 200–300 m of marine mudstone of the Clearwater Formation (Albian), resulting from the Boreal Sea transgression southwards along the Western Interior Seaway (Leckie & Smith 1992).

McMurray Formation strata are informally divided into three successions consisting of fluvial lower, estuarine middle and estuarine–marine upper intervals (Carrigy 1959; Hein *et al.* 2013). The Athabasca (Fig. 6) and Cold Lake oil sand deposits (Fig. 1) accumulated along a broad palaeovalley trunk segment of an Early Cretaceous continent-wide drainage that flowed northwards towards the Boreal Sea at the northern margin of the Mesozoic Laurasia continental mass (Benyon *et al.* 2014, 2016; Blum & Pecha 2014). The lowermost strata of the McMurray Formation accumulated along small fluvial channels characterized by sediment fills with detrital zircon having age dates that suggest a constrained palaeodrainage area emanating from the adjacent Canadian Shield. This accounts for the pervasive distribution of 0.2–2.5 cm diameter quartzite pebbles in beds of the lower McMurray Formation, presumably derived from conglomeratic beds in the Proterozoic

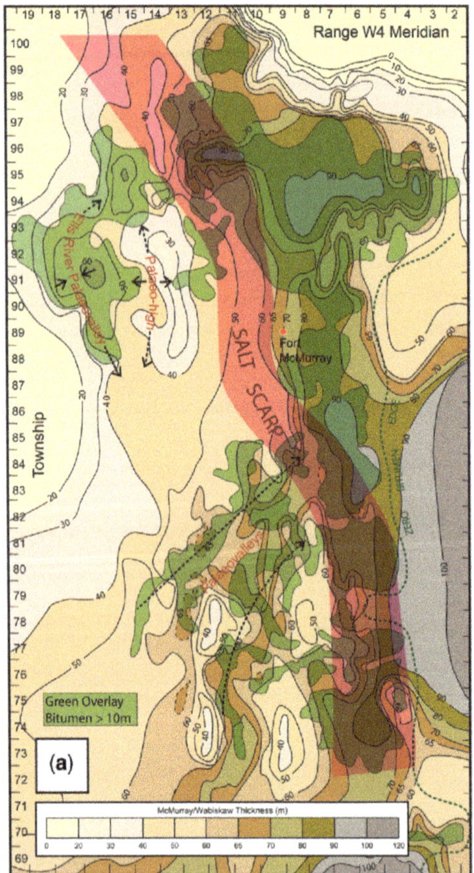

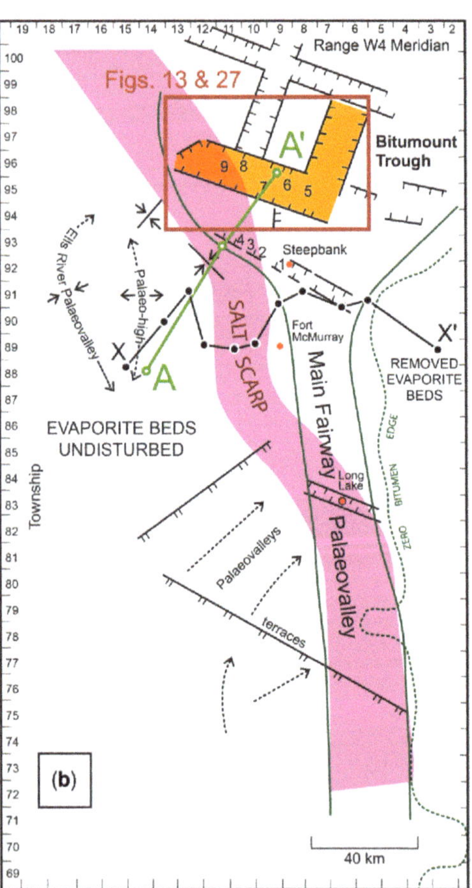

Fig. 6. Geology of the Lower Cretaceous Athabasca Oil Sands, northeastern Alberta. (**a**) Thickness of the McMurray Formation (Aptian) and overlying Wabiskaw Member, Clearwater Formation (Albian) with overlay of bitumen sands >10 m. The 300 km long salt scarp in the underlying Prairie Evaporite (Middle Devonian) separates salt removal areas to the east from undisturbed salt beds to the west. (**b**) Salt dissolution collapse trends on the sub-Cretaceous unconformity surface were covered by strata of the McMurray Formation (Aptian), accumulated as the Athabasca deposit. Mines are located along McMurray Formation bituminous sand fills of tens of kilometre long troughs on the sub-Cretaceous unconformity surface. The Steepbank district mines are (1) Suncor Millennium, (2) Steepbank, (3) Base and (4) North. The Bitumount Trough mines are (5) Imperial Oil Kearl Lake, (6) Shell Canada Jackpine, (7) Shell Canada Muskeg River, (8) Syncrude Aurora North and (9) Canadian Natural Resources Horizon. Area of Figures 13 and 27 shown in box. Modified from Broughton (2013, 2015).

Athabasca basin in northern Saskatchewan (note that the Proterozoic Athabasca basin in northern Saskatchewan is not related to the Cretaceous Athabasca deposit in northern Alberta). These quartzite pebbles are widely distributed in the lower McMurray Formation strata, but are absent in the middle–upper interval strata. By contrast, the fluvial–estuarine deposits of the middle–upper McMurray Formation are characterized by relatively abundant Appalachian- and Grenville-derived detrital zircon grains transported into Western Canada along a continental-scale river system reaching as far as Appalachia in eastern North America (Benyon *et al.* 2014, 2016).

Orogeny and tectonic controls on regional patterns of salt dissolution

The Paleozoic stratigraphic succession accumulated across Western Canada and adjacent areas of the northern USA was structurally configured by Paleozoic and Mesozoic periods of continental-scale compressional and extensional tectonism, resulting in the foreland Alberta and intracratonic Williston basins as constituent sub-basins of the WCSB. These orogenic events were linked to the creation of the Late Paleozoic Pangaea supercontinent and subsequent Mesozoic fragmentation. Middle Jurassic to Early Cretaceous Columbian and Late Cretaceous to Early Paleogene Laramide tectonism broadly configured the Devonian Prairie Evaporite salt basin into regional sub-basins eastwards of the Rocky Mountain deformation front: (1) the deepening foreland Alberta basin with strata tilted to the SW, resulting from the uplift of northeastern Alberta, the subsequent site of the Athabasca Oil Sands deposit; and (2) the northern area of the intracratonic Williston basin in southern Saskatchewan and southwestern Manitoba, resulting in strata tilting to the SW towards the Central Montana Uplift. The Alberta and Williston basins were separated by the Bow Island Arch, a northern extension of the Sweetgrass Arch in northeastern Montana. These constituent basins were covered by a succession of Cretaceous strata several kilometres thick as sediments were shed from the rising Rocky Mountains during Columbian and Laramide tectonism, resulting in the overarching WCSB.

The western margin of the WCSB is now preserved in the Cordillera. It is uncertain to what extent the ancestral tectonism associated with the proto-Pacific margin of North America during the Paleozoic Antler orogenic events regionally deformed this area of western Alberta and eastern British Columbia. The WCSB was probably a distal foreland basin area during most of the Devonian to Carboniferous periods of this Paleozoic tectonism (Root 2001). The earliest deformation of the Alberta basin probably took place as the response of the southern Canadian Cordillera to Antler and Telsin orogenic events. Late Proterozoic extensional rifting controlled the accumulation of a 7 km thick lower Paleozoic stratigraphic succession (Bond & Kominz 1984). Tectonism associated with the Columbian orogeny commenced by the Middle Jurassic and continued into the Early Cretaceous, followed by Late Cretaceous to Eocene Laramide tectonism. As this Cordilleran deformation front advanced eastwards, the foreland Alberta basin axis and the tectonic load similarly migrated, resulting in the progressively uplifted terrains of the Rocky Mountains (Wright *et al.* 1994).

Paleozoic Antler and the Columbian–Laramide orogenic tectonism resulted in the deformation of the Precambrian basement underlying the foreland Alberta and intracratonic Williston constituent sub-basins of the WCSB. Movements by these underlying Precambrian blocks, individually and in concert, imposed a dominant regional structural gain oriented to the NW and NE. Similarly oriented sets of tens to hundreds of kilometres long lineaments resulted. Phanerozoic movements resulted in these near-vertical fault and fracture trends being variously reactivated and extended upward into the overlying Paleozoic and Mesozoic strata. These orthogonal fault–fracture lineament sets were the dominant regional control exerted on salt removal patterns within the strata of the Prairie Evaporite Formation (Broughton 2015).

Foreland Alberta basin

The Late Proterozoic–Phanerozoic Peace River Arch in western Alberta is located eastwards of the Lower Cretaceous Athabasca and parallel to the Peace River oil sand deposits (Fig. 1). It resulted in the prominent NE–SW-trending Skeena Arch in the Cordilleran interior, active since the Mesozoic, and divided the Intermontane Belt into major crustal blocks. The Mesozoic Skeena and Stikine arches in the Cordillera followed ancestral NE–SW tectonic trends in the craton (Lyatsky *et al.* 1999; Lyatsky & Pana 2003), resulting in innumerable fault trends that controlled the Cordilleran tectonic grain, with most orientations to the NW and NE. The Peace River Arch and the West Alberta Ridge have been two major high areas in the Alberta basin since the Cambrian Period (Fig. 3). The Peace River Arch and subsequent Peace River embayment largely controlled NW-trending fault alignments across northern Alberta. The crest of the Peace River Arch was an elliptical topographic ridge until earliest Mississippian time, when tectonic inversion resulted in it becoming part of the Peace River embayment. The north side of the graben was a positive element in the late Mississippian and again in the post-middle Cretaceous (Wright *et al.* 1994).

The sedimentary cover was largely controlled by orogenic and post-orogenic buckling of the craton, which resulted in faults tens to hundreds of kilometres long. These controlled the Phanerozoic movements of basement blocks, resulting in NW–SE- and NE–SW-oriented structural grains dominated by orthogonal fracture lineaments and faults across northern Alberta, including the basement underlying the Athabasca Oil Sands deposit (Broughton 2013, 2015). Numerous studies have been made of the lineament patterns at the surface throughout northern Alberta, including the area of the Athabasca deposit (Babcock & Sheldon 1976), and how these extend to deeper structural patterns in the Precambrian basement. The structural grain of the Precambrian basement below the Athabasca Oil Sands deposit has been interpreted by Lyatsky & Pana (2003) and Lyatsky *et al.* (2005) using gravity and magnetic maps.

Figure 7 illustrates examples of gravity and magnetic maps across northeastern Alberta, including the area underlying the Athabasca Oil Sands. The structural grain is characterized by regional lineaments paralleling the dominant NW-oriented regional direction and counter trends oriented to the NE. These lineament trends within the basement were responses to chains of differentially moved Precambrian blocks as the craton buckled during the Paleozoic and Mesozoic orogenic events. The displacements resulted in parallel sets of smaller,

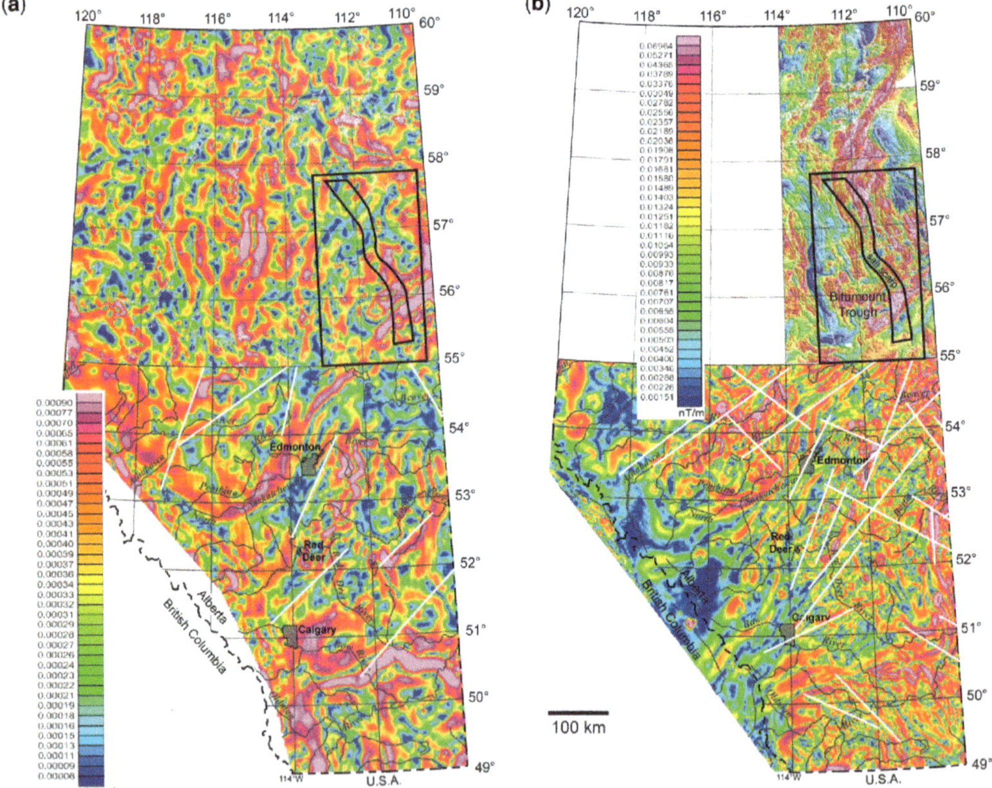

Fig. 7. Gravity and magnetic maps of Alberta, including the area of the Athabasca Oil Sands, displaying the structural grain oriented to the NW and NE, which was propagated upwards into Devonian strata as NW–SE and NE–SW lineament sets. (**a**) Bouguer gravity map of Alberta constructed with horizontal gradient. (**b**) Magnetic map showing southern Alberta with horizontal gradient and regional lineaments, adjoined northern Alberta constructed with a vertical shadowgram, measuring lateral variations in the rock density. The hundreds of kilometres long lineaments are breaks in the pattern trends and represent the distribution of ductile structural anomalies. The box outlines the Athabasca deposit area illustrated by Figure 6. The Bitumount Trough and the salt scarp trend follow the patterns imaged by the gravity and magnetic maps. Each map is a composite of images from different times (southern Alberta, 2005, 49–55° N; northern Alberta, 2003, 55–60° N). Modified from Lyatsky & Pana (2003) and Lyatsky *et al.* (2005).

several kilometre long, cross-cutting lineaments in the overlying Middle Devonian salt beds, similarly oriented to the NW and NE (Broughton 2015). Uncertainty remains about the style of lineament displacement, such as steep normal to vertical faults, linear fracture trends, or as en echelon fault arrays (Broughton 2015, 2016). The extent to which such movements included wrench or strike-slip adjustments is also uncertain. Nevertheless, these lineament patterns were responsible for the linear fracture-prone pathways along which dissolution fronts progressively advanced to the NW and NE within the salt beds. These salt removal collapse patterns were propagated upwards as near-vertical faults several kilometres long that cross-cut the overlying Middle–Upper Devonian and Cretaceous strata, resulting in differentially subsided orthogonal fault blocks varying in size from tens to 1000s of metres.

Intracratonic Williston basin

About 5 km of sediment accumulated at the intracratonic Williston basin depocentre in western North Dakota as it subsided from the Late Cambrian to the Tertiary (Figs 1, 8 & 9). The basin is bounded by the Central Montana Uplift on the west and exposures of Precambrian rocks to the east. A segment of the north–south-trending Trans-Hudson orogenic belt extends beneath the eastern Williston basin, representing an electrical conductivity zone in the

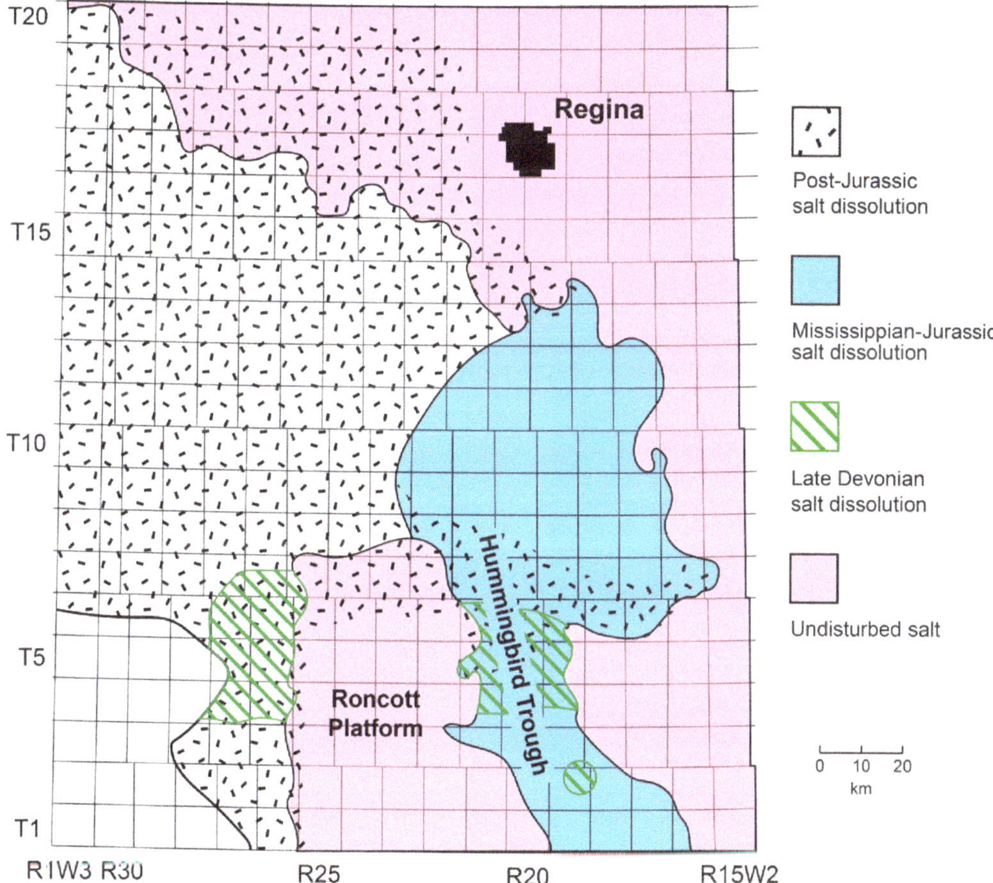

Fig. 8. Multi-stage salt dissolution patterns delineate the Roncott Platform and Hummingbird Trough in southern Saskatchewan, resulting in the removal of up to 200 m of section. The earliest known dissolution trend occurred during the Late Devonian along the incipient margins of the proto-Roncott Platform. Subsequent Late Paleozoic and Mesozoic salt removal finalized the configuration of the Roncott Platform as a salient of undisturbed salt beds in the northern Williston basin, bounded on the east by the Hummingbird Trough. Extensive salt removal patterns developed during the Cretaceous across southern Saskatchewan, removing most of the salt interval accumulated in the northern Williston basin. Modified from Holter (1969) and Broughton (1977, 1985).

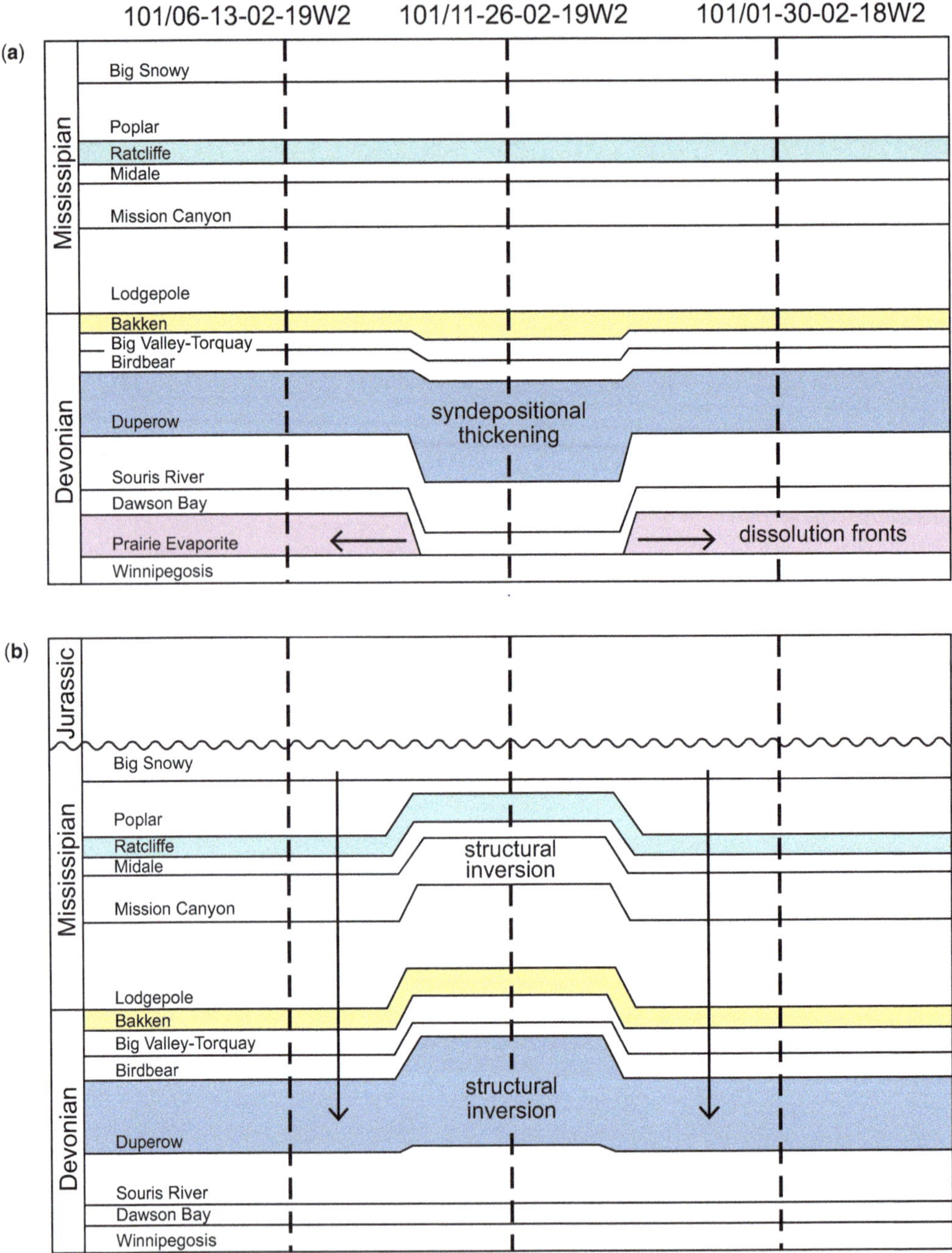

Fig. 9. Multi-staged salt dissolution events resulted in structural inversion sites within the Hummingbird Trough and elsewhere across southern Saskatchewan, which led to commercially attractive hydrocarbon traps. (**a**) Salt removal trends resulted in syndepositional accumulations in the overlying Devonian Souris River, Duperow and Bakken formations. (**b**) During the Jurassic and Cretaceous, structural inversions were formed as a result of additional salt removal peripheral to previously formed thickening of the Devonian beds related to salt collapse. Modified from Smith & Pullen (1967).

crystalline crust from the Black Hills Uplift in South Dakota northwards into Saskatchewan. The intrabasin architecture is dominated by the NW-trending Cedar Creek and Nesson anticlines and the Bowdoin Dome. Responses to movements by orthogonal basement blocks along the Trans-Hudson orogenic belt configured the basin (Kent 1974; Thomas 1974; Brown & Brown 1987). Rotational movements by basement blocks resulted in scissor-type wrench faults adjusting to compressional forces in the basin (Brown & Brown 1987). Kent (1974) interpreted the surface lineaments as evidence for the dominance of NW–SE and subordinate NE–SW structural trends, resulting from basement block movements, as tectono-stratigraphic controls on the salt dissolution patterns and distribution of sediments filling the basin (Fig. 9).

Regional trends in the hypogene removal of salt

Deep basin water flows were mostly along permeable Devonian strata in the foreland Alberta and intracratonic Williston constituent basins of the WCSB. A 1000 km long and 100–150 km wide dissolution trend developed along the eastern margin of the Prairie Evaporite salt basin, extending from northeastern Alberta to southeastern Saskatchewan and into southwestern Manitoba (Figs 10, 11 & 12). Dissolution stages significantly impacted the Prairie Evaporite salt basin in the Cordilleran foreland during the Middle Jurassic–Early Cretaceous (Columbian) tectonism and to a lesser, but uncertain, extent during the Late Cretaceous–Paleogene (Laramide). Compressional tectonism and the hydraulic head, charged in the mountainous terrains to the west and SW, provided the drive for the water flows up-structure to the NE towards topographically low terrains (Garven & Freeze 1984; Garven 1989). Basin aquifer water flows were regionally up-structure to the NE as the basins deepened and the strata tilted downwards to the SW. Such flows mixed with influxes of surface-charged groundwater (Stoakes *et al.* 2014*a*). Alternative models suggest that most salt dissolution patterns resulted from localized compaction-driven vertical water movements within the Devonian strata as a result of sediment loading (Connolly *et al.* 1990; Bachu & Underschultz 1993, Bachu *et al.* 1993, Bachu 1995).

Tectono-stratigraphic and palaeogeographical models, such as those by Garven (1989), interpret the regional controls on deep-seated aquifer water flows in western North America as responses to continental-scale tectonism. However, it remains uncertain to what extent the deep-seated aquifer flows obtained sufficient hydraulic head to have basin-wide flows up-structure for hundreds of kilometres, in contrast with the more localized and incremental movements. Both manners of flow would have included compaction-driven vertical flows resulting from sediment loading (Bachu 1995, 1999). It is also uncertain how much these basin water flows to the NE towards the basin margin mixed with influxes of glacial meltwater flow into the subsurface and meteoric charged groundwater (Stoakes *et al.* 2014*a*). All of these various water sources came into contact with shallow buried Devonian strata, including the Prairie Evaporite salt beds.

Late Devonian meteoric charged groundwater flows occurred in southern Saskatchewan, impacting the shallowly buried salt beds. The dissolution of more deeply buried salt beds occurring throughout the Paleozoic as Antler tectonism resulted in ancestral movements of the Williston basin. These salt removal events delineated the eastern and western margins of the proto-Roncott Platform and incipient subsidence of the proto-Hummingbird Trough. Subsequent dissolution events during the Carboniferous late Antler tectonism resulted in more widespread salt removal patterns along the Hummingbird Trough (Figs 8, 9 & 10). Comparable Paleozoic salt removal stages have not been recognized in northeastern Alberta, possibly because stratigraphic evidence recorded by post-Devonian Paleozoic strata had been eroded during the Columbian tectonism that uplifted northeastern Alberta as the basin deepened and tilted to the SW.

Three areas have been selected to illustrate the hypogene evaporite karst dissolution processes and patterns that resulted from the removal of 100–150 m of halite and anhydrite beds along the eastern basin margin extending from northeastern Alberta to southeastern Saskatchewan into southwestern Manitoba, and up to 200 m across southern Saskatchewan (Irvine *et al.* 1978; McTavish & Vigrass 1987; Anderson & Knapp 1993; Ford 1997; Broughton 1979, 2013, 2015, 2016). These dissolution trends are: (1) the northeastern area of Alberta underlying the Athabasca Oil Sands deposit, which was emplaced over a 300 km long segment of the 1000 km long dissolution trend along the eastern margin of the WCSB; (2) the Roncott Platform–Hummingbird Trough structural complex in the southern Saskatchewan area of the northern Williston basin; and (3) the collapse structures across the potash mining district of south-central Saskatchewan and beyond, which were reactivated during the Pleistocene by strongly pressured subglacial meltwater flows into the subsurface coming into contact with the salt beds.

Area 1. Athabasca Oil Sands, Alberta foreland basin

The Athabasca Oil Sands deposit accumulated above a 300 km long segment of the 1000 km long

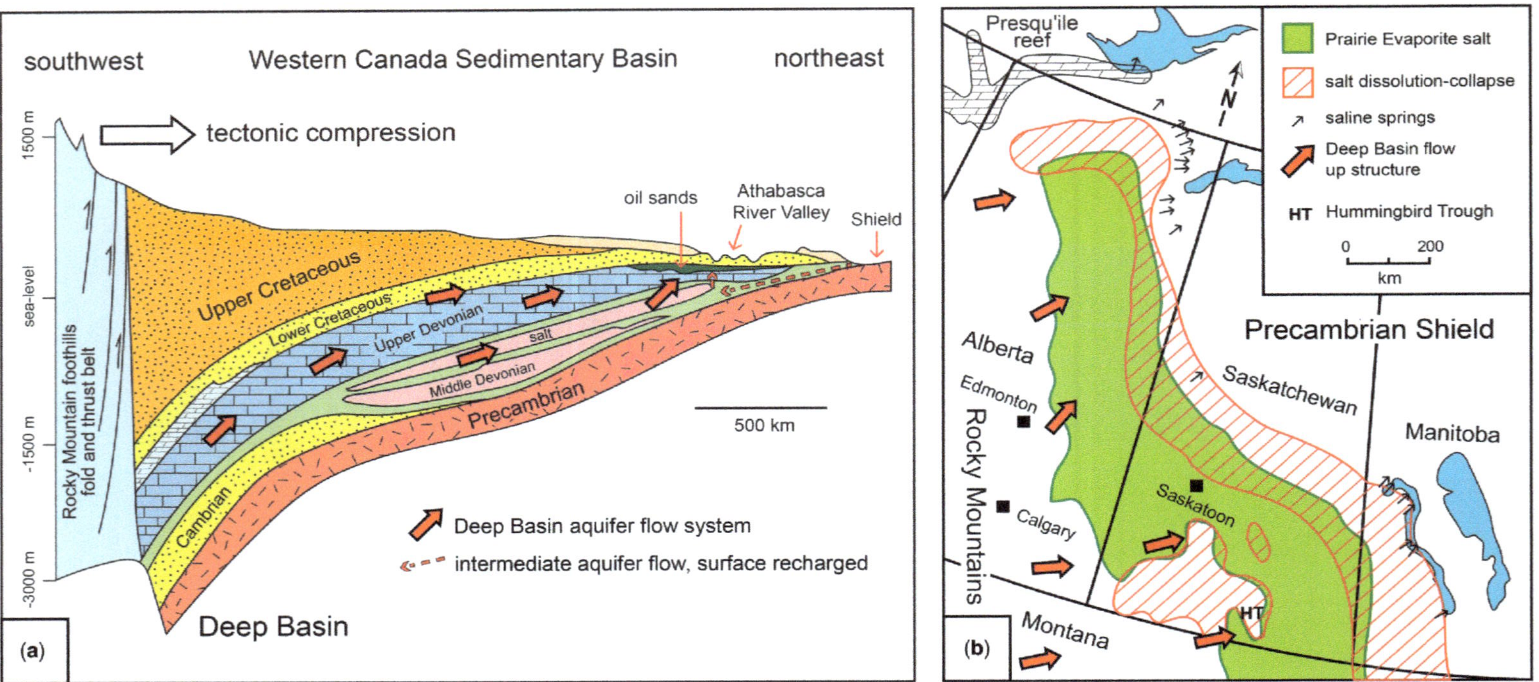

Fig. 10. Deep basin aquifer water flows in the foreland Alberta and Williston constituent basins of the Western Canada Sedimentary Basin. (**a**) Schematic profile of the foreland Alberta basin. Mountainous terrains to the west resulting from the Columbian and Laramide orogenic tectonism provided the hydraulic head for deep aquifer water flows up-structure to the NE as the basins deepened and the strata tilted to the SW. Incremental to basin-wide aquifer flows, combined with compaction-driven vertical flows resulting from basin loading, migrated up-dip into overlying salt beds of the Prairie Evaporite Formation. The resulting salt dissolution trends collapsed the overlying strata and configured the Devonian palaeotopography with troughs tens of kilometres long, subsequently filled and covered over by Lower Cretaceous strata as the Athabasca Oil Sands. (**b**) Water flows along the eastern margin of the Prairie Evaporite salt basin, extending across the Alberta and Williston constituent basins of the Western Canada Sedimentary Basin, resulted in a 1000 km long and 150 km wide dissolution trend. Extensive salt removal also occurred across the southern Saskatchewan area of the northern Williston basin. During the Pleistocene, these water flows to the NE were reversed by an influx of strongly pressured subglacial meltwaters into the subsurface, responding to the hydraulic head of the 1.5–2 km thick Laurentide ice sheet. Modified from Garven (1989).

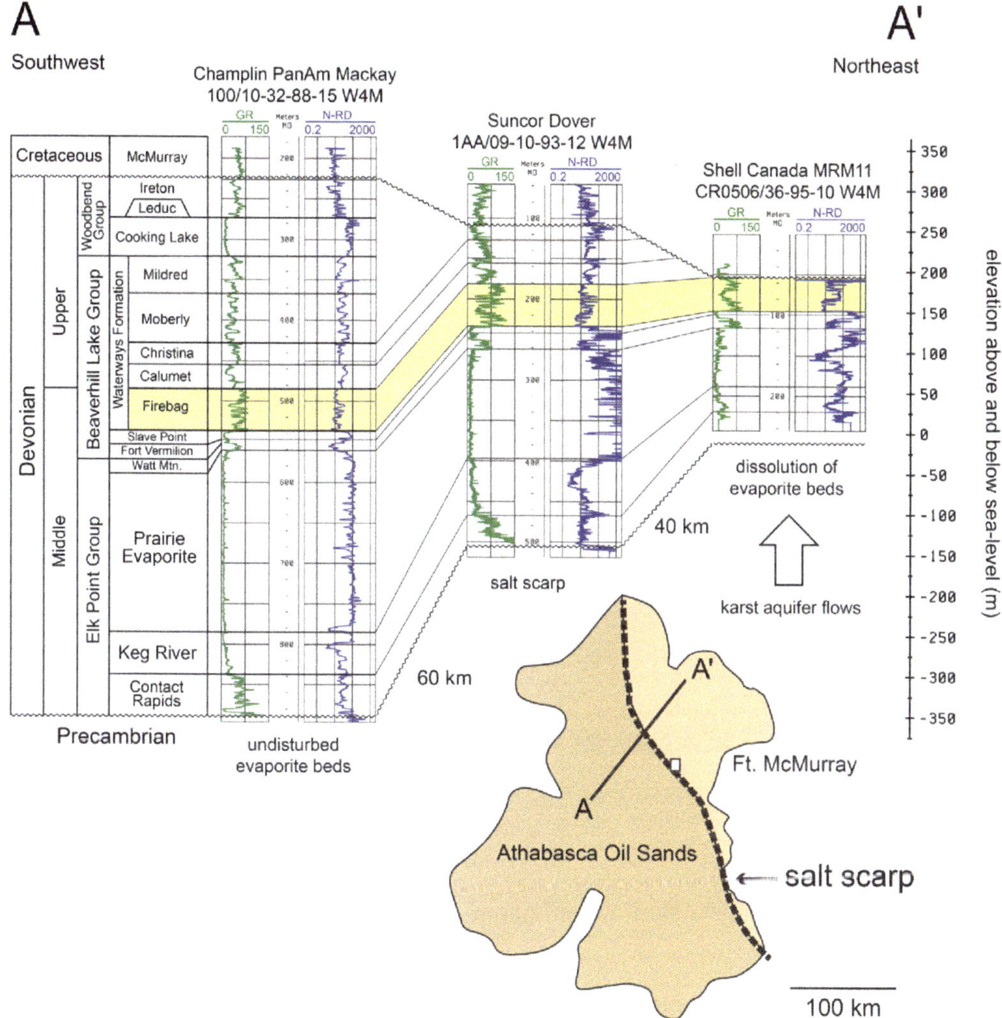

Fig. 11. Cross-section A–A′ of the Athabasca Oil Sands deposit illustrating the eastwards erosion of Devonian strata as the foreland Alberta basin deepened and the uplifted strata in northeastern Alberta tilted to the SW during the Middle Jurassic–Early Cretaceous Columbian orogeny. Concurrent subsurface dissolution of the Prairie Evaporite halite and anhydrite beds occurred 200 m below the uplifted Devonian limestone palaeotopography. Trace of A–A′ located on Figure 6. Well logs are gamma ray (GR) and neutron density (N-RD) log profiles. Modified from Broughton (2013, 2015).

dissolution trend that extends along the eastern updip margin of the Prairie Evaporite Formation salt basin. A 10–20 km wide trend of partially removed salt beds, known as the salt scarp, demarcates the westwards limit of the regional dissolution front, separating large areas of mostly complete salt removal to the east from mostly undisturbed salt beds to the west (Fig. 1). This salt scarp trend below the eastern and northeastern areas of the Athabasca deposit also underlies the eastern margin of the Cold Lake Oil Sands deposit to the south (Figs 1 & 6). Dissolution fronts advanced along cross-cutting lineament trends oriented to the NW and NE that dissected the salt scarp. These fracture–fault lineaments in the salt beds originated as vertical structures propagated upwards by movement between the underlying Precambrian blocks, mostly during the Columbian orogeny, but also to an unknown extent during the earlier Paleozoic Antler tectonism. As these linear dissolution trends widened with additional salt removal, the reticulate dissolution patterns merged into larger salt removal areas, resulting in migration of the scarp westwards and collapse of the overlying strata, thus producing extensive breccia deposits,

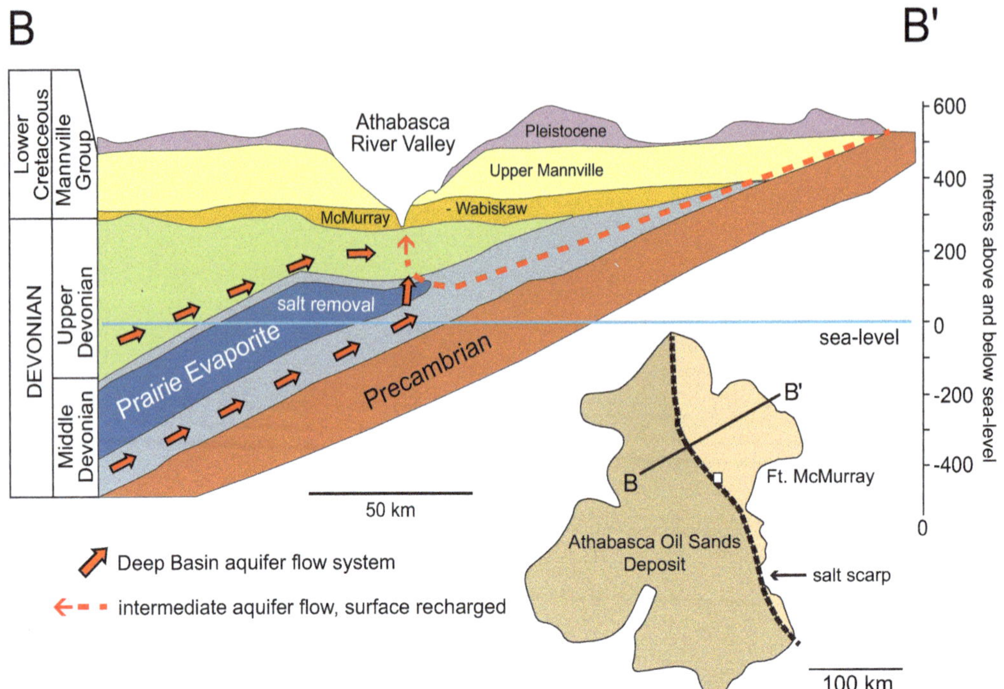

Fig. 12. Schematic section B–B′ across the northern Athabasca Oil Sands deposit showing the Devonian unconformity surface and trend of the Middle Devonian salt scarp 200 m below. The salt scarp underlying the Athabasca Oil Sands deposit formed as a 300 km segment of the 1000 km long salt dissolution trend extending along the eastern margin of the Prairie Evaporite salt basin across Western Canada. Salt dissolution trends developed as regional basin waters flowed up-structure to the NE, combined with compaction-driven vertical flows as the basin filled and strata tilted to the SW. Flows into the overlying salt beds advanced along orthogonal sets of cross-cutting NW- and NE-oriented fracture–fault lineaments that were propagated upwards into the Devonian strata as responses to the movements between underlying Precambrian blocks during orogenic tectonism. Influxes of meteoric charged groundwater further stimulated dissolution along the shallowly buried salt scarp.

in contrast with the gradual subsidence elsewhere (Broughton 2013).

Multiple dissolution stages removed halite and anhydrite beds in the Middle Devonian strata below the pre-Cretaceous unconformity that floors the Athabasca deposit. The dissolution trends may have started by the Late Devonian, but evidence for this occurrence in northeastern Alberta is scant. Widespread salt removal occurred during the Middle Jurassic to the Early Cretaceous Columbian orogeny, resulting in a mosaic of differentially subsided orthogonal Devonian fault blocks as the basin strata tilted to the SW (Bachu 1995, 1999; Anfort *et al.* 2001; Broughton 2013, 2015; Hein *et al.* 2013; Schneider & Grobe 2013). The collapse of the Middle–Upper Devonian limestone strata, overlying the salt removal areas only 175–250 m below, occurred prior to and concurrent with the accumulation of the McMurray Formation (Aptian) strata on this karstic limestone palaeotopography (Fig. 11). Salt dissolution continues to the present day, but at a much subdued rate. Surface erosion of the Devonian and post-Devonian Paleozoic strata resulted in the progressive down-cutting removal of the stratigraphic section towards the NE as the basin tilted to the SW. The concurrent removal of salt beds in the subsurface resulted in bed dip reversal (Figs 11 & 12). Strata of the Middle–Upper Devonian Waterways Formation members subcrop across large areas subsequently covered over by Cretaceous strata of the McMurray Formation. The thickness of the eroded Devonian and post-Devonian Paleozoic strata is uncertain, but was probably 1–1.5 km.

Broughton (2015) interpreted the configuration of the palaeotopography between underlying salt removals interacting with the concurrent down-cutting surface erosion of the Devonian strata during the Columbian uplift of the area. Salt dissolution patterns propagated upwards to the overlying pre-Cretaceous palaeotopography resulted in subsidence troughs tens of kilometres long flooring the Athabasca deposit (Figs 12 & 13). Salt dissolution

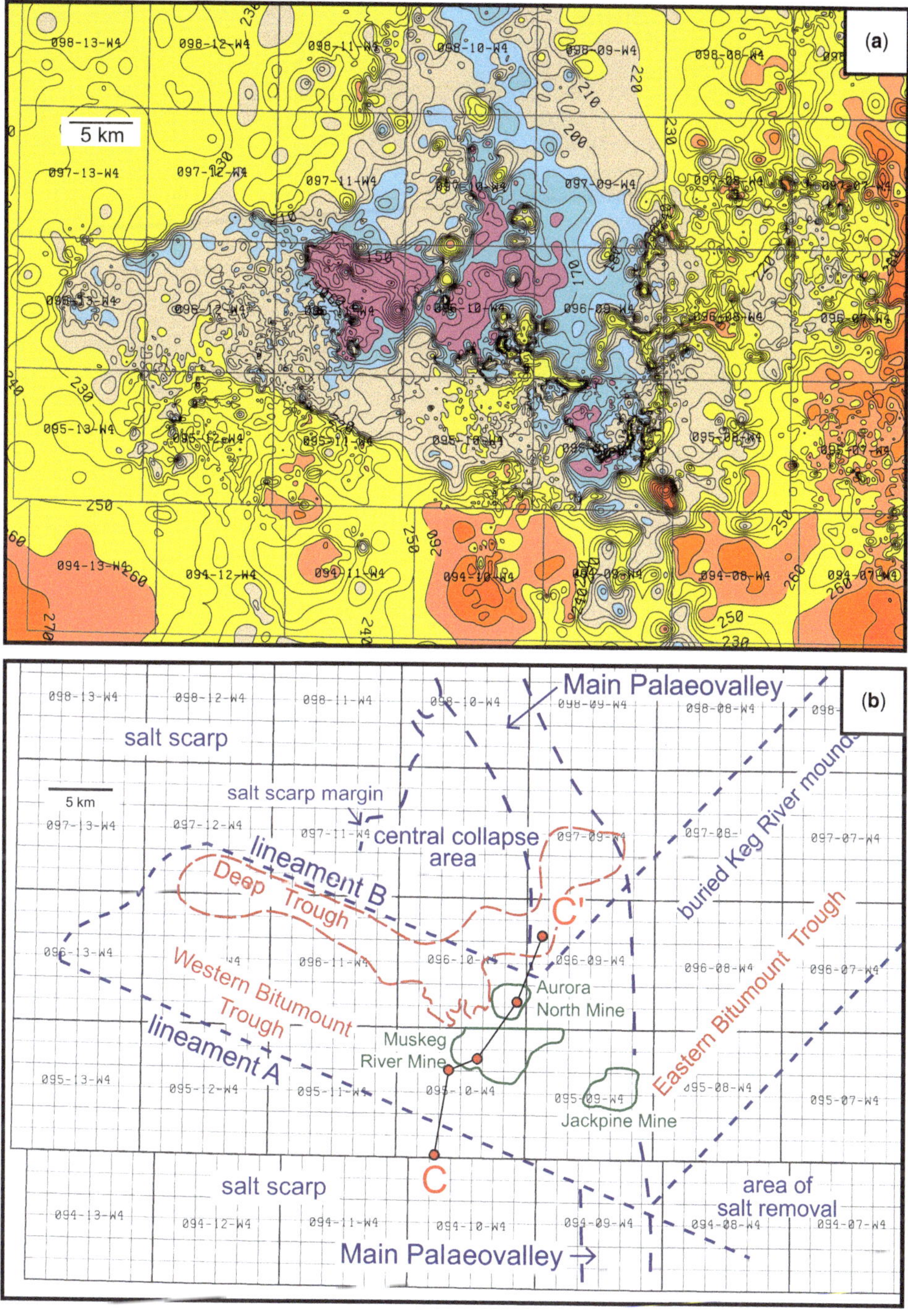

Fig. 13. Structure of the sub-Cretaceous unconformity surface of the Bitumount Trough area, northern Athabasca Oil Sands deposit. (**a**) Map of the Middle–Upper Devonian unconformity surface with up to 100 m of relief along the eastern margin of the salt scarp. Elevations above sea-level with 10 m contour intervals. (**b**) The sub-Cretaceous structure of the northern Athabasca deposit was dominated by the 50 km long V-shaped Bitumount Trough, including the deep trough structure along the northern margin, and the adjacent Central Collapse. Locations are shown of the Jackpine, Muskeg River and Aurora North mines. Area located on Figure 6b (box). Trace of cross-section C–C′ (Fig. 16). Modified from Broughton (2013, 2015).

trends mostly developed during the Middle–Late Jurassic to the Early Cretaceous Columbian orogeny, but post-orogenic dissolution trends continued into the Aptian concurrent with deposition of the McMurray Formation (Broughton 2013, 2015, 2016). More than 100 m of salt removal with only 175–250 m of overlying strata resulted in sharply defined collapse structures at the scale of tens of kilometres on this sub-Cretaceous unconformity surface. This is an unusually low 1:2 ratio of removed salt to thickness of the overlying strata (Broughton 2013). Biohermal build-ups and pinnacle reefs multiple kilometres long and oriented to the NW and NE were parallel to structural lineaments along the margins of displaced Precambrian fault blocks only 200–300 m below. Prominent structural ridges on the sub-Cretaceous unconformity surface mirror these underlying Keg River trends, resulting in sub-parallel valley and ridge trends to the NW and SE (Mahood *et al.* 2012; Broughton 2013, 2015, 2016). In contrast with these salt removal patterns, the Prairie Evaporite strata accumulated in central Alberta to the SW of the Athabasca deposit area preserve the complete 150–200 m salt interval, unaffected by salt dissolution.

Regional salt removal trends underlying the Athabasca deposit are the salt scarp and overlying main palaeovalley trunk, the Bitumount Trough and the Central Collapse (Figs 6 & 13).

Salt scarp and main palaeovalley trunk. The north–NW-oriented Prairie Evaporite Formation salt scarp, 200 m below McMurray Formation strata, is a 300 km long segment of the 1000 km long dissolution trend developed along the eastern margin of the evaporite basin (Figs 1 & 6). It represents the westernmost limit of the advance by the composite of dissolution fronts below the length of the Athabasca deposit and separates the 100–150 km wide salt removal areas to the east from mostly undisturbed salt beds to the west. A 175–250 m interval of fragmented and differentially subsided fault blocks, consisting of Middle–Upper Devonian and Lower Cretaceous strata, overlies the 10–20 km wide salt scarp and salt removal areas to the east. The salt scarp has been dissected by cross-cutting NW–SE- and NE–SW-oriented fracture–fault lineaments that lengthened and widened as the dissolution fronts advanced along these linear trends. Broughton (2016) refined the interpretation of these lineament trends, propagated upwards to the Devonian unconformity structural surface, as arrays of en echelon offsets consistent with progressively advancing dissolution collapse fronts. This often results in closely spaced parallel faults separating tilted fault blocks and Devonian breccia zones that are only tens of metres wide (Fig. 14). The northward trending main palaeovalley on the sub-Cretaceous

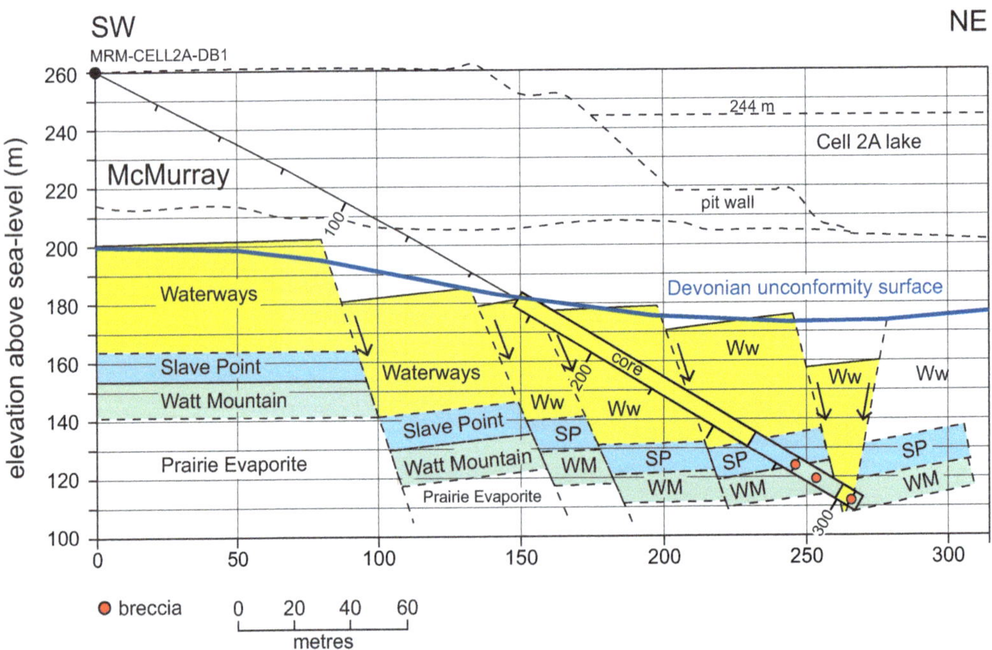

Fig. 14. Illustration of en echelon fault blocks of Middle–Upper Devonian strata that floor the Muskeg River Mine site in the Western Bitumount Trough. These fault patterns often consist of tilted 20–50 m wide blocks that responded to underlying salt dissolution fronts advancing along linear trends. Modified from Stoakes *et al.* (2014*a*, *b*).

unconformity surface parallels the eastern edge of the underlying scarp. It represents the structural offset fronting the approximately 100 m of relief along the eastern salt scarp margin (Figs 6 & 13).

Fracture, fault and joint patterns across the sub-Cretaceous palaeotopography resulted in lineaments oriented to the NW and NE that cross-cut and fragment the Middle–Upper Devonian limestone beds flooring the McMurray Formation deposits (Fig. 15). The fracture and fault patterns resulted from the removal of the underlying salt beds prior to and during McMurray deposition and the subsequent oil migration pathways into the overlying McMurray sand deposits during the Late Cretaceous to Early Paleogene. These pre-Aptian lattice-like joint patterns, often 2–3 m deep and 1–2 m wide at the surface and infilled with a green argillaceous clay palaeosol, a regolith that resulted from prolonged weathering of the Devonian Waterways Formation carbonate beds, but widened during Laramide tectonism.

Partial to complete removal of the halite–anhydrite succession along the length of the salt scarp was a westward progression of dissolution fronts impacting highly soluble halite and the partial preservation of less soluble anhydrite. Figure 16 illustrates this dissolution zoning westwards: (1) areas to the east with complete removal of all halite and anhydrite beds; (2) complete removal of the halite and partial removal of the anhydrite, preserving only anhydrite; (3) partial removal of the halite and preservation of anhydrite; and (4) unaffected areas to the west preserving the entire evaporite bed interval. This progressive dissolution resulted in NW-oriented regional trends with differing residual mineral compositions parallel to the eastern salt basin margin.

Bitumount Trough. The 50 km long V-shaped Bitumount Trough (Figs 6b & 13) is the largest salt dissolution collapse-related depression on the sub-Cretaceous unconformity surface (Flach & Mossop 1985; Hein *et al.* 2000, 2001; Hein & Cotterill 2006; Broughton 2013, 2015; Schneider & Grobe 2013; Schneider *et al.* 2014). It resulted from the collapse of Middle–Upper Devonian strata above a salt removal area between pairs of orthogonally cross-cutting parallel lineaments. During deposition of the lower and middle intervals of the McMurray Formation, the trough filled as a response to the underlying differential subsidence of the Devonian fault block mosaic along the trough floor. Significant pulses of subsidence occurred during and towards the end of the lower interval of the McMurray Formation at the site of the Bitumount Trough, but generally not elsewhere along the salt scarp (Broughton 2013, 2015). By contrast, collapses during the middle interval were more prevalent along the length of the underlying scarp, particularly to the north and to the south of the Bitumount Trough. Figure 17 illustrates examples of tilted, rotated and differentially subsided Devonian–McMurray fault blocks, each up to 1000 m long, exposed in mine cuts overlying the salt scarp. Portions of the trough subsided prior to the accumulation of the Cretaceous sediments, but most of the subsidence was concurrent with the deposition of the McMurray Formation strata, resulting in 20 km long syndepositional trends. Salt removal below the trough continued after the accumulation of the McMurray Formation fill, although to a significantly lesser extent, during the accumulation of the overlying Clearwater Formation strata. Dissolution subsidence may have continued during the Late Cretaceous and Tertiary, but the stratigraphic interval recording evidence for this has been eroded by Pleistocene glaciation.

Central Collapse. The Central Collapse area of the sub-Cretaceous unconformity surface developed along the northern side of the Bitumount Trough. This regional structural low resulted from embayment of the salt scarp that realigned the position of the westwards advancing scarp to a more northwesterly direction (Figs 13 & 18). The regional depression developed mostly during the deposition of the middle–upper interval of the McMurray Formation as the underlying dissolution fronts advanced northwards of the previously formed Bitumount Trough (Broughton 2016). The Bitumount Trough and adjacent Central Collapse area together collectively formed a regional-scale structural funnel-like area extending across the northern Athabasca deposit. As a result, the northern Athabasca deposit area was characterized by increasingly unstable Devonian substrate that included terraced structures, consisting of chains of aligned fault blocks of Devonian and lower McMurray strata, which stair step-down northwards into the Central Collapse (Fig. 18). These salt collapse structures exerted significant control on the depositional trends of the middle–upper McMurray Formation.

Area 2. Roncott Platform–Hummingbird Trough, northern Williston basin

Multi-stage salt dissolution impacted most of the southern Saskatchewan area of the northern Williston basin and its ancestral constituent structures (Figs 1 & 10). The earliest dissolution patterns probably resulted from meteoric charged groundwater coming into contact with the shallow buried salt beds. Deeper burial of the salt beds followed and these were impacted by aquifer water flows northwards up-structure along ancestral structures of the basin. Elevated terrains to the south, such as the Montana Uplift, were recharge areas.

Fig. 15. Fracture, fault and joint patterns on the sub-Cretaceous palaeotopography with orientations to the NW and NE, cross-cutting Middle–Upper Devonian strata overlying the salt scarp. The fracture and fault patterns resulted from the dissolution of salt beds prior to and during McMurray deposition. These pre-Aptian lattice-like joint patterns, often with 2–3 m deep and 1–2 m wide surface channels, were infilled with a palaeosol consisting of greenish grey argillaceous clay resulting from prolonged weathering of the Devonian carbonate beds. (**a**) Example of Devonian topographic terrain with post-Devonian overburden removed to allow quarrying of the Waterways limestone at the Hammerstone Quarry located adjacent to the Muskeg River Mine. (**b**) Several metre-deep and wide joint channels with palaeosol fill removed to facilitate mining and processing of the limestone bed. (**c**) Example of fragmented limestone beds along the wall of a 3 m deep excavated joint channel filled with green clay. (d) Example of a green palaeosol fill along a narrow joint offsetting a larger filled channel with clay fill removed. Images courtesy of Hammerstone Corporation, Calgary, Alberta, Canada.

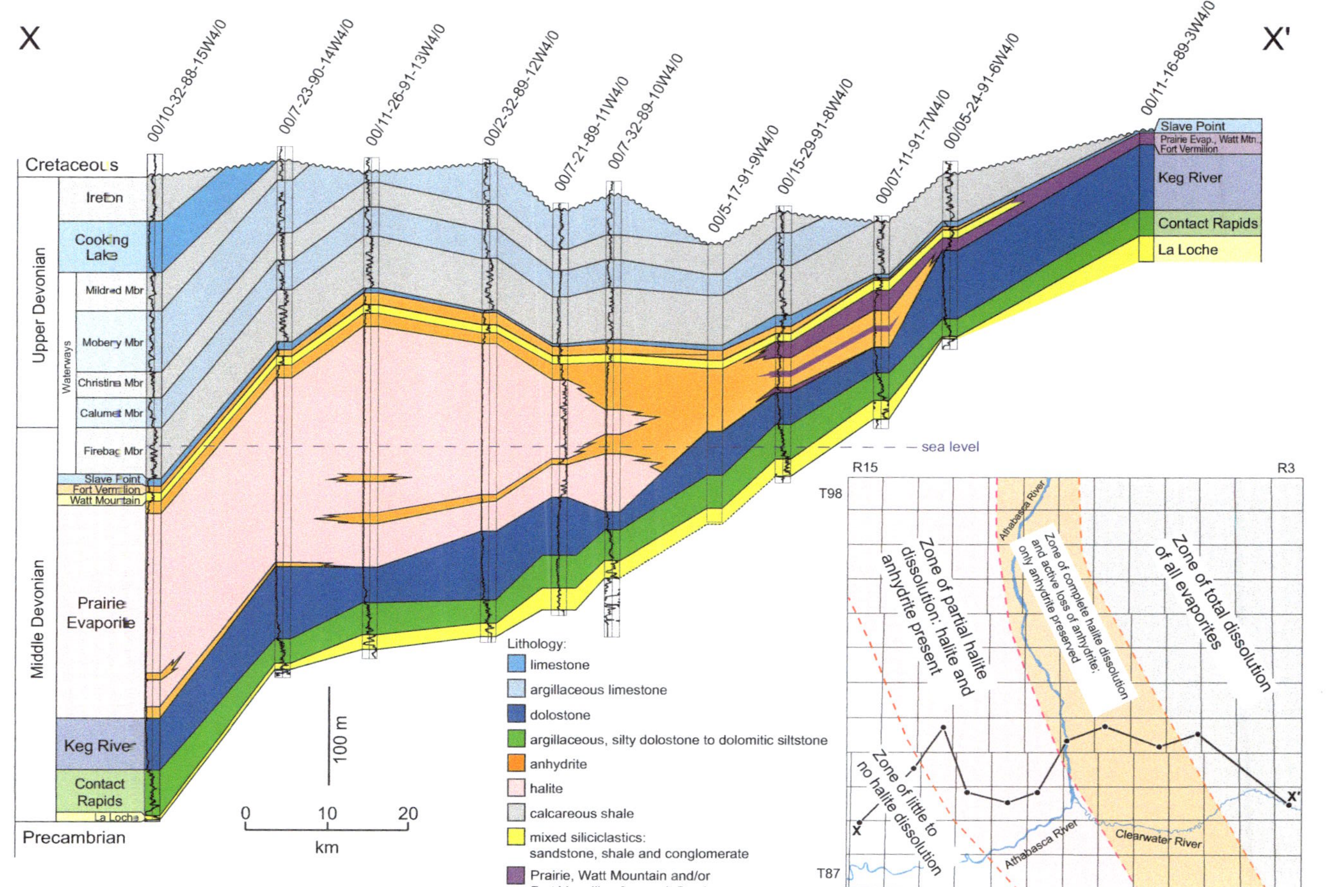

Fig. 16. Cross-section of Middle Devonian strata below the northern Athabasca Oil Sands illustrating progressive removal westward of highly soluble halite and less soluble anhydrite beds, resulting in NW-oriented regional mineral solubility trends. Trace of cross-section X–X′ located on Figure 6b. Modified from Schneider & Grobe (2013) and Schneider *et al.* (2014).

(a)
middle McMurray strata
upper bed-lower McMurray
tilted lower McMurray strata
rotated fault block
5 m
(b)
(c)
(d)
middle McMurray
strata

Fig. 17.

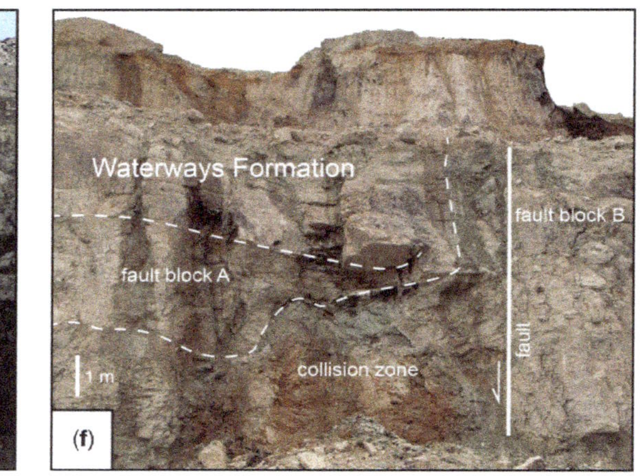

Fig. 17. Differentially subsided and rotated fault blocks responding to removal of underlying Prairie Evaporite salt beds and exposed by mining operations located above the salt scarp. (**a**) 1000 m long tilted and rotated fault block consisting of lower McMurray beds excavated at the Muskeg River Mine, Bitumount Trough. Erosion and reactivation surfaces indicate two collapse events prior to being covered by middle McMurray beds. Modified from Broughton (2013, 2015). (**b**) Collapse of an orthogonal-shaped fault block uncovered in the Muskeg River Mine, consisting of upper beds of the lower McMurray interval collapsed against lower beds of the interval. (**c**) Two adjacent collapsed lower McMurray fault blocks exposed in the Muskeg River Mine. Modified from Barton & Siebel (2016) and Barton *et al.* (2017). (**d**) Collapsed kilometre-long fault block consisting of middle McMurray strata exposed in the Millennium Mine, Steepbank district. The hundreds of metres long fault block broke during the collapse as the segments collided. The smaller middle block rotated to a near-vertical position as it was crushed between the two larger colliding blocks, resulting in a vertical breccia pipe (box). Modified from Broughton (2013, 2015). (**e**) Illustration of collisional impact between lower McMurray fault blocks on removal of the underlying salt beds, exposed along a bench cut in the Muskeg River Mine. (**f**) Collision zone between two adjacent Upper Devonian fault blocks, consisting of Moberly Member limestone beds, Waterways Formation, exposed at the Hammerstone Quarry near the Muskeg River Mine. Courtesy of G. Kozdial, Hammerstone Corporation, Calgary, Alberta, Canada.

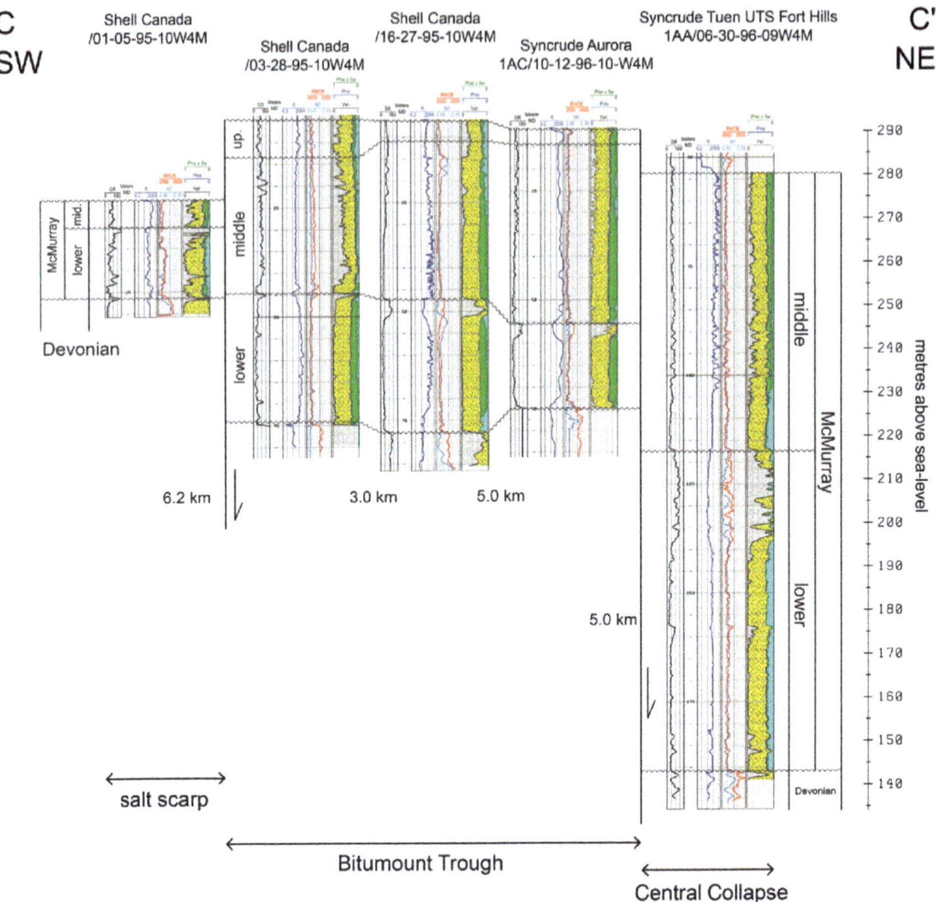

Fig. 18. Structural cross-section C–C′ of the Bitumount Trough area, northern Athabasca deposit. The SW–NE trace extends from an area above the salt scarp that was not significantly impacted by the removal of the underlying salt, northwards into areas of significant, but partial, salt removal below the western Bitumount Trough and into the Central Collapse, where most of the underlying salt section had been removed. Elevation in metres above sea-level. Location of trace on Figure 13b. Petrophysical analyses of the well logs consist of the percentages of rock and fluid types – sand (yellow), shale (grey), hydrocarbons (green) and water (blue) – calculated from the gamma ray (GR), resistivity (R), density (RHOB) and porosity (NP) log profiles. Modified from Broughton (2016).

The Roncott Platform in southern Saskatchewan consists of undisturbed salt beds (Figs 1 & 8). The 150 km long Hummingbird Trough developed along the eastern margin of the salt platform as a multi-stage collapse structure (DeMille *et al.* 1964; Smith & Pullen 1967; Holter 1969; Broughton 1977, 1985; McTavish & Vigrass 1987). The earliest known dissolution stage impacting this area of southern Saskatchewan was during the Late Devonian, resulting in the partial delineation of the western and eastern margins of the proto-Roncott Platform. Extensive salt removal phases during the Mississippian to Jurassic periods further defined the Roncott Platform, together with Cretaceous salt removal patterns across most of southern Saskatchewan (Fig. 8). Widespread salt removal during the Cretaceous occurred across most of southern Saskatchewan beyond the area of the Roncott Platform (Fig. 1). More subdued salt dissolution subsidence along the margins of the Roncott Platform occurred during the Late Cretaceous Maastrichtian and into the Early Paleogene, resulting in the accumulation of commercially important lignite coal basins of the Paleocene Ravenscrag Formation (Broughton 1977, 1985).

Salt removal patterns along the Hummingbird Trough (Figs 8 & 9) resulted in structural closures that trapped oil migration into the area (Wright *et al.* 1994). Complete removal of the salt beds during Cretaceous Laramide tectonism resulted in structural inversion of these earlier formed salt collapse structures. The Hummingbird and Lake Alma oil pools, which produce 40° API light oil, are examples of traps located within the southern trough that are related to salt collapse (Smith & Pullen 1967). The density of oil is an indication of the quality measured by the American Petroleum Institute with water as the standard (10° API). The higher the number above that of water, the better the quality of the oil.

The Hummingbird pool resulted from multiple salt removal collapses with 50 m of closure. The stacked traps occurred at intersections of basement-controlled salt removal trends, oriented to the NW and NE, at overlying levels of the Devonian Nisku and Mississippian Radcliffe formations (Fig. 9). Removal of the salt beds along these basement-controlled fault–fracture lineaments continued below each of the stacked reservoirs throughout the Paleozoic from the Late Devonian to the Mississippian, resulting in structural inversion traps. Many other oilfields across southeastern Saskatchewan, such as the Innes field (Twp 7–8, Rge 10 W2M), are located at the intersections of salt collapse induced conjugate NW and NE lineament sets. At the Innes field, for example, oil was stratigraphically trapped in Mississippian Frobisher carbonate beds. Initial salt dissolution collapse occurred during the Mississippian, resulting in an elongate trough that filled with carbonate beds. Subsequent pre-Triassic dissolution along an adjacent NW trend resulted in a 1.5 km wide structural low updip from the reservoir, thereby preventing further migration by the hydrocarbons (Crabtree 1982; Wright *et al.* 1994).

Area 3. Potash sub-basin, south-central Saskatchewan: impact by subglacial meltwaters

The upper interval beds of the Prairie Evaporite Formation accumulated sylvite and carnallite deposits at locations near the city of Saskatoon in south-central Saskatchewan. These commercially attractive potash accumulations are distinct from the halite–anhydrite dominated salt beds distributed elsewhere (Fuzesy 1982). This potash mining district is characterized by 100 m high brine-filled dissolution chimneys that cross-cut the Prairie Evaporite halite–anhydrite–gypsum and sylvite–carnallite beds. The roofs of many of these brine-filled dissolution chimneys have collapsed, resulting in the structural configuration of the overlying Devonian and Cretaceous strata as differentially subsided fault blocks. Seismic surveys, usually three-dimensional arrays, have been extensively used within the potash mining district to avoid the mine tunnels coming into contact with these brine-filled chimneys or collapsed structures leached of ore. Figure 19 illustrates an example of a depth-migrated three-dimensional

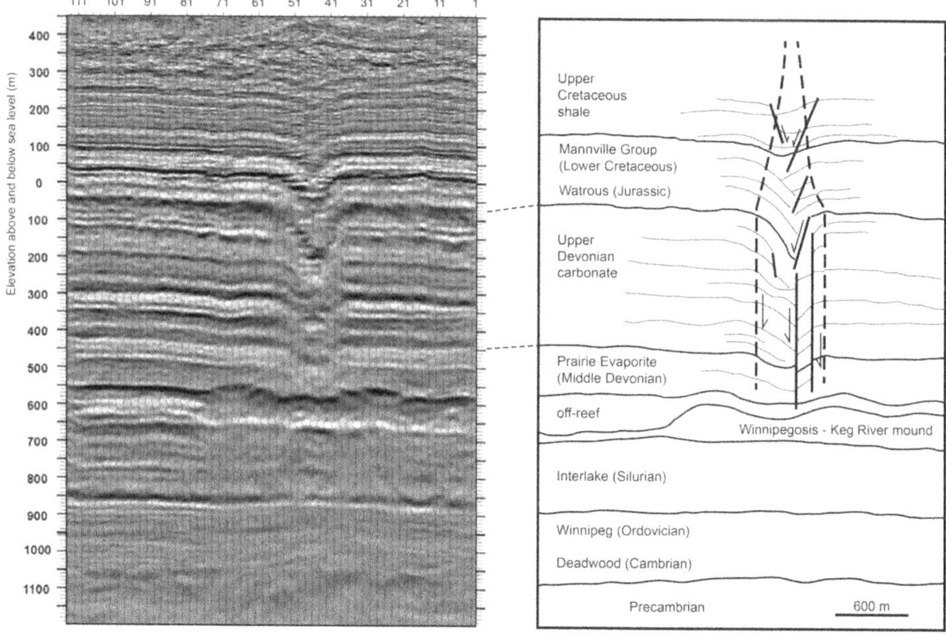

Fig. 19. Seismic profile at a site in the potash mining district, south-central Saskatchewan, showing a 100 m high salt dissolution chimney that collapsed and configured the overlying strata (Nemeth *et al.* 2002). Modified from Broughton (2017).

seismic profile image of a collapsed 100 m high chimney (Nemeth *et al.* 2002). The 500 m interval of overlying Upper Devonian and Cretaceous strata was cross-cut by faults that extended upwards from the level of salt removal, resulting in differentially subsided and rotated fault blocks similar to the fault block collapse patterns exposed by mining operations in the Athabasca Oil Sands (Fig. 17).

Strongly pressured subglacial meltwaters, driven by the hydraulic head resulting from a 1.5–2 km thick Laurentide ice sheet, invaded the regional aquifer system across southern Saskatchewan. Influxes of these freshwater flows coming into contact with the Prairie Evaporite salt beds stimulated further dissolution, resulting in large Pleistocene collapse structures located between the area of the Roncott Platform–Hummingbird Trough and the potash mining district to the north. Examples are the Saskatoon, Rosetown and Regina lows on the Quaternary topography (Christiansen 1967; Christiansen & Sauer 2002). These collapse structures were invariably multi-stage, consisting of older Paleozoic and Mesozoic structures rejuvenated during the Pleistocene, resulting in tens to hundreds of square kilometres of topographic lows on the subglacial topography and partially filled by Late Pleistocene and Holocene surficial deposits.

For example, the Saskatoon Low is a 40 km long and 25 km wide depression south of the city with 200 m of elevation relief on the Cretaceous structure (Fig. 20). The Saskatoon Low resulted from the removal of a 200 m thick salt bed interval (Christiansen 1967). The progressive removal impacted: (1) the Upper Cretaceous Lea Park, Judith River and Bearpaw formations of the Montana Group; (2) the Early and Middle Pleistocene Mennon, Dundurn and Warman formations of the Sutherland Group; and (3) the Late Pleistocene Floral, Battleford and Haultain formations of the Saskatoon Group. The initial collapse followed the deposition of the Ardkenneth Member of the Cretaceous Bearpaw Formation, but was prior to the middle Pleistocene glaciation (pre-Illinoian). This was followed by Pleistocene dissolution collapses during the Late Wisconsinan glaciation. These Pleistocene (Wisconsinan age) collapses resulted in a 16 km long and 8 km wide structure with 70 m of elevation relief superimposed on the pre-glacial collapse structure.

A similar structural collapse resulting from the removal of salt beds by influxes of subglacial meltwater is the Regina Low, located north of the Roncott Platform. This collapse reconfigured the structure of the Upper Cretaceous Bearpaw Formation and overlying Pleistocene deposits, resulting in 125–175 m of elevation relief on these structural surfaces (Christiansen & Sauer 2002), which is consistent with the estimated pre-dissolution thickness of the Prairie Evaporite salt beds. There were several

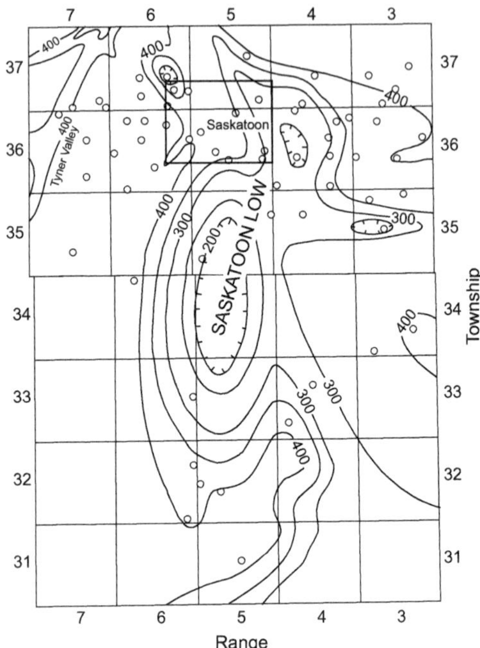

Fig. 20. Structure map of the Lea Park Formation, Upper Colorado Group, showing the Saskatoon Low, a regional collapse structure located in southern Saskatchewan. Modified from Christiansen (1967).

phases to the formation of this collapse structure. About 54 m of collapse occurred during the Early Pleistocene at the eastern end of the structure. Subsequent collapses of 55 and 85 m occurred to the north during the Middle Pleistocene (pre-Illinoian). The final collapse of 125 m was to the NE during the Late Pleistocene. Another site with significant salt collapse is the 360 km^2 Rosetown Low, with 110 m of structural closure, located to the east of the city of Saskatoon. It may have been a post-Bearpaw Formation (Late Cretaceous), pre-Pleistocene collapse (DeMille *et al.* 1964).

It is noteworthy that some crater lakes in southern Saskatchewan, such as Howe Lake, have been interpreted as blowout structures originating as salt dissolution collapses, but rejuvenated as blowouts resulting from the build-up of extreme pressure by invading subglacial meltwaters (Christiansen 1971; Christiansen *et al.* 1982).

Glacial loading and regional salt dissolution: geochemical perspective of aquifer flow reversals

Influxes of strongly pressured subglacial meltwaters into the subsurface coming into contact with shallowly buried Prairie Evaporite beds rejuvenated salt removal along the 1000 km long dissolution trend previously formed along the eastern margin of the Prairie Evaporite salt basin. This salt basin margin trend is largely coincidental with the eastern margins of the adjoining Alberta and Williston sub-basins of the WCSB (Grasby *et al.* 2000; Grasby & Chen 2005; Gue 2012; Cowie *et al.* 2014*a*, *b*, 2015). These strongly pressured subglacial meltwater flows into the subsurface were possible because of the hydraulic head consistent with a 1.5–2 km thick Laurentide ice sheet during the glacial maximum (*c.* 25–40 ka) and possibly, to some extent, impacted by proglacial meltwaters during retreat of the ice sheet (*c.* 8–12 ka). The subglacial meltwater flows into the Devonian strata encountered deep-seated regional basin flows up-structure to the NE. The hydraulic head behind these glacial meltwaters was sufficient to halt and, to some extent, reverse the regional water flows up-structure to the NE towards the eastern margins of both the Alberta and Williston basins (Figs 10 & 21).

Grasby *et al.* (2000) interpreted the strength of the hydraulic head of subglacial meltwaters as dependent on whether the flows were dominated by hydrostatic or lithostatic pressure. Free water under lithostatic load at the base of a 1.5 km thick ice sheet would have a pressure equivalent to a

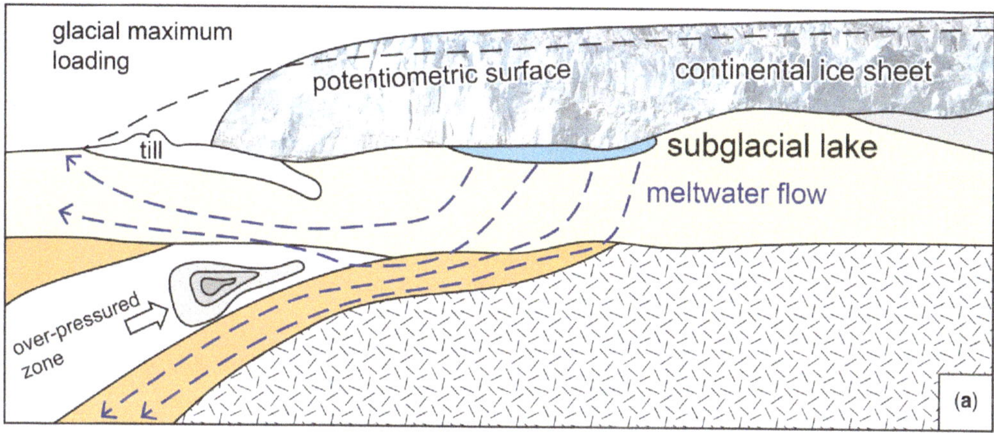

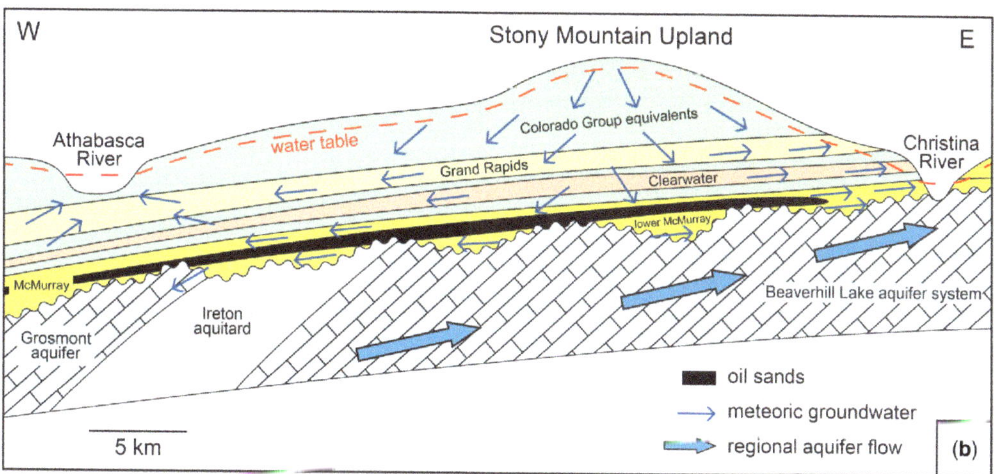

Fig. 21. Schematic illustration of the Quaternary water flows into Cretaceous and Devonian strata at the Athabasca Oil Sands deposit. (**a**) Strongly pressured subglacial meltwater flows were driven into the subsurface by the hydraulic head of a 1.5–2 km thick Laurentide ice sheet. Meltwater flows to the SW along Devonian strata were sufficiently pressured to reverse the direction of the regional aquifer water flows up-structure to the NE. The pre-glacial flow system was reasserted on retreat of the ice sheet. Modified from Person *et al.* (2007). (**b**) Holocene meteoric charged groundwater flows along river valleys into Cretaceous strata overlying the bitumen aquiclude were often counter to the regional aquifer flows in Devonian strata below the aquiclude. Modified from Barson *et al.* (2001).

1400 m thick water column. The thickness of the ice sheet may have been as much as 2 km across this area of Western Canada, but less than the 3 km high bulge over the proto-Hudson Bay (Dyke *et al.* 2002; Fisher *et al.* 2009; Margold *et al.* 2015). The thickness of such a relatively porous ice sheet would have resulted in a hydraulic head sufficient to reverse subsurface water flows to the NE towards the basin margin established prior to the onset of glaciation. Provided the modern regional aquifer water flows of the WCSB were similar to the pre-Quaternary system, there were two flow reversals. The first responded to glacial loading and the second to the reassertion of the earlier system following glacial retreat (Fig. 21a). At present, Devonian deep basin water flows up-structure to the NE are maintained and are often discharged as saline springs, in contrast with the counter flows of meteoric water along overlying permeable Cretaceous beds (Fig. 21b) (Barson *et al.* 2001; Wells & Price 2015).

The impact by these Pleistocene salt dissolution events was widespread across southern Saskatchewan, where many Paleozoic and Cretaceous collapse structures were rejuvenated by fresh glacial meltwaters (Area 3). As a dissolution process, the influx of glacial meltwaters may have laterally extended into the shallow subsurface for up to 100–300 km along the eastern margin of the salt basin. As late-stage collapse structures, however, they were only locally significant and were secondary to the more widespread salt removal trends that collapsed the overlying strata for tens to hundreds of kilometres during the Middle Jurassic to Early Cretaceous Columbian tectonism in the foreland Alberta basin and Paleozoic–Mesozoic structural configuration across the northern Williston basin.

Numerous studies of the water geochemistry associated with this 1000 km long dissolution trend (Fig. 1) have interpreted the impact of glacial meltwaters that mixed with Devonian formation waters. These studies measured the total dissolved solids (TDS) and stable isotope ratios. The stable light and heavy $\delta^{18}O$ (H_2O) isotope ratios indicate the extent to which fresh glacial meltwaters mixed with the Devonian saline formation water flows up-structure to the NE.

Williston basin

Recharge areas for the Williston basin, such as the Montana Uplift and the Black Hills, resulted in NE-directed aquifer water flows towards topographically lower areas in southern Saskatchewan and southwestern Manitoba, proximal to the erosional edge of mostly Devonian strata on the Canadian Shield (Grasby & Chen 2005). A persistent increase in dissolved solids is observed from the basin margin in southwestern Manitoba and southeastern Saskatchewan downdip to the SW towards the basin centre. The TDS measurements also have values that become progressively lower towards the modern day recharge zones in eastern Montana and western North Dakota (Grasby *et al.* 2000). A broad range of TDS values has been measured from 20 g l^{-1} to as much as 300 g l^{-1}, dominated by the Na–Cl associated with Devonian formation waters sampled from carbonate strata across the northern Williston basin. Shallow groundwater samples from water wells into the meteoric charged water table have <10 000 mg l^{-1} compared with saline seeps and springs discharged along the basin margin with 10 000–60 000 mg l^{-1} and formation waters of deeply buried Devonian strata to the SW with TDS >35 000 mg l^{-1} (Fig. 22). The highest TDS measurements are from water samples collected from Devonian strata in the north-central Williston basin overlying the Prairie Evaporite Formation salt beds (Fig. 22).

Salinity data indicate a chemical boundary between saline and fresh formation waters across the northern Williston basin areas of southern Saskatchewan and southwestern Manitoba. This partition resulted from the influx of glacial meltwaters that spread southwestward into bedrock for distances as great as 300 km long in the shallowly buried strata from the basin edge onlap with the Canadian Shield. This resulted in a geochemical and hydrological divide between Williston basin deep aquifer water flows up-structure to the NE and the southwestward flows downdip into the subsurface by strongly pressured glacial meltwaters (Betcher *et al.* 1995). The limits of these influxes by meltwaters into the subsurface may be approximated by the 2000 mg l^{-1} boundary (Betcher *et al.* 1995; Grasby & Betcher 2000; Grasby *et al.* 2000; Grasby & Chen 2005). Geochemical evidence suggests that this invasion by fresh glacial waters extended as far as areas currently characterized by high salinity, i.e. TDS values >100 000 mg l^{-1}. The position of this high >100 000 mg l^{-1} salinity boundary with freshwater in Devonian strata is located *c.* 300 km to the SW of the basin edge.

Grasby *et al.* (2000) and Grasby & Chen (2005) measured Na/Cl and Br/Cl ratios to interpret the mixture of multiple water sources combined as meteoric charged groundwater, glacial meltwater, connate formation water and deep-seated basin aquifer water flows (Fig. 22b). A progressive shift of Na/Cl molar and Br/Cl ratios from the deep basin brines altering the seawater chemistry occurs to the NE near the basin margin in waters that include dissolved halite. This suggests that the chemistry of the brines that discharged as saline seeps and springs at the surface along the eastern margin of the WCSB were only partially related to deep basin formation fluids and had mixed sources,

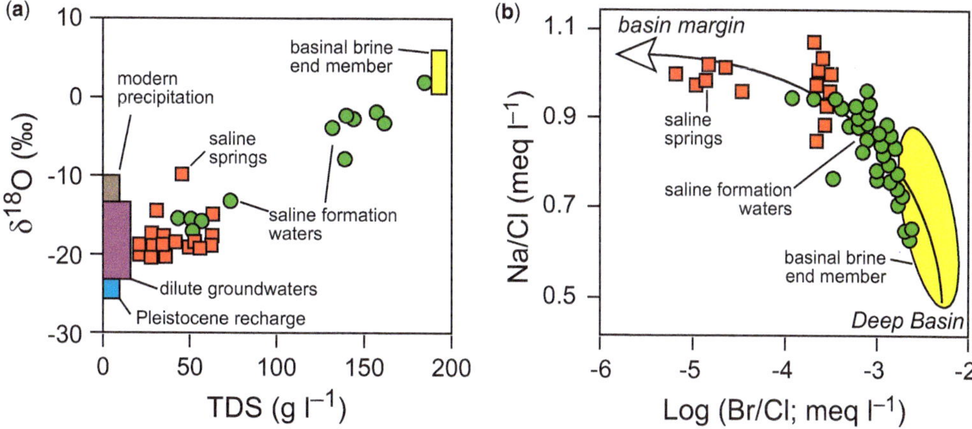

Fig. 22. Geochemistry of waters from mixed sources, such as deep basin brines, glacial meltwaters and saline surface springs of the Athabasca deposit area. (**a**) $\delta^{18}O$ vs. total dissolved solids (TDS). (**b**) Na^{+}/Cl^{+} vs. Br^{-}/Cl^{-}. A progressive shift in ion ratios occurs from deep basin waters to brines discharged along the basin margin. Modified from Grasby *et al.* (2000) and Grasby & Chen (2005).

including glacial meltwaters that came into contact with shallowly buried Prairie Evaporite salt beds. The geochemistry indicates that glacial meltwaters invaded Devonian carbonate aquifers, diluting the salinity, but also altering the composition of the TDS. This resulted in significantly diluted brines with higher Na/Cl ratios, but lower Br/Cl ratios, indicative of dissolved solids sourced from the Prairie Evaporite Formation salt beds. Halite deposits have a different Br ratio from that of residual connate seawater because Br is excluded from the crystal structure during halite precipitation as the seawater evaporates. The Br/Cl ratio in halite is generally >3000, but can be an order of magnitude higher, depending on the re-dissolution/re-precipitation cycles. The deep basin brines within the Devonian strata have Na/Cl molar ratios <0.8, in contrast with surface spring brines with Na/Cl molar ratios close to 1.0, and are indicative of halite dissolution. Measurements of numerous saline spring water samples show increased salinity, but low Br/Cl ratios (Fig. 22b). The Br/Cl ratios in spring waters are generally consistent with a halite dissolution source of Na and Cl ions mixed with lower TDS formation waters. These values contrast with very low Br/Cl ratios, <200, in meteoric waters (Gue *et al.* 2015).

Stable isotope values for spring waters show dissolved halite, indicative of freshwater contact with the underlying Prairie Evaporite Formation salt beds, as measured by $\delta^{18}O$ ratios in formation waters that migrated up-structure to the NE towards the basin margin in southeastern Saskatchewan and southwestern Manitoba. Water samples from sites up-structure to the NE were altered and mixed with influxes of freshwater that have end-member values of about −22‰ (Grasby *et al.* 2000). The $\delta^{18}O$ in water samples from oilfields have a dilutive mixing of saline formation and freshwater, resulting in values from −3 to +6‰ in the deeper areas of the basin. By contrast, brine springs discharged along the northeastern Williston basin margin in Manitoba, as well as in the Athabasca deposit area near Fort McMurray, have low $\delta^{18}O$ values, such as −20‰ in Manitoba and −25‰ in Alberta. The $\delta^{18}O$ measurements in Devonian strata formation waters indicate that they mostly represent residual evaporated seawater.

Alberta basin

The measurements of Gue *et al.* (2015) of spring waters have $\delta^{18}O$ (H_2O) values indicating that the proportion of glacial meltwater ranges from 39% to as much as 75%. Stable isotope ratio measurements of $\delta^{18}O$ (H_2O) in saline spring and surface seeps along the Athabasca and other river valleys dissecting the terrain of the Athabasca deposit indicate sourcing from glacial meltwaters mixed with Devonian aquifer waters, together with dissolved solids sourced from contact with Prairie Evaporite salt beds (Grasby & Chen 2005; Gue 2012). The isotope ratios and TDS measurements for these waters, consisting of mixed glacial, Devonian aquifer and dissolved Prairie Evaporite beds, are similar to the water samples collected from Devonian strata below the bitumen mine floors in areas overlying the salt scarp.

Subglacial meltwater flows into the subsurface, down-structure to the SW along permeable

Cretaceous and Devonian strata, encountered and mixed with the regional flow to the NE in topographically low areas, such as below the proto-Athabasca river valley. The chemistries of the formation waters sampled from strata overlying the salt scarp indicate mixing with glacial meltwaters. The TDS in McMurray Formation pore waters vary widely from near-freshwater (240 mg l^{-1}) to brines with measurements up to 280 g l^{-1} (Cowie 2013; Cowie *et al.* 2014*a*, *b*, 2015). Increased salinity values in water samples collected from beds of the McMurray Formation resulted in a roughly linear trend extending from Twp. 78, Rge. 4 to the NW into Twp. 100, Rge. 10, W4M (Fig. 23). This trend of elevated salinity values (>25 000 mg l^{-1}) in beds of the McMurray Formation follows the underlying Prairie Evaporite salt scarp only 200 m below (Cowie 2013; Cowie *et al.* 2015).

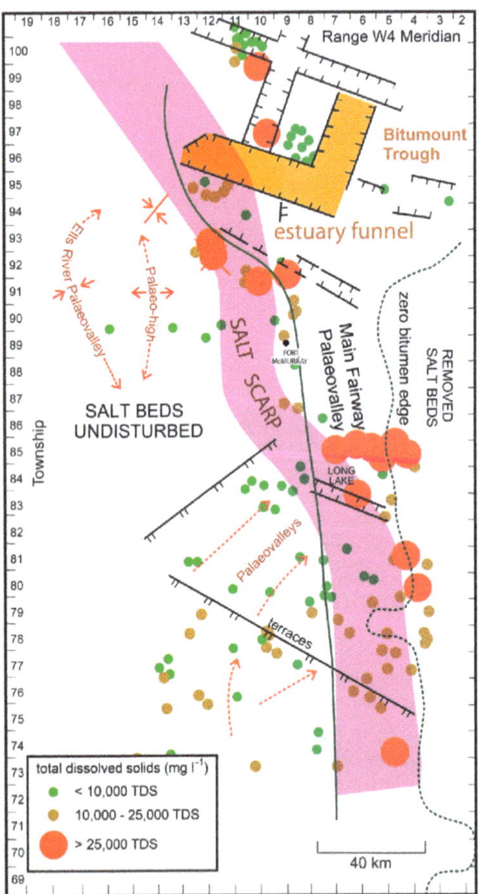

Fig. 23. Distribution of total dissolved solids (TDS) in formation waters sampled from McMurray Formation strata. A pattern of increased TDS occurs at sites overlying the Prairie Evaporite salt scarp. Modified from Cowie (2013) and Cowie *et al.* (2015).

Karstic conduits cross-cutting Devonian and Cretaceous strata were more developed and effective in the areas overlying the salt scarp and eastward into salt removal areas. They provided the connectivity between the Devonian and overlying McMurray Formation hydraulic systems. Broughton (2013) and Cowie (2013) interpreted karstic collapse features, such as breccia pipe–sinkhole complexes and dissolution-related faults within the Middle–Upper Devonian and Lower Cretaceous strata, as vertical migration pathways for saline water flows from Devonian beds below the Prairie Evaporite into beds within and above the salt beds and upwards into the overlying McMurray Formation beds. Some of these flows discharged as surface saline seeps and springs (Gibson *et al.* 2013). Vertically oriented water migration pathways also formed along the margins of Keg River Formation reefs, where onlapped lower Prairie Evaporite dolostone beds were de-dolomitized. The resulting diagenetic permeability pathways enabled water to flow upwards into the overlying salt beds and along fault–fracture trends into the overlying Upper Devonian and McMurray beds during Columbian tectonism. Some of these migration pathways may have been reactivated during glacial loading (Grasby & Chen 2005).

Areas overlying the salt scarp contrast with areas to the west, where the evaporite beds were not significantly impacted by regional dissolution trends. Nevertheless, extensive areas of the Middle Devonian salt beds underlying the broad expanse of the Athabasca Oil Sands deposit sourced dissolved solids into the overlying McMurray Formation beds to an uncertain extent, in addition to the more pervasive residual marine-sourced connate pore waters. Measurements of formation pore water samples collected west of the underlying salt scarp have increased salinity values, consistent with the widespread increased salinity of Keg River Formation waters (Fig. 24), but the impact of water flows up-section into the overlying Prairie Evaporite Formation salt beds is uncertain. The Devonian palaeotopography westwards of the underlying salt scarp is characterized by valleys tens to hundreds of kilometres long trending to the NE and jointing with the main palaeovalley. These secondary palaeovalley tributary trends on the pre-Cretaceous unconformity surface were subsequently filled by unusually thick McMurray Formation beds, resulting in favourable sites for steam-assisted gravity drainage (SAGD) bitumen recovery operations to the west of the main palaeovalley bituminous sand fairway. At these sites, steam is injected into the bituminous sands by horizontally emplaced drill pipes, allowing heated bitumen to flow to the surface.

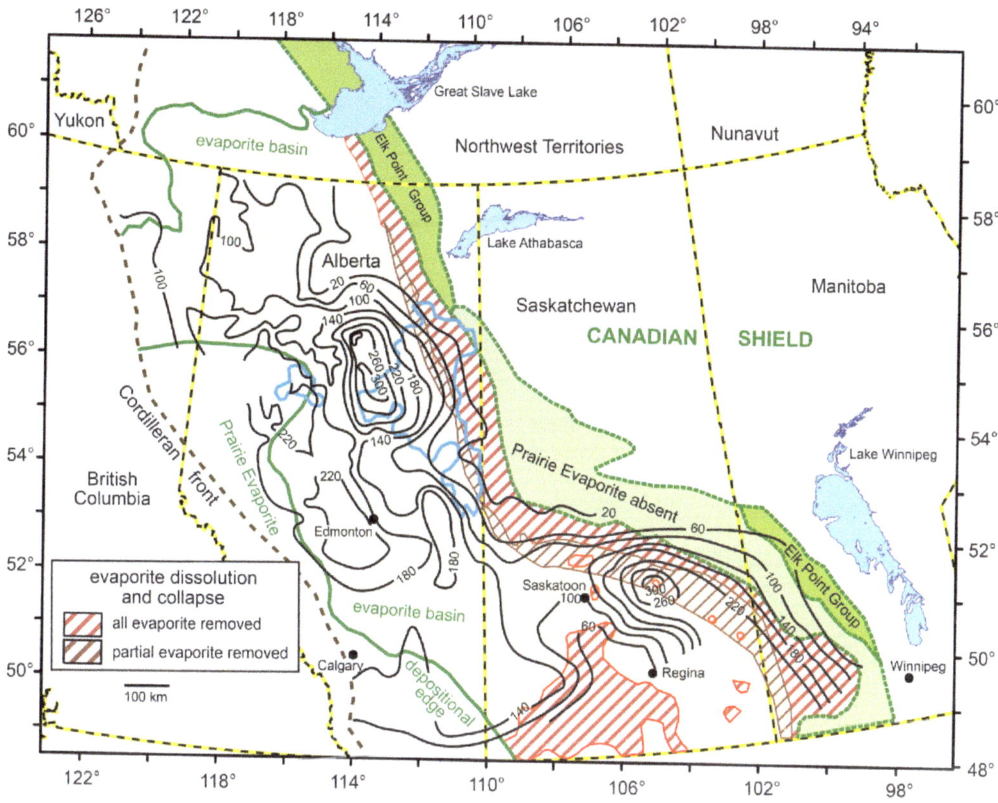

Fig. 24. Distribution of total dissolved solids (TDS) in the Devonian formation waters sampled across Western Canada. Overlay on map of the Prairie Evaporite salt basin with dissolution trends. Contour interval is 40 g l^{-1} TDS. Modified from Grasby & Chen (2005).

The structural configuration, but limited geochemical evidence, suggests that these palaeovalley trends underlying the Athabasca deposits west of the salt scarp resulted from early stage, but regionally important, salt dissolution trends.

Impact of evaporite karst collapse on the overlying Athabasca Oil Sands

The Athabasca Oil Sands is the largest and commercially the most important of the three Cretaceous bituminous sand deposits in northern Alberta (Fig. 1). The 45 000 km^2 deposit accumulated as a 200 billion cubic metre or a trillion barrel hydrocarbon resource (Head *et al.* 2003; Higley *et al.* 2009; Berbesi *et al.* 2012; Adams *et al.* 2013). Late Cretaceous–Paleogene sediments shed from the Cordillera during Laramide tectonism contributed to the deepening burial of organic-rich Upper Devonian–Lower Mississippian Exshaw Formation hydrocarbon source rocks accumulated in the foreland deep basin area of western Alberta. Hydrocarbons migrated eastwards up-structure towards and into unconsolidated sand deposits of the McMurray Formation (Aptian) and, if present, the overlying, but thinner, Wabiskaw Member basal sand beds of the Clearwater Formation (Albian) (Fig. 5). Marine mudstones of the Clearwater Formation, hundreds of metres thick, covered the McMurray Formation strata hosting these sand reservoirs, sealing the traps. Shallow burial during the Paleogene allowed biodegradation by meteoric charged groundwater, resulting in highly viscous bitumen (6.5–9° API) cementing the unconsolidated sand deposits of the McMurray Formation.

Pre-Cretaceous

The overarching Devonian salt karstification resulted in a pre-Cretaceous Devonian limestone palaeotopographic terrain characterized by a reticulate mosaic of differentially subsided orthogonal fault blocks varying in size from tens to thousands of metres long. Smaller structures included breccia pipes and sinkholes up to 100 m deep and solution-enlarged near-surface joints (Broughton 2013). The removal of tens of metres thick halite–anhydrite beds during the Middle Jurassic to Early Cretaceous Columbian orogeny collapsed the 200 m interval of overlying Middle–Upper Devonian strata prior to, but also concurrent with Aptian deposition of McMurray Formation strata. The salt removal patterns resulted in several kilometre long troughs on the pre-Aptian palaeotopography, which followed the structural grain resulting from older and concurrent movements by Precambrian blocks only 200 m below the salt beds. Dissolution trends within the highly soluble halite beds rapidly advanced along cross-cutting sets of fracture–fault lineaments, oriented to the NW and NE, which were propagated upwards from movements between basement blocks. Dissolution fronts advancing along these cross-cutting lineaments within the salt beds. They subsequently coalesced as linear trends that resulted in troughs on the palaeotopography. This subsurface evaporite karstification process eventually resulted in widening and coalescing dissolution trends into larger areas of complete salt removal.

McMurray Formation

The distribution of the McMurray Formation fluvial and fluvial–estuarine sand bar complexes, accumulated as the Athabasca deposit, responded to a tectono-stratigraphic architecture controlled to a significant extent by the removal of salt beds in the underlying Prairie Evaporite strata only 200 m below. Post-orogenic salt dissolution of the evaporite strata continued concurrent with the deposition of the McMurray Formation, which accumulated on the pre-Cretaceous karstic limestone palaeotopography. An unusually low 1:2 thickness ratio of removed salt to overlying strata at the hundreds of metre vertical scale extended as trends tens of kilometres long. This architecture resulted in syndepositional trends in the McMurray Formation up to 20 km long (Fig. 25).

Lower McMurray Formation. Prior to and during the deposition of the lower McMurray braided river channel deposits, the underlying salt dissolution patterns were propagated upwards and resulted in a reticulate Devonian unconformity surface structure characterized by differentially subsided fault blocks. Braided river channels and overlying anastomosing river floodplain deposits of the lower McMurray Formation filled the lows of this valley and ridge terrain. The lower McMurray Formation braided river channel network was stabilized as a lattice-like pattern draped onto this reticulate mosaic of kilometre-scale differentially subsided limestone fault blocks. Deposits tens of metres thick accumulated as valley fills in such areas as the Bitumount Trough and Central Collapse, but only sparsely elsewhere. This palaeogeography was favourable for the aggradation of sand deposits that constrained lateral accretion beyond the Devonian bedrock valley lows. As a result, the underlying salt removal patterns exerted a significant control on the distribution and geometry of the 2–3 km long sand reservoirs, consisting of gravelly sand channel fills oriented to the NW and NE (Fig. 26). These lower McMurray Formation deposits and protruding Devonian highs were subsequently covered over by fluvial–estuarine deposits of the middle McMurray Formation.

The northern margin of the Bitumount Trough collapsed during the deposition of the lower McMurray Formation, responding to dissolution along a 20 km long lineament that dissected the underlying salt scarp. This dissolution trend resulted in the collapse of a linear chain of Middle–Upper Devonian fault blocks, forming the Deep Trough structure along the northern margin of an extensive upper Keg River Formation reef complex underlying the western Bitumount Trough. As a result, an economically significant 20 km long syndepositional sand trend accumulated along the northern margin of the Bitumount Trough as a 40–70 m thick sand deposit (Broughton 2015, 2016). This overly thick lower McMurray sand complex trend was emplaced in an elevated position above the salt scarp, which enabled the accumulation of oil flow into the area, in contrast with structurally lower water-saturated sands to the east of the salt scarp (Fig. 26). The Muskeg River, Aurora North and Horizon mines are located along this tens of metres thick lower McMurray Formation bituminous sand trend.

Middle–upper McMurray Formation. Regionally significant salt dissolution fronts advanced along cross-cutting lineaments north of the Bitumount Trough during the deposition of the overlying middle–upper McMurray strata. This widespread removal of salt beds resulted in the Central Collapse and a shift more to the NW of the salt scarp orientation (Broughton 2015). Structural configuration of the salt removal area across the northern Athabasca deposit resulted in a semi- to relatively unstable depositional surface along which fluvial–estuarine fairways transported sand. These multiple channel belts paralleled the underlying dominant NW- and subordinate NE-oriented structural grain imposed by the salt removal patterns. Although the Bitumount Trough was mostly a lower McMurray Formation structural event, the adjoining Central Collapse to the north developed as an early middle interval structure. Collectively, they formed a funnel-like structural area on the sub-Cretaceous–lower

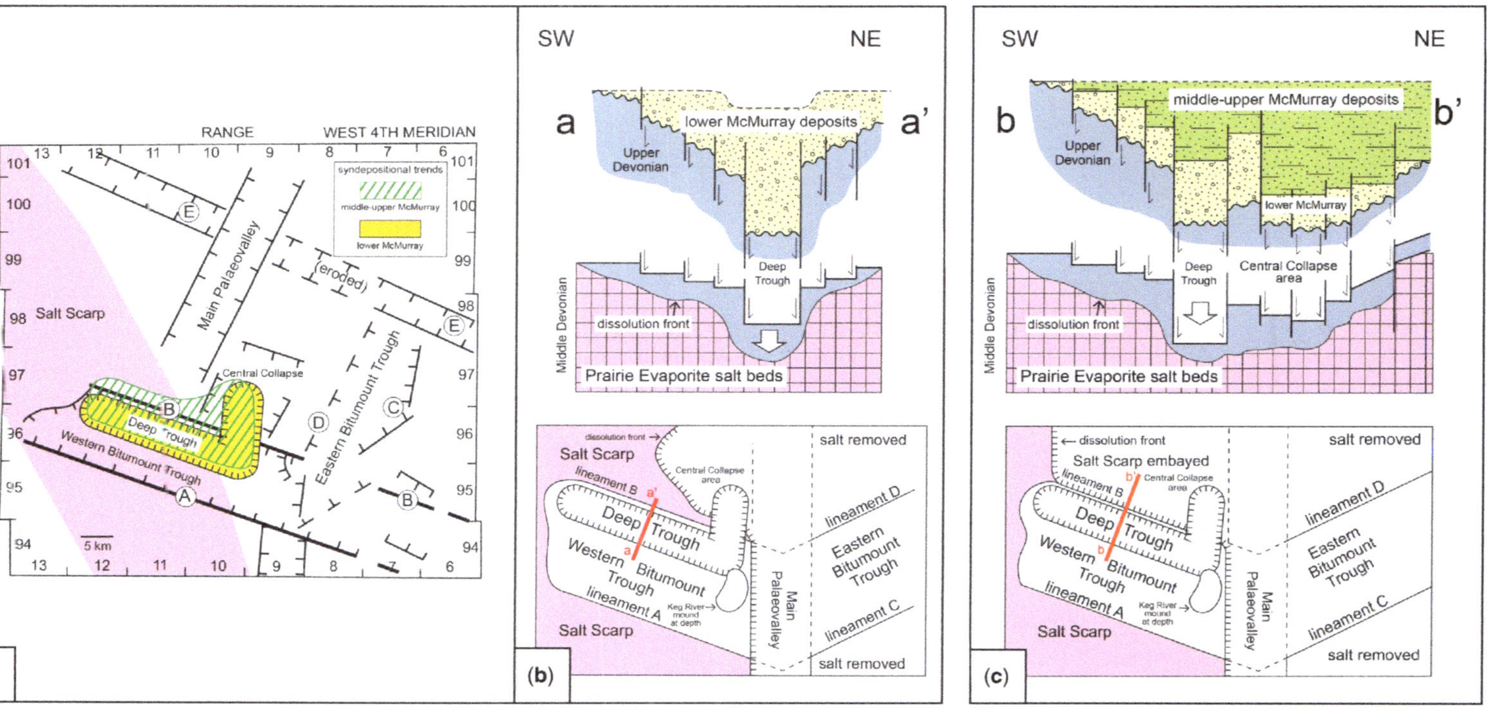

Fig. 25. Syndepositional trends in the McMurray Formation, northern Athabasca Oil Sands deposit. (**a**) Distribution of two 20 km long parallel trends developed along the northern margin of the western Bitumount Trough. These are a trend in the lower interval (yellow area) aligned above a salt dissolution lineament underlying the northern margin of the western Bitumount Trough. The dissolution front subsequently migrated northwards resulting in an overlying parallel, but partially offset to the north, trend in the middle–upper McMurray interval (green hatched area) along the southern margin of the Central Collapse. (**b**) Illustration of the 20 km long lower interval sand trend accumulated above a collapsed linear chain of Devonian fault blocks along the northern margin of the western Bitumount Trough. (**c**) Northwards migration of dissolution fronts during deposition of the middle–upper intervals that resulted in the Central Collapse and concurrent aggradation of the second 20 km long sand complex. The Muskeg River, Aurora North and Horizon mines are located along these two trends.

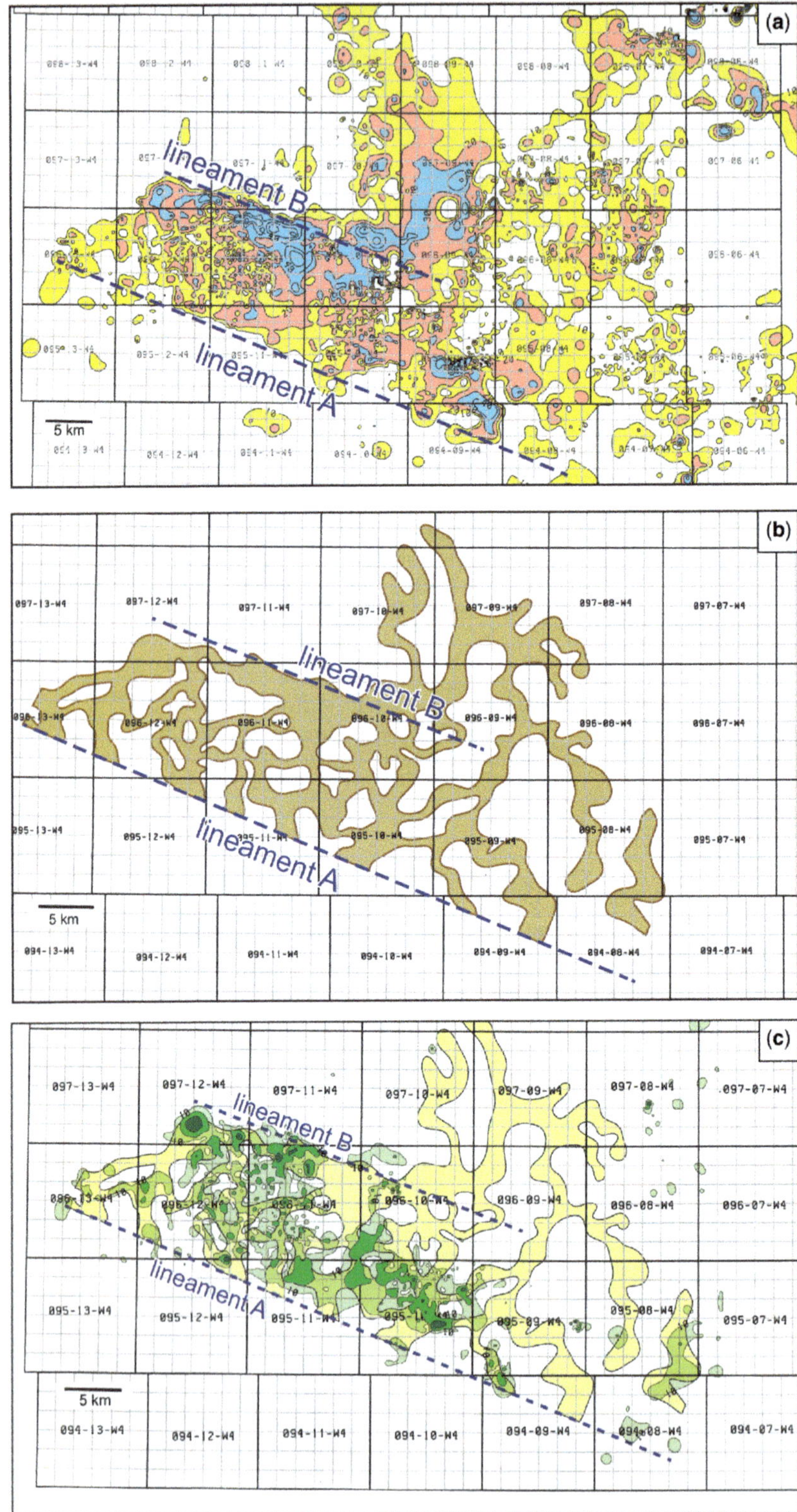
(a)
lineament B
lineament A
5 km
(b)
097-13-W4
097-12-W4
097-11-W4
097-09-W4
097-08-W4
097-07-W4
lineament B
096-12-W4
096-11-W4
096-09-W4
096-08-W4
096-07-W4
095-13-W4
095-12-W4
095-10-W4
095-09-W4
095-07-W4
lineament A
5 km
094-13-W4
094-12-W4
094-11-W4
094-10-W4
094-09-W4
094-08-W4
094-07-W4
(c)
097-13-W4
097-12-W4
097-11-W4
097-09-W4
097-08-W4
097-07-W4
lineament B
096-13-W4
096-12-W4
096-10-W4
096-09-W4
096-08-W4
096-07-W4
lineament A
095-13-W4
095-12-W4
095-11-W4
095-09-W4
095-08-W4
095-07-W4
5 km
094-13-W4
094-12-W4
094-11-W4
094-10-W4
094-09-W4
094-08-W4
094-07-W4

McMurray depositional surface of the northern Athabasca deposit that floored the fluvial–estuarine middle McMurray channels flowing northwards towards the Boreal Sea. Incremental advances northwards by the dissolution trends underlying this northern Bitumount Trough–Central Collapse funnel-like structural complex resulted in bench-like terraces, consisting of NE-oriented alignments of fault blocks that stepped down northwards into the Central Collapse. Some of these terraced step-downs paralleled multiple kilometre long SW–NE-trending ridges on the sub-Cretaceous unconformity surface resulting from the removal of the thick salt beds accumulated along the margins of linearly elongated upper Keg River reef trends (Fig. 13).

The thickness of the middle McMurray Formation sand deposits distributed across the northern Athabasca deposit varied from <10 m to as much as 80 m across the structural funnel, which is geographically comparable with a lower estuary. Giant sand complexes accumulated as stacked fluvial–tidal point bars along multiple modestly curvilinear tens of kilometre long parallel channel fairways (Broughton 2016). These fluvial–tidal point bar fairways contrast with the highly sinuous meandering channel belts of the upper estuary area associated with the southern Athabasca deposit. Figure 27 illustrates the distribution and geometric configuration of these tens of square kilometre lower estuary sand deposits in the northern Athabasca deposit as mostly large detached sand complexes. This map was constructed with a high cut-off grade map parameter, such as >10% bitumen weight. Other maps constructed with more inclusive bitumen quality map criteria, such as >6% bitumen weight, result in a continuous bitumen sand platform across almost all of the northern Athabasca deposit. These detached sand bar complexes tens of square kilometres in size each have a well-defined elongated linear geometric shape comparable with the giant fluvial point bars of the Mississippi River. They are typically 2–5 km long and 20–40 m thick. They are distributed as collinearly aligned deposits along each of the parallel fluvial–estuarine fairways. This distribution pattern resulted from the main fairway channel belt to the south of the funnel branching northwards as it dispersed onto an unstable floor. Multiple 10–30 km long sub-parallel fluvial–tidal channel fairways aligned parallel to the NW-oriented structural trends imposed by the underlying salt dissolution patterns only 200 m below.

The Central Collapse area of the northern Athabasca deposit accumulated collinear alignments of 1–5 km long detached sand complexes, each consisting of stacked point bars accumulated over terraced step-down aggradation sites distributed along the fluvial–tidal fairways. These step-down sites accumulated multiple kilometre long sand deposits extending over tens of square kilometres, many of which are commercially attractive bituminous sand deposits. For example, the tens of kilometre long Firebag mega-channel fairway from beyond the SE corner of the map area (Figs 27 & 28) transported sand across the eastern Bitumount Trough and northwards to the site of the step-down terrace that floors the Jackpine Mine deposit. The fairway continues to the NW towards the next significant step-down at the site of the Aurora North Mine site. In addition, an upper McMurray Formation syndepositional sand trend, 20 km long, developed parallel to a similar trend in the lower interval (Figs 25 & 26). This linkage between the distribution of large bitumen-saturated sand deposits and the underlying salt removal patterns is characteristic of the northern Athabasca Oil Sands deposit.

Point bar aggradations on unstable karst: reservoir sand deposits at the Jackpine and Muskeg River–Aurora North mines

The salt dissolution collapse that impacted the regional tectono-stratigraphic architecture of the middle McMurray Formation across the lower estuary-like funnel underpinning the northern Athabasca deposit differs from that of the point bar deposits accumulated along the upper estuary channel belts of the southern Athabasca deposit. This is consistent with the increasing instability northwards of the estuary floor as a result of the underlying salt dissolution

Fig. 26. Thickness maps of the lower McMurray sand beds distributed in the Bitumount Trough and Central Collapse areas. (**a**) The braided river channel network passes upwards as an anastomosing system. The channel organization drapes the reticulate Upper Devonian karstic palaeotopography, a mosaic consisting of differentially subsided orthogonal limestone fault blocks, each at the scale of several kilometres. A 20 km long syndepositional sand trend up to 60 m thick accumulated along the northern margin of the western Bitumount Trough. Map criteria standard map cut-offs, such as porosity >25% and shale volume <10% with 10 m contour intervals. (**b**) Sand trends distributed as a lattice-like anastomosing channel network with segments oriented to the NW and NE. The channel pattern resulted from the drape over the reticulated mosaic of differentially subsided Devonian limestone fault blocks flooring the 50 km long trough. Bitumen-saturated sand complexes are distributed within the western trough emplaced above the salt scarp, in contrast with the water-saturated sand beds to the east overlying the salt removal areas. (**c**) Thickness of bitumen sand deposits (BWB>10%) latterly constrained by the lattice-like distribution of the channels and sand fills (Vsh<10%).

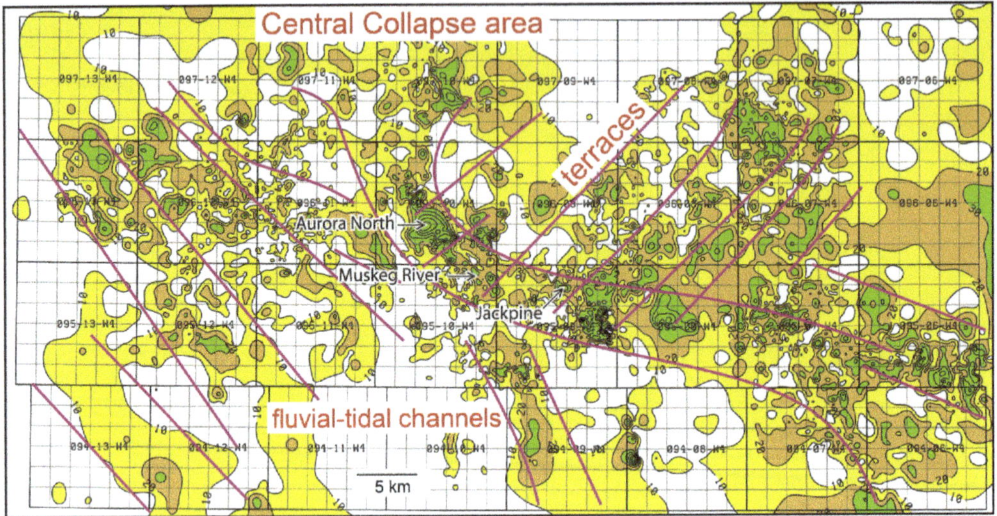

Fig. 27. Distribution of 3–5 km long stacked point bar complexes, >60 m thick, of middle McMurray interval across the northern Athabasca deposit. Aggradation sites of the point bar sand complexes, subsequently bitumen-saturated, were extensively controlled by the underlying salt removal trends as the fluvial–tidal fairways transported sand northwards along the unstable structural funnel floor towards the Boreal Sea. The aggradation sites were where the sand transporting the mega-channels cascaded over structural step-downs, consisting of aligned differentially subsided fault blocks, into the Central Collapse. Large commercially attractive bitumen sand deposits accumulated at these step-down sites, such as the Jackpine, Muskeg River and Aurora North mines. The map area, located on Figure 6b, illustrates the distribution of bitumen with >10% bulk weight of the rock mass, which is the average ore grade processed at most bitumen mines. Contour interval, 10 m. Modified from Broughton (2016).

trends, which impacted the fluvial fairways transporting sand and mud northwards towards the Boreal Sea. The sand transport fairway along the main palaeovalley trunk branched outwards across the widened structural funnel into numerous sub-parallel aligned curvilinear fairways, following the underlying salt removal trends. There was as much as 30–40 m of local relief in the underlying sub-Cretaceous and lower McMurray Formation structures, allowing stabilized aggradation along these step-down terraces with banking against the stepped high back walls.

The classic reference area for large, several kilometre long point bar deposits accumulated above a stable substrate is the thermal SAGD bitumen extraction operation at Long Lake (Smith *et al.* 2009; Hubbard *et al.* 2011). The fluvial channel belt at this upper estuary location developed over a relatively stable Devonian limestone floor that allowed highly sinuous channels to laterally migrate, forming multiple kilometre long meander loops. This floor stability allowed multiple kilometre long point bars to accumulate in a palaeogeography comparable with the highly sinuous lower Mississippi River channel belt as it approaches the Gulf of Mexico (Holbrook 2014). Figure 29 illustrates an example of this complex overlapping channel organization by the middle McMurray channel belt and shows a three-dimensional seismic image across a 12 km wide area near the Long Lake SAGD operation. The several kilometre long and tens of metres thick point bar sand complexes within this fluvial-dominated upper estuary are distributed along variously overlapping and complexly cross-cutting sinuous fluvial channels. Lateral accretion deposits are characteristic. The channel belt was geographically constrained by the 10–12 km width of the floodplain along the main palaeovalley fairway, which followed the structurally quiescent eastern margin of the underlying salt scarp.

Two examples follow to illustrate the impact of subsidence induced by episodic salt dissolution on floor instability, resulting in episodic aggradation architecture of the tens of metres thick point bar sand complexes characteristic of the northern Athabasca deposit. These accommodation spaces provided for unusually thick and commercially attractive giant bituminous sand deposits.

Example 1: Jackpine Mine

The Jackpine Mine is the site of one of the larger bitumen sand deposits of the Athabasca Oil Sands. The 25 km^2 deposit is 8–10 km long, 3–4 km wide and 10–40 m thick (Figs 13, 30 & 31). The deposit consists of several multiple kilometre long stacked

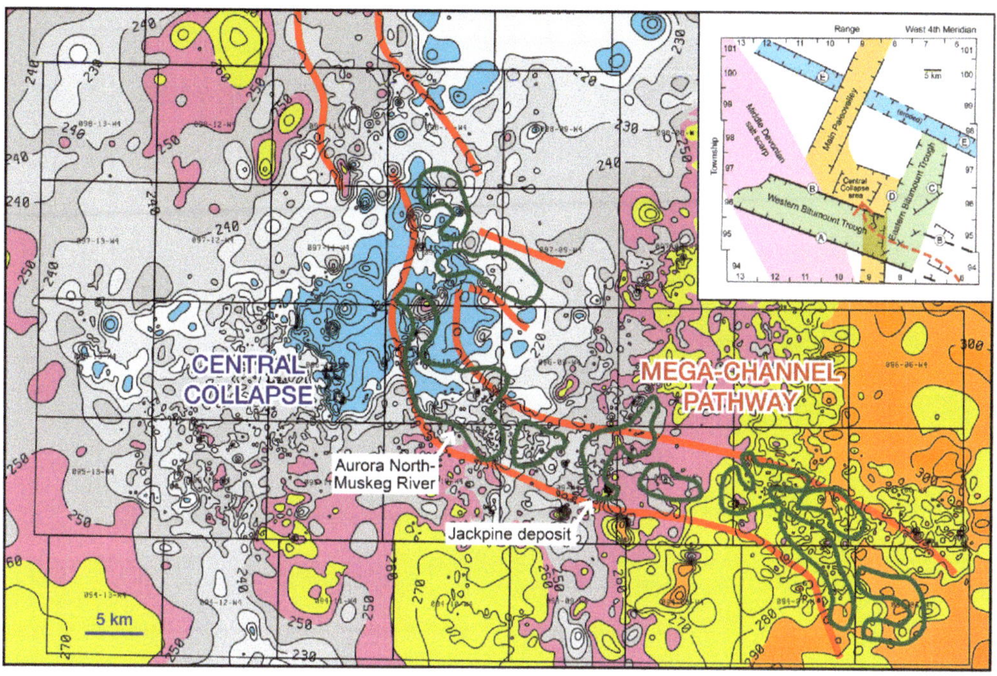

Fig. 28. The Firebag fairway, one of the largest mega-channels, transported middle McMurray sediment, mostly sand, across the Bitumount Trough and over terraced step downs into the Central Collapse towards the Boreal Sea. Stacked fluvial–tidal point bars accumulated at accommodation sites above structural step-downs, consisting of differentially subsided chains of aligned Devonian and lower McMurray fault blocks. Aggradations of collinearly aligned sand complexes, several kilometres long and tens of metres thick, were distributed along the fairways at these terraced step-down sites. Oil migration into the area accumulated in these sand stratigraphic traps, which were tens of square kilometres in area, and subsequently degraded into bituminous sand complexes. The structure of the sub-Cretaceous unconformity surface is shown with 10 m contour intervals with elevations above sea-level. Modified from Broughton (2015).

fluvial–tidal point bars of the middle McMurray Formation interval accumulated over a step-down terrace into the Central Collapse. The modified T-shaped deposit consists of multiple stacked sand bars oriented SW–NE, which accumulated above the site of terraced step-downs, and the SE orientation of a perpendicularly attached narrower and thinner sand bar, positioned over the upper back-step (Fig. 30a). Structural cross-section D–D′ illustrates this deposit (Figs 30d & 31a). The deposit is one of collinearly aligned sand aggradation sites along the Firebag mega-channel fairway that transported sand and mud northwards towards the Boreal Sea (Fig. 28). The stacked point bars, elongated SW–NE, accumulated parallel to the underlying orientation of the step-down terrace, which consisted of a NE-oriented chain of Devonian–lower McMurray fault blocks parallel to the underlying margin of a several kilometre long upper Keg River Formation reef-like biohermal mound trend (Figs 30b & 31a).

At this site, early movements by differentially subsided Devonian fault blocks occurred during the deposition of lower McMurray beds. Similar collapses related to salt removal were widely prevalent elsewhere during the deposition of the lower McMurray interval and are characteristic of the northern Athabasca deposit (Broughton 2013, 2015). At the Jackpine Mine site, the collapse of the Devonian strata resulted in accommodation space for the accumulation of up to 75 m of lower McMurray Formation beds (Fig. 31b). Salt removal continued during the deposition of middle McMurray interval strata, resulting in the collapse-rotation of Devonian–lower McMurray fault blocks. The resulting structural step-down retained sufficient accommodation space to constrain lateral accretion to the NW by the point bars. The fault blocks of these terraced step-downs continued to tilt with the additional removal of the underlying salt beds, resulting in relatively steep, near-vertical, apparent bed dips (Broughton 2016). Tide-impacted sand movements partially reversed fluvial transport to the NW, resulting in the sand accumulation banking against the high back wall of the step-down, further

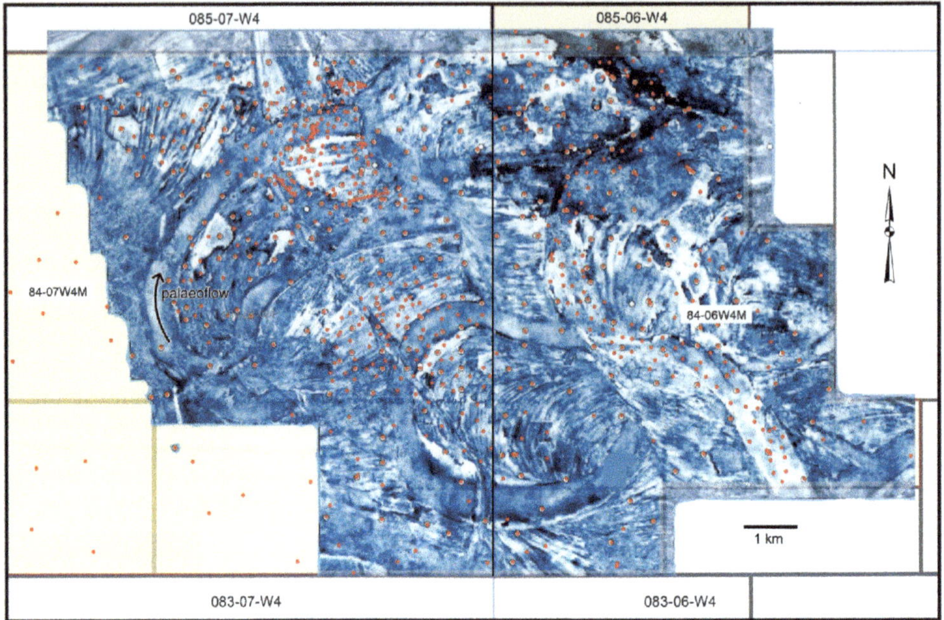

Fig. 29. Representative fluvial channel belt architecture for middle McMurray palaeogeography of the upper estuary of the southern Athabasca deposit. Seismic image of a 12 km wide segment of the channel belt near the Nexen SAGD project at Long Lake (Twp. 84, Rge. 6-7W4M), located on Figure 6. The fluvial channel belt architecture is characterized by overlaps of meander loops several kilometres wide of highly sinuous rivers developed along this section of the main palaeovalley fairway. Seismic time slice images of meander loops 5–7 km in diameter of 400–500 m wide channels with 3–5 km long point bars. Lateral accretion surfaces of fluvial point bars result in inclined heterolithic strata, such as scrolling patterns along the curved bend portions of channel meander loops. The image is a horizontal amplitude time slice on the three-dimensional pre-stack *c.* 12 m below the datum at the top of the middle McMurray point bar complex. Modified from Smith *et al.* (2009) and Hubbard *et al.* (2011).

enhancing vertical aggradation. Protrusions by Upper Devonian fault blocks were eventually covered, culminating in the T-shaped geometry of the deposit (Figs 30 & 31). The tens of metres thick sand bars were stabilized at these bench-like aggradation sites and fines were transported further to the NW. As a result, mud- and silt-dominated heterolithic deposits accumulated between the larger sand aggradation sites along the fairway, such as this deposit at the Jackpine Mine and the next terraced step-down sand aggradation site, such as the location of the Muskeg River–Aurora North mines.

Example 2: Muskeg River–Aurora North mines

An extensive bituminous sand complex accumulated along the northern margin of the Bitumount Trough (Muskeg River Mine site) above structural step-downs induced by salt dissolution collapse into the southern margin of the Central Collapse (Aurora North Mine site). A NW-oriented 20 km long point bar sand complex accumulated in northern Twp. 95, Rge. 10 and southern Twp. 96, Rge. 10, West 4th Meridian. The Muskeg River Mine is at the SE end and the Aurora North Mine is at the NW end of this extensive sand deposit (Figs 13 & 28). The Muskeg River Mine site overlies the elevated structure at the level of the sub-Cretaceous unconformity surface that overprints a 10 km diameter mound complex of the upper Keg River Formation below the salt scarp. Staged removal of the overly thickened salt beds along the northern margin of this large Keg River mound resulted in episodic sand aggradation during the deposition of the middle McMurray Formation interval, thickening northwards onto Devonian fault block step-downs. The deposit overlies unstable substrate at the eastern end of the Deep Trough structure (Fig. 13), which resulted from the collapse of the northern margin of the Bitumount Trough concurrent with the deposition of lower McMurray Formation strata. During the middle McMurray interval, the underlying salt dissolution fronts migrated northwards into the area underlying the southern Central Collapse (Fig. 25). The Devonian strata at the eastern end of the Deep Trough subsided by 20–30 m during middle McMurray deposition, following an earlier lower McMurray collapse of up to 70 m. The

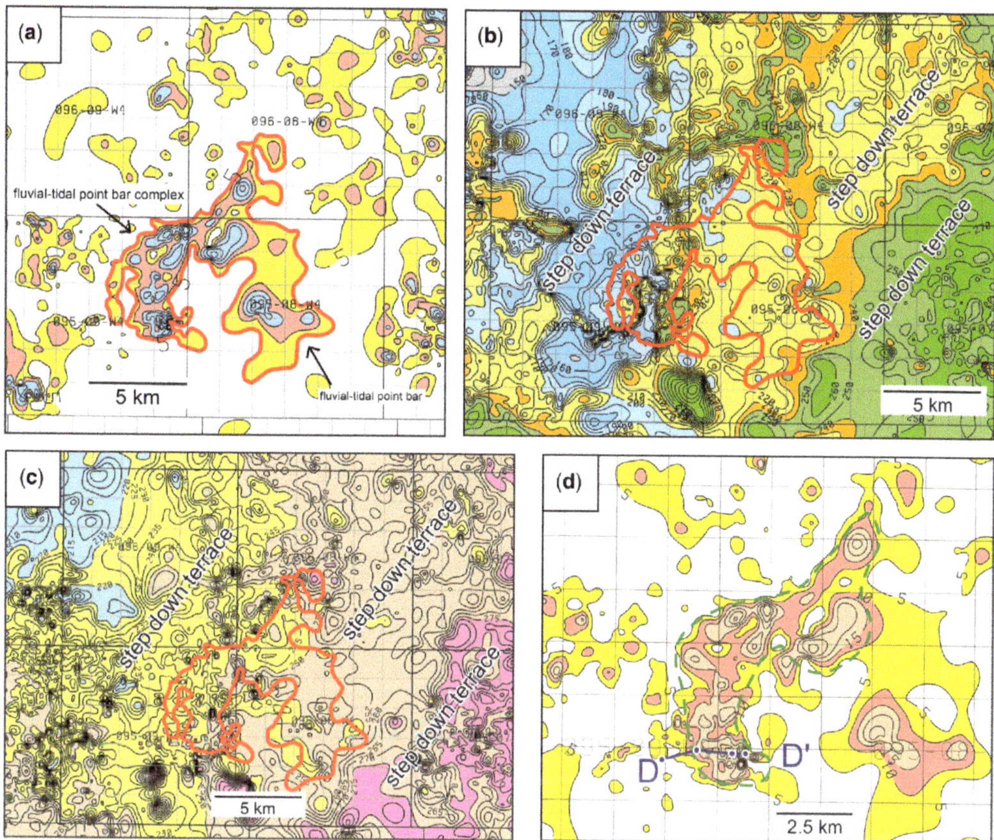

Fig. 30. Middle McMurray stacked point bar complex at the Jackpine Mine site, an example of aggradation extending along a SW-NE-oriented terraced stair step-down into the Central Collapse and associated tidal banking against the back-step high wall. (**a**) Distribution and thickness of very clean sand beds (BWB>14%), illustrating the NE–SW-oriented sand complexes elongated parallel to the underlying patterns of salt removal. Contour interval 5 m. (**b**) Overlay of the Jackpine point bar complex on the sub-Cretaceous structure, illustrating the SW-NE-oriented sand bar emplacement constrained along a step-down terrace. The structure and dimensions of the underlying structural terrace stabilized aggradation and constrained forward lateral migration northwards into the Central Collapse. Contour interval 20 m with elevation in metres above sea-level. (**c**) Orientation of the Jackpine Mine sand bar complex accumulated on lower McMurray structure, consistent with the step-down structure resulting from removal of the underlying salt. (**d**) Larger scale thickness map of the middle McMurray sand complex (BWB>14%) with trace of cross-section D–D' (Figure 31). Contour interval, 5 m. Modified from Broughton (2016).

removal of additional tens of metres thick salt intervals occurred northwards into the southern Central Collapse at the site of the Aurora North Mine. The increased accommodation space resulted in at least a 50% thicker sand deposit along the step-down site at the Aurora North Mine, northwards of the comparatively more stable area underlying the Muskeg River Mine site.

The McMurray Formation strata consist of a 20 m thick lower, 40 m thick middle and a relatively thin 5–10 m thick upper interval at the site of the Aurora North Mine. Bench cuts within the mine are oriented east–west, exposing the lower, middle and upper intervals of the McMurray Formation strata, and result in 80–100 m deep open pit excavations down to the underlying Devonian limestone (Fig. 32). These 3–4 km long bench cuts with 6–18 m high walls excavate into the central area of the 40–60 m thick complex, consisting of overlying point bars. Similar step-cut benches transect the southern end of the sand complex at the adjacent Muskeg River Mine, where a complete stratigraphic interval of the McMurray Formation is also preserved, although thinner.

The stacked point bar complexes of the commercially important middle McMurray interval are

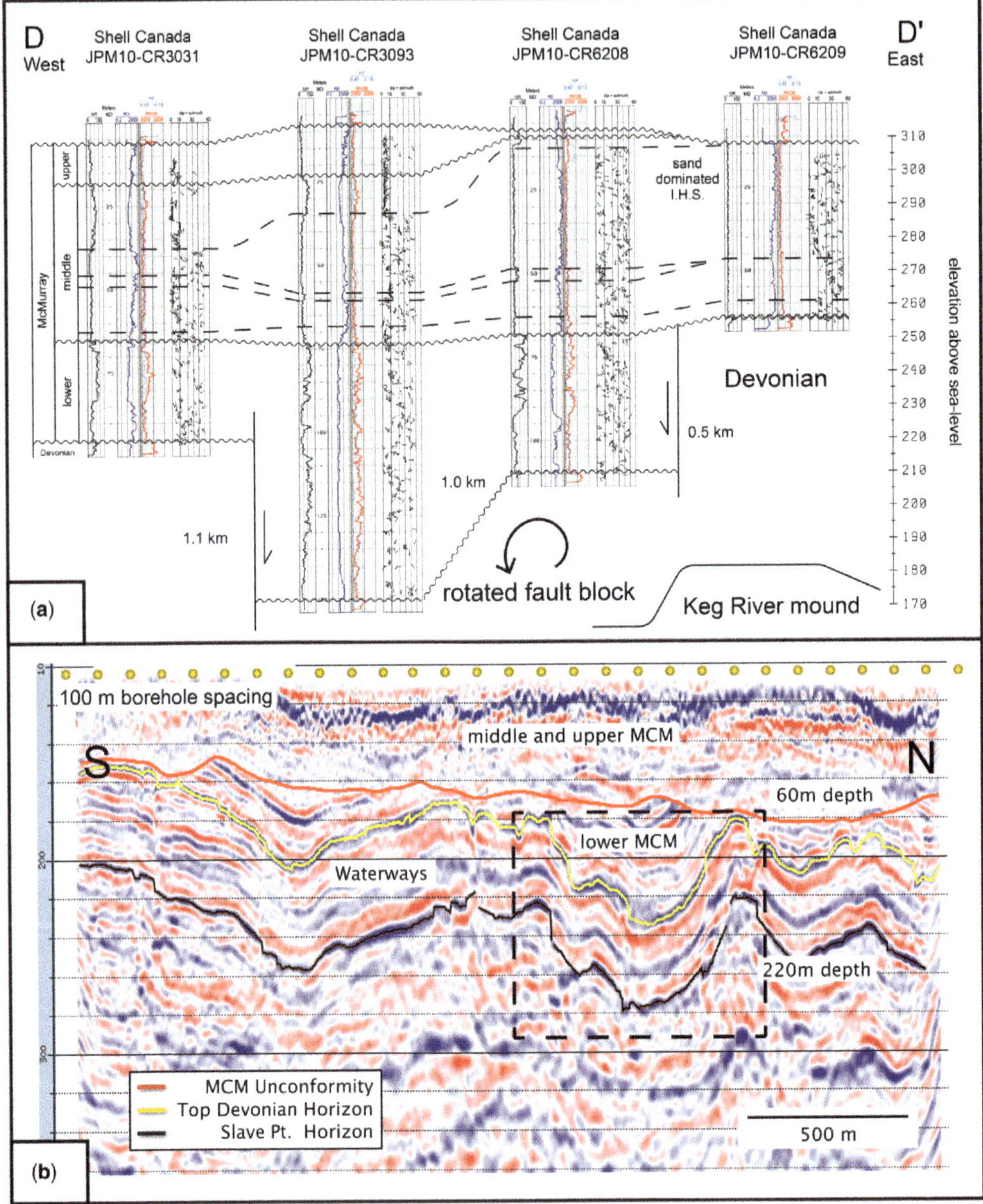

Fig. 31. Profile of the middle McMurray point bar complex at the Jackpine Mine. (**a**) Cross-section D–D′ illustrating aggradation of stacked point bars overlying a structural step-down consisting of collapsed Upper Devonian–lower McMurray fault blocks. This example is located along the northern margin of the Bitumount Trough, facing into the Central Collapse. The sand bars draped over a NE-oriented Devonian high that continued to subside during the deposition of lower McMurray strata, resulting in a syndepositionally thickened interval. The overlying stacked middle McMurray point bars were covered by overly thick sandy inclined heterolithic strata with steep to near-vertical bed dips responding to continued rotation by the underlying fault blocks as the sand transport fairway avalanched over the upper terrace step. Modified from Broughton (2016). (**b**) Seismic imaging along a south–north trace at the Jackpine Mine site, illustrating collapse subsidence of the lower McMurray interval (dashed box) and resulting terraced step-down. Modified from Stoakes *et al.* (2014*a*).

Fig. 32. Bench cuts at the Aurora North Mine, oriented east–west, excavating the central area of a large middle McMurray point bar sand complex that accumulated above a Devonian structural terrace step-down. Fault block tilt and collapses provided accommodation space for the 65 m thick McMurray Formation strata. The two or three middle McMurray stacked point bars were each separated by an erosion surface extending across the width of the sand complex. Channels with coeval flows northwards cut into the accretion beds of each point bar. The well log illustrates a petrophysical profile for the deposit consisting of percentages of rock and fluid: shale (grey), sand (yellow), hydrocarbon (mostly bitumen) (green) and water (blue) and the percentage of bitumen in the rock (0–20%). These are calculated from the well logs: gamma ray (GR), resistivity (R), porosity (NP) and density (RHOB). I.H.S., inclined heterolithic strata.

characterized by internal erosion surfaces. The Aurora North Mine bench cuts of the middle interval expose an internal sand complex architecture consisting of two, and variously preserved remnants of a third, overlying point bars each separated by an unconformity surface. Each of these intervals represents a partially preserved point bar terminated by a planar erosion surface that extends laterally for the multiple kilometre width of the sand deposit (Fig. 33). These two or three stacked sandy point bars, each 10–20 m thick, have four or five north-oriented channels cuts.

The erosion surfaces on each point bar exposed at the Aurora North Mine were responses to multi-stage differential subsidence by the underlying Devonian strata. This structural framework favoured vertical aggradation and constrained lateral accretion along relatively narrow areas responding to instability in the substrate. As a result, the giant sand complex evolved as a composite of tens of kilometres long and several kilometre wide aggradations cut into by a channel network established prior to and truncated by the successive complex-wide erosion events. Subsidence of the Devonian substrate stalled and then ceased, inviting erosional conditions, followed by renewed subsidence and reassertion of the northwards flowing channel system. Several of these subsidence–erosion cycles resulted in multiple kilometre long point bars stacked into a larger sand complex as intermittent aggradation responded to the available accommodation space created by underlying pulses of salt removal.

Breccia pipes and sinkholes at the Athabasca deposit

Regional salt removal trends, tens to hundreds of kilometres long, characterize the Prairie Evaporite halite-dominated salt deposit. These subsurface karstic processes also resulted in more localized structures such as breccia pipes and sinkholes distributed throughout the northern Athabasca deposit on the pre-Cretaceous karstic palaeotopography and within McMurray Formation strata. Some sinkholes were Quaternary structures on the subglacial topography. Thousands of sinkholes developed across the pre-Cretaceous karstic unconformity surface as the Middle–Upper Devonian limestone and calcareous mudstone. Most resulted from underlying breccia pipe collapses occurring during the Middle Jurassic to Early Cretaceous Columbian tectonism that deepened the Alberta foreland basin to the SW and exposed uplifted areas to the NE to erosion and karstification, concurrent with the removal of salt beds in the subsurface 200 m below. The uppermost

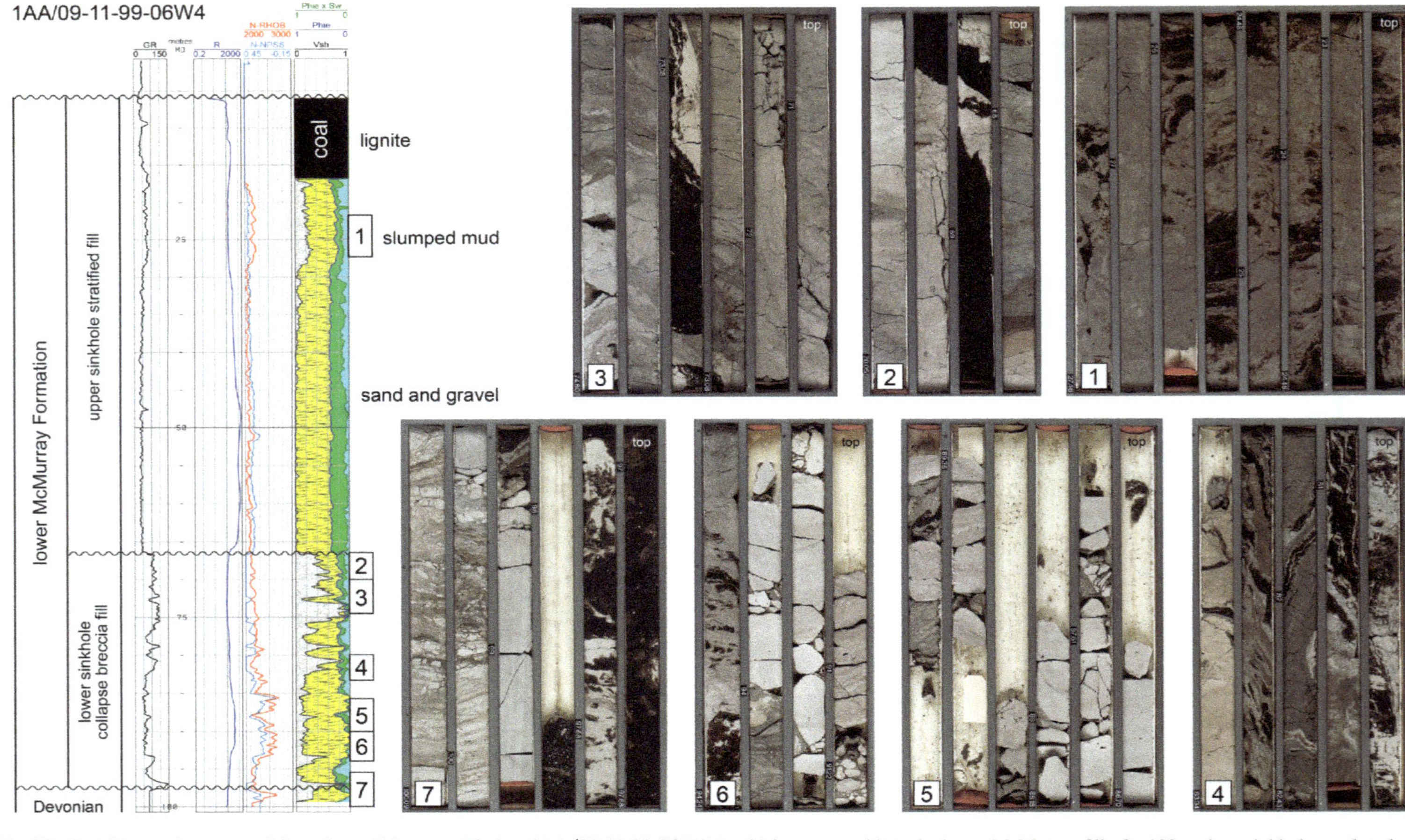

Fig. 33. Cored intervals recovered from the well Synenco Firebag 1AA/09-11-99-06W4M, which penetrated into the lower McMurray fill of a 100 m deep sinkhole emplaced on Middle Devonian palaeotopography (Fig. 34) NE of the Bitumount Trough. The lower sinkhole fill consists of collapsed Middle Devonian wall rock mixed with chaotic bed fragments of lower McMurray strata. The upper sinkhole fill consists of stratified lower McMurray gravelly sand, overlain by slumped mud beds and an uppermost accumulation of a 15 m thick lignite coal bed. Each core box slat is 75 cm. Modified from Broughton (2013).

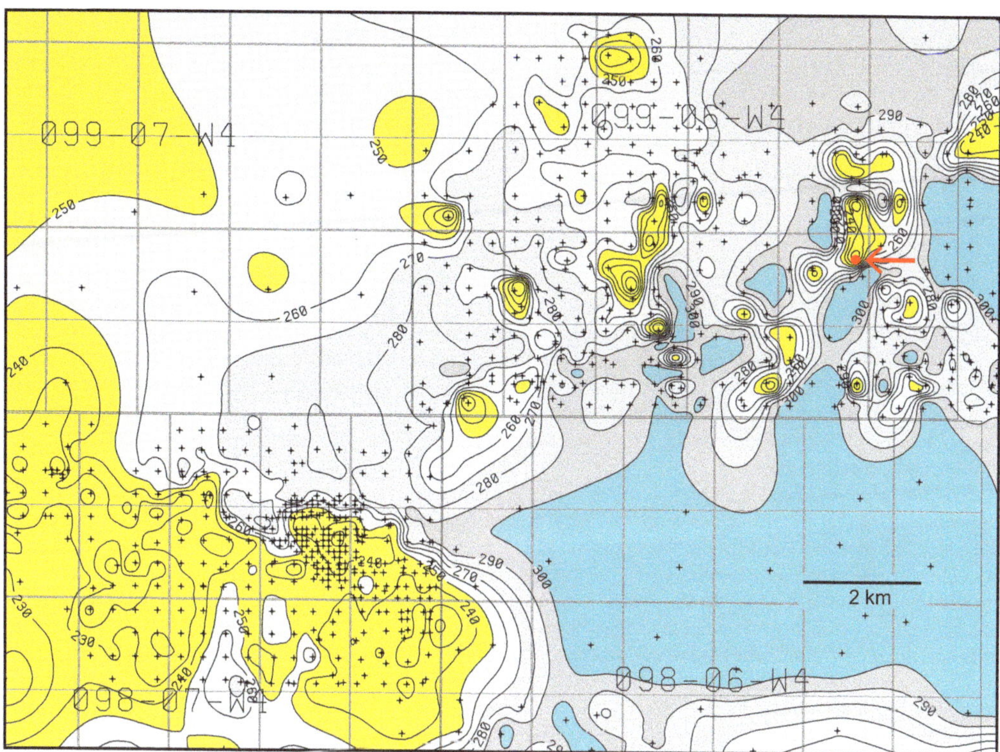

Fig. 34. Sub-Cretaceous map of the Devonian unconformity surface topography at the northeastern Athabasca deposit. Middle Devonian limestone sinkholes up to 100 m deep sustained open voids until subsequently filled and covered over by fluvial deposits of the lower McMurray Formation. Multiple breccia pipe–sinkhole complexes were distributed along 2–5 km long sinkhole valleys oriented to the NE, following the underlying salt removal trends and collapse of the Middle Devonian strata. Contour interval, 10 m; elevation above sea-level. Location (arrow) of the sinkhole at 1AA/09-11-99-06W4M (Fig. 33). Modified from Broughton (2013).

reaches of the resulting collapse breccia pipes, consisting of chaotic intervals with Middle–Upper Devonian limestone blocks, may or may not have extended to the surface (Broughton 2013).

Pre-Cretaceous

Sinkholes on the Devonian limestone topography, some with tens of metres deep open voids (Figs 33 & 34), are recognized across areas overlying the salt scarp and salt removal areas to the east (Broughton 2013). They are seldom observed to the west of the underlying salt scarp. The sinkholes developed as upper reaches of breccia pipes that extended upwards from salt removal zones within the Prairie Evaporite strata, cross-cutting overlying Middle–Upper Devonian strata, and to the surface. The lower reaches of the sinkholes filled with blocks of collapsed limestone wall rock as the floors intermittently collapsed the on additional removal of salt. The rigidity of the limestone wall rock allowed these sinkholes to remain open until the voids, often tens of metres deep, were infilled and covered over by lower McMurray Formation sediments (Fig. 33). The sinkholes often aligned along NW–SE- and NE–SW-oriented trends, which parallel the similar cross-cutting lineament pattern in the underlying salt beds along which the dissolution fronts rapidly advanced. Some of the sinkhole trends on the pre-Cretaceous topography amalgamated into 3–4 km long sinkhole valleys on the Middle Devonian palaeotopography during the prolonged pre-Cretaceous depositional hiatus (Fig. 34). Some of these aligned sinkholes and sinkhole valleys parallel the underlying salt removal trends along the near-linear margins of elongated upper Keg River reefs and chains of biohermal mounds.

Lower Cretaceous

Most breccia pipes and sinkholes observed on the Devonian palaeotopography remained inactive after being covered by Lower Cretaceous strata. Nevertheless, many of these breccia pipe–sinkholes were reactivated or initiated collapses during deposition of the McMurray Formation strata. These collapse structures extended upward to cross-cut lower or lower–middle McMurray Formation beds (Figs 35, 36 & 37). Most McMurray Formation sinkholes have been observed within the lower interval, in contrast with the middle–upper interval (Broughton 2013). Many smaller sinkholes may have formed during the accumulation of McMurray beds, but these were subject to almost immediate collapse because their unstable walls consisted of unconsolidated sand. Other sinkholes formed as post-burial structures, such as gravity-driven sags over the subsiding or tilted margins of the underlying Middle–Upper Devonian fault blocks, resulting in V-shaped McMurray bed deformations.

Breccia pipe–sinkhole complexes often have vertical profiles that record cataclysmic collapse or subsidence-sagged bed deformation that responded to decreasing amounts of removal of the underlying salt. For example, Figure 36 illustrates a cored interval that penetrated into a breccia pipe–sinkhole complex located along the western margin of the Muskeg River Mine. The cored interval penetrating the collapse structure consists of chaotic collapsed blocks of the breccia pipe that transitioned upwards into a sagged bed type sinkhole (Figs 33 & 36).

Fills of deeper sinkholes often have a persistent stratigraphy. The tens of metres deep pits formed on the Devonian palaeotopography and reactivated during lower McMurray deposition often have a lower fill interval consisting of chaotic breccia blocks of limestone wall rock with often vertical bed filaments or fragmented beds of lower McMurray sand and silt (Fig. 33). The middle fill, often 10–30 m thick, consists of stratified sand. The overlying upper interval consists of mud, often slumped, and an uppermost interval of peat (lignite). The thickest known accumulations of lignite coal up to 20 m in the McMurray Formation constitute the uppermost fills of these Devonian limestone sinkholes. This stratigraphic organization was a response to a decreasing rate of subsidence that probably resulted from the rapid removal of highly soluble halite, followed by the slower dissolution of anhydrite beds. By contrast, other breccia pipe–sinkholes collapses during the lower McMurray Formation interval were cataclysmic (Fig. 35) and probably represent the removal of only the underlying halite.

Broughton (2016, 2017) concluded that most, perhaps nearly all, of the larger McMurray Formation sinkholes, 10–25 m across, were emplaced against near-vertical normal fault planes between adjacent differentially subsided fault blocks (Fig. 37a). These sinkholes were positioned along the upper reaches of the block-to-block collision zones that extend downwards into underlying breccia

Fig. 35. Cored intervals recovered from two breccia pipe–sinkhole collapses located in the Bitumount Trough. The breccia consists of fragmented Waterways Formation (Middle–Upper Devonian) limestone mixed with and overlain by breccia clasts derived from the fragmented lower McMurray lignite, siltstone and mudstone. Each core box slat is 75 cm.

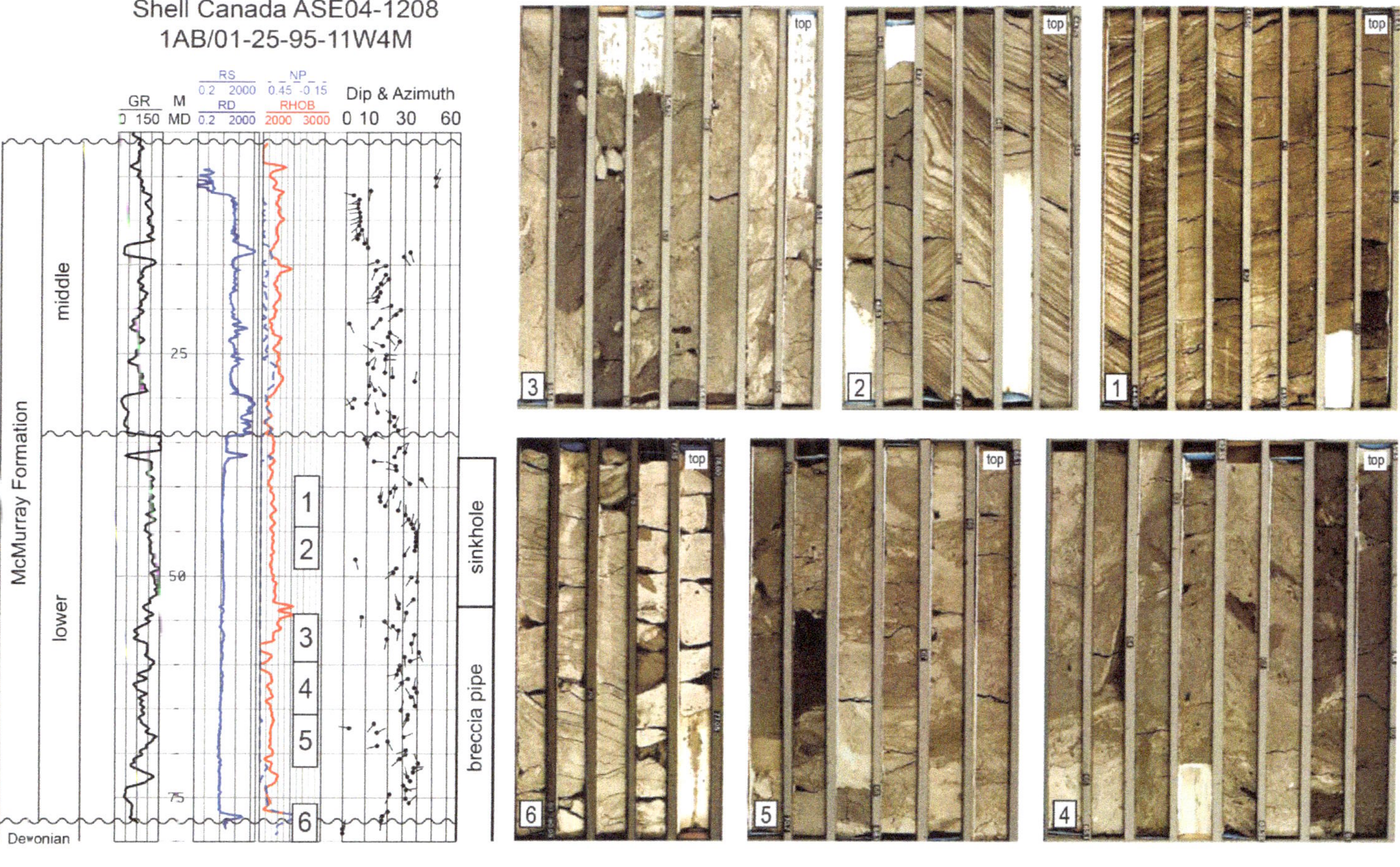

Fig. 36. Cored interval of a breccia pipe–sinkhole complex penetrated by the well at 1AB/01-25-95-11W4M, located on the west side of the Muskeg River Mine. The chaotic collapse breccia of the Devonian pipe extends upwards and cross-cuts a sinkhole with lower McMurray beds. The breccia pipe–sinkhole complex was overlain by undisturbed middle McMurray beds. Each core box slat is 75 cm. The well log illustrates a profile consisting of a gamma ray (GR), resistivity shallow (RS) and deep (RD) readings, neutron porosity (NP) and density (RHOB). The dip meter records bed dips and their direction.

Fig. 37. McMurray Formation sinkholes developed along the upper collision zone between colliding fault blocks and overlying breccia pipes. (**a**) Pair of lower McMurray sinkholes emplaced along block-to-block fault planes exposed along bench cut in the Aurora North Mine. The thickened kaolin fills of the sinkholes, compared with the surrounding beds, result from fills concurrent with subsidence. Modified from Broughton (2013). (**b**) Devonian–lower McMurray sinkhole collapse structure excavated in the Millennium Mine, Steepbank district. (**c**) Post-burial deformation of middle McMurray strata resulting in a sagged bed type sinkhole, excavated in the Millennium Mine. The sinkhole was emplaced against the fault plane between two adjacent differentially subsided fault blocks.

pipes (Fig. 38), although the vertical displacement interval can be tens of metres. The U- to V-shaped sagged bed deformation form of the sinkholes varied with the extent of lateral compression resulting from such block-to-block collisions that accompanied the gravity-driven vertical collapses towards the underlying salt removal voids (Figs 37 & 38). These compressive forces along the block-to-block collision zone often splintered and compartmentalized the sinkholes with internal faults and metre-scale breccia zones (Broughton 2016, 2017).

Sparsely distributed erosional remnants of overlying Clearwater Formation (Albian) mudstone beds sometimes record larger diameter sagged bed structures up to 100–150 m across. These 100 m scale synclines were emplaced above differentially subsided McMurray/Clearwater fault blocks (Broughton 2013, 2015). It is uncertain as to the extent to which these larger bed sags up to 100–150 m across were related to underlying salt removal or to compressional tectonism, or both. Glacial erosion of Late Cretaceous and Tertiary strata overlying the salt scarp may have obliterated the evidence of many sinkhole structures throughout the area.

Subglacial

Subglacial meltwater flows along permeable pathways in Devonian strata stimulated the additional dissolution of the salt scarp underlying the Athabasca Oil Sands (Cowie 2013; Cowie *et al.* 2014*a*, *b*). The weight of the Laurentide ice sheet and the freshwater flows rejuvenated the dissolution and reactivation of multiple kilometre long lineaments that controlled the earlier dissolution patterns in the salt beds and structural trends in the overlying Cretaceous strata. As a result, many linear vertical fracture trends extended upward to the subglacial surface as responses to glacial loading and unloading in the area of the Athabasca deposit (Anderson & Hinds 1997). Evidence for salt removal during the Pleistocene in northeastern Alberta, such as the area of the Athabasca Oil Sands deposit, was limited to changes in the water chemistry and the formation of lineaments and aligned sinkholes on the subglacial topography as responses to glacial loading and unloading by the 1.5–2 km thick Laurentide ice sheet. These influxes of meltwaters coming into contact with the salt beds did not result in regional-scale

Fig. 38. McMurray bed deformations aligned along the collision zone between adjacent fault blocks in the Muskeg River Mine, responding to the removal of underlying salt beds. (**a**) V-shaped offsets of twisted beds along the upper block-to-block collision zone. (**b**) Breccia pipe formed along the lower block-to-block collision zone.

Pleistocene collapse structures, comparable with the tens to hundreds of square kilometres collapses with up to 150 m of closure distributed across southern Saskatchewan.

The Quaternary topography in northeastern Alberta has sinkholes up to 130 m in diameter aligned along multiple kilometres long lineaments oriented NW–SE- and NE–SW (Haug *et al.* 2009, 2014; Broughton 2013). These are mostly distributed as kettle lakes on the subglacial topography. Figure 39 illustrates LiDAR images of two intersecting lineaments with aligned sinkholes, north of the Aurora North Mine site. Sinkholes are often exposed along river valley cuts into McMurray Formation beds and underlying Devonian strata (Fig. 40a). Some of these partially exhumed Devonian–McMurray sinkholes are interpreted as definitive post-bitumen collapses because of the near-vertical oil–water contacts indicative of bed adjustments subsequent to Laramide oil migration into the area and Paleogene degradation into bitumen (Fig. 40b). These contrast with most sinkholes that developed prior to, or concurrent with, the deposition of the McMurray Formation and covered over by undisturbed beds.

Conclusions

Dissolution trends in the Middle Devonian Prairie Evaporite Formation strata accumulated across Western Canada resulted in the largest known hypogene halite karst collapse and subsidence

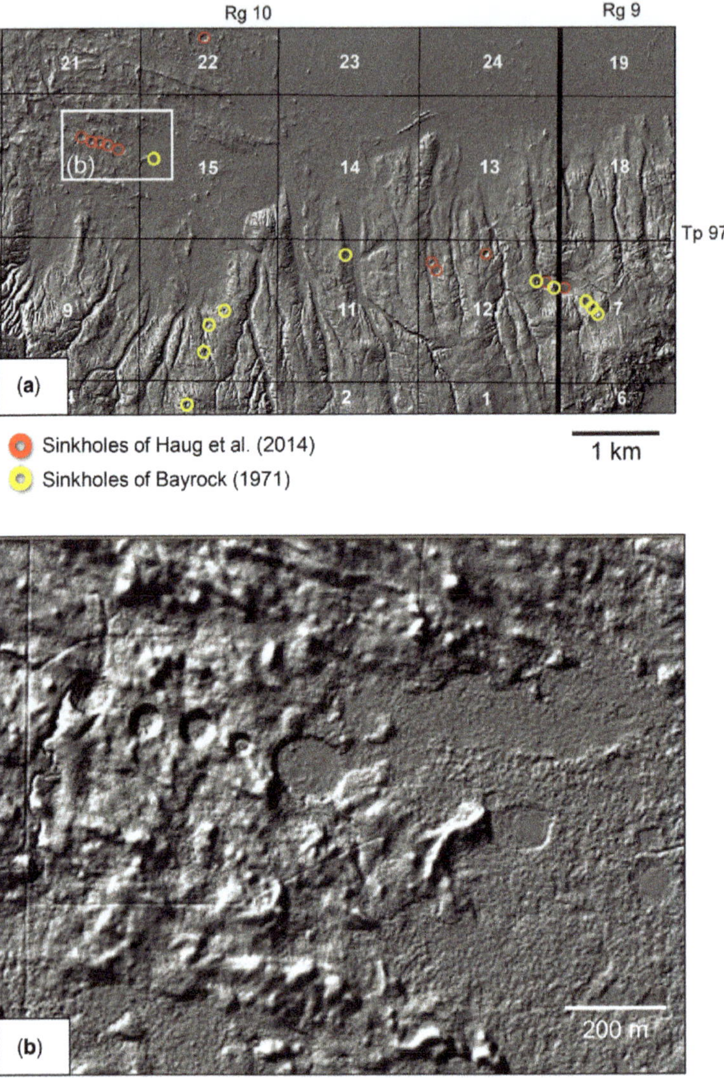

Fig. 39. LiDAR imaging of sinkholes on Holocene topography north of the Aurora North Mine. (**a**) Sinkholes aligned along NW–SE- and NE–SW-oriented subglacial lineaments. (**b**) Enlarged area illustrating linear alignment of the sinkholes up to 130 m in diameter. The surface lineaments, along which the sinkholes aligned, resulted from vertical fracture trends emanating upwards from salt removal zones responding to glacial loading and unloading. Modified from Bayrock (1971), Haug *et al.* (2009, 2014) and Broughton (2013).

structures. Continental-scale tectonism exerted significant regional controls on the development of the Alberta foreland and Williston intra-cratonic basins as constituents of the WCSB. Multiple sourced water movements into the salt beds of the Prairie Evaporite resulted in dissolution trends largely controlled by the Middle Jurassic to Early Cretaceous Columbian tectonism and post-orogenic Aptian structure in northeastern Alberta. Paleozoic Antler and Cretaceous Laramide tectonism impacted the Prairie Evaporite salt beds distributed across the southern Saskatchewan area of the northern Williston basin. The Columbian tectonism, which resulted in the deepening of the Alberta foreland and uplift of the northeastern Alberta basin. Erosion of post-Devonian and pre-Aptian strata was concurrent with dissolution trends within the shallowly buried Prairie Evaporite salt beds. By contrast, the dissolution trends across southern Saskatchewan occurred throughout the Paleozoic, particularly

Fig. 40. Sinkholes exposed along the Muskeg River near the site of the Muskeg River Mine. (**a**) Sinkhole with collapsed lower McMurray bituminous sand beds and underlying Devonian Waterways Formation limestone and calcareous mudstone. (**b**) A Pleistocene sinkhole collapse structure filled with tilted McMurray bitumen cemented sand beds with a vertical oil–water contact, responding to glacial loading and unloading. Images courtesy of Frank Stoakes, Calgary, Albert, Canada.

during the Late Paleozoic as the intracratonic Williston basin subsided.

The earliest known salt removal trends occurred in southern Saskatchewan during the Late Devonian along the margins of the proto-Roncott Platform. The dissolution events became more prevalent during the Carboniferous Period. Extensive salt removal trends occurred across the southern Saskatchewan area of the northern Williston basin during the Cretaceous as the underlying craton subsided and basin waters flowed up-structure to the NE. Salt removal patterns developed in northern Alberta during the Middle Jurassic to Early Cretaceous Columbian orogeny as the Alberta foreland basin subsided and the strata tilted to the SW. A 1000 km long and 150 km wide salt removal trend developed from NE Alberta to SE Saskatchewan and into SW Manitoba.

Compressional tectonism and the hydraulic head in the mountainous terrains to the west and SW provided the drive for deep basin water flows up-structure to the NE along permeable Devonian strata. These water flows were along permeable Keg River and other Devonian strata up-section into the overlying Prairie Evaporite salt beds, and variously mixed with influxes of surface-charged groundwater along the salt basin margin. Basin-wide flows may have flushed large areas or incrementally pushed the formation waters up-structure to the NE as they mixed with more localized compaction-driven vertical flows responding to sedimentary loading. It is uncertain whether the flows of aquifer water up-structure to the NE were largely basin-wide for distances of tens to hundreds of kilometres, were lesser incremental flows that did not flush the entire system, or were mixtures of both. Alternative models suggest that most salt dissolution patterns resulted only from compaction-driven vertical water movement within the Devonian strata during sediment loading. All of these hypogene salt karstification processes may have contributed to the widespread dissolution patterns in the halite salt basin, but the relative contribution from each remains uncertain.

Water flowing along the Keg River and other Devonian strata into the overlying Prairie Evaporite Formation salt beds resulted in rapidly advancing dissolution fronts that followed lineaments oriented NW–SE and NE–SW. These originated as fracture and fault patterns propagated 200–300 m upwards into the overlying Devonian strata as responses to adjustments between Precambrian fault blocks during the Paleozoic Antler and Mesozoic Columbian–Laramide orogenic events that warped the underlying basement. The resulting tens of kilometre long collapse troughs on the pre-Cretaceous unconformity palaeotopography, oriented to the NW and NE, followed dissolution trends that removed up to 200 m of halite–anhydrite across southern Saskatchewan and 100–150 m across northeastern Alberta.

Late-stage dissolution events occurred during the Pleistocene as subglacial meltwaters flowed into the subsurface down-structure to the SW and came into contact with shallowly buried Prairie Evaporite salt beds. The strong hydraulic head responding to the 1.5–2 km thick Laurentide ice sheet resulted in strongly pressured glacial meltwaters flowing down-structure to the SW, halting and locally reversing the regional basin water flows up-structure. The direction of the pre-glacial deep basin aquifer flows to the NE was reasserted on the withdrawal of the ice sheet. Dissolution of the salt beds in the subsurface continue at present, resulting in saline water seeps to the surface at topographic lows along the eastern margin of the evaporite basin.

The Middle Devonian halite karst collapse extending across Western Canada exerted significant controls on the distribution of oilfields in northern Alberta and southern Saskatchewan. Salt removal patterns exerted partial to extensive controls on the distribution of sand reservoirs in the largest known oil sand deposit in northeastern Alberta. The unusually low 1:2 thickness ratio of removed halite salt beds to overlying strata at the Athabasca Oil Sands resulted in collapse subsidence structures responding to the removal of a 100 m section or more of salt beds across areas covered by only 200–250 m of overlying strata. The mosaic of differentially subsided Middle–Upper Devonian fault blocks controlled the distribution and alignment of McMurray Formation point bars accumulated across the northern area of the Athabasca deposit, which subsequently trapped oil migration into the area. In southern Saskatchewan, many oilfields developed as structural traps related to salt dissolution, such as those formed in the Hummingbird Trough.

References

Adams, J., Larter, S., Bennett, B., Huang, H., Westrich, J. & Van Kruisdijk, C. 2013. Dynamic interplay of oil mining, charge timing, and biodegradation in forming the Alberta oilsands: insights from geologic modeling and biogeochemistry. *In*: Hein, F., Leckie, D., Larter, S. & Suter, J. (eds) *Heavy-Oil and Oil-Sand Petroleum Systems in Alberta and Beyond*. American Association of Petroleum Geologists, Studies in Geology, **64**, 23–102.

Anderson, N. & Hinds, R. 1997. Glacial loading and unloading: a possible cause of rock salt dissolution in the Western Canada Basin. *Carbonates and Evaporites*, **12**, 43–52.

Anderson, N. & Knapp, R. 1993. An overview of some of the large scale mechanisms of salt dissolution in western Canada. *Geophysics*, **58**, 1375–1387.

Anfort, S., Bachu, S. & Bentley, L. 2001. Regional-scale hydrogeology of the Upper Devonian sedimentary succession, south-central Alberta basin, Canada. *American Association of Petroleum Geologists Bulletin*, **85**, 637–660.

Babcock, E. & Sheldon, L. 1976. Structural significance of lineaments visible on aerial photos of the Athabasca oil sands area near Fort MacKay, Alberta. *Bulletin of Canadian Petroleum Geology*, **24**, 457–470.

Bachu, S. 1995. Synthesis and model of formation-water flow, Alberta basin, Canada. *American Association of Petroleum Geologists Bulletin*, **79**, 1159–1178.

BACHU, S. 1999. Flow systems in the Alberta basin: patterns, types and driving mechanisms. *Bulletin of Canadian Petroleum Geology*, **47**, 455–474.

BACHU, S. & UNDERSCHULTZ, J. 1993. Hydrogeology of formation waters, northeastern Alberta. *American Association of Petroleum Geologists Bulletin*, **77**, 1745–1768.

BACHU, S., UNDERSCHULTZ, J., HITCHON, B. & COTTERILL, D. 1993. Regional-scale subsurface hydrogeology in northeast Alberta. *Alberta Geological Survey Bulletin*, **61**, 1–44.

BARSON, D., BACHU, S. & ESSLINGER, P. 2001. Flow systems in the Mannville Group in the east-central Athabasca area and implications for steam-assisted gravity drainage (SAGD) operations for *in situ* bitumen production. *Bulletin of Canadian Petroleum Geology*, **49**, 376–392.

BARTON, M. & SIEBEL, C. 2016. The architecture and variability of valley fill deposits within the Cretaceous McMurray Formation, Shell Albian Sands Lease, northeast Alberta. *Bulletin of Canadian Petroleum Geology*, **64**, 166–198.

BARTON, M., PORTER, I., O'BRYNE, C. & MAHOOD, R. 2017. Impact of the Prairie Evaporite dissolution collapse on McMurray stratigraphy and depositional patterns, Shell Albian Sands Lease 13, northeast Alberta. *Bulletin of Canadian Petroleum Geology*, **65**, 175–199.

BAYROCK, L. 1971. *Surficial Geology, Bitumount*. Research Council of Alberta Map, **NTS 74E**.

BEBOUT, D. & MAIKLEM, W. 1973. Ancient anhydrite facies and environments, Middle Devonian Elk Point Basin, Alberta. *Bulletin of Canadian Petroleum Geology*, **21**, 287–243.

BENYON, C., LEIER, A., LECKIE, D., WEBB, A., HUBBARD, S. & GEHRELS, G. 2014. Provenance of the Cretaceous Athabasca Oil Sands, Canada: implications for continental-scale sediment transport. *Journal of Sedimentary Research*, **84**, 136–143.

BENYON, C., LEIER, A., LECKIE, D., HUBBARD, S. & GEHRELS, G. 2016. Sandstone provenance and insights into the paleogeography of the McMurray Formation from detrital zircon geochronology, Athabasca Oil Sands, Canada. *American Association of Petroleum Geologists Bulletin*, **100**, 269–287.

BERBESI, L., DE PRIMIO, R., ANKA, Z., HORSFIELD, B. & HIGLEY, D. 2012. Source rock contributions to the Lower Cretaceous heavy oil accumulations in Alberta: a basin modelling study. *American Association of Petroleum Geologists Bulletin*, **96**, 1211–1234.

BETCHER, R., GROVE, G. & PUPP, C. 1995. *Groundwater in Manitoba: Hydrogeology, Quality Concerns, Management*. National Hydrology Research Institute, Saskatoon, Contribution **CS-93017**.

BLUM, M. & PECHA, M. 2014. Mid-Cretaceous to Paleocene North American drainage reorganization from detrital zircons. *Geology*, **42**, 607–610.

BOND, G. & KOMINZ, M. 1984. Construction of tectonic subsidence curves for the early Paleozoic miogeocline, southern Canadian Rocky Mountains – implications for subsidence mechanisms, age of breakup, and crustal thinning. *Geological Society of America Bulletin*, **95**, 155–173.

BROUGHTON, P.L. 1977. Origin of coal basins by salt solution. *Nature*, **270**, 420–423.

BROUGHTON, P.L. 1979. *Origin of coal basin by salt solution tectonics in western Canada*. PhD thesis, University of Cambridge, Cambridge, UK.

BROUGHTON, P.L. 1985. Geology and resources of the Saskatchewan coalfields. *In*: PATCHING, T. (ed.) *Coal in Canada*. Canadian Institute of Mining and Metallurgy, Special Volume, **31**, 87–99.

BROUGHTON, P.L. 2013. Devonian salt dissolution-collapse breccia flooring the Cretaceous Athabasca oil sands deposit and development of lower McMurray Formation sinkholes, northern Alberta basin, Western Canada. *Sedimentary Geology*, **283**, 57–82.

BROUGHTON, P.L. 2015. Syndepositional architecture of the northern Athabasca Oil Sands deposit, northeastern Alberta. *Canadian Journal of Earth Sciences*, **52**, 21–50.

BROUGHTON, P.L. 2016. Alignment of fluvio-tidal sand bars in the middle McMurray Formation: implications for structural architecture of the Lower Cretaceous Athabasca Oil Sands deposit, northern Alberta. *Canadian Journal of Earth Sciences*, **53**, 896–930.

BROUGHTON, P.L. 2017. Breccia pipe and sinkhole linked fluidized beds and debris flows in the Athabasca Oil Sands: dynamics of evaporite karst collapse-induced fault block collisions. *Bulletin of Canadian Petroleum Geology*, **65**, 200–234.

BROWN, D. & BROWN, D. 1987. Wrench-style deformation and paleostructural influence on sedimentation in and around a cratonic basin. *In*: LONGMAN, M. (ed.) *Williston Basin – Anatomy of a Cratonic Oil Province*. Rocky Mountain Association of Geologists, Denver, 57–70.

CARRIGY, M. 1959. *Geology of the McMurray Formation, Part III: General Geology of the McMurray Area*. Alberta Geological Survey, Memoirs, **1**.

CHRISTIANSEN, E. 1967. Collapse structures near Saskatoon, Saskatchewan. *Canadian Journal of Earth Sciences*, **4**, 757–767.

CHRISTIANSEN, E. 1971. Geology of the Crater Lake collapse structure in southeast Saskatchewan. *Canadian Journal of Earth Sciences*, **8**, 1505–1513.

CHRISTIANSEN, E. & SAUER, E. 2002. Stratigraphy and structure of Pleistocene collapse in the Regina Low, Saskatchewan, Canada. *Canadian Journal of Earth Sciences*, **39**, 1411–1423.

CHRISTIANSEN, E., GENDZWILL, D. & MENELEY, W. 1982. Howe Lake: a hydrodynamic blowout structure. *Canadian Journal of Earth Sciences*, **19**, 1122–1139.

CHRISTOPHER, J. 1974. *Upper Jurassic Vanguard and Lower Cretaceous Mannville Groups of southwestern Saskatchewan*. Saskatchewan Department of Mineral Resources, Geological Report, **151**.

CONNOLLY, C., WALTER, L., BAADSGAARD, H. & LONGSTAFFE, F. 1990. Origin and evolution of formation waters, Alberta basin, Western Canada Sedimentary Basin: II, isotope systematic and water mixing. *Applied Geochemistry*, **5**, 397–413,

COWIE, B. 2013. *Stable isotope and geochemical investigations into the hydrogeology and biochemistry of oil sands reservoir systems in northeastern Alberta, Canada*. PhD thesis, University of Calgary.

COWIE, B., JAMES, B., NIGHTINGALE, M. & MAYER, B. 2014*a*. Determination of the stable isotope composition and total dissolved solids of Athabasca oil sands reservoir porewater: part 1. A new tool for aqueous fluid characterization in oil sands reservoirs. *American*

Association of Petroleum Geologists Bulletin, **98**, 2131–2141.

Cowie, B., James, B., Nightingale, M. & Mayer, B. 2014*b*. Determination of the stable isotope composition and total dissolved solids of Athabasca oil sands reservoir porewater: part 2. Characterization of McMurray Formation waters in the Suncor-Firebag field. *American Association of Petroleum Geologists Bulletin*, **98**, 2143–2160.

Cowie, B., James, B. & Mayer, B. 2015. Distribution of total dissolved solids in McMurray Formation water in the Athabasca oil sands region, Alberta, Canada: implications for regional hydrogeology and resource development. *American Association of Petroleum Geologists Bulletin*, **99**, 77–90.

Crabtree, H. 1982. Lithological types, depositional environment and reservoir properties of the Mississippian Frobisher Beds, Innes Field, southeastern Saskatchewan. *In*: Christopher, J. & Kaldi, J. (eds) *Fourth Williston Basin Symposium*. Saskatchewan Geological Society, Special Publications, **6**, 203–210.

DeMille, G., Shouldice, J. & Nelson, H. 1964. Collapse structures related to evaporites of the Prairie Formation, Saskatchewan. *Geological Society of America Bulletin*, **75**, 307–316.

Dyke, A., Andrews, J., Clark, P., England, J., Miller, G., Shaw, J. & Veillette, J. 2002. The Laurentide and Innuitian ice sheets during the last glacial maximum. *Quaternary Science Reviews*, **21**, 9–31.

Fisher, T., Waterson, N., Lowell, T. & Hajdas, I. 2009. Deglaciation ages and meltwater routing in the Fort McMurray region, northeastern Alberta and northwestern Saskatchewan, Canada. *Quaternary Science Reviews*, **28**, 1608–1624.

Flach, P. & Mossop, G. 1985. Depositional environments of Lower Cretaceous McMurray Formation, Athabasca oil sands, Alberta. *American Association of Petroleum Geologists Bulletin*, **69**, 1195–1207.

Ford, D. 1997. Principal features of evaporite karst in Canada. *Carbonates and Evaporites*, **12**, 15–23.

Fuzesy, A. 1982. *Potash in Saskatchewan*. Saskatchewan Energy and Mines Report, **181**.

Garven, G. 1989. A hydrogeological model for the formation of the giant oil sands deposits of the Western Canada Sedimentary Basin. *American Journal of Science*, **289**, 105–166.

Garven, G. & Freeze, R. 1984. Theoretical analysis of the role of groundwater flow in the genesis of stratabound ore deposits, 2, Quantitative results. *American Journal of Science*, **284**, 1125–1174.

Gibson, J., Fennell, J., Birks, S., Yi, Y., Moncur, M., Hansen, B. & Jasechko, S. 2013. Evidence of discharging saline formation water to the Athabasca River in the oil sands mining region, northern Alberta. *Canadian Journal of Earth Sciences*, **50**, 1244–1257.

Grasby, S. & Betcher, R. 2000. Pleistocene recharge and flow reversal in the Williston basin, central North America. *Journal of Geochemical Exploration*, **69–70**, 403–407.

Grasby, S. & Chen, Z. 2005. Subglacial recharge in the Western Canada sedimentary basin – impact of Pleistocene glaciations on basin hydrodynamics. *Geological Society of America Bulletin*, **117**, 500–514.

Grasby, S., Osadetz, K., Betcher, R. & Render, F. 2000. Reversal of the regional-scale flow system of the Williston Basin in response to Pleistocene glaciation. *Geology*, **7**, 635–638.

Grobe, M. 2000. *Distribution and Thickness of Salt within the Devonian Elk Pont Group, Western Canada Sedimentary Basin*. Alberta Geological Survey, Earth Sciences Report, **2000-02**.

Gue, A. 2012. *Geochemistry of saline springs in the Athabasca oil sands region and their impact on the Clearwater and Athabasca Rivers*. MSc thesis, University of Calgary.

Gue, A., Mayer, B. & Grasby, S. 2015. Origin and geochemistry of saline spring waters in the Athabasca oil sands region, Alberta, Canada. *Applied Geochemistry*, **61**, 132–145.

Haug, K., Chao, D., Hein, F. & Greene, P. 2009. Mapping surficial features in the Athabasca oil sands area using airborne LIDAR. *Abstracts and Proceedings*, Gussow Geoscience Conference, Canadian Society of Petroleum Geologists, 5–7 October 2009, Banff, Alberta.

Haug, K., Greene, P. & Mei, S. 2014. *Geological Characterization of the Lower Cretaceous Shale in the Athabasca Oil Sands Area, Townships 87–99, Ranges 1–13, West of the Fourth Meridian*. Alberta Geological Survey, Open File Report, **2014-04**.

Head, I., Jones, D. & Larter, S. 2003. Biological activity in the deep subsurface and the origin of heavy oil. *Nature*, **426**, 344–352.

Hein, F. & Cotterill, D. 2006. The Athabasca oil sands – a regional geological perspective, Fort McMurray area, Alberta, Canada. *Natural Resources Research*, **15**, 85–102.

Hein, F., Cotterill, D. & Berhane, H. 2000. *Atlas of Lithofacies of the McMurray Formation, Athabasca Oil Sands Deposit, Northeastern Alberta: Surface and Subsurface*. Alberta Geological Survey, Earth Science Report, **200-07**.

Hein, F., Langenberg, C., Kidston, C., Berhane, H. & Berezniuk, T. 2001. *A Comprehensive Field Guide for Facies Characterization of the Athabasca oil Sands, Northeast Alberta*. Alberta Geological Survey, Special Report, **13**.

Hein, F., Dolby, G. & Fairgrieve, B. 2013. A regional geologic framework for the Athabasca oil sands, northeastern Alberta, Canada. *In*: Hein, F., Leckie, D., Larter, S. & Suter, J. (eds) *Heavy-Oil and Oil-Sand Petroleum Systems in Alberta and Beyond*. American Association of Petroleum Geologists, Studies in Geology, **64**, 207–250.

Higley, D., Lewan, M., Roberts, L. & Henry, M. 2009. Timing and petroleum sources for the Lower Cretaceous Mannville Group oil sands of northern Alberta based on 4-D modelling. *American Association of Petroleum Geologists Bulletin*, **93**, 203–230.

Holbrook, J. 2014. Connectivity within and between channel-belt reservoirs: a Mississippi River analog for the McMurray Oil Sands. *Abstracts and Proceedings*, Oil Sands and Heavy Oil Symposium: a Local to Global Multidisciplinary Collaboration, 14–16 October 2014, Calgary Alberta, Canadian Society of Petroleum Geologists and American Association of Petroleum Geologists.

HOLTER, M. 1969. *Middle Devonian Prairie Evaporite of Saskatchewan*. Saskatchewan Department of Mineral Resources Report, **123**.

HUBBARD, S., SMITH, D., NIELSEN, H., LECKIE, D., FUSTIC, M., SPENCER, R. & BLOOM, L. 2011. Seismic geomorphology and sedimentology of a tidally influenced river deposit, Lower Cretaceous Athabasca oil sands, Alberta, Canada. *American Association of Petroleum Geologists Bulletin*, **95**, 1123–1145.

IRVINE, J., BROUGHTON, P.L. & WHITAKER, S. 1978. *Coal Resources of Southern Saskatchewan: a Model for Evaluation Methodology*. Geological Survey of Canada Economic Geology Report, **30**; Saskatchewan Department of Mineral Resources Report, **209**; Saskatchewan Research Council Report, **20**, vol. 1, microfiche; vol. 2, atlas.

KENT, D. 1974. Relationship between hydrocarbon accumulations and basement structural elements in the northern Williston basin. *In*: PARSLOW, G. (ed.) *Fuels: a Geological Appraisal*. Saskatchewan Geological Society, Special Publications, **2**, 63–80.

KREBS, W. & MACQUEEN, R. 1984. Sequence of diagenetic and mineralization events, Pine Point lead-zinc property, Northwest Territories, Canada. *Bulletin of Canadian Petroleum Geology*, **32**, 434–644.

LECKIE, D. & SMITH, D. 1992. Regional setting, evolution, and depositional cycles of the Western Canada foreland basin. *In*: MACQUEEN, R. & LECKIE, D. (eds) *Foreland Basins and Fold Belts*. American Association of Petroleum Geologists, Memoirs, **55**, 9–46.

LEFEVER, J. & LEFEVER, R. 2005. *Salts in the Williston Basin, North Dakota*. North Dakota Geological Survey, Report of Investigations, **103**.

LYATSKY, H. & PANA, D. 2003. *Catalogue of Selected Regional Gravity and Magnetic Maps of Northern Alberta*. Alberta Geological Survey, Special Report, **56**.

LYATSKY, H., FRIEDMAN, G. & LAYATSKY, V. 1999. *Principles of Practical Tectonic Analysis of Cratonic Regions with Particular Reference to Western North America*. Lecture Notes in Earth Sciences, **84**. Springer, Berlin.

LYATSKY, H., PANA, D. & GROBE, M. 2005. *Basement Structure in Central and Southern Alberta: Insights from Gravity and Magnetic Maps*. Alberta Geological Survey, Special Report, **72**.

MAHOOD, R., VERHOEF, M. & STOKES, F. 2012. Paleozoic stratigraphic framework beneath the Muskeg River Mine (Twp 95, Rge 9-10): controls and constraints on present day hydrogeology. *Abstracts and Proceedings, 2012 GeoConvention*, Canadian Society of Petroleum Geologists, 14–18 May 2012, Calgary, Alberta.

MAIKLEM, W. 1971. Evaporative drawdown – a mechanism for water-level lowering and diagenesis in the Elk Point basin. *Bulletin of Canadian Petroleum Geology*, **19**, 485–501.

MARGOLD, M., STOKES, C. & CLARK, C. 2015. Ice streams in the Laurentide ice sheet: identification, characteristics and comparison to modern ice sheets. *Earth-Science Reviews*, **143**, 117–146.

MCTAVISH, G. & VIGRASS, L. 1987. Salt dissolution and tectonics, south-central Saskatchewan. *In*: CARLSON, G. & CHRISTOPHER, J. (eds) *Proceedings, Fifth International Williston Basin Symposium*. Saskatchewan Geological Society, Special Publications, 9, 157–168.

MEIJER DREES, N. 1986. Evaporitic deposits of western Canada. *Geological Survey of Canada, Papers*, **85-20**.

MEIJER DREES, N. 1994. Devonian Elk Point Group of the Western Canada Sedimentary Basin. *In*: MOSSOP, G. & SHETSEN, I. (eds) *Geological Atlas of the Western Canada Sedimentary Basin*. Canadian Society of Petroleum Geologists/Alberta Research Council, Calgary, Alberta, 129–147.

NEMETH, B., DANYLUK, T., & PRUGGER, A. 2002. Benefits of 3D poststack depth migration: case study from the potash belt of Saskatchewan. *Abstracts and Proceedings*, 2002 CSEG Geophysics Convention, Canadian Society of Exploration Geophysicists, 12–14 May, Calgary, Alberta.

PERSON, M., MCINTOSH, J., BENSE, V. & REMENDA, V. 2007. Pleistocene hydrogeology of North America: the role of ice sheets in reorganizing groundwater flow systems. *Reviews of Geophysics*, **45**, 1–28.

QING, H. 1998. Petrography and geochemistry of early-stage, fine- and medium-crystalline dolomites in Middle Devonian Presqu'ile barrier at Pine Point. *Sedimentology*, **45**, 433–446.

QING, H. & MOUNTJOY, E. 1992. Large-scale fluid flow in the Middle Devonian Presqu'ile barrier, Western Canada Sedimentary Basin. *Geology*, **20**, 903–906.

QING, H. & MOUNTJOY, E. 1994. Origin of dissolution vugs, caverns, and breccias in the Middle Devonian Presqu'ile barrier, host of Pine Point Mississippi Valley-type deposits. *Economic Geology*, **89**, 858–876.

RHODES, D., LANTOS, E., LANTOS, J., WEBB, R. & OWENS, D. 1984. Pine Point orebodies and their relationship to the stratigraphy, structure, dolomitization, and karstification of the Middle Devonian barrier complex. *Economic Geology*, **79**, 991–1055.

ROOT, K. 2001. Devonian Antler fold and thrust belt and foreland basin development in the southern Canadian Cordillera: implications for the Western Canada Sedimentary Basin. *Bulletin of Canadian Petroleum Geology*, **49**, 7–36.

SCHNEIDER, C. & GROBE, M. 2013. *Regional Cross-Sections of Devonian Stratigraphy in Northeastern Alberta (NTS 74D, E)*. Alberta Geological Survey, Open File Report, **2013-05**.

SCHNEIDER, C., MEI, S., HAUG, K. & GROBE, M. 2014. *The Sub-Cretaceous Unconformity and the Devonian Subcrop in the Athabasca Oil Sands Area, Townships 87–99, Ranges 1–13, West of Fourth Meridian*. Alberta Geological Survey, Open File Report, **2014-07**.

SKALL, H. 1975. The paleoenvironment of the Pine Point lead-zinc district. *Economic Geology*, **70**, 22–45.

SMITH, D. & PULLEN, J. 1967. Hummingbird structure of southeast Saskatchewan. *Bulletin of Canadian Petroleum Geology*, **15**, 468–482.

SMITH, D., HUBBARD, S., LECKIE, D. & FUSTIC, M. 2009. Counter point bar deposits: lithofacies and reservoir significance in the meandering modern Peace River and ancient McMurray Formation, Alberta, Canada. *Sedimentology*, **56**, 1655–1669.

STOAKES, F., VERHOEF, M. & MAHOOD, R. 2014*a*. Multiphase solution removal of the Prairie Evaporite Formation in northeast Alberta (implications for the oil sands mining community). *Abstracts and Proceedings, 2014*

GeoConvention, Canadian Society of Petroleum Geologists, 12–16 May, Calgary, Alberta.

Stoakes, F., Mahood, R., Verhoef, M. & Gomez, A. 2014*b*. The Devonian caprock succession beneath the oil sands in northeastern Alberta; core data from shallow sub-horizontal boreholes. *Abstracts and Proceedings, 2014 GeoConvention*, Canadian Society of Petroleum Geologists, 12–16 May, Calgary, Alberta.

Thomas, G. 1974. Lineament block tectonics: Williston-Blood Creek basins. *American Association of Petroleum Geologists Bulletin*, **58**, 1305–1322.

Wells, C. & Price, J. 2015. The hydrogeologic connectivity of a low-flow saline-spring fen peatland within the Athabasca oil sands region, Canada. *Hydrogeology Journal*, **23**, 1799–1816.

Wright, G., McMechan, M., Potter, D. & Holter, M. 1994. Structure and architecture of the Western Canada Sedimentary Basin. *In*: Mossop, G. & Shetsen, I. (eds) *Geological Atlas of the Western Canada Sedimentary Basin*. Canadian Society of Petroleum Geologists/Alberta Research Council, Calgary, Alberta, 24–40.

Tafoni and honeycomb structures as indicators of ascending fluid flow and hypogene karstification

ALEXANDER KLIMCHOUK

Institute of Geological Sciences, National Academy of Sciences of Ukraine, 55-B Gonchara Street, Kiev, 01054, Ukraine

klim@speleogenesis.info

Abstract: Tafoni and honeycombs remain some of the most enigmatic and puzzling geomorphological phenomena. Their globally widespread occurrence across a wide range of lithologies and environmental conditions suggests that their formation is determined by a factor that overarches variations in these conditions and weathering processes. Based on a study of tafoni and honeycombs in the Crimean Piedmont, this paper demonstrates that the primary factor in their formation is the pre-exposure alteration of rocks along fractures and karst conduits as a result of fluid–rock interactions. Such alteration is commonly induced by ascending flow and is related to hypogene karstification. The morphological expression of cavernous features through removal of the alterite can occur under subsurface conditions, but is more frequent on exposure to atmospheric weathering. The local or regional characteristics of a weathering system are irrelevant or of only secondary importance in determining the localization and morphology of cavernous features. New features do not form on rock surfaces where the alteration zone was totally denuded or never present. The proposed model resolves major issues inherent to previous interpretations of cavernous features. It has general applicability and important implications for geodynamic and palaeohydrogeological reconstructions. Typical tafoni and honeycombs may indicate past events of ascending flow and the potential presence of hypogene karst systems.

Variously shaped and arranged cavities and hollows on the exposed surfaces of rocks, ranging in size from several millimetres to several metres, are collectively termed cavernous weathering or cavernous rock decay features. They are a global phenomenon and occur under a number of different environmental conditions. There are many descriptive terms used for cavernous features on rock surfaces, of which tafoni, alveoli and honeycomb are the most popular, but the terminology is highly inconsistent and entangled (see Groom *et al.* 2015 for a detailed review). In agreement with a suggestion by Robinson & Williams (1994), there has been a tendency in recent decades to use tafoni (singular tafone) for larger features (decimetre- to metre-scale) and alveoli (individual forms) and honeycomb (groups of forms) for smaller features (centimetre- to decimetre-scale). The terms honeycomb and honeycomb weathering are often interchanged with alveoli and alveolar weathering. Other terms used for similar structures include stonelace, lacework, boxwork, fretwork, fretting and pitting. Larger tafoni are also called rock shelters, rock overhangs, alcoves or abri. Groom *et al.* (2015) proposed using tafoni as the non-scalar universal term denoting all kinds of cavernous features.

Turkington (2004) describes tafoni as hollows of a spherical or elliptical shape with arch-shaped entrances, concave inner walls, overhanging visors and gently sloping debris-covered floors. Whereas tafoni may occur separately and even isolated, individual cells (alveoli) in honeycomb structures are closely spaced, separated by narrow, delicate walls and arranged into characteristic intricate patterns. Although these patterns most commonly resemble honeycombs, other pattern styles such as boxwork, fretwork and lacework, or complex mixes of all of these, are also common. Despite the variability in size, morphology and patterns, tafoni and honeycomb features have a recognizable appearance, maintained across a wide range of lithologies and environmental conditions.

Cavernous decay features are most commonly found on subvertical outcrops and on boulders, but occur much less frequently at sub-horizontal outcrops. They are found in a wide range of rocks, including sandstone, limestone, quartzite, granite, greywacke, dolerite, rhyolite, greenschist, conglomerate and tuff. They have been described from regions in various climatic zones and environments on all of the continents, from deserts to humid areas, from coasts and lowlands to high mountains, and even on Mars (e.g. Rodriguez-Navarro 1998).

Studies of the features and processes of cavernous decay have been a prominent part of geomorphology for more than 100 years, with a large body of literature on the subject, particularly in such fields as weathering research and sandstone landscapes. Notable works and recent reviews include, among

From: Parise, M., Gabrovsek, F., Kaufmann, G. & Ravbar, N. (eds) 2018. *Advances in Karst Research: Theory, Fieldwork and Applications*. Geological Society, London, Special Publications, **466**, 79–105.
First published online November 28, 2017, https://doi.org/10.1144/SP466.11

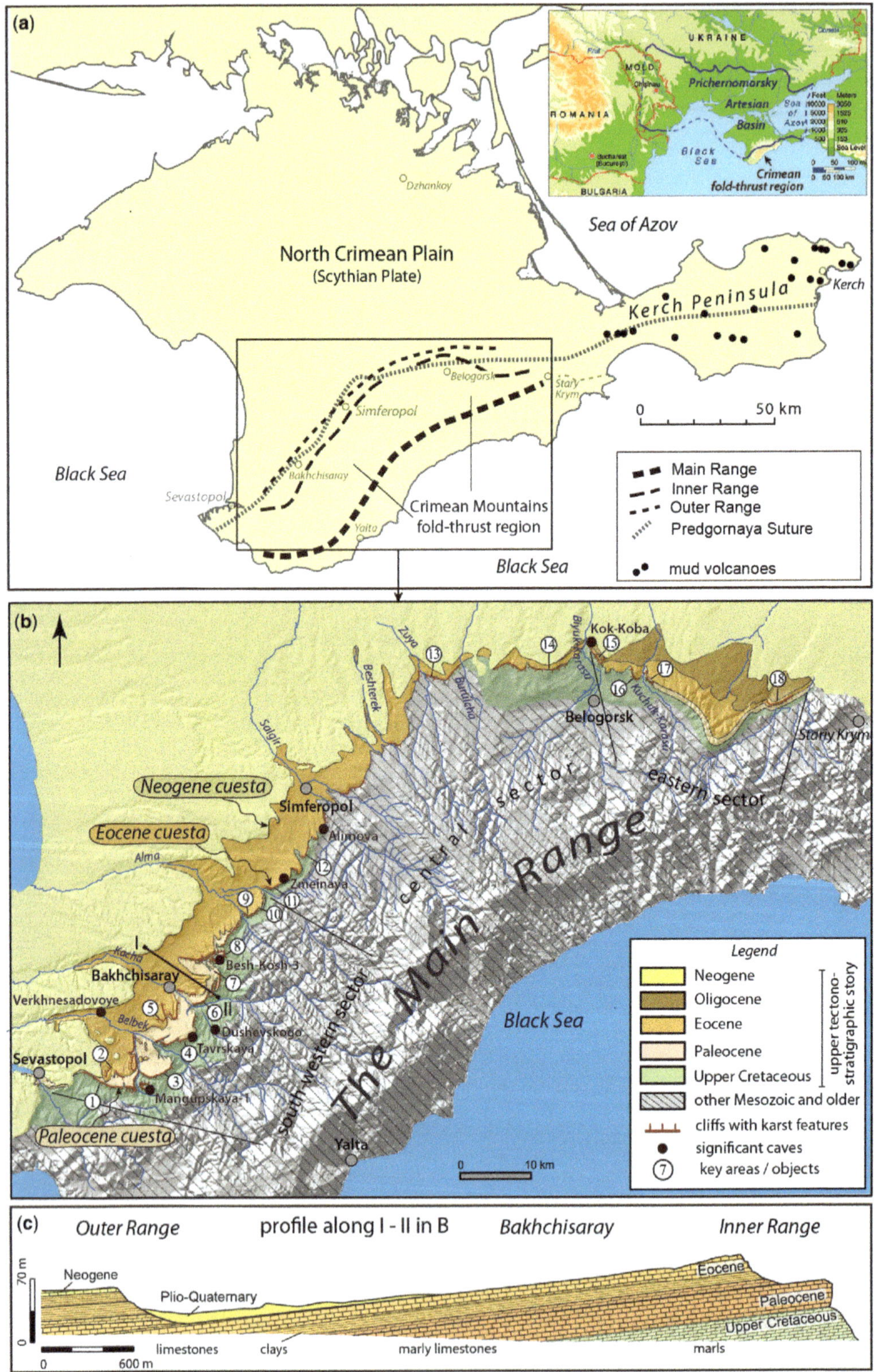
(a)
UKRAINE
ROMANIA
BULGARIA
Prichernomorsky
Artesian
Basin
Sea of Azov
Black Sea
Crimean
fold-thrust region
Dzhankoy
Sea of Azov
North Crimean Plain
(Scythian Plate)
Kerch Peninsula
Kerch
Belogorsk
Stary Krym
Simferopol
Bakhchisaray
Sevastopol
Yalta
Black Sea
Crimean Mountains
fold-thrust region
0 50 km
Main Range
Inner Range
Outer Range
Predgornaya Suture
mud volcanoes
Black Sea
(b)
Zuya
Beshterek
Burulcha
Biyuk-Karasu
Kok-Koba
Kuchuk-Karasu
Belogorsk
Stary Krym
Salgir
Neogene cuesta
Simferopol
Eocene cuesta
Alimova
Alma
Zmeinaya
central sector
eastern sector
The Main Range
Kacha
Bakhchisaray
Besh-Kosh-3
Verkhnesadovoye
Belbek
Dushevskogo
Tavrskaya
Sevastopol
Mangupskaya-1
Paleocene cuesta
south-western sector
Yalta
Black Sea
0 10 km
Legend
Neogene
Oligocene
Eocene
Paleocene
Upper Cretaceous
other Mesozoic and older
upper tectono-stratigraphic story
cliffs with karst features
significant caves
key areas / objects
(c)
Outer Range
profile along I - II in B
Bakhchisaray
Inner Range
70 m
0
Neogene
Plio-Quaternary
Eocene
Paleocene
Upper Cretaceous
limestones
clays
marly limestones
marls
0 600 m

others, Martini (1978), Mustoe (1983), Young & Young (1992), Goudie & Viles (1997), Huinink *et al.* (2004), McBride & Picard (2004), Turkington (2004), Turkington & Phillips (2004), Turkington & Paradise (2005), Viles (2005), Brandmeier *et al.* (2011), Paradise (2013), Groom *et al.* (2015) and Siedel (2015). However, despite vigorous research efforts, tafoni and honeycomb structures remain one of the most enigmatic and puzzling geomorphological phenomenon.

This paper shows that the major problem in studies of tafoni and honeycomb features is that they have always been constrained by the weathering paradigm, i.e. the action of external agencies (the weathering system) on exposed rock surfaces was *a priori* considered as the principal factor in the formation of these features. However, the occurrence of these features across a wide range of lithologies and environmental conditions, either regional or local, or both, strongly suggests that their formation is determined by a factor that overarches variations in these conditions and weathering processes. This study argues that this factor is the pre-exposure alteration of the wall rocks of fluid-conducting discontinuities (such as fractures and karst conduits) as a result of fluid–rock interactions. Such alteration occurs mainly, although not exclusively, in deep-seated settings due to the action of rising flow. Using the Crimean Piedmont as a prime example, it is demonstrated here that the rock alteration induced by focused rising fluid flow and the subsequent formation of tafoni and honeycombing are associated with hypogene karstification. The study in the Crimean Piedmont is mainly based on field data (geomorphological mapping, speleological investigations and the region-wide systematic documentation of cavernous features in cliffs) and a regional geological analysis. A new conceptual model for the formation of cavernous features is proposed and is shown to be applicable to other regions and a wide range of lithologies.

Tafoni and honeycomb structures as components of the hypogene karst system in the Crimean Piedmont (southern Ukraine)

Regional background

Geological and geomorphological settings. The Crimean Piedmont is part of the Crimean Mountains located in the south of Crimea, a large peninsula in the North Black Sea. The region lies at the margin of the Prichernomorsky artesian basin, along a geodynamically active regional suture zone that separates the fold–thrust structure of the Crimean Mountains (the Alpine fold–thrust belt) from the Scythian plate to the north (Fig. 1a). The Mesozoic Predgornaya Suture, formed as a result of the collision of the palaeo-terrain of the Mountainous Crimea with Eurasia in Jurassic to Early Cretaceous times, underlies the Piedmont along its whole extension and continues to the east through the Kerch Peninsula as the North Kerch retro-thrust zone (Yudin 2011). The Piedmont stretches for *c.* 130 km as an arched belt 5–15 km in width (Fig. 1b) and includes the Inner (the Second) and the Outer (the Third) ranges of the Crimean Mountains (Fig. 1c), as well as the inter-range lowlands. It is separated from the Main Range by longitudinal erosional depressions.

The upper tectonostratigraphic level of the Piedmont consists of alternating marly–clayey and carbonate rocks of the Late Cretaceous, Paleogene and Neogene, which were moderately deformed during reactivation of the suture in the Cenozoic. It forms a homocline dipping at 5–15° to the NW (in the southwestern sector) and north (in the central and eastern sectors) (Fig. 1b, c). Further in these directions, under the North Crimean Plain, the Cretaceous–Paleogene strata plunge steeply to a depth of 600–1000 m. The Outer Range, consisting of Neogene carbonates, is gentler than the Inner Range. The distinct limestone beds of Paleocene (Upper Danian) and Eocene (Ypresian and Lutetian) strata comprise the dip slope of the more prominent Inner Range, whereas the south- and SE-facing front slopes are typically steep, resulting in a prominent cuesta relief with limestone cliffs (Fig. 2). In the southwestern sector of the Inner Range, the Paleocene and Eocene limestone beds, separated by a marly–clayey sequence, constitute two distinct cuesta ridges. Major transverse valleys of rivers that originate in the Main Range cross the Inner Range, dividing it into a series of large cuesta massifs. In places in the southwestern sector, the combination of secondary longitudinal and transverse entrenchments dismember the ridges into separated mesas and buttes with inclined flat tops (Fig. 2c). In the central sector, the Paleocene carbonates are missing

Fig. 1. Location and main orographic and geological features of the Crimean Piedmont. (**a**) Outline of the Crimean Peninsula showing the relations between the major tectonic structures and the location of the ridges of the Crimean Mountains. (**b**) Shaded relief map of the Crimean Mountains with the overlay showing the formations of the upper tectonostratigraphic layer (contours of the geological units from Yudin 2011). Numbers in circles indicate key sites and objects of the Crimean Piedmont referred to in the text: 1, Kara-Koba; 2, Eski-Kermen; 3, Chardakly; 4, Kurushljuk; 5, Belokamennoye; 6, Kachi-Kalyon; 7, Chufut-Kale; 8, Besh-Kosh; 9, Skalistoye; 10, Bakla; 11, Malynove; 12, Tash-Dzhargan; 13, Rusakovka; 14, Sarak-Kaya; 15, Ak-Kaya; 16, Ailanma; 17, Burunduk-Kaya; 18, Bor-Kaya. (**c**) Geological section along line I–II in part (b). Modified from Klimchouk *et al.* (2013*b*).

Fig. 2. Cuesta-like landscape of the Inner Range of the Crimean Piedmont: (a–c) southwestern sector; (d) central sector; and (e) eastern sector. (**a**) Side view of the double cuesta in the Skalistoye area (9 in Fig. 1b). (**b**) Front view of the Paleocene cuesta in the Kachi-Kalyon area (6 in Fig. 1b). (**c**) View of the Paleocene cuesta back slope and the Churuk Su valley east of Bakhchisaray (7 in Fig. 1b). (**d**) Front view of Eocene cuesta in the Sarak-Kaya massif (14 in Fig. 1b). (**e**) Front view of the Ak-Kaya massif (15 in Fig. 1b).

in the cross-section and the Inner Range consists of only Eocene limestones.

Fracture distribution in the homocline sequence of the upper tectonostratigraphic level is uneven. Individual karst conduits and linear segments of cliffs are controlled by large subvertical fractures with north–south, west–east and oblique orientations. In cross-section, the fractures normally extend vertically for 10–40 m across several lithostratigraphic units, but often terminate upwards at certain stratigraphic boundaries. Also common in the Inner Range, particularly within the upper part of the Eocene limestones, are pinnate fractures inclined at 25–40°, associated with major subvertical cross-formational disruptions. Although fracturing is intense in the cliffs and the crest areas of the cuesta massifs, it becomes less intense at short distances away from the cliffs, as evidenced by observations in numerous quarries located on the dip slopes. Detailed studies in key areas (Tymokhina *et al.* 2011, 2012; Klimchouk *et al.* 2013*b*) show that the longitudinal and transverse valleys that limit and cross the Inner Range are controlled by linear karstified fracture zones, 100–400 m wide, beyond which the rocks remain largely intact. Such a distribution is best described by a concept of fracture corridors, structural features recognized in many regions (Singh *et al.* 2008; Bush 2010; Questiaux *et al.* 2010; Ogata *et al.* 2014). Peripheral fractures enlarged by solution of such karstified corridors are observed at the cliffs and crests of the cuesta massifs. Their presence causes the detachment of blocks and the dominant toppling mechanism of the cliff retreat.

With respect to the regional hydrogeology, the Crimean Piedmont is situated on the southern edge of the North Crimean artesian basin, a part of the larger Prichernomorsky artesian basin that extends through continental Ukraine in the North Black Sea region (see inset in Fig. 1a). In the present geomorphological situation, the Inner and Outer ranges constitute a marginal recharge area of the basin. The limestone beds are now completely drained in the crest of the Inner Range, but they contain groundwater and become confined with a plunge of the strata to the NW and north. There are flowing artesian wells from the Paleocene and Eocene aquifers in the lower dip slope, the inter-range lowland and the Outer Range. Some wells and rising springs are characterized by increased concentrations of dissolved gases and the degassing of N, CO_2, CH and H_2S. In places, the groundwater has increased concentrations of He, Ra and Hg vapour (Lushchik *et al.* 1981, 1988). The presence of cross-formational faults and fracture corridors, as well as hydraulic head relationships, allow vertical hydraulic communication in the confined zone of the upper tectonostratigraphic levels and localized upwelling leakage to shallower aquifers from deeper groundwater systems.

Evolution of the landscape and hydrogeological conditions. The Lower Cretaceous–Paleogene cover originally extended to the south over the present day Main Range. Tectonic compression since the early Oligocene may have promoted early phases of rising flow in the Piedmont area. Uplifts in the late Miocene (Sarmatian) exhumed the Upper Jurassic limestone massifs that now form the majority of the Main Range and formed the homocline of the northern slope (Lysenko 1976, 2002) and the marginal recharge area for the adjacent artesian basin (Klimchouk *et al.* 2013*b*). Subsequent tectonic phases in the late Sarmatian to Pliocene enhanced fracturing of the Lower Cretaceous–Paleogene cover above the Predgornaya Suture, where a marginal discharge zone established along the present day Piedmont. In this zone, groundwater flow in stratiform aquifers of the homocline mixed with rising deep fluids, promoting hypogene karstification in fractured zones (Klimchouk *et al.* 2013*a*, *b*, 2017). Continued erosion on the northern slope of the Main Range promoted the increase in recharge and flow in the homocline aquifer system, but later caused the orographic and hydraulic separation of the Main and Inner ranges. By the middle Pleistocene, the geological and hydraulic continuity of the homocline was completely breached in the southwestern sector, but was partly preserved in the central and eastern sectors of the mountain region.

U/Th dates from samples of phreatic calcite (417–263 ka) and the base parts of stalagmites (130–125 ka) in caves of the Inner Range provide constraints on the timing of the transition from phreatic to vadose conditions, which corresponds to the expression of the Paleocene cuesta in the landscape (Klimchouk *et al.* 2011). These data suggest that in the southwestern sector of the Piedmont, the Paleocene cuesta escarpment had formed at the end of the Middle Pleistocene (between 263 and 130 ka), significantly later than previously thought (Late Pliocene–Early Pleistocene, e.g. Nikishin *et al.* 2006; Vakhrushev 2010).

Hypogene karst in the Crimean Piedmont. Until recently, karst was considered to be scarcely developed in the Crimean Piedmont as a result of unfavourable geological and climatic conditions. Superficial karstic forms are rare in the Inner Range and include subdued karren features in the barren crests of the cuestas and a few small dolines on the dip slopes. In contrast to this is the occurrence of caves with well-developed phreatic morphology. Although these caves display no functional relationship with the landscape, they have been interpreted to be of epigenic origin (Dublyansky & Lomaev 1980; Dushevsky 1987).

In contrast to the scarcity of classical superficial karst features such as dolines and karren, the Inner Range is characterized by an abundance of various conspicuous geomorphological features, including rock amphitheatres, protruding rocks (bastions) and pinnacles (rock idols) along the massif edges, rock shelters, stratiform notches, niches, subvertical sculptured indentations (half-conduits) and various forms of cavernous weathering (e.g. tafoni and honeycomb structures) in the cliffs. Following conventional geomorphological views, all these features have been loosely attributed to atmospheric weathering and differential erosion on exposed surfaces.

Systematic studies carried out by the Ukrainian Institute of Speleology and Karstology since 2006 have firmly established the hypogene origin of caves in the region (Klimchouk *et al.* 2009, 2011, 2013*a*, *b*, 2017; Amelichev *et al.* 2011; Klimchouk & Timokhina 2011; Tymokhina *et al.* 2011, 2012; Dublyansky *et al.* 2014). The regional model of hypogene speleogenesis (Fig. 3a) links different types of cavities occurring in different lithostratigraphic units into integral, complexly structured, but functionally united, void–conduit systems formed by the rising flow of deep fluids along cross-formational discontinuities, interacting with lateral flow along stratiform permeability features, such as aquifers and bedding planes (Klimchouk *et al.* 2012, 2013*a*, *b*, 2017). The basic elements of the hypogene void–conduit systems, which constitute their spatial framework, are karst rifts formed along major subvertical fractures arranged in fracture zones (corridors). When surface denudation/erosion reached the limestone beds, the presence of hypogenically karstified, rift-dominated zones in the Piedmont guided erosional incision and the formation of cuesta cliffs (Fig. 3b). Further retreat of the cliffs occurred via block detachment and toppling and continuously exposed the morphology of the walls of the karstified fractures remaining on the periphery of the incised fracture corridors.

Limestone cliffs and karst rifts. The front and side (in the transverse valleys) scarps of the cuesta massifs in the Inner Range are characterized by subvertical limestone cliffs and steep lower slopes (Fig. 2). These major scarps reach up to 250 m in vertical extent, whereas the cliffs are up to 100 m high (Fig. 2a, b). To the east of Simferopol in the central sector, the Inner Range has a relatively low prominence in the landscape and the front cliffs are only 5–20 m high (Fig. 2d). In the southwestern sector and in places in the eastern sector, there are many deeply incised ravines and pocket valleys on the back slope of the cuestas, with steep to vertical heads and 10–50 m high side cliffs (Fig. 2c).

With several hundred kilometres of limestone cliffs, the Inner Range presents an excellent opportunity to examine the conspicuous cavernous features of diverse morphology abundantly displayed here, particularly rock shelters, tafoni and honeycomb structures. In addition, there are numerous elongated

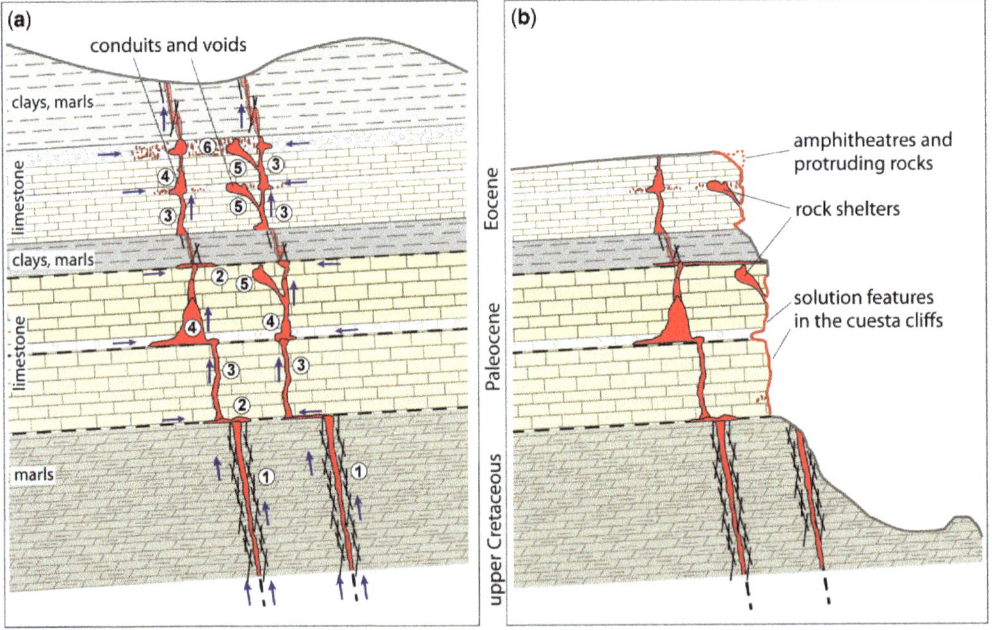

Fig. 3. Conceptual model of hypogene speleogenesis in the Crimean Piedmont. (**a**) Structure and function of void–conduit systems at the main stage of hypogene speleogenesis. Numbers in circles indicate morphogenetic components: 1, rift-type conduits along faults and damage zones in the Maastrichtian marls; 2, bedding partings enlarged by solution; 3, rift-type conduits along subvertical fractures in limestones; 4, large passages with abundant convection dissolution speleomorphs formed along intersections of subvertical fracture conduits with bedding-parallel high-permeability elements, or by the merging of closely spaced rifts; 5, blind chambers formed along inclined pinnate fractures; 6, zones of touching-vug porosity rimming major conduits and voids. (**b**) Exposure of hypogene solution morphology in cuesta cliffs in the present day geomorphic situation. Modified from Klimchouk *et al.* (2013*b*).

sculptured indentations, most commonly vertical and subvertical, that can be best described as wall half-conduits. Stratiform niches that form vertically stacked complexes are also frequent. Some cliffs demonstrate complex arrays of recessed forms constituting interconnected patterns of half- and true conduits, the morphology of which suggests rising flow (Fig. 2d).

Most of the vertical cliffs retreat through the failure and toppling of blocks, detached from the massif along karstified fractures (Fig. 4). This process maintains the verticality of cliffs and continuously exposes relatively fresh walls of karstified fractures with an abundant and diverse cavernous morphology. This mode of cliff retreat is evidenced by the obvious speleogenetic character of solution features in the cliff faces and by the presence on the lower slopes of fallen rock boulders retaining fragments of similarly sculptured walls (Fig. 2).

The lines of the cliffs are determined by combinations of individual faces of karstified fractures. The cliff-forming fractures can belong to a single

Fig. 4. Karst rifts in the cuesta cliffs of the Inner Range. (**a**) Rift co-planar to the cliff face. (**b–f**) Rifts occurring parallel and next to the cliff faces. Note (b) and (e) accessed via side window formed due the breach of the partition separating the rifts from the cliffs. (**g**) Rift normal to the cliff face and accessed from the back slope surface at the top. (**h**) Top part of a rift in Tavrskaya cave. Parts (a), (b), (d–f) Eocene limestones; parts (c), (g) and (h) Paleocene limestones. All photos from the southwestern sector (circled numbers in Fig. 1b: parts (a, b), 5, (c) 1, (d) 2, (e) 11, (g) 6, (h) 4; (f) Bakhchisaray).

(a)
degrading tafoni
rift
shallow notches
unroofed rock shelters
fresh fracture surface
inclined rock shelters
footslope
(b)
degrading tafoni
degrading notch
rift
fresh fracture surface
inclined rock shelters
(c)
(d)
(e)
(f)
(g)
(h)

dominant set, occurring en echelon so that the cliff line shifts back and forth or stepwise along its extension, or to fractures that are sub-perpendicular or diagonal to each other so that the cliff line zigzags. Both cases may produce situations where elementary faces in the cliff continue laterally in the massif as the walls of karstified fractures (Fig. 4a). Subvertical fractures that are more than a few metres in height and length, enlarged by solution to a width >10 cm, are termed karst rifts (Klimchouk *et al.* 2017; Fig. 4). Comparing the morphologies of the walls exposed in the cliffs with those in the intact rifts gives important insights into the relative roles of speleogenetic and external weathering processes in sculpturing the walls.

Rifts that can be penetrated by humans are numerous in the Inner Range. They are accessible either laterally from the cliff face or vertically from the cuesta top surface in the adjacent crest areas (Klimchouk *et al.* 2013*b*, 2017). Rifts that occur parallel to the cliff face and next to it, separated from the cliff by thin (3–10 m) partitions, are sometimes accessible via windows formed by a partial breach of the partitions to the valley side (Fig. 3b, e). The height dimensions of individual rifts vary from several metres to 40–60 m and their length varies from a few tens to a few hundreds of metres. The widths of rifts are also variable, but rarely exceed 2 m. Most rifts have lateral terminations pinching out more or less gradually into narrow, almost unchanged fractures.

Rifts are generally slot-like and fissure-like in cross-section, but in many rifts there are symmetrically or asymmetrically widened intervals (swells) corresponding to sub-horizontal niches developed in particular beds or along some bedding planes. Because such intervals are stratigraphically controlled, the widened space (sometimes the only part of a rift that can be passed by humans) may look like a horizontal or slightly inclined passage with narrower vertical slots extending down and up. In fact, almost all the common caves documented in the region correspond to such situations (Klimchouk *et al.* 2013*b*, 2017).

For the rifts and the adjacent cliff faces in the Eocene limestones, corrugated surfaces are common, with sub-horizontal ridges and notches controlled by lithostratigraphic variations and a wavelength ranging from 1 to 2 m (Fig. 3a, b). Notches are often complicated by various hollows, such as pockets, niches, tafoni and small side conduits. Rifts in the Upper Danian sequence of the Paleocene cuesta, which form most of the vertical cliffs in the southwestern sector, display less bedding control over the solutional morphology, although in places it is prominent. Rifts and the adjacent cliff faces in the Lower–Middle Danian marly limestones and marls, which form the foot of the cuesta front slope in many escarpments of the southwestern and eastern sectors, also display corrugated surfaces with numerous pockets, but the wavelength is commonly <1 m.

Klimchouk *et al.* (2013*b*, 2017) demonstrated that karst rifts were formed through the widening of cross-formational fractures by upwelling deep flow, in which the dissolution capacity was locally enhanced due to mixing with ambient groundwater conducted by more permeable beds and bedding planes.

Cavernous decay features in the Piedmont cliffs

Rock shelters. Rock shelters are metre- to decametre-scale cavities in vertical and steeply inclined cliffs, in which the dimensions of the opening are greater than their normal-to-cliff extension into the rock. Synonyms for this term include rock overhangs, alcoves and abri, but such cavities are also often considered to be large tafoni. In this study, rock shelters are differentiated from what is designated here as tafoni because the characteristic size, distribution and proposed origin of rock shelters are different from those inferred for tafoni.

There are thousands of rock shelters in the cliffs of the Paleocene and Eocene cuestas in the Inner Range. They are found in cliffs with varied expositions and heights and occur in different geomorphological and environmental situations (Figs 5 & 6): in the main front of the cuesta massifs, in large transverse valleys, in headless valleys and narrow ravines dissecting the cuesta back slopes, and in open and forested escarpments. They occur in all textural and structural varieties of Paleocene and Eocene limestones and marls exposed in the cliffs, at different heights above the base.

Despite the variability of these conditions, many rock shelters display some persistent characteristics, allowing their classification into several types. The most conspicuous are features that have their floor

Fig. 5. Typical occurrences and morphologies of rock shelters in the Inner Range of the Crimean Piedmont. (**a**, **b**) Variety of rock shelters in the Eocene limestone cliffs in the Tekme-Tash valley, Eski-Kermen (2 in Fig. 1b). Photo (b) is a continuation of the cliff shown in photo (a) (white arrows indicate matching points). Larger inclined rock shelters occur at the base of the cliff and a number of other features are scattered at various heights (some upper features are unroofed). The cliff in part (b) also shows degrading tafoni and notches in the cliff. (**c–h**) Varying morphologies of rock shelters: (e) at the contact of the Maastrichtian marls with the Eocene limestones (the eastern sector; 15 in Fig. 1b); (g) in the Paleogene limestones (3 in Fig. 1b); other features in the Eocene limestones. All photos except part (e) are from the southwestern sector of the Piedmont.

Fig. 6. Ceiling solution cupolas, nested alcoves and rising chimneys in rock shelters. (**d**, **e**) Rock shelters in which solution cupolas are truncated by the denudation surface. The cupola in part (e) (Tash-Dzhargan; 12 in Fig. 1b) holds exactly the same lithostratigraphic position as similar cupolas in the nearby Zmeinaya cave. All the features are in Eocene limestones; part (g) is in the central sector of the Piedmont (13 in Fig. 1b); all other photos are from the southwestern sector.

inclined outwards at 25–40° and a concave vault that frequently rises substantially above the level of the visor (Figs 5b–f, h & 6a, e, f). Such rock shelters often have a gentler slope or a sub-horizontal bench at the rear (Fig. 5e, f). They are typically 10–30 m wide (although sometimes up to 60–80 m wide) and 3–15 m high, extending into the massif for 10–30 m relative to the cliff face. In the cliffs of the Eocene cuesta in the SW sector of the Piedmont, inclined rock shelters of remarkably similar morphology often occur in series along straight cliff segments, with a regular distance of 80–100 m between individual features that are at the same height above the base (Fig. 5h).

Other common forms of rock shelters are simple spherical or oval (aligned to bedding) recesses in the cliff walls (Fig. 4a), with openings of up to 10–12 m and indentations up to 5–7 m (typically 2–3 m). Some features have a horizontal base controlled by the contacts between different rock units (Fig. 6b, g).

Many rock shelters have nested alcoves at the rear (Fig. 6f), oval or spherical cupolas in the vaulted ceiling or elsewhere (Fig. 6a–e), rising chimneys in the ceiling (Fig. 6g) and half-channels in the overhanging visors. These features closely resemble convection-related speleogens found in hypogene caves. In some rock shelters lying at shallow depth beneath the back surface of the cuestas, the lowering denudation surface intercepts some high chimneys or cupolas, creating windows and bridges (Fig. 6d, e). Sometimes the rock shelters have honeycomb structures inside, often concentrated in the upper parts of the rear walls and at the ceilings.

Most of the rock shelters do not display obvious signs of fracture control of their occurrence and morphology, but it is likely that guiding fractures have been completely worked out by the progressive development of the features. Some inclined shelters do show traces of guiding inclined fractures in their rear or side walls. Fractures at the same angles are widely observed in the cliffs and often control characteristic facets in the denuded ledges of the cuesta massifs. Other rock shelters are controlled by vertical fractures normal to the cliff, traces of which are visible at the rear and ceiling (Fig. 6b).

Tafoni. In this work, the term tafoni is conventionally reserved to denote decimetre- to metre-scale cavities, i.e. features that are larger than honeycomb cells, but smaller than rock shelters. In the Inner Range, tafoni occur in groups and are commonly closely spaced within the tafoni-bearing cliff faces (Fig. 7a, b, f, g). They are found only in the Eocene limestone cliffs, mainly in the SW sector of the Inner Range.

Tafoni are characterized by smooth concave inner surfaces. Typical dimensions are within 10–50 cm, but occasionally large 1–2 m sized features occur. Smaller features are often nested in larger ones. In closely spaced groups, individual cavities often almost coalesce, with only thin (5–10 cm) walls remaining between adjacent features, so that the surface resembles a larger scale version of the honeycomb structures. Interestingly, the morphology and patterns of tafoni differ substantially between cliffs oriented along the dip (SE–NW in the southwestern sector of the Piedmont) and those oriented along the strike or at oblique directions (SW–NE). In the former case, the individual cavities tend to have near-circular shapes and are densely packed in stratiform series corresponding to certain beds and notches developed along them (Fig. 7a, b). In the latter case, the cavities are generally less densely distributed and tend to be elongated along the beds (Fig. 7g).

The most outstanding display of tafoni (Fig. 7a, b, f, g) is located near the village of Belokamennoye. In this locality, a long NW-trending cliff with abundant, well-shaped and distinct tafoni (Fig. 7a, b) continues to the NW as a wall of an intact karst rift (Fig. 4a). Geomorphological observations suggest that the cliff face was exposed relatively recently (on a timescale of several hundred to a few thousand years) by the detachment of large blocks along the rift and their toppling towards the valley. Further to the NW along this escarpment, another rift of the same set is accessible via a side window (Fig. 7c; see also Fig. 5b) that opened very recently (within several tens of years; see also Fig. 4b) via breakdown. It is therefore possible to contrast and compare the morphology of the karstified fracture walls prior (in the rifts) and after (in the cliff) exposure to external weathering. In both cases the walls are corrugated, with stratigraphically controlled sub-horizontal niches and ridges. The undulation is smoother in the rift, but more distinct and of higher amplitude in the cliff. Importantly, incipient, still largely latent, structures of future tafoni are easily recognizable on the rift wall (indicated by black arrows in Fig. 7c). They are filled with friable material, which is easily removed by weathering when the rift wall is exposed as a cliff.

An intermediate situation between the state of cryptic tafoni in the karst rift and the well-formed tafoni in the cliff is documented in another locality (Eski-Kermen), where a similarly oriented cliff face has been recently exposed, probably not more than a few hundred years ago, based on geomorphological relationships. At this location, incipient tafoni are recessed at 10–20 cm (Fig. 7d). Granular disintegration and flaking on the inner surfaces is an active ongoing process and 5–10 cm thick accumulations of the loose weathering products are present on the floor of cavities and the outcrop shelves.

The major cliffs in the Eski Kermen area were exposed tens of thousands of years ago. These cliffs demonstrate only barely discernible traces of tafoni because they have almost completely degraded under the action of weathering and surface washing (Fig. 5a, b). This indicates that tafoni are relatively short-lived features. When the wall of a karst rift is exposed as a cliff, a stage of the prograde development (several hundred to several thousand years) then passes into a stage of retrograde development,

Fig. 7. Tafoni in the Inner Range of the Crimean Piedmont. All photos show outcrops of the Eocene limestones: (**a–c**, **f**, **g**) Belokamennoye site (5 in Fig. 1); (**d**) Eski-Kermen site (2 in Fig. 1); (**e**) Malynove site (11 in Fig. 1). Black arrows in parts (c) and (e) point to latent tafoni; white arrows in part (e) point to the alterite in the near-wall zone of the karst rift. Photo in (e) by E. Tymokhina.

during which tafoni features degrade and eventually obliterate under the action of weathering.

These observations strongly suggest that the near-wall zone of the host rock in karst rifts is considerably altered, characterized by the irregular propagation of a porous, often friable rock. This creates a template for the formation of tafoni, which readily takes place when the wall surface is exposed to the action of vigorous denudation agencies, such as those of external weathering. The geometry of tafoni is predefined by the configuration of the alteration front at the depth of the rock, but not by the weathering mechanisms at the exposure stage. This is further illustrated by Figure 7e, which shows a karst rift normal to the cliff, where the cliff face cuts across the altered zone on the right-hand wall. Weathering attacks the altered zone from the side, exposing the contact between the alterite and the unchanged rock before the alterite is removed, showing the alteration zone in the cross-section. A tafoni-like pocket, still filled with a porous alterite, is clearly visible and the alterite crust extends into the depth of the rift along the right-hand wall.

Honeycomb structures. Honeycomb structures are characterized by closely spaced, centimetre- to decimetre-scale hollows, separated by thin ribs and walls and arranged into characteristic patterns. Honeycomb structures occur widely in the Crimean Piedmont, in both the Inner and Outer ranges, in all varieties of carbonate rocks exposed in the cuesta escarpments. They occur in cliffs of varied orientations and in different environmental situations. Honeycombing commonly covers the inner surfaces of various concave features in cliffs (niches, half-conduits and rock shelters; Fig. 8a–f), but often spreads across large faces of recently exposed karstified fractures (Figs 4c, 8g, h & 9d–h). In the concave features, honeycombing tends to concentrate in the upper parts of the inner surfaces. In several instances, honeycomb structures are found in close association with caves (Fig. 8i). They also often occur on certain sides of rock blocks and boulders scattered beneath the cliffs, but, in most

Fig. 8. Typical occurrences of honeycomb structures in the Inner Range of the Crimean Piedmont: (**a–f**) in the concave forms in the cliffs; (**g**, **h**) in the relatively flat cliff faces; and (**i**) in the cave walls. Parts (a–f) show Paleocene limestones; the other photos show the Eocene limestones. Parts (e) and (f) are from the eastern sector (16 and 17 in Fig. 1b), parts (h) and (i) are from the central sector (14 in Fig. 1b). Other photos are from the southwestern sector (parts (**a–c**) 4 and parts (d) and (g) 10 in Fig. 1b). Large arrow in part (e) indicates the now-exposed plane of the major cliff-forming fracture.

Fig. 9. Variations in the patterns of honeycomb structures in the Inner Range of the Crimean Piedmont. **(d–i)** Honeycombing on exposed fracture walls (obvious remnants of the fracture planes in parts (g) and (i) are indicated by large arrows). Small black arrows point to patches where the basement honeycomb-bearing zone has been removed and the surface of the unchanged rock is exposed. **(a–c)** Eocene limestones; other photos, Paleocene limestones. Part (c) is from the eastern sector of the Piedmont (16 in Fig. 1b); other photos are from the southwestern sector (parts (a, b) 10, parts (d–f) and (i) 4, and parts (g, h) 1 in Fig. 1b).

instances, it is evident that honeycombing was imposed on them prior to the failure. The obvious independence of honeycombing with respect to the landscape and microclimatic conditions suggests that the weathering system is not the determinant in the formation of these structures. Honeycomb structures are best expressed on surfaces protected from surface runoff and show clear signs of the obliteration of features when they are directly exposed to water.

The patterns of honeycomb structures are varied (Figs 8 & 9). Although patterns resembling honeycomb are the most common, other pattern styles – such as spongework, lacework, boxwork and fretwork – also frequently occur. The sizes of individual cells may vary randomly and irregularly, but may change sharply or gradually (directionally) from one area to another (Fig. 9a, b). The geometries and sizes of the ribs and walls in the honeycomb structures also vary between the honeycombing localities, as well as within individual patches. The walls may have the same thickness along the entire height, or thicken towards the base. In some honeycombing surfaces the walls vary only slightly in width, whereas in others the variations are substantial. On the general background of a regular wall pattern, there may be patches of botryoidal or tufa-like masses of the same material. In some places, the ribs and walls line macro-fractures (Fig. 8a). In many cases, flat tabular sheets of the hardened material, the same as that of the ribs and walls, form the outer surface of the honeycomb-bearing zone, clearly indicating the wall of the major cliff-forming fracture (Figs 8e & 9h).

In addition to the variations in the geometry of the pattern, the honeycomb structures vary according to the relative volume of the hollows and the material that separates them. At one extreme, the cavities occupy most of the volume of the honeycomb-bearing zone, in which the only proxy of the original rock mass is the skeleton consisting of delicate ribs and walls (framework-dominated honeycombing; Fig. 9a–c). At the other extreme, the cavities are separated by more massive walls or are isolated in the rock mass, although still densely packed across the surface (matrix-dominated honeycombing; Fig. 9g–i). Transitions from one extreme to another may occur within a single honeycombing surface.

An important trait of honeycomb structures in the Inner Range is that they develop within a 5–50 cm thick zone of the obviously altered rock, which commonly has a distinct, although not necessarily even, inner boundary with the unchanged rock substrate. The ribs and walls of honeycomb cells commonly consist of more dense, hardened material than the surrounding alterite. The same material lines clearly recognizable fractures occurring within the honeycombed surface. Based on the analyses of occasional samples and field acid tests, the wall and rib material is calcite, although a systematic study of the mineralogy of honeycomb structures is still to be carried out. Several isotope analyses of the material from the walls and ribs of honeycomb structures in the Paleocene cuesta gave $\delta^{18}O$ and $\delta^{13}C$ values shifted towards more positive values (by up to 3 and 1‰, respectively) compared with the unchanged rock (Dublyansky *et al.* 2014).

The growth of cells at depth stops when they reach the boundary of the altered zone with the substrate and further development is dominated by the destruction and lowering of the walls and protrusions in the honeycomb structures, leading to their complete obliteration. This is illustrated by the photos in Figure 9a, b, g and i, where the basement of the honeycomb-bearing zone is exposed by denudation in certain patches within the honeycombed surfaces. In many cases, especially in the Paleocene limestones, the honeycomb-bearing zone is distinct and consists of tufa-like, porous, irregularly indurated material (Figs 9g, i & 10).

In a few places, the fragmentation of rock outcrops by fresh fractures or small, old stone quarries have made the tufa-like honeycomb-bearing zone available for observation in cross-section (Fig. 10). The boundary between the altered and unaltered rock is clearly visible and is accentuated by the dark coloration of the alterite. This coloration is caused by the penetration of cryptoendoliths (Golubic *et al.* 1981) – cyanobacteria that invade and inhabit the porous rock. The distinct boundary of the cryptoendolith invasion marks the difference in porosity between the alterite and the unchanged rock. Figure 10 also shows that the development of honeycombing is preconditioned by the presence of the alterite and does not occur in the unchanged rock.

Discussion: a new paradigm for the origin of cavernous decay features

Cavernous decay phenomena: the crisis of the weathering paradigm

Tafoni and honeycomb structures are still some of the most enigmatic and puzzling geomorphological phenomena, despite vigorous research efforts over more than a century. They are generally considered to form by selective weathering on exposure to atmospheric conditions. Assorted processes have been invoked to take part in their formation, including aeolian deflation, weathering through insolation, frost action, wetting and drying, chemical and salt weathering, and biological decay. Case hardening and/or core softening were thought by some workers to play a major part. Salt weathering is most commonly invoked. Although all these processes can contribute to granular disintegration, flaking, the resultant rock decay and the formation of caverning, none of them alone, or in combination with other processes, has been shown to satisfactory explain the major traits of localization, development, distribution, morphology and patterns of tafoni and honeycomb structures. Turkington (2004, p. 128) admits: 'Considerable literature has accumulated on the nature of both tafoni and alveoli, but as more information has been presented, their possible origins, rather than being clarified, seem to have become more confused'. Brandmeier *et al.* (2011, p. 840) write: 'While it is largely accepted that lithology, the presence of salts, climate and exposition form a complex system of weathering conditions, the major processes controlling tafoni-type-weathering are poorly understood.'. A thorough review by Groom *et al.* (2015, p. 10) concluded that 'Notwithstanding substantial academic attention, scientific understanding of the principal driving mechanisms

Fig. 10. Cross-sections of the altered, honeycomb-bearing zone in the Paleocene limestones, the southwestern sector of the Piedmont: (**a**) a split along fresh fractures in the Kurushljuk massif (4 in Fig. 1); and (**b**) an artificial cut in a small stone quarry in the Besh-Kosh massif (8 in Fig. 1b). The altered zone is darkened due to the invasion of cryptoendoliths, which highlights the limits of the porous altered zone. Note the distinct boundary with the unchanged rock.

for cavernous decay remains yet to be discovered'. These and similar acknowledgements in other publications attest the obvious explanatory crisis in cavernous decay studies.

In searching for a way out, recent work has invoked the polygenetic nature of cavernous phenomenon, which is thought to be '... the result of multiple, if not simultaneous, decay processes varying case to case.' (Groom *et al.* 2015). Other workers invoke self-organization characterized by positive feedback as a response to dynamic instability within the weathering system, independent of the particular decay processes (Turkington & Phillips 2004). Although these thoughts are certainly sensible, they add no clarity to the problem of the obviously weak dependence of cavernous phenomena on the characteristics of the weathering system.

That the formation of cavernous features is not directly dependent on lithology, the local environmental conditions and weathering mechanisms is further illustrated by the following opposing facts:

- the same features can develop on different rocks, such as limestone and sandstone, exposed under drastically different local environmental conditions, such as in open cliff faces and in caves (compare Fig. 11a–d), and in rocks of the same composition, but of different age and diagenetic maturity (compare Fig. 11e, f);
- in the case of honeycombing, different patterns can develop on a seemingly homogenous rock, in the same lithostratigraphic unit, in adjacent parts of the same cliff face;
- features in the same region, in cliffs characterized by the same aspect and geomorphological position, display a fresh appearance in one outcrop and clear signs of degradation in another (compare Figs 7a, b & 5b).

The contradictions of the weathering paradigm with regard to the origin of cavernous features further show up when weathering processes are invoked to explain the formation of features of contrasting scales and styles, such as rock shelters and honeycomb structures, especially considering that the latter are often found nested in the former (Fig. 8d–f). Why and how the operation of the weathering system may produce such drastically different features at the same place is not explained.

Fig. 11. A comparison of tafoni and honeycomb structures in different rocks and environmental conditions. (**a**, **b**) Nested tafoni in a larger cavity: part (a) in the Eocene limestone, Piedmont Crimea, Ukraine and part (b) in the Carboniferous arkose sandstone, Bohemia, Czech Republic. (**c**, **d**) Honeycomb structures: part (c) in the Eocene limestone, Piedmont Crimea, Ukraine and part (d) in the Cretaceous arkose sandstone, Western Caucasus, Russia. (**e**, **f**) Tafoni-like cavities: part (e) in the Eocene limestone, Piedmont Crimea, Ukraine and part (f) in the Cretaceous limestone, Owl Mountains, Texas, USA. Note that part (f) shows tafoni-like cavities in a cliff face, most of which represent the fracture wall that was only recently partially exposed by a block failure. This wall extends to the right into the fracture (the exposure limit is outlined by the dotted line and indicated by the arrow), but it can be clearly seen that tafoni are already present there. Photograph in part (b) by V. Suchy, photograph in part (d) by A. Ostapenko and photograph in part (e) by K. Stafford (reproduced by permission); other photos by the author.

The major problem in studies of cavernous decay features is that they have always been constrained by the weathering paradigm – that is, the action of external agencies (the weathering system) on exposed rock surfaces was *a priori* considered to be the principal factor in the formation of these features. A central postulate of the weathering paradigm with regard to the cavernous features is that the main control of their localization, morphology and patterns is thought to be in the weathering conditions and that the processes determined by these conditions are paramount for the formation of the cavernous features. However, the occurrence of similar features across wide ranges of lithologies and environmental conditions, either regional or local, or both, strongly suggests that their formation is determined by a factor that overarches the variations in these conditions and the weathering processes.

Lessons from the Crimean Piedmont

The extended cliffs of the Eocene, Paleocene and (in places) Upper Cretaceous carbonate rocks in the Inner Range cuestas of the Crimean Piedmont abundantly display various cavernous features, including rock shelters, notches, sculptured channel-like indentations, tafoni and honeycomb structures. These features commonly occur either in close proximity to each other, or they overlap one another, forming a spatially and, presumably, genetically related assemblage that constitutes the most conspicuous geomorphological trait of the Inner Range cliffs.

Referring to evident signs of granular disintegration and flaking in cavernous forms and following a deeply rooted geomorphological tradition, previous studies loosely attributed the origin of these features to various weathering processes (Dushevsky 1987; Dushevsky *et al.* 1979; Dushevsky & Kuznetsov 1991). In explaining rock shelters, stress release effects were also invoked (Blaga & Popov 2009), similar to a more elaborated model later suggested by Bruthans *et al.* (2014). However, neither these processes alone nor their combinations can adequately explain the specific morphologies and singularities of the distribution and localization of rock shelters and other cavernous features. The study from the Crimean Piedmont confirms one of the greatest issues of the weathering paradigm with regard to cavernous features, well recognized in the literature on cavernous decay – that is, the morphology and distribution of the cavernous features in the cliffs do not show appreciable controls by geomorphological, microclimatic and other environmental conditions, despite their wide variation. Although there is a distinct lithological/stratigraphic control over the localization of some features at the site scale, the same forms often occur across very different lithologies. These characteristics obviously contradict the relations that would be expected of features whose formation is primarily controlled by the weathering system.

The features of the cavernous assemblage, however, are strongly structurally controlled. They are all ultimately related to karstified fractures, the most enlarged of which (conventionally >10 cm in width) are termed karst rifts in this study. Karst rifts and caves in the region have been shown (Klimchouk *et al.* 2012, 2013*b*, 2017) to be of hypogenic origin, i.e. to have formed by the rising flow of deep fluids along cross-formational discontinuities, interacting with lateral flow along stratiform permeability features such as aquifers and bedding planes. It has also been shown that karstified fractures directly control most of the cliffs faces in the Inner Range (Klimchouk *et al.* 2013*b*, 2017). It is therefore natural to assume that the major (and apparently multi-phase) hypogene stage of the fluid history could have pronounced effects on the properties of the host rocks around karst rifts and on the morphology of cliffs formed along them.

The data from the Crimean Piedmont demonstrate that these effects were crucial to the formation of cavernous features in the cuesta cliffs. First and most obvious is that the dissolutional hollows of various geometries in the walls of karstified fractures become external features after their exposure in cliffs. Among these forms, sub-horizontal notches, spherical and oval pockets, and simple rock shelters are the most common cliff features that have their direct analogies in the documented karst rifts and caves in the region (Klimchouk *et al.* 2013*b*, 2017). The same is true for ceiling cupolas in the rock shelters (Fig. 6c–e). In some locations (e.g. Fig. 6e) it is documented that they hold exactly the same lithostratigraphic position as similar cupolas in the nearby true caves. Klimchouk *et al.* (2012) demonstrated that various side alcoves, notches and lateral conduits in the fracture-controlled hypogene rifts and caves formed as a result of the mixing corrosion effect, where locally increased inflow of the ambient groundwater occurred along enhanced matrix porosity zones or permeable bedding planes.

A specific origin is inferred for the inclined rock shelters (Klimchouk *et al.* 2013*b*, 2017). Their formation was guided by inclined pinnate fractures associated with major subvertical fractures (Fig. 12). Where part of the rising flow deviated from the main rift conduit along the pinnate fractures, natural convection cells established to form inclined blind chambers. Where the upper parts of pinnate fractures intercepted high-permeability beds in the limestone aquifer, dissolution could be enhanced as a result of mixing. The model shown in Figure 12 is also corroborated by the frequent occurrences of patches of cemented breccia, attached

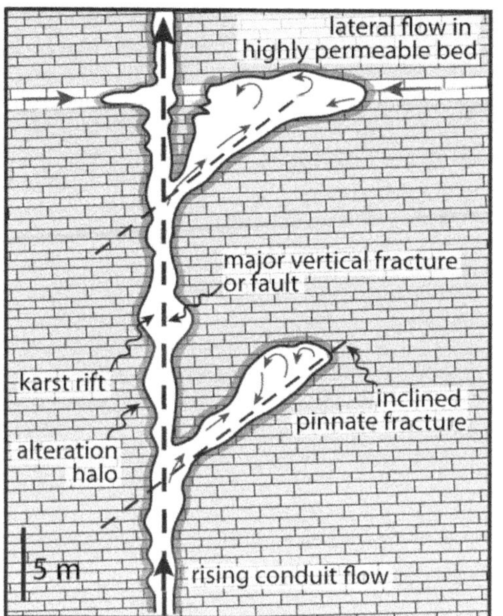

Fig. 12. Conceptual model of the formation of inclined chambers along pinnate fractures (from Klimchouk *et al.* 2013*b*). When a cliff is formed along the karst rift in the course of cuesta front slope retreat by block toppling, such chambers become inclined rock shelters.

to inclined surfaces of rock shelters at their mouth and sometimes to the vertical cliff walls, consisting of unsorted angular rock shatter. Such accumulations apparently formed before the exposure of the cliffs, at a time when karst rifts were opened at the top of the cuestas, trapping out the weathering shatter (Klimchouk *et al.* 2017). Direct modern analogues to this situation are documented in several places.

In addition to apparent speleogenetic morphologies, another important effect imposed by focused rising fluid flow and hypogene karstification is the alteration of the wall rock in karst rifts. This study provides strong field evidence that tafoni, as defined in this paper, devèlop in a cryptic form in the wall rocks of karst rifts, as pockets of the altered zone (Fig. 13b). Thus the formation of tafoni is predefined by the internal geometry of the alteration front (the boundary between the altered, friable material and the unchanged rock) and the features become morphologically expressed when the wall surface is exposed to the action of weathering processes in a cliff, which removes the alterite (Fig. 13c).

The alteration process in hypogene karst rifts is influenced by the interactions between the focused rising flow along them and the ambient groundwater in the host carbonate sequences. The unchanged Eocene limestones have a matrix porosity generally varying within 5–30% (Lygina 2010), with considerable cyclic variations between adjacent beds in some units. This causes lateral matrix flow to concentrate in more porous/permeable beds so that such beds supply a greater inflow to the walls of karst rifts, imposing a hydrostratigraphic control over the amount of interaction and dissolution/alteration effects along the vertical profile. This is expressed in the formation of sub-horizontal notches and tafoni in these beds (Fig. 7a–d). The deeper progression of the alterite pockets along conductive beds (the localization of cryptic tafoni) is caused by an even greater inflow from local systems of larger, connected pores.

Honeycomb structures occur on all the carbonate rocks exposed in the cuesta cliffs, but the diversity in their patterns do not demonstrate any regular affinity to specific varieties of rocks. The most important characteristic of honeycomb structures is that they

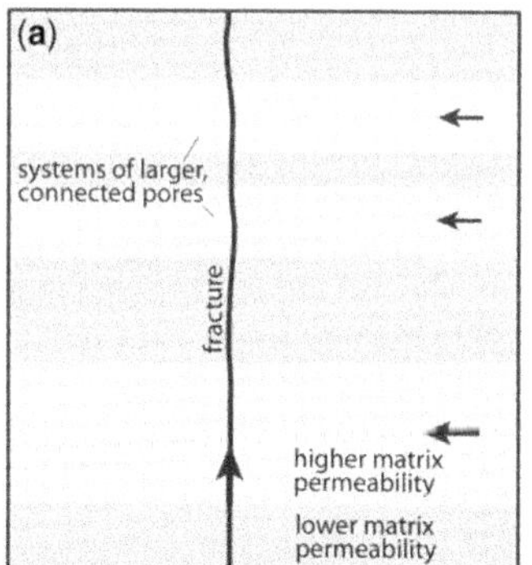

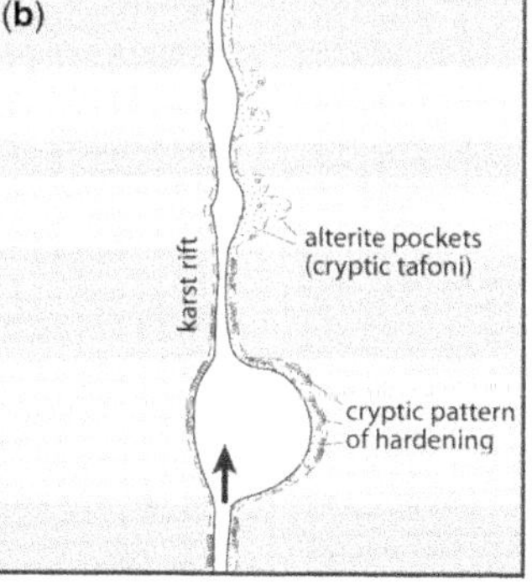

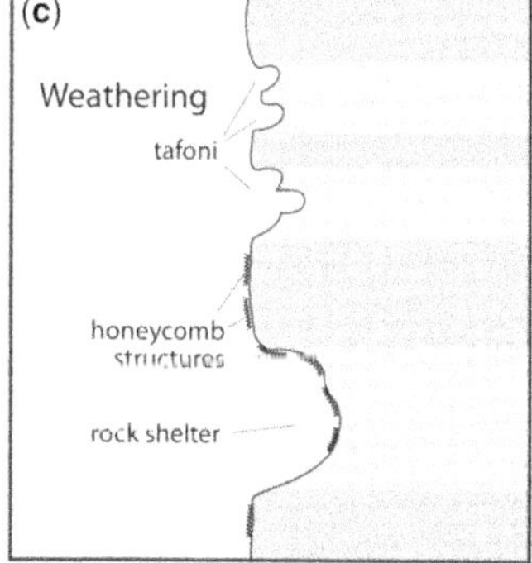

Fig. 13. Conceptual model of the formation of simple rock shelters, tafoni and honeycomb structures in the Crimean Piedmont by hypogene karstification and denudational cleaning. (**a**) Initial conditions; (**b**) multi-phase karstification though removal by dissolution and alteration; and (**c**) denudation of the alteration front and cryptic structures after exposure of the wall of a karst rift.

develop within a 5–50 cm thick zone of altered, significantly more porous rock, which commonly has a distinct boundary with the substrate. The ribs and walls of honeycomb cells consist of more dense, hardened material than the surrounding alterite. The fact that the same material often lines the fractures crossing the honeycombed surface suggests that the framework of the features is the result of differential fluid-induced precipitation/cementation within the altered zone by percolating fluids. This is supported by the more positive $\delta^{18}O$ and $\delta^{13}C$ values in the walls and ribs than in the unchanged rock. Such isotopic alteration cannot be produced by external weathering, but suggests a link to fluids enriched in both ^{18}O and ^{13}C isotopes (Dublyansky *et al.* 2014). The only waters in Crimea with the stable isotope composition required to drive the observed shift in the honeycomb material are found in a deep (>800 m) thermal aquifer in the north of the peninsula and in deeply rooted mud volcanoes in the Kerch Peninsula, to the east of the Piedmont (Fig. 1a). Klimchouk *et al.* (2013*a*, *b*) inferred that, in the past, most probably during the Middle Miocene (the time of the peak activity of the mud volcanoes in the Kerch Peninsula and the adjacent part of the Caucasus; Shniukov *et al.* 2006), similar deeply derived fluids ascended along cross-formational fracture zones in the Crimean Piedmont.

It is suggested here that the development of honeycomb structures is preconditioned by the formation of a three-dimensional framework of a hardened material within the altered zone of hypogenic void–conduit systems (Fig. 13b). As the cryptic hardened structure forms within the already existing alteration zone, its formation post-dates or is concomitant with the phase of the preferential creation of porosity. This cryptic pattern is then exhumed by a vigorous denudation agency to form honeycomb structures (commonly at the exposure).

As the formation of tafoni and honeycomb structures is pre-disposed by the fluid-induced alteration of the walls of conducting structures, these features only form where the exposed surface is in the altered zone. This explains the fact that cavernous features are not found on the apparently long-exposed surfaces of fractures formed after the emergence of the cuestas in the landscape – that is, where the altered zone is originally lacking. Equally, new features do not form on long-exposed surfaces where the alteration zone and the associated pre-existing caverns have been obliterated by denudation (Fig. 5b). Observations in the Crimean Piedmont suggest that the cavernous features are relatively short-lived, with typical lifespans between several hundred to several thousand years.

The extremely rich occurrence of cavernous features in the Crimean Piedmont is determined by the favourable geological, geodynamic and geomorphological factors. First, the position of the region along the margins of the large artesian basin and along the junction of two major tectonic structures provided favourable conditions for the cross-formational rise of deep fluids during periods of geodynamic activation, their interaction with groundwater flow in the stratified homocline and hypogene karstification. Second, the cuesta escarpments of the Inner Range have emerged in the landscape relatively recently, at the end of the Middle Pleistocene (Klimchouk *et al.* 2011). The entrenchment of valleys and the formation of the cliffs were guided by the presence of hypogenically karstified, rift-dominated fracture corridors. Further retreat of the cliffs via block failure and toppling continuously opens the walls of the remaining karstified fractures that bear solution-induced forms and the altered zones, exposing them to weathering. Which particular weathering mechanisms are involved in the emergence of the cavernous features pre-determined by the alteration structures is not relevant.

Ghost rock karstification, metasomatism and hypogene karstification

The mechanism of the formation of tafoni and honeycomb structures inferred for the Crimean Piedmont corresponds to the process known as ghost rock karstification. The concept of ghost rock karstification has been developed and publicized internationally during the last two decades by Belgium researchers (e.g. Vergari & Quinif 1997; Vergari 1998; Bruxelles *et al.* 2009; Quinif & Bruxelles 2011; Dubois *et al.* 2014, and references cited therein) to describe a two-stage process of cave formation by (1) chemical weathering of rocks by groundwater flow focused in permeable paths, such as joints and bedding planes, in a slow flow environment, and (2) subsequent removal of the porous and weakened altered rock (the ghost rock) by fast-flowing groundwater after a base level drop, mainly in the vadose zone. The first stage of the process is also termed rock phantomization. In the petrology and ore geology literature, the term ghost rock is often used to denote the alteration product, particularly if some elements of the texture and structure of the parent rock are retained in it. The removal of the alterite may result in substantial caves, often in maze systems of passages, commonly displaying a typical phreatic morphology, including spherical niches and cupolas (Dubois *et al.* 2014). In the context of karst, this phenomenon has also been described in France, Italy, the UK and other countries.

The publications by the Belgian group attribute ghost rock karstification to weathering that propagates with the downward infiltration from the surface, i.e. to epigene processes. However, the substantial isovolumetric wall rock alteration of carbonates, which requires a slow flow environment and specific conditions of water–rock interactions, although not excluded in epigene settings, is more liable to occur in the deep subsurface due to the action of hypogene fluids. Ghost rock phenomena in association with hypogene karstification have been reported from France (Audra 2017), Macedonia (Temovsky 2017), Hungary (Klimchouk 2017) and the Crimean Piedmont (Klimchouk *et al.* 2017 and this paper).

Klimchouk (2017) argued that the removal of the alterite and the creation/increase of a cave-type space can occur in different environments and through different processes. It may occur under hypogene conditions through dissolution induced by changes in pressure and temperature or in the composition of fluids. It may also occur through mechanical removal (erosion) when flow rates increase dramatically following an initial breach of the major upper confining unit, also under hypogene conditions. Publications on ghost rock karstification emphasize the mechanical removal of the alterite in the water-table and under vadose conditions, which is indeed an efficient mechanism. When cavities still filled with alterite are exposed to the surface, especially in subvertical faces in structurally controlled cliffs, they can be quickly cleaned out by external weathering agencies and display various conspicuous morphologies, including tafoni and honeycomb structures.

Dubois *et al.* (2014) strongly differentiate between traditional karstification by total removal and ghost rock karstification by separating chemical dissolution and mechanical erosion into two distinct stages. Although this contraposition was probably useful for introducing the concept of ghost rock karstification, the processes of normal karstification (dominated by removal by dissolution) and ghost rock karstification (with incomplete dissolution and the subsequent removal of the alterite) are closely related and interspersed, particularly in hypogene settings.

The alteration of the fracture wall rock by moving fluids, when accompanied by substantial changes in chemical composition, falls within the domain of infiltration metasomatism, a process well-known in petrology, ore geology and sedimentology. Substantial chemical changes in the alterite are reported in studies on ghost rock karstification (e.g. Dubois *et al.* 2014) and therefore ghost rock karstification could be considered as metasomatic karstification, i.e. karstification through metasomatism (Klimchouk 2017).

Metasomatism, a fluid-induced mass transfer process of re-equilibration in a rock involving a change in chemical composition (mineral replacement), is widely recognized to play a major part in the formation of the Earth's continental and oceanic crust and the lithospheric mantle (e.g. Korzhinskii 1953; Pospelov 1973; Zharikov *et al.* 1998, 2007; Harlov & Austrheim 2013; Yardley & Bodnar 2014). The metasomatic modification of rocks by moving fluids (infiltration metasomatism) mainly occurs in the upper crust, where the fluid fluxes are largest (Yardley 2013). Yardley & Bodnar (2014) point out that any fluid that moves across the boundaries between contrasting lithologies, or between regions with different temperatures and pressures, may cause metasomatism in the wall rocks, but they stress that the process is especially liable to occur under increased temperatures and pressures.

Rock transformations in metasomatic zones and the resultant metasomatites (alterites) vary greatly depending on the physico-chemical conditions (the compositions of the parent rocks and metasomatizing fluids and the pressure–temperature conditions). Among the metasomatites that occur in carbonate rocks, the most common are structurally controlled hydrothermal dolomites (Warren 2000; Davies & Smith 2006), scarns (at the contact with a silicate rock or magmatic melt) and argillisites (clay mineral assemblages, formed through the alteration of the host rock by low-temperature hydrothermal fluids around veins and fractures; Zharikov *et al.* 2007). Replacement by iron ores and silicification (replacement by chalcedony, chert or quartz) are other common metasomatic processes in carbonates.

Both metasomatism and karstification occur through a fluid-aided dissolution–transport–precipitation process, which underpins their likely intimate relationship Ezhov *et al.* (1992). Pospelov (1973) introduced the concept of extended metasomatism, in which the operation of zones of dissolution, transport and precipitation is stretched in time or space, or both. According to Pospelov, dissolution in extended metasomatism may lead to the development of macropores, cavities or even chambers. He considered that volume disequilibrium may develop locally, resulting in the formation of cavities (intersomatism). Further developing the ideas of Pospelov, Ezhov *et al.* (1992) asserted that karst is a component of a more general process of metasomatism.

In discussing the relationships between metasomatism and hypogene karstification, Klimchouk (2017) considered that spatial and temporal variations in pressure–temperature conditions and the compositions of rocks and metasomatizing fluids can lead to a breach in the balance between dissolution and precipitation and cause a local prevalence of dissolution, i.e. speleogenesis (Pospelov's extended metasomatism). The metasomatic increase in porosity in the rock matrix around a conductive fracture (the alteration halo) may serve to focus fluid flow and removal by dissolution and, in turn, speleogenesis may facilitate alteration of the rock by increasing fluid flow and changing the conditions and intensity of water–rock interactions. The degree of individualization of these mechanisms and their relative roles depend on the spatial and temporal variation in the pressure–temperature conditions and the composition of the rocks and metasomatizing fluids. Because rising deep fluids are characterized by variable flow rates and composition during the evolution of flow regimes and systems, the relationships between alteration and proper speleogenesis are complex.

The Buda Thermal Karst (Hungary) contains remarkable examples of hypogene caves in which development was guided by distinct zones of strong alteration along fractures. Here, large rift-like passages in some caves (e.g. the Pál-völgy cave system) were formed along fractures with up to metre-thick zones of friable, often silicified, alterite still preserved in the distal parts (the author's own observations). This corresponds to a scenario in which a fluid flow phase with dominant isovolumetric alteration was followed by a phase with the dominant creation of open cavities. In the Piedmont Crimea, the relationships between the creation of open spaces and alteration of the wall rock are more complex. Many rock shelters that were definitely open space chambers before their exposure in cliffs have distinctly altered zones (friable or tufa-like rock and honeycomb structures) on the inner surfaces. Similarly, most karst rifts have alteration zones, the formation of which post-dates the major solutional enlargement of fractures. By contrast, most of the documented common hypogene caves in the region lack obvious signs of significant physical alteration in their walls, although distinct haloes of isotopic alteration (carbon and oxygen) are found in at least some of them (Dublyansky *et al.* 2014). It is not clear whether these caves developed along altered precursor fractures, with the removal of the alterite by subsequent speleogenetic enlargement, or solely through dissolution. Enlarged fractures with argillisites and strongly altered wall rocks are found side by side with open hypogene conduits with no obvious sign of wall rock alteration, guided by fractures of the same set. This suggests that some fluid phases may selectively affect existing fractures and conduits and not influence other fractures in the same or other sets. This is in agreement with the view that transverse flow paths can be individualized in hypogene speleogenesis, operating largely independently of each other (Klimchouk *et al.* 2016). Specific events of ascending fluid flow may not involve all flow paths.

Further evidence of the alteration that accompanied or followed major speleogenetic phases in hypogene caves is the common occurrence of boxwork and corrosion residue (Northup *et al.* 2000) or speleosols (Spilde *et al.* 2003, 2009; DuChene *et al.* 2017). Klimchouk (2017) suggested that boxwork is the result of later selective dissolution in the altered zone of the cave walls, leading to the partial exhumation of a three-dimensional framework of denser material (veins) precipitated in fractures and microfractures within this zone during the preceding alteration phase. In effect, the formation of boxwork is the same process as that inferred here for honeycomb structures. The formation of the corrosion residue, an oxide-rich iron–manganese layer that develops on the walls of many hypogene caves, is strongly influenced by microbiological activity (Northup *et al.* 2000; Spilde *et al.* 2003), which is favoured by the presence of porous and chemically enriched alteration zones in the cave wall rocks.

The features of ghost rock karstification, tafoni and honeycomb structures are remarkable manifestations of the link between metasomatism (rock alteration) and hypogene karstification (or between two aspects of the same general process of extended metasomatism in the sense of Pospelov). The recognition of the genetic affinity of these phenomena (and of their common fluid-aided nature) helps to resolve many long-lasting interpretive problems regarding the origin of the cavernous decay features and improves the methodology of further research.

Origin of cavernous features through fluid-induced rock alteration and subsequent removal of the alterite

A conceptual model for the origin of cavernous features through fluid-induced rock alteration and subsequent removal of the alterite is briefly summarized and generalized here. The primary factor in the origin of cavernous features on exposed surfaces is the pre-exposure (i.e. in the subsurface) alteration of the host rocks along fluid-conducting discontinuities (such as fractures and karst conduits) as a result of fluid–rock interactions. It is argued here that such alteration occurs mainly in deep-seated settings and is commonly induced by rising flow. The morphological expression of cavernous features on rock surfaces through removal of the alterite (or its friable parts) can occur under subsurface conditions, but this happens more often on exposure to atmospheric weathering and erosion. The local or regional characteristics of the weathering system and specific weathering processes are irrelevant or have only secondary importance in determining the localization and morphology of cavernous features.

Depending on the rock composition, pressure–temperature conditions, fluid chemistry and dynamics, and fluid history, the wall rock alteration may precede, be concomitant with, or follow the enlargement of an initial conducting discontinuity that allows some focusing of the fluid flow. The alteration zone in the wall rock is characterized by a more or less distinct boundary at the rear (the alteration front), which may have a complex, but generally smooth, geometry, with deeper embayments and intervening salients developed as a result of the initial variations in the permeability of the rock matrix and hence variations in the interaction between the pore fluids and the flow in the main conduit. The geometry of the alteration front attained by the time of exposure to the action of vigorous denudation agencies determines the future morphology of larger cavernous features, such as tafoni, in a similar manner to the cavities created by ghost rock karstification.

The character of the alteration may change with a change in the fluid composition or conditions of fluid–rock interactions. Within the porous and friable bulk mass of the alteration zone, a three-dimensional pattern of more dense (hardened) material can be created through a dissolution–reprecipitation process. The mechanisms of this process depend on many factors, can be varied and require additional research. One possible mechanism is preferential precipitation in larger pores and fractures. According to Putnis (2002), the dissolution–reprecipitation mechanism allows fluids supersaturated with respect to certain phases to be transported through the microporous matrix without crystallization until a larger pore volume or fracture is encountered. In this way, hardened areas can be created along fractures or swarms of larger pores. By contrast, Laubach *et al.* (2009) showed that the quartz cement in sandstones typically seals microfractures, but merely lines or bridges larger fractures. Sealing the microporosity around larger fractures or systems of larger interconnected pores is another way of forming hardened structures. Preferential precipitation in some areas within the altered zone leads to the formation of the cryptically structured heterogeneity in the bulk porous alterite, pre-determining the diverse and intricate patterns of the future honeycomb structures.

The tafoni and honeycomb features are expressed morphologically after the partial or complete removal of the bulk mass of the porous and weakened alterite in the wall rock. Removal of the alterite can occur in different environments and through different processes, either speleogenetic processes or external weathering after the fracture (karst rift) wall is exposed to the surface in escarpments. Which particular weathering processes are involved is not crucial in the formation of the cavernous features; their exposure and appearance are only slightly controlled by the local environmental conditions.

On exposure, the evolution of cavernous features under the action of weathering proceeds in two stages: (1) prograde development, when the features are progressively cleaned from the bulk mass of the alterite; and (2) retrograde development, when the exposed features are destroyed by weathering and denudation. Cavernous decay features are relatively short-lived and are ultimately obliterated by weathering. New features are generated through the exposure of other fractures with altered wall rocks, but they do not form on rock surfaces where the alteration zone was totally denuded or never present.

This model allows the discrimination of two types of cavernous decay feature: (1) features in which the morphology is controlled by the geometry of the alteration front (commonly larger tafoni-like cavities); and (2) features in which the morphology is determined by the three-dimensional pattern of hardened structures within the alteration zone (honeycomb-like features). These two types of feature can be combined where a three-dimensional pattern of hardened structures is formed at the rear of the alteration zone, so that the honeycomb features are displayed on the walls of a tafoni-like cavity after removal of the alterite.

This proposed model is in good agreement with the major characteristic traits of the cavernous decay features described in different rocks in various locations worldwide, which were difficult to explain within the traditional weathering paradigm.

- Preferential development on subvertical surfaces, e.g. in cliffs. This is mainly because cliffs are commonly structurally controlled – that is, they are formed by block retreat along subvertical fractures, which are often affected by karstification and alteration. The features in cliffs are also less subject to obliteration by surface runoff.
- Preferential occurrence in coastal and arid areas. Structurally controlled cliffs occur widely in coastal areas, where their faces are frequently renewed by wave undercutting at the base and block failure. The cliff faces and fallen blocks are also subject to a high-energy weathering system, which promotes the opening of altered fractures and washing out of the alterite. In arid areas, the exposed wall rock morphologies are preserved for longer. According to the proposed model, the development of cavernous decay features is independent of climate, but their preservation is less favoured in humid conditions.
- The occurrence of honeycomb/boxwork features on the walls of larger tafoni-like and normal hypogene cavities.
- The occurrence of honeycomb/boxwork features within a distinct zone in the wall rock, beneath which they do not develop.
- The coexistence of degrading and well-articulated features in close proximity, in the same lithology and under the same environmental conditions.

Many studies have addressed cavernous decay phenomena on historical buildings and other artificial structures, interpreting tafoni and honeycombing on stone blocks as evidence of the high rate of their development. Although the latter is true, the hard question within the weathering paradigm is why these features develop on one stone block, but are absent on the adjacent block. The proposed model offers a simple answer: caverning develops on historical timescales only on those stone blocks in which the exposed side represents the altered wall rock of a fracture. It was common practice in ancient stone mining to use natural fractures to split blocks. Blocks with honeycombed sides were sometimes used intentionally because these patterns are often visually appealing.

This example illustrates one methodological problem of tafoni/honeycombing research, obvious in the light of the proposed concept. Measurements of denudation rates and rock properties on surfaces with cavernous features, as well as sampling for various analyses, can be misleading when interpreted as representative of the original rock. The methodology of further research should account for, and focus on revealing, physical and chemical alteration and heterogeneities in the near-surface zone of exposed rocks and fracture/cavity wall rocks in subsurface environments.

Some implications

Tafoni and honeycomb structures in sandstones

This study used the example of the Crimean Piedmont to support and illustrate a new approach to interpreting tafoni and honeycomb structures and to demonstrate their genetic affinity with hypogene karstification. Although the features in this region develop in carbonate rocks, tafoni and honeycomb structures elsewhere develop in many other lithologies, most notably in sandstones. As noted by Young & Young (1992, p. 71), ‘cavernous weathering is probably as characteristic a feature of sandstones as fluting is of limestones, but it is certainly not uniformly or universally developed’.

In the light of the proposed model, the wider occurrence of cavernous features in sandstones than in limestones is as expected. For the fracture-aligned alteration processes, the interaction between conduit fluids and pore fluids is important, but the amount and distribution of matrix flow, and hence the amount of the interaction, depend on the matrix porosity and permeability. Limestones are more reactive rocks, in which the original matrix porosity is quickly reduced with burial as a result of cementation, so that diagenetically mature (telogenetic) limestones tend to have a low matrix porosity and permeability. Although hypogene karstification is widespread in limestones, it may not be accompanied by significant wall rock alteration when started in rocks of high diagenetic maturity and low matrix porosity. Hence cavernous decay features are less expected to form in telogenetic limestones on exposure, although the alteration may occur earlier in the diagenetic history.

By contrast, sandstones are less reactive rocks. They commonly retain much of their primary porosity on burial and can generally transmit much higher flows through the matrix. Hypogene karstification in sandstones can be assumed to have a limited ability to produce voluminous caves (although they may form under high pressure–temperature conditions and under the action of fluids of specific compositions), but fracture-controlled, rift-like conduits with well-developed alteration haloes may be a common result. This assumption is supported by the wide occurrence in many sandstone landscapes worldwide of deep slot-like canyons and various conspicuous forms of differential weathering, including tafoni and honeycomb structures. Their interpretation has so far been constrained by the erosion/weathering paradigm; the role of fluid-induced alteration and hypogene karstification along focused flow paths has not been considered, with notable exceptions of recent work by Suchý & Filip (2016) and Suchý *et al.* (2017), which demonstrated the subsurface hypogene origin of remarkable tafoni-like cavities in Carboniferous arkose sandstones in central and western Bohemia (Czech Republic). The proposed model calls for and opens exciting possibilities for revisiting a number of conspicuous features of sandstone geomorphology in many regions worldwide.

Tafoni and honeycomb structures as indications of rising fluid flow and hypogene karstification

The proposed model has important implications for reconstructing the geodynamic and palaeo-hydrogeological history of regions and the identification of hypogene karst. Although the author admits that some caverning may occur through mechanisms that do not involve pre-exposure fluid-induced alteration along focused flow paths, and that some alteration may also be related to downward infiltration, the massive and widespread development of typical tafoni and honeycomb structures should be regarded as a strong indication that exposed formations could once have been subjected to significant ascending fluid flow. As hypogene karstification is related to focused rising flow, the presence of tafoni and honeycomb structures in carbonates and sandstones may indicate the potential presence of hypogene karst systems in a region.

A cursory review of the global distribution of areas renowned for the wide occurrence of tafoni and honeycomb structures supports these assumptions. Most of these structures are located in geodynamic and hydrogeological settings that favour, or favoured in the past, the ascending migration of basinal and/or deep endogenous fluids, such as subduction zones, regions of extensional geodynamic regimes (rifting), regions of asthenospheric rise (including hotspots), large-scale fault zones or marginal recharge areas of large artesian basins. Not surprisingly, many of these regions are known to host recognized examples of hypogene karst.

Conclusions

The study of tafoni and honeycomb structures in the Crimean Piedmont has shown that these features are related to hypogene karstification and that their formation is pre-determined by the alteration of host rocks along fractures and karst conduits by focused rising flow. The morphological expression of cavernous features occurs through removal of the alterite, either by speleogenetic agencies in subsurface conditions or (more frequently) on exposure to atmospheric weathering.

These findings are generalized to propose a new conceptual model for the formation of cavernous decay features, which provides an explanation for their characteristic traits and resolves the major issues inherent to previous interpretations bounded by the weathering paradigm. The pre-exposure alteration of the wall rocks of fluid-conducting discontinuities as a result of fluid–rock interactions is the major factor overarching variations in environmental conditions and weathering processes. The local or regional characteristics of the weathering system are irrelevant or of secondary importance in determining the localization and morphology of cavernous features. Typical tafoni and honeycomb structures may indicate past events of ascending flow and the potential presence of hypogene karst systems.

The author thanks colleagues at the Ukrainian Institute of Speleology and Karstology, particularly G. Amelichev, E. Tymokhina, V. Naumenko, S. Tokarev and B. Vakhrushev, for their assistance in the field studies. This study was partially supported by grant 0110U002248 of the Ministry of Education and Science of Ukraine.

References

Amelichev, G.N., Klimchouk, A.B. & Timokhina, E.I. 2011. Speleogenesis in Cretaceous and Eocene deposits of the valleys of Zuya and Burulcha rivers (eastern part of the Crimean Piedmont). *Speleology and Karstology (Simferopol)*, **7**, 52–64 [in Russian].

Audra, P. 2017. Hypogene caves in France. *In*: Klimchouk, A., Palmer, A., De Waele, J., Auler, A. & Audra, P. (eds) *Hypogene Karst Regions and Caves of the World*. Springer International, Berlin, 61–84.

Blaga, N.N. & Popov, A.V. 2009. Some aspects of morphogenesis of rock shelters of the Inner Range of the Crimean Mountains. *Cultura Narodov Prichernomorya (Simferopol)*, **155**, 7–9 [in Russian].

Brandmeier, M., Kuhlemann, J., Krumrei, I., Kappler, A. & Kubik, P.W. 2011. New challenges for tafoni research. A new approach to understand processes and weathering rates. *Earth Surface Processes and Landforms*, **36**, 839–852.

Bruthans, J., Soukup, J. *et al.* 2014. Sandstone landforms shaped by negative feedback between stress and erosion. *Nature Geoscience*, **7**, 597–601.

Bruxelles, L., Quinif, Y. & Wiénin, M. 2009. How can ghost rocks help in karst development? *In*: White, W.B. (ed.) *Proceedings of 15th International Congress of Speleology, Speleogenesis*. National Speleological Society, Kerrville, TX, USA, 814–818.

Bush, I. 2010. An integrated approach to fracture characterization. *Oil Review Middle East*, **2**, 88–91.

Davies, G.R. & Smith, L.B., Jr. 2006. Structurally controlled hydrothermal dolomite reservoir facies: an overview. *American Association of Petroleum Geologists Bulletin*, **90**, 1641–1690.

Dublyansky, V.N. & Lomaev, A.A. 1980. *Karst Caves of Ukraine*. Naukova Dumka, Kiev [in Russian].

Dublyansky, Y.V., Klimchouk, A.B., Spötl, C., Timokhina, E.I. & Amelichev, G.N. 2014. Isotope wallrock alteration associated with hypogene karst of the Crimean Piedmont, Ukraine. *Chemical Geology*, **377**, 31–44.

Dubois, C., Quinif, Y., Baele, J.M., Barriquand, L., Bini, A., Bruxelles, L. & Maire, R. 2014. The process of ghost-rock karstification and its role in the formation of cave systems. *Earth-Science Reviews*, **131**, 116–148.

DuChene, H., Palmer, A. *et al.* 2017. Hypogene speleogenesis in the Guadalupe Mountains, New Mexico and Texas, USA. *In*: Klimchouk, A., Palmer, A., De Waele, J., Auler, A. & Audra, P. (eds) *Hypogene Karst Regions and Caves of the World*. Springer International, Berlin, 511–530.

Dushevsky, V.P. 1987. Speleological knowledge on the Piedmont Crimea karst area [abstract]. Paper presented at the Problems of Studies, Ecology and Protection of Caves Conference, Institute of Geological Sciences, Academy of Sciences of the UkrSSR, 27–29 October 1987, Kiev, Ukraine [in Russian].

Dushevsky, V.P. & Kuznetsov, A.G. 1991. Singularities of development of karst in low-mountain areas of the cuesta relief. Paper presented at the Studies and Utilization of Karst in Western Caucasus Conference, 1–3 October 1991, Sochi, Russia [in Russian].

Dushevsky, V.P., Kljukin, A.A. & Soldatov, Y.V. 1979. Conditions and growth rate of denudational cavitics in cliffs of the Crimean cuestas [abstract]. Paper presented at the Karst Sredney Asii I gornyh stran Conference, 9–10 November 1979, Tashkent, Uzbekistan, 49–50 [in Russian].

Ezhov, Y.A., Lysenin, G.P., Andreychouk, V.N. & Dublyansky, V.N. 1992. *Karst in the Earth's Crust*. Sibirskoye otdeleniye Instituta geologii, Novosibirsk [in Russian].

GOLUBIC, S., FRIEDMANN, E.I. & SCHNEIDER, J. 1981. The lithobiontic ecological niche, with special reference to microorganisms. *Journal of Sedimentology and Petrology*, **51**, 475–478.

GOUDIE, A.S. & VILES, H.A. 1997. *Salt Weathering Hazards*. Wiley, New York.

GROOM, K.M., ALLEN, C.D., MOL, L., PARADISE, T.R. & HALL, K. 2015. Defining tafoni: re-examining terminological ambiguity for cavernous rock decay phenomena. *Progress in Physical Geography*, **39**, 775–793.

HARLOV, D.E. & AUSTRHEIM, H. (eds) 2013. *Metasomatism and the Chemical Transformation of Rock*. Springer, Berlin.

HUININK, H.P., PEL, L. & KOPINGA, K. 2004. Simulating the growth of tafoni. *Earth Surface Processes and Landforms*, **29**, 1225–1233.

KLIMCHOUK, A.B. 2017. Types and settings of hypogene karst. *In*: KLIMCHOUK, A., PALMER, A., DE WAELE, J., AULER, A. & AUDRA, P. (eds) *Hypogene Karst Regions and Caves of the World*. Springer International Publishing AG, Cham, 1–42.

KLIMCHOUK, A.B. & TIMOKHINA, E.I. 2011. Morphogenetic analysis of Tavrskaya cave (Inner Range of the Crimean Piedmont). *Speleology and Karstology (Simferopol)*, **6**, 36–52 [in Russian].

KLIMCHOUK, A., TYMOKHINA, E. & AMELICHEV, G. 2009. Hypogene speleogenesis in the Piedmont Crimea Range. *In*: KLIMCHOUK, A. & FORD, D. (eds) *Hypogene Speleogenesis and Karst Hydrogeology of Artesian Basins*. Ukrainian Institute of Speleology and Karstology, Special Papers, **1**, 159–171.

KLIMCHOUK, A.B., TYMOKHINA, E.I., AMELICHEV, G.M., DUBLYANSKY, YU.V. & STAUBWASSER, M. 2011. U/Th dating of speleothems of karst caves in the southwest part of the Inner Range of the Mountainous Crimea and determination of relief age and its development dynamics. *Speleology and Karstology (Simferopol)*, **7**, 29–39 [in Russian].

KLIMCHOUK, A., TYMOKHINA, E. & AMELICHEV, G. 2012. Speleogenetic effects of interaction between deeply derived fracture-conduit flow and intrastratal matrix flow in hypogene karst settings. *International Journal of Speleology*, **41**, 37–55.

KLIMCHOUK, A.B., AMELICHEV, G.N., TIMOKHINA, E.I. & TOKAREV, S.V. 2013*a*. Hypogenic speleogenesis in the Crimean fore-mountains (the Black Sea region, South Ukraine) and its role in the regional geomorphology. *In*: FILIPPI, M. & BOSÁK, P. (eds) *Proceedings of the 16th International Congress of Speleology*, 21–28 July, Brno. Czech Speleological Society, Praha, **3**, 364–365.

KLIMCHOUK, A.B., TYMOKHINA, E.I., AMELICHEV, G.M., DUBLYANSKY, YU.V. & SPÖTL, C. 2013*b*. *Hypogene Karst of Crimean Piedmont and its Geomorphological Role*. DIP, Simferopol [in Russian].

KLIMCHOUK, A., AULER, A.S., BEZERRA, F.H., CAZARIN, C.L., BALSAMO, F. & DUBLYANSKY, Y. 2016. Hypogenic origin, geologic controls and functional organization of a giant cave system in Precambrian carbonates, Brazil. *Geomorphology*, **253**, 385–405.

KLIMCHOUK, A., AMELICHEV, G., TYMOKHINA, E. & DUBLYANSKY, Y. 2017. Hypogene speleogenesis in the Crimean Piedmont, the Crimea Peninsula. *In*: KLIMCHOUK, A., PALMER, A., DE WAELE, J., AULER, A. & AUDRA, P. (eds) *Hypogene Karst Regions and Caves of the World*. Springer International, Berlin, 407–430.

KORZHINSKII, D.S. 1953. Infiltration metasomatism at the presence of temperature gradient and contact metasomatic leaching. *Zapiski Vsesoyuznogo Mineralogicheskogo. Obshchestva (Russia)*, **282**, 161–172 [in Russian].

LAUBACH, S.E., OLSON, J.E. & EICHHUBL, P. 2009. Fracture diagenesis and producibility in tight gas sandstones. *In*: CARR, T., D'AGOSTINO, T., AMBROSE, W., PASHIN, J. & ROSEN, N.C. (eds) *Unconventional Energy Resources: Making the Unconventional Conventional*, 29th Annual GCSSEPM Foundation Bob F. Perkins Conference. Gulf Coast Section of the Society of Economic Paleontologists and Mineralogists Foundation, Houston, Texas, 483–499.

LUSHCHIK, A.V., MOROZOV, V.I. & MELESHIN, V.P. 1981. *Groundwaters of Karst Platform Regions in the South of Ukraine*. Naukova Dumka, Kiev [in Russian].

LUSHCHIK, A.V., LISICHENKO, G.V. & YAKOVLEV, E.O. 1988. *Formation of the Groundwater Regime in Areas of Active Geodynamic Processes*. Naukova Dumka, Kiev [in Russian].

LYGINA, E.A. 2010. *Danian and Eocene platforms of the Crimea: build-up and conditions of formation*. PhD thesis, Moscow University [in Russian].

LYSENKO, N.I. 1976. New data on the Miocene planation surface in mountainous Crimea. *Geomorphologiia*, **1**, 86–90 [in Russian].

LYSENKO, N.I. 2002. About new finding of Lower Cretaceous sediments in the Crimean Yayla. *Priroda*, **1**, 2–4 [in Russian].

MARTINI, I.P. 1978. Tafoni weathering, with examples from Tuscany, Italy. *Zeitschrift für Geomorphologie*, **22**, 44–67.

MCBRIDE, E.F. & PICARD, M.D. 2004. Origin of honeycombs and related weathering forms in Oligocene Macigno sandstone, Tuscan coast near Livorno, Italy. *Earth Surface Processes and Landforms*, **29**, 713–735.

MUSTOE, G.E. 1983. Origin of honeycomb weathering. *GSA Bulletin*, **93**, 108–115.

NIKISHIN, A.M., ALEKSEEV, A.S. & BARABOSHKIN, E.Y. 2006. *Geological History of the Bakhchisaray Region of Crimea*. Moscow State University, Moscow [in Russian].

NORTHUP, D.E., DAHM, C.N. ET AL. 2000. Evidence for geomicrobiological interactions in Guadalupe caves. *Journal of Cave and Karst Studies*, **62**, 80–90.

OGATA, K., SENGER, K., BRAATHEN, A. & TVERANGER, J. 2014. Fracture corridors as seal-bypass systems in siliciclastic reservoir–cap rock successions: field-based insights from the Jurassic Entrada Formation (SE Utah, USA). *Journal of Structural Geology*, **66**, 162–187.

PARADISE, T.R. 2013. Tafoni and other rock basins. *In*: SHRODER, J. (ed.) *Treatise on Geomorphology*. Academic Press, San Diego, CA, 111–126.

POSPELOV, G.L. 1973. *Paradoxes, Physicochemical Nature and Mechanisms of Metasomatism*. Nauka, Novosibirsk [in Russian].

PUTNIS, A. 2002. Mineral replacement reactions: from macroscopic observations to microscopic mechanisms. *Mineralogical Magazine*, **66**, 689–708.

QUESTIAUX, J.-M., COUPLES, G.D. & NICOLAS, R. 2010. Fractured reservoirs with fracture corridors. *Geophysical Prospecting*, **58**, 279–295.

QUINIF, Y. & BRUXELLES, L. 2011. L'altération de type 'fantôme de roche': processus, évolution et implications pour la karstification. *Géomorphologie*, **4**, 349–358.

ROBINSON, D.A. & WILLIAMS, R.B.G. 1994. Sandstone weathering and landforms in Britain and Europe. *In*: ROBINSON, D.A. & WILLIAMS, R.B.G. (eds) *Rock Weathering and Landforms Evolution*. Wiley, Chichester, 371–391.

RODRIGUEZ-NAVARRO, C. 1998. Evidence of honeycomb weathering on Mars. *Geophysical Research Letters*, **25**, 3249–3252.

SHNIUKOV, E.F., SHEREMETIEV, V.M., MASLAKOV, N.A., KUTNII, V.A., GUSAKOV, I.N. & TROFIMOV, V.V. 2006. *Mud Volcanous of the Kerch-Taman' Region*. Glavmedia, Krasnodar [in Russian].

SIEDEL, H. 2015. Cavernous weathering features. *In*: HARGITAI, H. & KERESZTURI, A. (eds) *Encyclopedia of Planetary Landforms*. Springer, Berlin, 231–236.

SINGH, S.K., ABU-HABBIEL, H., KHAN, B., AKBAR, M., ETCHECOPAR, A. & MONTARON, B. 2008. Mapping fracture corridors in naturally fractured reservoirs: an example from Middle East carbonates. *First Break*, **26**, 109–113.

SPILDE, M.N., BOSTON, P.J. & NORTHUP, D.E. 2003. Subterranean soil development. *Journal of Cave and Karst Studies*, **65**, 188.

SPILDE, M.N., KOOSER, A., BOSTON, P.J. & NORTHUP, D.E. 2009. Speleosol: a subterranean soil. *In*: *Proceedings of 15th International Congress of Speleology, Speleogenesis*. National Speleological Society, Kerrville, TX, USA, 338–344.

SUCHÝ, V. & FILIP, J. 2016. Uvahy o puvodu konkavnich tvaru na nekterych lokalitach karbonskych arkoz ve strednich av zapadnich Cechach [Thoughts about the origin of convex shapes in some locations of Carboniferous arkose in the middle and western Bohemia]. *Zpravy o geologickych vyzkumech*, **49**, 61–67.

SUCHÝ, V., SÝKOROVÁ, I., ZACHARIÁŠ, J., FILIP, J., MACHOVIČ, V. & LAPČÁK, L. 2017. Hypogene features in sandstones: an example from Carboniferous basins of central-western Bohemia, Czech Republic. *In*: KLIMCHOUK, A., PALMER, A., DE WAELE, J., AULER, A. & AUDRA, P. (eds) *Hypogene Karst Regions and Caves of the World*. Springer International, Berlin, 313–328.

TEMOVSKY, M. 2017. Hypogene karst In Macedonia. *In*: KLIMCHOUK, A., PALMER, A., DE WAELE, J, AULER, A. & AUDRA, P. (eds) *Hypogene Karst Regions and Caves of the World*. Springer International, Berlin, 241–256.

TURKINGTON, A.V. 2004. Cavernous weathering. *In*: GOUDIE, A.S. (ed.) *Encyclopedia of Geomorphology*. Vol. 1. Routledge, London, 128–130.

TURKINGTON, A.V. & PARADISE, T.R. 2005. Sandstone weathering: a century of research and innovation. *Geomorphology*, **67**, 229–253.

TURKINGTON, A.V. & PHILLIPS, J.D. 2004. Cavernous weathering, dynamical instability and self-organization. *Earth Surface Processes and Landforms*, **29**, 665–675.

TYMOKHINA, E.I., KLIMCHOUK, A.B. & AMELICHEV, G.N. 2011. Geomorphology and speleogenesis of the marginal southwestern part of the Eocene cuesta of the Inner Range of the Crimea Mountains. *Speleology and Karstology (Simferopol)*, **7**, 40–51 [in Russian].

TYMOKHINA, E.I., KLIMCHOUK, A.B. & AMELICHEV, G.N. 2012. Role of hypogene karst in the geomorphogenesis of the Inner Range of the Crimean Mountains. *Speleology and Karstology (Simferopol)*, **9**, 38–51 [in Russian]

VAKHRUSHEV, B.O. 2010. Crimean Mountains. *In*: VAKHRUSHEV, B.O., KOVALCHUK, I.P. & KOMLEV, O.O. (eds) *Landscape of Ukraine*. Slovo, Kiev, 432–485 [in Russian].

VERGARI, A. 1998. Nouveau regard sur la spéléogénèse: le 'pseudo-endokarst' du Tournaisis (Hainaut, Belgique). *Karstologia*, **31**, 12–18.

VERGARI, A. & QUINIF, Y. 1997. Les paléokarsts du Hainaut (Belgique). *Geodinamica Acta*, **10**, 175–187.

VILES, H. 2005. Self-organized or disorganized? Towards a general explanation of cavernous weathering. *Earth Surface Processes and Landforms*, **30**, 1471–1473.

WARREN, J. 2000. Dolomite: occurrence, evolution and economically important associations. *Earth-Science Reviews*, **52**, 1–81.

YARDLEY, B.W.D. 2013. The chemical composition of metasomatic fluids in the crust. *In*: HARLOW, D.E. & AUSTRHEIM, H. (eds) *Metasomatism and the Chemical Transformation of Rock*. Springer, Berlin, 17–52.

YARDLEY, B.W. & BODNAR, R.J. 2014. Fluids in the continental crust. *Geochemical Perspectives*, **3**, 1–127.

YOUNG, R. & YOUNG, A. 1992. *Sandstone Landforms*. Springer, Heidelberg.

YUDIN, V.V. 2011. *Geodynamics of the Crimea*. DIP, Simferopol [in Russian].

ZHARIKOV, V., PERTSEV, N., RUSINOV, V., CALLEGARI, E. & FETTES, D.J. 2007. Metasomatism and metasomatic rocks. *In*: FETTES, D. & DESMONS, J. (eds) *Metamorphic Rocks: a Classification and Glossary of Terms*. Cambridge University Press, Cambridge, 58–68.

ZHARIKOV, V.A., RUSINOV, V.L. & MARAKUSHEV, A.A. 1998. *Metasomatism and Metasomatic Rocks*. Nauchnyy Mir, Moscow [in Russian].

Relations between surface and underground karst forms inferred from terrestrial laser scanning

ALEKSANDAR S. PETROVIĆ[1]*, JELENA ĆALIĆ[2], ALEKSANDRA SPALEVIĆ[2] & MARKO PANTIĆ[3]

[1]*University of Belgrade, Faculty of Geography, Studentski trg 3/3, Belgrade 11000, Serbia*

[2]*Geographical Institute 'Jovan Cvijić' of the Serbian Academy of Sciences and Arts, Đure Jaksica 9, Belgrade 11000, Serbia*

[3]*Vekom Geo, Trebinjska 24, Belgrade 11000, Serbia*

**Correspondence: apetrovic@gef.bg.ac.rs*

Abstract: This paper details methods that contribute towards solving the problem of the spatial relations between surface and underground karst morphology – relationships that are often unclear. The karst landforms studied in this context are karst valleys, through-caves and natural bridges. Two study sites are situated in the Carpatho-Balkan Mountains of eastern Serbia: the dry valley of the Radovanska Reka River on Mt Kučaj (together with the Pećura through-cave) and the Zamna Cave in the wider area of the Danube Gorge. The caves and the closest adjoining parts of the valleys were measured in detail using the terrestrial laser scanning method. The data obtained showed that some previous measurements at these locations, performed with classical traditional instruments, are insufficiently accurate and may lead to wrong conclusions.

Surface and underground karst relief in contact karst and fluviokarst settings shows characteristic features, such as karst valleys, through-caves and natural bridges. Karst valleys are linear forms of karst surface relief that occur either in the areas of eroded caprock from karst-levelled surfaces, where they occur as lined-up dolines (e.g. 'dry valleys'), or in areas of contact karst and/or fluviokarst (e.g. 'blind valleys'; cf. Ford & Williams 2007). Within the study of the evolution and morphological types of karst valleys of the Carpatho-Balkan Mountains (Carpatho-Balkanides) of eastern Serbia (Petrović 2015), particular focus was given to collapse valleys. By the term 'collapse valleys' we consider those parts of karst valleys formed by cave roof collapses. Apart from being relatively rare in comparison with other types of karst valleys, their peculiar characteristic is that they often host natural bridges – remnants of former cave roofs. Numerous natural bridges are a specific trait of karst in northeastern Serbia (Ambert & Nicod 1981; Gavrilović 1998).

Valleys formed by cave roof collapse are mentioned in several studies on the morphological evolution of karst (Cvijić 1918; Gams 1974, 2004), karst canyons (Nicod 1997), the morphogenesis of particular river basins (Petrović 1969, 1974) or the formation of natural bridges (Ćalić-Ljubojević 2000; Petrović & Carević 2015). The initial phase for the development of this type of karst valley comprises the segments of water courses known as 'by-pass' segments (in the sense of Nicod 1997). These by-passes are mostly short, through-caves. By thinning of the roof and further enlargement of a cave passage, which typically occur simultaneously with the karst surface lowering, the ceiling becomes unstable, which leads to the formation of daylight holes. Further collapse of the ceiling and enlargement of daylight holes leads to the formation of collapse valleys, with or without natural bridges. In order to understand the genesis and development of collapse valleys, it is important to follow the development of both surface and underground karst morphology. For this reason, the aim of the present research is to analyse the relationship between surface and underground karst in through-caves and collapse valleys at different stages of evolution. By the term through-caves, we consider the caves passable from a river sinking point to a spring (Jennings 1971, 1985; Audra & Palmer 2015; De Waele *et al.* 2017; Jouves *et al.* 2017). This study follows a morphological approach to the subject, without applying any engineering approach, neither geophysical/geotechnical (as in Waltham & Fookes 2003 or Kaufmann 2014), nor related to geohazard issues (e.g. Parise & Lollino 2011; Gutiérrez *et al.* 2014). It is important to distinguish the valleys formed by cave roof collapse from the phenomenon of the unroofed (denuded) caves, studied by Knez

From: Parise, M., Gabrovsek, F., Kaufmann, G. & Ravbar, N. (eds) 2018. *Advances in Karst Research: Theory, Fieldwork and Applications*. Geological Society, London, Special Publications, **466**, 107–120.
First published online January 31, 2018, https://doi.org/10.1144/SP466.23

& Slabe (1999, 2002), as well as Mihevc *et al.* (1998). In both cases, there is the process of karst surface lowering (karst denudation), but the differences are significant. The unroofed caves are fossilized, i.e. filled with sediments, so they are presently inactive compared with other types of karst conduits. Collapse valleys are, on the other hand, formed in hydrologically active settings, with constant washout of sediments in fluviokarst.

Terrestrial laser scanning (TLS) techniques were chosen as presently the most appropriate for determining the relationship between the surface and the underground. The use of TLS in geomorphological studies is becoming ever more frequent and affordable because of technological advances. It is particularly useful in studying surface morphology in conditions of difficult access (Siart *et al.* 2013; Andrić & Bonacci 2014) or for more proficient geomorphological mapping (Tilly *et al.* 2016). TLS is also very useful in speleological explorations because, with the improvements in technology, the dimensions of the instruments are decreasing, while robustness and scanning precision are increasing. To date, these techniques have been used in caves for archaeological (Lerma *et al.* 2010; Garcìa-Puchol *et al.* 2013), geological (Grussenmeyer *et al.* 2010; Jaillet *et al.* 2011) and geomorphological studies (Roncat *et al.* 2011; Fabbri *et al.* 2017), as well as for producing 3D models and detailed maps of caves (Westerman *et al.* 2003; Zlot & Bosse 2014).

In this study, we applied TLS surveying in two caves and their immediate surroundings. The first case study is a valley situated in the drainage area of the Radovanska Reka River (Mt Kučaj; Fig. 1a), where the first, initial evolutionary phase is described: a through-cave stretching below the karst valley. The second case study is Zamna Cave, a through-cave situated exactly over the river bed of the Zamna River (Fig. 1b). This is an example of the mature phase of

Fig. 1. Study sites: (**a**) location of the sites within Central Europe and eastern Serbia; (**b**) detailed map of the two sites and rivers (1, Radovanska Reka River site; 2, Zamna Cave site in the Danube Gorge).

cave evolution, in which daylight holes appear on the cave ceiling.

Study sites

The basis for the development of karst relief in the Carpatho-Balkanides of Serbia is Jurassic and Cretaceous carbonate platforms (Grubić & Jankićević 1973). Subsequent tectonic activity led to subsidence of parts of these platforms, and formation of basins and graben (Zeremski 1991). Therefore the karst relief of the Carpatho-Balkanides of Serbia developed on relatively isolated block structures. Also, particular sectors of the carbonate platforms are divided by deep systems of valleys (rivers Resava, Svrljiški Timok, Nišava, etc.), which contribute to the isolation of the karst units.

Apart from faulting, there is a significant thrust tectonic activity (Anđelković & Nikolić 1980). Within regional thrusting, high and extensive karst mountains have been formed (Suva Planina, Kučaj, Beljanica), while at the same time some of the areas have been covered with impermeable rocks. Another means of development of covered karst were Neogene lacustrine transgressions from the Pannonian Basin, as well as from the Dacian Basin (Milić 1970). These transgressions, during which sea-level was rising to the present 850 m a.s.l. (Gavrilović 1975; Krstić *et al.* 2012), caused the deposition of large quantities of sediments. After the withdrawal of the Neogene lakes and seas from both sides of the mountain chain, karst at the higher elevations started to develop under the strong influence of contact with the deposited sediments. In the parts of the mountains that were covered with Neogene sediments, karst started (or continued?) to develop only after the sediments were washed away. In cases where limestone thrusts were fragmented, karst continued to develop within smaller units (e.g. Mt Starica, Vratna River Basin, Zamna River Basin).

Radovanska Reka River

The valley of the Radovanska Reka River is situated on Mt Kučaj (Fig. 1b–1), hosting the largest karst unit of the Serbian Carpatho-Balkanides, with an area of 491 km^2 (Petrović 2015). Radovanska Reka has incised the valley at the contact between the Upper Jurassic and Lower Cretaceous limestones on the western side, and the Paleozoic schists on the east (Fig. 2) (Veselinović *et al.* 1970). The limestones are thrust over the schists from the east (Anđelković & Nikolić 1980).

Žljebura gorge is a part of the Radovanska Reka Valley cut in limestones (Fig. 2). It is 2 km long and 340 m deep. Owing to its position along the contact, the valley is asymmetrical. The right valley side is higher and steeper, cut in the rim of the karst-levelled surface Bele Vode. Analysing the karstic characteristics of the valley bottom, it is possible to single out several types of karst valleys (Petrović 2015).

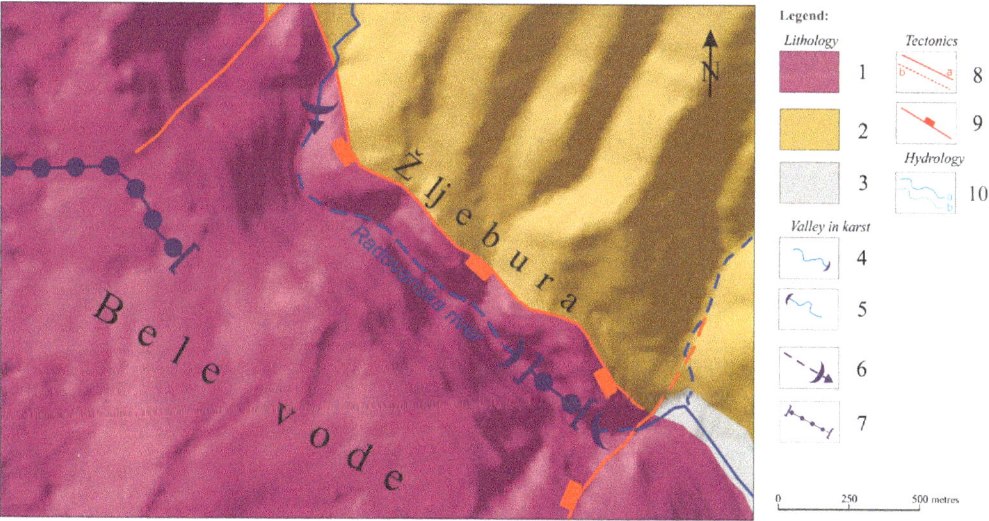

Fig. 2. Map of Radovanska Reka River Valley. 1, Carbonate rocks; 2, metamorphic rocks; 3, Quaternary deposits; 4, blind valley; 5, pocket valley; 6, semi-blind valley; 7, hanging valley with dolines on valley bottom; 8, fault (a, observed; b, covered); 9, vertical fault, relative movement of blocks; 10, river (a, with perennial flow; b, with seasonal flow).

Downstream from the contact between limestones and impermeable rocks, there is a half-blind valley, ending in a ponor (inflow) fissure on the left side of the river bed. During low and medium discharge, all of the water from the river sinks underground into the fissure, so that the downstream part of the river bed is dry. During high-water conditions, the river passes a small ridge in the river bed and continues downstream to the through-cave called Pećura. This part is the blind valley of the Radovanska Reka. Part of the valley above the cave remained at a higher, hanging position, both on this upstream end, and at the opposite, downstream end above the karst spring. The last part of the valley is a pocket valley, starting from the karst spring at the foothills of the 70 m high escarpment. This karst spring is the outflow of the Radovanska Reka, but also drains the waters that dispersedly percolate from the Bele Vode karst plateau.

Zamna River

In the furthest northeastern part of the Carpatho-Balkanides, which belong to the western rim of the Dacian Basin, there are three smaller karst units. They are cut by the short right tributaries of the Danube. The Zamna River, as well as its tributary Udubašnica, flows through this karst unit.

Zamna karst (Figs 1b-2 & 3) is formed in reef limestones, surrounded by conglomerates and sandstones from all sides (Kalinić *et al.* 1976). Zamna karst belongs to the type 'contact karst' and, in more detail, to the sub-type 'isolated karst with concentrated outflow' (Gams 1974).

The rivers Zamna and Udubašnica have deeply cut into limestones and incised steep-sided gorges (Fig. 3). Valleys and mouths of their former tributaries are now hanging at about 70 m on the sides of the gorges. The valley of the Zamna River ends blindly immediately before it meets the Udubašnica River. Just before the confluence with the Udubašnica, the Zamna flows through a 155 m long through-cave (Petrović 1971), which we call the Zamna Cave. At the beginning, as well as at the end, of the cave, there are daylight holes wider than 10 m.

Methodology

Geomorphological mapping

Geomorphological mapping and basic structural-geological observations were carried out in order to determine the settings of collapse-valley development. Geomorphological maps, as well as basic structural-geological sketches, were prepared using the previously designed Geomorphological Information System (GmIS) of karst units in the Carpatho-Balkanides of Serbia (Petrović 2015). GmIS is basically a GIS based on geomorphological data, criteria and principles (Letal 2005). GmIS is based on a 10 m resolution digital elevation model (DEM) derived by digitizing elevation data from topographic maps at a 1:25 000 scale. For preparation of the DEM, as well as the entire GmIS, we used ArcMap, ESRI™ software for geospatial data processing. GmIS consists of several layers: lithological composition including a layer for lithological types of carbonates, tectonic pattern, hydrography,

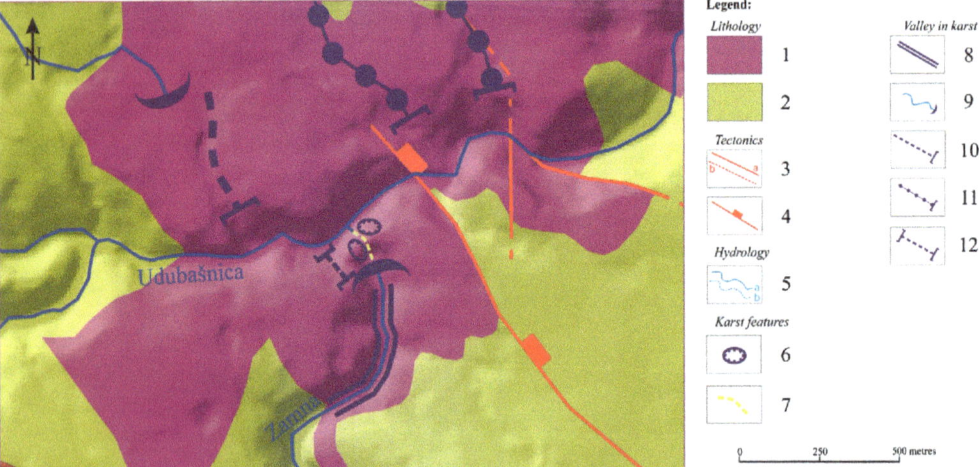

Fig. 3. Map of the Zamna River Valley. 1, Carbonate rock; 2, marl, shale and sandstone; 3, fault (a, observed; b, covered); 4, vertical fault, relative movement of blocks; 5, river (a, with perennial flow; b, with seasonal flow); 6, daylight hole; 7, cave passage; 8, gorge or canyon; 9, blind valley; 10, hanging valley; 11, hanging valley with dolines on valley bottom; 12, both sides hanging valley.

karst landforms (with a particular layer for karst valleys), slope and vegetation.

Data acquisition

Technical details. For detailed field measurement we used a Leica Nova MS50 (Fig. 4a), a combination of the surveying total station and a laser scanner, which gave us the opportunity to collect discrete points together with detailed high-precision scans, and even high-resolution images depending on the situation in the field. The angular accuracy of this instrument is 1 inch, and the accuracy of the distance measurements is 1 mm + 1.5 ppm, which enables measurement of a large number of points in short time with millimetre accuracy. Point scanning is automatically performed on the whole field of view 360° × 270°, or in the smaller user-defined area.

The Leica Nova MS50 electro-optical distance measurement system is based on wave form digitizing technology (WFD). WFD is a specific type of time-of-flight measurement system that sends out a signal and detects and stores a digital signal vector. The WFD-based electronic distance meter sends out signals. When a signal is sent out, a small part of each pulse gets directly guided on to

Fig. 4. (**a**) Field measurement with Leica Nova MS50; (**b**) scanning in the very hush conditions; (**c**) surface scanning in Radovanska Reka Valley; (**d**) surface scanning in Zamna River Valley.

a photo detector and serves as internal calibration measurement. This pulse is called the start pulse. The rest of the pulse is emitted out of the telescope and reflected by the object. The returning pulse is called the stop pulse. For every measurement, an internal calibration measurement is done.

In addition to such technology, this instrument features:

- automatic target aiming, which enables measurement of the centre of the reflector in low light, even in conditions without light, which was very convenient for referencing of the instrument and scans;
- integrated cameras, which are easy and fast to use for measurements, and for taking pictures or panoramic images.

Further, the factor that recommended this instrument in our field measurements is its size of 248 × 228 × 351 mm and weight of 7.6 kg. The physical characteristics of the instrument allow easy transport, and IP65 standard allows its use even in very hush (wet, dusty) underground conditions (Fig. 4b).

Surface measurement. All measurements were positioned using two points, 3D coordinates determined using the GNSS RTK rover, receiving corrections from the Reference network. These two points were used for the first setup of the instrument, one as a station and the other as orientation. After the first setup was finished, all further setup points were measured as foresight.

When talking about surface measurements, two types can be distinguished. The first is single-point measurement of the points on the surface (traditional surveying method), which is more or less manual, or at least the measured point must be chosen manually (Fig. 4c). Scanning is the preferred method but is not suitable in areas of dense vegetation or where the natural surface is not visible owing to natural obstacles, such as grass, bush and dense trees (Fig. 4d). Points to be measured are determined in the field by the instrument operator, as a function of their ability to reach it and to safely use the measuring rod.

If the scanning is used, the working process is straightforward: after setup is made, automatic scanning of the whole visible area (full dome scan) is performed. We considered that the resolution of more than 100 points/m^2 is sufficient for the realistic representation of the terrain and its main characteristics. This process takes *c.* 15 min. Usually 10 000–40 000 points are taken during this process.

Underground measurements. Stations for the instrument setup were determined as foresight from the surface, which ensures that the surface and underground measurements are in the same coordinate systems and the whole representation of the cave and surface is consistent.

Setup is performed using the known back-sight method. The instrument is placed on the tripod previously measured as a foresight, and orientation is a reflector on previous station. Since the instrument has automatic target aiming, this process is independent of the light conditions, and the instrument will automatically find the centre of the reflector, even in the dark. During orientation, as a control distance measurement between the instrument, both position and orientation are taken.

After setup, scanning of the whole visible area is performed, using the same resolution as for surface measurements. The average distance between stations for underground scanning depends on the complexity of the underground object.

Limitations during underground scanning are that:

- the instrument will not measure if the distance to the object is less than 1.8 m;
- the instrument must be placed on the tripod, so measurements of a small object must be taken from the side;
- the instrument will not measure points below the station position;
- in narrow long caves, more station positions are needed, which can be time consuming.

On the other hand, the advantages are obvious:

- representation of the object is realistic, owing to the large number of points taken;
- the measurement process is simple and the accuracy of the results is good;
- all points are consistent – there is no need for office registration of the measured points.

TLS data processing

Data processing of the field measurements is performed in several stages.

- Completeness of the data is checked in the field in the instrument cloud viewer, which integrates all measurements performed.
- The instrument integrated viewer is limited to a relatively small number of points (around 20 000). In order to perform full-scale inspection, data are imported to Leica Cyclone, in which all measured points are shown (Fig. 5).
- After measurements are checked, usually it is necessary to perform data cleaning, deleting the points that are not of interest. These points are deleted manually because they are clearly visible as 'noise'. This is also performed in Cyclone using limit box functionality, which enables only part of the whole project to be seen and keeps only necessary points. After the cleaning is performed, points are exported to be used in 3DReshaper.

Fig. 5. Cave represented as cloud of points.

- In 3DReshaper, models of the cave and surface are created. After triangulation model inspection, this model is exported back to Cyclone.
- In Cyclone, on the model, we have created longitudinal cross-sections automatically, and performed all measurements (Fig. 6).

Results

Radovanska Reka

The relationship between surface and underground morphology in the Radovanska Reka gorge was not the subject of any previous study. The Pećura Cave, extending downstream from the ponor of the Radovanska Reka, is a 155 m-long through-cave (Petrović *et al.* 1978). At the entrance, there is a fissure that swallows the average waters (Fig. 7a). During maximum water levels, when the river flows through the cave, sediments are deposited in the narrowest segment of the passage, thus limiting accessibility.

Above the cave, for a length of 200 m, the former river bed is preserved. This stretch is a palaeovalley which on both sides has the character of a hanging valley. It begins 40 m above the cave entrance

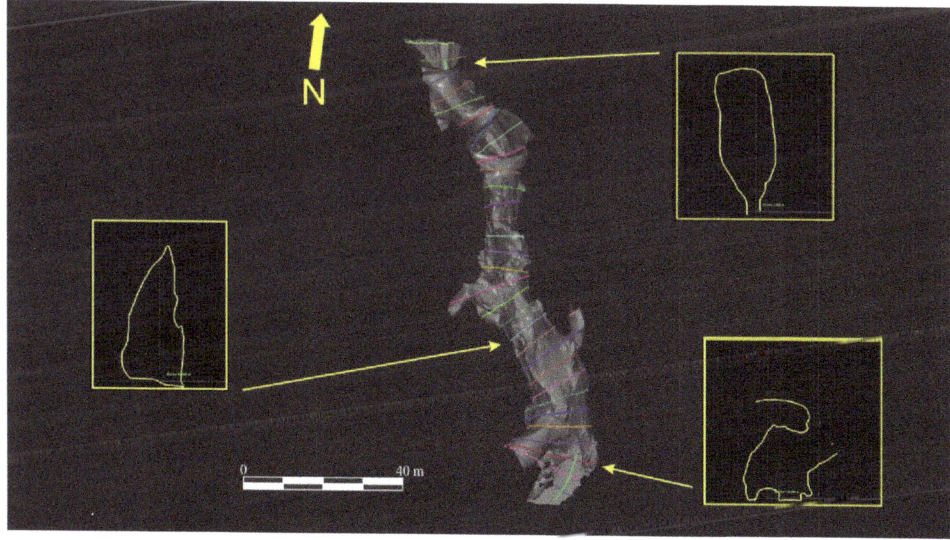

Fig. 6. Model of the Zamna Cave with a characteristic profile.

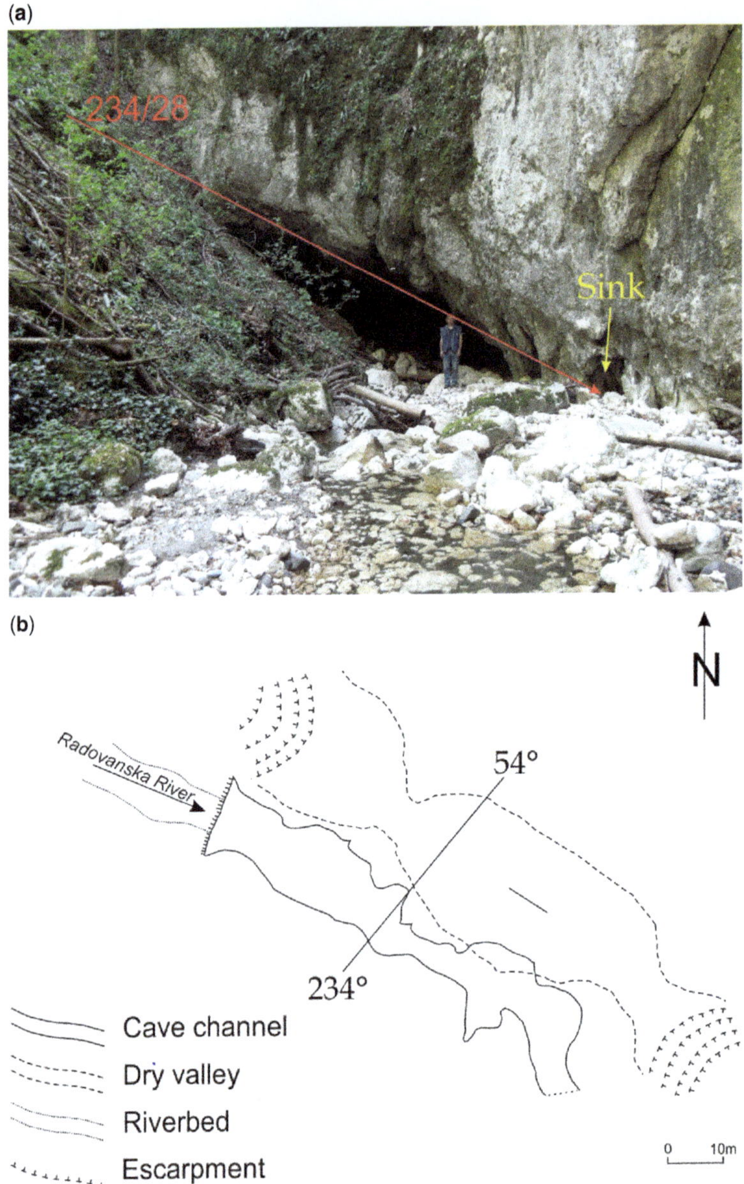

Fig. 7. (**a**) Entrance to Pećura Cave; (**b**) plan.

(Fig. 7a). The end of the valley is a 70 m-high rock face above the spring of Radovanska Reka. Between these two escarpments, there is a doline formed at the bottom of the valley (Fig. 7b). The doline is 70 m long and 36 m wide, while its maximum depth is 16 m. The strike of the doline along the longer axis is 300/120.

Mapping and scanning of the Radovanska Reka Valley showed that the hanging valley is positioned parallel to the cave passage (Fig. 7c), but is offset to the SW (20–30 m horizontally and about 8 m vertically; Fig. 7d). This offset is due to the existence of a significant tension joint dipping 234/32, which guided the development of both the cave passage and the dry valley. The entrance to the cave shows a clear cross-section of this passage (Fig. 7a). The above-mentioned ponor on the right side, by the cave entrance, is positioned at the same tension joint.

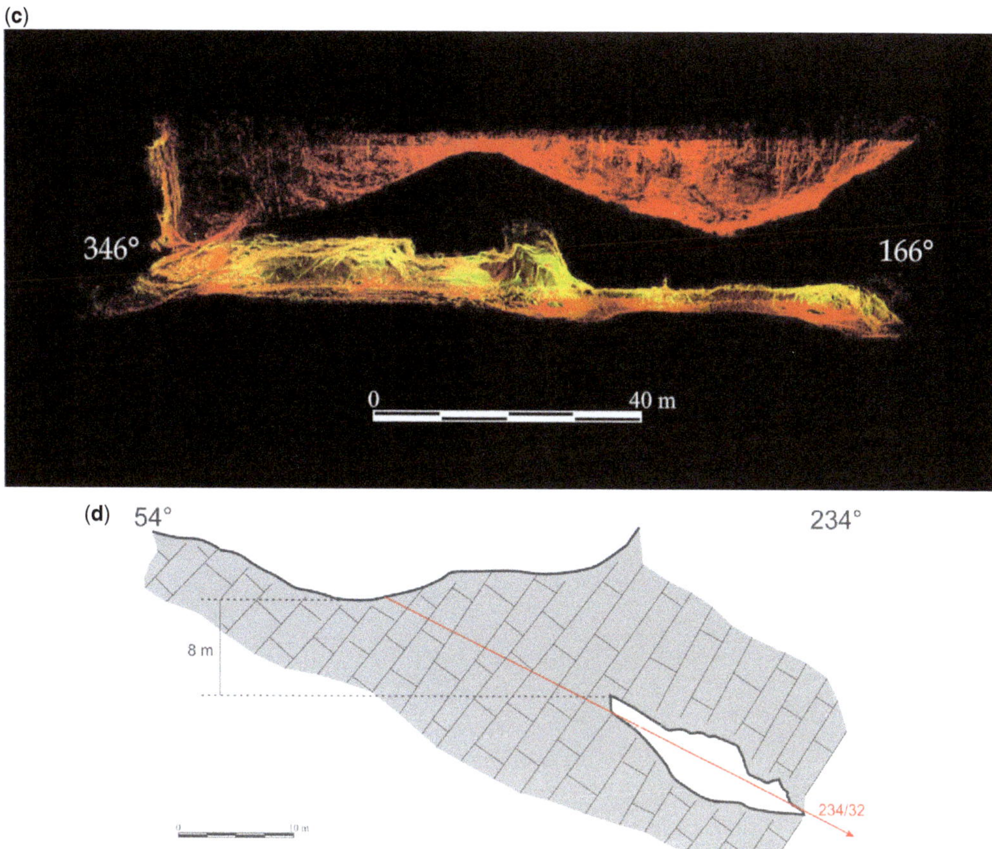

Fig. 7. (*Continued*) (**c**) scan and (**d**) profile of relations between surface and underground karst.

Zamna River

The Zamna Cave (Fig. 8a) was formed by the bypass of the river through the narrow limestone ridge between the valleys of Zamna River and its tributary Udubašnica. The analysis of GmIS and the field observations pointed out that part of the palaeovalley is visible on the ridge, west of the through-cave. The bottom of this 120 m-long palaeovalley is situated at 298 m a.s.l., i.e. 40 m above the bottom of the present Zamna River Valley (at the entrance to the cave). This points to significant incision of the Zamna River below the level of 300 m a.s.l, which influenced the dimensions of the cave, but also to the formation of a 1200 m-long and 30–40 m-deep gorge just upstream from it (Fig. 3).

After compiling the model of the cave, based on the results of TLS, precise data on the cave dimensions were obtained (Fig. 8a). The Zamna Cave is 142 m long, a maximum of 31 m high (26 m average) and up to 12 m wide. The cave generally strikes 346/166, following a large joint. Two large daylight holes are present close to the entrance/exit of the cave (Fig. 8b, c). These are the spots where the bottoms of former dolines have cut down to the cave ceiling, causing collapse (Figs 8b, c & 9). The remnants of the ceiling are still visible on the cave floor (Fig. 8c).

Discussion and conclusions

Through-caves, daylight holes and natural bridges are specific forms of karst relief. All of them may be linked to some of the evolutionary phases of collapse-valley formation. Although not exclusively, these forms are mostly linked to contact karst and its evolution (Gams 1974, 1986, 2004). Analysing the spatial distribution of collapse valleys and through-caves in the karst of the Carpatho-Balkanides of Serbia, it was realized that they occur immediately at the contacts of karst units (especially those of smaller areas) and impermeable Neogene marine or lacustrine sediments (Petrović 2015). Similarly,

Fig 8. **(a)** Zamna Cave; daylight holes at the **(b)** beginning and **(c)** end of cave.

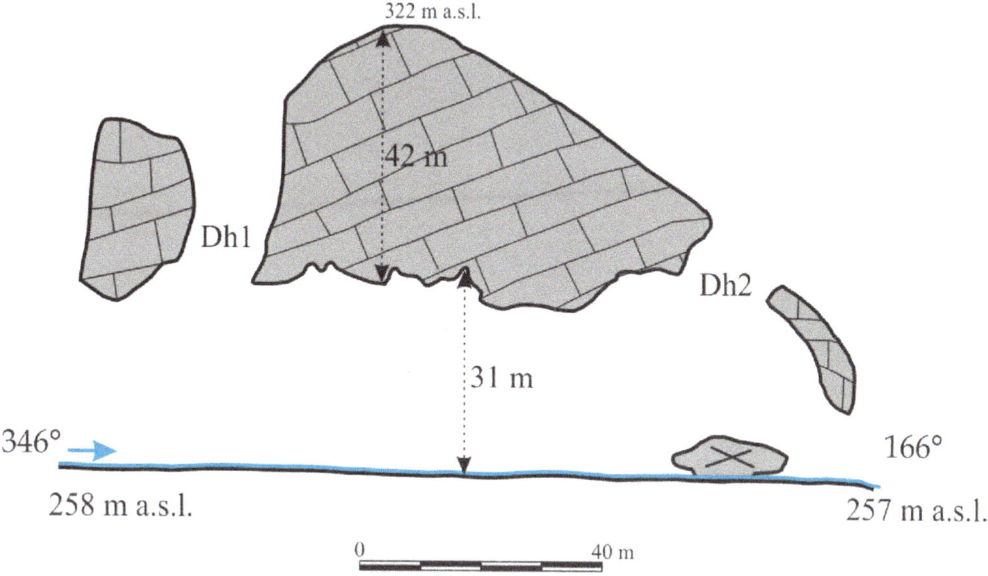

Fig. 9. Zamna Cave longitudinal profile.

through-caves are often linked to stripe karst (as defined by Lauritzen 2001 and later studied by Ćalić 2008).

Through-caves may be formed by widening of the underground passages situated right below the former valley bottom (Petrović 1971; Petrović *et al.* 1978; Bočić 2003). In other cases, the direction of cave passages does not overlap with the direction of the dry valley (Ćalić 2008). By TLS of the Pećura Cave and the Zamna through-cave, we tried to establish the mutual relation between the positions of the dry valleys and caves, as the conditions for the development of collapse valleys.

Pećura Cave in the valley of the Radovanska Reka River was selected for scanning thanks to the fact that GmIS analysis showed a high probability that the cave passage is positioned exactly below the dry part of the valley. Moreover, after comparison between the cave map drawn by standard speleological procedure and the topographical maps at a scale of 1:25 000, we expected the thickness of the cave roof to be extremely small.

The scanning results, however, showed the parallel positions of the palaeovalley and the cave passage, but with a certain horizontal offset. The former valley of the Radovanska Reka and the through-cave are developed along a significant tension joint, the dip of which causes the horizontal offset. The active ponor of the river also points to the same structure, as well as the largest chamber in the cave. Its development is directed, along the joint, towards a doline in the palaeovalley. This can be seen in the profile generated from the model (Fig. 6a), and the field observations point to that as well (stronger air circulation). However, the deepest part of the doline is not at that exact place, but directed to the lower segment of the passage. The thickness of the roof (8 m) is much larger than expected. Stability of this roof could theoretically be threatened with further increase in the passage height, but this process is considerably slowed down by the fact that the majority of the water sinks to the ponor in front of the cave.

Having these points in mind, it is less probable that the roof would collapse in the 'near' future, during further development of the cave and the doline. Without the prerequisites necessary for a collapse-doline formation above the cavity (in the sense of Šušteršič 2002), we will probably not have the second phase of collapse-valley evolution in the short term.

In the published literature, Zamna Cave was mentioned as an example of the second phase in natural bridge development, as a through-cave with daylight holes. The text and drawings by Petrović (1971) show a dry valley stretching exactly above the cave (Fig. 10). Therefore it was necessary to perform TLS and determine the exact positions.

Before scanning, the analysis of GmIS pointed to the possibility that relations in the field could not correspond to the sketch by Petrović (1971). Geomorphological mapping showed that part of the palaeovalley is situated on the ridge between the present blind valley of the Zamna and the valley of its tributary Udubašnica. The new longitudinal profile, created on the basis of TLS data, indicates that the

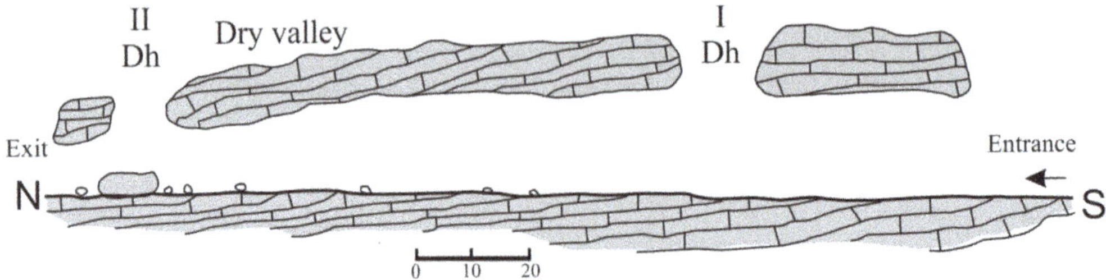

Fig. 10. Profile of Zamna Cave (according to Petrović 1971).

cave is formed between the formal 'normal' valleys of these two rivers. The highest point on the ridge above the cave is at 332 m a.s.l. (Fig. 9), while on the topographical map at 1:25 000 scale, the elevation of that point is mistakenly written as 302 m a.s.l. Most probably, this difference led previous authors to position the dry valley above the cave. Similar cases of mapping of caves whose development does not underlie the surface landforms (such as dry valleys) have been reported elsewhere, in other karst settings (Ford 1964; Cafaro *et al.* 2016).

Analysis of the longitudinal profile, the series of cross-sections obtained from the model (Fig. 11) and the surface above the cave indicates that we can theoretically expect further collapses in the relatively near future, but this issue has not yet been analysed from engineering geology and geotechnical standpoints (as in Parise & Lollino 2011; Fazio *et al.* 2017). Widening of daylight holes at the beginning and at the end of the cave could potentially lead to the development of a collapse valley. The middle part of the cave, bearing in mind the great thickness of the roof in that segment (up to 42 m), will probably remain as a large natural bridge.

The use of TLS and creation of the model of surface and underground karst relief have enabled comparative analysis and conclusions about their mutual relations. These two examples show that

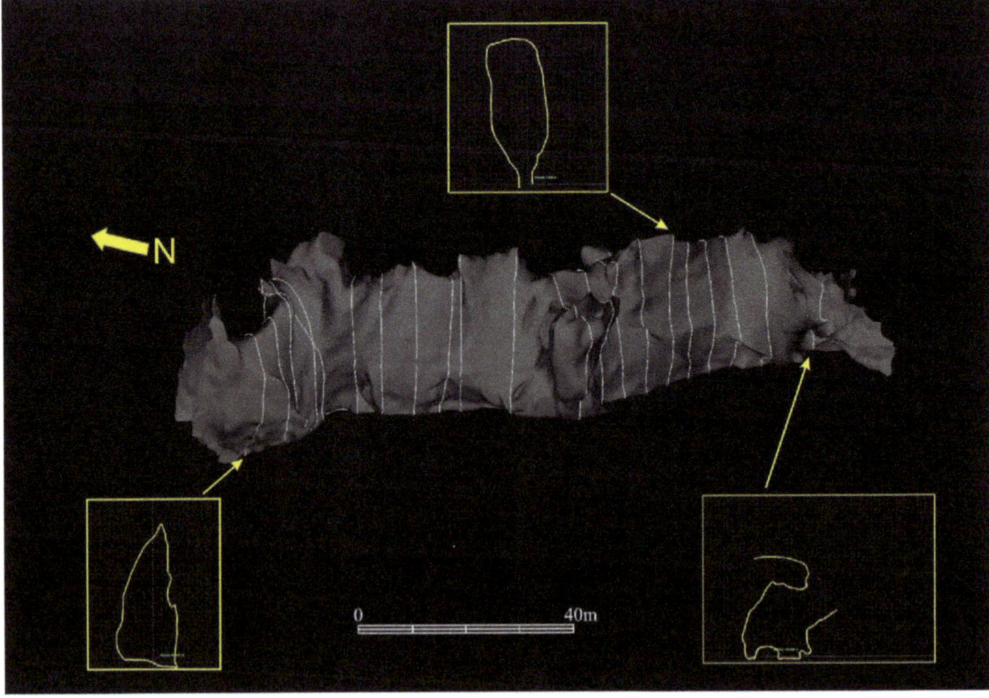

Fig. 11. Zamna Cave model with the series of cross-section lines.

some previous hypotheses were wrong owing to insufficiently accurate input information (from fieldwork and topographical maps). In the case of Zamna, there was a problem with thick vegetation that prevented the free positioning of the instruments. This could be overcome in future by the usage of airborne LiDAR (Weishampel *et al.* 2011), but it would considerably increase the costs of mapping. As future perspectives, we plan to perform the scanning of a few other locations in the Carpatho-Balkanides of eastern Serbia, which will make a crucial contribution to the issue of collapse-valley evolution.

References

Ambert, P. & Nicod, J. 1981. Sur quelques karsts de Serbie, au voisinage du Danube. Leurs rapports avec l'évolution du bassin pannonien. *Revue Géographique de l'Est*, **21**, 235–249.

Anđelković, M. & Nikolić, P. 1980. *Tektonika Karpato-balkanida Jugoslavije [Tectonics of the Carpatho-Balkan Mountains]*. Univerzitet u Beogradu, Monografija, **20** [in Serbian].

Andrić, I. & Bonacci, O. 2014. Morphological study of Red lake in Dinaric karst based on terrestrial laser scanning and sonar systems. *Acta Carsologica*, **43/2**, 229–239.

Audra, P. & Palmer, A.N. 2015. Research frontiers in speleogenesis. Dominant processes, hydrogeological conditions and resulting cave patterns. *Acta Carsologica*, **44**, 315–348.

Bočić, N. 2003. Relation between karst and fluviokarst relief on the Slunj Plateau (Croatia). *Acta Carsologica*, **32/2**, 137–146.

Cafaro, S., Gueguen, E., Parise, M. & Schiattarella, M. 2016. Morphometric analysis of karst features of the Alburni Mts, Southern Apennines, Italy. *Geografia Fisica e Dinamica Quaternaria*, **39**, 21–128.

Ćalić, J. 2008. Kontaktne i strukturne odlike karsta Dževrinske grede [Contact karst and structural characteristics of the ridge Dzevrinska Greda]. *Geografski institut 'Jovan Cvijić' SANU*, **72**, 1–156 [in Serbian].

Ćalić-Ljubojević, J. 2000. Natural bridges on the Vratna River (Eastern Serbia) as the last remnants of a former cave. *Acta Carsologica*, **29/2**, 241–248.

Cvijić, J. 1918. Hydrographie souterraine et evolution morphologique du karst. *Revue de Géographie Alpine*, Grenoble, **6-4**, 375–426.

De Waele, J., Carbone, C., Sanna, L., Vattano, M., Galli, E., Sauro, F. & Forti, P. 2017. Secondary minerals from salt caves in the Atacama Desert (Chile): a hyperarid and hypersaline environment with potential analogies to the Martian subsurface. *International Journal of Speleology*, **46**, 51–66.

Fabbri, S., Sauro, F., Santagata, T., Rossi, G. & De Waele, J. 2017. High-resolution 3-D mapping using terrestrial laser scanning as a tool for geomorphological and speleogenetical studies in caves: an example from the Lessini Mountains (North Italy). *Geomorphology*, **280**, 16–29.

Fazio, N.L., Perrotti, M., Lollino, P., Parise, M., Vattano, M., Madonia, G. & Di Maggio, C. 2017. A three-dimensional back analysis of the collapse of an underground cavity in soft rocks. *Engineering Geology*, **238**, 301–311.

Ford, D.C. 1964. Origin of closed depressions in the central Mendip Hills, Somerset, England. Paper presented at Karst Symposium, 20th International Geographical Congress, 1964, England.

Ford, D.C. & Williams, P.W. 2007. *Karst Hydrogeology and Geomorphology*. John Wiley and Sons, Chichester.

Gams, I. 1974. *Kras*. Slovenska matica, Ljubljana.

Gams, I. 1986. Kontaktni fluviokras. *Acta Carsologica*, **14-15**, 71–88.

Gams, I. 2004. *Kras v Sloveniji v prostoru in času*. Založba ZRC, ZRC SAZU, Ljubljana.

Garcìa-Puchol, O., McClure, S.B., Blasco Senabre, J., Cotina Villa, F. & Porcelli, V. 2013. Increasing contextual information by merging existing archaeological data with state of the art laser scanning in the prehistoric funerary deposit of Pastora Cave, Eastern Spain. *Journal of Archaeological Science*, **40**, 1593–1601.

Gavrilović, D. 1975. Kras Karpato-balkanskih planina Jugoslavije [Karst of the Carpatho-Balkanides of Serbia]. *Bulletin of the Serbian Geographical Society*, **LV/2**, 3–28 [in Serbian].

Gavrilović, D. 1998. Natural bridges – phenomenom of the fluviokarst in Eastern Serbia. *Nature Protection*, **48-49**, 25–32 [summary in English].

Grubić, A. & Jankićević, J. 1973. Karbonatna platforma u gornjoj juri i donjoj kredi Istočne Srbije [Upper Jurassic and Lower Cretaceous carbonate platform in Eastern Serbia]. *Zapisnici Srpskog geološkog društva za 1972. godinu*, Srpsko geološko društvo, Beograd, 73–85 [in Serbian].

Grussenmeyer, P., Landes, T., Alby, E. & Carozza, L. 2010. High resolution 3D recording and modelling of the Bronze Age Cave 'Les Fraux' in Perigord (France). *International Archives of Photogrammetry, Remote Sensing and Spatial Information Sciences*, **38**, 262–267.

Gutiérrez, F., Parise, M., DeWaele, J. & Jourde, H. 2014. A review on natural and human-induced geohazards and impacts in karst. *Earth Science Reviews*, **138**, 61–88.

Jaillet, S., Sadier, B., Hajri, S., Ployon, E. & Delannoy, J.J. 2011. A 3D analysis of underground karst: lasergrammetry applications in Orgnac's cave (Ardèche, France). *Géomorphologie: relief, processus, environnement*, **4**, 379–394.

Jennings, J.N. 1971. Karst. *In*: *An Introduction to Systematic Geomorphology*. MIT Press, **7**.

Jennings, J.N. 1985. *Karst Geomorphology*. Blackwell, Oxford.

Jouves, J., Viseur, S., Arfib, B., Baudement, C., Camus, H., Collon, P. & Guglielmi, Y. 2017. Speleogenesis, geometry, and topology of caves: A quantitative study of 3D karst conduits. *Geomorphology*, **298**, 86–106.

Kalinić, M., Đorđević, M. et al. 1976. *Tumač za list Bor [Interpretation for Bor sheet of the Basic Geological Map 1:100 000]*. Savezni geološki zavod, Beograd [in Serbian].

Kaufmann, G. 2014. Geophysical mapping of solution and collapse sinkholes. *Journal of Applied Geophysics*, **111**, 271–288.

Knez, M. & Slabe, T. 1999. Unroofed caves and recognising them in karst relief (discovered during motorway construction at Kozina, South Slovenia). *Acta Carsologica*, **28**, 103–112.

Knez, M. & Slabe, T. 2002. Unroofed caves are an important feature of karst surfaces: examples from the classical Karst. *Zeitschrift für Geomorphologie NF*, **46**, 181–191.

Krstić, N., Savić, L. & Jovanović, G. 2012. The Neogene Lakes on the Balkan Land. *Annales Géologiques de la Péninsule Balkanique*, **73**, 37–60.

Lauritzen, S.E. 2001. Marble stripe karst of the Scandinavian Caledonides: an end-member in the contact karst spectrum. *Acta Carsologica*, **30**, 47–79.

Lerma, J.L., Navarro, S., Cabrelles, M. & Villaverde, V. 2010. Terrestrial laser scanning and close range photogrammetry for 3D archaeological documentation: the Upper Palaeolithic Cave of Parpalló as a case study. *Journal of Archaeological Science*, **37**, 499–507.

Letal, A. 2005. *Aplikace GIS v geomorfologicke mapove tvorbe.* Disertacni prace, Prirodovedecka Fakulta, University Karlovy, Prague.

Mihevc, A., Slabe, T. & Šebela, S. 1998. Denuded caves. *Acta Carsologica*, **27**, 165–174.

Milić, Č. 1970. Osnovne karakteristike geomorfološke evolucije krečnjačkih terena u Istočnoj Srbiji [Basic characteristics of geomorphological evolution of limestone areas in Eastern Serbia; in Serbian]. *Journal of the Geographical Institute 'Jovan Cvijić' SASA*, **23**, 33–50.

Nicod, J. 1997. Les canyons karstiques 'Nouvelles approches de problèmes géomorphologiques classiques' (spécialement dans les domaines méditerranéens et tropicaux). *Quaternaire*, **8**, 71–89.

Parise, M. & Lollino, P. 2011. A preliminary analysis of failure mechanisms in karst and man-made underground caves in Southern Italy. *Geomorphology*, **134**, 132–143.

Petrović, D. 1969. Reljef u slivu Vratne. *Zbornik radova PMF, Beograd*, **16**, 7–25.

Petrović, D. 1971. Pećina na reci Zamni. *Zbornik radova PMF, Beograd*, **18**, 15–25.

Petrović, D. 1974. Morfogeneza doline Zamne. *Zbornik radova PMF, Beograd*, **21**, 5–18.

Petrović, D., Gavrilović, D. & Lješević, M. 1978. Pećine u dolini Radovanske reke [Caves in the valley of Radovanska Reka]. *Zbornik radova Geografskog instituta PMF*, **25**, 13–23 [in Serbian].

Petrović, S.A. 2015. *Influence of modifiers on the polymorphism of karst valleys of the Serbian Carpatho-Balkanides* [in Serbian]. PhD thesis, Geography Faculty, Belgrade University.

Petrović, S.A. & Carević, I. 2015. Geological influence on the formation of Samar natural bridge and collapse valley of Ravna river from the NE Kučaj mountains (Carpatho-Balkanides, Eastern Serbia). *Acta Carsologica*, **44**, 37–46.

Roncat, A., Dublyansky, Y., Spötl, C. & Dorninge, P. 2011. Full-3D surveying of caves: a case study of Marchenhohle (Austria). *In*: *Proceedings of the International Association for Mathematical Geosciences*, ISPRS Annals of the Photogrammetry, Remote Sensing and Spatial Information Sciences, Volume II-5/W1, XXIV International CIPA Symposium, 2–6 September 2011, Strasbourg, France.

Siart, C., Forbringer, M., Nowaczinski, E., Hecht, S. & Hofle, B. 2013. Fusion of multi-resolution surface (terrestrial laser scanning) and subsurface geodata (ERT, SRT) for karst landform investigation and geomorphometric quantification. *Earth Surface Processes and Landforms*, **38**, 1135–1147.

Šušteršič, F. 2002. Collapse dolines and deflector faults as indicator of karst flow corridors. *International Journal of Speleology*, **34**, 115–127.

Tilly, N., Kelterbaum, D. & Zeese, R. 2016. Geomorphological mapping with terrestrial laser scanning and UAV-based imagined. *The International Archives of the Photogrammetry, Remote Sensing and Spatial Information Sciences*, **XLI-B5**, 591–597.

Veselinović, M., Antonijević, I. *et al.* 1970. *Tumač za list Boljevac* [*Interpretation for Boljevac sheet of the Basic Geological Map 1:100 000*]. Savezni geološki zavod, Beograd [in Serbian].

Waltham, A.C. & Fookes, F.G. 2003. Engineering classification of karst ground conditions. *Quarterly Journal of Engineering Geology and Hydrogeology*, **36**, 101–118, https://doi.org/10.1144/1470-9236/2002-33

Weishampel, J.F., Hightower, J.N., Chase, A.F., Chase, D.Z. & Patrick, R.A. 2011. Detection and morphologic analysis of potential below-canopy cave openings in the karst landscape around the Maya polity of Caracol using airborne LiDAR. *Journal of Cave and Karst Studies*, **73**, 187–196.

Westerman, A.R., Pringle, J.K. & Hunter, G. 2003. Preliminary LiDAR survey results from Peak Cavern Vestibule, Derbyshire, UK. *Cave and Karst Science*, **30**, 129–130.

Zeremski, M. 1991. Planinski niz Krša i istočnoj Srbiji [Mountain range of Krs in Eastern Serbia]. *Zbornik radova Odbora za kras i speleologiju, SANU, Beograd*, **4**, 1–28 [in Serbian].

Zlot, R. & Bosse, M. 2014. Three-dimensional mobile mapping of caves. *Journal of Cave and Karst Studies*, **76**, 191–206.

A genetic classification of caves and its application in eastern Austria

P. OBERENDER & L. PLAN*

Natural History Museum Vienna, Cave and Karst Group, Museumsplatz 1/10, A-1070 Wien, Austria

**Correspondence: lukas.plan@univie.ac.at*

Abstract: Based on existing classifications of caves that often involve descriptive terms, a classification is presented that is based purely on genetic processes. An attribute key is developed that allows the classification of caves by means of cave maps, photographs and reports. This method is applied to a dataset of 6007 caves in a study area in eastern Austria. The area comprises diverse geological units of the Eastern Alps and the southern Bohemian Massif. A total of 94% of the caves could be classified with the surprising result that mechanical weathering and erosion caves are almost as common as solution caves even though the vast majority of caves are developed in carbonate rocks. Field checks confirmed the result and showed that the error is acceptable. The classified caves can also be used as indicator of natural phenomena like gravitational mass movements or vulnerable karst areas by decision-makers non-specialized in cave genesis.

Generally, a cave is defined as a natural underground opening in rock that is large enough for human entry (Ford & Williams 2007, p. 209). This purely phenomenological definition is not based on any genetic processes. Caves are complex features where mostly more than one process is involved in the often long-lasting history of development. It is surprising that there are only a few works on basic genetic classifications of caves compared with the mass of studies concerning special types of caves like karstic ones (e.g. Ford & Williams 2007; Palmer 2007; Klimchouk 2015) or lava tubes (Halliday 2004*d*). It is even more surprising that, to our knowledge, the existing classifications have not been applied to extensive cave datasets yet.

The need for and advantage of classifications in general can be questioned, especially in dealing with complex geomorphological features such as caves. It can be argued that all caves are polygenetic and a classification is not useful or necessary. However, we think that classifications are helpful in scientific communication where a specific vocabulary is needed. Furthermore, certain types of caves can provide information for the detection of natural phenomena that may have impacts on the surface or are even hazardous. This information can be important for planning infrastructure or defining land use areas. For example, a cave mainly formed owing to the dissolving action of water (i.e. a karstic cave) implies that a specific underground drainage system has to be expected and the regional hydrology may be especially vulnerable to contamination. A widened crack owing to the gravitational sliding of a rock mass (i.e. crevice cave) may be linked to a potentially hazardous mass movement.

In this paper we present a genetic classification of caves related to our study area which is a synthesis of existing approaches that focused on processes of cave development. In a further step this classification is applied to a dataset of 6007 caves in a study area with a wide variety of geological units and landscapes. It comprises the eastern part of the Eastern Alps and southernmost part of the Bohemian Massif. The approach was motivated by the Federal Government of Lower Austria, which uses these data for decision-making in land use planning. The statistical analysis of the classified caves gives a pattern of cave type distribution which in some parts leads to surprising results.

Previous concepts and classifications of caves

There are many possibilities for classifying caves, e.g. according to host rock, morphology or pattern. Here an overview of previous works, where cave forming processes are involved, is given.

After several attempts to find a universal theory for the formation of caves, Virlet was one of the first who noticed that several processes can lead to underground openings. In 1835 he published an article where he distinguished seven different reasons for the formation of caves.

To our knowledge the first comprehensive genetic classification of caves was given by Kraus (1894).

From: Parise, M., Gabrovsek, F., Kaufmann, G. & Ravbar, N. (eds) 2018. *Advances in Karst Research: Theory, Fieldwork and Applications*. Geological Society, London, Special Publications, **466**, 121–136.
First published online January 29, 2018, https://doi.org/10.1144/SP466.21

The entire structure of his 308-page monograph 'Höhlenkunde' (speleology) was based on the differentiation of genetic cave types in various lithologies. According to the development of the cave with respect to the formation of the host rock, he differentiated two main types: (1) 'ursprüngliche Höhlen' (primarily developed caves) like magmatic caves or holes formed during the growth of reef organisms; and (2) 'später gebildete Höhlen' (secondarily developed caves). Here, he differentiated erosional, corrosional, crevice caves, etc.

Trimmel (1968) in his monograph (also titled 'Höhlenkunde') also used the concept of primary and secondary caves and extended the classification of Kraus (1894). The nomenclature of the subtypes, however, mixed processes, the erosive medium (e.g. wind or water) and tectonic structures.

Bögli (1978) also used primary and secondary caves and subdivided the first class into exogenous and endogenous caves. The latter are formed by forces within the mountain comprising tectonic caves, open joint caves (i.e. crevice caves) and karstic caves. In the Anglo-American literature the concept of primary and secondary caves was generally not adopted. White (1988, p. 353) gave a cave classification table where he distinguished *caves formed mainly by chemical processes* (solution, lava and ice caves) and *caves formed mainly by mechanical processes* (aeolian, sea, tectonic, suffosion, erosion and talus caves). In addition to these types, he briefly gave the process and the 'optimum host rock' for each type.

Klimchouk (in Gunn 2004) gave a list of seven cave types distinguished by the process and partly by the host rock.

White & Culver (2005; reproduced in 2012) gave a classification scheme that was process based in the first level. For their further subdivision they used different criteria like the resulting form, erosive medium or host rock.

A comprehensive cave classification of Striebel (2005) was among the few in the German-speaking realm that did not consider the concept of primary and secondary caves. He classified caves based on the lithology and further differentiated according to the cave-forming processes. The nomenclature was unique and consistent.

Palmer (2007) used a mixture of host rock, shape and process in his introduction to cave geology. Bella & Gaál (2013) gave a comprehensive genetic classification of non-solution caves only, in which they distinguished 56 cave types. They differentiated between caves formed by endogenous processes (i.e. magmatic, volcanic and tectogene caves) that were further differentiated and exogenous processes. The caves related to exogenous forces were structured by the process and the medium, e.g. fluvial erosion, mechanical weathering and aeolian caves. Within this category they further differentiated by the shape of the cave or the genetic process.

Urban & Margielewski (2013) discussed caves originated by gravitational movements and differentiated three types of caves, suggesting two systems of nomenclature. One was based on a genetic approach that focused on the stages of slope evolution in relation to the position of the cave and the second was a geomechanical viewpoint considering the processes in detail.

Genetic classification and its application

The structure of the presented classification scheme is mainly based on White & Culver (2005), Striebel (2005) and Bella & Gaál (2013), which in our opinion were the most comprehensive and distinct in their nomenclature. The schemes are combined into a new scheme that includes the relevant types but is simple enough for the achieved classification based on the available documents of the caves (see below). We restrict the scheme to cave types that occur in the study area described below. Within the nomenclature we use mainly established terms from the works cited in the previous chapter.

As mentioned above, it is evident that each cave is polygenetic and a couple of processes have been active to form the present shape. On the other hand, for most caves it is possible to identify a dominant process, i.e. the process that was responsible for the biggest gain in volume. This approach was the basis for the presented nomenclature of the first-order genetic cave types (Figs 1 & 2). Three types are distinguished:

(1) *Solution caves* – mainly formed by the dissolving action (chemical enlargement) of underground water. The term *dissolution cave* would be more precise, but it is not that widely used (e.g. White & Culver 2005; Palmer 2007).
(2) *Mechanical weathering and erosion (MWE) cave* – on the basis of Striebel (2005) we introduce this term. It allows the grouping of several, often differently named, cave types (Klimchouk 2004; White & Culver 2005; Palmer 2007). Often these caves, or a group of them, are just called erosion caves but strictly speaking weathering is also important for their formation.
(3) *Deposition cave* – here we prefer a purely process-based term in contrast to the descriptive term *framework cave* used by Palmer (2007).

For the second order of the genetic cave type the scheme is based on processes, shapes and hydrological conditions in order to use existing terms. We simplify the scheme of White & Culver (2005) and subdivide *solution caves* into *epigean* and *hypogean*,

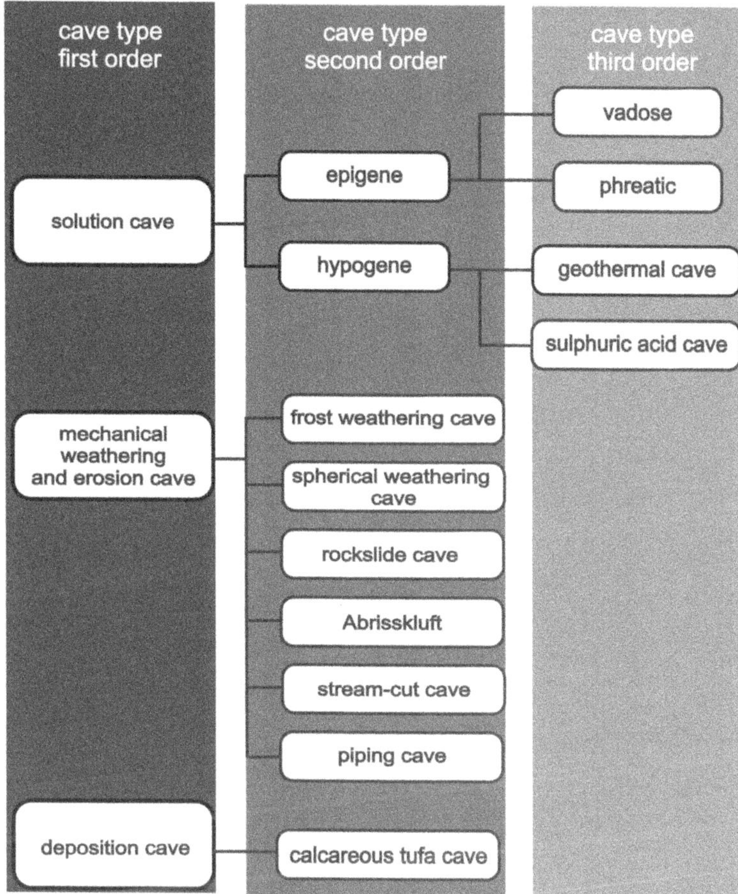

Fig. 1. Genetic cave classification scheme. Differentiation between cave types of first order, second order and third order. The first-order cave type is based on the process regime. The second-order cave type for the solution caves is based on the hydrological condition, while for the mechanical weathering and erosion caves it is also the process in the second order. The nomenclature of the second-order cave type of deposition caves is based on the name of the material. The differentiation within a third order is only used for solution caves and is also related to the hydrological condition and the agent.

which follows the main hydrogeological conditions described in Palmer (2007) and many others. Again, according to hydrology, epigean caves are further subdivided into *vadose* and *phreatic* cave development. Similar to White & Culver (2005), the hypogean caves are differentiated into *geothermal* and *sulphuric acid caves.*

Within the *MWE caves* we distinguish the following types:

(1) *Frost weathering cave* – these caves were called *rock shelters* (e.g. White & Culver 2005) and *shelter caves* (e.g. Palmer 2007) or '*Auswitterungshöhle*' in the German literature (Striebel 2005). From field measurements and observations, Oberender & Plan (2015) suggested that frost weathering is the dominant process and used the term suggested by Bella & Gaál (2013).

(2) *Spherical weathering cave* – following Striebel (2005) we use this term based on the process.

(3) *Rockslide cave* – this is an established term (e.g. Halliday 2004*c*) that is focused on a process in contrast to the widely used term *talus cave* (e.g. White & Culver 2005) or *boulder cave* (Urban & Margielewski 2013), which are descriptive terms that include different processes leading to caves between boulders.

(4) *Abrisskluft* – this German term is well established (Trimmel 1968; Bögli 1978) for crevices

Fig. 2. Examples for the various genetic cave types in the study area. (**a**) Active vadose canyon (Fledermausschacht, Mt Tonion), dark chert nodules project out of the walls; (**b**) epiphreatic tube with large scallops (Trockenes Loch, Schwarzenbach/Pielach); (**c**) calcareous tufa cave (Tuffsteinhöhle, Opponitz); (**d**) chaotic boulders form a rockslide cave (Thorbachgrabenhöhle, Scheibbs); (**e**) parallel walls of an Abrisskluft (Knochenkluft, Kleinzell); (**f**) vertical entrance into a piping cave in loess (Wasserschloß, Kirchberg/Wagram); (**g**) ascending passage with increasing dimensions of a frost weathering cave (Untere Traisenbacherhöhle, Kleinzell); (**h**) 30 m-wide and 9 m-deep stream-cut cave in Pleistocene conglomerates (Inselblock Felsdach, Erlauf river); (**i**) spherical weathering cave in granite (Kaltenberghöhle, Altmelon). All photographs by Lukas Plan except (c) by Heiner Thaler.

widened by gravitational sliding. We use this German term instead of *crevice cave* (e.g. Halliday 2004*a*; Palmer 2007) as Abrisskluft is focused only on mass movement-induced crevices and does not include stresses in the Earth's surface (i.e. tectonic cave).

(5) *Stream-cut cave* – following Klimchouk (2004) it is used for caves formed in bedrock at river-sides owing to the mechanical action of water streams.
(6) *Piping cave* – formed where loose soil or sediment subsides into zones of rapid underground water flow and is carried out to the surface at lower elevations (Halliday 2004*b*; Palmer 2007). Other authors termed them *suffosion (al) caves* (e.g. White & Culver 2005; Bella & Gaál 2013).

Among *deposition caves* only caves formed by stream-deposited tufa occur in the study area. We think that the term *framework cave* (e.g. used by Palmer 2007) is too general and *tufa cave* is also used for volcanic cave. Therefore, we decided to specify them as *calcareous tufa caves*.

Application of the classification

Quantitative research in geomorphology focuses on the functional relations between form, materials and Earth surface processes (Slaymaker *et al.* 2009, p. 5). Therefore, it is a common approach that different forms are characteristic for different processes (e.g. Dikau 2006). Often the basis of understanding involves empirical relationships between form element measurements and their relation to physically induced movements (Gregory & Lewin 2014, p. 221). The application of our classification is based on this relation between form and process. Therefore, we set up a key of characteristic attributes for each cave type.

Method

The classification scheme described above (Fig. 1) was applied based on the data available in the Austrian Cave Register. These are the location of the cave (coordinates), standardized cave maps (most often in plan-view and longitudinal section and sometimes cross-sections), morphological descriptions, often hydrogeological observations and photographs.

To be able to classify the caves based on these documents the morphological characteristics for each cave type have to be defined. The specific properties of the attributes for each cave type were determined and merged in a table to establish a classification key (see Appendix A). The six specific attributes were determined and derived from the following data basis:

(1) *overall cave morphology and pattern* – can be determined from the cave maps and descriptions;
(2) *host rock lithology* – based on descriptions and geological maps;
(3) *relation of the cave to the surface and local hydrology* – via the coordinates the cave entrances were plotted in a GIS containing topographic maps, orthophotos and hill-shade and slope maps derived from airborne laser scan data with 1 m resolution (Geoland 2017);
(4) *cave sediments* – from descriptions, cave maps and photographs;
(5) *shape of cross-sections* – from maps and photographs;
(6) *micro morphological features* – photographs, descriptions and maps.

As not all of this information is available for every cave, the attributes have different consistencies. The first three attributes are known for all caves. The last three are less consistent, as they are not known for all registered caves. Cave sediments should be shown on the cave map – at least a differentiation between fine-grained and coarse cobbles and boulders should be possible. Owing to the scale (normally between 1:100 and 1:500), cartographic generalization information gets lost. The attribute *shape of cross-section* is not very consistent because it is often missing on the cave maps. The attribute *micro morphological features* is the least consistent one. Only sometimes is it detectable on photographs, mentioned in the description or indicated on the map. Hence we have different reliabilities as to the identification of the cave types. Therefore, we marked uncertain cave types within each order. For example, it is possible that the first-order cave type is clearly identified while the second or third order has a higher uncertainty. For some caves all orders are uncertain. Caves which could not be classified at all are marked as of *unknown genesis*.

In addition, caves were investigated in the field to clarify their genesis, recheck the reliability of the attributes and reduce the error. In total some 750 caves in the study area were visited by the authors.

Study area

The area of investigation includes the Austrian provinces of Lower Austria and Vienna as well as some neighbouring parts of Upper Austria, Styria and Burgenland (23 666 km^2; Fig. 3). Most of the caves in this area were documented and registered by the Speleological Society of Vienna and Lower Austria.

The landscape is highly diverse, ranging from high-Alpine karst plateaux with summits up to 2277 m a.s.l. (Mt Hochschwab), hilly pre-Alpine regions and parts of the Bohemian Massif to extended lowlands down to 115 m a.s.l. at Lake Neusiedl.

The mean annual air temperatures and the precipitation range from 10°C and 500 mm a^{-1} in the

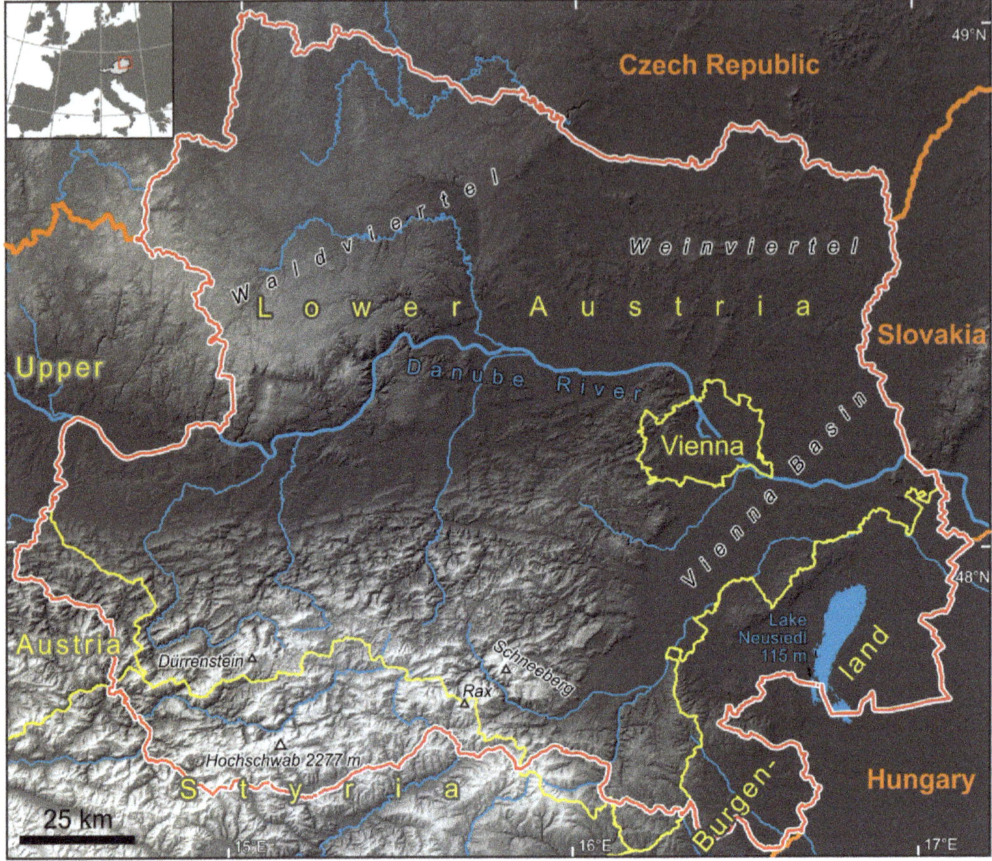

Fig. 3. Geographical overview of the study area (red outline) in the NE of Austria. *Source*: EU-DEM, Copernicus.

lower parts to slightly below zero and more than 2000 mm a^{-1} at the highest mountain tops (NÖ-Landesregierung 2017). In winter snow piles up to some metres in the mountains while there are rarely some decimetres of snow in the lower areas.

The study area comprises the Bohemian Massif at the southern margin of the European plate and the NE end of the Eastern Alps, whereas most of the Alpine tectonic superunits are present (Fig. 4; Wessely 2006; Schuster *et al.* 2014):

(1) The Bohemian Massif comprises highly metamorphic rocks of the Moldanubian and Moravian units. Stripes of marbles, containing few caves, are rare. During the Variscan Orogeny massive plutonites (mainly granites) intruded into the Moldanubian unit. These granites were exposed and spheroidal weathering left many characteristic rounded boulders (woolsacks) that coin the hilly landscape of the so-called *Waldviertel*, that reaches up to 1063 m a.s.l.

(2) The Molasse Zone and intramontane basins contain Tertiary, mainly clastic, sediments. The biggest basin is the Vienna Basin, a pull-apart structure with up to 6 km of Neogene sediments. At its margins Miocene limestones occur, which host some caves. The so-called Waschberg Zone within the Molasse Zone is formed by tectonically pushed up fragments of autochthonous sediments (including Jurassic limestones) deposited on the margins of the Bohemian Massif.

(3) At the northern margin of the Eastern Alps nappes of the Penninic superunit consist of Rhenodanubian Flysch. These fine-grained clastic sediments and marls were deposited from Jurassic to Paleogene times.

(4) The Helvetic unit is represented only by small tectonic slices within the Penninic unit. They consist of diverse, highly deformed sediments.

(5) The Northern Calcareous Alps (NCA) mainly consist of a stack of non-metamorphic Upper Austroalpine nappes. Mainly along Lower

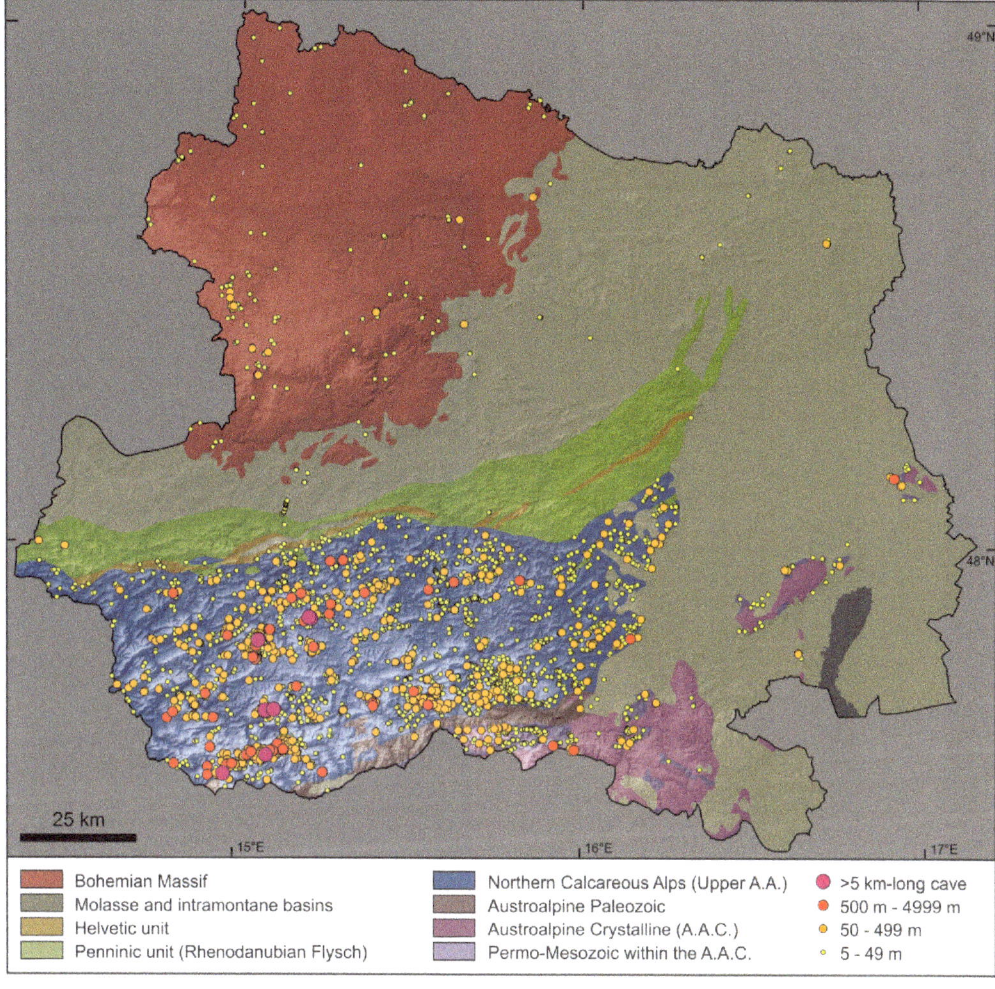

Fig. 4. The main tectonic units in the study area and investigated caves classified according to length; A.A., Austroalpine. *Source*: Geological Map of Austria 1:1 500 000 (in Schuster *et al.* 2014).

Triassic sandstones they were sheared off from their basement. The NCA are dominated by sequences of Middle and Upper Triassic limestones and dolomites that reach up to 3 km in thickness. Jurassic to Paleogene sediments are only of local relevance. Partly extensive karst plateaux are preserved in various elevations between the pre-Alpine areas in the northern and NE part of the NCA and the highest massifs like Hochschwab, Schneeberg, Rax, Schneealpe, Dürrenstein and others. Some 84% of the caves in the study area are located within the NCA.

(6) The Greywacke Zone is a small strip of the weakly metamorphic Paleozoic sedimentary rocks (mainly schists) of the Austroalpine. Carbonate rocks are rare and there are almost no caves within the Greywacke Zone in the study area.

(7) The Austroalpine Crystalline (metamorphic Upper and Lower Austroalpine) is characterized by smaller units of weakly metamorphic Permo-Mesozoic sediments – mainly Triassic carbonates that host several caves – between schists and gneisses. For simplicity, we also add the Hainburger Berge, a tiny part of the Tatricum, to this unit.

In the Pleistocene it was only during the heaviest glaciations that the Hochschwab was part of the Alpine ice stream network and some other massifs showed local glaciers (Van Husen 2000). In the

Alpine valleys and on the foothills massive fluviatile terraces have formed and are partly conglomerated. Loess (aeolian silt deposit) was deposited mainly in the Danube area.

Databases

In Austria caves longer than 5 m are registered. There are also some underground cavities in the register that turned out to be of artificial origin. These were excluded from the study. Especially in eastern Austria there has been a long tradition of cave documentation (Herrmann & Plan 2016) and the first systematic study, published in 1954 (Pirker & Trimmel), lists 697 caves for Lower Austria and adjacent areas. Meanwhile, by February 2017 a total of 6007 caves had been registered in the study area (Fig. 4). For most of the area the caves were systematically described in a five-volume monograph (Hartmann & Hartmann 2000) and an overview is given in Plan (2016) and Plan *et al.* (2016). All caves are included in the online GIS database *Spelix* by the Austrian Speleological Association. There, besides all basic information like cave name, length, depth, coordinates, exploration status, etc., all cave maps are accessible and for most caves photographs are available.

The caves in the study area are highly heterogeneous in their spatial distribution (Fig. 4) as well as in their morphometric parameters. The cumulative length of all caves is 320.6 km. Even though the Frauenmauer–Langstein–Höhlensystem is 41.4 km long and the Warwas–Glatzen–Höhlensystem is 756 m deep, the majority of caves in the area is small. Some 50% of the caves are only 5–12 m long. The average length is 53 m and average depth is 11 m. A total of 333 caves are longer than 100 m and 27 caves are longer than 1 km.

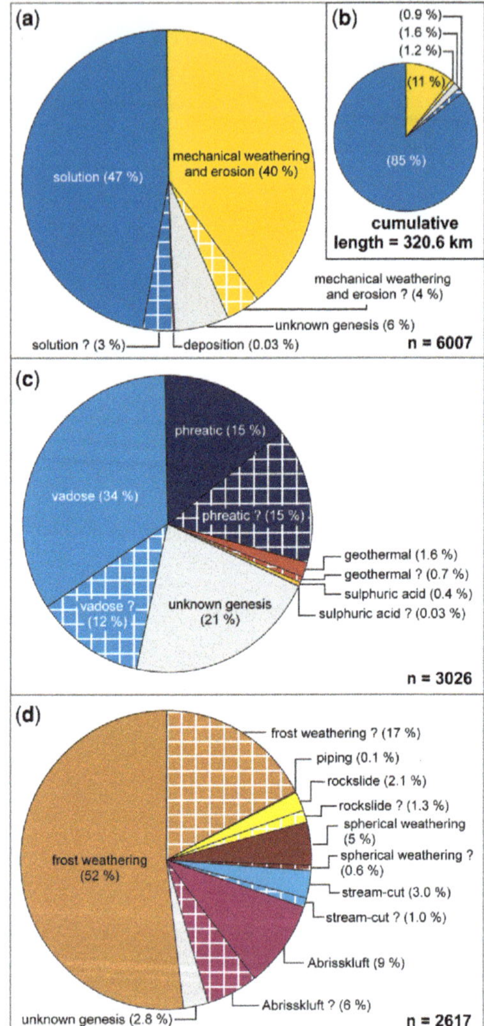

Fig. 5. Percentage frequency distribution of genetic cave types (cf. Fig. 1). (**a**) First-order cave types per cave; (**b**) first-order cave types per cave length; (**c**) third-order cave types of solution caves only (per cave); and (**d**) second-order cave types of MWE caves only (per cave). A question mark after the cave type indicates that the classification was uncertain.

Results and interpretation

The distribution of the first-order genetic cave types is quite balanced between solution caves (50% including uncertain ones) and MWE caves (44%; Fig. 5a). Only two caves are deposition caves, i.e. calcareous tufa caves. For 7% (3% of the solution and 4% of the MWE) of the first-order cave type the classification was uncertain. In all, 362 (6%) caves could not be classified. If the caves are weighted by their length (Fig. 5b), the vast majority of passages (86% including that of the uncertain ones) belong to caves that were classified as solution caves.

Of the 3026 solution caves the vast majority are epigean and only 2.7% have a hypogene origin. They are entirely located around the Vienna Basin (Plan & Spötl 2016). For the third-order cave type of solution caves the majority (46% including the uncertain ones) are vadose followed by phreatic caves (30%; Fig. 5c). However, the uncertainty within the phreatic caves is very high (49%) compared with vadose ones (26%). Concerning the distinction between vadose and phreatic in many polygenetic systems we also followed the principle that we looked for the process that created the most gain in volume, e.g. if a cave was dominated by vadose pits and canyons and only showed small initial phreatic tubes it was classified as vadose. Some 2.4% of the hypogean caves were classified as geothermal and only 0.4% were of sulphuric acid speleogenesis.

The distribution within the 2617 MWE caves was clearly dominated by frost weathering caves (69% including the uncertain ones; Fig. 5d). Some 15% were Abrissklüfte, 5.6% were classified as spherical weathering and 4.0% as stream-cut caves. Only 3.4% were rockslide caves and there were just three piping caves in loess (Pavuza & Plan 2011). Within the MWE caves for 2.8% the second-order subtype is unknown.

Figure 6 shows the number of first-order genetic cave types per cave length. The class of the 5–15 m-long caves is significantly dominated by MWE caves. Already among the 16–49 m-long caves there are twice as many solution caves as MWE caves. This trend is also visible in the further classes. It is also notable that caves that could not be classified decrease significantly with increasing length. Only for three caves longer than 158 m is the genesis unknown.

The caves longer than 500 m are solution caves, except for one. The longest MWE cave is the 634 m-long and 55 m-deep Otterkluft. This Abrisskluft in Triassic Dolomite is part of a system of seven slope-parallel crevices in which 1.3 km of passages were mapped. Other major Abrissklüfte have developed in the sandstones of the Penninic Flysch (the 97 m-long Windloch) and the Greywacke Zone (the 323 m-long Rabensteinhöhle). The longest frost weathering cave is the 95 m-long Wetzsteinloch in Hochschwab massif (NCA) that developed owing to the weathering of a fold core that is made up of volcanic tuff of the Middle Triassic Grafensteig Formation while the remaining fold consists of limestone of the same formation (Oberender & Plan 2015). The longest spherical weathering caves in Granite of the Bohemian Massif are Obere- and Untere Saubachlhöhle near Ysper, north of the Danube river. An active brook washed out the weathering products underground. For a distance of 135 and 160 m, respectively, the subsurface brook can be followed between the rounded granite boulders. The surveyed length of both caves is about 300 m each but not all passages are mapped yet (Plan & Oberender 2016). The longest rockslide cave is the 450 m-long Thorbachgrabenhöhle IX near Scheibbs at the northern margin of the NCA. Here Jurassic limestones slid down on marls and left a body of chaotic boulders. Generally, the penetration depth of stream-cut caves is only 5–10 m. The widest in the study area is the 80 m-wide Rettensteinerhöhle in the city of Waidhofen at the Ybbs River that developed in Pleistocene conglomerates.

The distribution of the first-order cave types with respect to entrance elevation shows a peak of MWE caves between 600 and 900 m a.s.l. (Fig. 7). Above 1500 m they become rare (fewer than 50 caves per 100 m interval). The caves that could not be classified are more or less proportional (10–20%) to the distribution of the MWE caves, and above 1800 m a.s.l. there are none. Among the solution caves there are two peaks visible: a less significant one around 650 m and a more significant one at around 1550 m a.s.l. This is analysed in more detail in Figure 8, where only the epigene solution caves are plotted against the entrance elevation. Here it is obvious that the lower peak is due to the phreatic caves and the upper one is a result of the abundance of vadose caves. The uncertain classification of the phreatic caves is *c.* 50% over the whole elevation range. For the vadose caves it is very high below 600 m a.s.l. and it becomes significantly less than 50% above 1500 m a.s.l.

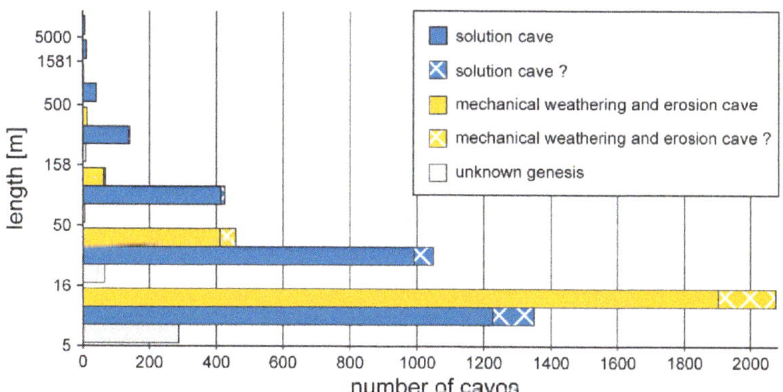

Fig. 6. Number of first-order genetic cave types classified by classes of cave lengths (logarithmic scale). A question mark after the cave type indicates that the classification was uncertain. The two deposition caves are neglected.

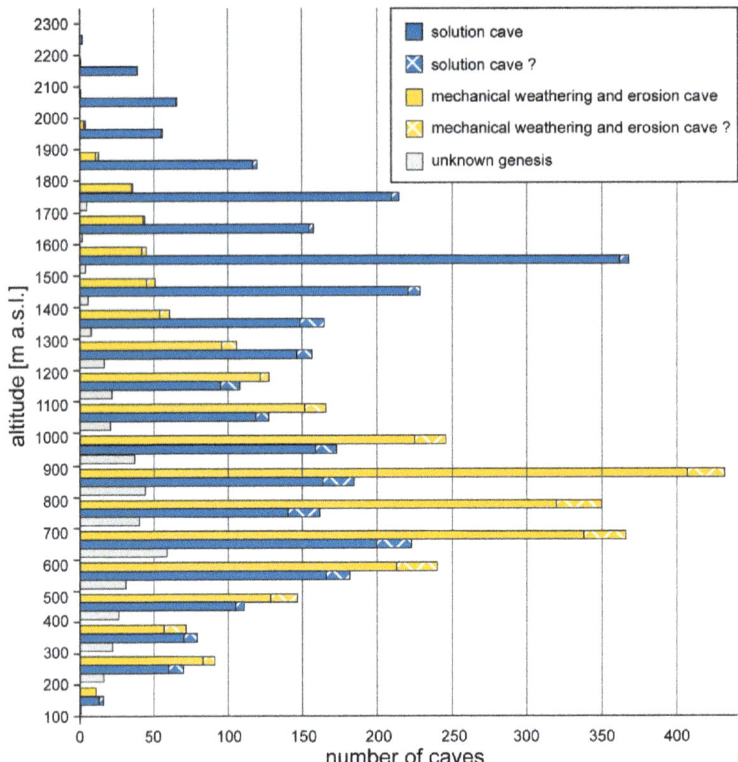

Fig. 7. Height-dependent distribution of first-order cave type. The elevations of the main entrances are regarded. The two deposition caves are neglected.

The regional distribution of the first-order caves types is shown in Figure 9. The analysed sub-areas are based on the division of the mountain ranges used for the Austrian Cave Register (Stummer & Plan 2002). These are mainly delineated following natural borders like rivers or thalwegs and geological aspects. For our analysis many smaller units of the Austrian Cave Register (in total 79 for the study area) with similar geomorphological and geological properties and similar cave type distribution were combined into 34 units.

As expected, MWE caves are dominant in the Bohemian Massif, in most parts of the Molasse Zone, and in the Rhenodanubian Flysch. Only in the NE are some caves in Jurassic and Miocene limestone probably of solutional origin. There, a clear distinction is difficult as they show significant anthropogenic modifications. In the Austroalpine Crystalline (SE part of the study area) in the two eastern areas solution caves are dominant while in the westernmost one (containing 297 caves) MWE caves are dominant. In most areas of the NCA solution caves are dominant, as expected. Surprisingly there are also some areas that are significantly dominated by MWE caves.

Discussion

Classification scheme

Modern approaches in geomorphology tend to identify and quantify processes (Schumm 1991). Therefore, our aim was to provide a classification scheme for caves based on processes and hydrological regimes only. We tried to avoid descriptive terms which are included in most previous concepts. To our knowledge, the only classification scheme that only used process-relevant terms was provided by Bella & Gaál (2013) and is limited to non-karstic caves. However, this scheme and others that consider only restricted groups of caves, e.g. gravitational caves (Urban & Margielewski 2013), are partly too complex for the achieved classification based on the data available. Also from an applied viewpoint it is not relevant that a cave is in the transition between Abrisskluft and rock slide cave (which is called the *dilatancy cave* by Urban & Margielewski 2013). For land use planning it is relevant that this cave indicates a gravitational mass movement. The presented scheme in combination with key attributes allows quantification of the dominant genetic

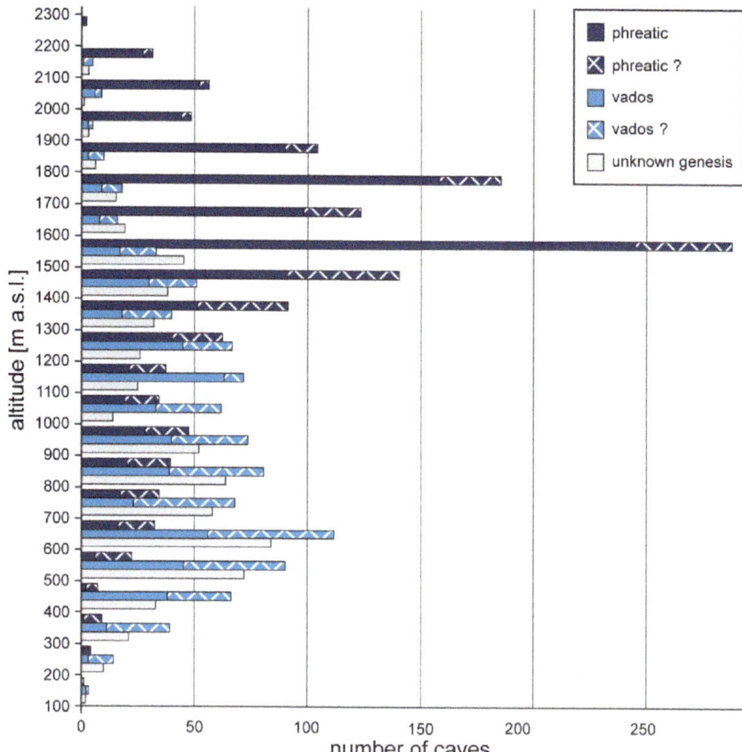

Fig. 8. Height-dependent distribution of third-order cave type of epigene solution caves.

process for a large number of caves if there is a solid database.

It is obvious that our classification scheme has the drawback that it is restricted to cave types that occur in the study area. However, especially among non-karstic caves, the understanding of processes is often influenced by local observations and studies (e.g. Halliday 2004*a*, p. 249) and a worldwide valid process based classification of caves is hardly achievable at the present time. Therefore, we suggest that the presented scheme should be modified and extended by experts who are able to identify attributes for specific genetic cave types occurring in their working area.

A genetic classification scheme is always related to the present knowledge of the processes. Therefore, further findings on details of speleogenesis will necessitate adoption of the classification scheme. For example, in our study area there are four caves in poorly consolidated very pure Oligocene to Miocene Quartz sandstones in the Molasse basin. From the cave morphology a genesis owing to piping is most likely but their relation to the present hydrological regime is ruling that out. Therefore, at the moment these caves are classified as being of unknown genesis as there are no detailed studies on their genesis yet.

Classification method

To achieve a preferably objective classification of a large number of caves mainly based on the available documents from the cave register, an attribute key was developed (see Appendix A). The six defined attributes (listed in the 'Method' section) differ in their consistency. While the first three attributes are available for all caves, attribute 4 (cave sediments) is only available for about half of the caves and attributes 5 and 6 could only be used for roughly one-third of the caves. Attribute 1 (overview and pattern of the cave) is the strongest attribute. Therefore, short caves that consist of single chambers or short passages only are difficult to classify and show the highest occurrence of unknown genesis (Fig. 6).

If more than one process is active during cave development it is more difficult to specify the dominant one in a small cave than in a big one. In particular, frost weathering is active in most cave entrance parts and often overprints the original morphology that was created by a different process (e.g. a short

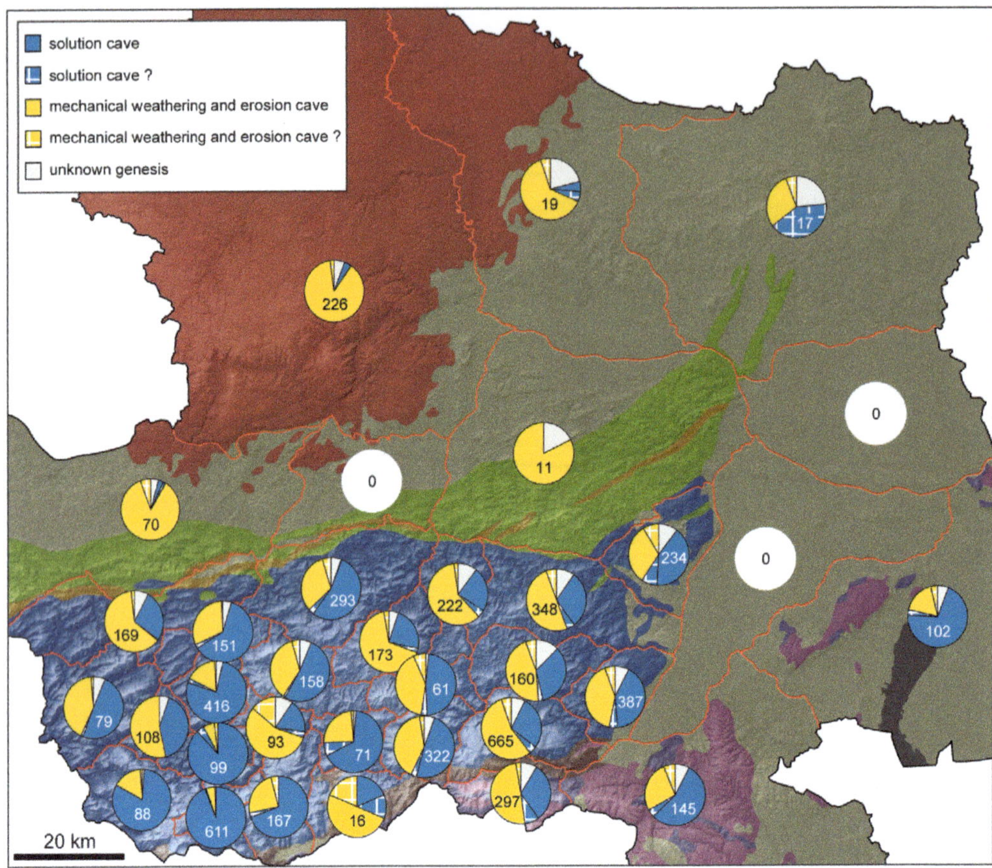

Fig. 9. Distribution of first-order cave types within geomorphically and geologically similar areas (red polygons). For explanation of geology see Figure 4. Numbers in the pie charts give the quantity of caves in each sub-area. Sub-areas without caves are marked by white circles. The two deposition caves are neglected. *Source*: Geological Map of Austria 1:1 500 000 (in Schuster *et al.* 2014).

section of a solution cave can look like an MWE cave).

Attribute 1 was derived from cave maps of various authors, some of them dating back to the 1950s. Depending on scale, skill or focus of the author, the maps differ in quality (e.g. Häuselmann 2007; Palmer 2007, p. 85). In particular, the scale determines the readability of the map and the degree of generalization. However, small caves especially, which are hard to interpret from the map, are usually drawn in big scales and extensive ones in smaller scales. Nevertheless, we are aware that a cave map is a generalization and interpretation of the nature and our method performs a second level of interpretation on that, which can lead to fuzzy results in some cases. It has to be mentioned that attribute 2 (host rock lithology) has a strong influence on the process itself and is therefore not completely independent of the other attributes.

Field checks of caves that have been already classified by the proposed method show that roughly for two-thirds the classification based on cave documents is in accordance with the field observations. Field checks mainly led to shifts between caves types attributed as 'uncertain' or 'certain'. For less than about 15% a different subtype was classified and only in a few cases was the first order cave type changed.

For the application of the classification (e.g. for land use planning), errors within the first-order cave type are for most cases more relevant than within the second or third order. However, mostly caves where difficulties occurred during the classification were checked in the field. Therefore, the sample was not random and the error estimation is probably too pessimistic.

Especially problematic are natural caves that suffered from human overprint as they were utilized for subsurface quarrying, mining, dwelling or as shelters during wars. Often these impacts took place a long time ago and natural processes have obscured them. This can lead to incorrect interpretations and for some caves it is not clear whether they are of artificial or natural origin.

Classification of caves in the study area

In the study area, most caves occur in the NCA (Fig. 4), which are dominated by carbonate rocks. Therefore, one might assume that solution caves are by far the dominant cave type in the study area. However, the classification shows that the distribution between MWE caves and solution caves is almost equal (44–50% respectively). Especially in some pre-Alpine areas in the north of the NCA (in Fig. 9 the areas where 169, 173, 222 and 348 caves were classified), but also in high-Alpine areas that reach up to 2076 m (Schneeberg and Rax 665 caves in Fig. 9), as many as two-thirds are MWE caves. Only for the Abrissklüfte, which are dominant in few areas (e.g. Mt Veitsch in the south of the NCA; 16 caves in Fig. 9), is the reason for this obvious, as they have formed in carbonates that overlie less competent sandstones of the Werfen Formation. Of course, the minimum length for caves documented and included in this study (i.e. 5 m) has an impact: if it were longer (e.g. 10 m) the proportion of solution caves would increase (Fig. 6) and if it were shorter the percentage of MWE would probably rise. However, so far there is no satisfactory explanation why in some areas frost weathering caves are dominant and there are so few solution caves, respectively. Neither the age of the carbonate host rock (mainly Middle v. Upper Triassic) nor the difference between limestone and dolomite or a different tectonic regime is significant. Concerning the elevational distribution also for the distinct peak of frost weathering caves between 600 and 900 m a.s.l. there is no explanation yet (see also discussion in Oberender & Plan 2015). The vertical distribution of vadose and phreatic caves is as expected. There are numerous vadose pits and canyons, especially on the elevated plateaux of the NCA. There, old phreatic caves are rare. This is in contrast to massifs in the central part of the NCA like Tennengebirge, Dachstein and Totes Gebirge, where several palaeo-phreatic cave levels can be identified (Spötl *et al.* 2016). In the study area, phreatic caves occur more often in lower areas and are sometimes related to present-day hydrology.

A similar classification approach was presented by Herrmann & Fischer (2013) that was focused on the genesis of solution caves. For the small high-Alpine mountain range Hochtor (summit 2369 m a.s.l.) west of the Hochschwab massif 252 caves were regarded. Some 69% of the caves were classified as vadose solution caves and further subdivided into invasion vadose and drawdown vadose canyon-shaft systems according to Ford & Williams (2007); 5% were phreatic caves. The other types are hard to compare with our classification. However, in general the pattern is similar to that of the central Hochschwab (661 caves in Fig. 9), which shows similar altitude but different morphology (there are extensive plateaux on Hochschwab) and different lithology (Hochschwab is dominated by Middle Triassic Wetterstein Formation and Hochtor by Upper Triassic Dachstein Formation).

Conclusions

The presented classification scheme is purely based on the dominant speleogenetic process and avoids descriptive terms. The specific attributes defined for each cave type allow the classification of caves based on the use of cave maps, photographs and reports. Therefore, it can be applied to a large number of caves within a reasonable time. A total of 94% of the 6007 caves in the study area in eastern Austria could be classified; 7% of them are uncertain. Short (<15 m) caves and caves with anthropogenic overprint are especially difficult to classify as characteristic morphological patterns are unclear and an overprint by frost weathering is common. The most surprising result is that 44% of the caves (including the uncertain ones) are mechanical weathering and erosion caves even though most of them developed in carbonate rocks. The solution caves account for 50%.

In addition to the scientific significance, the location of the classified caves provides a tool for land use planning in order to identify gravitational mass movements (Abrissklüfte and rockslide caves) or highly vulnerable karst aquifers (solution caves), for example. On the other hand, a frost weathering cave, even though it has developed in limestone, does not indicate karstification.

The classification was supported in the frame of the project NÖ-Höhlenkataster by the Federal Government of Lower Austria (Klemens Grösl). Harald Zeitlhofer, who developed the SPELIX-database, helped with data management. For discussion, we thank Harald Bauer, Yuri Dublyansky, Wilhelm Hartmann, Eckart Herrmann and Rudolf Pavuza. Philipp Häuselmann and an anonymous reviewer are thanked as well. Christa Pfarr improved spelling and grammar.

Appendix A

Attribute table of cave classification

Table A1. *Attribute table of cave classification*

	Attribute 1 Overall cave morphology and pattern	Attribute 2 Host rock lithology	Attribute 3 Relation of the cave to the surface and local hydrology	Attribute 4 Cave sediments	Attribute 5 Cross-sections	Attribute 6 Micro morphological features
Solution cave; epigen, vados	Mostly canyons (narrower than high), often meandering, rather branching than network pattern, vertical pits, continuous inclination in longitudinal section	Carbonates, sulphates	Often developed in areas with other karst (hydrological) features, sometimes entrances connected to sinkholes or springs	Allochthonous and autochthonous fluviatile sediments, breakdown debris, and diverse speleothems are common	Narrow and high, often shape	Dissolutional microforms, rather small scallops, differential solution
Solution cave; epigen, phreatic	Mix of horizontal passages and intertwined pits, rather continuous cross-sections of passages, cave pattern is often single conduits or network pattern, changes of inclination in longitudinal section	As above	Often developed in areas with other karst (hydrological) features, entrances sometimes connected to springs, in elevated areas hydrologically inactive	As above	Rahter isometric or elliptic shape, exception: paragenetic canyons	Dissolutional microforms, rather big scallops, paragenetic half tubes, Rinnenkarren
Solution cave; hypogen, hydrothermal	Branching and network pattern is common, abrupt changes in dimension and inclination, complex shape of rooms and passages, spherical chambers, blind galleries and chimneys	Carbonates	Often no connection to the surface and landscape evolution, often no natural entrance, often close to hydrothermal springs	(para) Autochthonous sediments, breakdown debris, rarely allochthonous coarse sediments, speleothems which are related to evaporation like coralloids and fristwork	Often isometric or narrow depending on faults	Dissolutional microforms especially cupolas, no scallops, rarely differential solution, feeder channels
Solution cave; hypogen, sulphuric acid	Branching and network pattern is common, abrupt changes in dimension and inclination, complex shape of rooms and passages, spherical chambers	Carbonates	Often no connection to the surface and landscape evolution, often no natural entrance, often close to a H_2S-bearing springs	(para) Autochthonous sediments, breakdown debris, rarely allochthonous coarse sediments, gypsum and other sulphates	Often isometric or narrow depending on faults, eventually sulphuric acid water table morphology	Dissolutional microforms, replacement pockets, especially cupolas, no scallops, rarely differentiated dissolution, feeder channels

Mechanical weathering and erosion cave; frost weathering cave	Often entrance wider than cave depth, often decrease in dimension, rarely branching, normally increase of declination or horizontal passages caused by sediments	Bedrocks	At the foot of cliffs or in the cliffs, natural entrances, often connected to fissures and faults with dripping water	Autochthonous sediments: from boulders and debris size to fine-grained sediments, rather evaporative speleothems	Various	Often rough and cracked surface
Mechanical weathering and erosion cave; spherical weathering cave	Often well-rounded boulders, sharp changes in dimensions, different levels are common, changes in inclination in longitudinal section	Granite, massive conglomerates	Anywhere in the landscape, sometimes connected to river beds, towers of rounded boulders, often many natural entrances	Autochthonous sediments, rarely allochthonous sediments	Separate boulders	Often smooth rock surface
Mechanical weathering and erosion cave; rockslide cave	(sub) Angular boulders in different dimensions, chaotic arrangement, sharp changes in dimensions, different levels are common, changes in inclination in longitudinal section	Any rock-fall material, boulders	Rock fan on the slope below a cliff, often many natural entrances	As above	As above	Often rough and cracked surface
Mechanical weathering and erosion cave; Abrisskluft	Straight parallel walls, rarely branching or network pattern, often different levels, ceiling often built by boulders or rock walls	Any competent rock	Crack often visible on the surface (laser scan), close to steep slopes, natural entrance	Autochthonous sediments mainly debris, rarely allochthonous sediments	Narrow, high with parallel cave walls	Often rough and cracked surface
Mechanical weathering and erosion cave; stream-cut cave	Variety of dimensions and extensiveness, mostly entrance wider than cave depth, rarely branching, rarely different levels, rather horizontal	Any bedrocks, sometimes under competent rock	Natural entrance, close to rivers or palaeo river beds	Allochthonous sediments, rarely autochthonous debris	Niches like cross-sections	Often smooth surface of walls, rough and cracked surface of the ceiling possible
Mechanical weathering and erosion cave; piping cave	Parallel walls, tubes and channels, short pits, rarely branched, small extension, continuous inclination, narrow profiles	Poorly consolidated sediments, e.g. loess and soil	Natural entrances, close to brim, fast changes of shape and dimension during precipitation events	Autochthonous sediments, often collapse of walls and ceiling	Various channels or breakdown cross-sections	Often straight plain walls
Deposition cave; calcareous tufa cave	Mainly single chambers with small extension	Carbonate tufa	Natural entrance, related to terraces in or close to streams	Carbonate tufa, allochthonous fine sediments	Triangular, rounded profiles	Structure of the carbonate tufa visible, sometimes structure of leaves or moss visible

References

Bella, P. & Gaál, L. 2013. Genetic types of non-solution caves. *In*: Filippi, M. & Bosák, P. (eds) *Proceedings 16th International Congress of Speleology*. Czech Speleological Society, Brno, **3**, 237–242.

Bögli, A. 1978. *Karsthydrographie und physische Speläologie*. Springer, Berlin.

Dikau, R. 2006. Oberflächenprozesse – ein altes oder ein neues Thema? *Geographica Helvetica*, **61**, 170–180.

Ford, D. & Williams, P. 2007. *Karst Hydrogeology and Geomorphology*. John Wiley and Sons, Chichester.

Geoland 2017. Kostenloses Geodatenportal der österreichischen Länder, http://www.geoland.at

Gregory, K.J. & Lewin, J. 2014. *The Basics of Geomorphology*. SAGE, London.

Halliday, W.R. 2004*a*. Crevice caves. *In*: Gunn, J. (ed.) *Encyclopedia of Caves and Karst Science*. Fitzroy Dearborn, New York, 249–252.

Halliday, W.R. 2004*b*. Piping caves and badlands Pseudokarst. *In*: Gunn, J. (ed.) *Encyclopedia of Caves and Karst Science*. Fitzroy Dearborn, New York, 589–593.

Halliday, W.R. 2004*c*. Talus caves. *In*: Gunn, J. (ed.) *Encyclopedia of Caves and Karst Science*. Fitzroy Dearborn, New York, 721–724.

Halliday, W.R. 2004*d*. Volcanic caves. *In*: Gunn, J. (ed.) *Encyclopedia of Caves and Karst Science*. Fitzroy Dearborn, New York, 760–764.

Hartmann, H. & Hartmann, W. 2000. Die Höhlen Niederösterreichs, Band 5. Die Höhle, Supplement 54.

Häuselmann, P. 2007. Sustainable mapping of caves. *NSS News*, **65**, 10–13.

Herrmann, E. & Fischer, F. 2013. Höhlen im Hochtor. *Die Höhle*, Supplement 59.

Herrmann, E. & Plan, L. 2016. Höhlendokumentation. *In*: Spötl, C., Plan, L. & Christian, E. (eds) *Höhlen und Karst in Österreich*. OÖ-Landesmuseum, Linz, 399–410.

Klimchouk, A. 2004. Caves. *In*: Gunn, J. (ed.) *Encyclopedia of Caves and Karst Science*. Fitzroy Dearborn, New York, 203–205.

Klimchouk, A. 2015. The karst paradigm: changes, trends and perspectives. *Acta Carsologica*, **44**, 289–313, https://doi.org/10.3986/ac.v44i3.2996

Kraus, F. 1894. Höhlenkunde. – Facsimile reprint 2009, *Die Höhle*, Supplement 56.

NÖ-Landesregierung 2017. Wasserstandsnachrichten und Hochwasserprognosen Niederösterreich, http://www.noel.gv.at/ExterneSeiten/Wasserstand/htm/wndcms.htm

Oberender, P. & Plan, L. 2015. Cave development by frost weathering. *Geomorphology*, **229**, 73–84, https://doi.org/10.1016/j.geomorph.2014.07.031

Palmer, A.N. 2007. *Cave Geology*. Cave Books, Dayton, OH.

Pavuza, R. & Plan, L. 2011. Loess caves of Austria – a preview. *Cadernos Lab. Xeolóxico de Laxe Coruña*, **37**, 65–72.

Pirker, R. & Trimmel, H. 1954. *Karst und Höhlen in Niederösterreich und Wien*. Jugend und Volk, Wien.

Plan, L. 2016. Hochschwab. *In*: Spötl, C., Plan, L. & Christian, E. (eds) *Höhlen und Karst in Österreich*. OÖ-Landesmuseum, Linz, 645–660.

Plan, L. & Oberender, P. 2016. Nicht-Karsthöhlen. *In*: Spötl, C., Plan, L. & Christian, E. (eds) *Höhlen und Karst in Österreich*. OÖ-Landesmuseum, Linz, 61–72.

Plan, L. & Spötl, C. 2016. Hypogene Karsthöhlen. *In*: Spötl, C., Plan, L. & Christian, E. (eds) *Höhlen und Karst in Österreich*. OÖ-Landesmuseum, Linz, 49–60.

Plan, L., Hartmann, H. & Hartmann, W. 2016. Kalkalpen-Ostabschnitt. *In*: Spötl, C., Plan, L. & Christian, E. (eds) *Höhlen und Karst in Österreich*. OÖ-Landesmuseum, Linz, 645–660.

Schumm, S.A. 1991. *To Interpret the Earth. Ten Ways to be Wrong*. Cambridge University Press, Cambridge.

Schuster, R., Daurer, A., Krenmayr, H.G., Linner, M., Mandl, G., Pestal, G. & Reitner, J. 2014. *Rocky Austria – The Geology of Austria – Brief and Colourful*. Geological Survey of Austria, Vienna.

Slaymaker, O., Spencer, T. & Embleton-Hamann, C. (eds) 2009. *Geomorphology and Global Environmental Change*. Cambridge University Press, Cambridge.

Spötl, C., Plan, L. & Christian, E. 2016. *Höhlen und Karst in Österreich*. OÖ-Landesmuseum, Linz.

Striebel, T. 2005. Höhlenbildung in 'nicht verkarstungsfähigen' Gesteinen: welche Formen sind Karstformen? *Laichinger Höhlenfreund*, **40**, 31–52.

Stummer, G. & Plan, L. 2002. *Speldok-Austria – Handbuch zum Österreichischen Höhlenverzeichnis. Speldok-10*. Verband Österreichischer Höhlenforscher, Wien.

Trimmel, H. 1968. *Höhlenkunde*. Vieweg und Sohn, Braunschweig.

Urban, J. & Margielewski, W. 2013. Types of non-karst caves in Polish Outer Carpathians – historical review and perspectives. *In*: Filippi, M. & Bosák, P. (eds) *Proceedings 16th International Congress of Speleology*, **3**. Czech Speleological Society, Brno, 314–319.

Van Husen, D. 2000. Geological processes during the Quaternary. *Mitteilungen der Österreichischen Geologischen Gesellschaft*, **92**, 135–156.

Virlet, T. 1835. Observations faites en Franche-Comté, sur les cavernes et la théorie de leur formation. *Bulletin de la Société géologique*, **6**, 154–164.

Wessely, G. 2006. *Geologie der Österreichischen Bundesländer: Niederösterreich*. Geological Survey of Austria, Vienna.

White, W.B. 1988. *Geomorphology and Hydrology of Karst Terrains*. Oxford University Press, Oxford.

White, W.B. & Culver, D.C. 2005. Definition of Cave. *In*: Culver, D.C. & White, W.B. (eds) *Encyclopedia of Caves*. Elsevier Academic Press, Amsterdam, 81–85.

White, W.B. & Culver, D.C. 2012. Definition of Cave. *In*: Culver, D.C. & White, W.B. (eds) *Encyclopedia of Caves*. 2nd edn. Elsevier Academic Press, Amsterdam, 103–107.

Surface landforms and speleological investigation for a better understanding of karst hydrogeological processes: a history of research in southeastern Italy

M. PARISE[1,2]* & L. BENEDETTO[3]

[1]*Department of Earth and Environmental Sciences, University Aldo Moro, Via Orabona 4, 70125, Bari, Italy*

[2]*National Research Council, Institute of Research for Geo-Hydrological Protection, Bari, Italy*

[3]*Gruppo Archeo Speleologico Pugliese, Gioia del Colle, Bari, Italy*

**Correspondence: m.parise@ba.irpi.cnr.it*

Abstract: Small-size karst landforms may potentially provide very useful information to fully understand the behaviour of karst systems and their dynamics. In this chapter we demonstrate the need to pay attention to such features. 'Inghiottitoio della Masseria Rotolo', located in a remarkable karst area of southern Italy, has in recent years become the most controversial and discussed speleological site in Apulia. Even though it has been known for several decades, recently excavation work has allowed cavers to enter a huge karst system, eventually reaching the water table. The total depth of the cave is now 324 m, making it the deepest in the region. This chapter summarizes the history of discoveries at the site, starting from the description of the polje, also including information about the link between toponymy and karst. The works carried out at the swallet site are then described to emphasize the importance of the often neglected small-size karst features. In fact, when carefully observed and studied, these might be able to shed new light and greatly increase our knowledge about karst. The final part of the chapter deals with the cave system and provides an outline of the ongoing research.

Karst landscapes may continuously offer new and interesting opportunities to find, explore, survey and document sites previously unknown. There are many possible situations leading to such circumstances: advancement of a quarry face, which may bring to light a 'new' cave; excavations for building foundations, leading to possible access below underground; natural (or anthropogenic) unclogging of swallets or sinkholes, opening passages previously inaccessible to exploration, etc. At the same time, even known caves may host unexplored spaces (Putiska *et al.* 2014; Zvab Rozic *et al.* 2015; Despain *et al.* 2016), hidden behind a pile of breakdown deposits, concealed by calcite flowstones or simply not accessible owing to the narrowness of passages.

In particular, when caves were originally discovered and explored several decades ago, when survey techniques and the procedures for correctly locating their position were not so precise and well established, a re-visitation might be interesting, with the potential to shed new light on the knowledge so far acquired (Zhu *et al.* 2014; Fabri *et al.* 2015).

In this chapter, we describe a swallet which has in recent years 'become' a real cave, in Apulia (southeastern Italy), and now represents the deepest karst system in the region and opens a direct window on the deep water table of central Apulia. Apart from the importance of the cave in the framework of the karst features of this sector of Apulia, the history presented here demonstrates the relevance of historical documentation in karst (Shaw 1988, 2004, 2005), and the need to take into account even its small features (which are often neglected) to properly recognize the main properties of the terrains and acquire better knowledge of karst.

Study area: the Canale di Pirro polje

Geological and morphological features

The area object of this study is located in central Apulia (SE Italy), precisely within one of the main karst landscapes of the region, a polje called *Canale di Pirro* (Figs 1 & 2). This polje, showing an overall length of some 12 km, extends from the surroundings of the towns of Gioia del Colle and Putignano towards the east (Anelli 1957; Parise 2006), creating a very elongated landform, clearly bounded on both the northern and the southern sides by tectonically controlled ridges (Fig. 3). The polje is located in the so-called Low Murge, one of the main karst sub-regions of Apulia (Sauro 1991; Parise 2011),

From: Parise, M., Gabrovsek, F., Kaufmann, G. & Ravbar, N. (eds) 2018. *Advances in Karst Research: Theory, Fieldwork and Applications*. Geological Society, London, Special Publications, **466**, 137–153.
First published online January 25, 2018, https://doi.org/10.1144/SP466.25

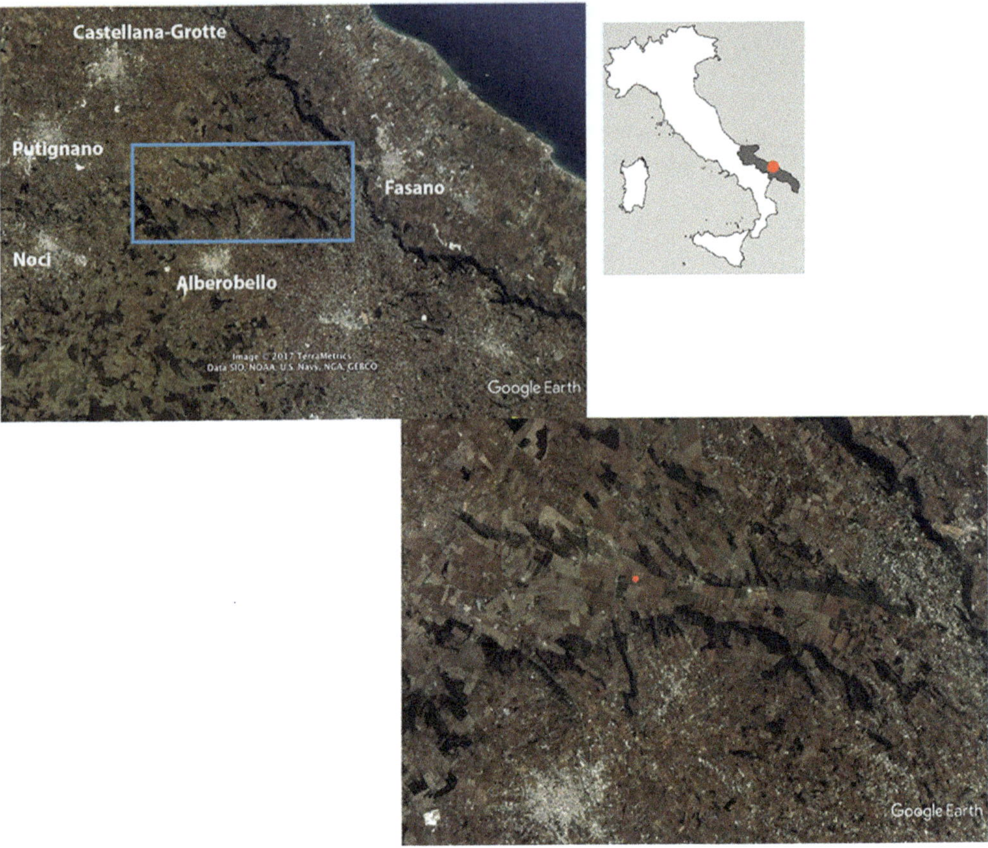

Fig. 1. Location of Canale di Pirro, the polje where the cave object of this study is located (after Google Earth). The blue rectangle marks the polje, which is enlarged in the image below, where Inghiottitoio di Masseria Rotolo is marked by the red dot. In the enlargement, the polje field is clearly delimited by the two approximately west–east-oriented green lines, showing the forested polje borders.

characterized by the direct outcrop of Cretaceous limestones, and reaches on its eastern side the Murgia escarpment (Figs 1 & 2), that is the main NW–SE tectonic line separating inland Murge from the Adriatic coastline (Di Geronimo 1970; Bruno *et al.* 1995). Locally, a thin layer of residual deposits (terra rossa) covers the limestone bedrock. At some sites, such as in depressions or at the ridge foothills, the terra rossa deposit might reach the maximum thickness of a few metres.

From a structural standpoint, the study area is characterized by high-angle, mainly normal faults, strongly affecting the regional-scale morphology. The most important fault systems are represented by normal, NNW–SSE-, west–east- and NW–SE-trending faults (Apenninic faults), affecting the Cretaceous bedrock (Funiciello *et al.* 1991; Doglioni *et al.* 1994; Pieri *et al.* 1997). Apenninic faults are also responsible for the main slopes bordering both the Adriatic and the Ionian sides of the Apulian peninsula (e.g. Neboit 1975; Baldassarre *et al.* 1978; Ciaranfi *et al.* 1988; Delle Rose & Parise 2003). Recent meso-structural analyses by Di Bucci *et al.* (2009) ascribe Apenninic faults to middle and late Pleistocene extensional tectonics. Normal, east–west-trending faults mark the boundary between the Murge and the Salento highlands (*Soglia Messapica line*), and are best represented by the polje of Canale di Pirro.

The above fault systems locally result in the development of a karst landscape consisting of morpho-structural ridges bounding elongated depressions, as also morphologically expressed further south, in the Salento karst (Tozzi 1993; Bosellini & Parente 1994; Gambini & Tozzi 1996; Gil *et al.* 2013; Pepe & Parise 2014).

Dolines and other depression-like landscapes, hosting temporary lakes or marshlands, or

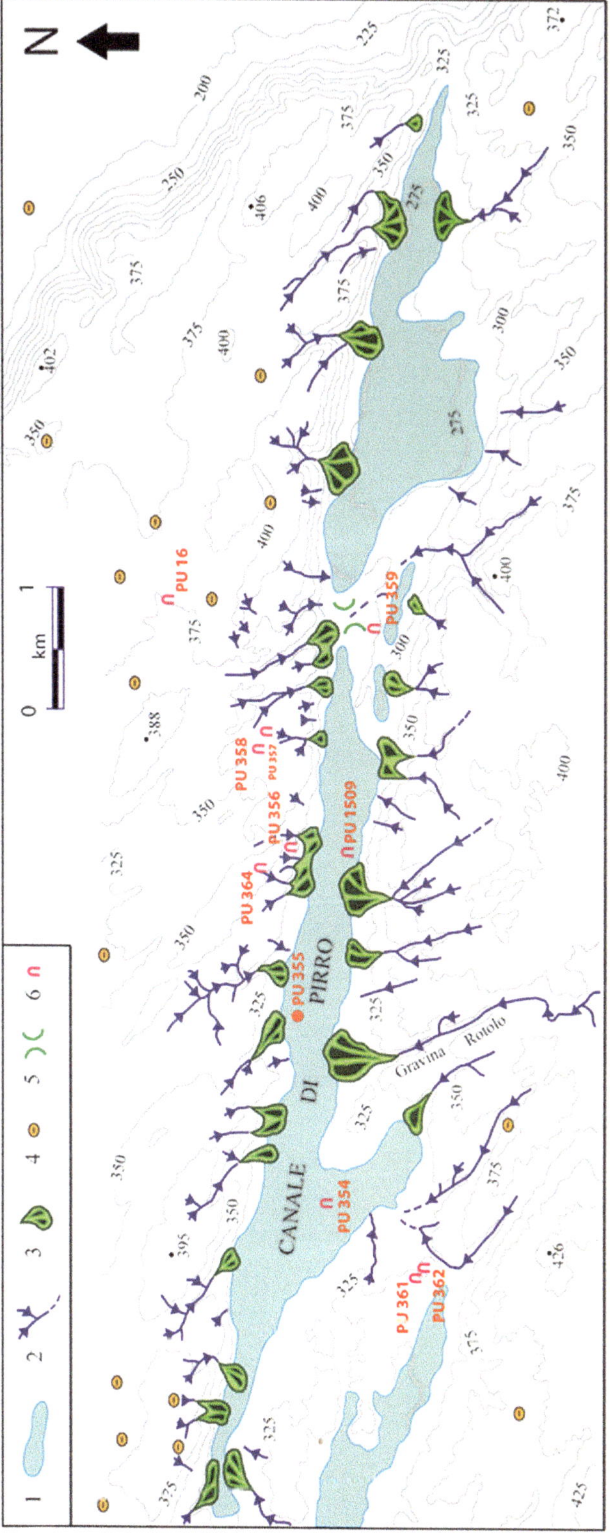

Fig. 2. Geomorphological map of Canale di Pirro (modified after Parise 2006). Key: 1, polje floor; 2, karst valley (dashed where less evident); 3, alluvial fan; 4, doline; 5, morphological saddle; 6, cave (label PU and the numbers indicate the cadastral number of the caves in the Regional Cadastre of Natural Caves of Apulia). The red dot marks the location of the Masseria Rotolo swallow hole. The pattern of the contour lines at the right sector of the map highlights the escarpment separating the Murge plateau from the Adriatic plain. Contour interval 25 m.

Fig. 3. Views of Canale di Pirro: (**a**) northern flank; and (**b**) overall view looking westward.

developing over larger territories such as poljes, are extremely important in terms of land management in all karst areas (Gams 1978, 2005; Pavicic *et al.* 2002; Dogan 2003; Nicod 2003; Bonacci 2004; Gracia *et al.* 2014); the karst of Apulia is not an exception to this (Lopez *et al.* 2009; Parise 2006, 2009). The lowlands are often flooded in the aftermath of the main and/or most intense rainstorms, a situation which has caused historically heavy economic losses throughout the region (Parise 2003; Andriani & Walsh 2009; Cotecchia & Scuro 2010; Martinotti *et al.* 2017). The effects of the floods are often exacerbated by land mismanagement, typically carried out without proper knowledge of the karst landforms and their hydrological functioning (Calò & Parise 2006; Parise & Gunn 2007; De Waele *et al.* 2011). Among these actions, clogging of natural swallow holes and other adsorption points, and expansion of urbanized areas above fragile karst lands often result in obstruction of the natural drainage and surface runoff during and after heavy rainstorms, leading to significant economic losses (Šušteršic & Šušteršic 2003; Breg 2007; Gutierrez *et al.* 2014).

Several swallow holes are present in the Canale di Pirro polje, the biggest one being the Gravaglione (Fig. 4), located in its central sector (Fig. 2). Gravaglione and the surrounding area become periodically flooded, in consequence of the main rainstorm events hitting the polje.

Fig. 4. Gravaglione, the main swallow hole within Canale di Pirro, seen in (**a**) normal condition and (**b**) soon after a heavy rainstorm.

Historical setting and karst toponymy

Toponymy – the attribution of names to a geographical place or entity – often provides very interesting insights into the history, geography and sometimes the geology of a specific site. Apulia is a region which, owing to its orography and geographical configuration (a very elongated peninsula, surrounded by seas and connecting the outer sectors of the Southern Italian Apennines to the very easternmost tip of Italy), has seen very different civilizations inhabit its territory, giving rise to a variety of languages and local dialects. This diversity is also reflected in the terms used to designate karst features, which present a great variety of names, with significant differences, moving from the Gargano (the northern karst sub-region), to Murge (in central Apulia) and eventually to Salento (the southernmost part). A study dedicated to karst terminology in Apulia showed that the various karst features (swallets, different types of valleys, dolines, etc.) are indicated by many different terms, each one having its own history, etymology and motivations (Parise *et al.* 2003).

Starting from this, but most of all from the long history of this region of Italy, the issue of the origin of the toponym *Canale di Pirro* cannot be disregarded. For a long time, this name was attributed to Pyrrhus, the king of Epirus, who was in Apulia to fight against the Romans in support of Taranto during the third century BC (Trisciuzzi 1989, and references therein). It has been historically documented, however, that Pyrrhus was not in the Canale di Pirro area during his wars with the Roman army, so it is difficult to attribute the name of the polje to such a historical circumstance. The toponym is more likely to be a corruption of the original Canale delle Pile (from the latin '*de pile*', through the later '*de pilo*', until '*de piro*'). *Pila* (plural *pile*) is a term used to designate wide-open cisterns for the collection of water, also equipped with channels to facilitate the collection and flow of water within the structure. Several *pile* (troughs, wells) are still present in the Canale di Pirro (Fig. 5), built in the local limestone rock with dry-stone wall technique, and showing widths of 8–9 m, with depths ranging from 4 to 6 m. The technique used consists of rock blocks put together without any mortar, the same as for the dry-stone walls that are widespread in several karst countries of the Mediterranean Basin (Neboit 1975; Parise 2012). This type of well-cistern, or hydric structure, is quite common in several other parts of Apulia, typically located in small basins or depressions at the sites where rainwater is most likely to remain after a heavy rainstorm (Fig. 6).

Actually, the history of the Canale di Pirro is strictly linked to water. Many ancient maps of the area show the presence of a water course, named *Cana*, within the polje. The first cartographic documents depicting this river date back to the sixteenth century. There is no consistent documentation throughout history, since the water course seems to be present on some maps, but not in others. Even when present, its course is diverse, from a short water course located near the coastline to a very long river extending for the whole length of the Canale. The name (*Cana flumen*) is not always present either. Some scholars, for all these reasons, identify it as an 'elusive' river.

Historiographic research carried out by Sisto (2006) demonstrated that many ancient maps documented the presence of the river. Its disappearance from the surface is attributed to deforestation and agricultural practices that resulted in changes in land use and the original flow of water. Among the ancient maps in the historical cartography, those produced by the cartographer Giacomo Gastaldi (Fig. 7) are particularly significant (Gastaldi 1561, 1567).

Beside the maps by Gastaldi, Carlone & Angelini (1986) re-published another interesting document, probably dated between the end of the sixteenth and the beginning of the seventeenth century: the

Fig. 5. Water trough (*pila*) at the valley floor of Canale di Pirro.

Fig. 6. Water troughs (*pile*) in sites of the Murge karst area, central Apulia: (**a**) general view of a site with presence of two water troughs; and (**b**) detail.

map is dedicated to the defensive structures located along the coasts in the Kingdom of Naples, but also extends inland to cover several miles from the coastline. In this document the Cana river is shown clearly, with a length that makes it the second most important water course in the region, following the river Ofanto, which flows in Campania and northern Apulia.

Flood problems in the Canale di Pirro polje

In karst, water courses may not be permanent, but flow only during the rainy season, or soon after the main rainstorms. The karst landscape is also characterized by changing features in the forms of infiltration sites that might be present at different locations from year to year, as a consequence of changes of both natural and anthropogenic origin (White 1988, 2002; Ford & Williams 2007).

In this regard, as concerns Canale di Pirro, a local scholar, Martellotta (2006), citing Notarnicola (1933), describes nomadic swallow holes in the area of our study, the positions of which keep changing according to the different rainstorms and the arrival of variable amounts of rainfall and runoff water from the polje flanks. This testifies to the dynamic nature of the karst systems within Canale di Pirro, and how they accommodate local clogging and closure of swallets caused by anthropogenic

Fig. 7. Map of Apulia (after Gastaldi 1561). Note the very long Cana river (*cana f.* in the map) flowing towards the east, and reaching the Adriatic Sea south of Monopoli.

actions and sediment blockage. Among these, for instance, during the 1990s the loss of some swallets and caves was documented, including the Grotticella del Canale di Pirro (PU 1509 in Fig. 2 and Table 1), owing to agricultural practices to establish a vineyard.

The possibility of intense and clustered rainstorms is not a rare situation in these lands, and this can also be proved through historical documents. A sector within the Canale di Pirro was used at the turn of the fifteenth century as a stud to breed horses for the Venetian Republic: called 'la Cavallerizza', this breeding farm was destroyed in 1506 when a summer rainstorm caused the formation of a huge lake in the *Canale*, killing almost all of the animals (only four stallions and nine foals survived; Notarnicola 1933; Corsi 2006).

Flooded areas can also be observed today after heavy rainstorms, and wide sectors of Canale di Pirro become lakes which require several hours or days to be absorbed underground (Fig. 4): during the most intense storms, both of the flat sectors of the polje, separated by the intervening morphological saddle, are flooded, and the adsorption of water in the subsoil occurs as a function of the amount of rainfall and the conditions of the terrain before the rainstorm (dry, partially saturated, etc.). On the other hand, this part of Apulia is well known for a number of historical floods that hit the main towns: the nearby town of Castellana-Grotte, located some 10 km apart, is the best-known example, where at the turn of the twentieth century, 10 or more severe flood events were recorded, the strongest one in 1893 claiming four casualties (Ce.Ri.Ca. 1996;

Table 1. *Caves in the Canale di Pirro*

No.	Name	Municipality	Length (m)	Depth (m)
PU 16	Grave Santa Lucia	Monopoli	175	250
PU 354	Gravaglione	Alberobello	25	7
PU 355	**Inghiottitoio Mass. Rotolo**	**Monopoli**	**1649** (5)	**324** (3)
PU 356	Caverna dei Buoi	Monopoli	7	0.5
PU 357	Grotta delle Spine a Paretano	Monopoli	8	1
PU 358	Grotta dei Suini	Monopoli	14	0.5
PU 359	Caverna di Marzalossa	Alberobello	14	0
PU 360	Inghiottitoio di Micele	Martina Franca	4.5	7
PU 361	Caverna Grande di Lama Grotta	Alberobello	22	1
PU 362	Cavernetta di Lama Grotta	Alberobello	8.5	1
PU 363	Grotta del Cane	Martina Franca	4	5
PU 364	Grotta del Busine	Monopoli	11.5	0.5
PU 1509	Grotticella del Canale di Pirro	Alberobello	8	10

Data extracted from the Regional Cadastre of Natural Caves, managed by the Apulian Speleological Federation; see http://www.fspuglia.it/catastogrotte.htm. In the *Inghiottitoio di Masseria Rotolo* row, the numbers between brackets in the 'length' and 'depth' columns indicate the morphometric data before the recent exploration of the cave (see Fig. 8).

Parise 2003). At the lowest part of town, a lake formed on that occasion, reaching 6 m in depth, and boats had to be brought from the Adriatic coast to save people from the upper storeys of buildings. Such a response to high rainfall was due to the closure of a large number of swallow holes during previous decades in the process of urban expansion, as well as the practice of covering the local roads with asphalt, many of which were the remains of the ancient hydrographic network (Parise 2003). Dry for most of the year, on the occasion of rainfall this hydrographic network becomes active, allowing water to move downslope towards the main swallets.

Re-discovering Inghiottitoio di Masseria Rotolo

Main features of the cave system

In the framework so far described, it is not surprising that many swallow holes are present within the area occupied by Canale di Pirro. As mentioned, the main one is Gravaglione (once again, a name which has as its root a meaning associated with the idea of depth, of underground, from the Latin *grava* and the Messapian *graba*; see Parise *et al.* 2003, and also Palagiano 1965; Parise 2011). Located in the central sector of the polje, it consists of an elongated depression that collects most of the surface runoff from the head of the catchment. Since 1969 Gravaglione has been included (cadastral number PU 354) in the Regional Cadastre of Natural Caves (Table 1), managed by the Federazione Speleologica Pugliese (Apulian Speleological Federation; http://www.fspuglia.it/catastogrotte.htm).

In addition to Gravaglione, several other minor caves and swallets are known (see Table 1; Orofino 1965; Parise 1999), but many of them have been destroyed by human action (wide sectors of the polje are currently intensely cultivated with vineyards). Two of these swallets (Inghiottitoio di Masseria Rotolo, and Grotticella del Canale di Pirro) were also included in the Cadastre, with numbers 355 and 1509, respectively. Grotticella del Canale di Pirro was later destroyed to create a vineyard on the southern side of the polje, and its precise location has been lost.

As for the other cave, this is the site to which we are going to dedicate the rest of this chapter. *Inghiottitoio di Masseria Rotolo* was originally a very slight swallet, which many cavers considered too small and insignificant a feature to include in the Regional Cadastre (the official rules for including a cave in the Apulian cadastre require a spatial development of 5 m).

The swallet was registered following observations by a local teacher, Giovanni Longo, and was described with the following words, accompanied by the sketch shown in Figure 8 (Longo 1969, pp. 21–22):

> a 3 m-deep doline, with two narrow absorbing points which walls are covered by thick vegetation. The collector channel, developing toward E, reaches a length of 90 meters, with low slope, and collects the rainfall water.

In recent years, the present owner of the land where the swallet is located has been curious to observe the huge amount of water that infiltrates underground at the site after heavy rainstorms (Fig. 9). Being convinced that, to absorb such a quantity of water, wide spaces had to be present behind his land, he allowed local cavers to attempt to dig the site.

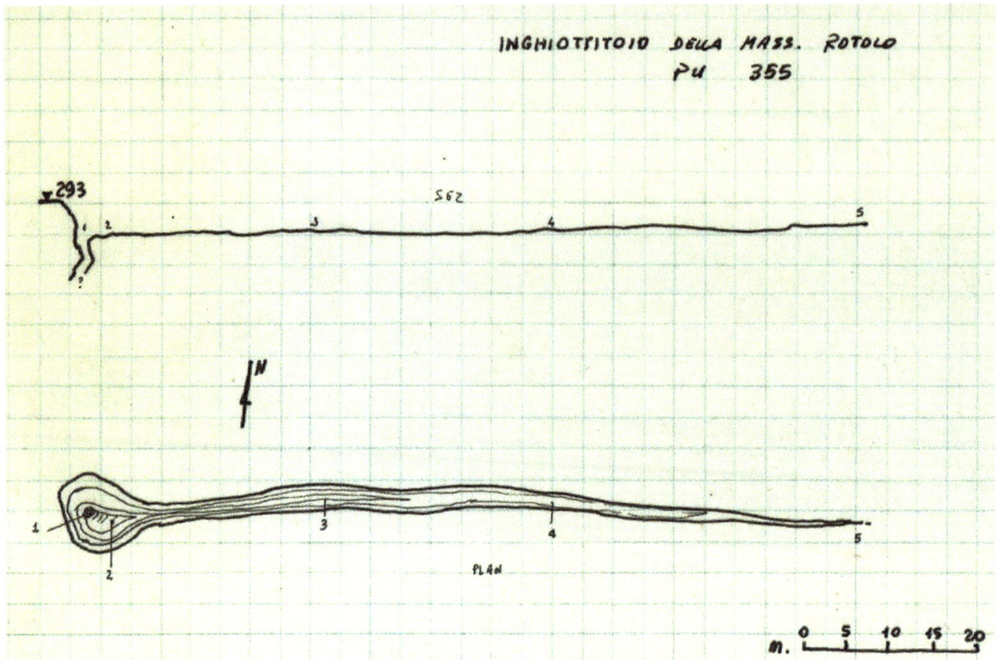

Fig. 8. The original survey of the Masseria Rotolo swallow hole, as registered in 1969 in the Cadastre of Natural Caves of Apulia.

In May 2012, the work started (Fig. 10). Much time, patience and strength was required for the work, together with a great deal of attention to safety. Air flow was detected, but the passage looked too narrow to allow any access, so that further months of hard work were required. Eventually, in September 2012, a passage wide enough to allow human ingress was found at the base of what had become a 8 m-deep shaft. After putting in place cement rings to reinforce the entrance shaft, the first exploration began.

From the start, it was evident to the explorers that they were entering a complex and remarkable system (Figs 11 & 12): a first series of vertical shafts brought them rapidly to a depth of about 100 m below the ground surface (Fig. 13a–c). From here, a sub-horizontal level started, which also included lateral branches (Fig. 13d), whilst at several locations hanging deposits related past levels of karstification, now suspended above the main cave level (Fig. 13f).

The main sub-horizontal level, at an average depth of −110 m below the ground surface, develops for a few hundred metres, before becoming a meandering, canyon-like feature. Some parts of the cave are highly decorated by speleothems (Fig. 13g), with active dripwater feeding the many stalactites and stalagmites present. Through small jumps and some shafts, the cave system deepens, until reaching what is without any doubt the most impressive feature at Grotta Rotolo: the so-called Pozzo dei Veneti, a 20 m-wide and over 100 m-deep circular shaft, leading down to the water table at a depth of 264 m below the surface. The explorations initially stopped here, in the final lake at the bottom of the pit. However, soon the first attempt to explore by cave diving the site was organized: during the spring of 2013, cave diver Luca Pedrali was able to dive 48 m underwater, exploring new and wide spaces of the karst system (Fig. 13e). A second attempt, carried out a year later (spring 2014), brought the same explorer well beyond the point reached previously, to a maximum depth of 324 m below the surface (60 m below the lake level at the bottom of the pit). There, the exploration had to stop for safety reasons. Nevertheless, the documentation shows very large flooded galleries at the site, with massive columns and speleothems, once again testifying to the impressive dimensions and remarkable richness of this cave system.

On the opposite side of Pozzo dei Veneti, a huge breakdown area probably marks the sector of the cave with the clearest evidence of tectonics controlling the cave development, with a fault face and passages extending to the NE probably marking another part of the cave system where explorations are still in progress.

Fig. 9. Pictures of the Grave Rotolo swallow hole, before starting the excavation works (photos courtesy of GASP). (**a**,**b**) The swallow hole during a heavy rainstorm, collecting and absorbing a great amount of rainfall. (**c**) The excavated channel to facilitate water infiltration at the site.

Future perspectives

The new data derived from the explorations at Inghiottitoio di Masseria Rotolo offer remarkable opportunities for the study of the Apulian karst, with particular reference to past and recent evolution of the specific karst system, to its relationships with sea-level stands (Rudnicki 1980; Dini *et al.* 2000; Mastronuzzi & Sansò 2002; De Waele & Parise 2013) and to the quantity and quality of karst aquifers in the region as well (Stevanovic 2015, and contributions therein). These latter, in particular, are very important in Apulia, a region where water availability is among the main problematic issues, further amplified by huge touristic pressure during the summer season, and with serious problems of saline intrusion from both the Ionian and Adriatic sides over several decades (Cotecchia 1977; Tulipano *et al.* 1990; Tulipano & Fidelibus 2002).

Actually, Inghiottitoio di Masseria Rotolo represents one of only two sites in Apulia where the development of an accessible cave allows the water table

Fig. 10. Excavation works at Grave Rotolo (April through September, 2012; photos courtesy of GASP): (**a**) the initial condition at the site; (**b**) the works to remove the big blocks of rock; (**c**) the installation of concrete rings to protect the entrance; (**d**) a view of the entrance through the series of concrete rings.

to be reached directly (the second site, located in Salento, is a much less developed system, consisting of a series of shafts leading down to a depth of some 70 m). Based upon such circumstances, Apulia Region funded a project aimed at exploring the scientific, hydrogeological and karstic potential of Inghiottitoio di Masseria Rotolo. The project, started in the first half of 2017, will develop through acquisition of data dealing with the geology, morphology and structural geology in the Canale di Pirro and within the caves located in the area, sampling, analysing and monitoring the water table in the cave system, and understanding the boundaries of the catchment feeding the system, its areal extension and the spatial and temporal response to meteoric recharge.

The survey of the cave system will be carried out in fine detail. The main goal is to define with precision the altitude (depth) of the main sub-horizontal sectors of the system, in an attempt to correlate them with the sequence of marine terraces identified at the surface along the Adriatic coast, and/or to evaluate the connection of low-gradient cave stretches with stratigraphic or tectonic features. From a hydrogeological standpoint, the scheduled research will include the positioning of a multi-parameter probe to monitor water table oscillations for at least 12 months. These data will be compared with the rainfall registered by a rain gauge established near the cave access, with the goal of assessing the time response of the system to different types of rainfall recharge (from low-intensity and long rainfall, to short and intense downpours). Water samples and analyses are programmed for the whole hydrological year, with sampling in both the wet and dry seasons, following the standardized procedure by the Regional Agency for Environmental Protection.

Another part of the research programme concerns biospeleology. Previous studies in karst areas of Apulia have shown that some aquatic hypogeous species that are highly sensitive to changes in the hydrological cycle and to habitat modifications caused by anthropogenic activities can be used as indicators of environmental quality (Masciopinto *et al.* 2006). For example, *Spelaeomysis bottazzii*, a stygobiont omnivore that has largely increased in number and size with respect to other smaller species, and is less resistant to water quality modifications, has been observed to be increasing in

Fig. 11. Plan view of the Inghiottitoio di Masseria Rotolo cave system (survey 2012–15, by GASP). Colours mark the different sectors in the cave system, to facilitate comparison between plan and profile (Fig. 12) of the cave.

number in the Nardò district. *Spelaeomysis bottazzii* is eurytherm and tolerant to a wide range of salinities, and its increase could lead in the near future to the extinction of other stygobionts, thus causing a decline in the hypogeous aquatic fauna biodiversity (Masciopinto *et al.* 2006; Inguscio *et al.* 2009).

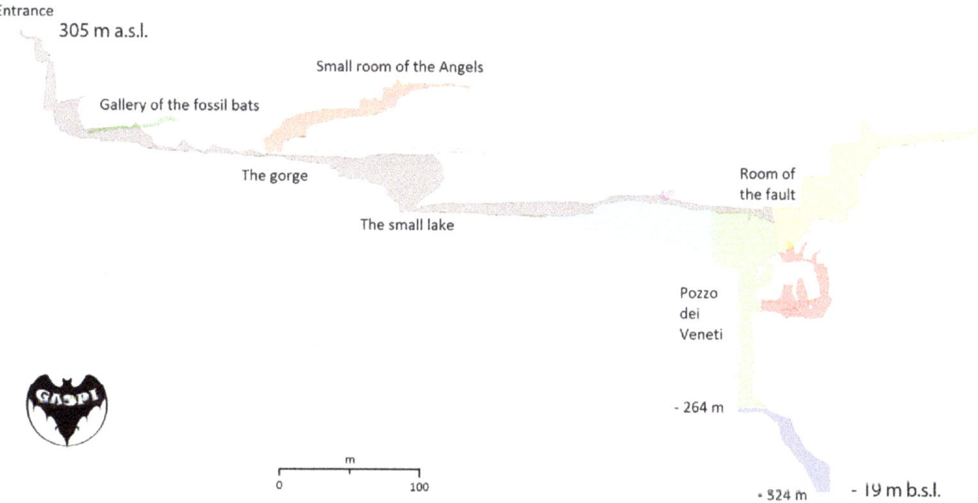

Fig. 12. Main profile of the Inghiottitoio di Masseria Rotolo cave system (survey 2012–15, by GASP). Colours mark the different sectors in the cave system, to facilitate comparison between plan (Fig. 11) and profile of the cave.

Fig. 13. Views within the Grave Rotolo karst system (photos courtesy of GASP): (**a–c**) the first sequence of shafts, at the entrance of the cave system, going down rapidly to about 100 m below the ground surface; (**d**) sub-horizontal passages, in one of the lateral branches (gallery of the fossil bats, in Figs 11 & 12); (**e**) underwater exploration, below the lake at the bottom of Pozzo dei Veneti; (**f**) a natural 'bridge' within one of the main horizontal passages of the karst system, testifying ancient base levels; and (**g**) one of the most speleothem-rich rooms in the karst system.

Further, the cave system, and the Canale di Pirro area at the surface, will be the subject of geological–structural and morphological surveys, aimed at defining the main structural elements conditioning the development of the cave system, and its later evolution as well. In particular, all of the breakdown deposits and the areas of greater tectonic disturbance within the cave system will be mapped and analysed to identify the sectors mostly controlled by tectonics. At the surface, these data will be compared with the distribution of dolines and karst depressions, identified by means of an integration of techniques, including field surveys, interpretations of multi-temporal aerial photographs and semi-automated methods to detect doline-like features from air photographs and orthophotos.

Integration of classical geomorphological karst data, surveyed at the ground surface, with those deriving from direct surveys and analysis in caves (Ford & Ewers 1978; Šebela 1998; Cucchi *et al.* 2001; De Waele *et al.* 2003; Iurilli *et al.* 2009; Pepe & Parise 2014) should be particularly appropriate, in order to achieve a deeper knowledge of the evolution of the karst system and the water circulation pattern.

Conclusions

Inghiottitoio di Masseria Rotolo is an open window on the deep Apulian water table. Together with another cave in Salento, it represents the only possibility in Apulia of moving within a cave system until the water level is reached. This allows great opportunities to check the quality and quantity of groundwater resources. Notwithstanding the difficulties owing to the complexity of the cave system, and to logistical problems in transporting and installing scientific instruments and measurement tools, the research that has started and is scheduled for the coming months will provide a unique opportunity to verify the possibility of an effective co-operation among cavers and scientists.

The results will be of great use to regional and local administrators in a land where paucity of water is a recurrent problem and where, in recent years, several difficult and costly solutions have been proposed (including, among others, the withdrawal and transport of water from the famous Blue Eye spring in Albania, across the Adriatic Sea).

Apart from the future outcomes of the research, the present chapter demonstrates the importance of taking into account even small surface karst landforms. Since their location is typically not casual, observation of the dynamics of such features in different situations can be important for a full understanding of the functioning of karst systems, and can help in defining optimum management actions in fragile karst environments. For instance, these goals have been pursued and achieved in other karst systems where, after speleological survey, research has been carried out in order to investigate several fields, such as palaeoclimatology, surface landscape morphology, water chemistry, hydrogeology, mineralogy and the impact and disturbance of human pressure on show caves (Genty & Deflandre 1998; Buecher 1999; Onac & Vereş 2003; Luetscher *et al.* 2005; Toomey 2009; De Waele *et al.* 2012; Liu & Brancelj 2014; Novas *et al.* 2017).

References

Andriani, G.F. & Walsh, N. 2009. An example of the effects of anthropogenic changes on natural environment in the Apulian karst (southern Italy). *Environmental Geology*, **58**, 313–325.

Anelli, F. 1957. Guide to the excursion II. Bari-Alberobello-Selva di Fasano-Castellana Grotte-Bari. *Proceedings of the XVII Italian Congress on Geography*, 23–29 April 1957, Bari, 69–120 [in Italian].

Baldassarre, G., Boenzi, F. et al. 1978. Dati preliminari sulla neotettonica dei fogli 148 (Vasto), 154 (Larino), 188 (Gravina di Puglia), 201 (Matera), 202 (Taranto) e 203 (Brindisi). Pubbl. no. 155, C.N.R. *Progetto Finalizzato Geodinamica*, 35–67.

Bonacci, O. 2004. Poljes. *In*: Gunn, J. (ed.) *Encyclopedia of Caves and Karst Science*. Fitzroy, Dearborn, MI, 599–600.

Bosellini, A. & Parente, M. 1994. The Apulia Platform margin in the Salento peninsula (southern Italy). *Giornale di Geologia*, **56**, 167–177.

Breg, M. 2007. Degradation of dolines on Logasko polje (Slovenia). *Acta Carsologica*, **36**, 223–231.

Bruno, G., Del Gaudio, V., Mascia, U. & Ruina, G. 1995. Numerical analysis of morphology in relation to coastline variations and karstic phenomena in the southeastern Murge (Apulia, Italy). *Geomorphology*, **12**, 313–322.

Buecher, R.H. 1999. Microclimate study of Kartchner Caverns, Arizona. *Journal of Caves and Karst Studies*, **61**, 108–120.

Calò, F. & Parise, M. 2006. Evaluating the human disturbance to karst environments in southern Italy. *Acta Carsologica*, **35**, 47–56.

Carlone, G. & Angelini, G. 1986. *Castelli e fortificazioni in Puglia. Visite alle difese marittime nell'età del Viceregno spagnolo*. Capone Ed., Cavallino.

Ce.Ri.Ca. (Centro Ricerche Castellanese) 1996. *Le inondazioni a Castellana*. Amministraz. Comunale di Castellana-Grotte.

Ciaranfi, N., Pieri, P. & Ricchetti, G. 1988. Note alla Carta geologica delle Murge e del Salento (Puglia centromeridionale). *Memorie della Società Geologica Italiana*, **41**, 449–460.

Corsi, P. 2006. Aspetti economici e sociali nella Murgia sud-orientale in età medievale. *In*: Martellotta, A. (ed.) *Dal Canale di Pirro al Canale delle Pile, tra storia e geografia*. Proceedings of the Seminar, Alberobello, 11 July 1997. Corpus Scriptorum Alberobellensium, 49–63.

Cotecchia, V. 1977. Studi e ricerche sulle acque sotterranee e sull'intrusione marina in Puglia (Penisola Salentina). *Quaderni CNR IRSA*, **20**, 1–466.

Cotecchia, V. & Scuro, M. 2010. Portrait of a coastal karst aquifer: the city of Bari. *Aqua Mundi*, **1**, 187–196.

Cucchi, F., Marinetti, E., Potleca, M. & Zini, L. 2001. Influence of geostructural conditions on the speleogenesis of the Trieste Karst (Italy). *Geologica Belgica*, **4**, 241–250.

Delle Rose, M. & Parise, M. 2003. Il condizionamento di fattori geologico-strutturali e idrogeologici nella speleogenesi di grotte costiere del Salento. *Proceedings of the 19th National Congress of Speleology*, Bologna, **1**, 27–36.

Despain, J.D., Tobin, B.W. & Stock, G.M 2016. Geomorphology and paleohydrology of Hurricane Crawl Cave, Sequoia National Park, California. *Journal of Cave and Karst Studies*, **78**, 72–84.

De Waele, J. & Parise, M. 2013. Discussion on the article 'Coastal and inland karst morphologies driven by sea level stands: a GIS based method for their evaluation'. *Earth Surface Processes and Landforms*, **38**, 902–907.

De Waele, J., Plan, L. & Audra, P. 2003. Recent developments in surface and subsurface karst geomorphology. An introduction. *Geomorphology*, **106**, 1–8.

De Waele, J., Gutiérrez, F., Parise, M. & Plan, L. 2011. Geomorphology and natural hazards in karst areas: a review. *Geomorphology*, **134**, 1–8.

De Waele, J., Ferrarese, F., Granger, D.E. & Sauro, F. 2012. Landscape evolution in the Tacchi area (Central-East Sardinia) based on karst and fluvial morphology and age of cave sediments. *Geografia Fisica e Dinamica Quaternaria*, **35**, 119–127.

Di Bucci, D., Coccia, S. *et al.* 2009. Late Quaternary deformation of the southern Adriatic foreland (southern Apulia) from mesostructural data: preliminary results. *Italian Journal of Geosciences*, **128**, 33–46.

Di Geronimo, I. 1970. Geomorphology of the Adriatic slope of SE Murge (Ostuni area, Brindisi). *Geologica Romana*, **9**, 47–58 [in Italian].

Dini, M., Mastronuzzi, G. & Sansò, P. 2000. The effects of relative sea-level changes on the coastal morphology of southern Apulia (Italy) during the Holocene. *In*: Slaymaker, O. (ed.) *Geomorphology, Human Activity, and Global Environmental Changes*. John Wiley & Sons, New York, 43–66.

Dogan, U. 2003. Sariot polje, central Taurus, (Turkey): a border polje developed at the contact of karstic and non-karstic lithologies. *Cave and Karst Science*, **30**, 117–124.

Doglioni, C., Mongelli, F. & Pieri, P. 1994. The Puglia uplift (SE Italy): an anomaly in the foreland of the Apenninic subduction due to buckling of a thick continental lithosphere. *Tectonics*, **13**, 1309–1321.

Fabri, F.P., Auler, A.S., Calux, A.S., Cassimiro, R. & Augustin, C.H.R.R. 2015. Cave morphology and controls on speleogenesis in quartzite: the example of the Itambè do Mato Dentro area in southeastern Brazil. *Acta Carsologica*, **44**, 23–35.

Ford, D.C. & Ewers, R.O. 1978. The development of limestone cave systems in the dimension of length and depth. *Canadian Journal of Earth Sciences*, **15**, 1783–1798.

Ford, D.C. & Williams, P. 2007. *Karst Hydrogeology and Geomorphology*. Wiley, Chichester.

Funiciello, R., Montone, P., Parotto, M., Salvini, F. & Tozzi, M. 1991. Geodynamic evolution of an intra-orogenic foreland: the Apulia case history (Italy). *Bollettino della Società Geologica Italiana*, **110**, 419–425.

Gambini, R. & Tozzi, M. 1996. Tertiary geodynamic evolution of the Southern Adria microplate. *Terra Nova*, **8**, 593–602.

Gams, I. 1978. The polje: the problem of definition. *Zeitschrift für Geomorphologie*, **22**, 170–181.

Gams, I. 2005. Tectonic impacts on poljes and minor basins (case studies of Dinaric Karst). *Acta Carsologica*, **34**, 25–41.

Gastaldi, G. 1561. La universale descrittione del mondo, descritta da Giacomo de' Castaldi piamontese. Matteo Pagano, Venezia.

Gastaldi, G. 1567. Il disegno della geografia moderna de tutta la provincia de la Italia.

Genty, D. & Deflandre, G. 1998. Drip flow variations under a stalactite of the Père Noël cave (Belgium): Evidence of seasonal variations and air pressure constraints. *Journal of Hydrology*, **211**, 208–232.

Gil, H., Pepe, M. *et al.* 2013. Sviluppo ed evoluzione di sprofondamenti in rocce solubili: un confronto tra il carso coperto del Bacino dell'Ebro (Spagna) e la Penisola Salentina (Italia). *Memorie Descrittive della Carta Geologica d'Italia*, **93**, 253–276.

Gracia, F.J., Geremia, F., Privitera, S. & Amore, C. 2014. The probable karst origin and evolution of the Vendicari coastal lake system (SE Sicily, Italy). *Acta Carsologica*, **43**, 215–228.

Gutierrez, F., Parise, M., De Waele, J. & Jourde, H. 2014. A review on natural and human-induced geohazards and impacts in karst. *Earth Science Reviews*, **138**, 61–88.

Inguscio, S., Rossi, E. & Parise, M. 2009. Biogeographical distribution of subterranean fauna in Apulia (Italy) in the context of the palaeo-geographic evolution of the area. *Proceedings 15th International Congress of Speleology*, Kerrville, TX, 19–26 July, **2**, 749–754.

Iurilli, V., Cacciapaglia, G., Selleri, G., Palmentola, G. & Mastronuzzi, G. 2009. Karst morphogenesis and tectonics in south-eastern Murge (Apulia, Italy). *Geografia Fisica e Dinamica Quaternaria*, **32**, 145–155.

Liu, W. & Brancelj, A. 2014. Hydrochemical response of a cave drip water to snowmelt water, a case study from Velika Pasica cave, central Slovenia. *Acta Carsologica*, **43**, 65–74.

Longo, G. 1969. *Osservazioni geomorfologiche sulla zona di Alberobello*. Arti Grafiche Angelini e Pace, Locorotondo.

Lopez, N., Spizzico, V. & Parise, M. 2009. Geomorphological, pedological, and hydrological characteristics of karst lakes at Conversano (Apulia, southern Italy) as a basis for environmental protection. *Environmental Geology*, **58**, 327–337.

Luetscher, M., Jeannin, P.-Y. & Haeberli, W. 2005. Ice caves as an indicator of winter climate evolution: a case study from the Jura Mountains. *The Holocene*, **15**, 982–993.

Martellotta, A. (ed.) 2006. Dal Canale di Pirro al Canale delle Pile, tra storia e geografia. *Proceedings of the Seminar*, Alberobello, 11 July 1997, Corpus Scriptorum Alberobellensium.

Martinotti, M.E., Pisano, L. *et al.* 2017. Landslides, floods and sinkholes in a karst environment: the 1–6 September 2014 Gargano event, southern Italy. *Natural Hazards and Earth System Sciences*, **17**, 467–480.

Masciopinto, C., Semeraro, F., La Mantia, R., Inguscio, S. & Rossi, E. 2006. Stygofauna abundance and distribution in the fissures and caves of the Nard. (Southern Italy) fractured aquifer subject to reclame water injections. *Geomicrobiology Journal*, **23**, 267–278.

Mastronuzzi, G. & Sansò, P. 2002. The morphogenetic effects of relative sea level changes on the coastal area of Apulia (Italy). *Proceedings of the Workshop MACRiVaLiMa*, Ostuni, 30–31 May, GI2S Coast, **1**, 29–34.

Neboit, R. 1975. *Plateaux et collines de Lucanie orientale et des Pouilles*. Étude morphologique. Librairie Honore Champion, Paris.

NICOD, J. 2003. A little contribution to the karst terminology: special or aberrant cases of poljes? *Acta Carsologica*, **32**, 29–39.

NOTARNICOLA, G. 1933. *La Cavallerizza della Serenissima in Puglia*. Gastone Bellini ed., Venezia.

NOVAS, N., GAZQUEZ, J.A., MAC LENNAN, J., GARCIA, R.M., FERNANDEZ-ROS, M. & MANZANO AGUGLIARO, F. 2017. A real-time underground environment monitoring system for sustainable tourism of caves. *Journal of Cleaner Production*, **142**, 2707–2721.

ONAC, B.P. & VEREŞ, D.Ş. 2003. Sequence of secondary phosphates deposition in a karst environment: evidence from Măgurici Cave (Romania). *European Journal of Mineralogy*, **15**, 741–745.

OROFINO, F. 1965. Interessante serie di grotte lungo la parete sinistra del Canale di Pirro. *L'Alabastro*, **5**, 10.

PALAGIANO, C. 1965. On the 'lame' and 'gravine' of Apulia. *Annali della Facoltà di Economia e Commercio*, University of Bari, **21**, 357–386 [in Italian].

PARISE, M. 1999. Morfologia carsica epigea nel territorio di Castellana-Grotte. *Itinerari Speleologici, series II*, **8**, 53–68.

PARISE, M. 2003. Flood history in the karst environment of Castellana-Grotte (Apulia, southern Italy). *Natural Hazards and Earth System Sciences*, **3**, 593–604.

PARISE, M. 2006. Geomorphology of the Canale di Pirro karst polje (Apulia, Southern Italy). *Zeitschrift für Geomorphologie N.F.*, **147**, 143–158.

PARISE, M. 2009. Lakes in the Apulian karst (Southern Italy): geology, karst morphology, and their role in the local history. *In*: MIRANDA, F.R. & BERNARD, L.M. (eds) *Lake Pollution Research Progress*. Nova Science, New York, 63–80.

PARISE, M. 2011. Surface and subsurface karst geomorphology in the Murge (Apulia, southern Italy). *Acta Carsologica*, **40**, 79–93.

PARISE, M. 2012. Management of water resources in karst environments, and negative effects of land use changes in the Murge area (Apulia). *Karst Development*, **2**, 16–20.

PARISE, M. & GUNN, J. (eds) 2007. *Natural and Anthropogenic Hazards in Karst Areas: Recognition, Analysis and Mitigation*. Geological Society, London, Special Publications, **279**, http://sp.lyellcollection.org/content/279/1

PARISE, M., FEDERICO, A., DELLE ROSE, M. & SAMMARCO, M. 2003. Karst terminology in Apulia (southern Italy). *Acta Carsologica*, **32**, 65–82.

PAVICIC, A., BENAMATIC, D., PEST, D. & MARASOVIC, M. 2002. Water reservoir within the karst field overburden: Gusic polje, Croatia. *Geologia Croatica*, **55**, 93–100.

PEPE, M. & PARISE, M. 2014. Structural control on development of karst landscape in the Salento Peninsula (Apulia, SE Italy). *Acta Carsologica*, **43**, 101–114.

PIERI, P., FESTA, V., MORETTI, M. & TROPEANO, M. 1997. Quaternary tectonic activity of the Murge area (Apulian foreland, southern Italy). *Annali di Geofisica*, **40**, 1395–1404.

PUTISKA, R., KUSNIRAK, D. *ET AL*. 2014. Integrated geophysical and geological investigations of karst structures in Komberek, Slovakia. *Journal of Cave and Karst Studies*, **76**, 155–163.

RUDNICKI, J. 1980. Karst in coastal areas – development of karst processes in the zone of mixing of fresh and saline water (with special reference to Apulia, Southern Italy). *Studia Geologica Polonica*, **65**, 9–59.

SAURO, U. 1991. A polygonal karst in Alte Murge (Puglia, Southern Italy). *Zeitschrift für Geomorphologie*, **35**, 207–223.

ŠEBELA, S. 1998. *Tectonic structure of Postojnska Jama cave system*. Založba ZRC, **18**. Ljubljana.

SHAW, T.R. 1988. Martel's visit to Mendip in 1904. Part of his international strategy? *Proceedings of the University of Bristol Speleological Society*, **18**, 278–291.

SHAW, T.R. 2004. Valvasor – a common error about his publications on Cerknisko Jezero, Slovenia. *Acta Carsologica*, **33**, 313–317.

SHAW, T.R. 2005. Skocjanske Jame, Slovenia, in 1891 – an alpine club excursion. *Acta Carsologica*, **34**, 236–260.

SISTO, P. 2006. Il Canale delle pile e il fiume Cana tra storia, 'letteratura' e leggenda. *In*: MARTELLOTTA, A. (ed.) *Dal Canale di Pirro al Canale delle Pile, tra storia e geografia*. Proceedings of the Seminar, Alberobello, 11 July 1997, Corpus Scriptorum Alberobellensium, 65–77.

STEVANOVIC, Z. (ed.) 2015. *Karst Aquifers – Characterization and Engineering*. Professional Practice in Earth Sciences. Springer, Berlin.

ŠUŠTERŠIC, F. & ŠUŠTERŠIC, S. 2003. Formation of the Cerknišèica and the flooding of Cerkniško polje. *Acta Carsologica*, **32**, 121–136.

TOOMEY, R.S. III. 2009. Geological monitoring of caves and associated landscapes. *In*: YOUNG, R. & NORBY, L. (eds) *Geological Monitoring*. Geological Society of America, Boulder, CO, 27–46.

TOZZI, M. 1993. Assetto tettonico dell'Avampaese Apulo meridionale (Murge meridionali –Salento) sulla base dei dati strutturali. *Geologica Romana*, **29**, 95–111.

TRISCIUZZI, A.S. 1989. Il Canale di Pirro. *Fasano*, **20**, 67–74.

TULIPANO, L. & FIDELIBUS, M.D. 2002. Mechanisms of groundwaters salinisation in a coastal karstic aquifer subject to over-exploitation. *Proceedings of the 17th SWIM*, Delft (The Netherlands), 39–49.

TULIPANO, L., COTECCHIA, V. & FIDELIBUS, M.D. 1990. An example of multitracing approach in the studies of karstic and coastal aquifers. *Proceedings of the International Symposium and Field Seminar 'Hydrogeologic Processes in Karst Terranes'*. IAHS, **207**. Antalya, Turkey, 381–389.

WHITE, W.B. 1988. *Geomorphology and Hydrology of Karst Terrains*. Oxford University Press, New York.

WHITE, W.B. 2002. Karst hydrology: recent developments and open questions. *Engineering Geology*, **65**, 85–105.

ZHU, J., TAYLOR, T.P., CURRENS, J.C. & CRAWFORD, M.M. 2014. Improved karst sinkhole mapping in Kentucky using LiDAR techniques: a pilot study in Floyds Fork Watershed. *Journal of Cave and Karst Studies*, **76**, 207–216.

ZVAB ROZIC, P., CAR, J. & ROZIC, B. 2015. Geological structure of the Divaca area and its influence on the speleogenesis and hydrogeology of Kacna Jama. *Acta Carsologica*, **44**, 153–168.

The Puerto Princesa Underground River (Palawan, Philippines): some peculiar features of a tropical, high-energy coastal karst system

GIOVANNI BADINO[1,2†], ANTONIO DE VIVO[1], PAOLO FORTI[1,3]* & LEONARDO PICCINI[1,4]

[1]*La Venta Associazione Geografica, Treviso, Italy*

[2]*Dipartimento di Fisica, University of Torino, 1-10125 Torino, Italy*

[3]*Italian Institute of Speleology, via Zamboni 67, I-40126 Bologna, Italy*

[4]*Department of Earth Science, University of Firenze, Italy*

**Correspondence: paolo.forti@unibo.it*

Abstract: The Puerto Princesa Underground River, amongst the largest caves of the Philippine Islands, is the most visited show cave in the country, even though it has undergone no tourism adaptation at all. Its scientific importance primarily relies on the fact that it is one of the largest known underground estuaries in the world, and the effect of tides is visible along more than 7 km of the cave length. The complex relationships between sea and freshwater influence not only the hydrodynamics of the system and the speleogenetic processes presently active, but also its climate and its ecosystem. The systematic exploration and research of this coastal karst system started some 40 years ago and have shown that the Puerto Princesa Underground River is one of the most important caves in the world with regard to many different scientific fields. Speleogenesis concerns the initial phreatic solution followed by vadose erosion with periodical marine invasion, and subsequent saline/freshwater-mixing processes during sea-level highstands. The hydrodynamic behaviour of the water flowing inside the cave is rather complex, being simultaneously controlled by allogenic recharge and tides. Speleothems abundantly occur with several forms, some of which have never been described before. Several minerals, some of which are very rare, are present, together with palaeontological remains exposed by differential corrosion on rock walls. Last but not least, two large populations of bats and swiftlets sustain a complex subterranean ecosystem.

In 1971, when the St Paul Subterranean River (now popularly known as Puerto Princesa Underground River: PPUR) was declared a National Park, only a small part of the cave was known, no map was available and no one knew what unique and amazing scientific discoveries were hidden inside this karst system. Since then, many changes have occurred, transforming this once virtually unknown cavity into what it is today: a fully accredited UNESCO World Heritage Site (1999), a highly protected National Geological Monument (2003), an ASEAN Heritage Site (2005) (Restificar *et al.* 2006) and, more recently, one of the New Seven Wonders of Nature (2012). Last but not least, PPUR is the most visited show cave in the Philippines, with some 300 000 visitors per year.

The outflow of the cave has always been known to local people, and the first explorers would certainly have been local people, probably entering the cave to search for drinkable water or swallows' nests. Some visible writing left by visitors to the first part of the underground river show dates back to 1930 and 1934, but the first documented speleological explorations were carried out by Australian cavers in 1980 and 1981 (Hayllar 1980, 1981). From 1989 to 1991 the first research project, co-ordinated by the Italian Speleological Society, took place and led to the discovery of large new lateral branches (Piccini & Rossi 1994), making this cave one of the most important in the Far East and, above all, giving a start to a new series of studies on hydrodynamics (Forti *et al.* 1993*b*), speleogenesis (Piccini & Rossi 1994; Piccini & Iandelli 2011) and cave ecosystems (Messana 1994; Sbordoni 2007).

Thanks to the explorations carried out by the Geographic Association La Venta in the last 25 years (Piccini & Rossi 1994; Piccini *et al.* 2007; De Vivo *et al.* 2009; De Vivo & Piccini 2013), the PPUR presently is a karst system over 34 km (Fig. 1) long developed inside the St Paul Dome, a

†Deceased 8 August 2017.

From: Parise, M., Gabrovsek, F., Kaufmann, G. & Ravbar, N. (eds) 2018. *Advances in Karst Research: Theory, Fieldwork and Applications*. Geological Society, London, Special Publications, **466**, 155–170.
First published online January 24, 2018, https://doi.org/10.1144/SP466.22

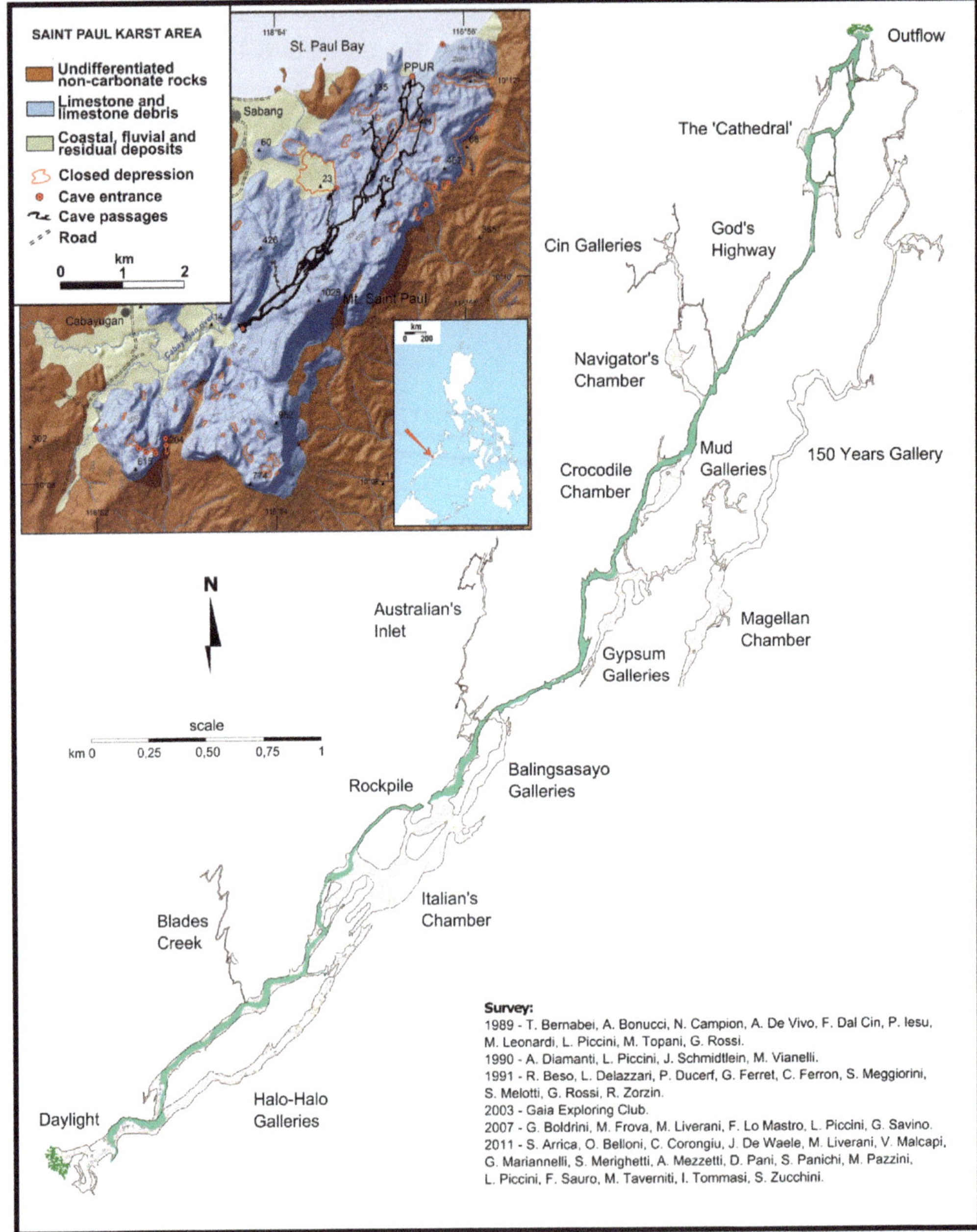

Fig. 1. Index map, geological sketch and present day extension of the PPUR (after De Vivo *et al.* 2013).

limestone massif slightly higher than 1000 m a.s.l., which flows directly into the sea with a huge, 7.5 km-long tunnel, along which tides propagate undisturbed for 7 km.

The cave is inhabited by massive colonies of swiftlets and bats, as well as many other terrestrial and aquatic species. All of these environmental features make the PPUR one of the richest underground ecosystems on the planet.

In 2000, the La Venta team went back to Palawan with the aim of filming a documentary on this extraordinary cave. Following that expedition, in

2007, a new exploration and research project was set up. This new project involves the fields of geology, hydrogeology, hydrodynamics, hydrochemistry, speleogenesis, epigean and hypogean morphogenesis, biology and palaeontology in the PPUR and in the whole karst area embraced by the National Park. It is currently being carried out in strict collaboration with important Italian and other universities and research institutions. Some of the results and data collected during these expeditions have already been published in international or local magazines (Badino 2010; Piccini & Iandelli 2011; De Vivo *et al.* 2013; De Vivo & Forti 2014*a*, *b*; Forti 2014; Coombes *et al.* 2015).

Finally, in 2015 a joint research project between Tagbalay Foundation (Philippines) and La Venta (Italy) was granted by the Philippines–Italy Debt for Development Swap Program (De Vivo *et al.* 2017*a*, *b*) in order to complete the exploration of and studies on PPUR, with special reference to safeguarding its ecosystem without compromising the tourism activities in Palawan.

Geological setting

The St Paul karst area covers *c.* 35 km^2 and is made from massive to roughly bedded limestone, light to dark grey in colour, with levels rich in fossils of Late Oligocene–Early Miocene ages, which represent the remains of an ancient coral reef, about 25–20 Ma old (Almasco *et al.* 2000). Such a rock formation, more than 400 m thick, lies over sedimentary (mudstones, sandstones and marls) and volcanic rocks, dating back to about 30 Ma before present (Oligocene) and lying on an older metamorphic basement.

The limestone outcrop is shaped as a NNE–SSW elongated, asymmetrical ridge sloping down to the west. From a structural standpoint, it consists of a multiple NW-dipping homoclinal ridges, bounded by NE–SW-oriented faults. Such lineaments have controlled both the general shape of the mountain and the karst landforms, influencing the alignment of dolines and the development of the major caves (Piccini & Iandelli 2011).

After having been drowned and covered by terrigenous sediments, the limestone reef gradually rose up and was exhumed again by erosion some millions of years ago (Piccini & Iandelli 2011). This process probably led to the formation of a wide and regular limestone ridge, less elevated than now, about 6–5 Ma BP (late Miocene). Rivers coming from the south crossed this mountain, probably forming gorges and through-caves of spectacular size, nowadays preserved as ancient tunnels in the southern sector of the ridge (De Vivo *et al.* 2009). Successively, a further tectonic uplift probably led to the present morphology and altitude (Piccini & Iandelli 2011).

At present, limestone makes up the upper part of the right (west) side of the Babuyan River valley, as the water course has deeply eroded the underlying impermeable formations. Such carbonate rocks form a continuous wall up to 500 m high, with the exception of the northern part of the ridge, where the limestone merges into the surrounding non-karstic landscape. On the other side of the range, carbonate rocks descend to the Cabayugan valley and to Sabang with steep slopes cut by rectilinear NE–SW-oriented fault-steps.

Climate

Palawan is located on the Intertropical Convergence Zone and therefore its average temperature is relatively high (*c.* 27°C) with daily, monthly and yearly excursions rarely exceeding ±5°C. The annual average relative humidity is high and almost constant, staying around 84%.

The climate is 'tropical wet and dry' (Aw in the Köppen climate classification): the wettest month is September, while the driest month is February. The average rainfall is relatively high (close to 2000 mm a^{-1}), and 95% of it falls during the wet (monsoon) period (from May to November), concentrated in short but intense rainstorms.

Consequently the PPUR underground climate is extremely stable (Badino 2010, 2013), being controlled by the general island climate and by the seawater, which during the dry season invades a large part of the cave every *c.* 12.5 h with more than 100 000 m^3 of seawater (Forti 2014). The flow regime within the PPUR dramatically changes from the dry to wet period (during floods the river discharge can increase from less than 0.2 m^3 s^{-1} to much more than 10 m^3 s^{-1} in only a few hours).

The very high external relative humidity, together with the low daily temperature fluctuation and the presence within the cave of a wide free-water surface, greatly inhibit the evaporation processes, which are limited to places where relatively strong air currents occur (up to 1.5 m s^{-1} displacing over 100 m^3 s^{-1}). Conversely, at some sites within the cave, it is sometimes possible to see active condensation processes with the development of large clouds, the genesis of which is induced by cold air currents coming from the upper levels of the cave. This phenomenon has mainly been observed when occasional rainstorms occurring in the dry season cause an abrupt and local decrease in the cave temperatures (Badino 2017). In turn, this results when relatively cool percolation water coming from the top of the mountain rapidly reaches the main cave passages. Such a phenomenon might also explain the fact that in the thermostable area of the cave the average temperature is almost 2°C lower than the external one (De Vivo & Forti 2017).

Speleogenesis

One of the most significant features of the PPUR is the fact that tides affect a large part of the cave, up to about 7 km from the outflow. This represents the current situation, while in the past the sea-level was usually lower than today, so that PPUR was not a marine cave (Piccini & Iandelli 2011). Indeed, despite the occurrence of corrosion produced by the mixing of freshwater with salty water, the speleogenesis of the PPUR is mainly due to solution by meteoric and continental water and to mechanical erosion by suspended load during floods. Only in its middle to lower part has mixing corrosion produced forms typical of coastal caves, such as waterline notches, spongework and lateral diverting conduits (Bunnell & Kovarik 2013). From this perspective, the system may be considered a classic example of an underground estuary.

Features that indicate former water levels are present along the PPUR up to 7 km upstream from the coastal outflow. In the last sector of the navigable path, where the ceiling of the main tunnel rises to 20 m or higher, there are two evident old corrosion notches, owing to persistent levels of water (Piccini & Iandelli 2011): the upper notch is at +11–12 m above the present mean sea-level (p.m.s.l.). The second notch is at +7–8 m above p.m.s.l. and it can be correlated with that visible on the coastal cliff at about 7 m a.s.l., which dates back to the last interglacial phase, that is about 120 ka BP (Maeda *et al.* 2004; Omura *et al.* 2004). Some of the lateral branches of the PPUR contain alluvial terraces consisting of sands and gravels, which are probably related to this lower high-stand notch.

The morphology of the PPUR active level is clearly adjusted to the current sea-level, but it should be taken into account that in the last 500 ka BP, the sea was mostly lower than it is today (mainly 50–60 m below p.m.s.l.). This implies that the PPUR has functioned mainly as a vadose through-cave affected by freshwater flow, with a substantial load of insoluble material, creating a subterranean canyon that was filled by the alluvial sediment now forming the current riverbed (Piccini & Iandelli 2011).

The PPUR cave profile shows several large passages at an elevation of mainly 50–80 m a.s.l. This level consists of large inactive tunnels parallel to the current river, containing thick alluvial deposits covered by flowstones, which in places have almost completely filled the conduits.

In the upstream sector of the cave, erosion features, such as canyon-shaped conduits, and large collapse chambers, can be found, indicating a long phase of vadose entrenchment. This ancient 'underground river' could reasonably date back to the Early Pleistocene (about 1 Ma BP), as suggested by the extrapolation of the recent low uplift rate of the coastal zone.

Several morphological features, such as the presence of corrosion notches at +12.4 m a.s.l. and the huge speleothems corroded and interbedded with alluvial deposits, suggest that the lower and presently active level passed through more than two sea-level high-stands, and might have been formed during most of the Middle–Late Pleistocene (Piccini & Iandelli 2011).

Hydrodynamics of the main river

The analysis of the hydrodynamics of the underground river was the first stage of the scientific research performed inside this cave. The study, started during the first expeditions (Forti *et al.* 1993*a*, *b*; Piccini & Rossi 1994), was continued later (De Vivo *et al.* 2013; Forti 2014).

As a matter of fact, the water dynamics within the cave changes dramatically from the dry season, with a discharge often lower than 200 l s^{-1}, to the flood periods, when it may reach over 15–20 m^3 s^{-1} soon after heavy rainstorms. The observations made over 20 years allowed the definition of two extreme scenarios. The first scenario, in high flood conditions, occurs when the discharge is some 10–15 m^3 s^{-1}: during this limited period (2–3 days maximum) the only water present inside the cave is fresh water. In such situations all of the salt and/or brackish water, potentially trapped even in the deep depressions existing at the bottom of the riverbed, is pushed out and no seawater may enter the cave, even during the highest high tides (Fig. 2).

During high flow conditions, the tides propagate within the cave just as a pressure wave, without any significant water displacement, the increase in seawater level being balanced by the amount of freshwater entering the system. Thus, tides occur at exactly the same time at the entrance and 5–6 km within the cave (Fig. 3), and the only flow direction is always directed outside.

During flood peaks, when the discharge can be higher than 15 m^3 s^{-1}, tides do not affect the cave, and the freshwater level, which may reach up to 2 m above the mean sea-level, completely masks the high and low tide fluctuation inside the cave.

The other extreme scenario corresponds to a very low discharge (below 0.2 m^3 s^{-1}) occurring during the dry season: in this circumstance the first 5–6 km of the cave is characterized by three distinct

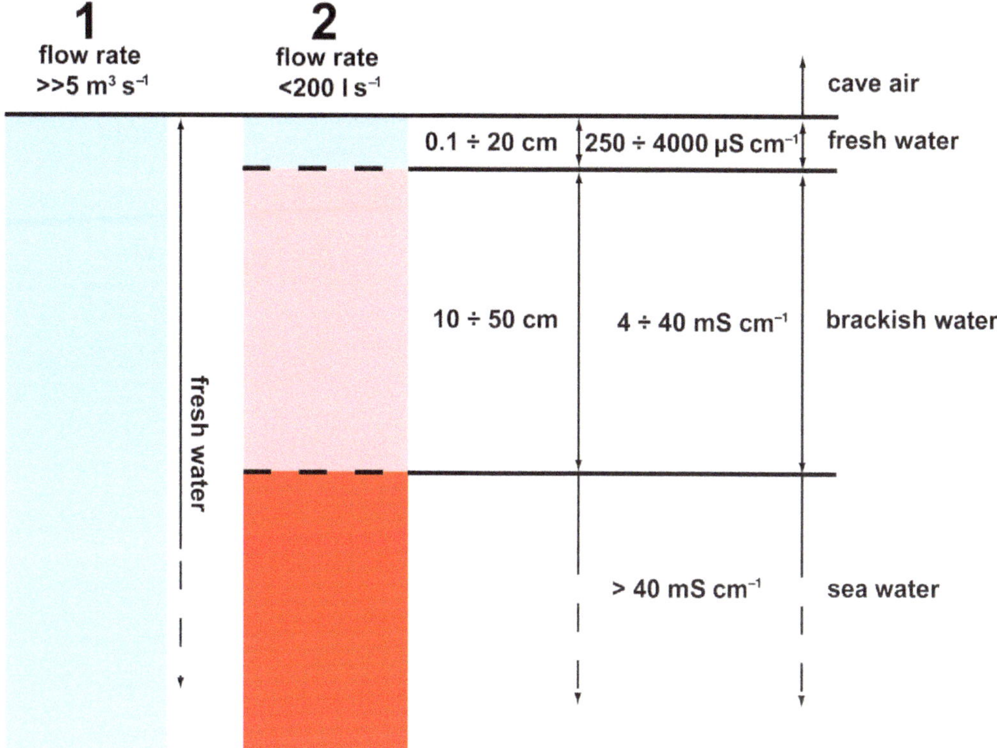

Fig. 2. Behaviour of the water flow in the first 5–6 km of the cave in two extreme scenarios: (**1**) during floods when flow rates exceeds 5 $m^3 s^{-1}$ (as observed in the final part of the 2000 expedition), with only freshwater present; and (**2**) during dry periods, with freshwater flow not exceeding 0.2 $m^3 s^{-1}$ (as observed during the 1991, 2011 and 2017 expeditions), with three water layers present with different saline contents.

water layers with different salinities: the upper layer consists of freshwater, the intermediate of brackish, and the lower one of seawater (Fig. 2, right). Owing to the absence of turbulence (laminar flow), these three water layers do not mix together and move rather independently under the influence of the tides currents. Decreasing the freshwater supply, propagation of the tides inside the cave is induced not only by up and down movements of water masses, but also by tide currents (back and forth horizontal movements of water masses), the contribution of which is inversely proportional to the amount of freshwater discharge.

The physical displacement of water masses by tide currents induces a certain delay in the propagation of the tide peaks, which progressively increases back-stream from the cave entrance towards the inflow. This delay is also partially reflected in the flow direction of a single or of all three water layers, which may locally revert four times a day (Fig. 4). The variation of the freshwater flow rate is also responsible for another peculiar phenomenon which has been observed along the main, partially flooded, tunnel, where well-developed scallops are present on the cave walls up to 2–3 m above the present mean sea-level.

The statistical analyses of scallop morphometry highlight that the scallop size increases with their elevation above water level, a fact that seems to be in contrast with the normal trend of such forms (Curl 1974). Normally, the dimension of the scallops is inversely proportional to the speed of the water flow that produces them. In the PPUR the reverse seems to occur: with increasing water discharge (e.g. rising level of water within the gallery), the flow speed decreases, as indicated by a larger scallop size (Fig. 5). This probably happens because during a flood, which can last some tens of hours to some days, the water speed is affected by the tidal fluctuation and consequently it decreases during high tides and accelerates during low tides. Furthermore the peaks of water discharge are normally registered during or soon after intense rainstorms (often corresponding in the tropics to marine storms) which

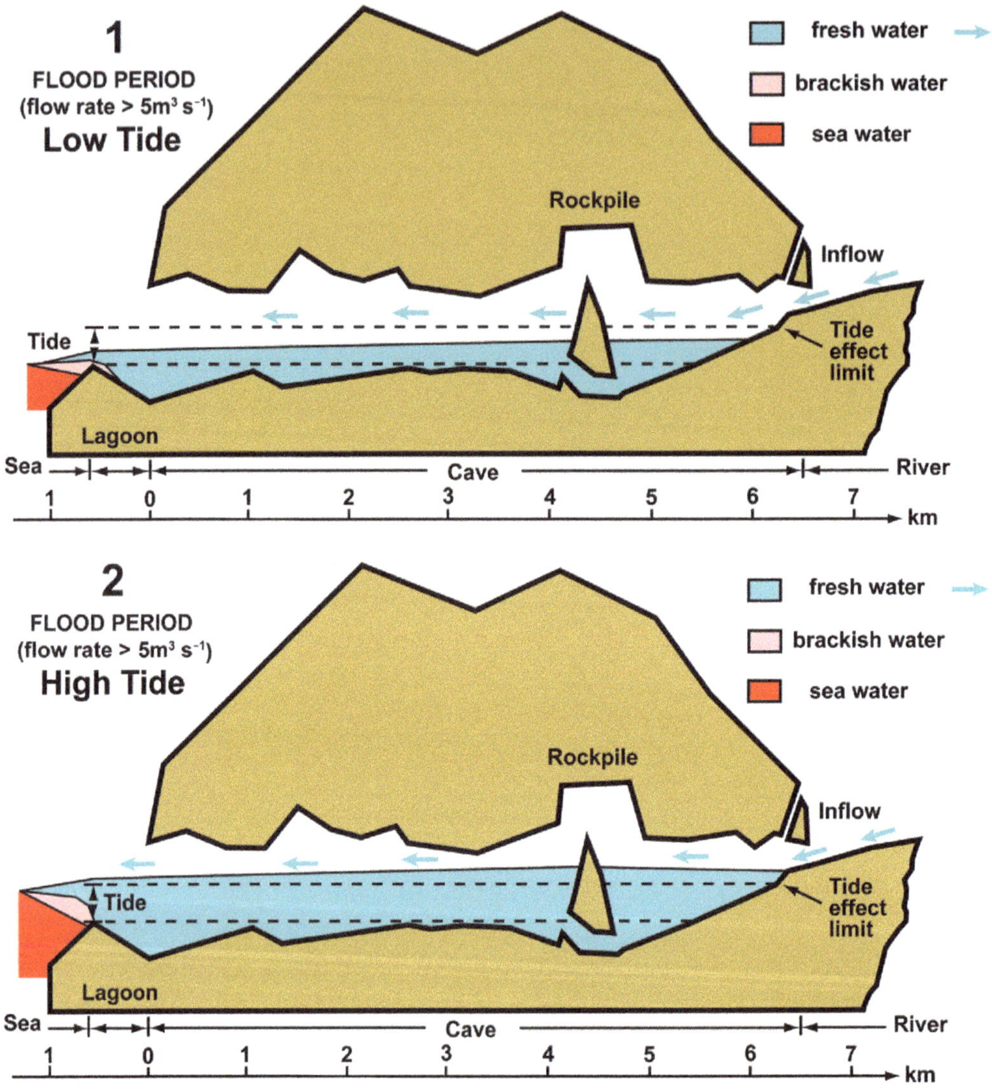

Fig. 3. Sketch of the PPUR system which undergoes tidal effects during a flood (discharge >5 $m^3 s^{-1}$): only freshwater is present inside, always flowing towards the outflow, and tide maximum occurs simultaneously at the cave entrance and at 5–6 km inside the system (after Forti 2014, modified).

can, at least partially, hinder the discharge from the cave outflow.

Other scientific interests

The explorations and studies carried out in the last 25 years have demonstrated that the scientific value of the PPUR goes beyond the speleogenesis and hydrodynamics topics. Among the other disciplines worthy of mention are those related to biology, mineralogy, palaeontology and meteorology, which will be briefly reported in the following sections.

Biology

In common caves the dominant factors controlling the subterranean ecosystems are the lack of light and the scarcity of food. This is definitely not the case in the PPUR, where visitors are impressed by the remarkable availability of trophic energy (Messana 1994; Sbordoni 2007). Indeed, PPUR is a

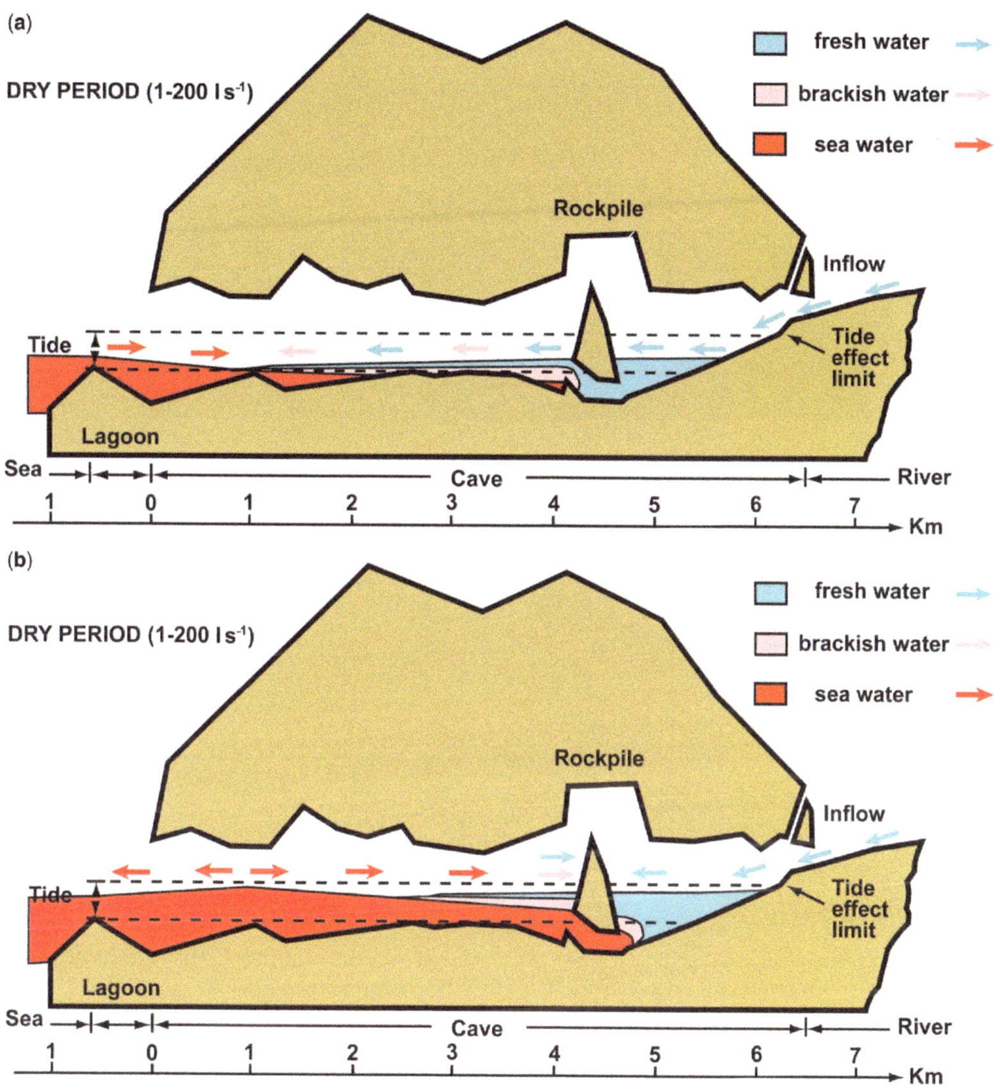

Fig. 4. Sketch of the PPUR system which undergoes tidal effects during a dry period (discharge <0.2 $m^3 s^{-1}$): three well-defined water layers are present moving back and forth driven by tidal currents (after Forti 2014, modified): (**a**) immediately after the LT peak, seawater enter the cave while back-tream brackish and freshwater are still flowing towards the cave entrance; (**b**) immediately after the HT peak saltwater goes out of the cave while back-stream salt, brackish and freshwater go in the opposite direction.

typical tropical cave with a rich fauna and large amounts of food available. In most cases, the former comprises poorly specialized animals with no evident signs of adaptation to cave life, such as the large populations of bats and swiftlets (*Collocalia* genus; Fig. 6(1)). These are the main producers of the huge amounts of guano that represents the main trophic resource of the cave, especially in its terrestrial environment.

Similar situations have been found, however, in many other caves in tropical areas. What makes a real difference at PPUR is that this cave behaves like a classic estuary. This means that high tide extends up to 5–6 km within the cave, whereas the freshwater flows as a sheet on the top surface of the salty water, except during floods. The two types of water do not mix much, and at the interface there is an accumulation of large bacterial populations.

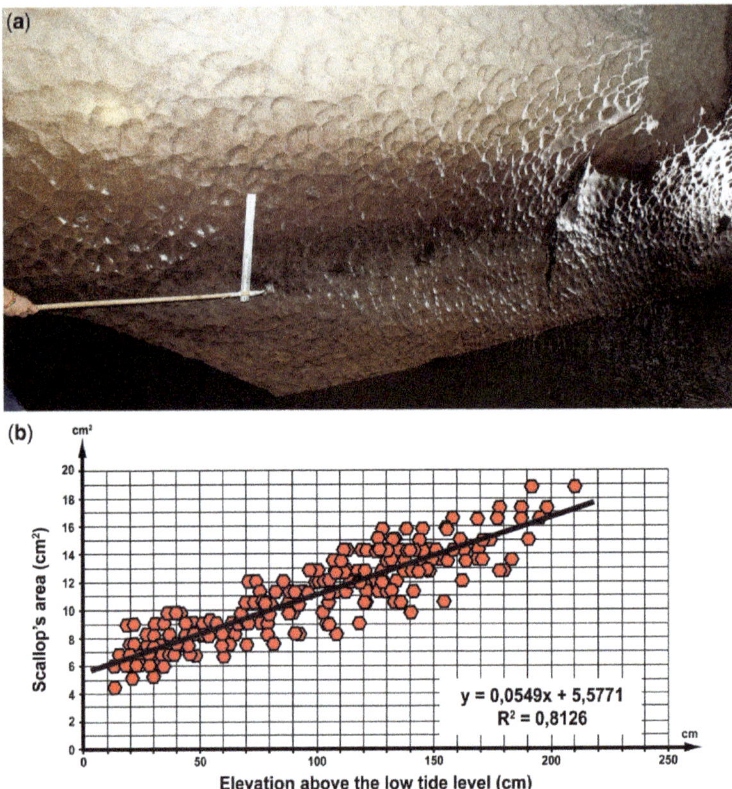

Fig. 5. (**a**) A general view of the scallops occurring on the wall of the PPUR main tunnel. (**b**) The relationship between scallop size and their elevation above water level (low tide).

The different distribution of the fauna during the day was clearly documented by some fish captured during the 2000 expedition a few hundred metres from the entrance of the cave (Sbordoni 2007). In 24 h, the net captured typical freshwater fish, such as a microphtalmic catfish (with clear signs of adaptation to cave life), as well as sea fish, like the indo-pacific blackmouth croaker *Atrobucca nibe*. This testifies that the catfish had descended towards the sea following the freshwater flow during a low tide, whereas the blackmouth croakers had entered the cave following the saltwater cone created by a high tide.

The terrestrial fauna is also quite rich, representing all the levels of the trophic chain. The presence of at least three species of snake is noteworthy; amongst them is the colubrid *Elaphe erythrura*, often found in the caves of SE Asia where it feeds mostly on bats (Fig. 6(2)). Finding a cobra a few kilometres inside the underground river was unexpected, as was the discovery of a population of pythons (*Python reticulatus*), well fed on bats and swiftlets. Many Thereuopoda individuals, large and voracious scutigeromorpha centipedes and a huge population of large migalomorph spiders have been found. Meanwhile, the walls are home to *Amblypygi* and the large cave crickets of the *Rhaphidophora* genus. The guano community includes a new species of the Leiodidae beetle *Phomaphaginus*, as well as a small *Tineidae lepidopter*.

In addition to these organisms 'little adapted to caves', the PPUR also hosts in its upper passages troglobiont organisms (Fig. 6(3)): in 2000 in a branch of the cave, far from the areas where bats stay and swiftlets nest, a small terrestrial isopod crustacean and a pseudo-scorpion, both blind and de-pigmented, were found (Sbordoni 2007).

No systematic study has been performed on the PPUR ecosystems, but this gap was partially filled during the scientific expeditions of 2016–17 (De Vivo *et al.* 2017*a*, *b*). This research, started in 2016, thanks to the grant by the Philippines–Italy Debt for Development Swap Program, has been mainly focused on the ecosystems existing in the different sectors of the cave (Agnelli *et al.* 2017).

Fig. 6. (**1**) Swiftlet's nest about 6 km from the PPUR entrance (photo G. Savino, La Venta); (**2**) the colubrid *Elaphe erythrura* (photo G. Savino, La Venta); (**3**) a crab found in a side branch (photo A. Romeo, La Venta); (**4**) the 'cave grass' in the 150 Years Gallery (photo A. Romeo, La Venta); (**5**) the black crusts in the Australian inlet (photo J.M. Calaforra, La Venta); and (**6**) the sirenian bones (photo N. Russo, La Venta).

It has been demonstrated that the PPUR hosts five distinct ecosystems (Fig. 7), each one characterized by the different nature and abundance of the trophic resources and the salinity level of the water. These features make this karst system an extraordinary natural laboratory to study the evolutionary processes and the ecology of hypogean environments.

A relatively high number of zoological samples was taken during the aforementioned two expeditions and they are now under examination by specialists around the world. The preliminary results clearly demonstrate that the PPUR hosts several new species of well-specialized cave fauna (Agnelli *et al.* 2017).

Further specific analyses are needed but it is fairly certain that the ecosystem of PPUR will become one of the most varied and complex underground ecosystems of our planet.

Speleothems and minerals

Until recently the PPUR was supposed to be a cave with only 'normal' speleothems (stalactites, stalagmites, flowstones, etc.) and cave minerals of local

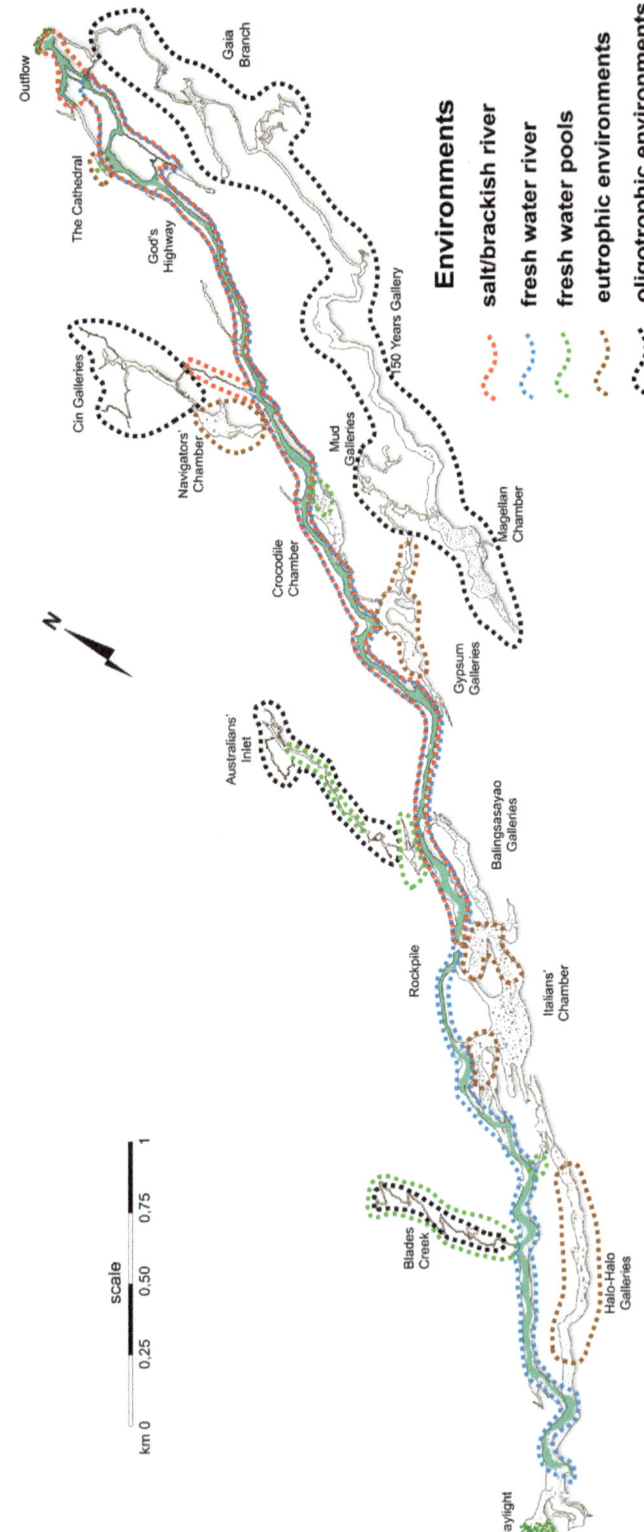

Fig. 7. The five ecosystems hosted by the Puerto Princesa Underground River.

or regional importance at most. Some of them have noticeable aesthetic interest and can easily be seen by tourists in the first part of the cave, like the gigantic stalagmites in the 'Cathedral' hall, not far from the cave entrance, or some large 'canopy bells', which are surely worth mentioning for their beauty and dimension. Recent studies, nevertheless, proved that the PPUR hosts rare, and sometimes aesthetically astonishing, speleothems. One of these is the 'calcite grass', an extremely rare kind of helictite, consisting of bended and twisted calcite mono-crystals growing from the floor and covering several tens of square metres of the 150 Years Gallery (Fig. 6 (4)). Their development is controlled by the strong air currents that characterize this area. Other rare speleothems, observed in the same cave area, are the triangular mono-crystal calcite micro-gours and the triangular mono-crystal calcite stalagmites.

Further, PPUR also hosts a new type of speleothem, presently restricted to the 'God's Highway' tunnel (see Fig. 1): the 'ribbed' drapery, observed during the 2011 La Venta expedition (Badino *et al.* 2017). This drapery, over 10 m high, is characterized by the presence, along its lateral sides, of several close-to-horizontal symmetrical ribs (Fig. 8).

The distance between the ribs ranges from a few decimetres to over 1 m, while their size ranges from a few centimetres to 20–30 cm wide. The increase in rib size corresponds to a larger deviation from horizontality: in fact the largest ones are clearly bent upward with a possible interference with the upper ribs, which may eventually be incorporated.

Finally, at least in the area close to the contact with the cave wall (Fig. 8(1)), the drapery is characterized by a higher number of smaller ribs, which progressively coalesce, forming a few, larger ribs. The vertical section of the drapery is characterized by zones of sudden enlargement, followed by a downward, fast progressive reduction until its normal thickness is re-established. The enlargements are always symmetrical, being of the same size and at the same level on both sides of the drapery.

The morphological study (Badino *et al.* 2017) allowed us to infer that the main factor guiding the

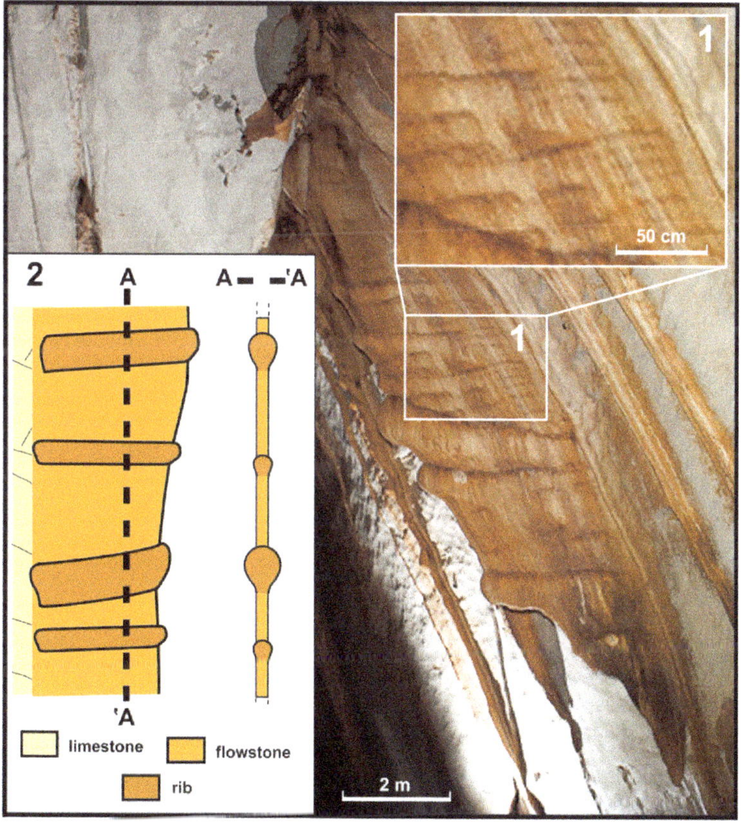

Fig. 8. The 'ribbed' drapery (photo N. Russo, La Venta): (**1**) magnification showing the high number of 'embryonic' ribs close to the cave wall; and (**2**) simplified sketch of the drapery (after Badino *et al.* 2017, modified).

development of the ribbed drapery is the particular climate of Palawan Island, which is characterized by a long dry season with occasional heavy rainstorms. The great variability of the feeding water allows the genesis of the close-to-horizontal symmetric embryonic ribs induced by the perturbation of the flow along the external rim of the drapery, and their progressive evolution owing to the water flow along both lateral sides. The alternation of dry and wet periods also justifies the possible union of two or more ribs to form a larger one, and is responsible for the slight upward deflection, more evident in the larger ribs.

The mineralogical studies (Forti *et al.* 1993*a*, *b*; Billi *et al.* 2013; De Vivo *et al.* 2013) have shown a complex situation, with the number of cave minerals much higher than previously supposed and even more important, with some of them never having been observed in other caves in the world.

Most of these minerals are concentrated within the thick 'black crusts', widespread along the main galleries of the PPUR (Fig. 6(5)). These crusts, often completely detached from the cave wall, were formed during periods of strong corrosion of the bedrock induced by the biogenic mineralization of guano. The same process has been active in all the other caves of the St Paul karst: in two of these cavities the process also allowed peculiar black cave pearls to be formed (Forti *et al.* 1993*a*, *b*). In general the crusts consist of several very thin layers of different colours: besides the far more dominant black ones, other layers are present from black to reddish, and from yellow to white. Each colour corresponds to one or more compounds. In the black layers most of the components are manganese-rich minerals, the reddish layers are characterized by iron-manganese minerals, while the white and yellow ones consist of gypsum and phosphates, many of which are amorphous.

Until now, 13 different minerals have been detected in the PPUR: calcite ($CaCO_3$), aragonite ($CaCO_3$), hydromagnesite [$Mg_5(CO_3)_4(OH)_2 \cdot 4(H_2O)$], gypsum ($CaSO_4 \cdot 2H_2O$), apatite [$Ca_5(PO_4)_3 \cdot$ (C,F, Cl,O,OH)], variscite [$AlPO_4 \cdot 2H_2O$)], strengite [(Fe, Al)$PO_4 \cdot 2H_2O$)], manganite [MnO(OH)], rhodochrosite ($MnCO_3$) pyrolusite (MnO_2), robertsite [$Ca_6\,Mn_9(PO_4)_9O_6(H_2O)_6 \cdot 3(H_2O)$], janggunite [$Mn_{5-x}$ $(Mn,Fe)_{1-x}O_8(OH)_6$] and serrabrancaite ($MnPO_4 \cdot 2H_2O$). The first 10 minerals were already known from the cave environment while the last three (robertsite, janggunite and serrabrancaite) are new cave minerals, being restricted to the St Paul karst only: in particular, serrabrancaite is very rare even outside caves, and it has been found so far only once in Brazil (Witzke *et al.* 2000). Thanks to these findings, the PPUR has to be considered among the most interesting caves in the world from a mineralogical point of view.

Palaeontology

The Palawan limestone hosts well-preserved fossils, which are highlighted along the PPUR walls by selective corrosion. Normally they consist of fragments of marine shells. In 2011, however, several partially articulated bones (ribs and dorsal vertebrae) exposed for *c.* 10 cm of their length were observed about 3 m above the river on the left wall of the 'God's Highway' (Fig. 6(6)). For conservation purposes, it was decided to take only pictures of the bones because sampling would surely result in destruction of the fossil (Forti *et al.* 2011). A detailed comparison with several available photos of closed related taxa, and with the fossil specimens present in many museums all over the world, has concluded that they belong to a sirenian. Apart from isolated and fragmentary findings in Java, Pakistan and Sri Lanka, nearly all of the known sirenian fossils from East Asia are from India (Bajpai & Domning 1997; Bajpai *et al.* 2010). Therefore, the specimen recovered on the Island of Palawan represents the first from the Philippines and the easternmost finding in the Asia region.

The partially articulated thoracic elements include a well-preserved vertebra with exposed neural spine and processes, partial centrum and complete articulation with the head of the ribs (Fig. 6(6)). Several fragments of ribs are preserved *in situ*, showing the flat and recurved shape typical of sirenians. The specimen is *c.* 60 cm wide, whereas the vertebra is 10 cm wide and 15 cm high, including the centrum. It is therefore possible to estimate a total length, for the individual, of 180 cm. This sirenian is coeval with the hosting rock, the St Paul Limestone, which is Oligo-Miocene in age. Therefore the fossil of the Underground River is extremely important because it pushes the limit of diffusion of these animals during the Neogene some 1500 km eastward. At the moment, using only the photos of the outcropping bones it is impossible to state if this sirenian belongs to an already known fossil genus or to a new one.

Meteorology

One of the many factors that make the PPUR so exceptional is that the underground meteorological phenomena are here at their most intense level: actually we can say that there is a permanent invisible 'underground storm' stirred up by the shape of its galleries (Badino 2010, 2013). The external river flowing through this cave carries the effects of all of the meteorological events that occur outside: precipitation peaks and temperature fluctuations. This is not exceptional in itself – around the world there are many through-caves of this type, but what makes this cave so unique is that it extends to the sea-level deep in the core of Mount St Paul.

As already mentioned, the flows of freshwater from upstream and of saltwater from downstream do not mix at all for long distances, with the former flowing over the latter. This locally creates pockets of cool air in contact with the warmer freshwater during and/or immediately after rainstorms and, seasonally, generates ascension flows of relatively warm and humid air. Micro-meteorological niches and extremely complicated ecosystems are thus created, each having their own 'seasons'.

The St Paul 'underground storm' is probably the warmest in the world, because the island of Palawan lies directly on the thermal equator. Therefore, the PPUR cave could have the maximum temperature possible for a cave whose energy balance depends mainly on the external water flow.

Furthermore, the galleries are crossed by an important airflow, with a flow rate up to 150 m^3 s^{-1}. The origin of air circulation inside caves is usually the temperature difference between inside and outside. The PPUR airflow is instead driven by a factor that is usually insignificant elsewhere: the variations in the humidity of the outside air.

On the one hand PPUR is very hot, while on the other, it opens in an area where the climate is 'super-oceanic', that is, characterized by an extremely small thermal excursion between day and night, and between summer and winter as well. The result is therefore that the temperature difference between the inside and the outside is always small, just a few degrees, and this is ineffective for creating an underground wind. In these conditions, variation in the outside air density owing to humidity variations can assume a decisive role. Therefore, one of the power supplies for the underground atmospheric dynamic is the humidity of the external air, a factor which in other places is usually negligible. A real 'storm' should also have some big clouds: in fact they are present in the furthest sectors of the upper branches. There, the air masses coming from the surface of the mountain meet the air and water flows of the underground river, which are some fractions of a degree warmer. This mixture produces condensation, forming stationary clouds.

A storm should be also accompanied by a great quantity of energy. From the measures taken, we now know that the power of the flow entering the cave is some 10–20 MW (Badino 2013), which is much more than the power entering most of the caves around the world (Huppert *et al.* 1991).

The ancient huge upper galleries, now abandoned by water, have different conditions. There, the underground wind is much weaker, clouds are absent or very thin and temperature excursions are more limited. These galleries are a sort of 'safe harbour', which has allowed fantastic concretions to grow and many types of ecological and meteorological niches to develop.

In order to obtain detailed data on the climate within the PPUR, during the expedition of November 2016 several automatic devices have been placed in the different branches of the karst system (Fig. 9) in order to monitor temperature values every 15 min. When all of the registered data have been analysed, these temperature records will allow us to define the PPUR microclimate, knowledge that is fundamental not only for the detailed definition of the cave meteorology, but also to fully understand its different ecosystems.

Safeguarding PPUR

Usually, the transformation of a wild cavity into a show cave causes a more or less pronounced loss of its pristine condition, which sometimes may become critical even for the survival of the cave itself (Huppert *et al.* 1991; Parise 2011). This should have been particularly true for the PPUR considering the fact that in less than 40 years the cave has become the most visited show cave in the Philippines and among the most visited caves in the world, with over 300 000 visitors in 2015.

Luckily, the PPUR was appointed as a Natural Park and World Heritage site before its first part was opened to large tourist flows. Therefore, safeguarding it was always the main concern for local and national authorities when deciding on its implementation. In all these years they have never taken a decision in conflict with conservation, and the capacity to leave the environment and the cave absolutely untouched.

Despite the huge numbers of tourists visiting the cave and the opportunities they have had to make changes in the name of progress and development, the authorities have not built a single cement step, hung a single steel footbridge or set up an electric line to light up the cave.

Among the unique features characterizing the PPUR, the ecological approach to the cave is probably the most amazing: this cave, even while it is amongst the most visited show caves in the world, is at the same time one of the least damaged. This is absolutely exceptional, and represents the true point of force of this tourist site, since the disturbance induced by human presence (North *et al.* 2009) has been kept to a minimum with proper policies.

Final remarks

This short, surely not exhaustive, outlook on the main peculiarities of the Puerto Princesa Underground River emphasizes that the explorations and research performed within the PPUR in the last 25 years have made it possible to disclose just

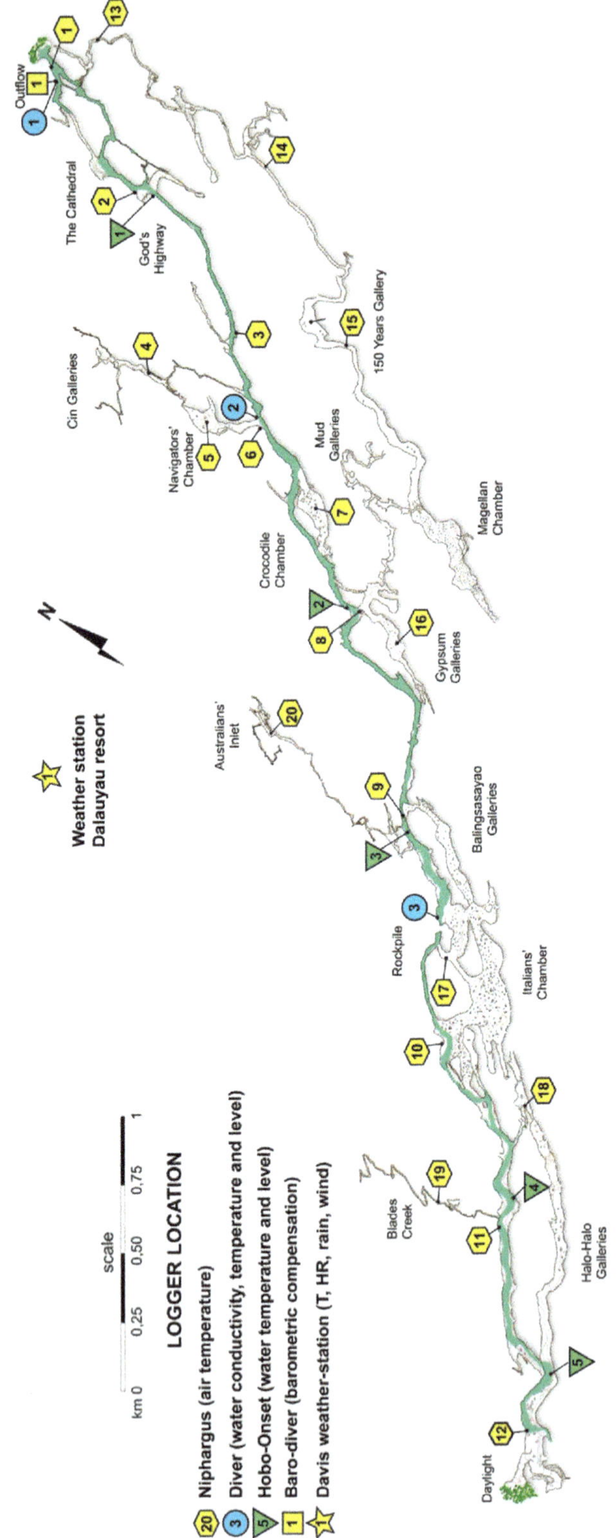

Fig. 9. PPUR plan with the location of the instruments positioned in November 2016.

some of the many features characterizing this astonishing cave.

Nevertheless, the large majority of exploration and research are still to be done. In the near future many other exciting discoveries await the cavers and scientists who dedicate their efforts to this amazing karst system.

The authors thanks the Municipality of Puerto Princesa and the Administration of the PPUR Park for the support given during the expeditions to the PPUR.

References

AGNELLI, P., CIARAMELLA, M. & VANNI, S. 2017. Biology. *In*: DE VIVO, A., FORTI, P. & PICCINI, L. (eds) . *Support for Sustainable Eco-Tourism in the Puerto Princesa Underground River Area, Palawan, Philippine. Report on the Second Expedition to Palawan.* La Venta, Treviso, 64–74.

ALMASCO, J.N., RODOLFO, K., FULLERB, M. & FROST, G. 2000. Paleomagnetism of Palawan, Philippines. *Journal of Asian Earth Sciences*, **18**, 369–389.

BADINO, G. 2010. Underground meteorology, what's the weather underground? *Acta Carsologica*, **39**, 427–448.

BADINO, G. 2013. Where are the hottest caves in the world? *Kur*, **20**, 43.

BADINO, G. 2017. Driving pressure of subterranean airflows: an analysis. *Proceedings 17th International Speleological Congress*, Sydney, **2**, 205–208.

BADINO, G., CALAFORRA, J.M., DE WAELE, J. & FORTI, P. 2017. The ribbed drapery of the Puerto Princesa Underground River (Palawan, Philippines): morphology and genesis. *International Journal of Speleology*, **46**(2), 93–97.

BAJPAI, S. & DOMNING, D. 1997. A new dugongine sirenian from the early Miocene of India. *Journal of Vertebrate Paleontology*, **17**, 219–228.

BAJPAI, S., DOMNING, D., DAS, D., VÈLEZ-JUARBE, J. & MISHRA, V. 2010. A new fossil sirenian (Mammalia, Dugonginae) from the Miocene of India. *Neues Jahrbuch fur Geologie und Palaontologie.*, **258**, 39–50.

BILLI, S., FORTI, P., GALLI, E. & ROSSI, A. 2013. Robertsite: un nuovo fosfato di grotta scoperto nella Tagusan Cave (Palawan – Filippine). *Congresso Nazionale di Speleologia, Trieste*, **2011**, 306–311.

BUNNELL, D.E. & KOVARIK, J.L. 2013. Littoral cave development on the Western U.S. Coast. *In*: LACE, M.J. & MYLROIE, J.E. (eds) *Coastal Karst Landforms*. Springer, New York.

COOMBES, M.A., LA MARCA, E.C., NAYLOR, L.A., PICCINI, L., DE WAELE, J. & SAURO, F. 2015. The influence of light attenuation on the biogeomorphology of a marine karst cave: a case study of Puerto Princesa Underground River, Palawan, the Philippines, *Geomorphology*, **229**, 125–133

CURL, R.L. 1974. Deducing flow velocity in cave conduits from scallops. *National Speleological Society Bulletin*, **36**(2), 1–5.

DE VIVO, A. & FORTI, P. (eds) 2014*a*. Puerto Princesa Underground River (Palawan Philippines): geological sketch, plan view, projected section and main points of interest (1991–2011). Map, supplement to *Kur*, **21**.

DE VIVO, A. & FORTI, P. 2014*b*. Puerto Princesa Underground River 2015: un nuovo salto di qualità delle attività di La Venta. *Kur*, **21**, 42–45.

DE VIVO, A. & FORTI, P. (eds) 2017. *Support for Sustainable Eco-Tourism in the Puerto Princesa Underground River Area, Palawan, Philippine*. Report on the first expedition to Palawan La Venta, Treviso.

DE VIVO, A. & PICCINI, L. (eds) 2013. *The River of Swallows*. La Venta, Treviso.

DE VIVO, A., PICCINI, L. & MECCHIA, M. 2009. Recent explorations in the St. Paul karst (Palawan, Philippines). *Proceedings 15th International Congress of Speleology*, Kerville, Texas, **3**, 1786–1792.

DE VIVO, A., PICCINI, L., FORTI, P. & BADINO, G. 2013. Some scientific features of the Puerto Princesa Underground River: one of the new 7 wonders of nature. *Proceedings International Congress of Speleology*, Brno, **3**, 35–41.

DE VIVO, A., FORTI, P. & PICCINI, L. 2017*a*. Support for sustainable eco-tourism in PPUR (Puerto Princesa Underground River) – project 2016–2017. *Proceedings 17th International Speleological Congress*, Sydney, **2**, 360–364.

DE VIVO, A., FORTI, P. & PICCINI, L. (eds) 2017*b*. *Support for Sustainable Eco-tourism in the Puerto Princesa Underground River Area, Palawan, Philippine*. Report on the second expedition to Palawan La Venta, Treviso.

FORTI, P. (ed.) 2014. Puerto Princesa Underground River (Palawan, Pilippines). Bathimetric, hydrochemical, hydrodynamical, and climatological data (1990–2011). Map, supplement to *Kur*, **21**.

FORTI, P., GORGONI, C., ROSSI, A. & PICCINI, L. 1993*a*. Studio Mineralogico e Genetico delle pisoliti nere della Lyon Cave (Palawan-Filippine). *Atti XVI Congresso Nazionale di Spelologia*, Udine, **1**, 59–72.

FORTI, P., PICCINI, L., ROSSI, G. & ZORZIN, R. 1993*b*. Note preliminari sull'idrodinamica del sistema carsico di St. Paul (Palawan, Filippine). *Bulletin Société Géographique de Liège*, **29**, 37–44.

FORTI, P., RUSSO, N., LO. & MASTRO, F. 2011. Laventino, il sirenide di Palawan – Laventino, Palawan's siren. *Kur*, **17**, 14–15.

HAYLLAR, T. 1980. A description of the St. Paul Cave, Palawan, Philippines. *The Journal of the Sydney Speleological Society*, **24**, 153–158.

HAYLLAR, T. 1981. Caving on Palawan. *The Journal of the Sydney Speleological Society*, **25**, 215–231.

HUPPERT, G.N., BURRI, E., CIGNA, A. & FORTI, P. 1991. Effects of tourist development on caves and karst. *In*: WILLIAMS, P. (ed.) *Karst Terrains*. Environmental Changes and Human Impact, *Catena* suppl., **25**, 251–268.

MAEDA, Y., SIRINGANA, F., OMURAD, A., BERDINA, R., HOSONOD, Y., ATSUMIB, S. & NAKAMURAC, T. 2004. Higher-than-present Holocene mean sea levels in Ilocos, Palawan and Samar, Philippines. *Quaternary International*, **115–116**, 15–26.

MESSANA, G. 1994. Biologia. *In*: PICCINI, L. & ROSSI, G. (eds) *Le esplorazioni speleologiche italiane nell'Isola di Palawan*. Speleologia, **31**, 57–60.

NORTH, L.A., VAN BEYNEN, P.E. & PARISE, M. 2009. Interregional comparison of karst disturbance: west-central

Florida and southeast Italy. *Journal of Environmental Management*, **90**, 1770–1781.

OMURA, A., MAEDA, Y., KAWANA, T., SIRINGAN, F. & BERDI, R. 2004. U-series dates of Pleistocene corals and their implications to the paleo-sea levels and the vertical displacement in the Central Philippines. *Quaternary International*, **115–116**, 3–13.

PARISE, M. 2011. Some considerations on show cave management issues in Southern Italy. *In*: VAN BEYNEN, P.E. (ed.) *Karst Management*. Springer, New York, 159–167.

PICCINI, L. & IANDELLI, N. 2011. Tectonic uplift, sea level changes and Plio-Pleistocene evolution of a costal karst system: the Mount Saint Paul. *Earth Surface Processes and Landforms*, **36**, 594–609.

PICCINI, L. & ROSSI, G. 1994. Le esplorazioni speleologiche italiane nell'Isola di Palawan. *Speleologia*, **31**, 5–61.

PICCINI, L., MECCHIA, M., BONUCCI, A. & LO MASTRO, F. 2007. Recent speleological explorations in the St. Paul Karst. Technical Notes, supplement to *Kur*, **9**.

RESTIFICAR, S., DAY, M. & URICH, P. 2006. Protection of karst in the Philippines. *Acta Carsologica*, **35**, 121–130.

SBORDONI, V. 2007. Life in caves. *Kur*, **9**, 14–15.

WITZKE, T., WEGNER, R., DOERING, T., PÖLLMANN, H. & SCHUCKMANN, W. 2000. Serrabrancaite, $MnPO_4 \cdot H_2O$, a new mineral from the Alto Serra Branca pegmatite, Pedra Lavrada, Paraiba, Brazil. *American Mineralogist*, **85**, 847–849.

Role of karst denudation on the accurate assessment of glacio-eustasy and tectonic uplift on carbonate coasts

JOHN E. MYLROIE* & JOAN R. MYLROIE

Department of Geosciences, Mississippi State University, Mississippi State, MS 39762, USA

**Correspondence: mylroie@geosci.msstate.edu*

Abstract: Quaternary glacio-eustasy has traditionally been determined in part by the examination of fossil coral reefs on carbonate islands and coasts uplifted by tectonics. These studies do not properly account for dissolutional denudation, which is cumulative, making higher and therefore older terraces exist at elevations far below their assumed depositional elevation. Karst pedestals (karrentische) on Guam reveal the extent of the denudation (*c.* 50 mm ka^{-1}) and demonstrate that theoretical denudation models can be accurately applied to eogenetic carbonates in tropical settings. Aeolian calcarenite islands such as the Bahamas have been used as tectonically stable sea-level calibrations for other islands, which may not be correct. Flank margin caves, forming in the distal margin of the freshwater lens within a carbonate island, are excellent sea-level indicators. Analysis of flank margin cave elevations indicates that the Bahamas have had past sea-level highstands >6 m, perhaps up to 15 m or more, for which no fossil coral data exist. Denudational removal of these older corals has biased the record to younger events and only flank margin caves remain as viable terrestrial signatures of these older sea-level highstands.

The elevation and age of coral reef terraces on tectonically uplifted islands and coasts has been the mainstay of studies of Quaternary sea-level. The two most famous locations are Barbados (see review by Ulrich & Schellmann 2006, and references cited therein) and the Huon Peninsula of New Guinea (see review by Chappell *et al.* 1996, and references cited therein). A limitation of these studies is that no account has been taken of the post-depositional denudation of these uplifted coral reefs. For example, Chappell *et al.* (1996, p. 231) provide a formula:

> $S = (H+z) - Ut$ with $U = (H^* - S^*)/t^*$ where t = age of a coral sample from height H above sea level, corrected for the palaeo water depth, z, in which the dated sample grew, and U is the tectonic uplift rate at the site. U is calculated from the height H^* of a reference terrace of age t^* at the same terrace transect, which formed when sea level was S^*. The reference terrace was the crest of the reef that formed during the climax of the Last Interglacial (equivalent to oxygen isotope stage 5e in deep sea cores). We used a value of $S^* = 5 = \pm 2$ m, on the basis of data from raised reefs at tectonically stable sites in many parts of the world.

Note that no denudation factor is included. The surface of the fossilized coral terrace is assumed to be the depositional surface, therefore the elevation of coral reef terraces as presented for Barbados and New Guinea is an underestimate that increases in error with terrace age, older terraces having experienced more denudational lowering than younger terraces. The formula of Chappell *et al.* (1996) is calibrated against Marine Isotope Stage (MIS) 5e fossil coral reefs from stable islands, which themselves have undergone denudation. As these stable islands are at the young end of the Quaternary timescale, the calibration error is small for very young rocks, but for the older and therefore higher reef terraces on uplifted islands and coasts the error is much greater and these higher reefs are certainly not correct sea-level indicators as a result of denudational lowering.

Karst denudation depends not only on climate, biogenic activity and rock purity, but also on the diagenetic stage of the host rock. For carbonate rocks, Choquette & Pray (1970, p. 215) defined 'the time of early burial as eogenetic, the time of deeper burial as mesogenetic, and the late stage associated with erosion of long-buried carbonates as telogenetic'. Karst denudation in carbonate rocks is a well-studied phenomenon in typical mid-latitude continental settings, where the rocks are diagenetically mature, or telogenetic. Telogenetic coastal carbonate rocks can also be found in specific locations – for example, the Adriatic Sea in the Mediterranean (Mylroie & Mylroie 2013) or Gotland Island in the Baltic Sea (Vacher & Mylroie 2002). However, the majority of coastal carbonate rocks globally are in the tropics and subtropics and are diagenetically immature, or eogenetic. The classic use of carbonate coasts to determine glacio-eustasy and tectonic uplift has been in tropical and subtropical eogenetic carbonates (Chappell 1974; Chappell *et al.* 1996; Ulrich & Schellmann 2006); however, the impact of denudation has not been fully appreciated. This paper examines karst denudation on eogenetic carbonate coasts

From: Parise, M., Gabrovsek, F., Kaufmann, G. & Ravbar, N. (eds) 2018. *Advances in Karst Research: Theory, Fieldwork and Applications*. Geological Society, London, Special Publications, **466**, 171–185.
First published online November 6, 2017, https://doi.org/10.1144/SP466.2

and how such denudation impacts the interpretation of tectonic uplift and glacio-eustatic sea-level change. This paper is not a review of coastal karren (the phytokarst of Folk *et al.* 1973). The readers interested in those specific coastal landforms are referred to the extensive detailed study of the topic by Taboroši & Kázmér (2013).

Eogenetic carbonate coasts

Eogenetic carbonate rocks are found proximal to their site of deposition and, as such, are excellent proxies for sea-level position at the time of deposition. Fossilized coral reefs and other shallow subtidal deposits are accurate indicators of lagoonal conditions, which are closely tied to sea-level. Other carbonate deposits, such as aeolian calcarenites, while not direct indicators of sea-level position, require carbonate platforms to be flooded and that beaches form on any remaining high ground, becoming the source of the aeolian allochems immediately adjacent to developing carbonate dunes (Carew & Mylroie 1995*a*, 1997).

Coastal carbonates of any diagenetic stage contain a freshwater lens, a water body that transfers meteoric recharge to the coast for discharge as diffuse flow. The freshwater lens thins at the coast, where the lens margin discharges to the sea (Fig. 1). The dissolution of carbonate rocks is especially favoured at this site as a result of the combination of three effects as the lens thins at its discharging margin: (1) the superposition of the vadose (lens top) mixing zone with the phreatic zone (lens base) mixing zone; (2) the superposition of organic decay horizons at the top and bottom of the lens; and (3) increased flow rates as the integrated lens discharge occurs across a declining cross-sectional area (Mylroie & Mylroie 2007; Mylroie 2013). Dissolutional caves are produced in large numbers at this site, called flank margin caves because they form under the flank of the enclosing land mass at the discharging margin of the freshwater lens (Fig. 1). Flank margin caves are excellent indicators of sea-level position at the time of their formation because the lens margin is intimately tied to sea-level position.

Flank margin caves form at specific horizons tied to sea level, along the entire discharging margin of the freshwater lens (Carew & Mylroie 1995*b*). This dissolution, and later exposure by sea-level change, creates voids along former coastlines in a pattern referred to as 'beads-on-a-string' (Figs 2 & 3) and the caves maintain a horizontal position despite both primary (Fig. 3c, d) and secondary structure (Fig. 2b). The top position of the freshwater lens, and hence the sea-level position, is considered the highest phreatic dissolutional surface in a given flank margin cave (i.e. the original cave ceiling in the absence of modification by collapse). The location of dissolutional voids at a constant horizon by exposure in road-cuts indicates an origin in the distal margin of a freshwater lens, i.e. at sea-level (Fig. 3a). For a discussion of sea caves and tafoni, other cave types common on carbonate coasts, and how to differentiate them from flank margin caves, see Waterstrat *et al.* (2010). Although best known from eogenetic carbonate rocks, flank margin caves are also found in coastal telogenetic carbonate rocks (Fig. 2b), as discussed in Mylroie & Mylroie (2013) and D'Angeli *et al.* (2015).

When large carbonate banks are exposed by sea-level fall, diffuse flow in the freshwater lens becomes inefficient (the discharge perimeter has increased

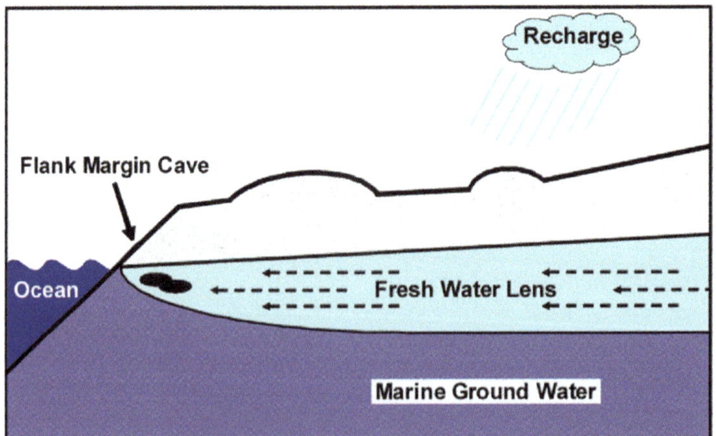

Fig. 1. Simple schematic diagram of a freshwater lens on a carbonate coast. The lens is vertically exaggerated here, but the image shows that the lens margin is a sea-level indicator. Flank margin caves preferentially develop in the lens margin.

Fig. 2. Flank margin caves exposed in the cliffs of uplifted carbonate islands. The repeated appearance of chambers at a constant horizon crossing primary and secondary structures is referred to as 'beads-on-a-string' and represents the position of the freshwater lens and hence the sea-level at the time of cave development. **(a)** Cave entrances on Isla de Mona, Puerto Rico in Mio-Pliocene reef carbonates. **(b)** Caves at Limestone Creek, South Island, New Zealand in Oligocene cold water carbonates; bedding is upper left to lower right and jointing is high-angle upper right to lower left, but the cave entrances are on a horizontal plane. **(c)** Cave entrances in uplifted Plio-Pleistocene reef carbonates, Tinian Island (telephone pole with transformer in left foreground for scale).

Fig. 3. Glacio-eustatic sea-level as indicated by flank margin caves in Bahamian aeolianites, produced by the last interglacial sea-level highstand (MIS 5e). (**a**) Small decimetre-sized voids (rock hammer in circle for scale) exposed in a road-cut in aeolianites, Eleuthera, cutting across the dipping foreset beds. (**b**) Flank margin caves exposed by Holocene coastal erosion, Rum Cay. (**c**) Dipping foreset beds in aeolianite, cut smoothly on a horizontal plane, Crown Cave, Cat Island. (**d**) Terra rossa palaeosol cut by a horizontal cave passage, Red Roof Cave, Long Island.

linearly with an increase in the size of the platform, but the meteoric catchment has increased by the square) and extensive conduit systems with turbulent flow develop within the lens to enable water discharge to the coast (Vacher & Mylroie 2002; Larson & Mylroie 2014). The >1000 km of mapped conduits known from Quintana Roo State in Mexico are an extraordinary example (Kambesis & Coke 2013).

Sea-level position is therefore determinable not only by the depositional facies of the coastal carbonate rocks, but also by the internal dissolutional megaporosity preserved within those very same carbonate rocks.

Karst denudation on carbonate coasts

Karst denudation, based on theoretical calculations (White 1984) and empirical relationships (Smith & Atkinson 1976), is a function of soil pCO_2, water surplus (precipitation minus evapotranspiration) and temperature (Fig. 4). Although $CaCO_3$ solubility is inversely proportional to temperature (the result both of kinetics and higher CO_2 solubility in colder water), the much higher soil pCO_2 levels of productive tropical soils, coupled with a significant water surplus, creates the highest rate of carbonate rock denudation in tropical areas (Brook *et al.* 1983). Empirical data on the denudation of eogenetic carbonates are rare, being focused primarily on coastal karren, bioerosion notches, and related landforms and dissolutional structures (e.g. Trudgill 1985; see also Taboroši & Kázmér 2013)

Vacher & Mylroie (2002, p. 183) defined eogenetic karst as 'the land surface evolving on, and the pore system developing in, rocks undergoing eogenetic, meteoric diagenesis'. Such rocks commonly have a primary depositional porosity of ≥30%; subsequent dissolution in the freshwater lens to produce touching vug porosity may create a system close to 50% porosity (Vacher & Mylroie 2002). Therefore the denudation of eogenetic carbonates can be expected to be more rapid than in telogenetic rocks because the mass to be removed for an equivalent

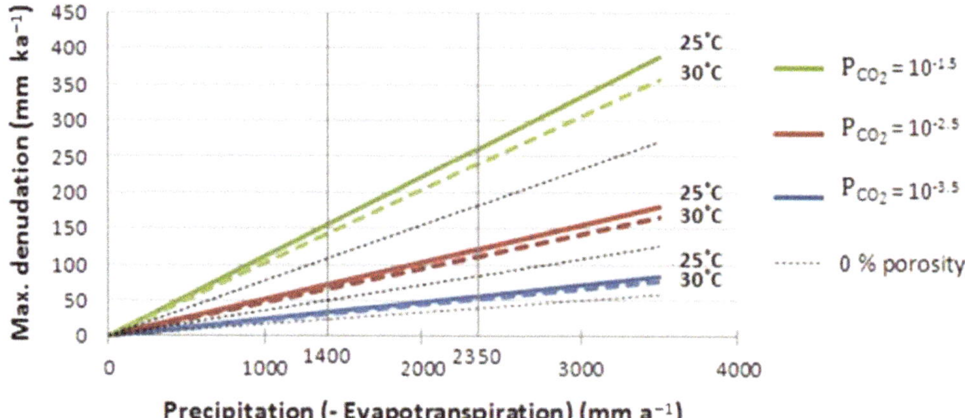

Fig. 4. Plot of karst denudation v. water excess (precipitation–evapotranspiration), with variations due to temperature and soil pCO_2 shown, assuming an aragonitic content and 30% porosity. Note that $CaCO_3$ solubility is inversely controlled by temperature and that increasing soil pCO_2 increases denudation. The Guam values were based on 2350 mm precipitation, 1400 mm evapotranspiration and a pCO_2 of $10^{-2.5}$. Soil pCO_2 values from White (1984), kinetic constants from Ford & Williams (2007) and plot from Miklavič (2011).

surface lowering can easily be a third less than for telogenetic carbonate rocks from continental interiors. Despite the higher biological production associated with low-latitude soils, which may increase soil pCO_2 and hence the dissolution rate, very young carbonate rocks usually have thin soils and pCO_2 production would be therefore diminished.

The majority of Cenozoic carbonate rocks were initially deposited as the $CaCO_3$ polymorph aragonite (e.g. green algae, corals and molluscs; the less abundant echinoderms and red algae typically precipitate high-Mg calcite), which is slightly more soluble than calcite. Aragonite inverts to its more stable polymorph calcite on a time span of several hundred thousand years (it can persist far longer under specific conditions). Late Quaternary eogenetic carbonates are commonly high in aragonite, so it can be expected these very young carbonates would denude more rapidly than older calcite carbonate rocks. Larson & Mylroie (2013), in a paper on the global CO_2 budget and carbonate deposition/dissolution cycles, argued that the aragonite v. calcite dissolution differential is a relatively minor denudation effect on a global scale; however, Miklavič *et al.* (2012) demonstrated that the effect has an important practical application when calibrating denudation rates on Guam.

Coral reef terraces

How much of an elevation error occurs in the evaluation of coral reef terraces if denudation is ignored? Larson & Mylroie (2013) used model calculations following the methodology of Gombert (2002) to estimate the denudational lowering of carbonate landscapes, which depends on climate and pCO_2. Their results indicate a 10–100 mm ka^{-1} rate of landscape lowering. These model data are consistent with the range of 9–60 mm ka^{-1} reported from a number of studies across a variety of areas by Ford & Williams (2007, their table 4.3). As their data are mostly telogenetic examples, for young eogenetic carbonate rocks the values may well be closer to the higher end of the range at 100 mm ka^{-1}. A fossil coral reef on a tectonically stable island from MIS 5e, 120 ka ago, accepting a minimum rate of 10 mm ka^{-1}, could have been denuded by 1.2 m. A much greater degree of lowering (12 m) would occur if the upper end of the denudation model is used; on tectonically uplifted islands, the degree of lowering would be greater on higher (and older) coral reef terraces.

Empirical evidence for the lowering rates of carbonate landscapes is seen in glaciated areas, where non-carbonate glacial erratics have protected underlying carbonates from meteoric precipitation and hence dissolution. These pedestals are called karrentische or 'karst tables' (Ford & Williams 2007), and their height is a measure of denudation since ice retreat after the Last Glacial Maximum. Insoluble components within the carbonates, such as quartz veins, will also protrude from the carbonate rock surface (Fig. 5). Figure 5 is instructive because it is from Arctic Norway, where ice retreated at about 8 ka and,

Fig. 5. (**a**) Karrentische from Arctic Norway and (**b**) a protruding quartz vein. Ice left this region north of the Arctic Circle about 8 ka before present. Despite the short time window, the lack of soil pCO_2 and the dense telogenetic nature of the host marble, a demonstrable surface lowering has occurred. This surface lowering would not be recognizable without these indicators.

being barren of soil, the denudation expressed is solely the outcome of meteoric precipitation, which has an unpolluted pH of 5.7 as a result of atmospheric CO_2 absorption ($pCO_2 = 10^{-3.5}$). The rock is dense telogenetic Cambro-Ordovician marble, the climate is poor in CO_2 and the time window is short, yet a demonstrable landscape lowering effect is obvious. A key point is that landscape lowering needs a frame of reference; the landscape lowering of Figure 5 would not be recognizable if not for the karrentische and protruding quartz vein.

The Guam example

Karrentische are not expected to be common on tropical islands where glaciation is not a viable mechanism. On Guam, in the west Pacific, a unique situation has created karrentische on a fossil coral reef terrace, which allows an empirical denudation rate to be established in a setting comparable with that of Barbados and the Huon Peninsula of New Guinea. The detailed setting is presented in Miklavič (2011) and summarized here. On northern Guam, in the area of Ritidian Point, the late Pleistocene Tarague Limestone forms a bench or terrace 3–8 m high. The terrace is made of fossil corals that are primarily aragonitic in composition, consistent with their young age of 126–132 ka based on U/Th measurements (Randall & Siegrist 1996). The elevation of some terrace areas above 6 m is a result of tectonic uplift since MIS 5e. This terrace abuts high cliffs (up to 170 m high) of Plio-Pleistocene Marianas Limestone, which is entirely calcite (Fig. 6a). The Tarague

Fig. 6. (**a**) Large cliffs of Plio-Pleistocene Marianas Limestone tower over an adjacent terrace of late Pleistocene Tarague Limestone, Ritidian Point, Guam. (**b**) The Tarague Limestone terrace a few dozen metres away from the Marianas Limestone cliff. The surface is *in situ* corals and coral rubble with a thin soil, and appears to casual observation to be a weathered depositional surface.

Limestone terrace is flat, with a thin soil and a coral rubble surface (Fig. 6b). At numerous locations on this terrace, increasing in number as the Marianas Limestone cliff is approached, are a series of karrentische (Fig. 7). The interpretation (Miklavič 2011; Miklavič *et al.* 2012) is that after the sea-level fell at the end of MIS 5e, calcitic boulders of Marianas Limestone fell from the cliff onto the Tarague Limestone terrace and acted in a manner similar to a glacial erratic in protecting the underlying aragonitic Tarague Limestone from denudation. As a result, karrentische developed, with pedestals commonly 2 m in height, with some examples reaching 5 m in height. The variation in height is attributed to the

Fig. 7. Three images of karrentische from Ritidian Point, Guam. The plain surrounding the pedestals and the pedestals themselves are aragonitic *c.* 125 ka late Pleistocene Tarague Limestone. The capping boulders are Plio-Pleistocene calcitic Marianas Limestone, which fell from the high cliff pictured in Figure 6a after MIS 5e. The Marianas Limestone boulders, when they fell from the adjacent cliff, rested on a flat terrace. Denudation has lowered that surface to create the karrentische pedestals.

random timing of when boulders fell from the cliff above, with higher pedestals forming from boulders that fell earlier than those on shorter pedestals. Unlike a deglaciation event, which in a local region is fairly synchronous in time, leading to karrentische with similar pedestal heights, the karrentische of the Tarague Limestone are the integration of the timing of the various boulder falls from the first exposure of the terrace until the present day. Both X-ray diffraction and Feigl solution tests indicate that the pedestal is aragonite and the capping boulder is calcite (Miklavič 2011).

The authors have personally visited this site and, while walking on the Tarague Limestone terrace, gained the impression that it is the weathered original depositional surface. It is flat with an irregular surface with patches of coral rubble and a thin soil. The terrace lacks dramatic pit and pinnacle topography and when the karrentische are first observed, the effect is stunning. Occasional limestone pinnacles are found and these are interpreted as karrentische in which the covering Marianas Limestone boulder has either eroded away or toppled off the pedestal, with subsequent dissolution creating the pinnacle shape (Miklavič 2011). If the toppling effect occurred more often with increasing pedestal age, it is possible that the oldest pedestals have lost their capping Marianas Limestone boulder cap and have been significantly denuded. In other words, the 5 m maximum pedestal height observed today may not date from the first exposure of the Tarague Limestone terrace, but some significant time after that exposure, the older pedestals, which should have been taller, having lost their boulder cap and eroded to shorter pinnacles.

The age of the pedestals is difficult to determine; the maximum age would be the formation of the Tarague Limestone terrace at *c.* 125 ka. Pedestal height would be proportional to the time of the fall of the capping boulder. Assuming that the highest pedestals represent boulder fall soon after the exposure of the Tarague terrace (with the discussed caveats regarding the loss of the boulder cap), estimates of the denudation rate can be made and compared with model calculations utilizing the White (1984) method (see Fig. 4). The greatest observed pedestal height is 5 m and, using this value as the end of MIS 5e at *c.* 120 ka, a minimum denudation value of 42 mm ka^{-1} is produced. Model calculations indicate a value of 50 mm ka^{-1} (Miklavič 2011), which is in the same range as the empirical data from the Tarague Limestone karrentische. The model values assume a thin soil and low soil pCO_2 values ($pCO_2 = 10^{-2.5}$), so the model rate would be a minimum. If boulder cap loss has biased the karrentische to younger examples, then the actual denudation rate would have been higher. The data also indicate that the small difference in solubility between calcitic and aragonitic eogenetic limestones can be a significant factor in the resulting expression of denudation.

In addition to the karrentische, the Marianas Limestone cliffs in the Ritidian area of Guam contain notches that are either fossil coastal bioerosion notches or dissolution features formed in the flank margin cave setting and exposed by cliff retreat (Miklavič 2011). Either interpretation establishes sea-level at the height of the notch. Differentiating the mechanism is not a crucial factor, but given that the Marianas Limestone cliff is clearly retreating (as shown by the accumulation of talus and boulders at the cliff base and on the Tarague Limestone terrace), it is likely that a bioerosion notch would have been removed and that the dissolutional voids (flank margin caves) formed in the freshwater lens behind the notch would have been exposed by cliff retreat. This issue of notch origin and interpretation is discussed in full in Waterstrat *et al.* (2010). The notch at Ritidian Point is a double notch, indicating either a pulse of tectonic uplift during MIS 5e or a double peak on MIS 5e itself, which has been suggested for the Pacific (Chappell 1974; Nunn 1999) and the Atlantic (Carew & Mylroie 1999) oceans. The two notches are 6.5 and 8 m above the current Tarague Limestone terrace (their current elevation with respect to current sea-level is higher, reflecting post-MIS 5e tectonic uplift) and, assuming that the notches represent a palaeo-sea-level, then the notches indicate a maximum of up to 6.5 m of denudation of the Tarague terrace if we assume that the corals were at wave base during terrace deposition and that the higher notch formed first (which is not necessarily true). The higher value for denudation based on the notches (*c.* 6.5–8 m) as opposed to the karrentische (*c.* 5 m) may suggest that the growing corals were not at wave base at the end of MIS 5e, or that the fall of the first boulders to produce the karrentische occurred after the Tarague terrace was exposed and when denudation had already been operating below the notches. The issue of boulder cap loss could mean the oldest (and presumably highest) karrentische are no longer present.

The Guam example demonstrates two things. First, models used to calculate karst denudation for eogenetic carbonates on tropical islands produce reliable results when compared with field examples and, second, the amount of denudation is significant and could change current thinking on sea-level positions obtained by using the flat tops of fossil coral reef terraces as true depositional surfaces. The degree to which these flat coral terraces look like actual depositional surfaces cannot be overemphasized; the observer must be mindful of the effects of denudation and that such denudation may not leave an obvious signal.

The Bahamas problem

Although tectonically uplifted islands have been valued in Quaternary sea-level studies because they raise datable marine facies into the subaerial environment, tectonically stable islands provide a data source in which the tectonic signal is eliminated and sea-level variation is solely due to glacio-eustasy. As noted earlier, such stable islands have been used to calibrate tectonic uplift on active islands. The Bahamian archipelago has attracted much scientific study addressing Quaternary sea-level, utilizing the glacio-eustatic highstands that left marine facies above modern sea-level. The work has focused primarily on MIS stage 5 and its substages 5e, 5c and 5a (for a thorough review of Bahamian geology, see Curran & White 1995). The subaerially exposed rocks of the Bahamas are all Holocene or mid- to late Pleistocene in age (Carew & Mylroie 1995*a*). The one exception is Mayaguana Island, which appears to be slightly tilted such that its north shore has risen up by a small amount, exposing marine rocks back to the Late Miocene (Kindler *et al.* 2011). The tilting is thought to be related to the island's proximity to faults associated with the North American–Caribbean plate boundary. Otherwise, all dated subtidal deposits in the Bahamas date to MIS 5e, in which sea-level was elevated above modern levels by *c.* 6 m from 124 to 115 ka (Thompson *et al.* 2011). There has been debate about subtidal or beach facies being present from earlier Quaternary glacio-eustatic sea-level highstands, such as MIS 7, 9 and 11 (Carew & Mylroie 1997; Mylroie 2008); MIS 9 and 11 appear to have been higher than current sea-level, so subtidal facies should be present from those earlier events, but none has been located. Such deposits may exist where they were entombed by later sediments, such that they escaped surface exposure and subsequent denudation.

Flank margin caves are indicators of sea-level position at the time of cave development. Carew & Mylroie (1995*b*) compared U/Th dates from fossil coral reefs around the Bahamas with the position and elevation of flank margin caves and found that the caves were consistent with an origin during MIS 5e. Stalagmites dated from Bahamian flank margin caves were all shown to be younger than MIS 5e, consistent with an origin during MIS 5e, but the sample number was extremely low and the results were considered to be inconclusive. The elevation of the flank margin caves was up to 7 m above sea-level, in agreement with the position of a freshwater lens floating on a sea-level 6 m above that of the modern day. The major problem with this interpretation is that the Bahamas are thought to be subsiding at 1–2 m per 100 ka (Carew & Mylroie 1995*a*, *b*), so caves formed at *c.* 120 ka should have subsided 1 m or more below their elevation of genesis, and perhaps should not have dissolutional ceilings at 7 m elevation unless the freshwater lens was higher above sea-level than expected (difficult in a coastal setting, which is where flank margin caves form). Could some of these caves be older than originally proposed?

Further work in the Bahamas, and in the geological extension of that archipelago SE to the Turks and Caicos islands, discovered flank margin caves with elevations higher than 7 m (up to 15 m or more) above modern sea-level. These caves also contain stalagmites with ages older than MIS 5e

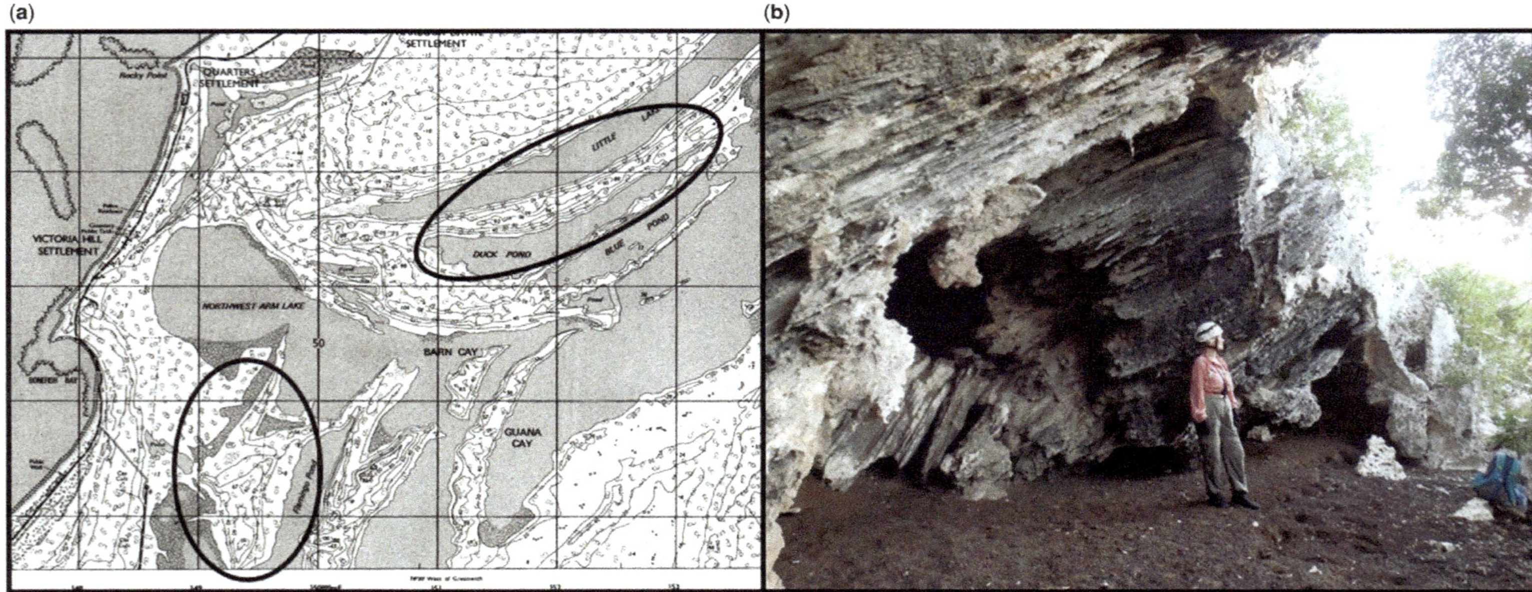

Fig. 8. Evidence of denudation of aeolian calcarenites in the Bahamas. (**a**) Older aeolian calcarenite dunes of the Bahamas developed in the Owls Hole Formation have a lumpy, irregular shape as a result of denudation (lower left oval); younger dunes of the MIS 5e Grotto Beach Formation (*c.* 120 ka) still retain their original depositional arcuate shape (upper right oval). Examples from San Salvador Island, Lands and Surveys topographic sheet 1; grid pattern is 1 km squares for scale. (**b**) Remnant dune, with foreset beds sticking up to the right, indicating extensive denudation, Cat Island, Bahamas. The flank margin cave once contained within the dune is now opened and partially destroyed by the denudation.

(e.g. 266 ka, Smart *et al.* 2008). If these caves were produced in the freshwater lens, either the Bahamas were not as tectonically stable as originally thought (as Mayaguana Island clearly demonstrates) or the caves formed during an earlier MIS, perhaps MIS 11, which may have been 10 m or more above modern sea-level. In either case, the problem is the location of the subtidal facies from these high sea-level events, regardless of the cause (tectonic v. glacio-eustatic). These data also failed to address the Bahamas platform subsidence question because MIS 11 is >400 ka back in time, meaning that subsidence should have been significant and cave passages formed at a +15 m highstand should not be at that elevation today, but 4–8 m lower.

The argument for the caves in the Turks and Caicos islands was that the proximity of those islands to the North American–Caribbean plate boundary had resulted in tectonic uplift, as also proposed for nearby Mayaguana Island (Kindler *et al.* 2008). This argument is more difficult to apply to Moores Island, located far to the north on Little Bahama Bank remote from any plate boundary, where there are caves with phreatic dissolutional surfaces above 12 m. Regardless of important questions about the degree of passive isostatic subsidence, active tectonics and Quaternary glacio-eustasy in the Bahamas, other questions remain. Where are the subtidal facies from those earlier events? Is the model for the elevation of flank margin cave formation relative to sea-level incorrect?

Quaternary glacio-eustatic sea-level highstands form only *c.* 10% of total Quaternary time. All Quaternary carbonate deposition as seen subaerially today must have formed in the relatively small time windows of glacio-eustatic sea-level highstands of *c.* 10 ka. During the other 90% of Quaternary time (*c.* 100 ka per event), the Bahamas were in a subaerial denudational environment. Quaternary marine facies are therefore a thin veneer in aeolianite islands such as the Bahamas and are vulnerable to complete removal by denudation. For uplifted coral reef islands such as Barbados, thick marine facies exist and aeolianites are rare or absent; denudation cannot entirely remove the marine facies. If carbonate denudation is accepted as occurring at the rate seen in

Fig. 9. Aeolian calcarenite dunes from Abaco Island, Bahamas, where denudation has removed the original asymmetrical depositional shape and created symmetrical shapes that resemble classic cone karst (from Walker *et al.* 2010). (**a**) Close up of truncated aeolian beds (machete 72 cm long for scale). (**b**) Cone-shaped hill (person in oval for scale), arrows show bedding dip, black arrow at the truncated end, talus collecting below the truncation. (**c**) Schematic diagram showing the evolution of the dunes into cone shapes by denudation.

Guam, it is possible that marine facies from MIS 11 and related events have been completely removed in the Bahamas. The examination of older aeolian calcarenites (carbonate dunes) in the Bahamas today reveals that they are highly modified from their original depositional configuration by denudation. Although younger dunes show the arcuate shape from when they first formed, older dunes are 'lumpy' and irregular (Fig. 8a). An examination of outcrops reveals the significant removal of bedforms, such that foreset beds stick up in the air with the remaining portion of the dune missing (Fig. 8b). Abaco Island has the most positive water budget of any Bahamian island (Whitaker & Smart 1997) and maximum denudation for the Bahamian archipelago would be expected there. Denudation on Abaco (Fig. 9) has altered the original upwind–downwind asymmetrical shape of aeolian calcarenite dunes into symmetrical features that resemble cone karst (Walker *et al.* 2010). The result is a much bigger denudational footprint than that provided by karrentische.

These observations explain two issues. First, such denudation may well have removed a thin (*c.* 10 ka long depositional window) marine facies overlay from events prior to MIS 5e in the Bahamian archipelago. The abundant MIS 5e fossil reefs survive today because they have only experienced 120 ka of denudation; however, the present tops of these exposed reefs are not their original depositional surface. Second, such extensive denudation would have removed smaller flank margin caves that formed during the earlier time, creating a population of flank margin caves biased to younger, and therefore better surviving, caves from MIS 5e. This latter fact could explain why only a few dozen caves in the Bahamas today, out of a database of >400 flank margin caves, appear at elevations higher than 7 m. Examples of such caves above 7 m include Sonny's Cave on Moores Island (+12 m), Hatchet Bay Cave

Fig. 10. Pleistocene aeolian calcarenite dunes with fossil wave-cut benches and fossil corals. (**a**) Rum Cay, Bahamas, showing flank margin caves exposed by slope retreat and a MIS 5e wave-cut bench. (**b**) The bench in part (a) showing fossil corals (circle) and terra rossa palaeosol remnants (arrows) that overlie the corals at this locality. Machete 55 cm long for scale. (**c**) Similar situation on Crooked Island, Bahamas; note that the bench has a terra rossa palaeosol surface, indicating it is pre-Holocene in origin. (**d**) Fossil corals (ovals) on the bench in part (c), where the overlying terra rossa palaeosol has been stripped by modern wave action. Machete 72 cm long for scale.

upper levels on Eleuthera Island (+12 m), Port Howe Cave on Cat Island (+15 m), Osprey Cave on Crooked Island (+18 m) and Conch Bar Cave on Middle Caicos Island (+20 m), a trend >800 km long through the archipelago from NW to SE, with increasing elevations to the SE.

The crucial point to be made is that flank margin caves, which form in the carbonate rock, may be a more robust and persistent indicator of sea-level position than fossil coral reefs and associated marine facies, or littoral caves and notches, which form on the rock and which may have been denuded from surface outcrops. For example, wave-cut benches containing fossil corals from MIS 5e can be found throughout the Bahamas (Fig. 10), incised in older aeolian calcarenite dunes. The back wall of these benches has erosionally retreated, such that flank margin caves on the inside of the dune are now exposed (Fig. 10a, see also Fig. 3b). These observations show that past sea-level highstands that formed flank margin caves above modern sea-level also created benches providing hard ground for coral reef development. The lack of such features from earlier highstands indicates that denudation has been significant and a large piece of the shallow marine rock record is missing. This situation is shown in Figure 11. The flank margin caves are indicators that sea-level highstands associated with MIS 9 and 11 (or perhaps earlier) were higher than the extant fossil coral record of the Bahamas displayed today. The observations suggest that platform subsidence may have been less in the late Quaternary than the long-term average derived from drillcores. The use of fossil corals to determine Quaternary sea-level in aeolian calcarenite islands may be biased by the selective removal of older units by denudation over time.

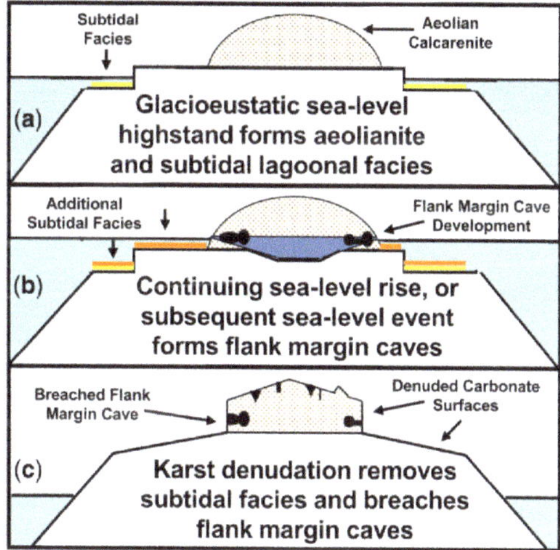

Fig. 11. Schematic model demonstrating how denudation may remove subtidal facies from pre-existing aeolian calcarenites in the Bahamas. (**a**) An initial sea-level highstand creates both subtidal facies in the lagoon and aeolian calcarenite dunes on the remaining subaerial platform. (**b**) Either continued sea-level advance on the initial sea-level highstand or a slightly higher second sea-level highstand create additional subtidal facies and place a freshwater lens in the dune, producing flank margin caves (note that the caves form without entrances). (**c**) Denudation during the glacio-eustatic sea-level lowstands (*c.* 10 times longer than the highstands in the Quaternary) strips off the thin veneer of subtidal facies and opens access by humans to the flank margin caves.

Conclusions

Karst processes must be taken into account when considering Quaternary sea-level change and the data provided by carbonate islands and coasts. Karst gives and karst takes away. Flank margin caves are a true gift, revealing the position of the freshwater lens and hence the sea-level position, even when other surface indicators have been removed by denudation and are not present. The dating of cave deposits, such as U/Th dating of stalagmites, may allow time constraints on when the caves developed. Karst denudation takes away surface evidence, such as bioerosion notches and marine facies, but also provides karrentische, which allows the independent calibration of denudation models. Karrentische are also a vivid reminder that flat coral reef terraces are probably not true depositional surfaces and that the terrace elevation needs to be corrected as a result of denudation. Most carbonate islands and coasts, especially those studied for sea-level purposes, consist of eogenetic carbonates and that condition must be taken into account when applying denudation models. The mixing dissolution and other phenomena that drive flank margin cave development are also unique to carbonate coasts. The speed at which these caves form make them reliable indicators of sea-level position, even when the sea-level is stable at a given elevation for only a few thousand years. Flank margin caves also persist through time, far longer than bioerosion notches, sea caves and other surface features used to assess sea-level position. Flank margin caves are a high-resolution, but long-lasting, indicator of sea-level position. The focus in the scientific literature on carbonate islands and coasts as sea-level indicators has almost solely used depositional features. Understanding and applying karst processes to these carbonate islands is a necessary and useful endeavour to more accurately determine Quaternary glacio-eustasy, subsidence rates and rates of tectonic uplift.

We thank the many colleagues who have worked with us on island karst processes over many decades, especially James Carew, John Jenson and Len Vacher. Former students have made major contributions, including, for this study, Erik Larson, Blaz Miklavič and Athena Owens Nagle. The Coastal Cave Survey, Michael Lace Executive Director, Nancy Albury of the Antiquities, Monuments and Museum Corporation of the National Museum of the Bahamas, and the Gerace Research Centre, San Salvador Island, Bahamas have been major supporters. Reviewers are thanked for comments that improved and strengthened the manuscript.

References

Brook, G.A., Folkoff, M.E. & Box, E.O. 1983. A world model of soil carbon dioxide. *Earth Surface Processes and Landforms*, **1**, 79–88. https://doi.org/10.1002/esp.3290080108

Carew, J.L. & Mylroie, J.E. 1995*a*. A stratigraphic and depositional model for the Bahama Islands. *In*: Curran, H.A. & White, B. (eds) *Terrestrial and Shallow Marine Geology of the Bahamas and Bermuda*. Geological Society of America, Special Papers, **300**, 5–31.

Carew, J.L. & Mylroie, J.E. 1995*b*. Quaternary tectonic stability of the Bahamian archipelago: evidence from fossil coral reefs and flank margin caves. *Quaternary Science Reviews*, **14**, 144–153.

Carew, J.L. & Mylroie, J.E. 1997. Geology of the Bahamas. *In*: Vacher, H.L. & Quinn, T.M. (eds) *Geology and Hydrogeology of Carbonate Islands*. Elsevier Science, Amsterdam, 91–139.

Carew, J.L. & Mylroie, J.E. 1999. A review of the last interglacial sea-level highstand (oxygen isotope substage 5e): duration, magnitude and variability from Bahamian Data. *In*: Curran, H.A. & Mylroie, J.E. (eds) *Proceedings of the Ninth Symposium on the Geology of the Bahamas and Other Carbonate Regions*. Bahamian Field Station, San Salvador Island, 14–21.

Chappell, J. 1974. Geology of coral terraces at Huon Peninsula, Papua New Guinea: a study of Quaternary tectonic movements and sea level changes. *Geological Society of America Bulletin*, **85**, 553–570.

Chappell, J., Omura, A., Esat, T., McCulloch, M., Pandolfi, J., Ota, Y. & Pillans, B. 1996. Reconciliation of late Quaternary sea levels derived from terraces at Huon Peninsula with deep sea oxygen isotope records. *Earth and Planetary Science Letters*, **41**, 227–236.

Choquette, P.W. & Pray, L.C. 1970. Geologic nomenclature and classification of porosity in sedimentary carbonates. *American Association of Petroleum Geologists Bulletin*, **54**, 207–250.

Curran, H.A. & White, B. (eds) 1995. *Terrestrial and Shallow Marine Geology of the Bahamas and Bermuda*. Geological Society of America, Special Papers, **300**.

D'Angeli, I.M., Sanna, L., Calzoni, C. & De Waele, J. 2015. Uplifted flank margin caves in telogenetic limestones in the Gulf of Orosei (Central-East Sardinia–Italy) and their palaeogeographic significance. *Geomorphology*, **231**, 202–211.

Folk, R.L., Roberts, H.H. & Moore, C.M. 1973. Black phytokarst from Hell, Cayman Islands, West Indies. *Geological Society of America Bulletin*, **84**, 2351–2360.

Ford, D.C. & Williams, P.W. 2007. *Karst Hydrology and Geomorphology*. Wiley, Chichester.

Gombert, P. 2002. Role of karstic dissolution in global carbon cycle. *Global and Planetary Change*, **33**, 177–184.

Kambesis, P.N. & Coke, J.G. 2013. Overview of the controls on eogenetic caves and karst development in Quintana Roo, Mexico. *In*: Lace, M.J. & Mylroie, J.E. (eds) *Coastal Karst Landforms*. Coastal Research Library, **5**. Springer, Dordrecht, 347–373.

Kindler, P., Godefroid, F. & Samankassou, E. 2008. Pre-Holocene island geology of the Caicos and Mayaguana (Bahamas) platforms: similarities and differences. *In*: Morgan, W.A. & Harris, P.M. (eds) *Developing Models and Analogs for Isolated Carbonate Platforms Holocene and Pleistocene Carbonates of Caicos Platform, British West Indies*. SEPM Core Workshop Notes, **22**, 211–220.

Kindler, P., Godefroid, F. *et al.* 2011. Discovery of Miocene to lower Pleistocene deposits on Mayaguana, Bahamas: evidence for recent active tectonism on the North American margin. *Geology*, **39**, 523–526.

Larson, E.B. & Mylroie, J.E. 2013. Quaternary glacial cycles: karst processes and the global CO_2 budget. *Acta Carsologica*, **42**, 197–202.

Larson, E.B. & Mylroie, J.E. 2014. A review of Whiting Formation in the Bahamas and new models. *Carbonates and Evaporites*, **29**, 337–347, https://doi.org/10.1007/s13146-014-0212-7

Miklavič, B. 2011. *Formation of geomorphic features as a response to sea-level change at Ritidian Point, Guam, Mariana Islands*. MSc thesis, Mississippi State University, http://sun.library.msstate.edu/ETD-db/theses/available/etd-03262011-212710/

Miklavič, B., Mylroie, J.E., Jenson, J.W., Randall, R.H., Banner, J.L. & Partin, J.W. 2012. Evidence of the sea-level change since MIS 5e on Guam, tropical west Pacific. *Studia UBB Geologia Special Issue. National Science Foundation Workshop Sea-Level Changes into the MIS 5e: From Observation to Prediction*. Studia Universitatas Babeş-Bolyai, Cluj, 30–32.

Mylroie, J.E. 2008. Late Quaternary sea level position: Bahamian carbonate deposition and dissolution cycles. *Quaternary International*, **183**, 61–75, https://doi.org/10.1016/j.quaint.2007.06.030

Mylroie, J.E. 2013. Coastal karst development in carbonate rocks. *In*: Lace, M.J. & Mylroie, J.E. (eds) *Coastal Karst Landforms*. Coastal Research Library, **5**. Springer, Dordrecht, 77–109.

Mylroie, J.E. & Mylroie, J.R. 2007. Development of the carbonate island karst model. *Journal of Cave and Karst Studies*, **69**, 59–75.

Mylroie, J.E. & Mylroie, J.R. 2013. Telogenetic limestones and island karst. *In*: Lace, M.J. & Mylroie, J.E. (eds) *Coastal Karst Landforms*. Coastal Research Library, **5**. Springer, Dordrecht, 375–393.

Nunn, P.D. 1999. *Environmental Change in the Pacific Basin*. Wiley, Chichester.

Randall, R.H. & Siegrist, H.G. 1996. *The Legacy of Tarague Embayment and its Inhabitants, Anderson AFB, Guam. Vol. III. Geology, Beaches, and Coral Reefs*. International Archaeology, Honolulu.

Smart, P.L., Moseley, G.M., Richards, D.A. & Whitaker, F.F. 2008. Past high sea-stands and platform stability: evidence from Conch Bar Cave, Middle Caicos. *In*: Morgan, W.A. & Harris, P.M. (eds) *Developing Models and Analogs for Isolated Carbonate Platforms-Holocene and Pleistocene Carbonates of Caicos Platform, British West Indies*. SEPM Core Workshop Notes, **22**, 203–210.

Smith, D. & Atkinson, T.C. 1976. Process, landforms and climate in limestone regions. *In*: Derbyshire, E. (ed.) *Geomorphology and Climate*. Wiley, Chichester.

Taboroši, D. & Kázmér, M. 2013. Erosional and depositional textures and structures in coastal karst landscape. *In*: Lace, M.J. & Mylroie, J.E. (eds) *Coastal Karst Landforms*. Coastal Research Library, **5**. Springer, Dordrecht, 15–57.

Thompson, W.G., Curran, H.A., Wilson, M.A. & White, B. 2011. Sea-level oscillations during the last interglacial highstand recorded by Bahamian corals. *Nature Geoscience*, **4**, 684–687.

Trudgill, S.T. 1985. *Limestone Geomorphology*. Longman, Harlow.

Ulrich, R. & Schellmann, G. 2006. Uplift history along the Clermont Nose traverse on the west coast of Barbados during the last 500 000 years – implications for paleo-sea level reconstructions. *Journal of Coastal Research*, **22**, 350–356.

Vacher, H.L. & Mylroie, J.E. 2002. Eogenetic karst from the perspective of an equivalent porous medium. *Carbonates and Evaporites*, **17**, 182–196.

Walker, L.N., Mylroie, J.E., Walker, A.D. & Mylroie, J.R. 2010. Symmetrical cone-shaped hills, Abaco Island, Bahamas. Karst or pseudokarst? *Journal of Cave and Karst Studies*, **72**, 137–149.

Waterstrat, W.J., Mylroie, J.E., Owen, A.M. & Mylroie, J.R. 2010. Coastal caves in Bahamian eolian calcarenites: differentiating between sea caves and flank margin caves using quantitative morphology. *Journal of Cave and Karst Studies*, **72**, 61–74.

Whitaker, F.F. & Smart, P.L. 1997. Hydrogeology of the Bahamian archipelago. *In*: Vacher, H.L. & Quinn, T. (eds) *Geology and Hydrogeology of Carbonate Islands*. Developments in Sedimentology, **54**. Elsevier, Amsterdam, 183–216.

White, W.B. 1984. Rate processes: chemical kinetics and karst landform development. *In*: La Fleur, R.G. (ed.) *Groundwater as a Geomorphic Agent*. Allen and Unwin, London, 227–248.

Arid hypogene karst in a multi-aquifer system: hydrogeology and speleogenesis of Ashalim Cave, Negev Desert, Israel

AMOS FRUMKIN* & BOAZ LANGFORD

Israel Cave Research Center, Institute of Earth Sciences, The Hebrew University of Jerusalem, Israel 91904

**Correspondence: amos.frumkin@mail.huji.ac.il*

Abstract: Ashalim maze cave, and neighbouring caves in the NW Negev Desert, Israel demonstrate hypogene karst features. These features are shown to have developed as a result of the mixing of two types of groundwater flowing in opposite directions within two tiers of Cretaceous rock aquifers. The stable isotope composition indicates that the lower Kurnub sandstone aquifer was recharged over far-field Nubian Sandstone outcrops in the vicinity of the Precambrian basement outcrops of the Sinai Desert, which belongs to the Afro-Arabian dome. The water flows northward and rises into the Judea carbonate aquifer through deep faults. A similar hydrogeological system is inferred for the speleogenetic period of Ashalim Cave. Dewatering of the cave occurred in the Pliocene due to regional uplift. This is indicated by the first vadose speleothems, dated to the late Pliocene (3.1 Ma). This was followed by surface denudation, which breached the cave and formed the present entrance.

Water scarcity is the main constraint on karstification in arid regions. The study of relict hypogene karst within desert regions will lead to a better understanding of the palaeohydrology and palaeoclimate of such regions. The present study tackles this issue, addressing the effects of converging tiered aquifers which are recharged in remote outcrops.

The Negev and Sinai deserts are part of the Saharo-Arabian Desert, the largest desert on Earth. Several hypogene features have been observed by our team in the Negev Desert, among which is Ashalim Cave, the largest known cave in the NW Negev (Fig. 1). The Cretaceous rock sequence in the Negev consists of Lower Cretaceous sandstones, carbonates and marls (Kurnub Group), Cenomanian–Turonian carbonates and marls (Judea Group) and Senonian chalk, chert and marls (Mount Scopus Group). Most of the Negev caves are within late Cretaceous, Cenomanian–Turonian carbonates. Eocene limestones (Avedat Group) contain additional karst features, such as unconfined chamber caves.

Ashalim Cave is located on the flank of Boqer Ridge, a moderate limestone hill 370 m above sea-level, drained through Nahal (wadi) Besor to the Mediterranean. This region of the Negev is relatively rich in small caves located within Turonian limestone of the upper Judea Group. The regional structure is marked by NE-trending anticlines, forming topographic ridges separated by synclinal valleys (Fig. 2).

Some of these structures are associated with faults, most of which are buried (Weinberger & Rosenthal 1994) (Fig. 2). In addition, the Zin transverse fault (Bartov *et al.* 1976), which is the northernmost east–west-trending fault of the Negev Desert, apparently runs in the subsurface not far from the cave.

Rainfall is highly variable, with a mean annual amount of *c.* 110 mm; the rainy period is October to May. The vegetation is concentrated mainly along ephemeral stream beds and consists of sparse steppe and desert plants of C3 and C4 photosynthesis pathways. Few ligneous perennial shrubs are scattered on the limestone hill above the cave (Fig. 3), but annual plants are abundant after rain storms.

Cave morphology

Ashalim Cave is a three-dimensional maze of interconnected passages and chambers with a total length of 570 m and a depth of 31 m. The cave's single entrance opens subvertically at a mild limestone slope without evidence of water flowing intensively into the cave, except for tiny flows from the immediate surroundings. Colluvial sediments have accumulated through the shaft-like entrance in recent times (Holocene?).

The cave entrance shows no genetic connection with the surface. It was apparently opened when subaerial denudation breached the ceiling of chamber A. This must have occurred >6000 years ago, because at this time the cave was already used by humans as a burial site (Cohen 1971; Yahalom-Mack *et al.* 2015). The entrance chamber (A) is the largest in the cave (Fig. 4). At its bottom, a smaller

From: Parise, M., Gabrovsek, F., Kaufmann, G. & Ravbar, N. (eds) 2018. *Advances in Karst Research: Theory, Fieldwork and Applications*. Geological Society, London, Special Publications, **466**, 187–200.
First published online November 6, 2017, https://doi.org/10.1144/SP466.3

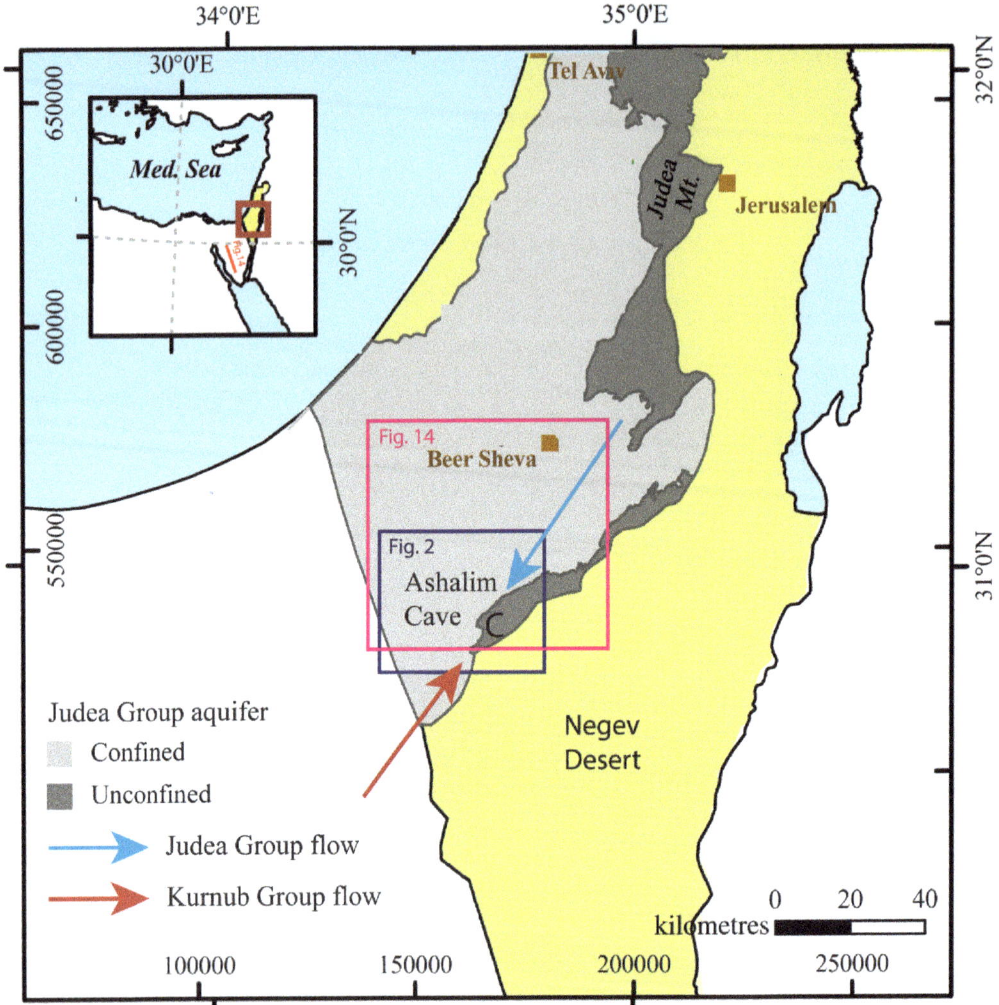

Fig. 1. Location map showing Ashalim Cave in the NW Negev Desert, Israel, at the confluence of the aquifers of the Judea Group and Kurnub Group. For stratigraphic sections see Figures 14 and 16.

passage (B) leads towards a speleothem-rich chamber (C), followed by narrow passages leading downwards to the lowermost level of the cave, where the main 110 m long lower passage (D–E–F) extends to the NW and SE (Fig. 5).

The inner parts of the cave can be reached from point Dc of this passage and include several medium-sized chambers (E, H, I, J, K and L) connected by narrow passages. The lower passage, D–F, developed along the bottom of a NW-trending oblique fracture that possibly acted as a fault. This fracture is the backbone of the three-dimensional maze of the cave.

The solutional micromorphology of the cave includes smooth walls, elliptical cross-sections of passages, cupolas (Fig. 6) and solution pockets in the walls and ceilings (Fig. 7). Corroded bubble trails are observed at some points along overhanging walls (Fig. 8). No evidence was found for fast-flowing water (such as scallops or clastic fluvial sediments).

The morphology of the cave indicates that water intruded the cave during speleogenesis through a fracture, which is hardly accessible at the bottom of the cave (bottom of passage D–F). From there the cave developed upwards through the oblique fracture, forming a sloping three-dimensional maze, divided to several tiers of sub-horizontal passages. The cross-section of the main passages is often elliptic or asymmetrical, reflecting the sloping guiding fracture (Fig. 9). The various levels of the

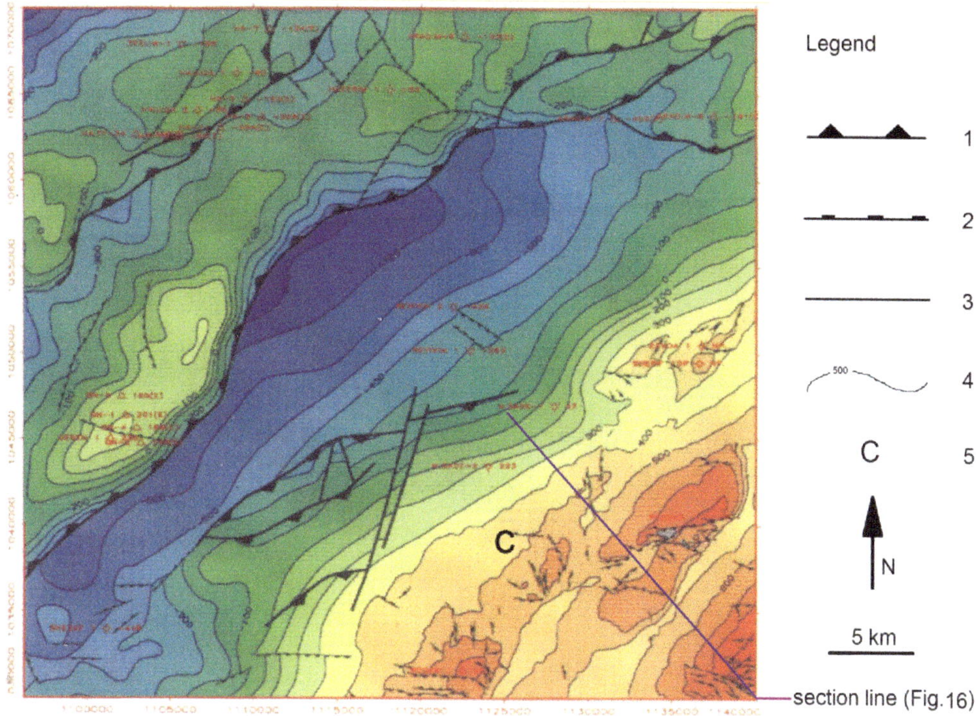

Fig. 2. Structural map (top Judea Group) showing the main regional Syrian Arc anticlines and faults in the vicinity of Ashalim Cave. Modified after map by Ofer Siman Tov, the Geophysical Institute of Israel. For location, see Figure 1. 1, Major fault; 2, medium fault; 3, small fault; 4, structural contour; and 5, Ashalim Cave.

cave are interconnected by smooth-walled shafts, either vertical or sloping (Fig. 10), which, having no vadose dissolution features, appear to have formed as rising conduits or feeders (Klimchouk 2013). Above the lowermost passages, the cave bifurcates in a fan-like fashion. Some of the passages interconnect in a complex three-dimensional fashion, forming a 'boneyard' and boulder-divided intricate voids. Water circulating through the cave can escape horizontally through tight outlets, or upwards through passage A.

Vadose speleothems (mainly flowstone) are observed at the entrance of the cave, half a metre below the present surface, indicating a previously thicker bedrock ceiling that has been eventually denuded. The walls and ceiling of the lowermost passage (Da) are covered by a thin gypsum crust (Fig. 11). Salt and gypsum 'flowers' are also common here. During the vadose stage of the cave, some of the rising shafts have served as downward routes for saturated vadose flow, which deposited calcite dripstones (Fig. 12).

The general morphology indicates that the cave initially formed in a confined, hypogene setting (e.g. Klimchouk 2013; Audra & Palmer 2015; De Waele *et al.* 2016 and references cited therein). Three gypsum samples have $\delta^{34}S_{SO_4}$ values of 14.4, 14.8 and 15 (std up to 0.5), indicating evaporative origin. The isotopic composition of the gypsum crust does not indicate the corrosion of limestone by sulphuric acid (e.g. Galdenzi & Maruoka 2003; Hose 2013; Palmer 2013; Naaman *et al.* 2014 and references cited therein). The voids started forming by the action of aggressive water; this aggressiveness could result from mixing corrosion or the cooling of thermal water. This process is no longer active in the cave and there are insufficient data on the composition of the groundwater currently present below the cave.

U–Pb and U–Th age dating of the vadose calcite speleothems show that they have been deposited on the bedrock wall and ceiling of the cave intermittently since the late Pliocene (3.1 Ma) (Vaks *et al.* 2013). The Pliocene age of initial deposition under vadose conditions suggests that late Neogene regional uplift, possibly associated with the deepening of the Dead Sea Rift (Guralnik *et al.* 2010; Matmon *et al.* 2014), caused the dewatering of groundwater from the cave, a process followed by the deposition of speleothems under vadose conditions.

Fig. 3. Ashalim Cave entrance. Photograph by Amos Frumkin.

Speleothem deposition lasted intermittently until the early–late Pleistocene (115 ka) (Vaks *et al.* 2010). There is no sign of speleothem deposition during the last 115 ka. The speleothems indicate increasing aridity since the relatively mild climate of the Pliocene, with ever-shortening moist episodes, ending at MIS 5.

Condensation corrosion by convective air flow has partly truncated some of the speleothems, exposing their internal stratification (Fig. 13). Ceiling collapse is common and its products are piled on the floors of most chambers and some passages. The chamber's ceilings and collapsed blocks are commonly smoothly corroded.

These types of maze cave, formed by hypogene dissolution, possibly assisted by the cooling of thermal water (Dublyansky 2013; Klimchouk 2013 and references cited therein), were typically formed in the Negev and Judean deserts under confined conditions during the Oligocene–early Miocene within Late Cretaceous massive limestones of the Shivta Formation (Frumkin & Fischhendler 2005; Frumkin *et al.* 2017*a*, *b*).

Groundwater flow

Two major aquifers are known in Cretaceous rocks of the NW Negev Desert: the Judea Group freshwater aquifer and the Kurnub Group brackish aquifer. Both aquifers are capped by poorly permeable layers of chalk and marls, resulting in

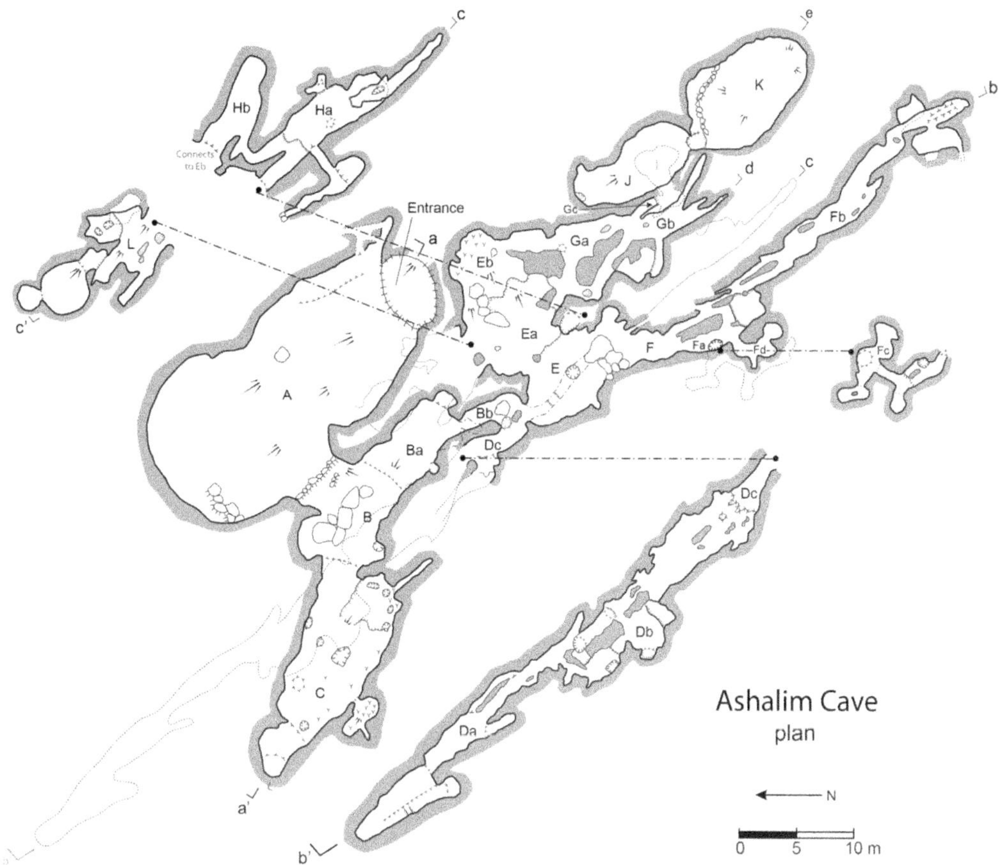

Fig. 4. Plan of Ashalim Cave. Surveyed by B. Langford, M. Ullman, N, Fishbein, L. Buchman, Israel Cave Research Center (2011).

confinement and artesian conditions mainly within synclines. These aquifers are clearly distinguished from each other by their hydrochemical properties (Kronfeld *et al.* 1993).

The Judea Group, composed of carbonate rocks, is recharged mainly in the western-central mountain range of Israel (Fig. 1) (e.g. Weinberger *et al.* 1994; Sheffer *et al.* 2010). The east Mediterranean precipitation that replenishes the Judea Group aquifer has a unique stable isotope signature, following the east Mediterranean meteoric water line, with d-excess values of *c.* 22‰ (Gat & Dansgaard 1972).

Conversely, most of the lower Cretaceous sandstone groundwater of the NE Negev Desert has d-excess values of *c.* 10‰, as well as low values of $\delta^{18}O_w$ and δD, similar to correlative aquifers in Egypt–Sinai (Gat & Issar 1974; Thorweihe & Heinl 2002). This is in agreement with the assumption that the Kurnub aquifer contains palaeowater recharged at the Nubian Sandstone outcrops of the Sinai Desert, downstream from the Precambrian basement outcrops (Fig. 14), which form part of the Afro-Arabian dome, south of the Negev Desert (Issar 1981; Rosenthal *et al.* 1998; Vengosh *et al.* 2007).

The above-mentioned studies suggest that the water of the sandstone aquifers flows gravitationally northward along the general regional dip, becoming confined under younger strata. The flow northward from Nubian Sandstone outcrops has been extensive since the rising and truncation of the Afro-Arabian dome during the Oligocene (Almond 1986; Avni *et al.* 2012; Bar *et al.* 2016).

The present water table in the Ashalim region is *c.* 400 m below the cave entrance. A groundwater anomaly was observed in this region, where the Nizana 1, Revivim, Mashabim and Ashalim wells, drilled into the Judea Group aquifer, encountered water with Kurnub Group aquifer properties, reaching up to 83% of this water type at Ashalim well (Fig. 15) (Rosenthal *et al.* 1998). The Kurnub-type

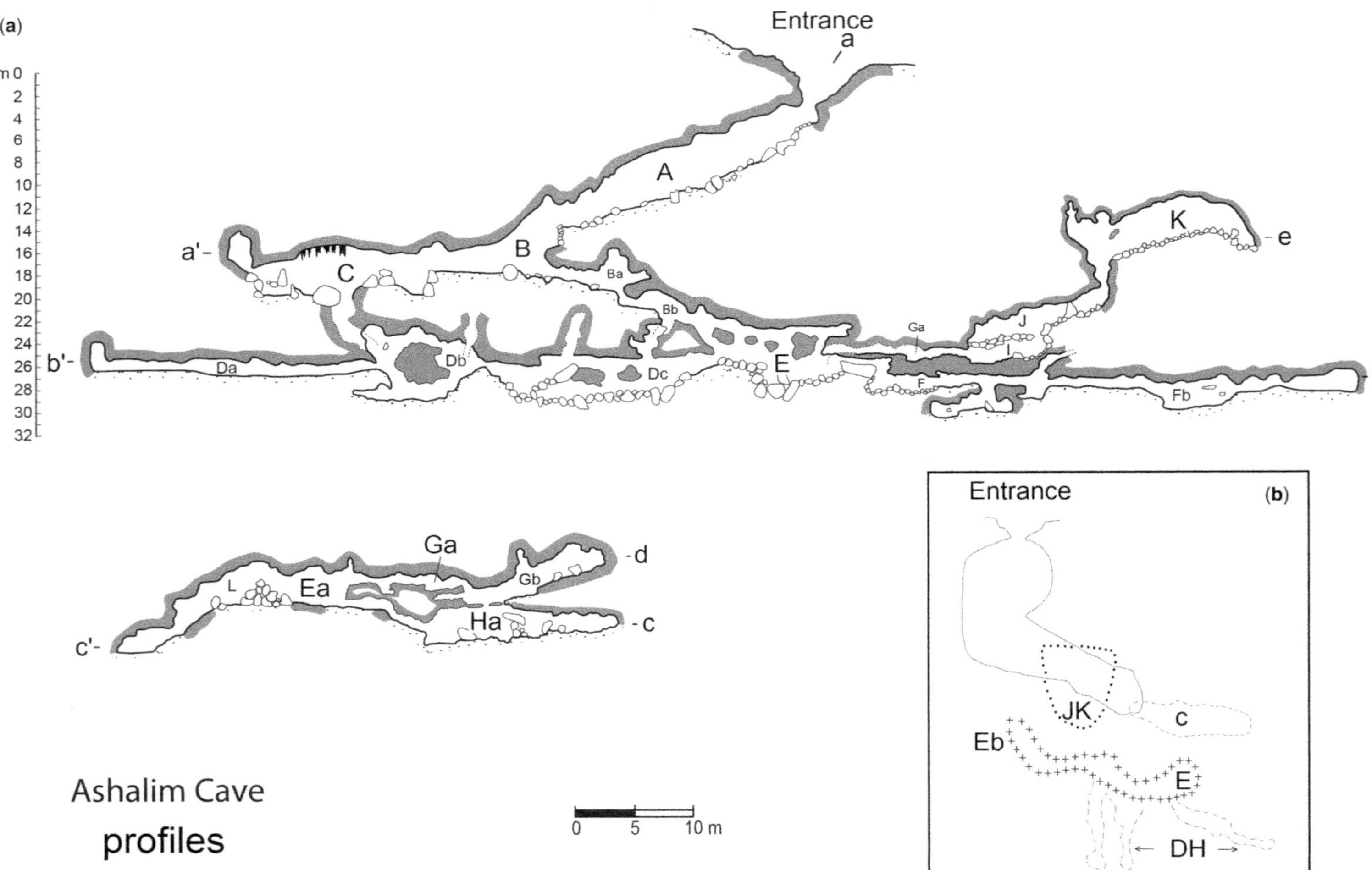

Fig. 5. Ashalim Cave profile. Surveyed by B. Langford, M. Ullman, N, Fishbein, L. Buchman, Israel Cave Research Center (2011). **(a)** Profiles along the main cave axis; **(b)** vertical projection of the cave perpendicular to its axis, view to azimuth 155°.

Fig. 6. Cupola at Ashalim Cave. Photograph by Amos Frumkin.

water gradually mixes with Judea-type water flowing in the Judea Group to the SW (Fig. 1). The contrasting water with intermediate mixing is best observed between Nizana 1 well, SW of Ashalim Cave, and Tel Shoqet 4 Well, NE of Ashalim Cave (Fig. 15).

Fig. 7. Solution pockets at Ashalim Cave. Locally resembling scallops, these are isolated and variable in size. Photograph by Amos Frumkin.

Fig. 8. Morphology indicating possible bubble trail at Ashalim Cave. Photograph by Amos Frumkin.

Fig. 9. Lower passage of Ashalim Cave. Cross-section determined by a sloping fracture. Photograph by Lihi Buchman.

Rosenthal *et al.* (1992), Kronfeld *et al.* (1993), Weinberger *et al.* (1994) and Weinberger & Rosenthal (1994) suggested that the anomalous properties of the groundwater, particularly the elevated temperatures, indicate transverse rising flow from the Kurnub Group into the overlying Judea Group, where the waters of the two aquifers mix. The upward flow is facilitated by the confined higher pressure heads measured in the Kurnub Group aquifer and the deep reverse faults revealed by lithological and structural analysis of the subsurface (Fig. 16) (Weinberger & Rosenthal 1998).

An increase in transmissivity was observed in the Judea Group aquifer between Revivim and Mashabim wells in the vicinity of Ashalim Cave (Fig. 15) (Weinberger & Rosenthal 1998). A geophysical study of wells across the aquifer, using resistivity, gamma ray, caliper and acoustic logs,

Fig. 10. A smooth-walled shaft at Ashalim Cave, probably formed by rising water. Photograph by Boaz Langford.

Fig. 11. Gypsum crust and evaporitic 'flowers' at the lower passage of Ashalim Cave. Note horizontal bottom line of flowers, indicating air stratification. Photograph by Lihi Buchman.

indicated that the Ashalim vicinity has more permeable fractured zones than most of the wells of the NW Negev Desert towards the Mediterranean coast (Laskow *et al.* 2011).

Discussion and conclusions

This paper presents evidence for past rising water along the presently known area of rising water, forming hypogene karst. This water must have been aggressive during cave formation. Analysis of the present hydrogeology suggests that the most probable palaeohydrological scenario during speleogenesis was similar to the present configuration: rising water from the Kurnub Group invaded the Judea Group under confined conditions, developing hypogene karst, as demonstrated by Ashalim Cave.

The palaeohydrogeological setting is assumed to have been generally similar to the present setting:

Fig. 12. Solutional v. depositional morphology at Ashalim Cave. Top-left: smooth walls and pockets with solutional morphology. Bottom-right: calcite speleothems deposited from down-flowing vadose water. Photograph by Lihi Buchman.

Fig. 13. Speleothems truncated by condensation corrosion. Photograph by Amos Frumkin.

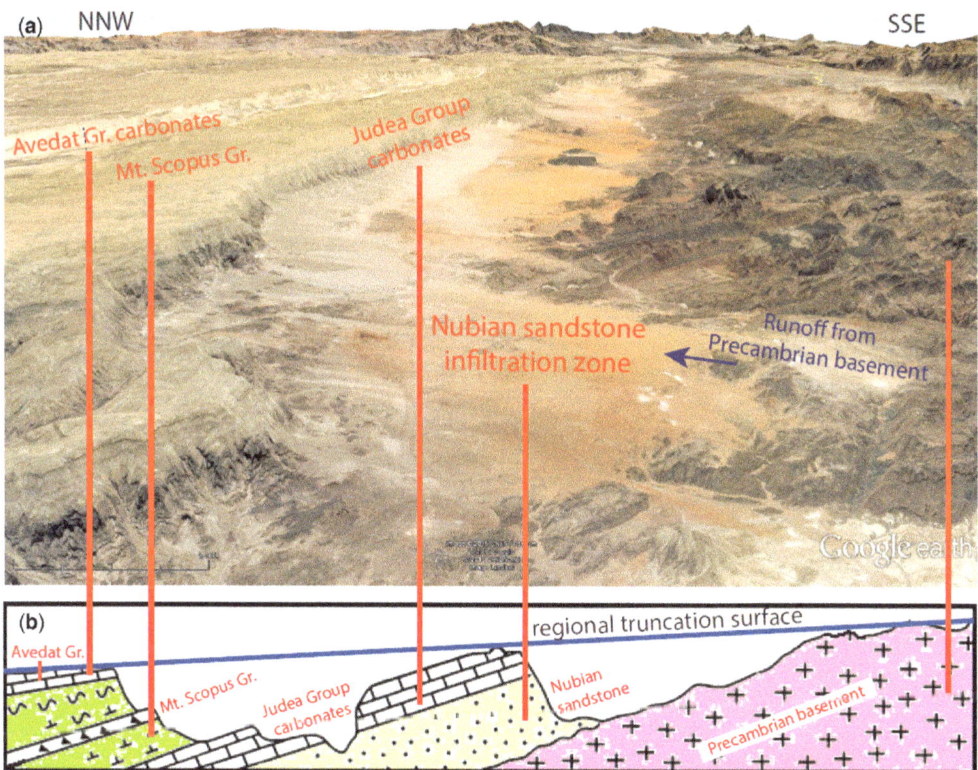

Fig. 14. Hydrogeological setting of central Sinai, *c.* 250 km SW of Ashalim Cave, as a possible recharge zone for the hypogenic water at Ashalim. Ashalim and most Negev caves are in the Judea Group, and some are in the Avedat Group. (**a**) Google Earth view of central Sinai showing an example of runoff from Precambrian basement flowing to Nubian Sandstone outcrops, where recharge of the aquifer takes place. Background courtesy of Google. (**b**) Schematic geological section of the area shown in part (a). Section length *c.* 200 km. Modified after Avni *et al.* (2012). For location, see inset in Figure 1.

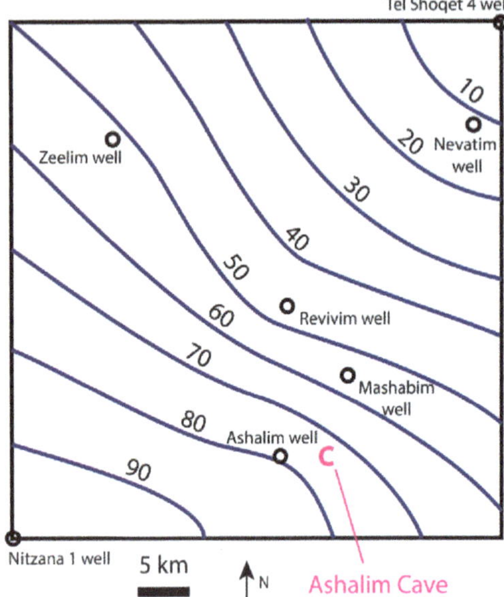

Fig. 15. Percentage of deep groundwater rising from the Nubian Sandstone into the modern Judea Group aquifer near Ashalim. The deep water rising near Nitzana 1 well is diluted towards the NE, where Judea Group groundwater dominates. Modified after Rosenthal *et al.* (1998). For location, see Figure 1.

Kurnub Group groundwater flowing northward from far-field recharge areas in the south and mixing with Judea Group groundwater, which flows southward from the Judea Mountains. If we assume that speleogenesis predated the formation of the Dead Sea Rift–Gulf of Aqaba depression, then the Kurnub Group groundwater palaeo-flow at Ashalim region could have been stronger than today due to:

(1) a larger recharge zone of Kurnub Group sandstone outcrops with gradients towards the NW Negev Desert, including regions around the present Gulf of Aqaba and the Red Sea, as well as the central Negev anticlines of Ramon and Hazera; and
(2) the Kurnub Group groundwater presently discharging into the Dead Sea Rift base level (Issar *et al.* 1972) would have flowed towards the Mediterranean, partly through the vicinity of Ashalim Cave.

The following processes may account for the increase in aggressiveness of the water within the Shivta Formation (Frumkin & Gvirtzman 2006; Auler 2013):

(1) mixing of the waters of the two aquifers;
(2) the addition of salt to 'normal' bicarbonate water, inducing undersaturation by the ionic strength effect;
(3) hydrogen sulphide degassing and the condensation of sulphuric acid just above the water table;
(4) the aggressiveness of Kurnub Group water flowing without being buffered through sandstone; and
(5) the cooling of hydrothermal waters;

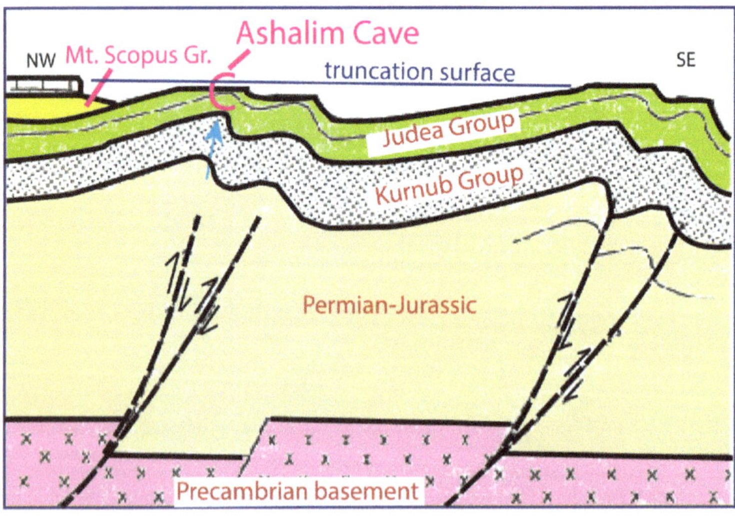

Fig. 16. Schematic cross-section of the Syrian Arc structures near Ashalim Cave and their relation with deep reverse faults, allowing upward flow of deep groundwater. Modified after Flexer *et al.* (2005). Arrow indicates possible palaeo-upwelling flow route to Ashalim Cave. For location, see Figure 2.

The Ashalim Cave therefore acts as a window to understanding past groundwater flow in a desert zone in which the groundwater flow was recharged a long distance away. The general flow in the Ashalim region today seems similar, although the groundwater level has dropped and some of the Kurnub Group waters currently flow eastward to the Dead Sea Rift.

The Ashalim Cave was surveyed by Boaz Langford, Micka Ullman, Vladimir Buslov, Nevo Fishbein and Lihi Buchman of the Israel Cave Research Center, The Hebrew University of Jerusalem. Alon Amrani of the Hebrew University analysed the sulphur isotopes of speleothems. Alexander Klimchouk and an anonymous reviewer considerably improved the manuscript with their remarks.

References

ALMOND, D.C. 1986. Geological evolution of the Afro-Arabian dome. *Tectonophysics*, **131**, 301–332.

AUDRA, Ph. & PALMER, A.N. 2015. Research frontiers in speleogenesis. Dominant processes, hydrogeological conditions and resulting cave patterns. *Acta Carsologica*, **44**, 315–348.

AULER, A S. 2013. Sources of water aggressiveness – the driving force of karstification. *In*: SHRODER, J. & FRUMKIN, A. (eds), *Treatise on Geomorphology*. Karst Geomorphology, **6**. Academic Press, San Diego, CA, 23–28.

AVNI, Y., SEGEV, A. & GINAT, H. 2012. Oligocene regional denudation of the northern Afar dome: pre-and synbreakup stages of the Afro-Arabian plate. *Geological Society of America Bulletin*, **124**, 1871–1897.

BAR, O., ZILBERMAN, E., FEINSTEIN, S., CALVO, R. & GVIRTZMAN, Z. 2016. The uplift history of the Arabian Plateau as inferred from geomorphologic analysis of its northwestern edge. *Tectonophysics*, **671**, 9–23.

BARTOV, Y., ARKIN, Y. & STEINITZ, G. 1976. The Zin fault: an example of Senonian faulting in northern Negev of Israel. *Israel Journal of Earth Sciences*, **25**, 40–44.

COHEN, R. 1971. A stalactite cave near Nahal Zalzal. *Hadashot Arkheologiyot*, **37**, 28 [in Hebrew].

DE WAELE, J., AUDRA, P. *ET AL.* 2016. Sulfuric acid speleogenesis (SAS) close to the water table: examples from southern France, Austria, and Sicily. *Geomorphology*, **253**, 452–467.

DUBLYANSKY, Y.V. 2013. Karstification by geothermal waters. *In*: SHRODER, J. & FRUMKIN, A. (eds) *Treatise on Geomorphology*. Karst Geomorphology, **6**. Academic Press, San Diego, CA, 57–71.

FLEXER, A., HIRSCH, F. & HALL, J.K. 2005. Tectonic evolution of Israel. *In*: HALL, J.K., KRASHENINNIKOV, V.A., HIRSCH, F., BENJAMINI, C. & FLEXER, A. (eds) *Geological Framework of the Levant*. Vol. **2**. The Levantine Basin and Israel. Historical Productions-Hall, Jerusalem, 523–537.

FRUMKIN, A. & FISCHHENDLER, I. 2005. Morphometry and distribution of isolated caves as a guide for phreatic and confined paleohydrological conditions. *Geomorphology*, **67**, 457–471.

FRUMKIN, A. & GVIRTZMAN, H. 2006. Cross-formational rising groundwater at an artesian karstic basin: the Ayalon Saline Anomaly, Israel. *Journal of Hydrology*, **318**, 316–333.

FRUMKIN, A., LANGFORD, B., LISKER, S. & AMRANI, A. (2017*a*) Hypogenic karst at the Arabian platform margins: implications for far-field groundwater systems: *Bulletin of the Geological Society of America*, https://doi.org/10.1130/B31694.1

FRUMKIN, A., LANGFORD, B. & PORAT, R. 2017*b*. The Judean desert – the major hypogene cave region of the southern Levant. *In*: KLIMCHOUK, A., PALMER, A., DE WAELE, J., AULER, A. & AUDRA, P. (eds) *Hypogene Karst Regions and Caves of the World*. Springer, Berlin, 463–477.

GALDENZI, S. & MARUOKA, T. 2003. Gypsum deposits in the Frasassi Caves, central Italy. *Journal of Cave and Karst Studies*, **65**, 111–125.

GAT, J.R. & DANSGAARD, W. 1972. Stable isotope survey of the fresh water occurrences in Israel and the northern Jordan Rift Valley. *Journal of Hydrology*, **16**, 177–211.

GAT, J.R. & ISSAR, A. 1974. Desert isotope hydrology: water sources of the Sinai Desert. *Geochimica et Cosmochimica Acta*, **38**, 1117–1131.

GURALNIK, B., MATMON, A., AVNI, Y. & FINK, D. 2010. ^{10}Be exposure ages of ancient desert pavements reveal Quaternary evolution of the Dead Sea drainage basin and rift margin tilting. *Earth and Planetary Science Letters*, **290**, 132–141.

HOSE, L.D. 2013. Karst geomorphology: sulfur karst processes. *In*: SHRODER, J. & FRUMKIN, A. (eds) *Treatise on Geomorphology*. Karst Geomorphology, **6**. Academic Press, San Diego, CA, 29–37.

ISSAR, A. 1981. The rate of flushing as a major factor in deciding the chemistry of water in fossil aquifers in southern Israel. *Journal of Hydrology*, **54**, 285–295.

ISSAR, A., BEIN, A. & MICHAELI, A. 1972. On the ancient water of the Upper Nubian Sandstone aquifer in central Sinai and southern Israel. *Journal of Hydrology*, **17**, 353–379.

KLIMCHOUK, A. 2013. Hypogene speleogenesis. *In*: SHRODER, J. & FRUMKIN, A. (eds) *Treatise on Geomorphology*. Karst Geomorphology, **6**. Academic Press, San Diego, CA, 220–240.

KRONFELD, J., ROSENTHAL, E., WEINBERGER, G., FLEXER, A. & BERKOWITZ, B. 1993. The interaction of two major old water bodies and its implication on the exploitation of groundwater in the multiple aquifer system of central and northern Negev, Israel. *Journal of Hydrology*, **143**, 169–190.

LASKOW, M., GENDLER, M., GOLDBERG, I., GVIRTZMAN, H. & FRUMKIN, A. 2011. Deep confined karst detection, analysis and paleo-hydrology reconstruction at a basin-wide scale using new geophysical interpretation of borehole logs. *Journal of Hydrology*, **406**, 158–169.

MATMON, A., FINK, D., DAVIS, M., NIEDERMANN, S. & ROOD, D. 2014. Unraveling rift margin evolution and escarpment development ages along the Dead Sea fault using cosmogenic burial ages. *Quaternary Research*, **82**, 281–295.

NAAMAN, I., DIMENTMAN, H. & FRUMKIN, A. 2014. Active hypogene speleogenesis in a regional karst aquifer. *In*: KLIMCHOUK, A., SASOWSKY, I.D., MYLROIE, J., ENGEL,

S.A. & Engel, A.S. (eds) *Hypogene Cave Morphologies*. Karst Water Institute, Special Publications, **18**, 73–74.

Palmer, A.N. 2013. Sulfuric acid caves: morphology and evolution. *In*: Shroder, J. & Frumkin, A. (eds) *Treatise on Geomorphology*. Karst Geomorphology, **6**. Academic Press, San Diego, CA, 241–257.

Rosenthal, E., Weinberger, G., Berkowitz, B., Flexer, A. & Kronfeld, J. 1992. The Nubian Sandstone aquifer in the western Negev, Israel: delineation of the hydrogeological model under conditions of scarce data. *Journal of Hydrology*, **132**, 107–135.

Rosenthal, E., Jones, B.P. & Weinberger, G. 1998. The chemical evolution of Kurnub Group paleowater in the Sinai–Negev Province – a mass balance approach. *Applied Geochemistry*, **13**, 553–569.

Sheffer, N., Dafny, E., Gvirtzman, H., Navon, S., Frumkin, A. & Morin, E. 2010. Hydrometeorological daily recharge assessment model (DREAM) for the Western Mountain Aquifer, Israel: model application and effects of temporal patterns. *Water Resources Research*, **46**, W05510.

Thorweihe, U. & Heinl, M. 2002. *Groundwater Resources of the Nubian Aquifer System, NE-Africa*. Technical University, Berlin.

Vaks, A., Bar-Matthews, M., Matthews, A., Ayalon, A. & Frumkin, A. 2010. Middle–Late Quaternary paleoclimate of northern Saharan–Arabian Desert: reconstruction from speleothems of Negev Desert, Israel. *Quaternary Science Reviews*, **29**, 1201–1211.

Vaks, A., Woodhead, J., *et al.* 2013. Pliocene–Pleistocene climate of the northern margin of Saharan–Arabian Desert recorded in speleothems from the Negev Desert, Israel. *Earth and Planetary Science Letters*, **368**, 88–100.

Vengosh, A., Hening, S., Ganor, J., Mayer, B., Weyhenmeyer, C.E., Bullen, T.D. & Paytan, A. 2007. New isotopic evidence for the origin of groundwater from the Nubian Sandstone Aquifer in the Negev, *Israel. Applied Geochemistry*, **22**, 1052–1073.

Weinberger, G. & Rosenthal, E. 1994. The fault pattern in the northern Negev and northern coastal plain of Israel and its hydrogeological implications for ground-water flow in the Judea Group aquifer. *Journal of Hydrology*, **155**, 103–124.

Weinberger, G. & Rosenthal, E. 1998. Reconstruction of natural groundwater flow paths in the multiple aquifer system of the northern Negev (Israel), based on lithological and structural evidence. *Hydrogeology Journal*, **6**, 421–440.

Weinberger, G., Rosenthal, E., Ben Zvi, A. & Zeitoun, D. G. 1994. The Yarkon–Taninim groundwater basin, Israel. Hydrogeology, case study and critical review. *Journal of Hydrology*, **161**, 227–255.

Yahalom-Mack, N., Langgut, D. *et al.* 2015. The earliest lead object in the Levant. *PloS One*, **10**, 1–14, https://doi.org/10.1371/journal.pone.0142948

Pliocene–Pleistocene palaeoclimate reconstruction from Ashalim Cave speleothems, Negev Desert, Israel

ANTON VAKS[1]*, MIRYAM BAR-MATTHEWS[1], AVNER AYALON[1], ALAN MATTHEWS[2] & AMOS FRUMKIN[2]

[1]*Geological Survey of Israel, 30 Malchei Israel Street, Jerusalem 9550161, Israel*

[2]*Institute of Earth Sciences, Hebrew University of Jerusalem, Jerusalem 9190401, Israel*

**Correspondence: antonv@gsi.gov.il*

Abstract: Speleothems from Ashalim Cave, located in the arid central Negev Desert, Israel, were used in a reconstruction of the palaeoclimate of the northern Saharan–Arabian desert margin. The sequence of speleothems is composed of three stratigraphic members: the yellow Pliocene Basal Member, the brown Early Pleistocene Intermediate Member and the thin Middle–Late Pleistocene Young Member. The age of the Basal Member is *c.* 3.1 Ma and the base of the Intermediate Member is 1.272 ± 0.018 Ma. Two last deposition periods of the Young Member occurred at 221–190 ka (Negev Humid Period (NHP) 2) and 134–114 ka (NHP-1), associated with interglacial marine isotopic stages 7.3–7.1 and 5.5, respectively. NHP-1 and -2 occurred when the African monsoon index was highest in the last 221 ka. The $\delta^{18}O$ values of the speleothems range between −6.9 and −11.2‰, 2–4‰ less than in the speleothems of central and northern Israel. This may indicate a remote southern tropical source of precipitation, although during NHP-1 and -2 the thickness of the Negev Desert speleothems decreases from north to south, showing a stronger northern Mediterranean source of moisture. The $\delta^{13}C$ values of the speleothems (3.5 to −8.5‰) show steppe to semi-desert C4 type vegetation. The $^{87}Sr/^{86}Sr$ ratios of the speleothems increased from *c.* 0.7078 in the Pliocene to 0.7082–0.7085 in the Pleistocene, indicating an increasing supply of desert dust and a decrease in host rock weathering.

Ashalim Cave is a karstic cave containing secondary carbonate cave deposits (speleothems). It is located in the central Negev Desert, which is part of the northern boundary of the Saharan–Arabian desert belt (Fig. 1a). Although this desert belt is the largest and one of the most arid regions in the world, its climatic history has been punctuated by numerous humid periods since its formation around *c.* 3 myr ago (deMenocal 1995, 2004; Vaks *et al.* 2010, 2013). The northern margin of the Saharan–Arabian desert belt currently receives precipitation predominantly from mid-latitude Atlantic–Mediterranean cyclones (Dayan 1986), whereas the southern margin derives most of its moisture from the African monsoon (Osborne *et al.* 2008; Fleitmann *et al.* 2011). Many factors affecting the palaeoclimate of this region are still not well understood.

Speleothems potentially provide one of the most valuable archives of palaeoclimatic data because of their ability to be dated with high precision and accuracy and the potential for high-resolution stable isotope records. Vadose speleothems (speleothems formed above groundwater level) grow in caves when water penetrates into the unsaturated zone and surface vegetation is present to supply the CO_2 necessary for limestone dissolution (Hendy 1971; Schwarcz 1986). They do not grow in arid/hyper-arid deserts where water and soil CO_2 are depleted (Holmgren *et al.* 1995; Fleitmann *et al.* 2003; Vaks *et al.* 2010). Thus speleothem deposition is an indication that seasons with positive effective precipitation (precipitation – (evaporation + runoff)) have occurred above the cave, leading to the seepage of water through the vadose zone.

Speleothem deposition was continuous in the Mediterranean climate areas of Israel (Soreq, Jerusalem and the Peqi'in caves; Fig. 1b, c) during the last 240 ka and still continues today (Frumkin *et al.* 2000; Bar-Matthews *et al.* 2003), whereas speleothem formation became more intermittent in the deserts towards the south of the country (Vaks *et al.* 2003, 2006, 2007, 2010, 2013; Lisker *et al.* 2010).

This paper reports the timing of humid periods from the speleothems record in Ashalim Cave, which is located in an arid desert with an annual mean precipitation of 100–120 mm. No speleothem deposition occurs today in caves to the south of the 300 mm annual mean isohyet (Fig. 1), but the existence of old speleothems in the cave shows that water seeped into the cave in the past. Ashalim Cave and adjacent caves span the transition zone between semi-arid steppe to the north and hyper-arid desert to the south (Vaks *et al.* 2007, 2010, 2013; Fig. 1b, c). The caves used in this study are all

From: PARISE, M., GABROVSEK, F., KAUFMANN, G. & RAVBAR, N. (eds) 2018. *Advances in Karst Research: Theory, Fieldwork and Applications*. Geological Society, London, Special Publications, **466**, 201–216.
First published online December 15, 2017, https://doi.org/10.1144/SP466.10

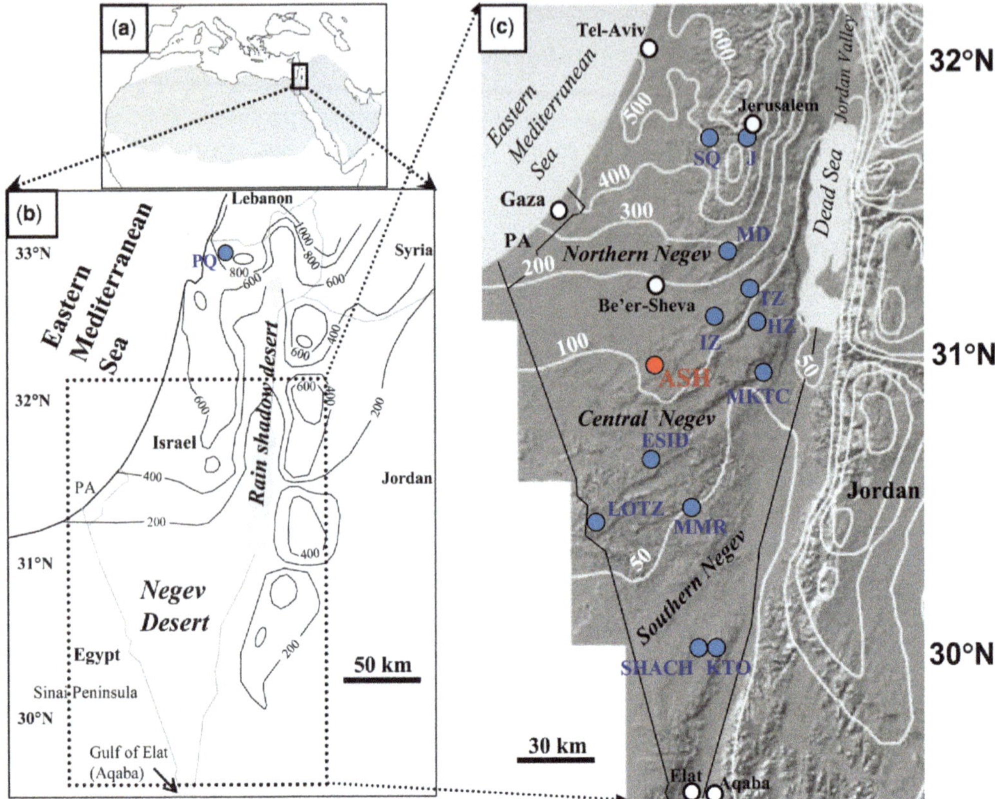

Fig. 1. Geographical location of the study area, annual precipitation and the location of Ashalim Cave relative to other major caves in which palaeoclimate studies have been performed. (**a**) Map indicating the extent of the Saharan–Arabian desert belt (grey shading). The rectangle marks the research area. (**b**) Annual rainfall map of Israel and adjacent lands: Palestinian Authority in Gaza (PA), northeastern Egypt, western Jordan, southwestern Syria and southern Lebanon. Isohyets are indicated by black lines. Peqi'in Cave (PQ) (Bar-Matthews *et al.* 2003) in northern Israel is marked. The dotted rectangle is enlarged in part (c). (**c**) The major research area showing the location of Ashalim Cave (ASH, red circle) relative to other studied caves (blue circles): ESID, Even-Sid mini-caves; HZ, Hol-Zakh Cave; IZ, Izzim Cave; J, Jerusalem Cave; KTO, Ktora Cracks LOTZ, Wadi Lotz Cave; MD, Ma'ale-Dragot Caves; MKTC, Makhtesh-ha-Qatan Cave; MMR, Ma'ale-ha-Meyshar Cave; SHACH, Shizafon mini-caves; SQ, Soreq Cave; TZ, Tzavoa Cave. Isohyets are shown by white lines.

located in the highlands, so each cave only has a limited recharge zone on the top of the hills above the cave, sampling only the direct rainfall/snowmelt there. No cave located beneath the stream channels (active or dry) was studied.

This paper summarizes the palaeoclimatic reconstruction from Ashalim Cave in the context of cave formation processes discussed elsewhere (Frumkin & Langford 2017). The paper focuses on two main topics:

(1) the reconstruction of the palaeoclimate of the arid area of Israel by:
  - determining the timing of speleothem growth (i.e. periods of positive effective precipitation) using U series chronology;
  - defining the origin of precipitation using spatial variations in the speleothem record along a north–south transect from the Mediterranean sub-humid zone in the north to the hyper-arid region in the south;
  - reconstructing the vegetation type above the cave by $\delta^{13}C$ values in the speleothems;

(2) understanding the links between the local palaeoclimatic history and regional/global palaeoclimatic changes in the Pliocene and Pleistocene.

Cave settings and climate

Ashalim Cave is located in the central Negev Desert (30° 56′ 36.2″ N, 34° 44′ 22.5″ E), 414 m above sea-level in karstified limestone rocks of the Turonian (Middle Cretaceous) Shivta Formation. The cave is 67 km SW of the coast of the Eastern Mediterranean Sea. The cave is a three-dimensional hypogene maze, 31 m deep and with a total length of 540 m (Fig. 2). The palaeo-hydrogeological setting during speleogenesis involved deep groundwater flowing northwards from recharge areas in the south, rising and mixing with waters from the Judea Group aquifer, which flowed southwards from the Judean Mountains (Frumkin & Langford 2017). The cave is richly decorated with vadose speleothems. The thickness of the speleothems varies from several centimetres to a few tens of centimetres. The soil above the cave is silicate loess originated mainly from aeolian dust (Crouvi *et al.* 2010) and the present day vegetation is composed of sparse xeric shrubs with <10% vegetation cover.

The northern and central Negev Desert receives most of its present day rainfall during the winter months from mid-latitude Atlantic–Mediterranean cyclones moving eastwards above the Eastern

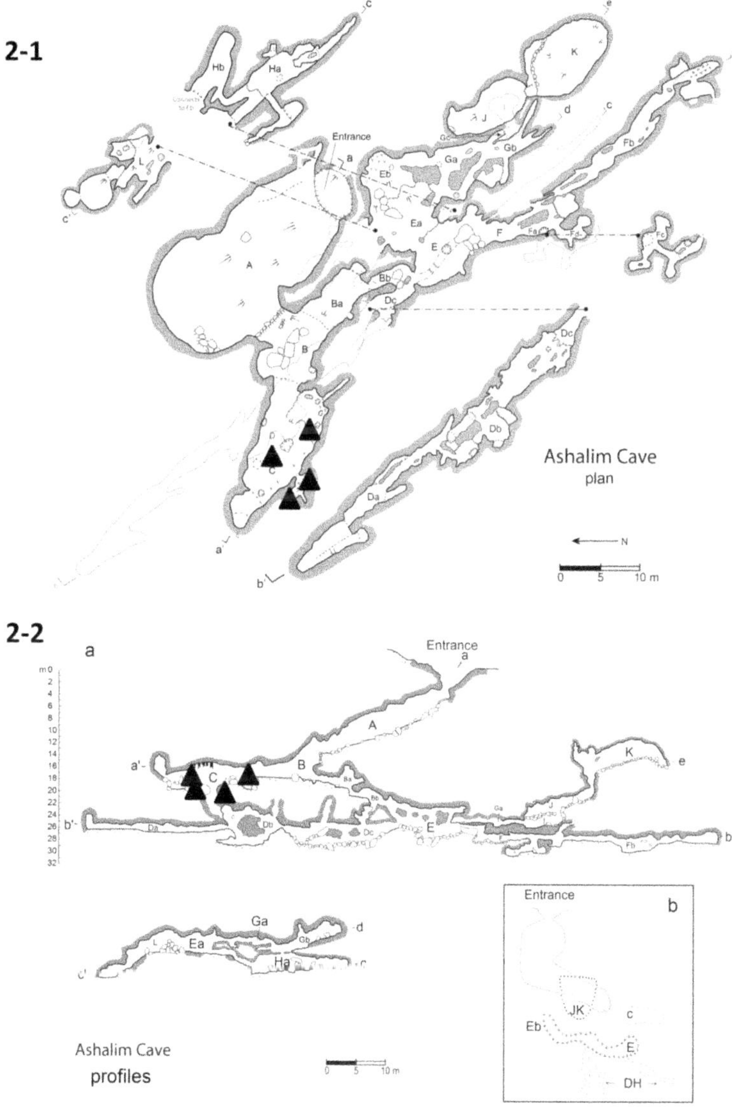

Fig. 2. 2–1. Map of Ashalim Cave showing the original location of the speleothems used for the study (marked as black triangles). 2–2. Vertical section of Ashalim Cave showing the original location of studied speleothems. Surveyed by B. Langford, M. Ullman, N. Fishbein and L. Buchmann (Frumkin & Langford 2017).

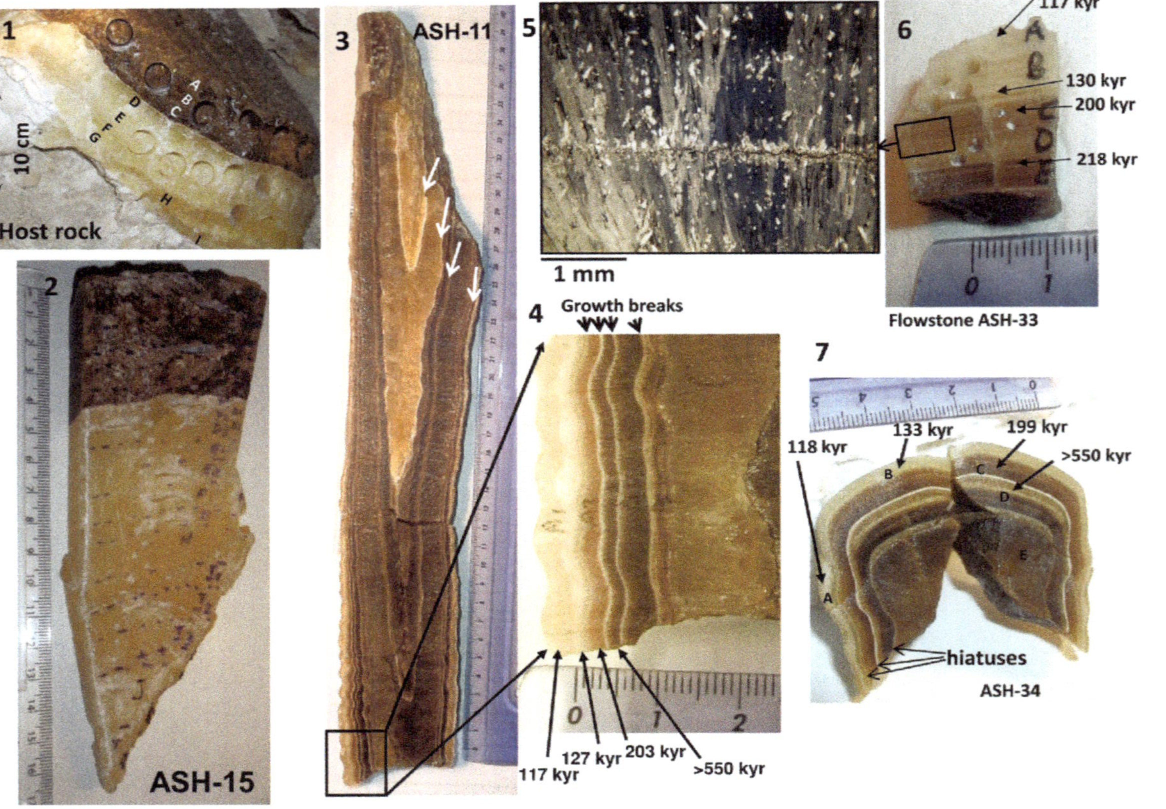

Fig. 3. Studied speleothems from Ashalim Cave. (**1**, **2**) Flowstone ASH-15 showing the massive yellow Basal Member (bottom section, layers D–I) composed of large (>2 cm) calcite crystals and the brown Intermediate Member (top section, A–C). The thin layer between the two stratigraphic members probably represents a growth break (hiatus). The slab in part (2) is the section in which isotopic profile was sampled. (**3**, **4**) Stalactite ASH-11. The Basal Member is yellow, the Intermediate Member is brown and the Young Member is composed of four thin layers; see enlargement in part (4) with U–Th ages. Numerous fine white layers divide between the thicker dark calcite layers; white arrows mark growth breaks. (**5**, **6**) Uppermost section of the Young Member in flowstone ASH-33. The magnification of layers C and D and the boundary between them under a petrographic microscope are shown in part (5). The entire section with U–Th ages is shown in part (6). (**7**) The Young Member in stalagmite ASH-34 with U–Th ages. The white layer between C and D is a long-term hiatus in growth.

Mediterranean Sea (Dayan 1986). Summers are hot and dry as a result of sinking air from the subtropical highs that develop over the Mediterranean Sea. The Eastern Mediterranean Sea is the major source of moisture for precipitation. As the air masses pass over the land, the supply of moisture and latent heat is dramatically reduced to the south and reduced more gradually to the east (Shay-El & Alpert 1991; Enzel *et al.* 2008). Thus the northern Sinai coastline defines the southern limit at which rain clouds can form (Zangvil & Druian 1990), resulting in a very sharp present day rainfall gradient that decreases from central Israel towards the Negev Desert in the south (Fig. 1b, c). Another source of precipitation is associated with synoptic systems bringing moisture from the tropical Atlantic Ocean. These systems approach the region from the SSW (Kahana *et al.* 2002) and mainly affect the southern parts of the Negev Desert. They are common at the beginning and end of the rainy season.

Methods

Five speleothems were sampled from chamber C at Ashalim Cave (Frumkin & Langford 2017), as shown in Figures 2 and 3. The samples were sectioned using a diamond saw to expose their internal structure and to identify and eliminate diagenetically altered samples (Bar-Matthews *et al.* 1997). The mineralogy and petrography were determined using a petrographic microscope, a Jeol 840 scanning electron microscope equipped with an Oxford ISIS energy-dispersive spectrometry system and a Philips PW 3020 X-ray diffractometer. The intervals of speleothem deposition were determined by counting the number of calcite layers divided by growth breaks (hiatuses).

Thirty-eight U–Th and five U–Pb ages were determined on five speleothems. Fifty-eight determinations of $\delta^{18}O$ and $\delta^{13}C$ and 20 determinations of the $^{87}Sr/^{86}Sr$ isotopic ratios were performed on the speleothems, the cave host rock and the overlying lithologies. The $^{87}Sr/^{86}Sr$ isotopic ratio was also measured in a loess soil sample from above the cave. Full descriptions of all procedures and techniques used in this study are given in Vaks *et al.* (2006, 2007, 2010, 2013).

Results

Petrography of the Ashalim vadose speleothems

The vadose speleothems of Ashalim Cave consist of low-Mg calcite stalagmites, stalactites and flowstones. Based on the stratigraphic position, layer thickness, colour and crystal size, the speleothems are divided into three main stratigraphic members: the Basal, Intermediate and the outermost thin Young members. The term 'Member' is used here because these major speleothem layers can be traced with the same ages among different caves. They define deposition under similar climatic conditions that prevailed in the central and southern parts of the Negev Desert. The oldest Basal Member is the thickest of the three stratigraphic members, varying from 5 to 25 cm in thickness and comprising *c.* 90% of the speleothem volume in the cave. It is composed of massive yellow calcite crystals (Fig. 3-1, 3-2), often showing continuous growth, suggesting deposition from continuously dripping water. Locally, the growth of stalactites is interrupted by thin layers of milky white or beige microcrystalline calcite or detrital material (Fig. 3). The Basal Member is terminated at its top by a <1 mm layer of microcrystalline calcite, evaporite minerals and reddish clays (Fig. 3-1, 3-2, 3-3). This terminal layer is interpreted as a hiatus (growth break) separating the Basal and Intermediate members (Vaks *et al.* 2013).

The Intermediate Member is stratigraphically younger than the Basal Member, 2–15 cm thick and comprises 9–10% of the speleothem volume in the caves. It is layered and is composed of alternating brown layers of calcite crystals (Fig. 3-1, 3-2, 3-3), with layer thicknesses of a few millimetres to 5 cm. Several layers <1 mm thick of milky white or beige microcrystalline calcite are found within the columnar crystalline structure, suggesting hiatuses in speleothem deposition (Vaks *et al.* 2013).

The Young Member (Fig. 3-4, 3-5, 3-6, 3-7) is 0.5–2 cm thick, comprises <1% of the speleothem volume and is completely missing in some speleothems (e.g. ASH-15). It is composed of thinly layered, light transmitting, columnar crystalline calcite (Fig. 3-5) of a brown or yellow colour, with multiple hiatuses, showing that the speleothems only grew during very short episodes. The hiatuses are variably coloured fine layers (<1 mm) (Fig. 3-4, 3-6, 3-7) composed of microcrystalline calcite, gypsum, halite, clays, quartz and hydroxides/oxides. This mineralogy is similar to the minerals forming a patina covering the surfaces of speleothems, suggesting that they formed due to condensation corrosion (Auler & Smart 2004; Dreybrodt *et al.* 2005) and/or to the accumulation of evaporites and detrital minerals, or both, that were washed into the cave during rare rain events and were then left on the speleothem surface after drying (Vaks *et al.* 2013).

U–Th and U–Pb chronology

The U–Th and U–Pb dating results are presented in Figures 4 and 5 and Tables 1 and 2. The U concentrations range between 1.9 and 19.7 ppm and the amounts of non-radiogenic Th are negligible.

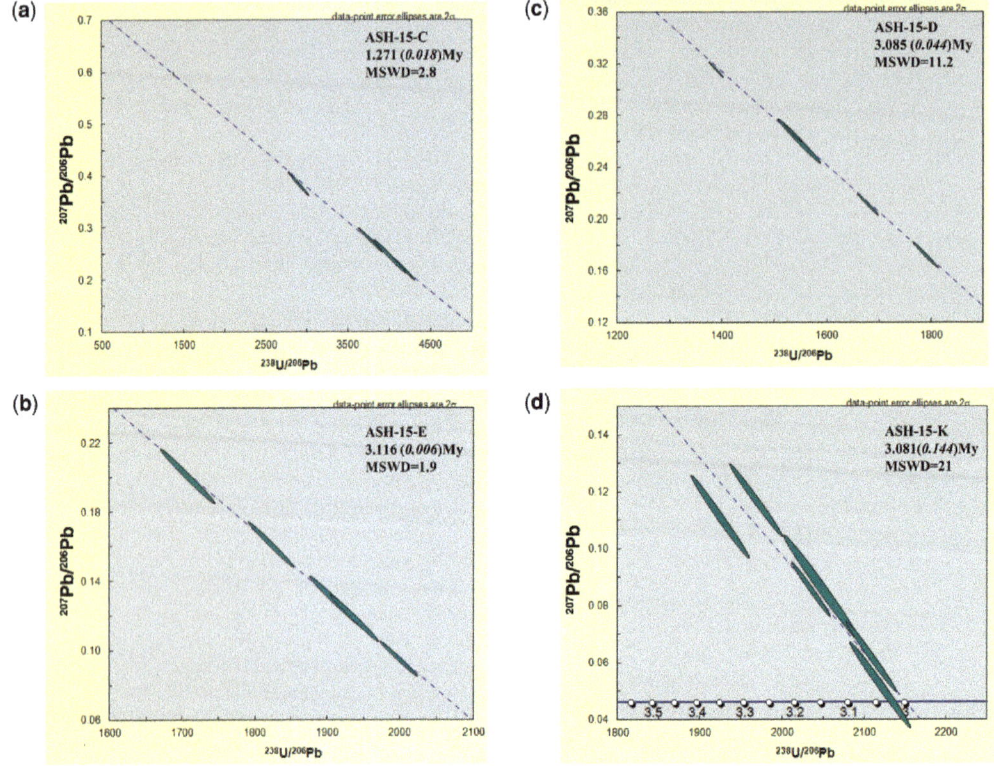

Fig. 4. U–Pb isochrons of four Ashalim Cave speleothem horizons with their ages and MSWD values. Analytical uncertainties are shown by ellipses (from Vaks *et al.* 2013).

U–Th ages determined on the layers of the Young Member show that speleothem deposition was highly discontinuous during the last 500 kyr. The upper layers of the Young Member grew in two periods of 221–190 ka and 134–114 ka, whereas the lower layers were older than 500 ka (Vaks *et al.* 2010). U–Pb ages were determined for the ASH-15 flowstone in two separate laboratories. The dating results obtained at the University of Melbourne (J. Woodhead, Fig. 4) show that the bottom layer C of the Intermediate Member is 1.272 ± 0.018 Ma, whereas the Basal Member gives ages of 3.085 ± 0.044, 3.116 ± 0.006 and 3.081 ± 0.144 Ma for layers D, E and K, respectively. Layer D was also dated at the University of Leeds (R.A. Cliff) to 3.005 ± 0.026 Ma (Vaks *et al.* 2013).

There is no speleothem growth between 185 and 135 ka and between 114 ka and the present day. For the older periods, the number of dates is not sufficient to determine the length of the older dry periods in this cave.

Most ages follow the stratigraphic order, with the outermost layers being the youngest. However, there are age reversals in some layers adjacent to hiatuses, with the most prominent example found in the flowstone layer ASH-33-E2.2 (Fig. 5b, Table 1). The proximity of these layers to hiatuses suggests that secondary U mobilization processes could have occurred and/or the deposition of secondary calcite. In some cases fragments of the host rock from the cave ceiling were identified, which could have resulted in older age measurements (Vaks *et al.* 2010).

$\delta^{18}O$ and $\delta^{13}C$ of speleothems and host rocks

δ^{18}O and δ^{13}C profiles of Ashalim Cave speleothems are plotted in Figure 6a (Vaks *et al.* 2010). The host rock δ^{18}O and δ^{13}C values are −6.1 to −5.8‰ and 0.35–1.0‰, respectively. δ^{18}O values of the Basal Member (ASH-15-D-K) vary from −9.7 to −11.2‰. The oldest layer of the Intermediate Member (ASH-15-C) shows δ^{18}O values from −9.3 to −9.5‰, whereas the younger layer ASH-15-A+B shows δ^{18}O values between −9.6 and −10‰. The average δ^{18}O values in the Young Member (speleothems ASH-11, -33 and -34) are usually higher by 1–2‰ than in the Intermediate and Basal members and are also more variable, fluctuating between −6.9

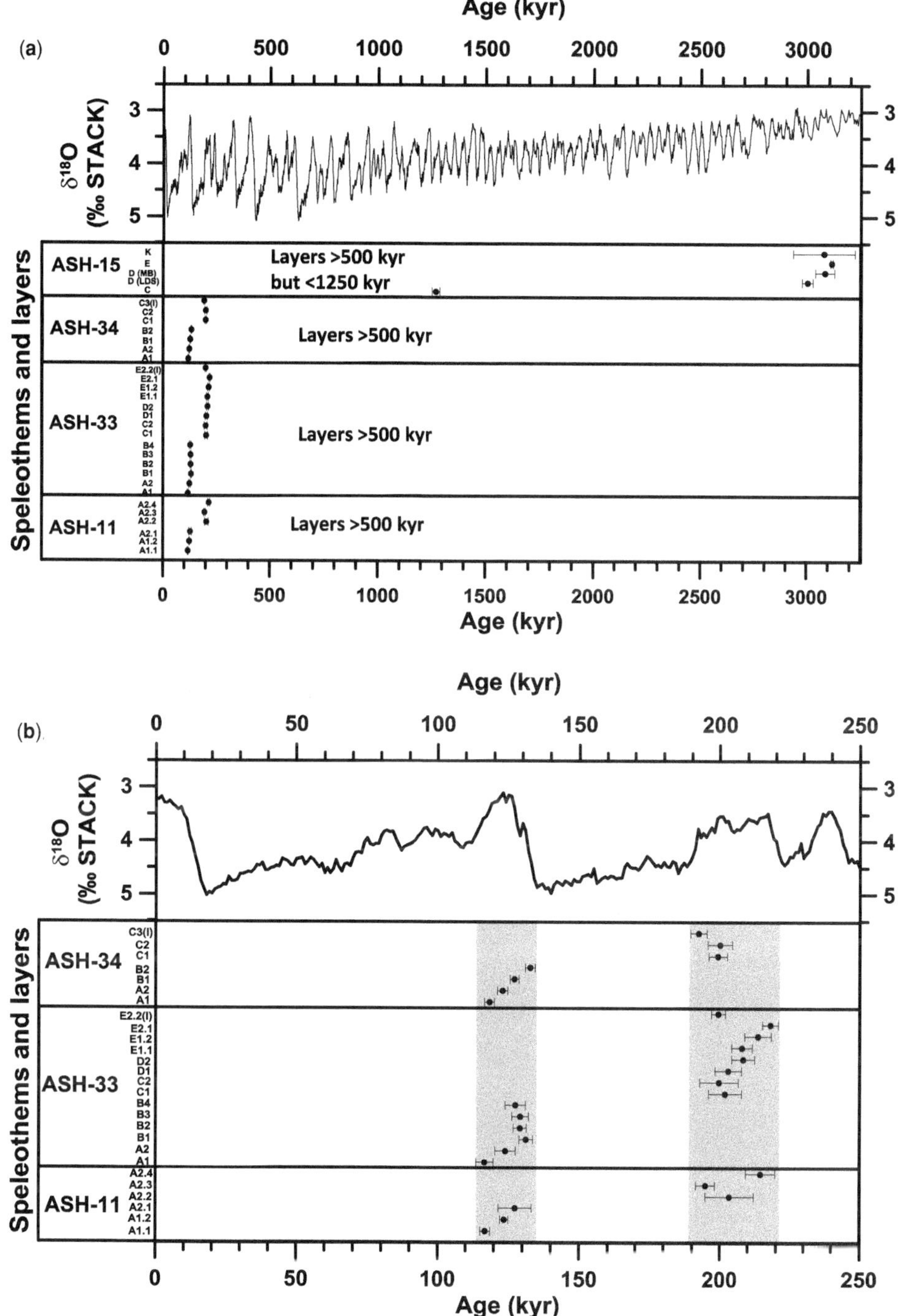

Fig. 5. (**a**) Periods of speleothem deposition in Ashalim Cave as determined by U–Th and U–Pb dating (Vaks *et al.* 2010, 2013). (**b**) Enlargement of the last 230 kyr. An oxygen benthic stack $\delta^{18}O$ record displaying the glacial and interglacial marine isotopic stages (Lisiecki & Raymo 2005) for the same period is shown above each plot for comparison. The horizontal axis marks the age (ka), whereas the vertical axis shows the individual speleothems and their layers. Ages are marked as black circles with horizontal error bars. The vertical grey rectangles in part (b) mark speleothem depositional periods.

Table 1. *U–Th ages of speleothems*

Sample description	Sub-sample number	Layer	Age (ka)	2σ	^{238}U (ppm)	$^{234}U/^{238}U$	2σ	$^{230}Th/^{234}U$	±2σ	^{230}Th (ppb)	^{232}Th (ppb)	$^{230}Th/^{232}Th$
Stalactite ASH-2	1	ASH-2-A	>500		8.416	0.99963	0.00151	0.99630	0.00522	0.13664	1.715	14 944
Stalactite (column) ASH-11	2	ASH-11-A1.1	116.9	1.8	3.891	1.03981	0.00095	0.66237	0.00275	0.0437	1.431	5761
	3	ASH-11-A1.2	123.6	1.4	4.055	1.02781	0.00078	0.68192	0.00208	0.0463	0.791	11 039
	4	ASH-11-A2.1	127.4	5.9	2.751	1.0434	0.00157	0.69474	0.00837	0.0325	0.953	6541
	5	ASH-11-A2.2	203.4	8.7	12.56	1.02947	0.00085	0.85177	0.00837	0.1795	1.113	30 363
	6	ASH-11-A2.3	194.8	3.4	12.56	1.03011	0.00083	0.83886	0.00254	0.1769	1.293	25 755
	7	ASH-11-A2.4	214.5	5.2	12.63	1.0271	0.00060	0.86654	0.00329	0.1833	1.328	25 979
	8	ASH-11-A3	>550*		11.92	1.01049	0.00122	1.03557	0.00207	0.2033	1.748	21 815
	9	ASH-11-A4	>550*		16.17	0.99917	0.00203	1.03400	0.00294	0.2723	2.384	21 428
Flowstone ASH-15	10	ASH-15-A+B	>550*		1.861	0.99939	0.00108	1.03532	0.00113	0.0314	2.394	2460
	11	ASH-15-C1	>550*		3.412	0.99925	0.00150	1.02507	0.00219	0.0570	0.766	13 956
Flowstone ASH-33	12	ASH-33-A1	116.7	3.1	3.298	1.03892	0.00209	0.66166	0.00471	0.03696	2.371	3013
	13	ASH-33-A2	124.0	3.6	3.371	1.02948	0.00191	0.68327	0.00513	0.03866	2.172	3508
	14	ASH-33-B1	131.2	2.4	3.583	1.02585	0.00103	0.70373	0.00322	0.04217	1.396	5805
	15	ASH-33-B2	129.1	2.3	4.156	1.02745	0.00088	0.69807	0.00320	0.04859	1.035	8969
	16	ASH-33-B3	129.2	2.9	4.161	1.02453	0.00087	0.69803	0.00406	0.04852	1.188	7780
	17	ASH-33-B4	127.5	3.6	3.634	1.02456	0.00138	0.69312	0.00506	0.04207	1.824	4447
	18	ASH-33-C1	201.9	6.0	9.854	1.0167	0.00080	0.84705	0.00419	0.13834	2.576	10 180
	19	ASH-33-C2	199.7	7.0	11.73	1.01826	0.00078	0.84413	0.00502	0.16434	2.306	13 498
	20	ASH-33-D1	203.0	4.7	11.53	1.01834	0.00088	0.84906	0.00323	0.16248	4.444	6934
	21	ASH-33-D2	208.3	4.1	11.00	1.01839	0.00117	0.85654	0.00256	0.15644	6.406	4629
	22	ASH-33-E1.1	207.9	3.6	12.37	1.01819	0.00061	0.85590	0.00238	0.17575	1.254	26 447
	23	ASH-33-E1.2	213.6	4.7	14.27	1.01996	0.00060	0.86394	0.00295	0.20495	2.627	14 713
	24	ASH-33-E2.1	218.0	2.8	13.29	1.01997	0.00073	0.86961	0.00161	0.19217	3.041	11 929
	25	ASH-33-E2.2(I)	199.5	2.5	15.59	1.02737	0.00110	0.84628	0.00226	0.22102	2.258	18 499
	26	ASH-33-E22(II)	199.7	4.2	14.63	1.02247	0.00081	0.84495	0.00295	0.20600	21.01	1850
	27	ASH-33-E22(III)	198.8	4.3	14.05	1.02368	0.00247	0.84368	0.00311	0.19785	5.700	6574
	28	ASH-33-F-up	>550		13.18	1.01930	0.00121	1.01496	0.00351	0.22235	2.589	16 219
Flowstone ASH-34	29	ASH-34-A1	118.4	1.7	3.805	1.04200	0.00078	0.66737	0.00264	0.04314	5.333	1526
	30	ASH-34-A2	122.9	1.9	3.626	1.05737	0.00108	0.68293	0.00272	0.04269	0.463	17 381
	31	ASH-34-B1	127.2	1.6	3.093	1.04600	0.00069	0.69434	0.00222	0.03662	1.510	4581
	32	ASH-34-B2	132.7	1.7	4.673	1.05782	0.00106	0.71139	0.00229	0.05732	1.542	7003
	33	ASH-34-C1	199.3	3.3	12.84	1.02963	0.00093	0.84575	0.00229	0.18229	10.75	3196
	34	ASH-34-C2	200.1	4.4	15.25	1.02501	0.00066	0.84605	0.00317	0.21560	14.47	2820
	35	ASH-34-C3(I)	192.3	2.9	19.68	1.02945	0.0007	0.83484	0.00222	0.27572	15.40	3370
	36	ASH-34-C3 (II)	188.4	6.9	13.99	1.02299	0.00095	0.82727	0.00557	0.19304	13.87	2636
	37	ASH-34-C3 (III)	194.6	4.3	15.19	1.02689	0.00231	0.83797	0.00296	0.21301	11.00	3667
	38	ASH-34-D	>550		16.45	1.01215	0.0013	0.99939	0.00431	0.27129	1.740	29 461

*Possible U loss.

Table 2. *U–Pb ages of speleothems*

Sample name	Weight (mg)	U (ppm)	Pb (ppm)	$Pb_{radiogenic}/Pb_{common}$	Sample (radiogenic + initial Pb) isotope ratios										Age assuming $^{234}U/^{238}U$ initial = 1 ± 0.1		Age assuming $^{234}U/^{238}U$ initial from measured disequilibrium			Young Member's $^{234}U/^{238}U$ initial	±2σ% (err)
					$^{238}U/^{206}Pb$	±2σ (%)	$^{207}Pb/^{206}Pb$	±2σ (%)	$^{204}Pb/^{206}Pb$	±2σ (%)	$^{238}U/^{204}Pb$	±2σ (%)	$^{206}Pb/^{204}Pb$	±2σ (%)	age (Ma)	95% confidence	age (Ma)	95% confidence	MSWD		
Melbourne data:																					
ASH-15-C	110	4.276339	0.001575	0.33	4184	2.50	0.221177	7.8	0.0169	6.2	2 47 427	8.7	59	6.2	1.288	0.04	**1.271**	**0.018**	2.8	1.047	0.0057
	130	3.271235	0.006419	0.11	1372	0.91	0.598244	0.5	0.0380	0.5	36 086	1.4	26.3	0.5							
	70	4.081154	0.001705	0.19	4000	3.93	0.245143	10.6	0.0200	7.6	1 99 517	11.5	50	7.6							
	90	3.849827	0.001781	0.19	3771	3.06	0.276776	7.0	0.0215	5.3	1 75 410	8.3	46.5	5.3							
	70	3.296282	0.00241	0.10	2893	3.52	0.386277	4.8	0.0285	3.7	1 01 370	7.2	35.0	3.7							
ASH-15-D	254	1.887356	0.002432	0.52	1389	0.81	0.316245	1.5	0.0182	1.8	76 295	2.6	54.9	1.8	3.103	0.058	**3.085**	**0.044**	11.2	1.047	0.0057
	129	1.569532	0.001681	0.68	1548	2.15	0.260713	5.3	0.0144	6.6	1 07 318	8.7	69	6.6							
	267	1.809399	0.001926	0.68	1536	0.89	0.266985	2.1	0.0149	2.6	1 03 030	3.5	67.1	2.6							
	267	1.799333	0.001568	0.99	1678	0.98	0.211816	3.2	0.0111	4.2	1 50 973	5.2	90.0	4.2							
	278	1.606193	0.001202	1.33	1789	1.13	0.172227	4.8	0.0085	6.7	2 11 143	7.8	118	6.7							
ASH-15-E	187	1.908065	0.001377	1.43	1823	1.44	0.161602	6.6	0.0078	9.4	2 34 643	10.8	129	9.4	3.134	0.038	**3.116**	**0.006**	1.9	1.047	0.0057
	225	2.361587	0.001305	3.50	1998	1.06	0.095048	9.0	0.0033	17.7	6 06 205	18.7	303	17.7							
	146	1.797302	0.001533	1.02	1708	1.82	0.201126	6.4	0.0104	8.4	1 63 742	10.2	96	8.4							
	174	1.879745	0.001193	2.05	1915	1.65	0.127174	10.1	0.0054	16.3	3 55 444	17.9	186	16.3							
	204	1.893589	0.001152	2.32	1938	1.41	0.118726	9.4	0.0048	15.7	4 01 005	17.1	207	15.7							
ASH-15-K	99	3.641304	0.002191	1.81	1924	1.50	0.111412	10.7	0.0057	13.9	3 37 505	15.4	175	13.9	3.098	0.148	**3.081**	**0.144**	21.0	1.047	0.0057
	102	2.99982	0.0016	3.79	2048	1.88	0.086013	18.0	0.0026	41.2	8 01 564	43.1	391	41.2							
	144	2.894629	0.00172	2.33	1969	1.32	0.117056	8.9	0.0048	14.8	4 08 486	16.1	207	14.8							
	137	3.011123	0.001377	11.24	2119	1.45	0.052263	23.6	0.0004	190.5	48 10 258	192.0	2270	190.5							
	189	2.591251	0.001218	7.73	2108	1.21	0.062107	16.4	0.0011	61.8	18 80 566	63.0	892	61.8							
	176	3.436165	0.001788	5.23	2035	0.95	0.085744	9.0	0.0022	24.8	9 40 603	25.7	462	24.8							
Leeds data:																					
ASH-15-D	231	2.20	0.0029	0.45	1388.00	0.64	0.3418	0.22	0.020461	0.08	67 756	0.60	48.87	0.16	3.022	0.045	**3.005**	**0.026**	0.0	1.0470	0.0057
	367	1.70	0.0059	0.13	735.00	2.00	0.5757	0.17	0.03721	0.15	19 750	0.30	26.87	0.15							

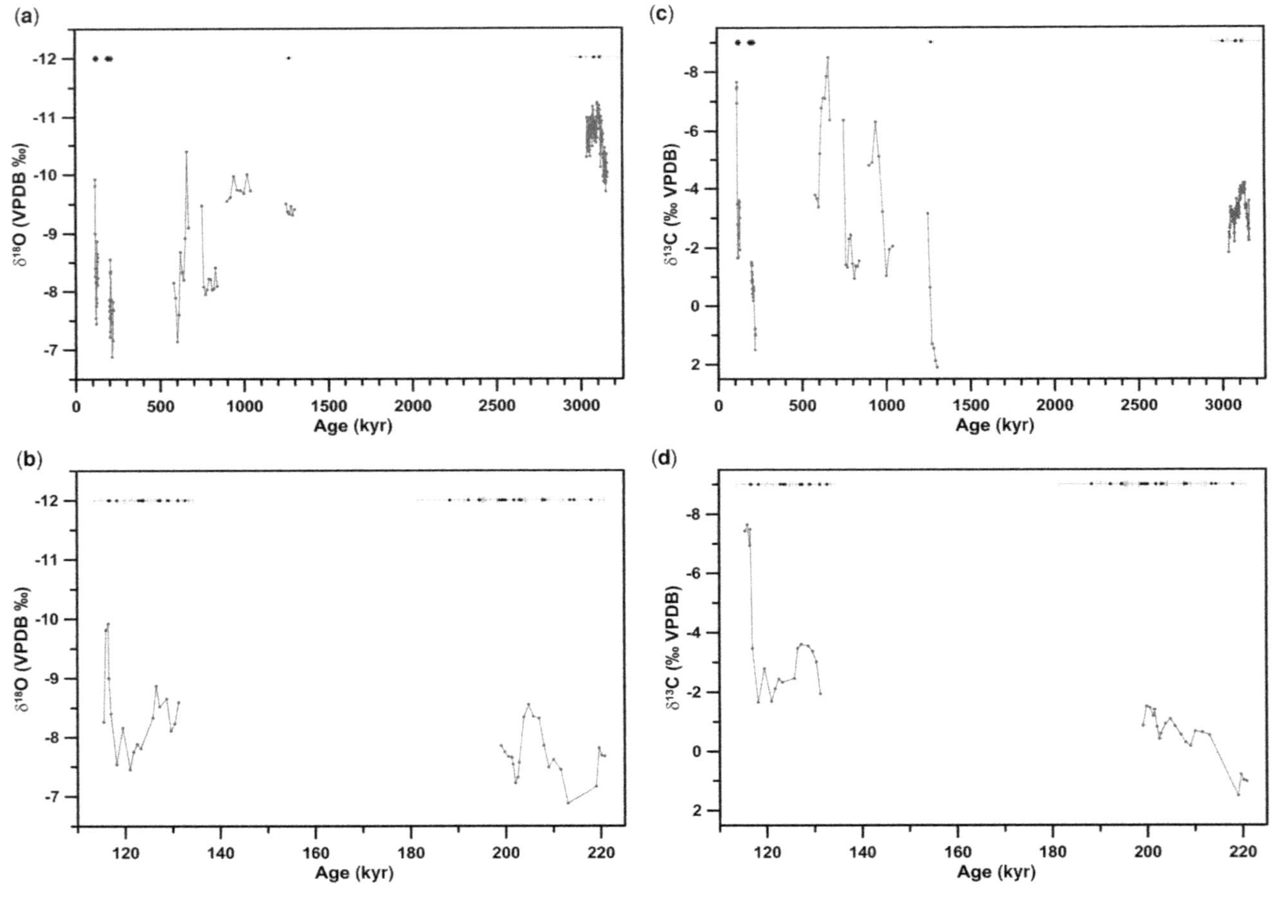

Fig. 6. Pliocene–present day (**a**) $\delta^{18}O$ and (**c**) $\delta^{13}C$ profiles in Ashalim Cave speleothems with (**b**, **d**) magnification of the last 230 kyr. U–Th and U–Pb ages with their uncertainties are given at the top of each plot. The duration of speleothem deposition periods older than 221 ka is arbitrary because of the uncertainties of U–Pb dating. The timing of speleothem layers between 1272 and 500 ka is based on their stratigraphic position in the speleothem sequence. The $\delta^{18}O$ and $\delta^{13}C$ values of the Young Member layers (<221 ka) are after Vaks *et al.* (2010).

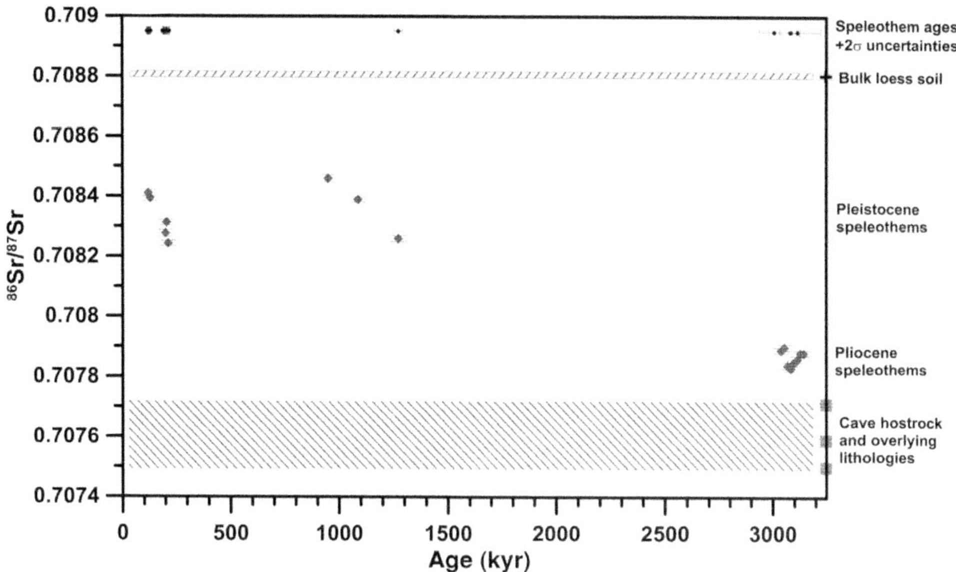

Fig. 7. $^{87}Sr/^{86}Sr$ ratios in speleothems (black diamonds), host rock/overlying rock formations (shaded rectangle in lower part of figure and the squares on the lower right) and dust-borne loess soil (shaded rectangle in the upper part of the figure and data point to the upper right). U–Th and U–Pb ages are shown on the top.

and −10.4‰ (Fig. 6b). $\delta^{13}C$ values in Ashalim speleothems range between 3.5 and −8.5‰. $\delta^{13}C$ values in Pleistocene speleothem layers often become more negative towards the end of the deposition interval (Vaks *et al.* 2010, 2013).

Sr isotopes

The $^{87}Sr/^{86}Sr$ ratios in speleothems, cave host rock/overlying rock formations and loess soil above the cave are presented in Figure 7. The $^{87}Sr/^{86}Sr$ ratios exhibit the following ranges: Basal Member, 0.7078–0.7079; Intermediate Member, 0.7083–0.7085; Young Member, 0.7082–0.7084; host rocks, 0.7075–0.7077; and loess soil, 0.7088 (Vaks *et al.* 2013). The Sr concentrations in speleothems are 252–326, 42–110 and 357–505 ppm in the Basal, Intermediate and Young members, respectively. In the cave host rocks and overlying formations, the Sr concentrations vary from 107 to 118 ppm in Turonian limestones to 950 ppm in Senonian chalks. The mean Sr concentration in the overlying loess soil is 333 ppm (Vaks *et al.* 2013).

Discussion

Timing of humid periods

Speleothems only form when meteoric water infiltrates into the cave, i.e. when the effective precipitation (or infiltration coefficient) during the rainy season is positive. Such periods are defined here as humid, whereas periods with no water infiltration (zero effective precipitation/infiltration coefficient) are defined as dry.

Vaks *et al.* (2010) defined four humid periods in the central–southern Negev Desert during the last 350 kyr, referred to as the Negev Humid Periods (NHPs): NHP-1 between 142 and 109 ka, NHP-2 between 220 and 190 ka, NHP-3 between 310 and 290 ka and NHP-4 between 350 and 310 ka. Only NHP-1 and NHP-2 are represented in Ashalim Cave speleothems, whereas NHP-3 and NHP-4 are missing, suggesting that NHP-1 and NHP-2 were the wettest periods during the last 350 kyr in the central Negev Desert.

The yellow Basal Member is of Pliocene age (*c.* 3.1 Ma) and the brown Intermediate Member is of Early–Middle Pleistocene age. The one age measured on the stratigraphically oldest layer (ASH-15-C of the Intermediate Member) is 1.272 ± 0.018 Ma (interglacial MIS-39). The counting of speleothem calcite layers in the Intermediate and Young members showed that 12–15 NHP speleothem deposition events occurred during the Pleistocene, half of them within the Intermediate Member and the rest within the Young Member (Vaks *et al.* 2013). At least nine of these deposition events have not yet been dated.

The $\delta^{18}O$ values of all speleothems vary between −6.9 and −11.2%. All dated speleothem deposition periods in the cave occurred during interglacial periods (MIS-5, MIS-7 and MIS-39) or during the even warmer conditions of the Pliocene. The $\delta^{18}O$ values of other undated Pleistocene deposition periods are similar, indicating that they also occurred during interglacial periods. Vaks *et al.* (2010) estimated the minimum annual precipitation needed for vadose speleothem deposition in Israel to be 300–350 mm a^{-1} during interglacial periods and 200–275 mm a^{-1} during glacial periods. Given the evidence that the Ashalim speleothems grew during interglacial periods, it is likely that the amount of precipitation during all the NHPs exceeded 300 mm a^{-1} (Vaks *et al.* 2010, 2013).

Source of atmospheric precipitation and vegetation

The speleothem sequences in caves located to the north of Ashalim Cave show a rapid increase in speleothem volume and quantity and in the duration of growth periods. Speleothem deposition in Tzavoa Cave in the northern Negev Desert, located on the present day 150 mm isohyet (Fig. 1c), occurred during most of the last 200 kyr, mainly during the last two glacial periods, but also during interglacial MIS-5a (84–77 ka), the MIS-6-5 deglaciation, MIS-5e (137–123 ka) and MIS-7a (200–190 ka). The dry periods in Tzavoa Cave occurred during the Younger Dryas, Holocene (13–0 ka), most of MIS-5b (92–85 ka) and MIS-5d (117–96 ka). The southernmost cave in which speleothem deposition occurred during the Holocene is Maale-Dragot Cave, located within the present day 280–300 mm isohyets (Vaks *et al.* 2006). In caves located in regions with an average precipitation of 500 mm a^{-1} (e.g. Soreq Cave and Jerusalem Cave) and northwards, speleothem deposition was continuous during the last 240 kyr (Frumkin *et al.* 2000; Bar-Matthews *et al.* 2003). Such an increase in the continuity of speleothem deposition and of speleothem volume in the northern direction is evidence of long-term humidity and the higher availability of high Atlantic–Mediterranean precipitation.

The speleothem deposition periods in the caves of the central Negev highlands (Even-Sid and Ma'a'le-ha-Meyshar caves, Fig. 1c) are similar to those in Ashalim Cave. There was no speleothem deposition during NHP-2 further to the south in the hyper-arid southern Negev Desert and very limited deposition during NHP-1, resulting in a single thin lamina (one layer 3 mm thick in the Maale-Ktora Cracks) (Vaks *et al.* 2010). The decrease in speleothem thickness in the southern Negev Desert indicates that the rainfall contribution from Atlantic–Mediterranean sources decreased to the south because of the rain shadow effect of the central Negev highlands (Amit *et al.* 2006). However, the thicknesses of the Basal and Intermediate members are similar in both Ashalim Cave and the Ktora Cracks in the southern Negev Desert. This similarity can be explained by a negligible rain shadow effect of the central Negev highlands during the Pliocene and early Pleistocene (Amit *et al.* 2006) due to the more southerly position of the Mediterranean coasts along the Sinai Peninsula and North Africa (Vaks *et al.* 2013) or a southern source of precipitation. It is important to emphasize that large lakes were also formed in this region during the Pliocene and Early Pleistocene humid periods, when >95% of the speleothem volume was deposited in the central and southern Negev caves (Avni *et al.* 2000, 2001; Ginat *et al.* 2003, Vaks *et al.* 2013).

The most intensive episodes of speleothem deposition during the last 350 kyr in the central Negev Desert (NHP-1 and NHP-2) coincide with highest northern hemisphere insolation (Berger 1978), African monsoon index highs of ≥51 (cal cm^{-2}) day^{-1} (Rossignol-Strick 1983), the formation of Mediterranean sapropels (Rossignol-Strick & Paterne 1999), speleothem deposition periods in Oman and Yemen (Fleitmann *et al.* 2003, 2011) and periods of the formation of lakes in southern Jordan (Rohling *et al.* 2002; Bar-Matthews 2014). In addition, the $\delta^{18}O$ of Ashalim and other central Negev speleothems are 2–4‰ lower than the $\delta^{18}O$ values of speleothems from central/northern Israel (Vaks *et al.* 2010). Taken together, this evidence may indicate a southern rain source, such as Red Sea trough or tropical plume events, as suggested in other studies (Waldmann *et al.* 2010). However, whereas the moisture source for speleothems in Oman and Yemen was associated with the African monsoon (Fleitmann *et al.* 2003, 2011), the Mediterranean sapropels formed during periods of both increased rainfall above the Mediterranean Sea and a higher influx of riverine water from the Nile and other north African rivers fed by the African monsoon (Rohling *et al.* 2015). Although a southern contribution of rainfall during the interglacial humid periods cannot be completely ruled out (Vaks *et al.* 2006), a significant southern rain source would have resulted in considerable speleothem deposition in the southern Negev Desert during the last 350 kyr. However, there is a decrease in speleothem thickness from north to the south during NHPs-1, -2 and -3, supporting an Atlantic–Mediterranean northern source. At present, there is no record of rainfall events exhibiting $\delta^{18}O_w$ values lower than −6‰ associated with the tropical Atlantic Ocean, whereas the strongest Atlantic–Mediterranean cyclones do deliver rain and snow with $\delta^{18}O_w$ values of −8.5‰ and lower. These cyclones bring cold air from Russia, causing snowfall in the low mountains of the Levant (Ayalon *et al.* 1998). Vaks *et al.* (2006) suggested that the lower $\delta^{18}O_{Cc}$ values of the Negev speleothems may be partly explained by Rayleigh isotopic fractionation associated with increasing distance from the Mediterranean Sea coast.

High $\delta^{13}C$ values in the Ashalim speleothems indicate that vegetation above the cave was of dry/hot C4 steppe. Most speleothem deposition periods in Ashalim Cave are characterized by high $\delta^{13}C$ values at the beginning (between +2 and −2‰) and a very sharp decrease towards the end of the growth periods (−6 to −8‰), i.e. an amplitude of the order of 4–5‰, most likely reflecting a change in vegetation during the wet event from almost pure C4 vegetation type to a mixed C3–C4 vegetation. This significant change in the vegetation type indicates that the wet event was long enough to cause a change in the vegetation cover and, as a result, in the $\delta^{13}C$ of soil CO_2.

Drying, increasing dust supply and decreasing weathering from the Pliocene to Pleistocene

Sr substitutes for Ca in the speleothem calcite lattice (Fairchild *et al.* 2001). Three major sources potentially supply Sr to the groundwater reaching the Ashalim cave system: cave host rock, dust-borne soil above the cave and sea spray (Ayalon *et al.* 1999; Bar-Matthews *et al.* 1999; Frumkin & Stein 2004). Sr contributions from sea spray ($^{87}Sr/^{86}Sr$ = 0.7092; Farrell *et al.* 1995) are highest near the coast, but decrease significantly inland. The contribution from sea spray in Soreq Cave is very low (Ayalon *et al.* 1999; Bar-Matthews *et al.* 1999) and the contribution of Sr from sea spray is most probably negligible in the Negev Desert. This is also evident in the $^{87}Sr/^{86}Sr$ ratios in present day Negev rainwater, which vary between 0.7079 and 0.7084, compared with 0.7078–0.7092 on the Israeli Mediterranean coast (Herut *et al.* 1993).

Soils in Israel are mainly composed of dust-borne silicates originating from two major sources: the distal Saharan–Arabian desert belt (Frumkin & Stein 2004) and the proximal northern Sinai Peninsula (Yaalon 1971; Yaalon & Ganor 1973; Yaalon & Dan 1974; Amit & Gerson 1986; Tsoar & Pye 1987). The latter includes wind-blown sediments of the Nile Delta marine shelf and sand fields originating from Nile sediments (Crouvi *et al.* 2008, 2010; Enzel *et al.* 2008, 2010). The Saharan dust source shows $^{87}Sr/^{86}Sr$ ratios of 0.7165–0.7200, whereas the Nile source displays an average value of 0.7070 (Krom *et al.* 1999). The $^{87}Sr/^{86}Sr$ ratio of the bulk loess soil above Ashalim Cave is 0.7088, whereas the range in the entire Negev Desert varies between 0.7084 and 0.7091 (Vaks *et al.* 2013). The $^{87}Sr/^{86}Sr$ ratios of silicate soils fall on a mixing line between Nilotic and Saharan end-members, closer to the Nilotic ratio. The similarity between the $^{87}Sr/^{86}Sr$ ratios of present day rainfall in the Negev Desert (0.7079–0.7084) (Herut *et al.* 1993) and the $^{87}Sr/^{86}Sr$ ratio of the Intermediate and Young speleothem members in the Negev Desert may suggest that the rainfall $^{87}Sr/^{86}Sr$ ratio may represent an important contribution of the water-soluble fraction of dust.

During the last *c.* 3.1 myr, the $^{87}Sr/^{86}Sr$ ratios in Ashalim Cave speleothems increased from 0.7078–0.7079 in the Basal Member to 0.7082–0.7085 in the Intermediate and Young members (Fig. 7). This Pliocene–Pleistocene increase in $^{87}Sr/^{86}Sr$ ratios is also seen in other caves of the Negev Desert. Although the amplitude of the $^{87}Sr/^{86}Sr$ changes are similar to that found in speleothems from the caves of sub-humid Mediterranean central Israel (0.7078–0.7085), the patterns of the variability in $^{87}Sr/^{86}Sr$ ratios is different. Variations of in Sr isotopic ratios of speleothems from the Soreq and Jerusalem caves in central Israel, as well as the Sr concentrations, roughly mimic the glacial–interglacial cyclicity, with the lowest $^{87}Sr/^{86}Sr$ ratio of *c.* 0.7075–0.7081 during the interglacial periods and *c.* 0.7080–0.7085 during the glacial periods (Ayalon *et al.* 1999; Bar-Matthews *et al.* 1999; Frumkin & Stein 2004). The $^{87}Sr/^{86}Sr$ ratios in Ashalim Cave increase from the Pliocene (0.7078–0.7079) to Pleistocene (0.7082–0.7085) and remain stable during the Pleistocene periods of speleothem deposition (Vaks *et al.* 2013).

The linear negative correlation between the 1/Sr and $^{87}Sr/^{86}Sr$ ratios in Soreq Cave speleothems (Ayalon *et al.* 1999) indicates that the Sr in the speleothems represents a mixture of carbonate host rock ($^{87}Sr/^{86}Sr$ = *c.* 0.7075) and an exogenic source (mainly dust, $^{87}Sr/^{86}Sr$ = *c.* 0.7087). Ayalon *et al.* (1999) and Bar-Matthews *et al.* (1999) interpreted periods of high $^{87}Sr/^{86}Sr$ ratios as periods with low chemical weathering of the soil and host rock above the cave and a relatively higher input of exogenic Sr, whereas periods with low $^{87}Sr/^{86}Sr$ ratios represent more intense chemical weathering. The latter explanation is supported by data from the Père Noël Cave in Belgium (Verheyden *et al.* 2000). Frumkin & Stein (2004) argued that the high $^{87}Sr/^{86}Sr$ ratios in Jerusalem Cave, 15 km east of Soreq Cave (Fig. 1c), are evidence of intensive fluxes of dust from the Saharan region to the Eastern Mediterranean Sea area during periods of increased Saharan hyper-aridity. Low $^{87}Sr/^{86}Sr$ ratios were interpreted as periods of reduced dust supply from the Sahara Desert (Green Sahara periods).

The relatively constant high $^{87}Sr/^{86}Sr$ values (0.7082–0.7085) of periods of Pleistocene speleothem deposition may be explained by a higher contribution of Sr from dust-borne soils ($^{87}Sr/^{86}Sr$ = 0.7088) and a lower contribution of Sr from the host rock ($^{87}Sr/^{86}Sr$ = 0.7075–0.7077). Only the $^{87}Sr/^{86}Sr$ ratios of the Basal Member show consistent values (0.7078–0.7079), which are similar to the interglacial speleothem values of central Israel (Fig. 7). The average increase in $^{87}Sr/^{86}Sr$ ratios by *c.* 0.0004 between the Pliocene and Pleistocene speleothems reflects a decrease in the contribution of Sr from the host rock and a greater contribution from an external source (dust). As in central Israel, this shift in the Negev Desert can be explained by an increase in the supply of dust from the Sahara and Sinai deserts (Frumkin & Stein 2004) and/or by a longer residence time of meteoric water in the vadose zone as a result of drier conditions (Verheyden *et al.* 2000) with less intensive chemical weathering of the host rock (Ayalon *et al.* 1999; Bar-Matthews *et al.* 1999). The speleothem $^{87}Sr/^{86}Sr$ values suggest that the effective precipitation in the Negev Desert during the Pliocene was much larger than that during the Pleistocene (Vaks *et al.* 2013).

Conclusions

During the last 3.1 myr, the speleothems of Ashalim Cave in the central Negev Desert recorded humid periods, equivalent to an annual precipitation of >300 mm in the present day desert. The speleothem deposition record of Ashalim Cave is more intermittent than that in caves further to the north, but similar to other caves of the central Negev Desert. The Ashalim Cave speleothems only grew during two periods during the last 350 kyr: 221–190 ka (NHP-2) and from 134 to 114 ka (NHP-1). These growth periods correspond to interglacial MIS-7(3–1) and Termination II + MIS-5(5–4), respectively, and are the wettest among the four NHPs. NHP-3 (285–300 ka) and NHP-4 (300–350 ka), found in other Negev caves, were not found in Ashalim Cave, showing that this cave is probably recording only the wettest NHPs. Earlier periods of deposition occurred in the Early–Middle Pleistocene, with one layer of this sequence dated to 1.272 ± 0.018 Ma, and during the Pliocene at *c.* 3.1 Ma. However, many Pleistocene humid periods in the cave speleothems have not yet been dated.

NHP-2 speleothems do not occur in the caves of the southern Negev Desert and the NHP-1 layer is much thinner. A decrease in speleothem volume and thickness from north to south in the Negev caves during these humid periods indicates that, as today, the Mediterranean Sea was the major source of atmospheric precipitation in Ashalim Cave. These decreases in volume and thickness from north to south are not seen in the older speleothems and therefore the source of rain is not clear in the older periods. The very low $\delta^{18}O$ values of the Ashalim speleothems during the last 220 kyr are more likely to be a result of Rayleigh isotopic fractionation associated with increasing distance from the Mediterranean Sea coast than to indicate a different source of rain.

The high $\delta^{13}C$ values in the Ashalim speleothems indicate that the vegetation was C4 steppe adapted to dry/hot conditions, whereas C3 Mediterranean vegetation was extremely rare. Relatively small fluctuations in the relatively high $\delta^{13}C$ values during the Pliocene indicate that the C4 vegetation cover was stable. The large variability in $\delta^{13}C$ values during the Pleistocene show large changes in vegetation type and density, in most cases with more C3 type plants towards the end of the wet period, implying that the wet periods were long enough to cause a change in vegetation.

During the last *c.* 3.1 myr, the $^{87}Sr/^{86}Sr$ ratios in the Ashalim Cave speleothems increased from 0.7078–0.7079 in the Pliocene to 0.7082–0.7085 in the Pleistocene. This trend is probably a result of drying, with an increasing supply of silicate dust and a decrease in the weathering intensity of the carbonate lithologies above the cave.

References

Amit, R. & Gerson, R. 1986. The evolution of Holocene Reg (gravelly) soils in the deserts; an example from Dead Sea region. *Catena*, **13**, 59–79.

Amit, R., Enzel, Y. & Sharon, D. 2006. Permanent Quaternary hyperaridity in the Negev, Israel, resulting from regional tectonics blocking Mediterranean frontal systems. *Geology*, **34**, 509–512.

Auler, A.S. & Smart, P.L. 2004. Rates of condensation corrosion in speleothems of semi-arid northeastern Brazil. *Speleogenesis and Evolution of Karst Aquifers*, **2**, http://speleogenesis.info/directory/karstbase/pdf/seka_pdf4497.pdf

Avni, Y., Bartov, Y., Garfunkel, Z. & Ginat, H. 2000. Evolution of the Paran drainage basin and its relation to the Plio-Pleistocene history of the Arava Rift western margin, Israel. *Israel Journal of Earth Sciences*, **49**, 215–238.

Avni, Y., Bartov, Y., Garfunkel, Z. & Ginat, H. 2001. The Arava formation – a Pliocene sequence in the Arava Valley and its western margin, southern Israel. *Israel Journal of Earth Sciences*, **50**, 101–120.

Ayalon, A., Bar-Matthews, M. & Sass, E. 1998. Rainfall–recharge relationships within a karstic terrain in the Eastern Mediterranean semi-arid region, Israel: $\delta^{18}O$ and δD characteristics. *Journal of Hydrology*, **207**, 18–31.

Ayalon, A., Bar-Matthews, M. & Kaufman, A. 1999. Petrography, strontium, barium and uranium isotope ratios in speleothems as paleoclimatic proxies: Soreq cave, Israel. *The Holocene*, **9**, 715–722.

Bar-Matthews, M. 2014. History of water in the Middle East and North Africa. *In*: Holland, H.D. & Turekian, K.K. (eds) *Treatise on Geochemistry*. 2nd edn. Elsevier, Oxford, 109–128.

Bar-Matthews, M., Ayalon, A. & Kaufman, A. 1997. Late Quaternary paleoclimate in the Eastern Mediterranean region from stable isotope analysis of speleothems at Soreq Cave, Israel. *Quaternary Research*, **47**, 155–168.

Bar-Matthews, M., Ayalon, A., Kaufman, A. & Wasserburg, G.J. 1999. The Eastern Mediterranean paleoclimate as a reflection of regional events: Soreq cave, Israel. *Earth and Planetary Science Letters*, **166**, 85–95.

Bar-Matthews, M., Ayalon, A., Gilmour, M., Matthews, A. & Hawkesworth, C.J. 2003. Sea–land oxygen isotopic relationships from planktonic foraminifera and speleothems in the Eastern Mediterranean region and their implication for paleorainfall during interglacial intervals. *Geochimica et Cosmochimica Acta*, **67**, 3181–3199.

Berger, A. 1978. Long-term variations of caloric insolation resulting from the Earth's orbital elements. *Quaternary Research*, **9**, 139–167.

Crouvi, O., Amit, R., Enzel, Y., Porat, N. & Sandler, A., 2008. Sand dunes as a major proximal dust source for late Pleistocene loess in the Negev Desert, Israel. *Quaternary Research*, **70**, 275–282.

Crouvi, O., Amit, R., Enzel, Y. & Gillespie, A.R. 2010. Active sand seas and the formation of desert loess. *Quaternary Science Reviews*, **29**, 2087–2098.

Dayan, U. 1986. Climatology of back trajectories from Israel based on synoptic analysis. *Journal of Climate and Applied Meteorology*, **25**, 591–595.

deMenocal, P.B. 1995. Plio-Pleistocene African climate. *Science*, **270**, 53–59.

deMenocal, P.B. 2004. African climate change and faunal evolution during the Pliocene–Pleistocene. *Earth and Planetary Science Letters*, **220**, 3–24.

Dreybrodt, W., Gabrovšek, F. & Perne, M. 2005. Condensation corrosion: a theoretical approach. *Acta Carsologica*, **34**, 317–348.

Enzel, Y., Amit, R., Dayan, U., Crouvi, O., Kahana, R., Ziv, B. & Sharon, D. 2008. The climatic and physiographic controls of the eastern Mediterranean over the late Pleistocene climates in the southern Levant and its neighboring deserts. *Global and Planetary Change*, **60**, 165–192.

Enzel, Y., Amit, R., Crouvi, O. & Porat, N. 2010. Abrasion-derived sediments under intensified winds at the latest Pleistocene leading edge of the advancing Sinai-Negev erg. *Quaternary Research*, **74**, 121–131.

Fairchild, I.J., Baker, A., Borsato, A., Frisia, S., Hinton, R.W., McDermott, F. & Tooth, A. 2001. Annual to sub-annual resolution of multiple trace-element trends in speleothems. *Journal of the Geological Society, London*, **158**, 831–841, https://doi.org/10.1144/jgs.158.5.831

Farrell, J.W., Clemens, S.C. & Gromet,P.L. 1995. Improved chronostratigraphic reference curve of late Neogene seawater $^{87}Sr/^{86}Sr$. *Geology*, **23**, 403–406.

Fleitmann, D., Burns, S.J., Neff, U., Mangini, A. & Matter, A. 2003. Changing moisture sources over the last 330,000 years in northern Oman from fluid-inclusion evidence in speleothems. *Quaternary Research*, **60**, 223–232.

Fleitmann, D., Burns, S.J. *et al.* 2011. Holocene and Pleistocene pluvial periods in Yemen, southern Arabia. *Quaternary Science Reviews*, **30**, 783–787.

Frumkin, A. & Langford, B. 2017. Arid hypogene karst in a multi-aquifer system: hydrogeology and speleogenesis of Ashalim Cave, Negev Desert, Israel. *In*: Parise, M., Gabrovsek, F., Kaufmann, G. & Ravbar, N. (eds) *Advances in Karst Research: Theory, Fieldwork and Applications*. Geological Society, London, Special Publications, **466**. First published online November 6, 2017, https://doi.org/10.1144/SP466.3

Frumkin, A. & Stein, M. 2004. The Sahara–East Mediterranean dust and climate connection revealed by strontium and uranium isotopes in a Jerusalem speleothem. *Earth and Planetary Science Letters*, **217**, 451–464.

Frumkin, A., Ford, D.C. & Schwarcz, H.P. 2000. Paleoclimate and vegetation of the last glacial cycles in Jerusalem from a speleothem record. *Global Biochemical Cycles*, **14**, 863–870.

Ginat, H., Zilberman, E. & Saragusti, I. 2003. Early Pleistocene lake deposits and Lower Palaeolithic finds in Nahal (wadi) Zihor, southern Negev Desert, Israel. *Quaternary Research*, **59**, 445–458.

Hendy, C.H. 1971. The isotopic geochemistry of speleothems – I. The calculation of the effects of different modes of formation on the isotopic composition of speleothems and their applicability as palaeoclimatic indicators. *Geochimica et Cosmochimica Acta*, **35**, 801–824.

Herut, B., Starinsky, A. & Katz, A. 1993. Strontium in rainwater from Israel: sources, isotopes and chemistry. *Earth and Planetary Science Letters*, **120**, 77–84.

Holmgren, K., Karlen, W. & Shaw, P.A. 1995. Paleoclimatic significance of the stable isotopic composition and petrology of a Late Pleistocene stalagmite from Botswana. *Quaternary Research*, **43**, 320–328.

Kahana, R., Ziv, B., Enzel, Y. & Dayan, U. 2002. Synoptic climatology of major floods in the Negev Desert, Israel. *International Journal of Climatology*, **22**, 867–882.

Krom, M.D., Cliff, R.A., Eijsink, L.M., Herut, B. & Chester, R. 1999. The characterisation of Saharan dusts and Nile particulate matter in surface sediments from the Levantine basin using Sr isotopes. *Marine Geology*, **155**, 319–330.

Lisiecki, L.E. & Raymo, M.E. 2005. A Pliocene–Pleistocene stack of 57 globally distributed benthic $\delta^{18}O$ records. *Paleoceanography*, **20**, PA1003.

Lisker, S., Vaks, A., Bar-Matthews, M., Porat, R. & Frumkin, A. 2010. Late Pleistocene palaeoclimatic and palaeoenvironmental reconstruction of the Dead Sea area (Israel), based on speleothems and cave stromatolites. *Quaternary Science Reviews*, **29**, 1201–1211.

Osborne, A.H., Vance, D., Rohling, E.J., Barton, N., Rogerson, M. & Fello, N. 2008. A humid corridor across the Sahara for the migration of early modern humans out of Africa 120,000 years ago. *Proceedings of the National Academy of Sciences of the United States of America*, **105**, 16444–16447.

Rohling, E.J., Cane, T.R. *et al.* 2002. African monsoon variability during the previous interglacial maximum. *Earth and Planetary Science Letters*, **202**, 61–75.

Rohling, E.J., Marino, G. & Grant, K.M. 2015. Mediterranean climate and oceanography, and the periodic development of anoxic events (sapropels). *Earth-Science Reviews*, **143**, 62–97.

Rossignol-Strick, M. 1983. African monsoons, an immediate climate response to orbital insolation. *Nature*, **304**, 46–49.

Rossignol-Strick, M. & Paterne, M. 1999. A synthetic pollen record of the eastern Mediterranean sapropels of the last 1 Ma: implications for the time-scale and formation of sapropels. *Marine Geology*, **153**, 221–237.

Schwarcz, H.P. 1986. Geochronology and isotopic geochemistry of speleothems. *In*: Fritz, P. & Fontes, J.C. (eds) *Handbook of Environmental Isotope Geochemistry*. Elsevier, Amsterdam, 271–300.

Shay-El, Y. & Alpert, P. 1991. A diagnostic study of winter diabatic heating in the Mediterranean in relation with cyclones. *Quarterly Journal of the Royal Meteorological Society*, **117**, 715–747.

Tsoar, H. & Pye, K. 1987. Dust transport and the question of desert loess formation. *Sedimentology*, **34**, 139–153.

Vaks, A., Bar-Matthews, M. *et al.* 2003. Paleoclimate reconstruction based on the timing of speleothem growth and oxygen and carbon isotope composition in a cave located in the rain shadow in Israel. *Quaternary Research*, **59**, 182–193.

Vaks, A., Bar-Matthews, M. *et al.* 2006. Paleoclimate and location of the border between the Mediterranean climate region and the Saharo-Arabian Desert as revealed by speleothems from the northern Negev Desert, Israel. *Earth and Planetary Science Letters*, **249**, 384–399.

Vaks, A., Bar-Matthews, M., Ayalon, A., Matthews, A., Halicz, L. & Frumkin, A. 2007. Desert speleothems reveal climatic window for African exodus of early modern humans. *Geology*, **35**, 831–834.

Vaks, A., Bar-Matthews, M., Matthews, A., Ayalon, A. & Frumkin, A. 2010. Middle–late Quaternary paleoclimate of northern margins of the Saharan–Arabian Desert: reconstruction from speleothems of Negev Desert, Israel. *Quaternary Science Reviews*, **29**, 2647–2662.

Vaks, A., Woodhead, J. *et al.* 2013. Pliocene–Pleistocene climate of the northern margin of Saharan–Arabian Desert recorded in speleothems from the Negev Desert, Israel. *Earth and Planetary Science Letters*, **368**, 88–100.

Verheyden, S., Keppens, E., Fairchild, I.J., McDermott, F. & Weis, D. 2000. Mg, Sr and Sr isotope geochemistry of a Belgian Holocene speleothem: implications for paleoclimate reconstructions. *Chemical Geology*, **169**, 131–144.

Waldmann, N., Torfstein, A. & Stein, M. 2010. Northward intrusions of low- and mid-latitude storms across the Saharo-Arabian belt during past interglacials. *Geology*, **38**, 567–570.

Yaalon, D.H. 1971. Soil forming processes in time and space. *In*: Yaalon, D.H. (ed.) *Paleopedology: Origin, Nature and Dating of Paleosols*. International Society of Soil Science and Israel Universities Press, Jerusalem, 29–39.

Yaalon, D.H. & Dan, J. 1974. Accumulation and distribution of loess-derived deposits in the semi-desert fringe areas of Israel. *Zeitschrift für Geomorphologie Supplementband*, **20**, 91–105.

Yaalon, D.H. & Ganor, E. 1973. The influence of dust on soils during the quaternary. *Soil Science*, **116**, 146–155.

Zangvil, A. & Druian, P. 1990. Upper air trough axis orientation and the spatial distribution of rainfall over Israel. *International Journal of Climatology*, **10**, 57–62.

Global distribution and use of water from karst aquifers

ZORAN STEVANOVIĆ

Faculty of Mining and Geology, Department of Hydrogeology, Centre for Karst Hydrogeology, Djušina St. 7, 11000 Belgrade, Serbia

zstev_2000@yahoo.co.uk

Abstract: Karst aquifers are some of the most important and well-used sources of water worldwide. The tapping of karst waters for use as drinking water has been important in the historical and economic development of many karst regions. Recent studies have found that karstified rocks and aquifer systems cover *c.* 15% of the Earth's ice-free land. The greatest area of karst outcrops (>1 × 10^6 km^2) is in Russia, the USA, China and Canada. In the Mediterranean basin, groundwater is generally more abundant in karst than in other aquifers and has been extensively exploited. Karst groundwater is also widely used in the Middle East, China, North America, and northern and eastern Africa and is of crucial importance for the sustainable development of tourism and the economy. Karst aquifers currently supply *c.* 10% of the global population with drinking water and, in some zones, they are the only water resource available. However, the share of karst aquifers in the global supply of water will decrease with the predicted increase in population, concentrated in urban areas, and improvements in treatment technologies for water from other sources.

Karst rocks and aquifer systems provide large springs and are widely used as a source of drinking water in many parts of the world. Aureli (2010) stated that 'groundwater in karst aquifers represents the most significant as well as the safest source of drinking water'. In their often cited book *Karst Hydrogeology and Geomorphology*, Ford & Williams (2007) estimated that surface and subsurface outcrops of potentially soluble karst rocks occupy *c.* 20% of the Earth's ice-free land. The first sketch map of carbonate rocks worldwide was published by Ford & Williams (1989). The World Karst Aquifer Map (WOKAM) project was established in 2012 to supplement the existing map of the groundwater resources of the world (WHYMAP) (Richts *et al.* 2011; Goldscheider *et al.* 2014). The draft WOKAM map, which shows carbonate rock and evaporite outcrops, was completed in 2017 (Chen *et al.* 2017).

Ford & Williams (2007) assumed that at least 20% of the global population largely depends on karst groundwater. Karst aquifers contribute to the water supply in regions where they have a wide extent: the Mediterranean basin, central and southeastern Europe (the Alps and Carpathian Mountains), the Near East and Middle East, northern and eastern Africa, SE Asia, the southern part of the USA and the Caribbean basin (Stevanović 2015).

Distribution of karst aquifers

The distribution of karst in the world was evaluated based on the map produced by the WOKAM project and its database, existing references and the opinions of local experts.

The WOKAM, which shows the distribution of karst aquifers worldwide, is based on the Global Lithological Map, which was assembled from 92 regional lithological maps of the highest available resolution (Hartmann & Moosdorf 2012). According to the generalization and reclassification of the Global Lithological Map, four lithological units and their subsequent aquifer systems were presented on the WOKAM: (1) igneous and metamorphic rocks; (2) non-carbonate sedimentary formations, including unconsolidated sediments, siliciclastic dominant sedimentary rocks and pyroclastic formations; (3) carbonate sedimentary rocks, which are further subdivided into continuous and discontinuous outcrops (when outcrops of karstified rocks are separated by non-karst rocks); and (4) evaporites. Only the last two groups belong to classical rock complexes resulting in two equivalent karst aquifer systems. Carbonate and evaporite rocks are further subdivided into continuous (>65% of the surface area of the lithostratigraphic unit) and discontinuous (15–65%). Areas formed by mixed carbonate and evaporite rocks (>15% of each rock type) are also displayed on the map (Goldscheider *et al.* 2014; Chen *et al.* 2017). This means that the actual total surface share of carbonate and evaporitic rock outcrops is smaller than the areas presented on the map and discussed in this paper. However, there are still some carbonate masses that have not been mapped due to their small scale or a smaller percentage of their contribution to the delineated rock complexes. There is thus a systematic standard error, which is always present in this type of global

From: Parise, M., Gabrovsek, F., Kaufmann, G. & Ravbar, N. (eds) 2018. *Advances in Karst Research: Theory, Fieldwork and Applications*. Geological Society, London, Special Publications, **466**, 217–236.
First published online January 4, 2018, https://doi.org/10.1144/SP466.17

analysis. The calculated surface area of karst outcrops is preliminary because the WOKAM project and its data are still undergoing evaluation.

The total area of ice-free land on the continents has been calculated as the land surface area of each UN Member State plus dependent territories (195 + 48) based on the information available (Food and Agriculture Organization 2016; Wikipedia 2016; WOKAM 2017). Although these three online sources show some differences, we have accepted a compiled figure of total ice-free land of 133 861 179 km^2.

According to the statistical data divided by continent and the map published by the University of Auckland (University of Auckland n.d.), the total karst coverage has been estimated to be 17.67×10^6 km^2. Although the WOKAM project is still not complete, the total karst surface area (continuous and discontinuous carbonate and evaporitic rocks) has been estimated to be *c.* 20×10^6 km^2, <15% of the total ice-free land surface area.

The largest area of karst is in Asia, with $>7.5 \times 10^6$ km^2, although Europe has the greatest percentage of karst at 21.6% of its territory. Of this percentage, 15.1% consists of continuous carbonate rocks (Chen *et al.* 2017).

The largest country in the world – the Russian Federation – is also the country with the most karst. Its total karst cover has been calculated at $>2.5 \times 10^6$ km^2, which represents *c.* 15% of its entire territory. It is followed by the USA, with *c.* 2×10^6 km^2 (21%) and China with 1.76×10^6 km^2 (18.5%; Guanghui Jiang pers. comm.). Canada, with $>1.5 \times 10^6$ km^2 of karst, is the last in the group of countries with $>1 \times 10^6$ km^2 of karst. Very large areas of karst can also be found in Mexico, Algeria, Egypt, Iran and Saudi Arabia.

Some other, much smaller, countries have a lithology dominated by karstified rocks. Karst cover of >80% is present in Cuba, Jamaica, Montenegro and several small, but almost pure karst (limestone) islands such as Malta, the Bahamas, Barbuda, the Cayman Islands and Antigua. About 25% of the UN Member countries have either no karst or only a negligible amount of karst.

Spain, France, Italy, Turkey and the countries of northern Africa are large Mediterranean countries with karst rocks extending over 25–35% of their territories. The Mediterranean basin and the Alpine orogenic belt in SE Europe are probably the richest regions in terms of karst features and waters (Fig. 1). These are areas of highly developed karst, with some classical karst regions, such as the Dinaric karst (Fig. 2), characterized by abundant water resources that are unequally distributed throughout the year due to a specific climate and high karstification rate (Bonacci 1987; Milanović 2005; Stevanović *et al.* 2016*a*). Some areas in the Dinaric karst, such as southern Montenegro, are characterized by intensive recharge with an average specific yield >0.04 m^3 s^{-1} km^{-2} (Radulović 2000). Apart from its abundance of water reserves, the Alpine system is also known for its karstic features and some of the deepest potholes and longest caves in the world. For instance, the WOKAM database contains the largest number of karst springs from the Dinaric karst: the density of springs in Croatia, Slovenia, Montenegro, and Bosnia and Herzegovina discharging >0.2 m^3 s^{-1} as a minimum is ≥ 1 spring per 2000 km^2 (Stevanović *et al.* 2016*b*). In this and the adjacent areas of the southeastern Mediterranean basin, which includes the Hellenides and Taurides mountain chains, 28 springs have a minimum discharge >2 m^3 s^{-1}, which is key evidence of the availability of water. The highest number of springs regularly discharging >2 m^3 s^{-1} are found in Bosnia and Herzegovina and Turkey (eight springs in each), followed by Montenegro (five springs).

Utilization of water from karst aquifers

A short history of the use of water from karst aquifers

Tapping structures are an ancient art, as old as the first human civilizations. Many remnants of intakes around large springs in karst have been found in ancient China, Babylon, Persia, Israel and Egypt. These springs were commonly used as central places around which to create settlements. In Jerusalem, for instance, the Gihon spring issues from Turonian limestones below the city walls, but to bring the water closer to citizens and to prevent the spring from falling into the hands of the enemy during the Assyrian siege, additional engineering works included the construction of a >500 m long tunnel, which was used to deliver water to the Siloam water pool (Frumkin & Shimron 2006).

Many past empires used water from karst springs simply because there was no alternative supply, but in some cases the principal factor for this decision was an awareness of the importance of karst springs and their high-quality water. To supply the city of Nineveh, near present day Mosul in Iraq, with water of better quality than that of the nearby Tigris River, the Assyrian emperor Sanherib, son of Sargon II (703–681 BC), constructed intake systems at the Khanis karst gravity spring (near Atrush, 16 km away). Spring water flowed to the city walls through one of the first aqueducts ever built (Reade 1978; Stevanović 2010*a*).

Another type of intake was necessary for supplying water to Persepolis, the historical capital of the Persian Kingdom. The city is surrounded by Sarvak Formation limestones and the excavation of a large

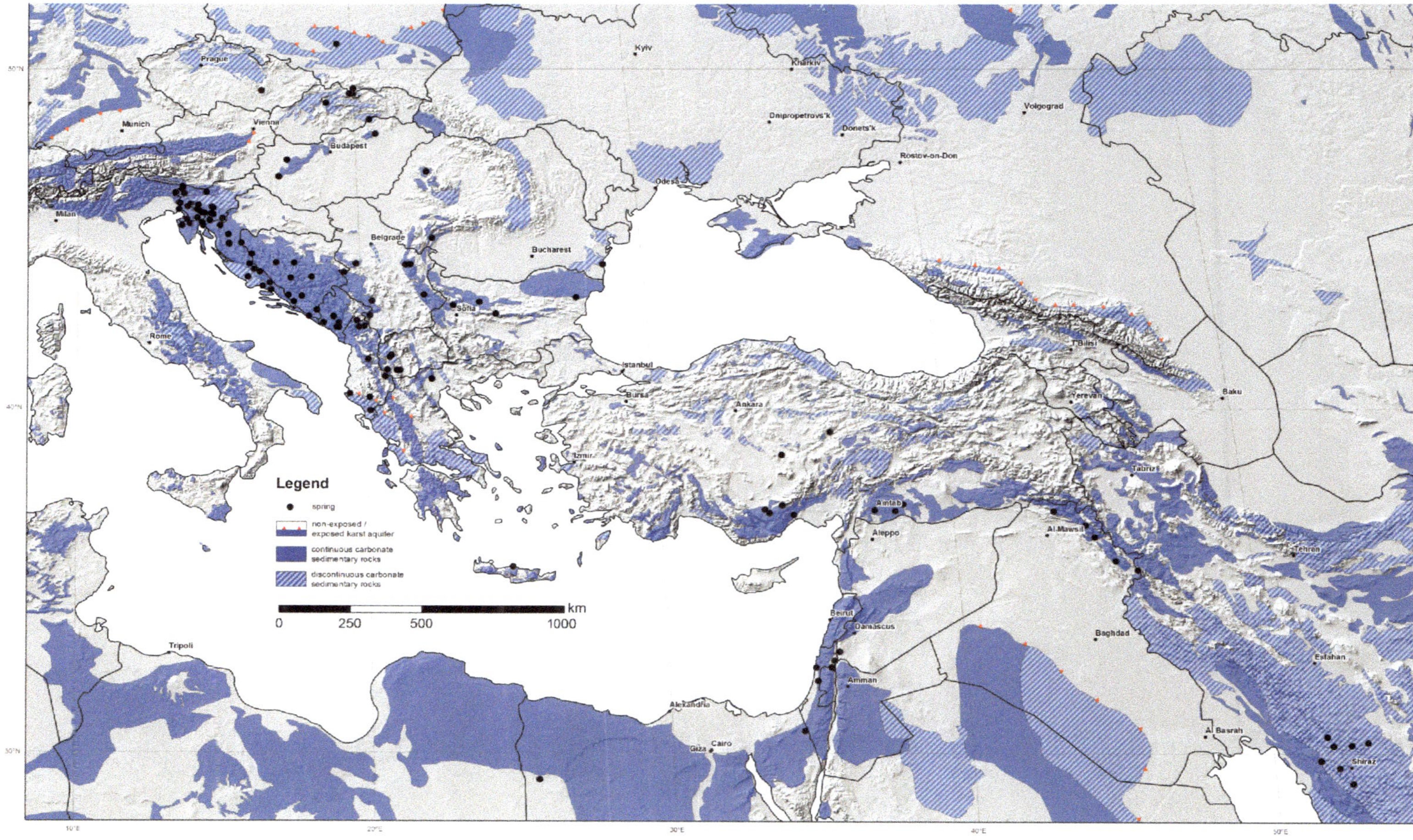

Fig. 1. Draft map of the distribution of karst aquifers in the eastern and southeastern parts of the Mediterranean basin as the outcome of the WOKAM project (Stevanović *et al.* 2016*b*; Chen *et al.* 2017; WOKAM 2017).

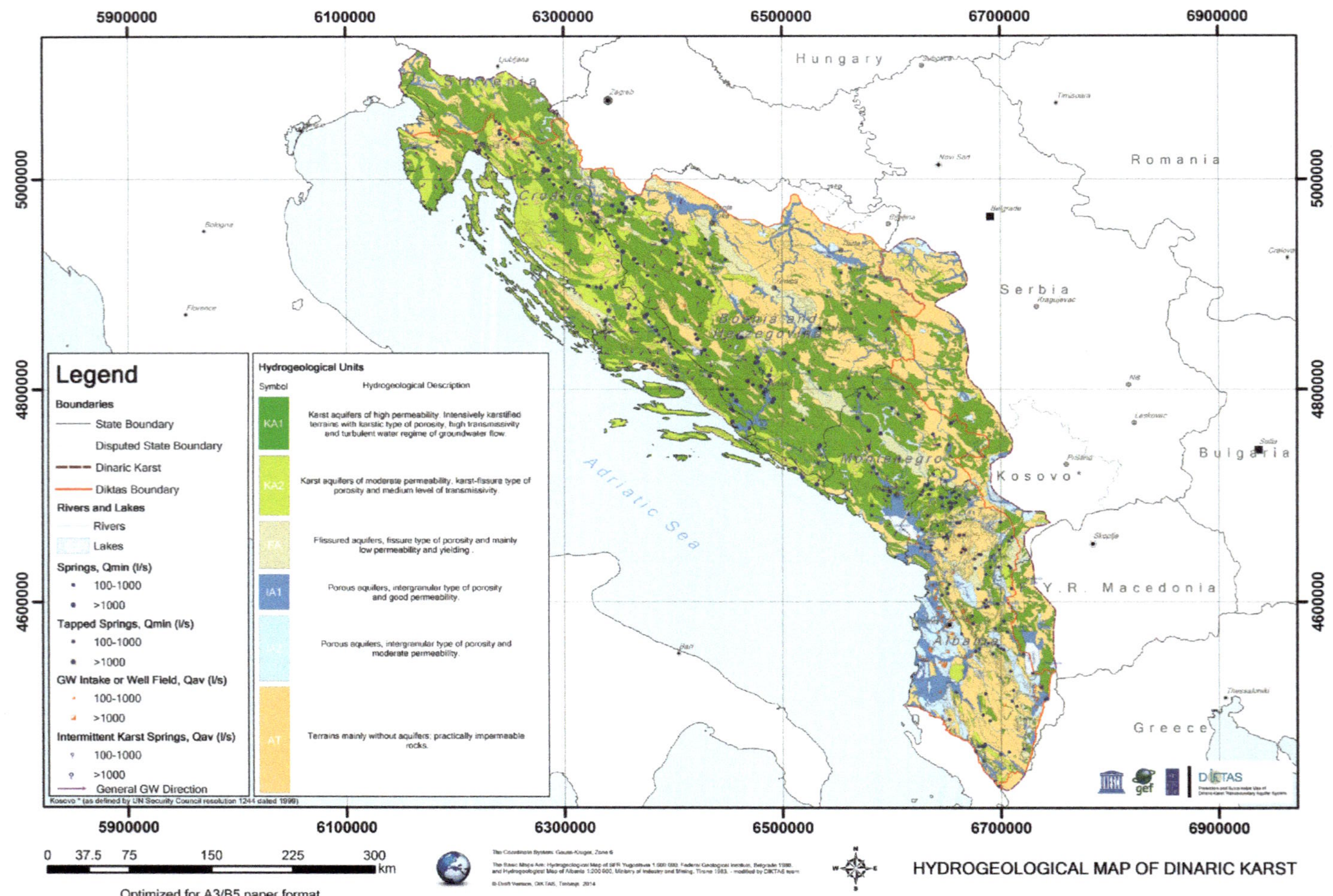

Fig. 2. Simplified hydrogeological map of the Dinaric karst as the outcome of the Dinaric Karst Transboundary Aquifer System Project (DIKTAS n.d.; Stevanović *et al.* 2016*a*).

Fig. 3. (**a**, **b**) Photographs of the stone well excavated to supply the historical city of Persepolis, near Shiraz, Iran, with water.

vertical hole in very hard rocks was found to be the optimum solution. A large square-shaped shaft ('the stone well') was attached to channels that delivered water to the city (Fig. 3a, b).

A typical example of the creation of cities around major springs in karst is found in the Adriatic part of the Mediterranean basin. All four major settlements established by the Romans (along today's Italian and Croatian shorelines) are linked to springs and depend on a spring water supply (Stevanović 2010*b*): Trieste (the Timavo springs); Rijeka (the Zvir group of springs); Split (the Jadro spring) (Fig. 4a, b); and Dubrovnik (the Šumet and Ombla springs).

The rounded arches systematically applied by the Romans enabled the construction of long aqueducts to transport water over long distances (Bono & Boni 1996). For instance, 11 long aqueducts delivered >13 $m^3 s^{-1}$ of spring water to the then city of Rome from distances of 16–91 km (Lombardi & Corazza 2008).

In the Medieval era, karst springs continued to be an important source of water for nearby cities. Large projects relating to water transport from remote areas were undertaken in Europe in the nineteenth century. The first 130 km long mountain pipeline was completed in 1873 to supply the city of Vienna (Drennig 1973). Long concrete tunnels and channels tap water from the Kaiserbrunn spring (Fig. 5a, b) and this water, along with that from other karst springs, ensures the supply of *c.* 1.7 million inhabitants of Vienna. The average amount of water distributed today is 1.40×10^8 $m^3 a^{-1}$. The quality of the water is excellent, generally requiring only chlorination, primarily for cleaning the distribution pipes.

The nineteenth century was also the period when chalk rocks in artesian basins became the main water sources for the cities of Paris and London. In addition to the confined aquifer, drinking water also came to Paris from several captured springs 100–150 km from the city and piped by aqueducts built between 1865 and 1893 (Margat *et al.* 2013).

Actual use of karst water

The intakes of springs are the dominant tapping structures applied in karst environments. Channelling gravity springs and diverting water over long distances is still a simpler solution than drilling many wells in a hard rock environment. However, in the case of some aquifers, especially less productive aquifers or those located in arid zones, with water at greater depths, digging followed by drilling is inevitable. Moreover, due to unstable regimes and the reduced discharge of karst springs in recession periods, many sources require a combined system: the delivery of water from a gravity spring during high water periods and the pumping of water during droughts.

Fig. 4. (**a**) Jadro spring near Split, Croatia. This spring has been used continuously since Roman times and supplied water to the Roman emperor Diocletian's palace (**b**) built between the third and fourth centuries AD (reconstruction by architect E. Hébrand). The spring intake was reconstructed in 1886.

Fig. 5. (**a**) An old sketch of the Kaiserbrunn spring (author's photo courtesy of the Kaiserbrunn Museum, Vienna, Austria) and (**b**) an inside view of this nineteenth century engineering masterpiece.

If we consider the number of dependent citizens, China is a major consumer of karst water. It is estimated that *c.* 150 million Chinese people use potable water from karst aquifers. According to Guanghui Jiang (pers. comm., 21 November 2016), the total number of consumers in northern China (including part of Beijing) is *c.* 50 million, whereas in the south, where karst outcrops have larger extents, there are *c.* 100 million consumers. Lu (2005) estimates that *c.* 80% of the large available karst groundwater flux of 12.5×10^9 m^3 a^{-1} in northern China is exploited.

The second-largest consumer of water from karstic aquifers is the USA. It has been estimated (KWI, Karst Waters Institute n.d.) that 40% of the groundwater used for drinking purposes comes from karst aquifers. However, the USA is a typical example of a situation where the expansion of urban settlements and industrialization resulted in the reduced use of karst aquifers. The assessment stating that *c.* 50 million people in the USA – mostly those from sparsely populated areas – actually depend on karst waters (Benjamin Tobin pers. comm., 6 December 2016) is more realistic. The largest city using water from a karst aquifer (the Edward aquifer) is San Antonio, with >1.5 million inhabitants. Eastern states such as Florida, Kentucky, Tennessee and West Virginia are also rich in karst aquifer waters and have many consumers.

An example of a small area of karst with a large number of consumers is India. Its densely populated areas located in karst comprise only 3% of the country's territory (about 80 000 km^2), which is spread over 106 administrative districts. However, carbonate rocks are also hidden under other types of rocks and soils and are being drilled to deeper levels to meet the water demands in stress areas (Dar *et al.* 2011). Although the total population in these districts is *c.* 160 million, the number of people directly or indirectly dependent on karst aquifers is lower. It is estimated that 46 million people in India use karst water for drinking purposes (Farooq Dar pers. comm., 3 December 2016).

More than 10 million people in Mexico, Iran, Indonesia, Russia, France, the Philippines, Turkey and Italy also use karst water. In Mexico, the largest karst area is the Yucatan Peninsula, where the platform Tertiary karst looks like Swiss cheese. The expansion of the tourist industry in the northern and northeastern coastal areas, coupled with intensive pumping (Merida, Cancun), has resulted in salt water intrusion deep inland.

The main karst aquifers in Iran are located in the south and provide the water supply of Shiraz, Kazeroon, Bushar and many other cities. Many islands in SE Asia, such as Java in Indonesia and the Palawan Province in the Philippines, heavily depend on karst water sources. Karst waters are also extensively used in certain parts of Vietnam, Laos, Malaysia, and northern and southern Thailand.

Both types of karstic aquifers – continuous unconfined and especially the discontinuous semi-confined and confined – have large extents in the southern parts of Russia, Ukraine and Moldavia, where large cities and industrial centres are located. For instance, *c.* 8 million Ukrainian residents consume potable water from karst aquifers (Alexander Klimchouk pers. comm., 28 November 2016).

In France, *c.* 60% of the population is supplied from groundwater and the proportion of use of alluvial and karst aquifers is about 50 : 50. Major cities in the south, such as Montpellier and Marseille, are exclusively supplied by karst water (Margat *et al.* 2013; Bakalowicz 2015).

Boni (1992) reported that there are *c.* 290 major springs in the central Italian Apennines, with a total mean discharge of 320 m^3 s^{-1}. Of these, 14 have a mean discharge between 5 and 18 m^3 s^{-1}. Rome receives 60% of its potable waters from the Peschiera karst source, whereas the Caposele spring is used for the regional waterworks of the Apulia region, including the city of Bari (Fiorillo & Guadagno 2012). Naples and most of the cities in southern Italy also traditionally use karst water for drinking purposes.

Thessaloniki and most of the Greek islands in the Ionian and Aegean seas use karst waters, the exception being those islands consisting of volcanic rocks. Karst occurs widely over southern Turkey and karst springs ensure the water supply of most tourist cities along the Mediterranean coast. Many reservoirs built across Turkey over the last 50 years have also captured karst waters. The Oymapinar reservoir has impounded one of the world's largest springs, Dumanli, which has a mean discharge of 50 m^3 s^{-1} (Günay 2010).

Many cities in Lebanon, Syria, Israel and Libya, especially those along the Mediterranean coast, have developed karstic sources to serve their populations. For instance, Figeh (Fijeh) and Barada springs, with a mean discharge of 4.5 and 5 m^3 s^{-1}, respectively, supply potable water to the Damascus area. Unfortunately, both springs have been exposed to contamination during the civil war in Syria and the city water supply has been interrupted several times. Using water as a weapon is especially dangerous in the case of vulnerable karst aquifers.

Benghazi in Libya used to obtain water from the Ain Zeina spring, but due to its high salinity this spring has been replaced by a newly constructed pipeline bringing water from the Sahara Desert and the rich Nubian sandstone aquifer.

The share of karst water exploitation is substantial in some countries, mostly due to the presence of very large springs. For instance, >90% of the population in Montenegro drinks water from karst aquifers. The most recent water supply project completed in this country involved the distribution of karst water to the Adriatic coastal area. Water from the inland karst spring Bolje Sestre is now delivered through a 40 km long pipeline to the coast, using <1 $m^3 s^{-1}$ of the 2.3 $m^3 s^{-1}$ minimum spring discharge (Radulović 2000; Stevanović 2010*b*). Six capitals in SE Europe use water from large karst springs for drinking purposes. In addition to Vienna and Rome, these are Sarajevo, Tirana, Skopje and Podgorica (Fig. 6a–c).

More than half the population in Bosnia and Herzegovina, Jordan, Austria, Slovakia and Albania also drink karst water. In Austria, the total renewable water resources per capita are >9000 $m^3 a^{-1}$ and the utilization rate is only 4.7% (Food and Agriculture Organization 2016). An even greater availability of water is present in Albania, where every citizen has *c.* 13 000 m^3 of water available per year, i.e. 35 m^3 each day (Food and Agriculture Organization 2016). There are *c.* 110 springs in the Albanian karst, with an average discharge >100 l s^{-1}. Of these, 17 have discharges >1 $m^3 s^{-1}$ (Eftimi 2010). There has been some discussion about the possible use of the largest of them, the Blue Eye in the Bistrica group of springs (mean discharge 18.4 $m^3 s^{-1}$), for an overseas supply to the Italian southern coast of the Adriatic.

Similarly, there is also low utilization of the renewable water resources in two other karst countries: Croatia (0.6%) and Bosnia and Herzegovina (0.9%) (Food and Agriculture Organization 2016). Some central European countries – those that share the Alpine orogenic belt, such as Switzerland, and the southern parts of Poland and Germany – dominantly use groundwater in their water supply systems (60–75%) and water from karst also plays an important part in such practices.

In Jamaica, where White Limestone is the dominant karstic formation, groundwater accounts for *c.* 84% of all the available water (Karanjac 2005). As a result of intensive irrigation (75% of the total water production), aquifers are over-exploited and, according to United Nations standards, Jamaica is near to being included in a group of water-stressed countries (1500 $m^3 a^{-1}$ per capita or 4 $m^3 day^{-1}$ per capita). The Dominican Republic is a little richer in karst aquifer water, with 2500 $m^3 a^{-1}$ available per capita, but both countries suffer from salt water intrusion and the contamination of groundwater by untreated industrial water, poor sewage management and tourism (Karanjac 2005). Several other purely karstic Caribbean island states also have no other option apart from karst groundwater (Robins 2013).

Like India, Brazil has a large number of consumers of karst water even though it has a relatively limited area of karst outcrops. In the large Brazilian territory, there are *c.* 622 municipalities with some karst, but not all their citizens are dependent on karst water. Karst water is used by *c.* 4.3 million people (Augusto Auler pers. comm., 15 November 2016).

According to Margat & Gun (2013), out of the 'top 10' groundwater abstraction countries, seven use karst waters to a considerable extent (the USA, Mexico, Saudi Arabia, Iran, Indonesia, China and Turkey). However, most of these waters are used for irrigation and it is well known that the irrigation portion represents 70% of all groundwater extraction globally (Zektser & Everett 2004; Fig. 7). In the majority of arid countries, the irrigation of arable land is the only way to ensure crop production.

In some countries with large areas of karst, viewed in relation to the total amount of water exploited, the percentage of water allocated for irrigation is as follows: Libya, 99%; Saudi Arabia, 97%; Algeria, 66%; and Jordan, 65% (Margat & Gun 2013; Food and Agriculture Organization 2016). Spain is also a large user of karst water. In the 1980s, *c.* 2×10^9 $m^3 a^{-1}$ was pumped from karst aquifers (40% of all pumped water) to irrigate *c.* 3000 km^2 of cultivated land (Sahuquillo 1986).

Some countries, despite a large area of karst aquifers, still have a great shortage of water. This is mostly the consequence of rare or erratic rainfall and insufficient recharge, but is sometimes also the result of uncontrolled extraction. Saudi Arabia, for instance, uses eight times more water than provided by its internal renewable water resources (Food and Agriculture Organization 2016). However, as a result of the specific behaviour of karst, there are broad, extended zones with limited water resources, even in karst areas with adequate rainfall and the replenishment of aquifers. This is seen in mountains high above the erosional bases that function exclusively as recharge zones. The shortage of water on their surface often results in the migration of local villagers, a reduction in livestock and limited crops (Stevanović 2015). Another problem is the exposure of drainage in littoral karst to salt water contamination.

Many reservoirs were built worldwide in the twentieth century to maintain karst river flows. Unfortunately, many dams were built at inappropriate sites or without proper research and feasibility studies, which resulted in significant water leakage and sometimes even complete abandonment (Milanović 2000). However, most of the projects were completed successfully, enabling the maintenance of the seasonal (base) flow, the generation of hydropower and, in many cases, the provision of potable water to local citizens (Fig. 8a–c).

Based on these data and the statistics available from the United Nations, I have calculated that groundwater from karst aquifers is probably currently satisfying the demands of *c.* 10% of the global

Fig. 6. (**a**) Mareza Spring supplying drinking water to Podgorica, the capital of Montenegro. (**b**) Vrelo Bosne supplying Sarajevo, the capital of Bosnia and Herzegovina. (**c**) The Izvarna spring intake room supplying 270 000 citizens of Craiova, Romania.

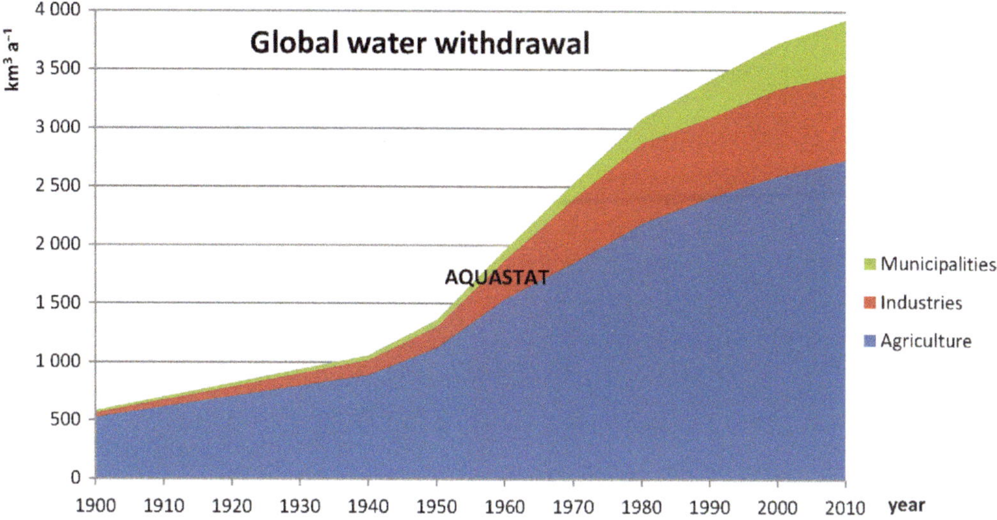

Fig.7. Global water withdrawal in the past 110 years. Sources of data: 2010, Food and Agriculture Organization AQUASTAT (www.fao.org/nr/water/aquastat/water_use/index.stm); 1900–2000, I.A. Shiklomanov.

population. This is much less than previously estimated, and is – among other facts – a consequence of the increase in the world's population. However, this evaluation still needs to be fine-tuned using more precise statistical data and the input of local experts.

Discussion: threats and opportunities

Pressures on karst water reserves

We live in an era when globalization is affecting many segments of everyday life, including natural resources and water supply. Fortunately, globalization also provides the widely applied concept of sustainable development and helps to increase awareness of the consequences of uncontrolled pollution. The question is whether sustainability is possible if water is already over-allocated for much of the year in many basins, especially in the arid parts of the world, and if the source basins are effectively closed to new users. The following discussion tries to evaluate sustainability and the prospect of further karst groundwater use from aspects of water reserves and availability. The larger problem of water quality and the easy pollution of vulnerable karst aquifers has been left aside because this has already been discussed elsewhere.

We are currently facing high population growth, especially in the developing world, and the expansion of large urban settlements requiring increasing amounts of water. Poor economic and sanitary conditions in many countries in the arid world, coupled with unstable political situations, are causing the migration of populations to an extent that has not been encountered previously in modern history. In 2015, the global population was estimated at 7.4 billion people (United Nations Development Program 2016; United Nations 2016); in the next 50 years there may be another two billion people on the planet. The Department of Economic and Social Affairs of the United Nations estimated that, in 2016, 54.5% of the world population lived in urban settlements (United Nations 2016). Urban areas are projected to house 60% of people globally by 2030 and one-third of the population will live in cities with at least 500 000 inhabitants (Fig. 9).

Are there proper solutions to improving the living conditions of poor people and enabling them to stay on their original land? And how can karst aquifers contribute to this process? According to the World Bank Group (2016), there are three overarching policy priorities that can help countries to find solutions to a water-secure and climate-resilient economy: (1) optimizing the use of water through better planning and incentives; (2) where appropriate, expanding water supply and availability; and (3) reducing the impact of extremes, variability and uncertainty.

The second priority has to be considered with caution. Historically, when supply is increased without corresponding management action, demand also increases to meet the new level of supply, resulting in a higher level of water dependence. Therefore these interventions must be accompanied by policies to promote water efficiency and improve water allocation across the sectors (World Bank 2016).

Fig. 8. Photographs of three dams restraining karst river waters. (**a**) Ravedis Dam on Cellina River, northern Italy; (**b**) the planned Bekhme dam site on the Great Zab River, Iraqi Kurdistan; and (**c**) Grančarevo on the Trebišnjica River, a major European sinking stream, Bosnia and Herzegovina.

With respect to karst aquifers, it is clear that they are not present everywhere and, even if they do exist, they are dependent on the climate regime and recharge. Along with the projections of the Intergovernmental Panel on Climate Change (2007), some newly implemented projects, such as the UNESCO GRAPHIC Project (2012; Treidel *et al.* 2012) or the European Union Project CCWaterS (2012; Stevanović *et al.* 2012) and their forecasts show that already vulnerable karst aquifers would be much

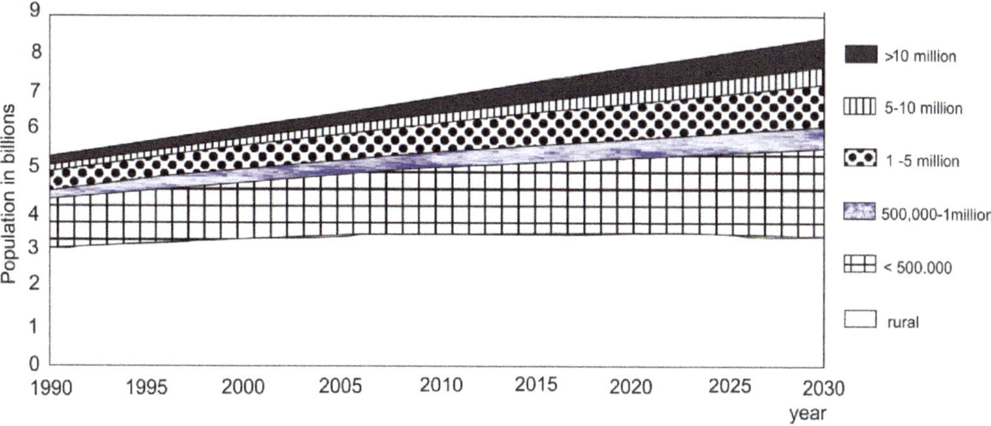

Fig. 9. World population by size class of settlement 1990–2030, according to the United Nations Department of Economic and Social Affairs (United Nations 2016).

more exposed to lowered discharge during recession periods. Although karst could be a good buffer against flooding because of its considerable effective infiltration capacity (Stevanović *et al.* 2015), a shortage of water during low water seasons and droughts would become even more frequent.

There is no general solution for combating a lack of water and maintaining groundwater reserves in karst aquifers; each particular case requires research and an appropriate local solution. In some cases intensive urbanization and industrialization is limiting the further development of karst aquifers; other solutions, although much more expensive (e.g. building new reservoirs, desalinization, water recycling and reuse or treatment by reverse osmosis), must be introduced into concerned water practice.

It is thus difficult, sometimes even impossible, to meet increased demands and provide sufficient water to the population in the developing world. A typical example of such a situation is the city of Sulaimani in northern Iraq (Iraqi Kurdistan Province). The local karst spring Sarchinar (discharge 0.6–7 m^3 s^{-1}; Figs 10 & 11) has been tapped for the municipal water supply and has successfully followed the expansion of the city from 46 000 inhabitants in 1957 to about 400 000 in the 1980s. Surface waters from the Dokan reservoir, 70 km away, were piped to over-bridge the water shortage during late summer months (Stevanović & Iurkiewicz 2009). Today the city has more than two million inhabitants and, along with water from the Sarchinar spring and Dokan reservoir, several thousand wells have been drilled, mostly illegally, in the city and its outskirts to support the domestic water supply (Salahaddin S. Ali pers. comm., 9 March 2016).

Some karst sources are exposed to intensive over-extraction. For instance, the largest Syrian spring and one of the largest in the Mediterranean karst, Ras el Ain, no longer exists because it dried out during the summer months due to forced pumping for the irrigation of cotton fields in the border area between Syria and Turkey.

However, in places where water is available, but inefficiently used by industry, agriculture or local waterworks, an adequate water policy and technical solutions could alleviate water-related problems and prevent negative environmental impacts. A preliminary water budget calculation at the global scale shows that only *c.* 1% of the total available karst groundwater resources (dynamic flux) is currently utilized. The problem is the distribution of water resources, which definitely does not correspond with the location of the major consumers. In addition, karst aquifers extend widely in arid regions, where the impacts of climate change and inefficient water use are much more present than in the developed world, which has an engineering tradition and a better awareness of the importance of the sustainable use of water and its protection against pollution.

Engineering control of karst aquifers as an option to maintain minimum flows

A high vulnerability to pollution, a direct hydraulic connection with seawater in coastal areas and the discharge of low springs during the recession periods are the major constraints faced by karst aquifer managers. Along with anthropogenic impacts, geological hazards are often present in karstic catchments (Milanović 2000; Parise & Gunn 2007; Parise *et al.* 2015) and may jeopardize efficient water use and protection from pollution. The phrase 'expect the unexpected' is commonly used by engineers dealing with karst and its properties.

In some regions where rainfall is sufficient and well balanced throughout the year – ensuring good

Fig. 10. A pumping station at the ascending karstic spring Sarchinar in the outskirts of Sulaimani city, Iraq.

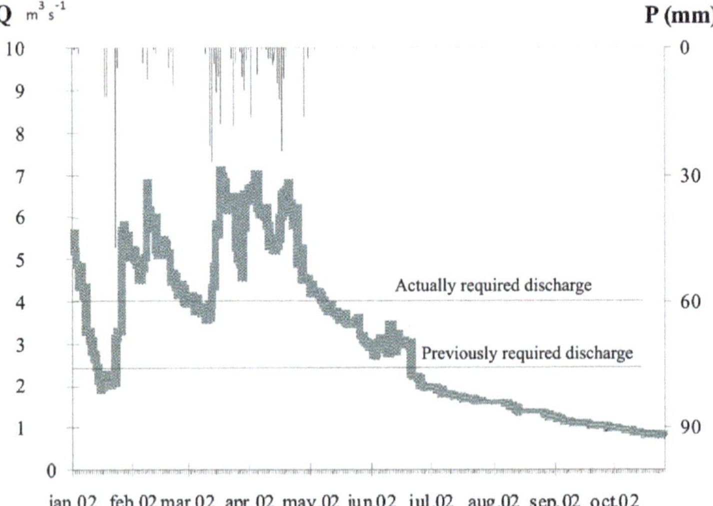

Fig. 11. Hydrograph of the Sarchinar spring for the year 2002 (modified from Stevanović & Iurkiewicz 2009). The average discharge in 2002 was 3.13 $m^3 s^{-1}$, while a discharge of *c.* 2.5 $m^3 s^{-1}$ was sufficient to ensure the domestic water supply of the city of Sulaimani. The period from June to October is regularly problematic as there is no rainfall in these months. The minimum needed to supply the city in 2017 has increased to 4 $m^3 s^{-1}$.

replenishment potential to well-karstified and permeable aquifers – it is often possible to regulate and manage flow by various engineering interventions (Stevanović 2015). Several interventions are possible; they can be classified as those applied directly in discharge zone and those applied in the wider catchment.

In the discharge zone, depending on the local hydrogeology, the following solutions are possible: (1) over-pumping of spring(s); (2) drilling of wells or other types of vertical shafts and pumping; (3) construction of a subsurface (underground) dam; and (4) artificial recharge of the aquifer.

The first two options are applied most often in engineering practice, simply as a result of certainty and cost. Drilling technology is known everywhere and, if an aquifer has a good storage capacity, large diameter vertical tube wells can easily be constructed to produce large yields. There are several successfully completed projects based on the utilization of groundwater from considerable storage in the deeper parts of karst aquifers, such as those implemented in Lez (Montpellier, France; Avias 1984), Bolje Sestre (the Montenegrin coast; Stevanović 2010*b*) and in several locations in China (Lu 1986).

There are many advantages to underground reservoirs relative to surface reservoirs (Stevanović 2010*a*):

- no problems with flooding of infrastructure, fertile land, monuments or compensation to reallocated people;
- no threat of dam collapse and major destruction downstream;
- no high evaporation rate as in surface reservoirs; and
- no negative impact on water quality as in open reservoirs (e.g. eutrophication, sedimentation).

However, with the exception of China, where several such projects have been implemented successfully (Yuan 1990), few such dams have been constructed in karst. The main reason for this is that, regardless of previous investigations, there is no 100% guarantee against leakage around the dam or plunges (Milanović 2000; Stevanović & Milanović 2015).

As regards the wider catchment, measures such as redirecting sinking streams and groundwater, building grout curtains or impermeable barriers, the construction of small reservoirs or subsurface dams in river beds, and closing or regulating the ponors (swallow holes) may considerably change the natural water regime and support the stabilization of discharge at the desired sites (springs).

Any type of intervention in aquifers requires available water resources. Otherwise, the applied intervention may result in only temporary or short-term effects and a negative environmental impact.

The example in Figure 12 shows a typical annual hydrograph of a karst spring under a normal continental type of rainfall distribution. The total annual rainfall is *c.* 800 mm and the spring discharge is in the range 0.38–1.2 m^3 s^{-1}. The ratio $Q_{max} : Q_{min}$ 1 : 3 indicates a relatively stable regime. However, if, for instance, water demands are 0.5 m^3 s^{-1} (including the ecological flow for a water-dependent ecosystem), water would have to be pumped for about two months (September–October). This loan from deeper aquifer storage can be compensated during the rainy period that follows (November–December). It is therefore important that the annual replenishment potential surpasses the amount of water pumped. The option of the temporary use of aquifer storage is helping not only to satisfy drinking water demands, but also to ensure ecological flow downstream. This is in line with the flexible concept of safe yield, which allows the temporary depletion

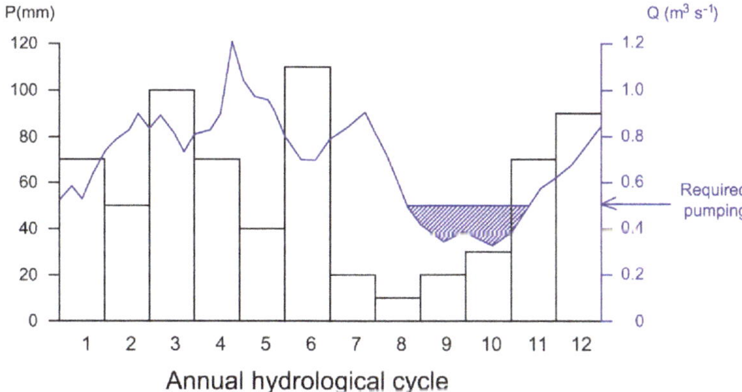

Fig. 12. Annual hydrograph of a karst spring with critical autumn yield (*Q*), although the sufficient water potential from precipitation (*P*) in the winter–spring months may compensate the amount of pumped water.

Fig. 13. Lez spring (photograph reproduced from the leaflet of the 43rd International Association of Hydrogeologists Congress).

of aquifer reserves (Custodio 1992; Burke & Moench 2000; Stevanović 2015).

One of the best, well-known examples of successful engineering regulations is the Lez spring, used to supply the city of Montpellier in France (Avias 1984; Fig. 13). By pumping groundwater since December 1982 at an average rate of 1.1 $m^3 s^{-1}$ using vertical shafts 400 m behind the spring, the city obtains sufficient water, while one pipe sends some of the water to the origin river bed to ensure ecological flow downstream. Therefore the natural minimum spring discharge regularly flows out from the spring orifice; artificial structures are visible and drawdown is >40 m below the spring site. Dörfliger *et al.* (2008) even noticed an improvement in the groundwater flow drainage of the saturated zone of the aquifer. This improvement might be the result of the notorious effect of the expulsion of clay out of karst conduits and their reactivation by pumping. The complete Lez project is thus fully sustainable and environmentally friendly.

Figure 14 shows the regulated karst spring Mrljiš as part of the regional water system in Bogovina, which supplies several towns in the Timok region of the Carpathian karst of eastern Serbia (Stevanović 2009, 2010*a*). The permanent vauclusian spring Mrljiš has a minimum natural discharge of 80 $l s^{-1}$, which is considered insufficient for a stable water supply. Because of this limit, hydrotechnicians proposed building a large dam. Several exploitation wells were drilled after a complex hydrogeological survey which, among other activities, included the drilling of exploratory boreholes and a pumping test of the spring's ascending karst channel. The capacities of wells during the pumping tests ranged from 50 to 110 $l s^{-1}$, producing a small drawdown of <2 m in the Mrljiš zone. As a result, a solution involving extraction wells that enabled a four-fold increase in the natural minimum spring flow of Mrljiš was approved and put into practice. One of the wells closest to Mrljiš (Fig. 14) only pumps during low water periods and for no more than 30 days per year; however, during pumping an overflow is diverted to the nearby river to ensure an unaffected streamflow for downstream consumers. Another downstream well, drilled to just sustain ecological flow, almost never activates due to the low impact of pumped aquifer waters on the hydrological system as a whole (Stevanović 2010*a*). The energy required for pumping can increase the cost of such solutions, but the cost is still negligible compared with the expense of large constructions such as dams, which require regular and expensive water treatment.

In general, there are two main factors that may go against such a technical solution: (1) insufficient water; and (2) disrespect of the ecology. A knowledge of the regional and local geological and hydrogeological properties is necessary to properly estimate and verify the groundwater reserves and the potential for replenishment. The aquifer geometry, i.e. the size of the catchment area, groundwater table and thickness of the saturated zone, permeability, storativity and water availability (reserves) have to be studied carefully and in great detail. A knowledge of the groundwater regime obtained from the systematic monitoring of groundwater quantity and

Fig. 14. Photograph of pumping well (B-5) located near the Mrljiš spring and river bank in Bogovina, Serbia.

quality and long-term pumping tests should be crucial verification indicators, but the collection and evaluation of other data, such as the climate, hydrology, morphology and vegetation, are also needed to ensure a successful regulation project.

Any attempt to change the natural hydrological regime of the springs is, by definition, opposed by environmentalists. The times when engineers used to change the world are far behind us; we live in an era dominated by a 'green' approach and sustainable development. It is thus not technically correct or environmentally friendly to reduce discharge or to completely dry out a spring by pumping and diverting its water to other locations and consumers. Figure 15 visualizes a compromise that was reached in the cases of Lez and Mrljiš. Maintenance of the natural spring flow (Q_s) is effected by pumping water from a drilled borehole (Q_b). In this way, both tasks are achieved: an ensured water supply (Q_{ws}) and a maintained ecological flow (Q_s).

Conclusions

Karst waters are an important natural resource with a high natural quality that often does not require expensive treatment. Climate change and human activities will increase the pressure on water

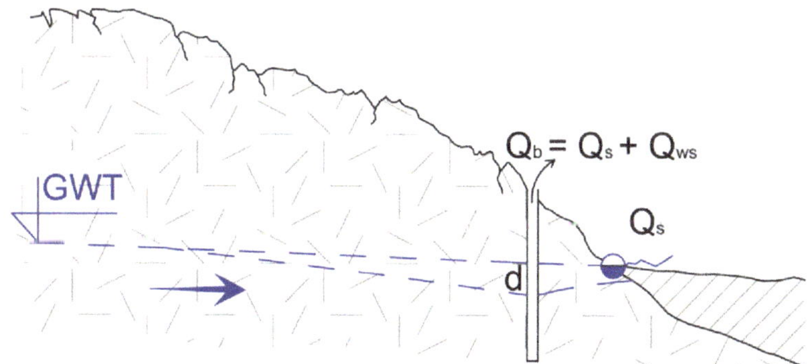

Fig. 15. A karst aquifer with a spring and nearby drilled borehole. When the discharge of the spring (Q_s) is insufficient to cover the water demands, the solution could be to pump the borehole (Q_b) with resultant drawdown (d). GWT, groundwater table. This results in an assured water supply (Q_{ws}).

resources globally, including karst aquifers, which are currently used for drinking water by *c.* 10% of the world's population. Many karst regions will experience strong natural and anthropogenic changes in the future. Water reserves in karst are abundant in many aquifers worldwide, but not everywhere and not always. Especially problematic are aquifers in arid regions, where water demands are highest due to population growth. Increasing demands for water for irrigation and domestic use will further increase the pressure on the number of karst aquifers. With further increases in population, its concentration in large cities and fast industrialization, few of the karst aquifers currently used will be able to respond, even if they are properly managed. The expected improvements in water treatment technology may provide an alternative in the use of other water sources with a poorer quality than karst water.

Engineering solutions for the control and maintenance of spring flows could be applied to many aquifer systems, especially those that are largely karstified and have a good storage capacity followed by adequate replenishment from rainfall. However, a complex and detailed investigation programme and a feasibility study are required before the implementation and success of such projects.

The author gratefully acknowledges data and comments provided by Guanghui Jiang, Farooq A. Dar, Benjamin Tobin, Alexander Klimchouk, Augusto Auler, Salahaddin S. Ali, Abe Springer and Jerome Perrin on karst water utilization. Joint work with members of the Advisory Board of the WOKAM project encouraged the author to embark on an extended inquiry into the global prospection of karst aquifers, work which will continue in the future.

References

Aureli, A. 2010. The UNESCO IHP's shared aquifer resources management global project. *AQUA Mundi*, **1**, 1–6.

Avias, J. 1984. Captage des sources karstiques avec pompage en periode d'etiage. L'example de la source du Lez. *In*: Burger, A. & Dubertret, L. (eds) *Hydrogeology of Karstic Terrains. Case Histories*. International Contributions to Hydrogeology, **1**. Heise, Hanover, 117–119.

Bakalowicz, M. 2015. Karst and karst groundwater resources in the Mediterranean. *Environmental Earth Sciences*, **74**, 5–14.

Bonacci, O. 1987. *Karst Hydrology With Special Reference to the Dinaric Karst*. Springer Series in Physical Environment, **2**. Springer, Berlin.

Boni, C. 1992. Karst hydrogeology in central Italy. *In*: Paloc, H. & Back, W. (eds) *Hydrogeology of Selected Karst Regions*. International Contributions to Hydrogeology, **13**. Heise, Hanover, 151–157.

Bono, P. & Boni, C. 1996. Water supply of Rome in antiquity and today. *Environmental Geology*, **27**, 126–134.

Burke, J.J. & Moench, H.M. 2000. *Groundwater and Society: Resources, Tensions and Opportunities. Themes in Groundwater Management for the Twenty-First Century*. Department of International Economic and Social Affairs, Statistical Office, United Nations, New York.

Chen, Z., Auler, A.S. *et al.* 2017. The World Karst Aquifer Mapping Project – concept, mapping procedure and map of Europe. *Hydrogeology Journal*, **25**, 771–785.

Custodio, E. 1992. *Hydrogeological and Hydrochemical Aspects of Aquifer Overexploitation*. International Association of Hydrogeologists, Selected Papers, **3**. Heise, Hanover, 3–27.

Dar, F.A., Perrin, J., Riotte, J., Gebauer, H.D., Narayana, A.C. & Shakeel, A. 2011. Karstification in the Cuddapah Sedimentary Basin, southern India: implications for groundwater resources. *Acta Carsologica*, **40**, 457–472.

DIKTAS (n.d.) GEF Project, 2011–2014, http://dinaric.iwlearn.org [last accessed 26 September 2016].

Dörfliger, N., Jourde, H. *et al.* 2008. Active water management resources of karstic water catchment: the example of Le Lez spring (Montpellier, South France). Paper presented at the 13th IWRA World Water Congress, 1–4 September 2008, Montpellier, France.

Drennig, A. 1973. *Die I. Wiener Hochquellenwasserleitung*. Magistrat der Stadt Wien, Abteilung 31 – Wasserwerke, Vienna.

Eftimi, R. 2010. Hydrogeological characteristics of Albania. *AQUA Mundi*, **1**, 79–92.

European Union Project CCWaterS 2012. Climate Changes and Water Supply, www.ccwaters.eu/ [last accessed 26 September 2016].

Fiorillo, F. & Guadagno, F.M. 2012. Long karst spring discharge time series and droughts occurrence in Southern Italy. *Environmental Earth Sciences*, **65**, 2273–2283.

Food and Agriculture Organization 2016. AQUASTAT, www.fao.org/nr/aquastat/ [last accessed 21 October 2016].

Ford, D. & Williams, P. 1989. *Karst Hydrogeology and Geomorphology*. 1st edn., Wiley, Chichester.

Ford, D. & Williams, P. 2007. *Karst Hydrogeology and Geomorphology*. 2nd edn., Wiley, Chichester.

Frumkin, A. & Shimron, A. 2006. Tunnel engineering in the Iron Age: geoarchaeology of the Siloam Tunnel, Jerusalem. *Journal of Archaeological Science*, **33**, 227–237.

Goldscheider, N. Chen, Zh. & the WOKAM Team 2014. The World Karst Aquifer Mapping Project – WOKAM. *In*: Kukurić, N., Stevanović, Z. & Krešić, N. (eds) *Proceedings of the DIKTAS Conference: Karst without Boundaries*, 11–15 June 2014, Trebinje, Bosnia and Herzegovina, Grafokomerc A.D., Trebinje.

Günay, G. 2010. Geological and hydrogeological properties of Turkish karst and major karstic springs. *In*: Kresic, N. & Stevanović, Z. (eds) *Groundwater Hydrology of Springs. Engineering, Theory, Management and Sustainability*. Elsevier, Amsterdam, 479–497.

Hartmann, J. & Moosdorf, N. 2012. The new global lithological map database GLiM: a representation of rock properties at the Earth surface. *Geochemistry, Geophysics, Geosystems*, **13**, Q12004.

Intergovernmental Panel on Climate Change 2007. www.ipcc.ch; www.ipcc.ch/pdf/technical-papers/climate-change-water-en.pdf [last accessed 12 August 2016].

Karanjac, J. 2005. Vulnerability of ground water in the karst of Jamaica. *In*: Stevanović, Z. & Milanović, P. (eds) *Water Resources and Environmental Problems in Karst. Proceedings of the International Conference KARST 2005*. University of Belgrade, Institute of Hydrogeology, Belgrade, 27–36.

Karst Waters Institute (n.d.) http://karstwaters.org/educational-resources/what-is-karst-and-why-is-it-important/ [last accessed 21 October 2016].

Lombardi, L. & Corazza, A. 2008. L'acqua e la città in epoca antica. *Memorie della Societa geologica italiana*, **LXXX**, 189–219.

Lu, Y. 1986. *Karst in China. Landscapes, Types, Rules [in Chinese]*. Special Edition of the Geological Publishing House, Beijing, **288**.

Lu, Y. 2005. Karst water resources and geo-ecology in typical regions of China. *In*: Stevanović, Z. & Milanović, P. (eds) *Water Resources and Environmental Problems in Karst. Proceedings of the International Conference KARST 2005*. University of Belgrade, Institute of Hydrogeology, Belgrade, 19–26.

Margat, J. & Gun van der, J. 2013. *Groundwater Around the World. A Geographic Synopsis*. CRC Press, London.

Margat, J., Pennequin, D. & Roux, J.C. 2013. History of French hydrogeology. *In*: Howden, N. & Mather, J. (eds) *History of Hydrogeology. International Contributions to Hydrogeology*, **28**. CRC Press, London, 59–99.

Milanović, P. 2000. *Geological Engineering in Karst*. Zebra, Belgrade.

Milanović, P. 2005. Water potential in southeastern Dinarides. *In*: Stevanović, Z. & Milanović, P. (eds) *Water Resources and Environmental Problems in Karst. Proceedings of the International Conference KARST 2005*. University of Belgrade, Institute of Hydrogeology, Belgrade, 249–257.

Parise, M. & Gunn, J. (eds) 2007. *Natural and Anthropogenic Hazards in Karst Areas: Recognition, Analysis and Mitigation*. Geological Society, London, Special Publications, **279**, http://sp.lyellcollection.org/content/279/1

Parise, M., Closson, D., Gutierrez, F. & Stevanović, Z. 2015. Anticipating and managing engineering problems in the complex karst environment. *Environmental Earth Sciences*, **74**, 7823–7835.

Radulović, M. 2000. Karst hydrogeology of Montenegro. *Geological Bulletin*, **XVIII**. Special issue of the Geological Survey of Montenegro, Podgorica.

Reade, J. 1978. Studies in Assyrian geography, part 1: Sennacherib and the waters of Ninveh. *Revue d'Assyriologie et d'archéologie orientale*, **72**, 47–72.

Richts, A., Struckmeier, W.F. & Zaepke, M. 2011. WHY-MAP and the groundwater resources of the world 1:25,000,000. *In*: Jones, J.A.A. (ed.) *Sustaining Groundwater Resources. A Critical Element in the Global Water Crisis*. International Year of Planet Earth Series. Springer, Dordrecht, 159–173.

Robins, N.S. 2013. Island hydrogeology: highlights from early experience in the West Indies and Bermuda. *In*: Howden, N. & Mather, J. (eds) *History of Hydrogeology*. International Contributions to Hydrogeology, **28**. CRC Press, London, 29–40.

Sahuquillo, A. 1986. Recursos hidraulicos en zonas kàrsticas. Experianca española. *In*: *Jornadas sobre el Karst en Euskadi*. San Sebastian, 341–363.

Shiklomanov, I.A. 2000. Appraisal and assessment of World water resources. *Water International*, **25**, 11–32.

Stevanović, Z. 2009. Karst groundwater use in the Carpathian–Balkan region. *In*: Paliwal, B. (ed.) *Global Groundwater Resources and Management*. Scientific Publishers, Jodhpur, 429–442.

Stevanović, Z. 2010*a*. Utilization and regulation of springs. *In*: Kresic, N. & Stevanović, Z. (eds) *Groundwater Hydrology of Springs. Engineering, Theory, Management and Sustainability*. Elsevier, Amsterdam, 339–388.

Stevanović, Z. 2010*b*. Major springs of southeastern Europe and their utilization. *In*: Kresic, N. & Stevanović, Z. (eds) *Groundwater Hydrology of Springs: Engineering, Theory, Management and Sustainability*, Elsevier, Amsterdam, 389–410.

Stevanović, Z. 2015. Characterization of karst aquifer. *In*: Stevanović, Z. (ed.) *Karst Aquifers – Characterization and Engineering*. Professional Practice in Earth Science Series. Springer, Heidelberg, 47–126.

Stevanović, Z. & Iurkiewicz, A. 2009. Groundwater management in northern Iraq. *Hydrogeology Journal*, **17**, 367–378.

Stevanović, Z. & Milanović, P. 2015. Engineering challenges in karst. *Acta Carsologica*, **44**, 381–399.

Stevanović, Z., Ristić-Vakanjac, V. & Milanović, S. (eds) 2012. *Climate Changes and Water Supply*. SE Europe Cooperation Programme Monograph. University of Belgrade, Belgrade.

Stevanović, Z., Ristić-Vakanjac, V., Milanović, S., Vasić, Lj., Petrović, B. & Čokorilo Ilić, M. 2015. Karstification depth and storativity as main factors of karst aquifer regimes: some examples from southern Alpine branches (SE Europe and Middle East). *Environmental Earth Science*, **74**, 227–240.

Stevanović, Z., Kukurić, N., Pekaš, Ž., Jolović, B., Pambuku, A. & Radojević, D. 2016*a*. Dinaric karst aquifer – one of the world's largest transboundary systems and an ideal location for applying innovative and integrated water management. *In*: Stevanović, Z., Krešić, N. & Kukurić, N. (eds) *Karst Without Boundaries*. CRC Press, London, 3–25.

Stevanović, Z., Goldscheider, N. & Chen, Z. & the WOKAM Team 2016*b*. WOKAM – the World Karst Aquifer Mapping Project, examples from south east Europe, Near and Middle East and Eastern Africa. *In*: Stevanović, Z., Krešić, N. & Kukurić, N. (eds) *Karst Without Boundaries*. CRC Press, London, 39–51.

Treidel, H., Martin-Bordes, J.J., Gurdak, J.J. (eds) 2012. *Climate Change Effects on Groundwater Resources: A Global Synthesis of Findings and Recommendations*, International Contributions to Hydrogeology, **27**. Taylor & Francis, London.

UNESCO GRAPHIC Project 2012. www.graphicnetwork.net [last accessed 26 September 2016].

United Nations 2016. The World's Population, www.unpopulation.org [last accessed 13 October 2016].

United Nations Department of Economic and Social Affairs, Population Division 2016. *The World's Cities in 2016 – Data Booklet*, **ST/ESA/SER.A/392**. United Nations, New York.

United Nations Development Program 2016. World Population in 2015, https://esa.un.org/unpd/wpp/Download/Standard/Population/ [last accessed 01 November 2016].

University of Auckland, n.d. World Map of Carbonate Rock Outcrops v3.0, http://web.env.auckland.ac.nz/our_research/karst/ [last accessed 6 November 2016].

Wikipedia 2016. Sovereign States and Dependencies by Area, https://en.wikipedia.org/wiki/List_of_sovereign_states_and_dependencies_by_area [last accessed 06 November 2016].

WOKAM – World Karst Aquifer Map 2017. https://www.bgr.bund.de/whymap/EN/Maps_Data/Wokam/whymap_ed2017_map_g.html?nn=1548136 [last accessed 5 December 2017].

World Bank Group 2016. *High and Dry: Climate Change, Water, and the Economy*. World Bank, Washington, DC. © World Bank. License: CC BY 3.0 IGO, https://openknowledge.worldbank.org/handle/10986/23665 [last accessed 09 October 2016].

Yuan, D. 1990. *Construction of Underground Dams on Subterranean Streams in South China Karst*. Institute of Karst Geology, Guilin.

Zektser, S.I. & Everett, G.L. 2004. *Groundwater Resources of the World and their Use*. IHP-VI, Series on Groundwater, **6**. UNESCO, Paris.

Characterization and hydraulic behaviour of the complex karst of the Kaibab Plateau and Grand Canyon National Park, USA

CASEY J. R. JONES[1], ABRAHAM E. SPRINGER[1]*, BENJAMIN W. TOBIN[2], SARAH J. ZAPPITELLO[2] & NATALIE A. JONES[2]

[1]*School of Earth Sciences and Environmental Sustainability, Northern Arizona University, NAU Box 4099, Flagstaff, AZ 86011, USA*

[2]*Grand Canyon National Park, National Park Service, 1824 South Thompson Street, Flagstaff, AZ, 86001, USA*

**Correspondence: abe.springer@nau.edu*

Abstract: The Kaibab Plateau and Grand Canyon National Park in the USA contain both shallow and deep karst systems, which interact in ways that are not well known, although recent studies have allowed better interpretations of this unique system. Detailed characterization of sinkholes and their distribution on the surface using geographical information system and LiDAR data can be used to relate the infiltration points to the overall hydrogeological system. Flow paths through the deep regional geological structure were delineated using non-toxic fluorescent dyes. The flow characteristics of the coupled aquifer system were evaluated using hydrograph recession curve analysis via discharge data from Roaring Springs, the sole source of the water supply for the Grand Canyon National Park. The interactions between these coupled surface and deep karst systems are complex and challenging to understand. Although the surface karst behaves in much the same way as karst in other similar regions, the deep karst has a base flow recession coefficient an order of magnitude lower than many other karst aquifers throughout the world. Dye trace analysis reveals rapid, conduit-dominated flow that demonstrates fracture connectivity along faults between the surface and deep karst. An understanding of this coupled karst system will better inform aquifer management and research in other complex karst systems.

Characterizing and understanding available water resources is an increasing priority for managers concerned with meeting both ecological and human needs in North America (Salcedo-Sanchez *et al.* 2013) and elsewhere in the world (Bakalowicz 2005; Hua *et al.* 2015; Szocs *et al.* 2015). The patterns of precipitation in the hydrological systems of the western USA are seasonal, with high precipitation in winter (as snowpack) and, in some regions, a strong summer monsoon, which requires significant aquifer storage to meet the requirement for water over extended periods of time. This seasonality of recharge results in a reliance on snowpack and groundwater systems to store water throughout the year, slowly releasing it during drier times. Research has shown that groundwater resources are the dominant control of low (base) flow in surface streams (Liu *et al.* 2008, 2012; Tobin & Schwartz 2012) and that they are essential sources of freshwater. Karst is present in many of these high elevation systems (Weary & Doctor 2014) and research from around the world suggests that high elevation karst aquifers provide an important amount of storage and are crucial to both ecosystems and humans (Han & Liu 2004; Karimi *et al.* 2005; Jemcov 2007; Mueller *et al.* 2017). Karst aquifers supply drinking water to 20–25% of the world's population via groundwater pumping and springs (Kresic & Stevanovic 2010).

High elevation groundwater systems with substantial hydraulic gradients have been found to provide significant groundwater storage and are relatively understudied due to their remote and often inaccessible locations (Clow *et al.* 2003). Karst aquifers in high elevation settings are equally underrepresented in the literature (Tobin & Schwartz 2016). Researchers have shown that these high elevation snow-dominated systems often have rapid conduit development (Faulkner 2009; Lauber *et al.* 2014), substantial amounts of diffuse recharge (Oraseanu & Mather 2000; Perrin *et al.* 2003; Goldscheider *et al.* 2007) and recharge to matrix storage through snowmelt (Tobin & Schwartz 2012). A changing climate may, however, promote more precipitation as rain rather than snowfall, changing the dynamics of snowmelt-dominated karst aquifers worldwide (Gremaud & Goldscheider 2010).

Karst aquifers are known to feed the largest springs in the world, making them essential not only for access to clean drinking and irrigation

From: Parise, M., Gabrovsek, F., Kaufmann, G. & Ravbar, N. (eds) 2018. *Advances in Karst Research: Theory, Fieldwork and Applications*. Geological Society, London, Special Publications, **466**, 237–260.
First published online November 6, 2017, https://doi.org/10.1144/SP466.5

water, but also as habitats for many spring-dependent species (Springer & Stevens 2009; Kresic & Stevanovic 2010) and unique cave-obligate species (Culver & Sket 2000). Groundwater systems provide a mechanism of control for changes in vegetative succession and fluvial geomorphic processes (Batz *et al.* 2016). To better protect these habitats, processes and water supplies, it is crucial to characterize the storage properties of karst aquifers (Scanlon *et al.* 2003).

Roaring Springs, a karst spring, is the sole source of water supplying the Grand Canyon National Park (GRCA) in the southwestern USA. The Redwall–Muav aquifer (R aquifer) of the Kaibab Plateau in the GRCA is part of an extremely complex hydrogeological system. The main aquifer unit is at a substantial depth below the surface of the plateau, with numerous overlying non-karst rock units, complex local and regional structural features, and a perched aquifer, all of which result in intricate recharge flow paths (Huntoon 1970, 1974, 1981; Beus 1990*a*). The Grand Canyon was originally designated as a National Monument in 1908 by President Theodore Roosevelt under the Antiquities Act as 'an object of unusual scientific interest'. The Grand Canyon became a National Park in 1919. Although neither of these descriptions specifically lists karst as one of the reasons for designating the Grand Canyon as an outstanding natural feature, the GRCA is actually the second largest karst region in the US National Park System, exceeded only by the Everglades National Park in Florida (Anderson 2000; Weary & Doctor 2014). Whereas other units of the US National Park System are well known for their karst phenomena, such as Mammoth Cave National Park in Kentucky, the GRCA contains >4000 km^2 of karst features (Weary & Doctor 2014). This includes surficial karst development, major cave development and the deeper R aquifer. Although karst systems are particularly difficult to quantify due to their heterogeneity and anisotropic dynamic nature, part of the GRCA's uniqueness is the depth of the main karst strata: the Redwall and Muav formations are buried >1000 m below the surface (Beus 1990*a*). Above this main aquifer is a smaller-scale limestone and sandstone perched aquifer (the Coconino or C aquifer) and various impermeable confining layers. The C aquifer and other stratigraphic layers affect the R aquifer in distinct ways that have not previously been determined in detail and examining their interconnectedness will lead to an improved understanding of the mechanisms of groundwater flow and storage in a multifaceted system.

The focus of this paper is on the northern rim of the Grand Canyon, located on the Kaibab Plateau in northern Arizona in the southwestern USA (Fig. 1). The karst aquifer in this region supplies numerous large springs that provide desert oasis habitats, drinking water and base flow to the Colorado River (Hart *et al.* 2002). Roaring Springs (Fig. 2) discharges from the R aquifer and supplies potable drinking water for all the residents of the National Park and more than six million yearly visitors to the Grand Canyon. Studies of the R aquifer north of the Grand Canyon, despite its ecological and economic importance, have been localized, resulting in limited datasets and making long-term evaluations challenging. Research in the Grand Canyon is also challenging as a result of difficult access and the extreme climatic conditions, but has enticed scientists for decades. Huntoon (1970, 1974, 1981, 2000) provided a valuable foundation of karst characterization in the Redwall and Muav formations, focusing on structural controls on the Kaibab Plateau, groundwater basin delineation and temporal karst development. Ross (2005), Brown (2011) and Schindel (2015) quantified R aquifer karstification, residence time and geochemical properties, but their analyses were limited by a lack of measurements of high-discharge events and long-term continuous data sampling.

Flow in karst aquifers can be turbulent in conduits and may have three components: intergranular porosity, fractures and conduit-dominated flow (Bonacci & Jelin 1988; Scanlon *et al.* 2003). Karst aquifers are typically challenging to study because of this heterogeneity (Goldscheider *et al.* 2007); in addition, the R aquifer of the GRCA is so deep that conventional methods of determining specific storage and transmissivity are difficult and expensive (Parise *et al.* 2015). However, recent studies and the application of a wider variety of field and geospatial techniques have been applied to better characterize the processes functioning in this karst aquifer system. This paper summarizes these new studies, which have been designed to provide: (1) more detailed characterization of the surficial karst on the Kaibab Plateau using geographical information system and LiDAR analyses; (2) an initial insight into flow paths from the surface to the springs using non-toxic fluorescent dye tracers; and (3) the use of hydrograph recession curves to analyse the recharge response of the R aquifer via Roaring Springs. The unique interactions of the shallow and deep karst systems in the GRCA may be able to better inform aquifer management decisions and research in other karst regions worldwide.

Study area

Located in northern Arizona in the southwestern USA (Fig. 1), the GRCA lies in an area with an arid to semi-arid environment characterized by

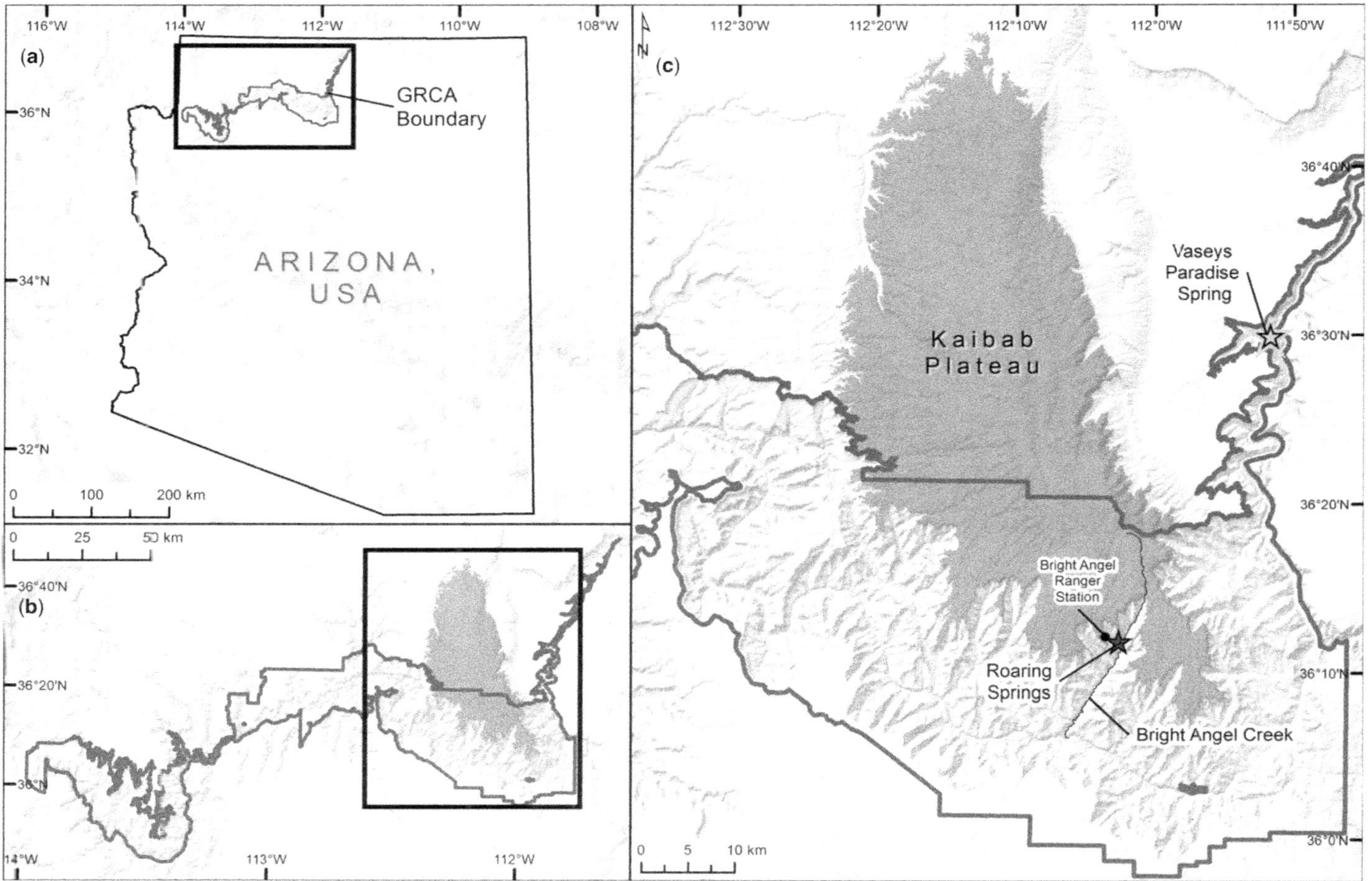

Fig. 1. (**a**) Location map showing the state border of Arizona and the boundary of the Grand Canyon National Park (GRCA). (**b**) Boundary of the GRCA and location of the study area. (**c**) Study area showing the locations of the Kaibab Plateau, the Bright Angel Ranger Station (precipitation data), Roaring Springs, Vaseys Paradise Spring and Bright Angel Creek.

Fig. 2. Photograph of Roaring Springs, Grand Canyon National Park discharging from several outlets (photograph taken by Abe Springer).

extreme vertical relief and stark climate gradients. Bisected by the Colorado River, the sheer canyon walls provide rare access to deep aquifers through springs gushing from caves in the cliff faces. These springs provide potable water, habitats and refuges hundreds of metres below the rim (Figs 2 & 3) in an otherwise desert environment. The Kaibab Plateau is an uplifted region bordered to the south by the North Rim of the Grand Canyon. The region covers *c.* 2460 km^2, existing within the bounds of both the GRCA and the adjoining Kaibab National Forest. Unlike the surrounding arid regions, the Kaibab Plateau is classified as a temperate forest climate, averaging 652 mm of precipitation per year (National Oceanic & Atmospheric Administration 2013) and reaching a maximum elevation of 2810 m a.s.l.

Climate

Precipitation in the Grand Canyon region is typically bimodal, with wet seasons occurring in both the summer and the winter. Winter precipitation is primarily fed by moisture originating in the north Pacific Ocean, which is transported eastward by polar and subtropical jet streams (Sheppard *et al.* 2002; Hereford 2007), whereas the summertime (July–September) precipitation is a result of the North American monsoon (Hereford 2007).

Since 1996, the southwestern USA has been in the midst of what scientists have called the 'early twenty-first century drought' (Cayan *et al.* 2010). Comprehensive climate records exist for the city of Flagstaff, Arizona, 130 km south of the Grand

Fig. 3. Photograph of Vaseys Paradise Spring, Grand Canyon National Park (photograph taken by Abe Springer).

Canyon. A 30% reduction in accumulated precipitation was observed in Flagstaff between 1996 and 2011 compared with the preceding 15 years (1981–96) (Hereford 2007). This reduction occurred mostly in winter precipitation, which is substantially influenced by large oceanic–atmospheric cycles that affect the winter temperatures and precipitation in the southwestern USA. The El Niño Southern Oscillation and the Pacific Decadal Oscillation (PDO) have been shown to affect weather patterns (Sheppard *et al.* 2002). Warm PDO phases have coincided with increased moisture in the southwestern USA, whereas cool phases in the PDO have coincided with drier conditions. The current 'early twenty-first century drought' in the SW has coincided with the latest cool phase of the PDO beginning in 1999 (Sheppard *et al.* 2002; Hereford 2007).

Recharge area, sources and mechanisms

The Kaibab Plateau is a classic representation of a snowmelt-dominated karst aquifer system. Snowmelt runoff and precipitation infiltrate the Kaibab Plateau rapidly via sinkholes, faults and fractures, and slowly through diffuse infiltration. Once in the subsurface, it travels hundreds of metres vertically and kilometres laterally through the karst system in the R aquifer (Brown 2011). Most precipitation (*c.* 60%) falls during the winter (November–March) as snow, which subsequently melts during spring (March–May) when low temperatures, minimal plant use and saturated conditions in the vadose zone allow more water to recharge the aquifer system. Roaring Springs primarily responds to recharge as a result of melting of the winter snowpack, with relatively little recharge to base flow occurring during the summer monsoon season (Ross 2005; Schindel 2015). Discharge from winter snowmelt peaks during late spring and decreases to base flow fed by the primary intergranular porosity during the summer monsoon season (Ross 2005) (Fig. 4).

Stratigraphy

The stratigraphy in the Grand Canyon is globally renowned and well-studied due to tremendous exposures resulting from a combination of the Kaibab Plateau uplift and downcutting by the Colorado River. Numerous studies have focused on quantifying the different stratigraphic layers and evolutionary

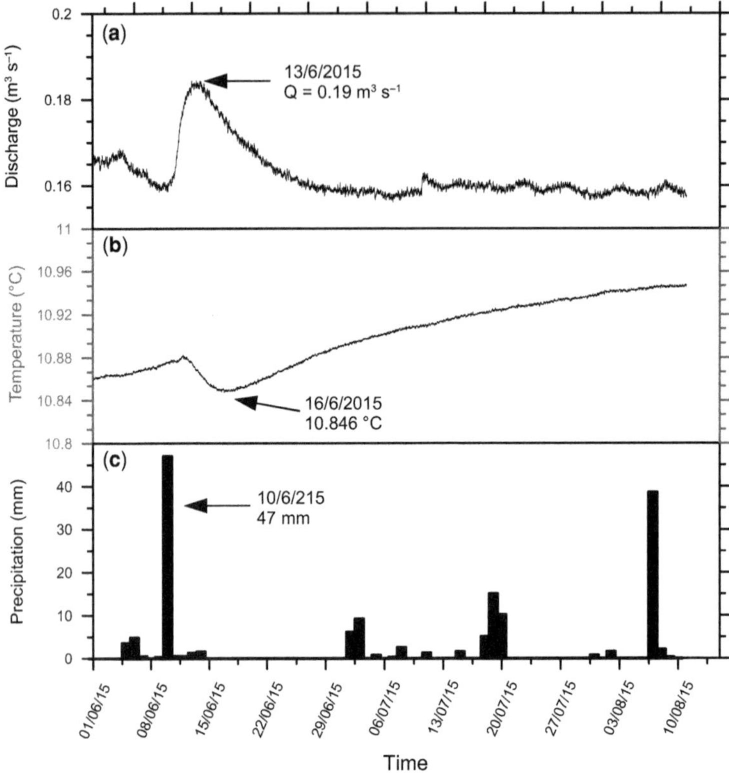

Fig. 4. (**a**) Discharge and (**b**) temperature response at Roaring Springs following a (**c**) 47 mm monsoonal precipitation event on 10 June 2015 over the Kaibab Plateau that drained to the spring.

history of the Kaibab Plateau over multiple decades (Huntoon 1974; Beus 1990*a*). The deep canyon and the regional, southward-dipping strata create two distinct aquifers with very different recharge areas and flow paths. Although more studies have focused on the aquifers of the southern rim of the canyon due to ease of access and the regional dependence on groundwater (Crossey *et al.* 2006), the aquifers of the Kaibab Plateau on the North Rim are less affected by human development and thus are less studied. Although the karstic R and C aquifers are the focus of this paper, the overlying and underlying units of the R and C aquifers are relevant for a comprehensive understanding of the hydrogeology (Fig. 5).

The R and C aquifers occur within Paleozoic strata of the canyon. Underlying this is a crystalline, Precambrian core and the sedimentary Grand Canyon Supergroup. These strata typically have very low porosity with minimal water storage. The Paleozoic sequence consists of sedimentary rocks, including sandstone, limestone and shale (Fig. 5). The Bright Angel Shale, a *c.* 100 m thick layer in this part of the Grand Canyon, acts as a regional aquitard, causing nearly all the groundwater to discharge at or above the shale (Huntoon 1974; Ross 2005). The composition of micaceous clay seals any secondary faulting or fracturing, further reducing the Bright Angel Shale's permeability (Huntoon 1974).

Overlying the Bright Angel Shale are strata that compose the R aquifer: the Muav, Temple Butte and Redwall formations (Fig. 5). These lower Paleozoic carbonates are well recognized in the Grand Canyon as steep, vertical cliffs and have low primary porosity unless fractured or karstified (Huntoon 1974). However, the large amount of dissolution and faulting in this aquifer causes it to be one of the largest stores of groundwater in this region.

The oldest layer in the R aquifer is the Muav Formation, a *c.* 100 m thick layer composed of laminated carbonates along with dolomitic and calcareous mudstone (Middleton & Elliot 1990) (Fig. 5). A complex intertonguing relationship characterizes the contact between the Muav Formation and the Bright Angel Shale. The Muav Formation is the base of the R aquifer and the majority of large springs below the Kaibab Plateau emerge at the contact between the Muav Formation and the Bright Angel Shale. Overlying the Muav Formation is the

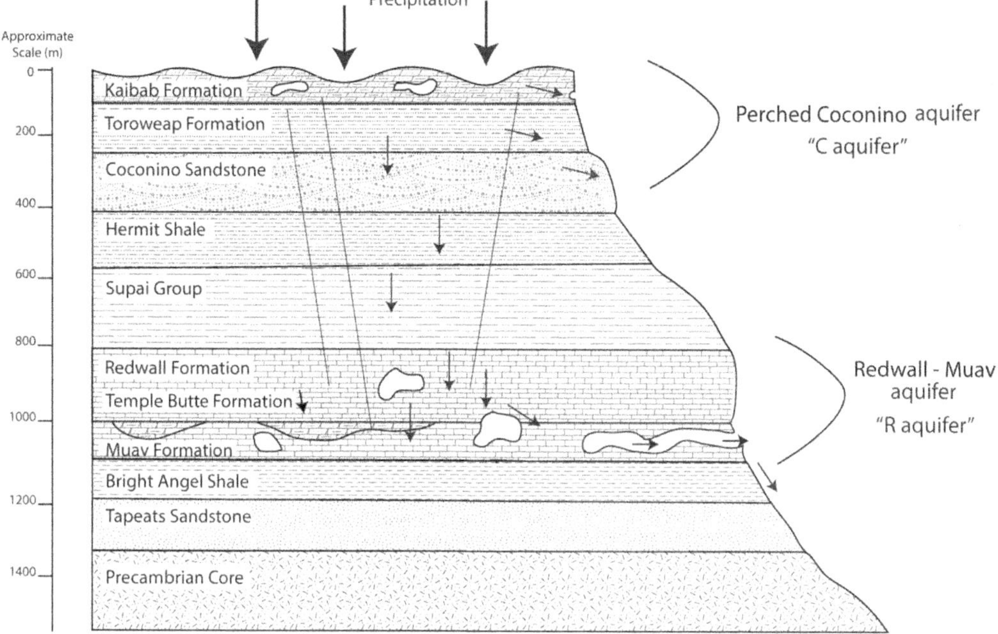

Fig. 5. Idealized conceptual profile oriented along the plane of a permeable fault zone cross-cutting all the strata and serving as a vertical hydraulic conduit. The permeable fault zone vertically connects an aerially extensive circulation system within the C aquifer. This circulation system drains to the fault and to horizontal karst conduits dissolved along the plane of the fault within the R aquifer.

Temple Butte Formation, which consists of dolostone lenses, most often <30 m thick (Beus 1990*b*). The Temple Butte Formation is negligible in the Kaibab Plateau region. The overlying Redwall Formation forms stark red cliffs stained by iron oxide from the overlying Supai Group. The Redwall is up to 250 m thick and consists predominantly of limestone (Beus 1990*a*). Large cavern development occurs throughout this formation (Huntoon 2000). Although there are a few hydrologically active unconfined caves, the Redwall Formation is primarily home to the majority of hydrologically inactive caves in the GRCA. Karstification throughout the formation has occurred along fractures parallel and sub-parallel to regional faults and fractures (Hill & Polyak 2010).

Huntoon (1974) has suggested that these caves and associated groundwater movement were driven by existing geological structures, with flow occurring along faults and fractures. Subsequently, Huntoon (2000) proposed that the dewatering of the aquifer and movement of springs from the Redwall Formation to the stratigraphically lower Muav Formation was due to the incision of the Grand Canyon and associated tributaries, such as Bright Angel Creek. The dissolution that forms the currently active karst springs and hydrologically inactive caves has been the result of a combination of epigenic and hypogenic waters, creating a unique karst system (Hill & Polyak 2010).

Between the R aquifer and the perched C aquifer are >300 m of sedimentary rock that act as a leaky aquitard. Although these formations are not water-bearing, they are important in the transport of water from the surface and the C aquifer to the R aquifer below. The Supai Group (Wescogame, Watahomigi and Manakacha formations and the Esplanade Sandstone) and the Hermit Shale compose the stratigraphic units between the R and C aquifers. The complex Supai Group is primarily composed of sandstone, mudstone and some limestone and dolomite, whereas the Hermit Shale consists of siltstones and mudstones (Blakey 1990). The rocks of these stratigraphic units have low permeability where undisturbed, but groundwater flows along localized faulting, vertical joints and bedding partings (Huntoon 1970).

The C aquifer occurs within the uppermost portion of the Grand Canyon stratigraphic sequence and consists of the Coconino Sandstone, the Toroweap Formation and the Kaibab Formation (Fig. 5). The C aquifer is a minor water-bearing unit, but small-scale springs do discharge on and near the North Rim and around the perimeter of the Kaibab Plateau (Huntoon 1970; Ross 2005). Where the underlying Hermit Shale is undisturbed, the C aquifer appears to have greater saturation and behaves as a perched aquifer. The Coconino Sandstone is *c.* 100 m thick and is a fine- to medium-grained cross-bedded quartz sandstone (Ross 2005). Overlying the Coconino Sandstone is the Toroweap Formation, which includes *c.* 120 m of non-cross-bedded sandstone, gypsum and limestone. The Kaibab Formation occurs at the surface of the Kaibab Plateau and is highly variable in thickness due to erosion. The formation is largely composed of gypsum, dolostone, chert and limestone (Huntoon 1970; Hopkins 1990). The Kaibab Formation is highly karstified with substantial dissolution along faults and fractures. The gypsum in the underlying Toroweap Formation is also highly susceptible to dissolution. The combination of karst features in the Toroweap and Kaibab formations has created an abundance of closed depressions throughout the region (Huntoon 1970).

Surface karst system

The near-surface karst system in the Kaibab and Toroweap formations of the C aquifer on the Kaibab Plateau is a defining feature of the plateau and the sinkholes provide the primary means of recharge to the underlying aquifers (Huntoon 1974, 2000). The unconfined nature of both the C and R aquifers on the plateau results in aquifer responses to storm events that are commonly flashy and variable, depending on the precipitation and the season (Huntoon 2000). These sinkholes recharge an underlying conduit system, both of which are structurally controlled. Limited geophysical evidence is available to map these conduits; however, the morphology of the sinkholes above these conduits can be indicative of conduit size and the ability to channel, store and discharge the incoming water (Panno *et al.* 2013). Geographical information system and LiDAR data for sinkholes were used to better quantify the surficial properties of the C aquifer and to relate the sinkholes as infiltration points to the overall hydrogeology of the system.

Surface karst methods

A sinkhole layer was created from 1 m resolution LiDAR data to characterize the sinkholes of the Kaibab Plateau (Fig. 6). Watershed Sciences was contracted by 3DiWest to fly LiDAR over the Kaibab Plateau for the Kaibab National Forest, Kaibab Ranger District to assess the habitat of the raptor species the northern goshawk. Ground returns with a resolution of 1 m were acquired. The elevation data were masked to an elevation of 2292 m and above to isolate the plateau surface. Depressions were delineated using a basin-fill function in ArcGIS 10.2 (ESRI 2014). The basin-fills were subtracted from the original topographic LiDAR data to create a depression layer. To reduce noise, this layer was smoothed using two iterations of the Focal Statistics-Mean function in ArcGIS (ESRI 2014) set to a 3 × 3 smoothing parameter. This smoothed layer was re-classified and filtered, resulting in sinkholes with a minimum depth of 0.1 m. The sinkhole size statistics table was then exported into Python Programming Language and filtered to eliminate all sinkholes smaller than four pixels in area (4 m^2) to remove artefacts and anthropogenic depressions (Fig. 7). A small sample of field observations and measurements was used to inform the parameters chosen to filter our model. However, additional field measurements would add value to, and increase the accuracy of, these parameters in future studies because field measurements remain a key component of sinkhole investigation to verify topographic data (Basso *et al.* 2013). Python Version 2.7.12 (Python Software Foundation 2016) was used to calculate the measures of sinkhole development, including the depression density and sinkhole area ratio (White 1988), in addition to general sinkhole population size statistics.

To assess similarities between the Kaibab Plateau sinkholes and other karst areas around the globe, sinkhole distributions were grouped by depth and plotted as a frequency–depth distribution using the equation:

$$n = N_o e^{-kd} \qquad (1)$$

where n is the number of sinkholes with a certain depth group, N_o is a constant representing the total number of sinkholes, k is a constant corresponding to the rate of attenuation of the number of sinkholes per depth and d is the depth of the sinkhole (Troester *et al.* 1984).

Surface karst results

The sinkhole analysis method delineated a total of 7457 sinkholes over the 1.45×10^3 m^2 area of the Kaibab Plateau, with volumes ranging from 0.40 m^3 to >1.4×10^6 m^3 (Table 1). The majority of the sinkholes were at the smaller end of the spectrum, typical of most surface karst (White 1988). A linear relationship ($r^2 = 0.68$) between the two-dimensional area of a sinkhole and its depth exists on the Kaibab Plateau (Fig. 8). A map of sinkhole density on the Kaibab Plateau is shown overlain by the mapped geological structures and reveals a correlation between sinkhole density and the presence of faults and fractures (Fig. 9).

A comparison of sinkhole frequency–depth distributions between the Kaibab Plateau and other karst regions shows that the Kaibab Plateau

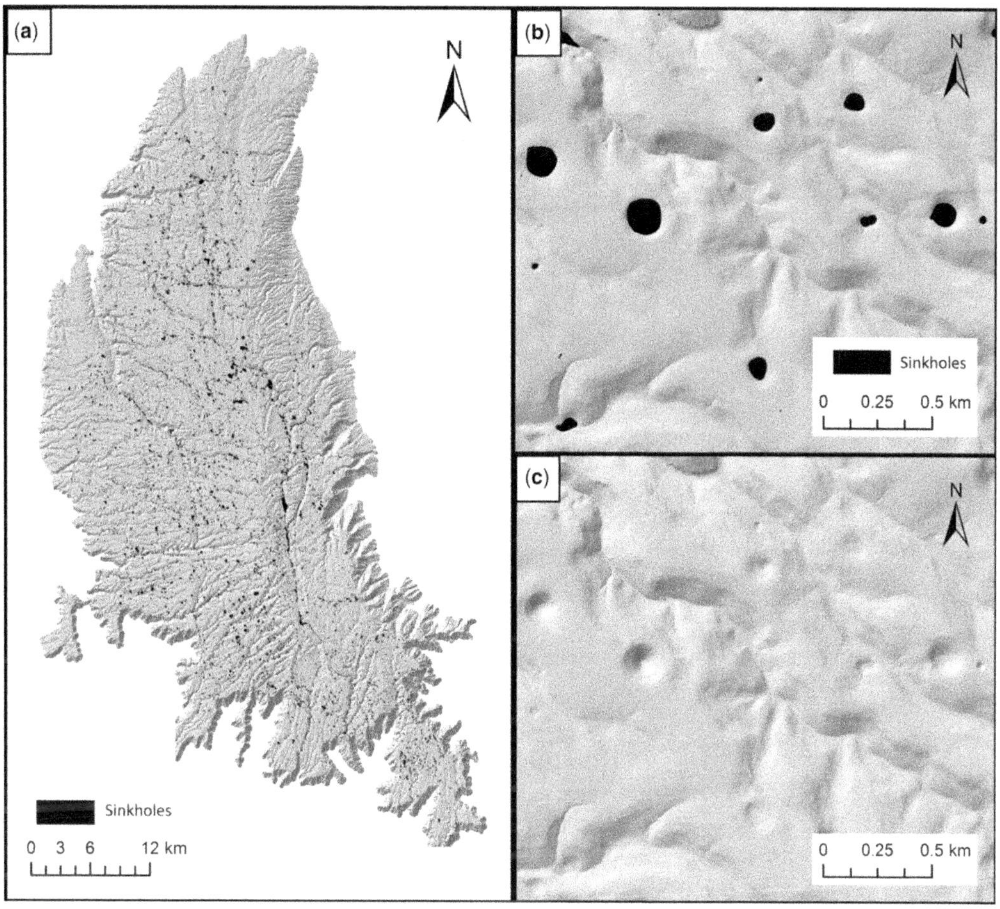

Fig. 6. (**a**) Hillshade layer of LiDAR elevation data of the Kaibab Plateau overlain with ArcGIS-automated sinkholes in black. (**b**) An enlarged section of the Kaibab Plateau LiDAR with Arc-GIS-automated sinkholes in black. (**c**) The same enlarged section of the Kaibab Plateau with the sinkhole layer removed.

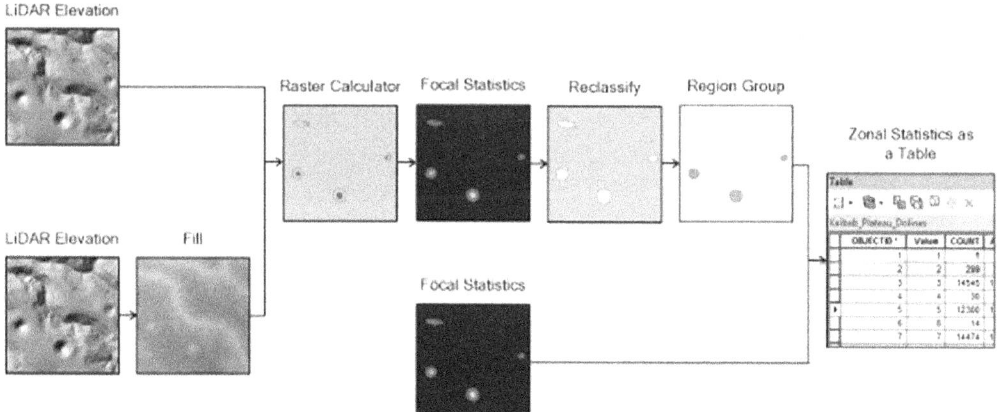

Fig. 7. Flow chart of ArcGIS functions used to identify sinkholes on the Kaibab Plateau and their measurements of area, depth and volume.

Table 1. *Statistical measurements for the Kaibab Plateau sinkholes*

	Depth (m)	Area (m^2)	Volume (m^3)
Minimum	0.10*	4*	0.40
Maximum	48.8	3.76×10^5	1.42×10^6
Mean	1.14	1.32×10^3	4.62×10^3

*Controlled value.

sinkhole depth distribution has a trend line equation of $n = 7460e^{-25d}$, which, when normalized, falls close to the trend line equation for the sinkhole plain of south-central Kentucky (Fig. 10, Table 2).

Surface karst discussion

The majority of the sinkholes were on the smaller end of the spectrum, typical of most surface karst (Table 1, Fig. 8; White 1988). Sinkholes plotted above the trend line in Figure 8 are likely to be steeper and more erosive, indicative of a larger conduit system capable of channelling, storing and discharging larger amounts of water. Sinkholes below the trend line are more likely to be shallow and wide, and may be closely tied to portions of the conduit system that have a lower drainage capacity. These characteristics could prove valuable in future vulnerability mapping of the Kaibab Plateau (Panno *et al.* 2013).

Figure 9 also illuminates a relationship between sinkholes and the location of possible active conduits near regions of major faults and fractures. A similar density model reported by Panno *et al.* (2008) suggests that sinkhole densities are higher in areas with prominent conduit systems. This interpretation is consistent with the conduit hypotheses developed by Huntoon (1974, 2000): dominant conduits exist in the proximity of faults and fractures on the plateau. Such conduits in highly fractured and faulted areas are hypothesized to provide direct connections between the surface and the shallow and deep karst systems of the plateau. Future investigations of sinkhole circularity and azimuth in relation to faults, fractures and joints would help to clarify the presence of structurally driven sinkhole formation on the Kaibab Plateau (Brinkmann *et al.* 2008; Basso *et al.* 2013).

The frequency–depth distribution trend line for the Kaibab Plateau is similar to the trend line of the sinkhole plain of south-central Kentucky (Fig. 10, Table 2) (Troester *et al.* 1984). The size and density characteristics may indicate that the overall geomorphology of the karst conduit system on the

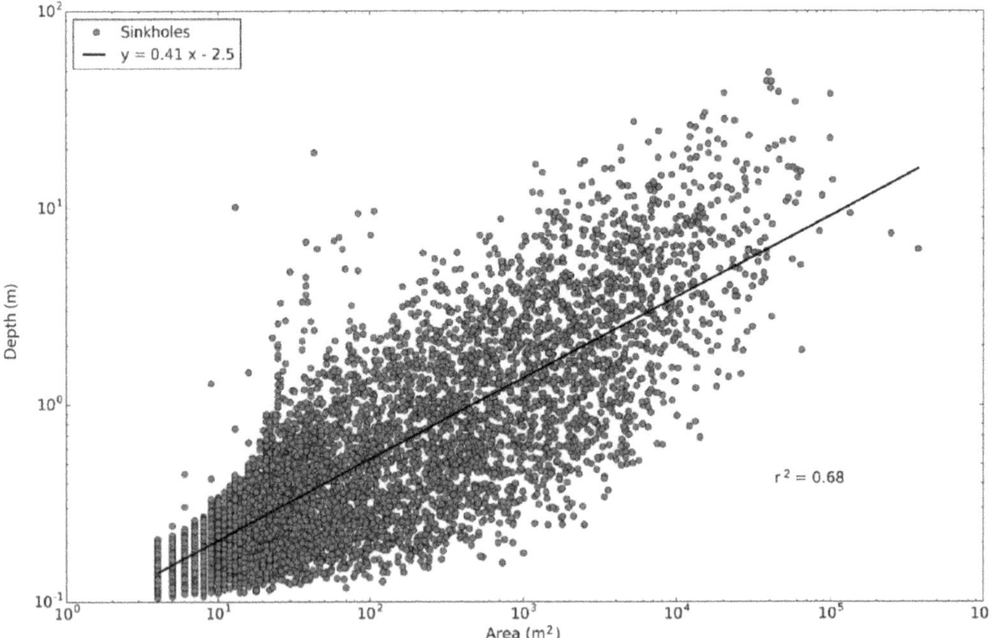

Fig. 8. Scatterplot of the positive relationship between the depth and area of sinkholes on the Kaibab Plateau. Note that both axes are on a logarithmic scale.

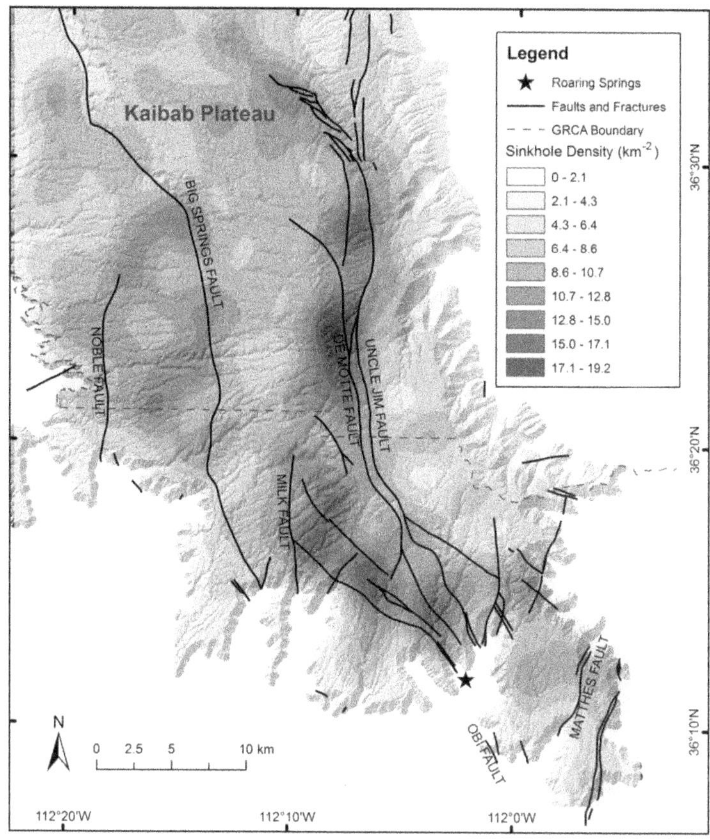

Fig. 9. Map of the Kaibab Plateau showing the density of sinkholes per km^2 overlain with known faults and fractures in the region. Note that the sinkhole density increases with proximity to faults.

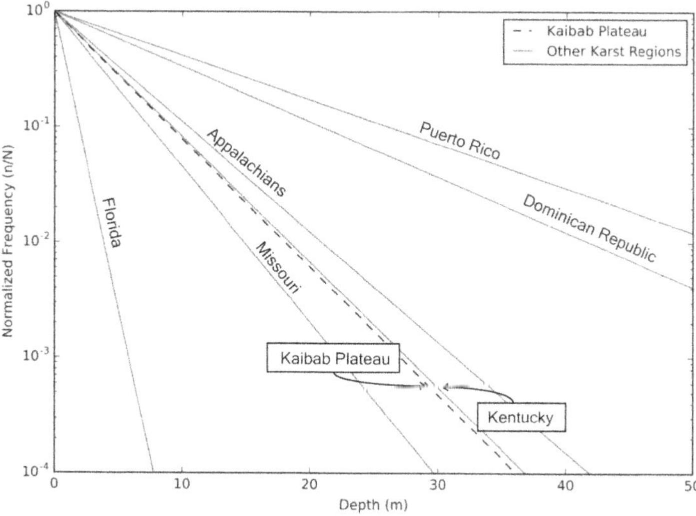

Fig. 10. Graph of the frequency–depth distribution exponential decay trend lines fitted to the karst regions in Troester *et al.* (1984) compared with those for the Kaibab Plateau sinkholes. Note that the slope of the Kaibab Plateau and the Kentucky trend lines are almost identical, indicating similar karst behaviour.

Table 2. *Properties of sinkhole populations and least-squares fitting coefficients for exponential depth distribution from equation (1)*

Karst region	Total number of sinkholes	Depression density (km^{-2})	Mean depth (m)	$D^e = 1/k$ (m)	N_o	k (m^{-1})	r^2
Temperate karst regions							
Appalachian mountains	5.16×10^3	1.25	7.8	4.48	1.26×10^4	0.22	0.99
Kentucky	8.30×10^2	5.41	5.4	4.02	8.92×10^2	0.25	0.99
Missouri	2.22×10^3	–	6.8	3.23	9.79×10^3	0.31	0.99
Florida	3.40×10^3	7.94	–	0.85	1.23×10^4	1.18	0.99
Kaibab Plateau	7.46×10^3	5.14	1.14	3.95	7.46×10^3	0.25	0.76
Tropical karst regions							
Puerto Rico	4.31×10^3	5.39	19	11.35	6.88×10^3	8.8×10^{-2}	0.99
Dominican Republic	7.21×10^3	5.71	23	8.93	6.92×10^4	0.11	0.99

N_o is a constant representing the total number of sinkholes, k is a constant corresponding to the rate of attenuation of the number of sinkholes per depth and r^2 is the fit of the modelled equation (1) to the actual sinkhole depth–frequency distribution (table adapted from Troester *et al.* 1984).

Kaibab Plateau is similar to that of temperate karst regions. The overall groundwater distribution patterns may follow similar trends, at least within the upper 90–100 m of the Kaibab Formation.

Deep karst system

The deep, karstic R aquifer of the Kaibab Plateau cannot be properly understood without acknowledging the interplay and dependence between it and the shallower components of this karst system. The R aquifer of the Kaibab Plateau is buried beneath a thick series of Paleozoic rocks and only outcrops to the east, west and south of the plateau, where it is bounded by deeply incised canyons. There are springs throughout the outcropping of the aquifer, indicating that recharge is probably not occurring in these locations. This suggests that all the water recharging the aquifer is via sinkholes in the overlying rocks. Recharge to the deep aquifer moves first through the perched karstic C aquifer. Groundwater then continues through non-karstic strata between the C aquifer and R aquifer via faults, joints and bedding partings, all of which alter the groundwater flow dynamics of the R aquifer. The interconnection between the two karst aquifers was interpreted through a dye tracer study and hydrograph analysis of discharge from the deep aquifer.

Dye tracer study

A qualitative tracer study was initiated in 2015 to determine the interconnection between sinkholes on the Kaibab Plateau and springs discharging from the R aquifer. This was the first dye trace study conducted in the region.

Dye injection methods. Four different non-toxic fluorescent dyes were injected into four different sinkholes over a two-year period (Fig. 11) to target those sinkholes whose recharge was theorized, based on research by Huntoon (1974), to contribute to Roaring Springs. The amount of dye was determined following the methodology of Worthington & Smart (2003). The injections were made just after snowmelt at two sinkholes in the GRCA in 2015 and just prior to snowmelt at two sinkholes adjacent to the GRCA in 2016 (Fig. 11). On 21–22 April 2015, 1.5 kg of phloxene B and 1.5 kg of sulforhodamine B dyes were flushed into the sinkholes using 3028 l of water each from water trucks, due to lack of flowing water at the time of injection. This lack of flowing water and the finite amount of water used for the injection probably resulted in the dyes being stranded in the shallow subsurface and they were not detected in the deep flow system.

As a result of the lack of positive results, access limitations, weather uncertainty and the remote nature of all the sites, it was not possible to conduct the 2016 injection during snowmelt. Instead, on 22 February 2016, 3 kg of eosin and 5 kg of uranine were buried in the snow within the sinkholes just prior to snowmelt and the subsequent snowmelt flushed the dyes into the groundwater system (Fig. 11). This method is not ideal because it increases the risk of exposing the dyes to increased photolytic decay and also increases the uncertainty of the flow rates, but it has the advantage of ensuring that there is enough water to push the tracer through the aquifer (Benischke *et al.* 2007). Flowing water generally only exists on the Kaibab Plateau during snowmelt, which is typically rapid, occurring over the course of days to weeks.

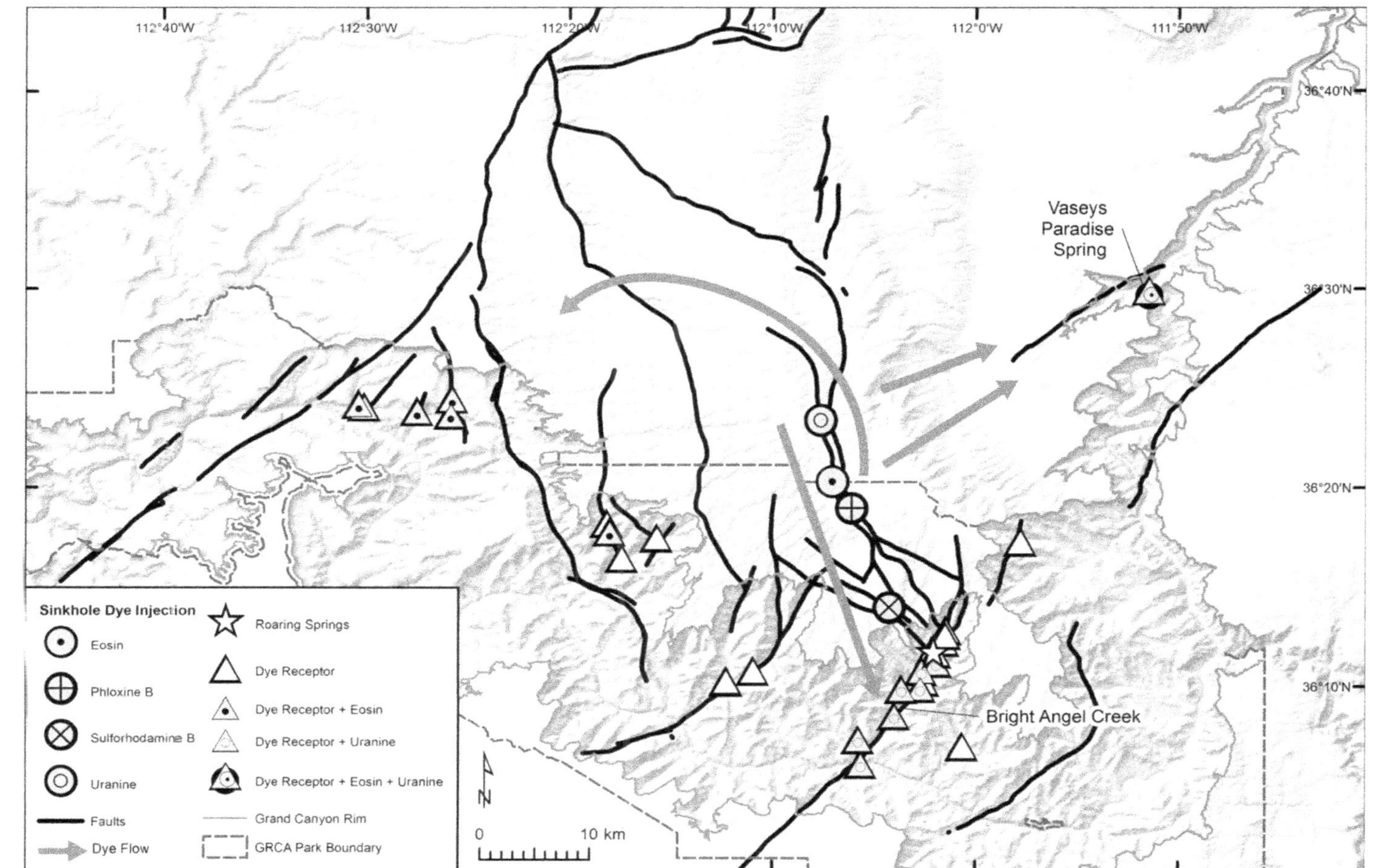

Fig. 11. Locations of the four dye injection sites and the 29 dye receptor sites. Phloxene B and sulforhodamine B were injected during April 2015; eosin and uranine were injected during February 2016. Eosin and uranine (filled triangles) were detected between February and July 2016. The potential generalized flow paths of the eosin and uranine dyes along faults are superimposed on the map. Phloxene B and sulforhodamine B had not been detected by July 2016.

Dye receptor methods. Passive dye monitoring occurred at 29 locations throughout the GRCA (Fig. 11). These monitoring locations were distributed to capture the discharge from more than 41 perennial springs; some locations were in creeks fed by multiple springs. Most of the springs were located at the base of the Muav Formation, with two occurring in the Redwall Formation. Although this methodology limits the amount of data it is possible to collect on the more detailed characteristics of the aquifer (Benischke *et al.* 2007), it was chosen due to a lack of background knowledge of the flow paths, the large number of springs potentially connected to recharge features on the Kaibab Plateau, and the difficulty in accessing and maintaining the monitoring sites.

The dye receptors consisted of nylon mesh packets containing activated charcoal and were assembled at the GRCA physical science laboratory following established protocols (Schindel *et al.* 2007). The receptors were collected and replaced at irregular intervals as allowed by the field conditions and the availability of personnel. Receptor placement and handling procedures were established to maximize the likelihood of recovering the dyes and to minimize the potential sources of error and cross-contamination of samples. After field collection, the dye receptors were rinsed, dried and shipped for analysis. The prepared receptors were analysed by Karst Works (San Antonio, TX, USA) following established methods (Schindel *et al.* 2007). The dyes used for the tracer test and other potential fluorescent substances in the system (from fire retardants and the water infrastructure) were analysed by the laboratory to verify their presence or absence in the receptors.

Methods of dye tracer spatial analysis. We assessed the relationships between positive dye locations, the regionally mapped geology and previous assumptions of flow paths using ArcGIS 10.2 (ESRI 2014) to determine the likely flow paths of the recovered dyes. The results from the dye tracer study were directly overlain on the assumed groundwater basins and flow paths using geological mapping data from Huntoon (1974). The assumed pathways were then modified based on the tracer results.

Results of dye tracer study. Based on the theoretical flow paths from Huntoon (1974), it was assumed that all of the injected dyes would discharge to Roaring Springs or to adjacent springs in Bright Angel Creek. However, eosin, the first dye detected, discharged at springs to the west and one to the east at Vaseys Paradise Spring, but not south near Bright Angel Creek as anticipated (Fig. 11). Eosin was detected between February and May 2016, with the initial detection occurring less than one month after injection. The dye persisted in the system for an additional two months before it was no longer detected (Table 3).

Uranine was detected between May and July 2016. The dye was detected in tributaries to Bright Angel Creek downstream of Roaring Springs and also to the east at Vaseys Paradise Spring (Fig. 11). Uranine injected in February 2016 arrived at these sites within three months of injection and was detected for an additional two months after the initial detection. The other two dyes injected in April 2015 were not detected at any of the dye receptor sites by July 2016.

Discussion of dye tracer study. When the surficial locations of the major faults are overlain on the dye detection results from the eosin test, the patterns that emerge can be related to the flow directions along faults described by Huntoon (1974). Karst is well known for complicated and, at times, unexpected conduit flow paths, and the fault-related structures on the Kaibab Plateau could be avenues for major conduit flow through the plateau. This interpretation may or may not be that simple when we

Table 3. *Dye tracer results for eosin and uranine injections*

Injection site	Injection elevation (m)	Amount of dye (kg)	Recovery site	Recovery elevation (m)	Distance between injection and recovery (km)	First detection after injection (months)
Eosin	2687	3	Deer Spring	835	35	<1
			Thunder River	980	31	<1
			Tapeats Spring	1125	28	<1
			Vaseys Paradise	895	29	<1
			Merlin Spring	1355	17	<1
Uranine	2657	5	Transept Creek	1300	24	2–3
			Ribbon Creek	1255	26	2–3
			Phantom Creek	940	30	2–3
			Wall Creek	1180	26	2–3
			Vaseys Paradise	895	27	2–3

consider the propagation of the surface expression of these faults through the various layers of limestone, sandstone and shale (Fig. 5). Shale typically deforms plastically along faults and therefore is considered to be a barrier to flow. Potential conduit flow along faults indicates that either the faulted shale is not acting as a barrier to flow, or that the flow could be predominantly along parallel and sub-parallel fractures related to the faults. Huntoon (1974) noted that the caves generally follow major fractures sub-parallel to larger regional faults. It is likely that more water follows these sub-parallel fractures than the faults themselves as a result of the prohibition of flow by deformation along the faults. In addition, because cave passage patterns in the region show strong fracture control (Fig. 12) highly related to the faults and fractures mapped on the surface, similar patterns may exist in intermediate geological units.

The results of the eosin dye tracer study suggest that the flow patterns follow the hypotheses proposed by Huntoon (1974); however, the locations of the detection of the uranine dye complicate these interpretations. The uranine dye was injected into the northernmost of two sinkholes associated with the same fault, whereas the eosin dye was injected into the southernmost of the two sinkholes. Although the eosin dye arrived at springs east and west, but not south, of Bright Angel Creek (Fig. 11), the uranine dye discharged into the tributaries feeding the creek south of Roaring Springs as well as to the east at Vaseys Paradise Spring. The contradictory arrival of uranine in Bright Angel Creek to the south suggests two possible explanations: (1) differences in horizontal and vertical flow paths from each sinkhole or (2) differences in the types, sources and timing of the water moving

Fig. 12. Photograph of a cave passage showing the linear morphology typical of fracture-controlled speleogenesis (photograph courtesy of Skye Salganek).

the dyes through the system, resulting in different arrival times. Although both dyes were injected during the same day, heterogeneity in snowmelt could have resulted in the delayed movement of uranine through the system.

Although a vertical connection between the C and R aquifers probably still occurs along these fractures, the downward movement of water along these vertical pathways and horizontal flow within the two aquifers are more complex than initially thought. The different flow paths and arrival times at the springs could suggest that the sinkhole where eosin was injected is more closely associated with major conduits, where the dye infiltrated rapidly and then moved quickly through the conduit system to the springs. Conversely, this explanation suggests that the sinkhole into which the uranine was injected is more removed from the conduit system, resulting in slower flow and/or a longer flow path.

Flow paths can vary depending on the intensity of precipitation, the source of water and the antecedent conditions, as has been observed in many other karst systems (Wong *et al.* 2012; Schwartz *et al.* 2013; Reisch & Toran 2014; Parise *et al.* 2015). It is likely that summer precipitation from intense, highly heterogeneous monsoon events results in very different flow patterns from the relatively slow and homogeneous infiltration associated with snowmelt. As a result of the later detection times associated with the uranine dye receptors (May–late July), it is possible that the dye was only partially transported vertically into the subsurface during snowmelt and was not mobilized until monsoonal moisture pushed it through the entire flow system. Conversely, the eosin dye was first detected at receptor sites within a month of injection, indicating that this dye was immediately transported vertically and horizontally through the system with only snowmelt.

Spring hydrograph analysis

Dye trace analyses alone are not sufficient to interpret the complexities of the geology overlying the deep karstic aquifer of the Kaibab Plateau. The spring discharge hydrographs in the deep karst aquifer are affected by the complex structure of the thick, overlying and partially karstified stratigraphy (Fig. 5). Spring hydrograph analysis allows the determination of aquifer water storage and temporal discharge distributions because the discharge can be directly measured in the field (Groves 2007; Fiorillo *et al.* 2012). Karstification leads to a hierarchical arrangement of conduit flow paths that converge and discharge at karst springs (Fiorillo 2014). In karst systems such as the GRCA, and other systems where springs are potable water sources, spring flow monitoring and forecasting are important for ensuring future water supplies (Ford & Williams 2007). Hydrograph analysis at Roaring Springs not only shows a connection between the R aquifer and surface recharge, but also aids in characterizing the vertical and horizontal flow patterns of the hydrogeological system. The discharge response accounts for flow through >1000 m of overlying lithostratigraphic units, including the perched C aquifer. Compared with the shallow karst aquifer system, the spring hydrographs of the deep karstic R aquifer indicate a complex, unknown history of transit time, flow paths and residence times. Analyses of spring hydrographs were used to quantify the dominant flow regimes from both summer monsoon storms and winter snowmelt events, the lag time of the spring response to precipitation and the qualitative amount of recharge contributing to the base flow for different hydrograph peaks. Although other large springs discharge from the R aquifer (e.g. the Vaseys Paradise Spring, Fig. 3), Roaring Springs (Fig. 2) was chosen for analysis because of its importance as the sole water source for visitor, residential and commercial use within the GRCA.

Stage–discharge relationship methods. Previous studies have attempted to collect complete discharge data (Ross 2005; Brown 2011; Schindel 2015), but the sites were inadequate for determining the total flow from Roaring Springs due to non-ideal transducer locations and the lack of high flow discharge measurements during the collection period. For this study, a transducer was relocated to capture all of the flow from Roaring Springs upstream of a distributary system in the cave near the spring's mouth. Because karst spring discharge can vary rapidly compared with other types of springs (Groves 2007), the stage and temperature were recorded at 15 minute intervals with an In-Situ Level TROLL 500 water level data logger (In-Situ, Fort Collins, CO, USA). Ten discrete discharges were measured, including the high flow after peak snowmelt, from November 2015 to August 2016, to create a stage–discharge relationship:

$$\log(Q) = 1.07 \times \log(H) - 0.264 \qquad (2)$$

where Q is the discharge in $m^3\ s^{-1}$ and H is the stage in metres. This equation showed a strong correlation between stage and discharge ($r^2 = 0.812$) and a rating curve was used to convert the stage data to continuous discharge values.

Hydrograph analysis methods. The shape of a discharge hydrograph and the timing of response varies with the type of precipitation, the intensity of precipitation, the flow paths and the drainage area. To assess the variability in transit time, groundwater recharge and precipitation response, the hydrographs generated from the transducer data at Roaring

Springs were evaluated for both summer monsoon events and winter snowmelt events. Precipitation was recorded at Bright Angel Ranger Station (COOPID 21001), the closest precipitation gauge to Roaring Springs on the Kaibab Plateau (Fig. 1).

Recession methods. The recession limb of each hydrograph extends from the discharge peak to the beginning of the next rise (Fiorillo 2014). Recession curves were analysed using a modified form of the equation of Maillet (1905), solved for the recession slope:

$$\alpha = \frac{\log\ (Q_n/Q_{n+1})}{0.4343 \times (t_{n+1} - t_n)} \quad (3)$$

where α is the recession coefficient in days^{-1}, Q is the discharge in m^3 s^{-1}, t is the time in days, 0.4343 is a constant conversion factor for relating Q and t in their respective units, and n corresponds to the microregime being evaluated. When the recession curve is plotted in semi-logarithmic space as log $[Q]$ v. t, α is the slope of the linear relationship between log $[Q]$ and time. The curve provides a characterization of aquifer drainage. Although Maillet's equation was initially constructed for homogeneous aquifers with high porosity, the equation is useful for comparing more complex karst aquifers (Kresic & Stevanovic 2010). The inverse of the recession coefficient, $1/\alpha$, determines the amount of time it would take for the aquifer to drain if the dominant microregime continued without any other recharge events (Tobin & Schwartz 2016). This value provides a method of comparing microregimes and different recession events.

Comparison of microregimes. Recession curves plotted using equation (3) display multiple straight-line components representing the different levels of aquifer porosity dominating groundwater flow. These different levels of aquifer porosity are referred to as microregimes (Bonacci & Jelin 1988; Kresic & Stevanovic 2010). Multiple microregimes are often observed as a result of the dissolution, structural patterns and conduit development occurring in karst. In the case of Roaring Springs, three microregimes have been identified from the recession curves. The initial steepest slope (α_1 days^{-1}) represents the tertiary porosity or the 'quick flow' through conduits and caves. The intermediate slope (α_2 days^{-1}) indicates the secondary porosity and is probably dominated by water discharging from fractures. The last slope (α_3 days^{-1}) is the flattest and probably represents water discharging slowly from the intergranular (matrix) porosity (Kresic & Stevanovic 2010). The base flow microregime (α_3 days^{-1}) provides a measure of aquifer storage during extended dry periods (Ford & Williams 2007). This is the most stable of the microregimes and is less dependent on surface precipitation patterns. It thus gives the best measure of the characteristics of the matrix porosity (Amit *et al.* 2002).

Regional precipitation patterns can provide long recession periods between precipitation events and flood events at karst springs, thus creating long recession curves for analysis. The Roaring Springs hydrograph data between February 2015 and August 2016 yielded four peaks suitable for recession curve analysis (Fig. 13). Recessions from monsoonal events (mn_1 and mn_2) were analysed and compared with snowmelt recessions (sw_1 and sw_2) (Fig. 13). Although hydrograph responses to the snowmelt events are much more complex, with multiple peaks, the final recession of the snowmelt season provided a method of comparing aquifer microregimes with the summer precipitation.

Storm response timing. The retardation time, or the time from the start of monsoonal precipitation to the spring response, describes the response time of the aquifer to precipitation. An initial increase in spring flow is the result of kinematic waves that pulse through the aquifer, at times 30% faster than the actual water (Ford & Williams 2007). The response of temperature changes at the spring also can be used to more accurately determine the retardation (lag) time of storm water events. The beginning of the decrease in recorded temperature and the bottom or base of the curve (pre-storm arrival) were analysed to determine the monsoonal retardation time rather than the peak discharge. The retardation time from winter snow, which often stays on the ground for days to months before melting and infiltrating the subsurface, is much more complicated and was not analysed.

Recession analysis. Recession curves for the monsoon events mn_1 and mn_2 had different microregime responses from those of the snowmelt events sw_1 and sw_2 (Fig. 14). The snowmelt responses are typified by large complex peaks, whereas the monsoon responses show only one recession slope and have a much lower discharge than the snowmelt responses. The recession curves for the monsoon event mn_1 and the snowmelt events sw_1 and sw_2 have three microregimes, whereas the recession curve for monsoon event mn_2 has only two microregimes. The three recession coefficients for monsoon event mn_1 are steep and about an order of magnitude different (Table 4). Monsoon event mn_2 has only two dominating microregimes, both of which have less steep slopes than mn_1. Monsoon event mn_2 was not as large or as steep as mn_1, which may indicate that the fracture (α_2 days^{-1}) and base flow (α_3 days^{-1}) microregimes are predominant.

The recession curves of the two snowmelt events (sw_1 and sw_2) also differ from each other (Fig. 14).

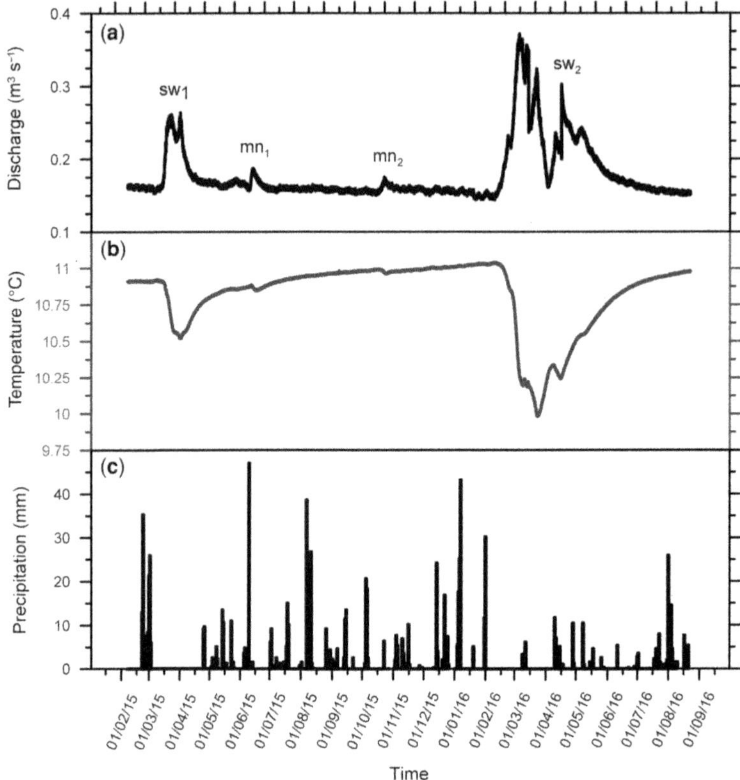

Fig. 13. (**a**) Discharge, (**b**) temperature and (**c**) total daily precipitation data for Roaring Springs cave from 6 February 2015–22 August 2016. sw_1 = early snowmelt recession in April–May 2015, mn_1 = summer monsoonal recession in June–October of 2015, mn_2 = autumn monsoonal recession in October–February 2015–16, sw_2 = snowmelt recession beginning in May 2016 and continuing until the end of data collection (22 August 2016).

Although the transducer was only active from the middle of winter in early February 2015, the response at the end of the 2014–15 winter season is apparent. This season was shorter and drier than the 2015–16 winter season and caused the discharge to appear to be similar to that of monsoon mn_1. The 2015–16 winter season, however, involved substantially more snowfall and subsequent snowmelt. Consequently, all three recession coefficients were much lower for snowmelt event sw_2 (Table 4).

Microregime analysis. The microregimes differed in length and intensity depending on whether the magnitude of various recharge events was sufficient to activate different flow paths in the system (conduit, fracture or intergranular flow). The monsoon recession curves were more influenced by lower magnitude precipitation events than the snowmelt recession curves. For monsoon event mn_1, the amount of time attributed to the conduit (α_1 days^{-1}) and fracture (α_2 days^{-1}) microregimes was relatively short, about six and five days, respectively; monsoon event mn_2 had only two microregimes (α_2 days^{-1}, fracture flow and α_3 days^{-1}, base flow). The fracture flow for monsoon event mn_2 lasted nearly four times as long as the same microregime in monsoon event mn_1 (Table 4). Snowmelt (sw_1), conduit (α_1 days^{-1}) and fracture (α_2 days^{-1}) flow dominated the system for about the first 11 days (2.1 and 8.6 days, respectively) (Fig. 14). For snowmelt (sw_2), however, conduit (α_1 days^{-1}) flow dominated the system for a full 20 days and fracture (α_2 days^{-1}) flow was dominant for 30 days (Table 4).

Storm response timing. The transit time of precipitation from the surface of the plateau to Roaring Springs can only be calculated for discrete monsoon events and not winter snowmelt. A significant rainfall event (mn_1) occurred on 10 June 2015, generating 47 mm of rain (Figs 4 & 13). This rainfall event followed several smaller summer monsoon precipitation events. Roaring Springs began rising one day after the peak and reached a maximum on 13

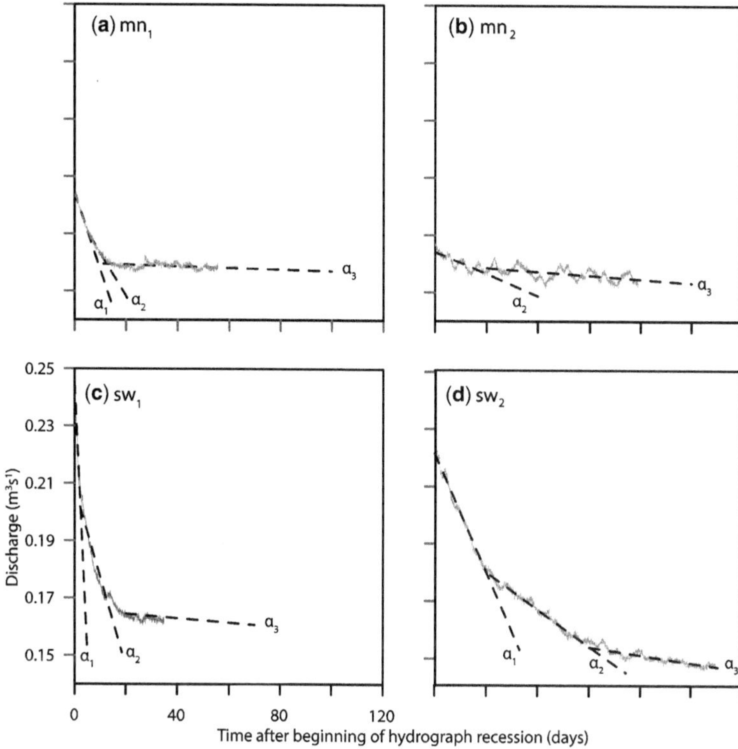

Fig. 14. Recession curves of all four peaks available from the Roaring Springs transducer. (**a**) sw_1 = early snowmelt recession in April–May 2015. (**b**) mn_1 = summer monsoonal recession in June–October 2015. (**c**) mn_2 = autumn monsoonal recession in October–February 2015–16. (**d**) sw_2 = snowmelt recession beginning in May 2016 and continuing until the end of data collection (22 August 2016).

June 2015, three days after the initial rainfall (Figs 4 & 13). The temperature of the spring reached a minimum on 16 June 2015, six days after the major rainfall event (Fig. 4). Monsoon mn_2 discharge and temperature peaks occurred 18 and 20 days after the previous four-day rain event, respectively (Fig. 13). These differences in hydrograph response are represented by the different microregimes

Table 4. *Alpha values (recession coefficients), dates of recessions, microregime duration in days and 1/α for each hydrograph peak and recession for Roaring Springs, 6 February 2015–22 August 2016*

Peak	Event	Recession dates	Alpha value (day^{-1})	Duration of microregime (days)	1/α (days)
mn_1	Monsoon	13 June 2015–8 August 2015	$\alpha_1 = 1.1 \times 10^{-2}$	5.8	97
			$\alpha_2 = 5.9 \times 10^{-3}$	5.3	1.7×10^2
			$\alpha_3 = 2.7 \times 10^{-4}$	45	3.7×10^3
mn_2	Monsoon	25 October 2015–11 January 2016	$\alpha_2 = 3.0 \times 10^{-3}$	20	3.3×10^2
			$\alpha_3 = 3.7 \times 10^{-4}$	59	2.7×10^3
sw_1	Snowmelt	2 April 2015–19 May 2015	$\alpha_1 = 8.6 \times 10^{-2}$	2.1	12
			$\alpha_2 = 2.2 \times 10^{-2}$	8.6	45
			$\alpha_3 = 3.3 \times 10^{-4}$	35	5.6×10^2
sw_2	Snowmelt	12 May 2016–22 August 2016	$\alpha_1 = 8.1 \times 10^{-3}$	20	1.2×10^2
			$\alpha_2 = 3.7 \times 10^{-3}$	30	2.7×10^2
			$\alpha_3 = 9.0 \times 10^{-4}$	53	1.1×10^3

(Fig. 14). The smaller magnitude precipitation event preceding monsoon event mn_2 was not of sufficient size to create a conduit (α_1 days^{-1}) flow response or was centred far enough away from the spring that the longer flow path resulted in a dampened response (Fig. 14).

Discussion of spring hydrograph analysis. The qualitative assessment of storm responses suggests that only snowmelt events recharge base flow, whereas the flashier monsoon events cause rapid infiltration through the conduit flow paths without recharging matrix storage. The base flow discharge after snowmelt is shown to increase (Fig. 13), whereas after a monsoon event the base flow discharge returns to its original pre-monsoon base flow discharge.

Differences in microregimes between hydrograph recession curves show how the R aquifer responds to different recharge events and can be compared with other karst aquifer systems (Table 5). Karst dolomite springs in the Judean and Galilee mountains of northern Israel (Amit *et al.* 2002), using the approach of Boussinesq (1904), had base flow (α_3 days^{-1}) recession coefficients about an order of magnitude higher than all the Roaring Springs base flow (α_3 days^{-1}) microregime coefficient, except for snowmelt event sw_1. A chalk spring from the same study, however, had a base flow (α_3 days^{-1}) microregime recession coefficient of 6×10^{-4}, which is similar to those for Roaring Springs. Amit *et al.* (2002) concluded that the local lithology has an important role and aquifers with lower permeability have a less steep base flow (α_3 days^{-1}) microregime. The quick flow microregimes of the system, which include both conduit (α_1 days^{-1}) and fracture (α_2 days^{-1}) flow, had variable recession coefficients, with a much higher value for the chalk spring (Amit *et al.* 2002). In the Kaweah river basin (California, USA), karst is present in marble bands (stripe karst) within the granitic to granodioritic host rock of the region (Tobin & Schwartz 2016). Only base flow (α_3 days^{-1}) recession coefficients were included in this study and the results were similar to the Israeli springs reported by Amit *et al.* (2002). The recession curve for the Crnojevića Spring in the Dinaric karst of Montenegro was analysed after the Cetinje polje flooded in 1986 (Bonacci 1993). The geology in this area consists largely of Mesozoic limestones, with some Quaternary dolomite (Bonacci 1993). The average base flow (α_3 days^{-1}) microregime coefficient for the R aquifer via Roaring Springs is 8.4×10^{-4}, which is similar only to the chalk spring from Amit *et al.* (2002). This suggests that the matrix controlling the discharge to Roaring Springs has very low permeability relative to most karst systems (Amit *et al.* 2002). The R aquifer is comparatively deep and overlain with both karst and non-karst features. The relatively flat-lying base flow recession curve could be controlled by either the very low porosity in the R aquifer, the low porosity of the overlying non-karst water-bearing units, such as the Coconino Sandstone, or a combination of water draining from the matrix of both the C and R aquifers.

Variability in precipitation also plays a major part in the variability of the aquifer response (Ford & Williams 2007; Schwartz *et al.* 2013). Precipitation events were highly variable for each hydrograph peak (Fig. 13). The precipitation preceding monsoon event mn_1 was shorter and higher in intensity, with 47 mm of rain in just one day (Fig. 4). Rain preceding monsoon event mn_2 was more temporally distributed, with 58 mm over four days. The location, intensity and antecedent conditions play a major part in defining the amount and type of recharge in the system (Schwartz *et al.* 2013). The perched C aquifer adds to the complications of aquifer response

Table 5. *Summary of seven microregime recession coefficients from previously published work*

Karst region	Geology	Conduit microregime (α_1 day^{-1})	Fracture microregime (α_2 day^{-1})	Intergranular porosity microregime (α_3 day^{-1})	Reference
Judean and Galilee mountains of Israel	Dolomite	2.83×10^{-2}–6.41×10^{-2} (average = 4.48×10^{-2})		3.3×10^{-3}–5.2×10^{-3} (average = 4.2×10^{-3})	Amit *et al.* (2002)
	Chalk	0.1052		6×10^{-4}	
Kaweah River Basin, California, USA	Karstified marble bands within granite to granodiorites	No data	No data	7×10^{-4}–1.27×10^{-2} (average = 5.9×10^{-3})	Tobin & Schwartz (2016)
Crnojevića Spring, Montenegro	Limestone with some dolomite	5.0×10^{-2}	2.7×10^{-2}	1.0×10^{-2}	Bonacci (1993)

with multiple storage types and transport pathways connecting it to the underlying R aquifer (Huntoon 1974).

Conclusions

The interpretations and analyses of the shallow and deep karst aquifers in this study show that the Kaibab Plateau is a highly developed and complex karst region. The surface expression of karst on the Kaibab Plateau resembles internationally recognized karst regions such as the sinkhole plain of south-central Kentucky, USA. The observable increase in sinkhole density in proximity to faults and fractures (or inferred faults and fractures) suggests a strong connection between these structural features and water infiltration into the karst system. Panno *et al.* (2008) have shown a significant relationship between the surface morphology of sinkholes and the underlying conduit system. This relationship in other regions suggests that the surface patterns of the Kaibab Plateau are representative of the shallow conduit system within the karstic portion of the C aquifer. This relationship may indicate that there is significant horizontal flow within the upper C aquifer.

Interpretations of the dye tracer study indicate the prevalence of significant conduit flow and a high level of connectivity between the shallow and deep aquifers. The relatively rapid flow rate of the eosin dye indicates that a large pulse of snowmelt pushed the dye vertically through almost 2000 m of strata in less than five to six weeks, while travelling >40 km horizontally. The rapid subsurface flow is highly indicative of conduit-dominated flow from sinkholes on the surface of the plateau to the springs. This rapid conduit-dominated flow, however, does not determine whether the horizontal flow is predominantly in the shallow C aquifer or the deeper R aquifer. There is a distinct possibility that horizontal flow could be occurring in conduits within either the shallow system, deep system, or both, with vertical flow occurring along fractures and faults in numerous locations.

The locations of detection and slower rate of flow for the uranine dye complicate the interpretation of flow paths. Uranine dye appeared counter-flow to the apparent direction of travel of the eosin dye and took an additional 8–12 weeks to travel through the system and be detected at springs. The apparent difference in flow paths could be due to different vertical flow paths through the system or to a shift in the regional groundwater divide depending on the amount of precipitation and the antecedent conditions. The longer travel time for the uranine dye indicates that the flow paths are either slower because of the conduit morphology or length, or activated by different precipitation regimes from some of the sinkholes on the plateau. The different flow paths are illustrative of the complicated heterogeneity associated with Grand Canyon karst, including vertical flow paths through non-karstic units and other variations between the sinkholes used for the dye injections and their relationship with the underlying conduits.

The possible differences in dye flow paths, arrival times and locations may be seen in the differences in hydrograph responses to snowmelt and monsoonal events. There are significant differences in the responses between the snowmelt and monsoonal storm events. These hydrograph variabilities indicate differences in the dominant flow paths during the recession limb. Snowmelt probably results in recharge and flow through conduits, faults and fractures, and recharge to the intergranular matrix, whereas the monsoonal events show the activation of primarily the conduit/fracture components of the aquifer, not the matrix. Differences in the travel times of these events are probably due to the heterogeneity of rainfall, the configuration of the flow path and the distance from the recharge source to the spring.

Once flow recesses to the base level, the recession coefficients (α mean = 0.00047) are an order of magnitude lower than those measured in other karst systems (i.e. Amit *et al.* 2002; Tobin & Schwartz 2016). The smaller recession coefficients of the Kaibab Plateau demonstrate that the base flow is emerging from matrix storage that has a substantially lower permeability than in most karst systems. This is consistent with the fact that, elsewhere in the Greater Grand Canyon region, the Coconino Sandstone is the significant, but low permeability, water-bearing unit in the C aquifer and releases water from storage that becomes an increasingly more important contributor to the base flow of the R aquifer. The well-indurated limestones that compose the R aquifer have negligible matrix permeability, but dissolution along fractures and bedding planes or partings can account for some of the contribution to base flow.

Our interpretations of the complex interactions of the shallow and deep karst aquifers of the Kaibab Plateau demonstrate how challenging it is to understand the coupled systems. The integration of multiple methods with dye tracers, geospatial analyses of the surface karst and hydrograph analyses allow the verification of aquifer models in a way that cannot be achieved with only one technique. These tools are vital in truly understanding karst aquifers and will allow planning of long-term impacts on water use from potential hazards. Through a combination of reinterpreting old data coupled with new analyses and data, the aquifer systems of the Kaibab Plateau appear to defy previous assumptions of flow path directions and storage patterns and encompass multiple aquifers and lithologies that are crucial in maintaining the flow of springs in the Grand Canyon. Some distinct flow patterns can be linked to major faults, caves and work hypothesized in the 1970s; however, some of those original hypotheses have been revised. Sinkhole characterization, dye trace analysis and hydrograph recession curve analysis provide an initial baseline of understanding of the shallow and deep karstic aquifer systems of the Kaibab Plateau and GRCA and can be used to develop strategies for protecting the Roaring Springs water supply as the sole source of water used in the GRCA. However, a considerable amount of work remains to be done to truly understand the dynamics of groundwater flow and the relationship between the shallow and deep karst aquifers in this region.

This research was conducted under NPS research permit #: GRCA-2016-SCI-0028. Essential funding was provided by the Geological Society of America, the National Park Service's Geoscientists-in-the-Parks Program and the Environmental Stewards Program. The authors thank Peter Huntoon for pre-submission editing and guidance, Lawrence Spangler for his helpful review and an anonymous reviewer.

References

Amit, H., Lyakhovsky, V. *et al.* 2002. Interpretation of spring recession curves. *Ground Water*, **40**, 543–551, https://doi.org/10.1111/j.1745-6584.2002.tb02539.x

Anderson, M.F. 2000. *Polishing the Jewel: An Administrative History of Grand Canyon National Park*. Grand Canyon Association, Monographs, **11**.

Bakalowicz, M. 2005. Karst groundwater: a challenge for new resources. *Hydrogeological Journal*, **13**, 148–160.

Basso, A., Bruno, E., Parise, M. & Pepe, M. 2013. Morphometric analysis of sinkholes in a karst coastal area of southern Apulia (Italy). *Environmental Earth Sciences*, **70**, 2545–2559.

Batz, N., Colombini, P. *et al.* 2016. Groundwater controls on biogeomorphic succession and river channel morphodynamics. *Journal of Geophysical Research: Earth Surface*, **121**, 1763–1785, https://doi.org/10.1002/2016JF004009

Benischke, R., Goldscheider, N. & Smart, C. 2007. Tracer techniques. *In*: Goldscheider, N. & Drew, D. (eds) *Methods in Karst Hydrogeology*. IAH International Contributions to Hydrogeology, **26**. CRC Press, Boca Raton, FL, 147–170.

Beus, S.S. 1990*a*. Redwall limestone and surprise canyon formation. *In*: Beus, S.S. & Morales, M. (eds) *Grand Canyon Geology*. Oxford University Press, New York, 119–145.

Beus, S.S. 1990*b*. Temple Butte Formation. *In*: Beus, S.S. & Morales, M. (eds) *Grand Canyon Geology*. Oxford University Press, New York, 107–117.

Blakey, R.C. 1990. Supai Group and Hermit Formation. *In*: Beus, S.S. & Morales, M. (eds) *Grand Canyon Geology*. Oxford University Press, New York, 147–182.

Bonacci, O. 1993. Karst springs hydrographs as indicators of karst aquifers. *Hydrological Sciences Journal*, **38**, 51–62, https://doi.org/10.1080/02626669309492639

Bonacci, O. & Jelin, J. 1988. Identification of a karst hydrological system in the Dinaric karst (Yugoslavia). *Hydrological Sciences Journal*, **33**, 483–497, https://doi.org/10.1080/02626668809491276

Boussinesq, J. 1904. Recherches théoriques sur l'écoulement des nappes d'eau infiltrées dans le sol et sur le debit des sources. *Journal de Mathématiques Pures et Appliquées*, **10**, 5–78.

Brinkmann, R., Parise, M. & Dye, D. 2008. Sinkhole distribution in a rapidly developing urban environment: Hillsborough County, Tampa Bay area, Florida. *Engineering Geology*, **99**, 169–184.

Brown, C.R. 2011. *Physical, geochemical, and isotopic analysis of R-aquifer springs, North Rim, Grand Canyon, Arizona*. Master's thesis, Northern Arizona University.

Cayan, D.R., Das, T., Pierce, D.W., Barnett, T.P., Tyree, M. & Gershunov, A. 2010. Future dryness in the southwest US and the hydrology of the early 21st century drought. *Proceedings of the National Academy of Sciences, United States of America*, **107**, 21271–21276, https://doi.org/10.1073/pnas.0912391107

Clow, D.W., Schrott, L. *et al.* 2003. Ground water occurrence and contributions to streamflow in an alpine catchment, Colorado Front Range. *Groundwater*, **41**, 937–950.

Crossey, L.J., Fischer, T.P. *et al.* 2006. Dissected hydrologic system at the Grand Canyon: interaction between deeply derived fluids and plateau aquifer waters in modern springs and travertine. *Geology*, **34**, 25–28, https://doi.org/10.1130/G22057.1

Culver, D.C. & Sket, B. 2000. Hotspots of subterranean biodiversity in caves and wells. *Journal of Cave and Karst Studies*, **62**, 11–17.

ESRI 2014. *ArcGIS Desktop: Release 10.2*. Environmental Systems Research Institute, Redlands, CA.

Faulkner, T.L. 2009. The endokarstic erosion of marble in cold climates: Corbel revisited. *Progress in Physical Geography*, **33**, 805–814, https://doi.org/10.1177/0309133309350266

Fiorillo, F. 2014. The recession of spring hydrographs, focused on karst aquifers. *Water Resources Management*, **28**, 1781–1805, https://doi.org/10.1007/s11269-014-0597-z

Fiorillo, F., Revellino, P. & Ventafridda, G. 2012. Karst aquifer draining during dry periods. *Journal of Cave and Karst Studies*, **74**, 148–156, https://doi.org/10.4311/2011JCKS0207

Ford, D. & Williams, P. 2007. *Karst Hydrogeology and Geomorphology*. Wiley, Chichester.

Goldscheider, N., Drew, D. & Worthington, S. 2007. Introduction. *In*: Goldscheider, N. & Drew, D. (eds) *Methods in Karst Hydrogeology*. IAH International Contributions to Hydrogeology, **26**. CRC Press, Boca Raton, FL, 1–8.

Gremaud, V. & Goldscheider, N. 2010. Geometry and drainage of a retreating glacier overlying and recharging a karst aquifer, Tsanfleuron-Sanetsch, Swiss Alps. *Acta Carsologica*, **39**, 289–300.

Groves, C. 2007. Hydrological methods. *In*: Goldscheider, N. & Drew, D. (eds) *Methods in Karst Hydrogeology*. IAH International Contributions to Hydrogeology, **26**. CRC Press, Boca Raton, FL, 45–64.

Han, G. & Liu, C.Q. 2004. Water geochemistry controlled by carbonate dissolution: a study of river waters draining from karst-dominated terrain, Guizhou Province, China. *Chemical Geology*, **204**, 1–21.

Hart, R.J., Rihs, J. *et al.* 2002. *Assessment of Spring Chemistry Along the South Rim of Grand Canyon in Grand Canyon National Park, Arizona.* US Geological Survey Fact Sheet, **096-02**.

Hereford, R. 2007. *Climate variation at Flagstaff, Arizona – 1950 to 2007.* Open-File Report 2007–1410, US Geological Survey, Reston, VA, 17.

Hill, C.A. & Polyak, V.J. 2010. Karst hydrology of Grand Canyon, Arizona, USA. *Journal of Hydrology*, **390**, 169–181.

Hopkins, R.L. 1990. Kaibab Formation. *In*: Beus, S.S. & Morales, M. (eds) *Grand Canyon Geology*. Oxford University Press, New York, 225–245.

Hua, S., Liang, J. *et al.* 2015. How to manage future groundwater resource of China under climate change and urbanization: an optimal stage investment design from modern portfolio theory. *Water Research*, **85**, 31–37.

Huntoon, P.W. 1970. *The hydro-mechanics of the ground water system in the southern portion of the Kaibab Plateau, Arizona.* PhD thesis, University of Arizona.

Huntoon, P.W. 1974. The post-Paleozoic structural geology of the eastern Grand Canyon, Arizona. *In*: Breed, W.J. & Roast, E.C. (eds) *Geology of the Grand Canyon*. Museum of Northern Arizona and Grand Canyon Natural History Association, Flagstaff, AZ, 82–115.

Huntoon, P.W. 1981. Fault controlled ground-water circulation under the Colorado River, Marble Canyon, Arizona. *Ground Water*, **19**, 20–27, https://doi.org/10.1111/j.1745-6584.1981.tb03433.x

Huntoon, P.W. 2000. Variability of karstic permeability between unconfined and confined aquifers, Grand Canyon region, Arizona. *Environmental & Engineering Geoscience*, **6**, 155–170, https://doi.org/10.2113/gseegeosci.6.2.155

Jemcov, I. 2007. Water supply potential and optimal exploitation capacity of karst aquifer systems. *Environmental Geology*, **51**, 767–773.

Karimi, H., Raeisi, E. & Bakalowicz, M. 2005. Characterising the main karst aquifers of the Alvand Basin, northwest of Zagros, Iran, by a hydrogeochemical approach. *Hydrogeology Journal*, **13**, 787–799.

Kresic, N. & Stevanovic, Z. 2010. *Groundwater Hydrology of Springs: Engineering, Theory, Management, and Sustainability.* Butterworth-Heinemann, Amsterdam.

Lauber, U., Ufrecht, W. & Goldscheider, N. 2014. Spatially resolved information on karst conduit flow from in-cave dye tracing. *Hydrology and Earth System Sciences*, **18**, 435–445.

Liu, F., Parmenter, R. *et al.* 2008. Seasonal and interannual variation of streamflow pathways and biogeochemical implications in semi-arid, forested catchments in Valles Caldera, New Mexico. *Ecohydrology*, **1**, 239–252.

Liu, F., Hunsaker, C. & Bales, R. 2012. Controls of streamflow generation in small catchments across the snow–rain transition in the southern Sierra Nevada, California. *Hydrological Processes*, **27**, 1959–1972, https://doi.org/10.1002/hyp.9304

Maillet, E. 1905. *Essai d'hydraulique souterraine et fluviale.* Hermann, Paris.

Middleton, L.T. & Elliot, D.K. 1990. Tonto Group. *In*: Beus, S.S. & Morales, M. (eds) *Grand Canyon Geology*. Oxford University Press, New York, 83–106.

Mueller, J.M., Lima, R.E. & Springer, A.E. 2017. Can environmental attributes influence protected area designation? A case study valuing preferences for springs in Grand Canyon National Park. *Land Use Policy*, **63**, 196–205, https://doi.org/10.1016/j.landusepol.2017.01.029

National Oceanic and Atmospheric Administration 2013. Bright Angel Ranger Station COOPID 21001, data for the period 1948–2013, www.ncdc.noaa.gov/cdo-web/datasets/GHCND/stations/GHCND:USC00021001/detail

Oraseanu, I. & Mather, J. 2000. Karst hydrogeology and origin of thermal waters in the Codru Moma. *Hydrogeology Journal*, **8**, 379–389, https://doi.org/10.1007/s100400000080

Panno, S.V., Angel, J.C., Nelson, D.O., Weibel, C.P., Luman, D.E. & Devera, J.A. 2008. *Sinkhole Distribution and Density of Renault Quadrangle, Monroe County, Illinois.* Illinois State Geological Survey, Champaign, IL.

Panno, S.V., Kelly, W.R., Angel, J.C. & Luman, D.E. 2013. Hydrogeologic and topographic controls on evolution of karst features in Illinois' sinkhole plain. *Carbonates and Evaporites*, **28**, 13–21, https://doi.org/10.1007/s13146-013-0157-2

Parise, M., Ravbar, N., Živanovic, V., Mikszewski, A., Kresic, N., Mádl-Szőnyi, J. & Kukuric, N. 2015. Hazards in karst and managing water resources quality. *In*: Stevanovic, Z. (ed.) *Karst Aquifers – Characterization and Engineering*. Professional Practice in Earth Sciences. Springer, Berlin, 601–687, https://doi.org/10.1007/978-3-319-12850-4_17

Perrin, J., Jeannin, P.-Y. & Zwahlen, F. 2003. Epikarst storage in a karst aquifer: a conceptual model based on isotopic data, Milandre test site, Switzerland. *Journal of Hydrology*, **279**, 106–124, https://doi.org/10.1016/S0022-1694(03)00171-9

Python Software Foundation 2016. Python Programming Language, Version 2.7.12, release date 2016, www.python.org

Reisch, C.E. & Toran, L. 2014. Characterizing snowmelt anomalies in hydrochemographs of a karst spring, Cumberland Valley, Pennsylvania (USA): evidence for multiple recharge pathways. *Environmental Earth Sciences*, **72**, 47–58.

Ross, L.E.V. 2005. *Interpretive three-dimensional numerical groundwater flow modeling, Roaring Springs, Grand Canyon, Arizona.* Master's thesis, Northern Arizona University.

Salcedo-Sanchez, E.R., Esteller, M.V. *et al.* 2013. Groundwater optimization model for sustainable management of the Valley of Puebla aquifer, Mexico. *Environmental Earth Sciences*, **70**, 337–351.

Scanlon, B.R., Mace, R.E. *et al.* 2003. Can we simulate regional groundwater flow in a karst system using equivalent porous media models? Case study, Barton Springs, Edwards aquifer, USA. *Journal of Hydrology*, **276**, 137–158, https://doi.org/10.1016/S0022-1694(03)00064-7

Schindel, G.M. 2015. *Determining groundwater residence times of the Kaibab Plateau, R-aquifer using temperature, Grand Canyon National Park, Arizona.* Master's thesis, Northern Arizona University.

SCHINDEL, G.M., JOHNSON, S., VENI, G., SCHNITZ, L. & SHADE, B.L. 2007. *Tracer Test Work Plan Kinney and Uvalde Counties.* Edwards Aquifer Authority, San Antonio, TX.

SCHWARTZ, B.F., SCHWINNING, S. *ET AL.* 2013. Using hydrogeochemical and ecohydrologic responses to understand epikarst process in semi-arid environments, Edwards Plateau, Texas, USA. *Acta Carsologica*, **42**, 315, https://doi.org/10.3986/ac.v42i2-3.670

SHEPPARD, A.C., COMRIE, G.D. *ET AL.* 2002. The climate of the US Southwest. *Climate Research*, **21**, 219–238.

SPRINGER, A.E. & STEVENS, L.E. 2009. Spheres of discharge of springs. *Hydrogeology Journal*, **17**, 83–93, https://doi.org/10.1007/s10040-008-0341-y

SZOCS, T., TÓTH, G. *ET AL.* 2015. Long-term impact of transboundary cooperation on groundwater management. *European Geologist*, **40**, 29–33, http://eurogeologists.eu/wp-content/uploads/2015/11/EGJ40_final_LR.pdf

TOBIN, B.W. & SCHWARTZ, B.F. 2012. Quantifying concentrated and diffuse recharge in two marble karst aquifers: Big Spring and Tufa Spring, Sequoia and Kings Canyon National Parks, California, USA. *Journal of Cave and Karst Studies*, **74**, 186–196.

TOBIN, B.W. & SCHWARTZ, B.F. 2016. Using periodic hydrologic and geochemical sampling with limited continuous monitoring to characterize remote karst aquifers in the Kaweah River Basin, California, USA. *Hydrological Processes*, **30**, 3361–3372.

TROESTER, J.W., WHITE, E.L. & WHITE, W.B. 1984. A comparison of sinkhole depth frequency distributions in temperate and tropical karst regions. *In*: BECK, B.F. (ed.) *Sinkholes: their Geology, Engineering, and Environmental Impact.* A.A. Balkema, Rotterdam, 65–73.

WEARY, D.J. & DOCTOR, D.H. 2014. *Karst in the United States: a Digital Map Compilation and Database.* US Geological Survey Open-File Report **2014–1156**, https://doi.org/10.3133/ofr20141156

WHITE, W.B. 1988. *Geomorphology and Hydrology of Karst Terrains.* 1st edn. Oxford University Press, Oxford.

WONG, C.I., MAHLER, B.J., MUSGROVE, M. & BANNER, J.L. 2012. Changes in sources and storage in a karst aquifer during a transition from drought to wet conditions. *Journal of Hydrology*, **468**, 159–172.

WORTHINGTON, S.R.H. & SMART, C.C. 2003. Empirical determination of tracer mass for sink to spring tests in karst. *In*: BECK, B.F. (ed.) *Sinkholes and the Engineering and Environmental Impacts of Karst.* Geotechnical Special Publication **122**. American Society of Civil Engineers, 287–298.

Vulnerability assessment and its validation: the Gömör-Torna Karst, Hungary and Slovakia

VERONIKA IVÁN & JUDIT MÁDL-SZŐNYI*

Department of Physical and Applied Geology, Faculty of Science, Eötvös Loránd University, 1/C Pázmány P. stny, Budapest-1117, Hungary

**Correspondence: szjudit@ludens.elte.hu*

Abstract: Reliability is the greatest concern in recent karst vulnerability assessments. Checking the validity of results and understanding the operation of karst systems are therefore the most promising areas for improvement. We report an application of the Slovene Approach to a test site in the Gömör-Torna Karst of Hungary and Slovakia. The resource vulnerability map provided an appropriate result and showed that the overlying sediments and thicker soil may effectively decrease the vulnerability of this unconfined area. The source vulnerability of the Kis-Tohonya Spring was evaluated for the purposes of protection zoning. The results of long-term hydrograph and recession curve analyses were helpful in understanding the functioning of the Kis-Tohonya Spring and the karst system. The importance of the different regimes contributing to the slow and fast flow of the spring under different hydrological conditions was shown. Ongoing research aims to feed these results back into the source vulnerability assessment of the spring.

Groundwater vulnerability maps support decision-making in the areas of environmental, land use and water management because they synthesize the relevant lithological, pedological, hydrogeological, meteorological, hydrological and geomorphological information in a visual and easy to understand way (Witkowski *et al.* 2007). However, the automatic application of a qualitative or semi-quantitative vulnerability method without an interpretation of the complex hydrogeological context can lead to questionable results. Therefore the validation of karst vulnerability maps using independent methods is a basic issue that should be addressed in conducting more reliable assessments. Hydrograph and chemograph analyses, tracer tests, isotope techniques, and analytical and numerical simulations can be used to achieve this (Neukum 2013).

The Gömör-Torna Karst is a transboundary aquifer straddling the border of Hungary and Slovakia at an elevation between 150 and 650 m a.s.l. (Fig. 1). As a result of its complex natural heritage – which includes karst formations, caves, sinkholes and swallow holes – the region is protected and lies within the Aggtelek National Park and the Slovak Karst National Park. Long-term protection of the groundwater, spring water and caves in the karst area is therefore also required. There are records of long-term daily observations (1964–93) of important springs in the Hungarian part of the system (Maucha 1998). A combined tracer test was carried out in the region in 2014 (Gruber *et al.* 2015*a*) involving swallow holes and springs from both the Slovakian and Hungarian sides.

Within the framework of this study, a source and resource vulnerability assessment was carried out using the Slovene Approach (Ravbar & Goldscheider 2007). This parametric semi-quantitative method is one of the most complex interpretations of the European Approach for Vulnerability and Risk Mapping for the Protection of Carbonate (Karstic) Aquifers recommended by the COST Action 620 Working Group (Daly *et al.* 2002; Zwahlen 2004). It is calibrated to the characteristics of the Slovene karst areas, but has been successfully applied to many different sites (Jimenez-Madrid *et al.* 2012; Zhang 2014). The area evaluated in this study is located in the SW of the Gömör-Torna Karst, partly in Hungary (Aggtelek Karst) and partly in Slovakia (Slovak Karst) (Fig. 1c). The 25.5 km^2 study area was delineated based on the recharge areas of the main springs, taking into account the availability of data. A source vulnerability assessment was completed for only one of the springs in the study area, the Kis-Tohonya Spring. The evaluation was validated based on hydrograph and recession curve analyses.

Hydrogeological environment

Topography, hydrology, climate, soils and vegetation

The study area is delineated by the EOV coordinates, where EOVX: 349 000–358 500, EOVY: 756 000–762 000. EOV is the Hungarian National Grid, a transverse Mercator projection in which the number

From: Parise, M., Gabrovsek, F., Kaufmann, G. & Ravbar, N. (eds) 2018. *Advances in Karst Research: Theory, Fieldwork and Applications*. Geological Society, London, Special Publications, **466**, 261–273.
First published online November 29, 2017, https://doi.org/10.1144/SP466.15

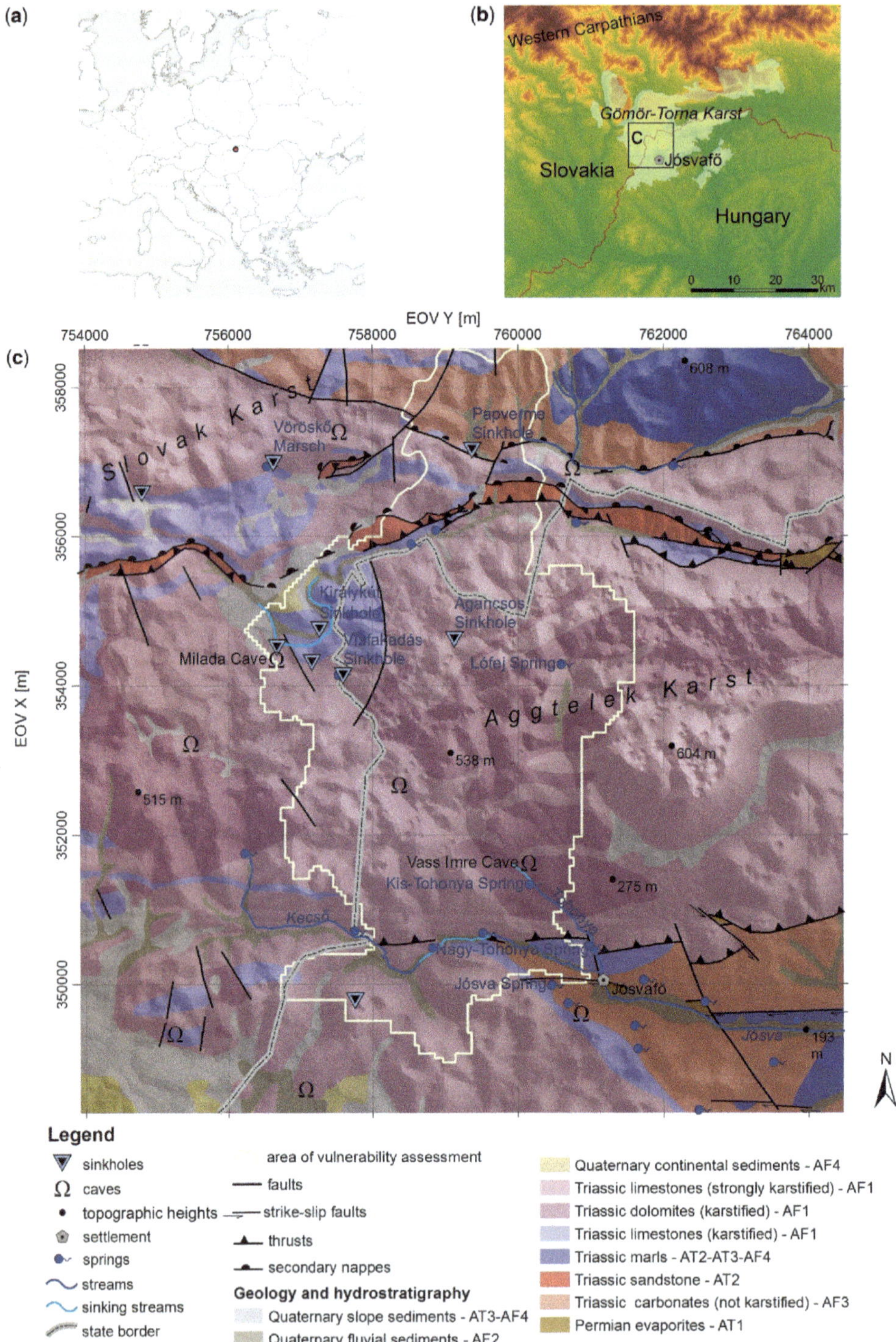

Fig. 1. (**a**) Location of the study area within Europe. (**b**) Location of the study area in the Gömör-Torna Karst, Slovakia and Hungary. (**c**) Geological map of the study area and its surroundings (Szentpétery & Less 2006; original scale: 1:25 000) with the addition of the hydrostratigraphic categories of different formations, shaded according to surface elevation.

refers to metres, and positive X indicates north and positive Y east. The highest elevation is located in the centre of the study area (538 m a.s.l.) and to the NE (608 m a.s.l.). The elevation decreases to the NW and the south to SE. The lowest elevation is 193 m a.s.l. in the Jósva valley. The area is characterized by dry valleys, with alignments of dolines (Zámbó 1998) (Fig. 1c). Two large caves (Milada and Vass Imre) are located in the study area and trend NW–SE across it. The springs are concentrated in the NW, east (Lófej Spring) and SE (Kis- and Nagy-Tohonya and Jósva springs) of the area. Sinkholes transport water from the surface into the karst, especially after precipitation or snowmelt. They are partly in Slovakia (Királykút, Papverme and Vöröskő) and partly in Hungary (Vízfakadás and Agancsos). The most important rivers in the study area are the Kecső Creek, which flows into the Tohonya Creek and later into the Jósva Creek at Jósvafő (Haviarová & Gruber 2006) (Fig. 1c).

The climate is cool and wet above 300 m a.s.l. Areas at lower elevations are characterized by a drier and more moderate climate. The annual mean temperature is 8°C in the highest regions and 8.5°C in the lower regions. The annual precipitation is between 640 and 700 mm (Dövényi 2010). The most common soil types in the area are black and brown rendzinas, together with brown forest soils in the valleys and hollows. Different transitional types of soil are also found depending on the slopes (Zámbó 1986, 1998; Tanács & Barta 2006) and these are complemented by the so-called red clay soil on both the Hungarian and Slovakian sides (Kiss 2012). The natural vegetation displays a good correlation with the bedrock, soil and water conditions. Hornbeam and oak, and in the valleys beech or mixed beech and oak, are typical. The meadows of the area are used for grazing (Tanács & Barta 2006).

Geology and hydrostratigraphy

From a geological point of view, the study area is part of the Aggtelek–Rudabánya Mountains in the Inner Western Carpathians (Szentpétery & Less 2006), but in the hydrogeological literature it is treated as equivalent to the southern part of the Gömör-Torna Karst (Fig. 1b). The Aggtelek and Rudabánya Mountains came into contact along the Darnó strike-slip fault in the Oligocene and Miocene. The geological evolution of the area was influenced by the Szilice facies area in the Triassic and Jurassic. The geological sequence evolved here on a continental crust and the area did not undergo later metamorphism. The stratigraphic units form the Szilice nappe system of the Aggtelek–Rudabánya Mountains (Szentpétery & Less 2006). The stratigraphic units were overthrust onto the lower tectonic sequences in the Middle Cretaceous.

The Perkupa Evaporite formed in a hypersaline environment in the Late Permian and Early Triassic (Fig. 1c, Table 1) and comprises the base of the Szilice nappe system. Deepening of the sedimentary

Table 1. *Important formations in the study area, their lithological characteristics and hydrostratigraphic category, and the derived ly score for the assessment using the Slovene Approach*

Formations in the study area (Less 1997)	Lithology	Hydrostratigraphic characteristics	Hydrostratigraphic category	ly score
Infilling of dry valleys, dolines, slope sediments, autochthon sediments, travertine (Q)	Slope sediments	Aquitard–poor aquifer	AT3–AF4	40
Fluvial sediments (Q)	Fluvial sediments	Good aquifer	AF2	10
Borsod Gravel (Q)	Continental sediments	Poor aquifer	AF4	10
Zlambach Marl (T3)	Marls	Aquitard	AT2	500
Szádvárborsa Limestone (T3)	Limestones (karstified)	Excellent aquifer	AF1	3
Wetterstein Limestone (T2–T3)	Limestones (strongly karstified)	Excellent aquifer	AF1	0.2
Wettersteini Dolomite (T2–T3)	Dolomites (karstified)	Excellent aquifer	AF1	1
Hallstatt Limestone (T3)	Limestones (karstified)	Excellent aquifer	AF1	1
Reifling Limestone (T2)	Limestones (karstified)	Excellent aquifer	AF1	1
Steinalm Limestone (T2)	Limestones (strongly karstified)	Excellent aquifer	AF1	0.2
Steinalmi Dolomite (T2)	Dolomites (karstified)	Excellent aquifer	AF1	
Gutenstein Formation (T2)	Limestones (karstified)	Excellent aquifer	AF1	1
Szinpetri Limestone (T1)	Carbonates (not karstified)	Medium aquifer	AF3	300
Szini Marl (T1)	Marls	Aquitard–poor aquifer	AT3–AF4	
Bódvaszilas Sandstone (T1)	Sandstones	Aquitard	AT2	300
Perkupa Evaporite (P–T1)	Evaporites	Excellent aquitard	AT1	

environment led to siliciclastic sedimentation with the deposition of the Bódvaszilas Sandstone. With transgression, sedimentation moved to the deeper part of the ramp, where the Szini Marl was deposited. Carbonate sedimentation became dominant from the Early Triassic, with deposition of the Szinpetri Limestone, which was followed by the bitumenic Gutenstein Limestone (250 m thick) and the Steinalm Limestone (200–400 m thick) in the Middle Triassic.

The dominantly Middle (and Upper) Triassic limestones of the Szilice nappe form the main karst aquifers of the area. They were formed on the European shelf of the Tethys Ocean. The shallow marine sedimentation in the Aggtelek series continued with the Wetterstein Limestone (1000 m thick). Caves up to 25 km in length mostly formed in the latter three limestone formations. The carbonate sedimentation ended with deep marine carbonate sedimentation (Hallstatt Limestone) and the Zlambach Marl. Quaternary sediments – such as gravel formations, various fluvial formations and doline infilling, slope sediments and travertine – were deposited above the Triassic formations. The formations have been evaluated and categorized based on their hydraulic conductivity (Szentpétery & Less 2006).

Hydrogeological connections and tracer tests

Most of the area consists of karstified and strongly karstified Triassic carbonates (Fig. 1c), constituting a good aquifer (Table 1). They are in complex tectonic relation with the aquitards (Permian evaporites, Triassic sandstones and marls). The typical karstic plateaus with sinkholes and dolines are separated by deep fluvial valleys, constituting smaller, connected, but nonetheless diverse hydrogeological units.

More than 40 tracer tests have been carried out on the Gömör-Torna Karst since 1952 (Maucha 1998). The connection from the Milada Cave to the Kis-Tohonya Spring was first detected in 1963 (Maucha 1998). A combined dye tracer test (uranine, Rhodamine, Tinopal, Phag and Eozine) was carried out in the study area to give a better understanding of the underground flow connections between the sinkholes and springs (Gruber *et al.* 2015*a*). The Kis-Tohonya Spring was also evaluated (Fig. 2, Table 2). The aim of the study was to detect and

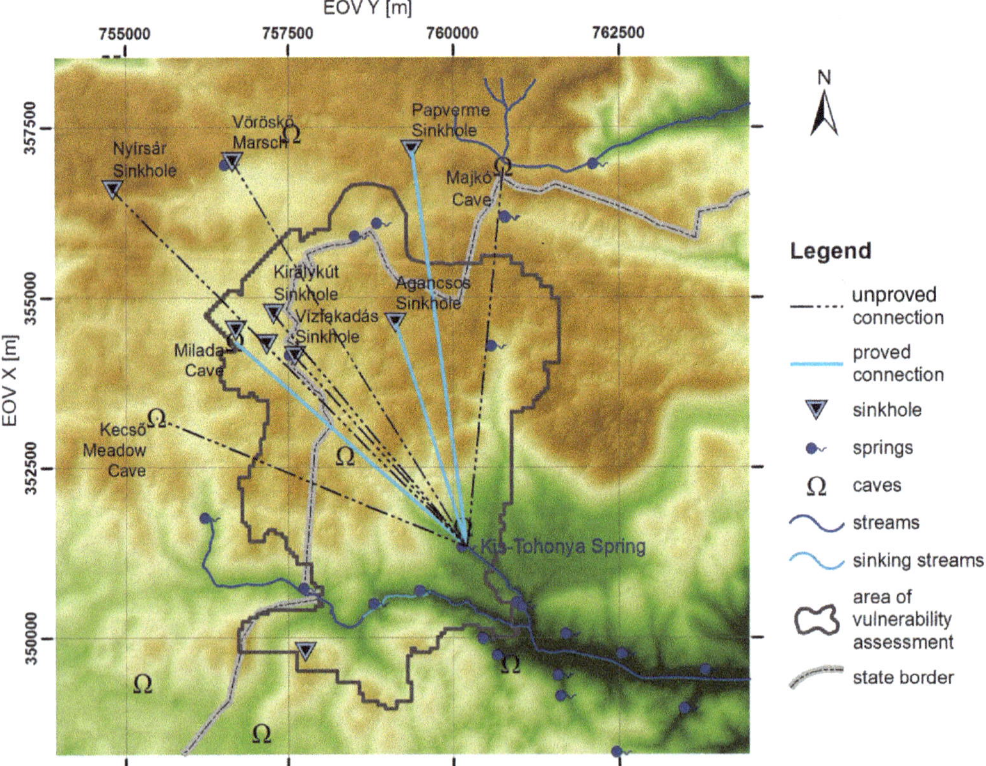

Fig. 2. Hydraulic connections (proved and unproved) between sinkholes, caves and Kis-Tohonya Spring based on the results of tracer experiments (Maucha 1998; Gruber *et al.* 2015*a*).

Table 2. *Basic and derived data of the tracer test between sinkholes, caves and the Kis-Tohonya Spring*

Location of tracer injection	Tracer	Amount of tracer	Distance (m)	Measured and calculated data			
				T_{app} (h)	V_{max} (m h^{-1})	T_{peak1} (h)	V_{peak1} (m h^{-1})
Papverme Sinkhole	Uranine	5 kg	6000	421	14.25	684	8.77
Agancsos Sinkhole	Rhodamine WT 20%	40 l (cca 15 kg)	3500	405	8.64	687	5.09
Vöröskő Marsh	Tinopal	7 kg	6700				
Vízfakadás Sinkhole	H40/1 Phag	8 l (10^{14}/ml pfu)	3800				
Királykút Sinkhole	Eozine	0.5 kg	4500				
Nyírsár Sinkhole	H10 Phag	8 l (10^{14}/ml pfu)	7500				
Kecső Meadow Cave	H40/1 Phag	8 l (10^{14}/ml pfu)	5100				
Majkó Cave	H40/1 Phag	8 l (10^{14}/ml pfu)	5600				

Basic data from Gruber *et al.* (2015*a*).

determine the hydrological connections between the sinkholes and springs in the karstified area, including measurements of the hydraulic parameters and more accurate mapping of the catchment area of the springs. Both Slovakian (Papverme Sinkhole, Királykút Sinkhole, Vöröskő Marsh, Kecső Meadow Cave and Majkó Cave) and Hungarian (Vízfakadás Sinkhole, Agancsos Sinkhole) locations were injected with the tracer (Fig. 2). The tracer study proved the connection between the Papverme and Agancsos sinkholes and the Kis-Tohonya Spring. There was no detectable arrival of the tracer from the other six injection points by the end of the observation period. The times to the first appearances of the tracers (17.5 and 16.8 days) and the first peaks (28.6 and 28.5 days) from the Papverme (6000 m) and Agancsos (3500 m) sinkholes were almost the same (Table 2, Fig. 3). The recoveries at the end of the monitoring period were 27.49 and 49.9% from the furthest and closest sinkholes, respectively. This tracer test also showed that the 48-day observation period was too short because the tracers remained in the system at significant concentrations.

Methods and data sources

The quantity and quality of the available and obtainable data determine the scale, approach and methods of vulnerability assessments. The source and resource vulnerability assessments for the test site on the Gömör-Torna Karst were carried out following the Slovene Approach (Ravbar & Goldscheider 2007). The reasons for this choice were the characteristics of the test site (classic karst with spatially and temporally heterogeneous diffuse and concentrated infiltration), the availability of data and the clear assessment scheme. Spring hydrograph and recession curve analyses were used for the validation.

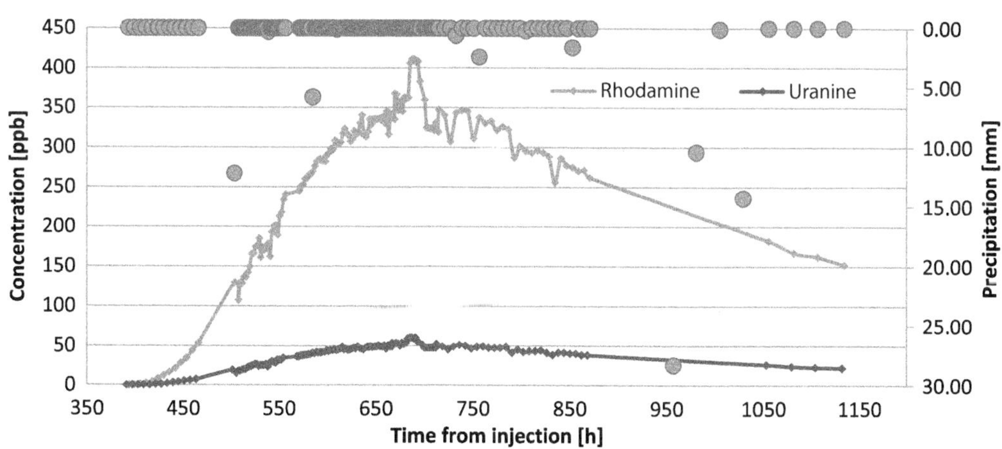

Fig. 3. Breakthrough curve of tracers arriving in the Kis-Tohonya Spring (uranine from Papverme; Rhodamine from Agancsos Sinkhole) and precipitation (data from Gruber *et al.* 2015*a*).

Resource and source vulnerability mapping with the Slovene Approach

The Slovene Approach is a semi-quantitative method for resource and source vulnerability assessment, hazard mapping and contamination risk analysis (Ravbar 2007; Ravbar & Goldscheider 2007) (Fig. 4). It is compatible with the conceptual framework of European COST Action 620 (Zwahlen 2004) and is calibrated to the characteristics of the Slovene karst aquifers. The resource vulnerability assessment is derived from the COP Method (Vias *et al.* 2006). The assessment of resource vulnerability considers mainly vertical percolation through the unsaturated zone towards the groundwater table. The O factor (overlying layers) determines the protection from contamination provided by the characteristics and thickness of the layers of the unsaturated zone in the case of diffuse infiltration (Fig. 4). The C factor (concentration of flow) is a modifier of the O factor and describes the potential for precipitation to bypass the unsaturated zone and to modify its protective function. It allows the consideration of two scenarios: (1) concentrated infiltration in the recharge area of swallow holes and sinking streams, and (2) diffuse autogenic recharge in the rest of the area. Hydrological variability is also integrated into the C factor assessment by considering the temporal variability (activity) of swallow holes. The P factor (precipitation) takes into account the annual number of rainy days and storm events. To determine source vulnerability, an additional factor, K (karst saturated zone), needs to be evaluated. This describes the lateral flow path within the saturated zone. The Slovene Approach has been successfully applied to many test sites (Ravbar & Goldscheider 2009; Jimenez-Madrid *et al.* 2012; Marín *et al.* 2014; Zhang 2014).

In the course of its recent application to the study area, a geoinformation database was constructed, processed and analysed, and the results were presented using ArcMap 10.2 software. The O map assessment started with the evaluation of soil parameters: texture, structure and thickness. These map layers were based on soil measurements taken using a hand auger at a density of 100 points km^{-2} (Zámbó 2003) on the Hungarian side of the test area and there were further detailed evaluations of the soils and red clays (Tanács & Barta 2006; Kiss 2012). The correlation between the geomorphological characteristics and soil thickness has been reported previously by Kiss (2012). The method reported in a study by Tanács & Barta (2006) was used in the interpolation of recent data points. The evaluation of the parameters relative to the lithology was based on the 1 : 100 000 geological map of the Gömör-Bükk area (Less *et al.* 2004), the 1 : 100 000 hydrogeological map of the Aggtelek–Rudabánya Mountains (Sásdi 2006), the 1 : 25 000 geological map of the Aggtelek–Rudabánya Mountains (Less *et al.* 1988) and its explanation by Szentpétery & Less (2006), and the environmental sensitivity map of the Aggtelek Karst and its surroundings (https://map.mfgi.hu/sajboderz25, Gyuricza 2004). Cross-sections of vertical electrical sounding measurements (Gruber *et al.* 2015*b*) were also used in the interpretation.

To evaluate every parameter of the C factor, the following information was used: detailed dynamic device mapping, aerial photographs, a cadaster of swallow holes, field observations and an environmental sensitivity map with Quaternary sediments (1 : 50 000). The assessment of the P factor was based on data from the meteorological station in the test area in Jósvafő. The calculations were based on a 35-year (1958–93) daily precipitation data series (Maucha 1998).

The Slovene Approach is a multiplicative method. The resource vulnerability map can be obtained by multiplication of the O, C and P values. The source vulnerability of the Kis-Tohonya Spring was analysed. The assessment of the additional K factor was based on the results of tracer tests (Gruber *et al.* 2015*a*) for the travel time and connection and contribution parameters, and on the geological map for information on the karst network parameter.

Applied validation methods

In any vulnerability assessment, the validation procedures should ensure that the conceptual understanding of the hydrogeological conditions is valid (Zwahlen 2004). Spring hydrograph and recession curve analyses are appropriate tools for understanding the general functioning of an aquifer by investigating the response of a spring to precipitation events. The most important springs (Lófej, Kis-Tohonya, Nagy-Tohonya and Jósva) have a long-term daily observational dataset for volume discharge from 1 January 1964 to 31 December 1993 (Maucha 1998). This long period is suitable for following the different scales of hydrological reaction in the system. As a first step, the discharge elevation of these springs was compared with the volume discharge values and temperature data (mean, minimum and maximum) to see the influence of different scales of gravity-driven flow systems (Mádl-Szőnyi & Tóth 2015). The variability of the spring discharge was evaluated based on the method used by Maillet (1905): $M = Q_{max}/Q_{min}$ (see Kresic & Stevanovic 2009), where M is the discharge variability coefficient, Q_{max} is the maximum discharge and Q_{min} is the minimum discharge.

The data for Kis-Tohonya Spring were also analysed using recession curve analysis. As a result

Factor	Subfactor				Characterization	Values	Calculation
O factor	O_S	Soils	Texture, structure, thickness		clayey soils >1 m	5	O score = $O_S + O_L$ O_L = Layer index value · cn Layer index = Σ(ly·m)
					clayey soils 0.5–1 m	2	
					clayey soils 0.2–0.5 m	1	
					clayey soils 0–0.2 m	0	
	O_L	Lithology	ly	Lithology and fracturation	strongly karstified limestones	0.2	
					karstified limestones	1-3	
					continental and fluvial sediments	10	
					slope sediments	40	
					not karstified carbonates	300	
					sandstones	300	
					marls	500	
			m	thickness of each layer	0-237 m	0-237	
			cn	confined conditions	confined	2	
					non-confined	1	
C factor	dh	Distance to swallow hole			≤ 10 m	0	C score = dh · ds · sv + tv for swallow hole recharge area; C score = sf · sv for the rest of the catchment
					(10-100 m]	0.2	
					(100-500 m]	0.4	
					(500-1000 m]	0.6	
					(1000-5000 m]	0.8	
	ds	Distance to sinking water body			≤ 10 m	0	
					(10-100 m]	0.5	
					> 100 m	0.75	
	tv	Temporal variability			> 100 d/y	0	
					(10-100 d/y]	0.1	
					≤ 10 d/y	0.25	
	sv	Slope and vegetation			≤ 8% on less permeable surface with less dense/dense vegetation	0.7/0.8	
					(8-31 %] on less permeable surface with less dense/dense vegetation	0.6/0.7	
					> 31% on less permeable surface with less dense/dense vegetation	0.5/0.6	
					≤ 8% on permeable surface with less dense/dense vegetation	1	
					(8-31 %] on permeable surface with less dense/dense vegetation	0.95/1	
					> 31% on permeable surface with less dense/dense vegetation	0.9/0.95	
	sf	Surface morphologic features			Developed karst with absent/permeable subsoil layers	0.25/0.5	
					Scarcely developed karst with absent/permeable subsoil layers	0.5/0.75	
					Fissured karst with absent/permeable subsoil layers	0.75	
					Absent karst features with absent/permeable subsoil layers	1	
P factor	rd	Rainy days			[0-10]	1	P score = rd · se
	se	Storm events			[0-1]	1	
K factor	t	Groundwater travel time			< 1 d	1	K score = t · n · r
					> 10 d	5	
	n	Information on karst network			Water conduit, clear evidence	1	
					Intermediate, unknown	3	
	r	Connection and contribution			< 1%, temporary, remote, not sure	5	
					> 10%, always, direct, sure	1	
Resource vulnerability	Resource score → Resource index				[0-0,5]	1 Extreme	Resource score = O score · C score · P score
					(0,5-1]	2 High	
					(1-2]	3 Moderate	
					(2-4]	4 Low	
					(4-15]	5 Very low	
Source vulnerability	Source score → Source index				1-2	2 High	Source score = Resource index + K index
					3	3 Moderate	
					> 4	4 Low	

Fig. 4. The categorization and values of the Slovene Approach (adapted from Ravbar & Goldscheider 2007) to the study area.

of the long-term nature of the observational data, it was possible to analyse many recession curves from different years within the observation period. An average recession curve was selected (Kresic 2012) with an envelope of minima. This is required to quantify the expected long-term minimum discharge. The equation for the falling limb of the hydrograph and the base flow are given in the Maillet (1905) equation. The recession coefficient α is a dimensionless parameter. It depends on the transmissivity and specific yield of the aquifer and is therefore a characteristic of the hydraulic condition of the aquifer. It is described by a straight line (where α is the slope) in the plot of a semi-logarithmic function. The recession plot was approximated using different recession curves (exponential functions) with different α values. These represent different micro-regimes of flow with different hydraulic behaviours. The flow rate for a spring at any given time after the recession can then be derived.

Results

Resource and source vulnerability assessment

The first parameter evaluated was the soil in the O factor. The Slovene Approach evaluates clayey soils as having lower protective values than loamy, silty soils as a result of the low water retention capacity of the latter as also due to the shrinkage of clay in dry conditions, which allows fast infiltration (Ravbar 2007). As the latter was not expected to occur with any frequency at the mostly densely vegetated test site, the assessed O_S scores for the typically clayey soils were: >1 m, 5; 0.5–1 m, 2; 0.2–0.5 m, 1; and 0–0.2 m, 0.

The lithological parameter was based on the modified ly scores (Table 1) according to the hydrostratigraphy. The thickness of the unsaturated zone was calculated by subtracting the estimated water table (Karst Survey Konzorcium 2014) from the ground surface. The resulting O map is shown in Figure 5a. Most of the area – the unconfined karst areas with a thin soil cover – has a low protective value, whereas the most vulnerable patches are the bare karst surfaces. A lower degree of vulnerability may be observed on aquitards.

To evaluate the two scenarios of infiltration conditions (concentrated and diffuse), the topographical recharge areas of sinkholes were delineated. The moderate vulnerability values on the C map are either from outside the recharge areas of sinkholes on dolomitic surfaces, or within the recharge area, but >1 km from the sinkholes. High and extreme C values appear in the immediate surroundings of the sinkholes and on strongly karstified areas where diffuse recharge is expected given the absence of significant subsoil layers. The finer pattern is a result of the different categories of slope (Fig. 5b).

According to the meteorological data series for Jósvafő, the calculated average annual number of rainy days is 5.16 and the number of storm events is 0.056. Thus, P = 1 over the whole area (Fig. 5c).

The resource vulnerability map is shown in Figure 5e. High and extreme vulnerability classes are dominant on karstified carbonates for both concentrated and diffuse infiltration. A more extensive patch of moderate vulnerability is observable in the south-central part of the test area, which is the far edge of the recharge area of sinking streams. Overlying sediments (such as the typical red clay of the area) or thicker soils may decrease the final result for the degree of vulnerability by one class.

The K factor was also mapped to assess the source vulnerability of the Kis-Tohonya Spring (Fig. 5d). The highest vulnerability was in the topographical recharge area of the sinkholes with a demonstrable and rapid connection to the Kis-Tohonya Spring, as proved by the tracer tests. Medium values were found for cases of proved slow or moderate hydraulic connection. The lowest K values were found in the area downstream from the spring, or divided from it by a valley. The resulting source vulnerability map is shown in Figure 5f. It is clear that the K parameter has the strongest influence on the source vulnerability map. Low K values may overwrite high resource vulnerability values, or refine the patterns. Nevertheless, the effect of overlying sediments or smaller lithological differences (dolomite–limestone) is still observable over most of the area.

Validation: spring hydrograph and recession curve analysis

The long-term observation data for the most important springs of the study area (Kis-Tohonya, Lófej, Nagy-Tohonya and Jósva) were used to calculate the characteristic discharge volume and the temperature data and their variability (Fig. 1, Table 3).

A strong correlation is observable if we evaluate these data as a function of the elevation of discharge, with the highest discharge volume at Jósva (Fig. 1c), which has the lowest elevation (218 m a.s.l.) of discharge. Nagy-Tohonya Spring is also characterized by a high discharge volume. These springs have the highest temperature. Lófej Spring discharges at the highest elevation (428 m a.s.l.) and is characterized by the lowest discharge volume and temperature. The variability of the discharge volume is highest for the Kis-Tohonya- and Lófej springs. By contrast, the maximum and minimum discharge volume is higher at Nagy-Tohonya and Jósva springs, respectively, which are located at lower elevations.

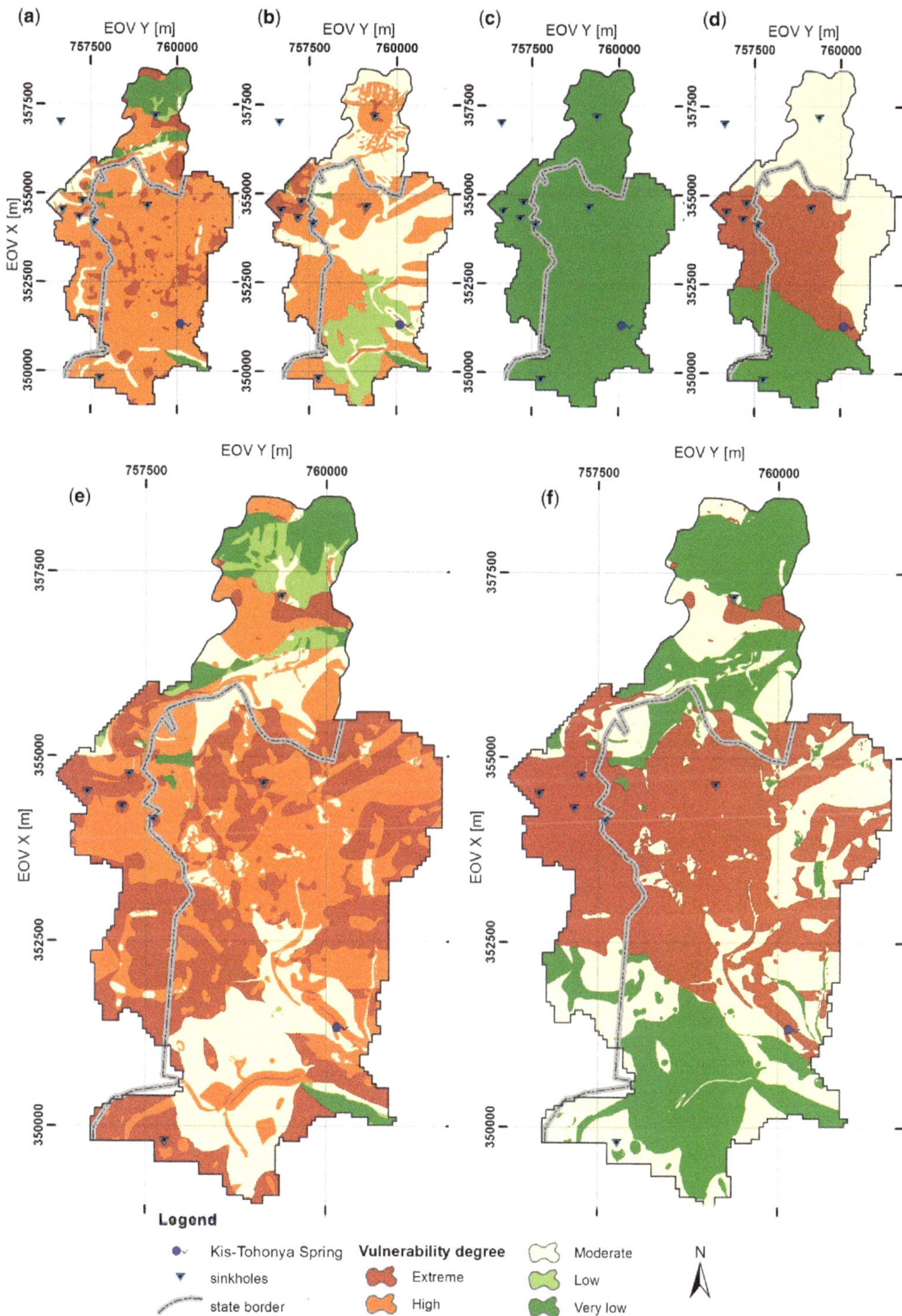

Fig. 5. Parameter maps for the Slovene Approach. (**a**) O map; (**b**) C map; (**c**) P map; (**d**) K map. (**e**) Resource vulnerability map using the Slovene Approach. (**f**) Source vulnerability map using the Slovene Approach for the Kis-Tohonya Spring.

Table 3. *Comparison of the elevation, discharge and temperature data of the most significant springs in the study area*

Spring	Elevation (m a.s.l.)	Q ($m^3 s^{-1}$)			Discharge variability coefficient (M)	T (°C)		
		Mean	Minimum	Maximum		Mean	Minimum	Maximum
Kis-Tohonya	258	0.0176[a]	0.0000[a]	0.4443[a]	38390.0[a]	9.7[d]	9.0[d]	10.2[d]
Lófej	428	0.0056[b]	0.0000[b]	0.1967[b]	16992.0[b]	9.0[d]	8.0[d]	9.8[d]
Nagy-Tohonya	218	0.1059[a]	0.0167[a]	1.1400[a]	68.4[a]	13.5[d]	11.5[d]	14.8[d]
Jósva	218	0.1663[c]	0.0633[c]	6.6667[c]	105.3[c]	13.0[d]	10.8[d]	14.2[d]

Code	Measurement		
	Period	Frequency	Length (years)
a	1964–93	Daily	20
b	1965–93	Daily	29
c	1974–93	Daily	20
d	1981–83	Weekly	3

The observation period and the length and the frequency of measurements are indicated by codes (data source Maucha 1998).

In the study area, the Kis-Tohonya Spring discharges at a medium elevation (258 m a.s.l.) and, among the monitored springs, is characterized by a medium discharge volume and temperature, with a high variability in discharge. These differences reflect the hierarchical flow systems connected to different springs. These indicate that the aquifer has a low storage capacity and a high degree of karstification above the 258 m a.s.l. elevation of the Kis-Tohonya Spring. These characteristics are associated with medium–high vulnerability.

Figure 6 shows the discharge hydrograph of the Kis-Tohonya Spring for half-year rainfall events. The time lag between the rainfall and the beginning of the increase in discharge is two days (18 July–20 July 1986), with values changing from 0.00450 to 0.00506 $m^3 s^{-1}$. The time within which the discharge reaches its maximum value is also two days (22 July 1986; 0.1229 $m^3 s^{-1}$). The time the discharge takes to fall is very long on the decreasing limb of the hydrograph. The volume discharge reaches its original value (0.00416 $m^3 s^{-1}$) within 81 days (by 9 December 1986). This hydrograph reflects the low level of accumulated storage compared with the peak volume discharge value. The peak was 27.3 times larger than the value of the base flow. This means that precipitation transmits through the system dominantly via well-developed

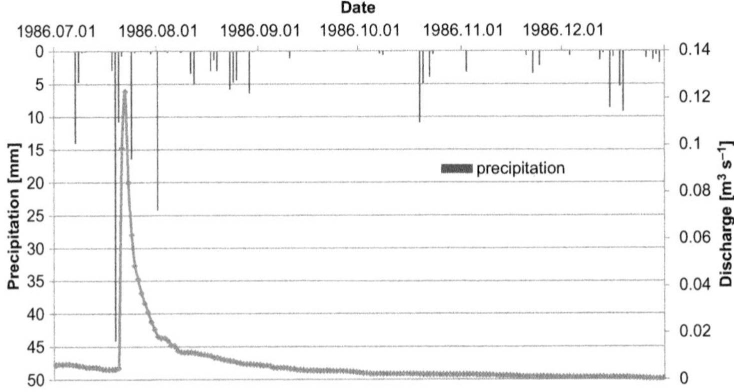

Fig. 6. Discharge hydrograph of the Kis-Tohonya Spring (daily observations, 30 June 1986–20 November 1986) (data source Maucha 1998).

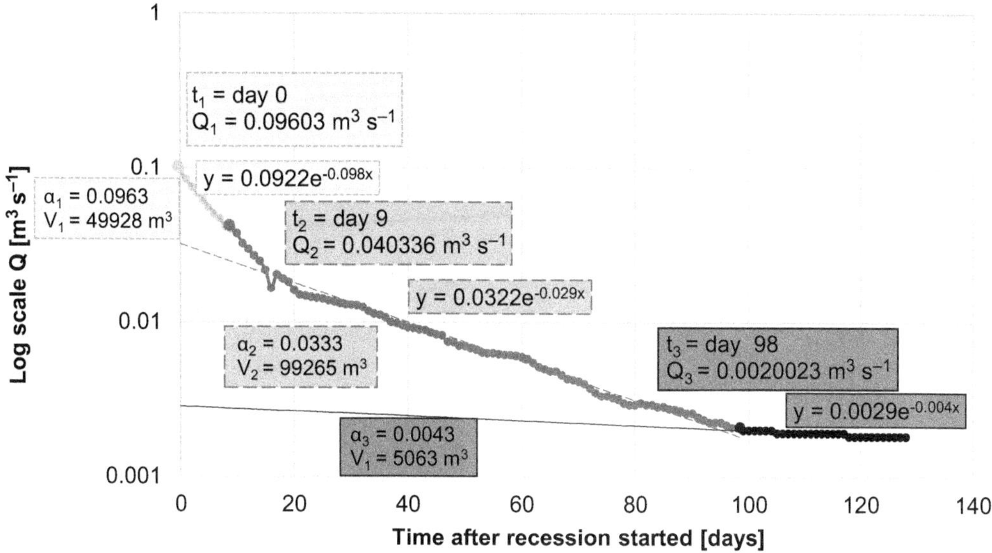

Fig. 7. Recession hydrograph for the Kis-Tohonya Spring (original data from Maucha 1998).

fractures, karst channels and discharges without a significant accumulation of groundwater in the matrix. In terms of vulnerability, this means there is less chance for the degradation and dilution for any possible contaminant.

The individual discharge hydrograph evaluated reflects the effect of only one heavy precipitation event on the volume discharge of the spring. To understand more about the reaction of the spring to precipitation events, the long-term observation data were analysed using recession curve evaluation.

The average recession curve was first selected and analysed. Three different micro-regimes were derived (Fig. 7) and characterized according to the time from the start of recession (0) to 9, 98 and 128 days, respectively, with characteristic threshold discharge volumes of 0.09603 (day 0), 0.040336, 0.0020023 and 0.001829 $m^3 s^{-1}$, respectively. The first micro-regime may be associated with the turbulent drainage of large fractures and conduits. The second transitional period indicates less turbulent flow with the contribution of smaller fractures and the rock matrix, whereas the third period may be interpreted as drainage of the rock matrix and smaller fractures (Kresic 2012). For the given period, 32.4% of the spring discharge volume can be sourced to the first micro-regime, 64.3% to the second and 3.3% to the third. This demonstrates the generally high vulnerability of the recharge area with concentrated infiltration points and also of the areas with diffuse infiltration without significant protection by overlying layers.

Summary and conclusions

The greatest concerns about recent karst vulnerability assessments relate to their validity and reliability (Iván & Mádl-Szőnyi 2017). Groundwater vulnerability maps can only provide an adequate background for decision-making about land use, water and environmental management if they provide relevant information. This study demonstrates the application of the Slovene Approach and compares the results with validation methods and other geological considerations.

The study area of the Gömör-Torna Karst (Hungary and Slovakia) consists of unconfined karst. The protective function of the overlying layers can be attributed to red clay, fluvial deposits and soil cover. This study proved the applicability of the Slovene Approach to the area and therefore, by implication, to other karst aquifers in Hungary and the Slovakian Karst. The available dataset, complemented by additional field data, provided the background for the compilation of the resource vulnerability map (Fig. 5e). Some smaller adaptations and changes were made in the course of the evaluation of soil parameters (for thin clayey soils) and ly scores (for those intermediate situations not specified in the assessment scheme, see Table 1 and Fig. 4). The resource vulnerability map shows the possibility of vertical flow as far down as the saturated zone. High and extreme vulnerability classes are dominant in karstified carbonates for both concentrated and diffuse infiltration. Moderate vulnerability is observed in the south-central part of the study area, which is the most distant part of the recharge area of sinking streams. Overlying sediments (such as the typical red clay) or thicker soils may decrease the final vulnerability index by one class. The source vulnerability of the Kis-Tohonya Spring was evaluated for protection zoning purposes.

The pattern of the source vulnerability map (Fig. 5f) reflects the first-order influence of the K factor. This factor determines the spatial distribution of very low, moderate and extreme vulnerability with regard to the Kis-Tohonya Spring. Based on this map, we suggest that the NW–SE central part of the study area is extremely vulnerable and that there are only limited areas with moderate vulnerability within it. The most vulnerable areas are the immediate surroundings of the swallow holes in the recharge area of the spring. The vulnerability of the north–NE part of the study area is moderate to low, representing slow or moderate hydraulic connections. The lowest K values were determined for the areas downstream from the spring and divided by a valley. The vulnerability here is very low to moderate. Low K values may modify finer patterns. However, the effect of overlying sediments or smaller lithological differences (dolomite–limestones) is still visible over most of the vulnerability map.

The geological map of the study area (Fig. 1c) shows thrust faults and secondary nappes (oriented west–east) in the northern part of the study area and to the north of it. Here, Triassic sandstones crop out and have an even greater extent north of this area. The Papverme Sinkhole is at the tectonic boundary of this larger Triassic sandstone block. In spite of this geological situation, the tracer test proved a direct connection 6000 m from the Papverme Sinkhole, indicating hydraulic communication with the Kis-Tohonya Spring. This example highlights the complexity of the problem of carrying out protection zoning in karstic areas.

The tracer test study was able to prove a connection between the Papverme and Agancsos sinkholes and the spring. There was no detectable tracer arrival in the 48-day observation period from the Vöröskő Marsh, Nyírsár Sinkhole, Majkó and Kecső Meadow caves, and the Királykútand and Vízfakadás sinkholes. The latter two sinkholes received the highest vulnerability rating on the source vulnerability map, which is not reflected in the results of the tracer test. A potential explanation for this might be found in the role of north–south-oriented tectonic elements and the Triassic marls between them (Fig. 1c), which may restrict flow towards the spring.

The results of hydrograph and recession curve analyses cannot be directly compared with the results of the source vulnerability mapping and assessment. However, a great deal of significant information of potential importance in the context of the vulnerability of the Kis-Tohonya Spring could be extracted. Long-term statistical data for the springs (Table 3) explain the position of the examined Kis-Tohonya Spring in relation to other springs and the flow systems of the area. A good correlation was found between the water table map derived from numerical simulations and decreasing topographic elevation. The discharge volume, the extent of temperature increases and the variability in the discharge volume decrease with elevation. This is in good agreement with the generalized findings for springs in Mádl-Szőnyi & Tóth (2015). In terms of this trend, the Kis-Tohonya Spring occupies an intermediate position. The spring hydrograph (Fig. 6) shows that the base flow of Kis-Tohonya Spring is lower than the peak, indicating an intense reaction of the spring to the given precipitation event. This also reflects the significant contribution of the fast flow component in the spring discharge volume after intense precipitation. The examined spring hydrograph shows a long recession period (81 days). To see the effect of more precipitation events on the spring hydrograph, an average recession curve analysis was carried out. Three different micro-regimes could be distinguished and characterized as contributing to the discharge volume of the spring. Research is ongoing to implement feedback from the validation results into the preliminary source vulnerability assessment to improve its reliability.

The authors acknowledge the help of Geogold Kárpátia Ltd in the provision of data and for allowing the VI to be involved in their research of the study area. Financial support of the Hungarian Scientific Research Fund (OTKA) NK 101356 is also acknowledged in the quest for a better understanding of the functioning of karst aquifers.

References

Daly, D., Dassargues, A. *et al.* 2002. Main concepts of the 'European approach' to karst-groundwater-vulnerability assessment and mapping. *Hydrogeology Journal*, **10**, 340–345.

Dövényi, Z. 2010. *Magyarország kistájainak katasztere [Inventory of Microregions in Hungary]*. MTA FKI, Budapest [in Hungarian].

Gruber, P., Haviarová, D., Balázs, I., Mátrahalmi, T., Serfőző, A. & Ambrus, M. 2015*a*. Víznyomjelzéses vizsgálatok a Haragistya–Szilice–Borzova karsztterületen [Tracer tests in the Haragistya–Silica–Silická Brezová karst area]. *Karsztfejlődés*, **20**, 63–79 [in Hungarian].

Gruber, P., Gaál, L., Balázs, I., Mátrahalmi, T., Serfőző, A. & Ambrus, M. 2015*b*. Geofizikai vizsgálatok a Haragistya–Szilice–Borzova karsztterületen (HU-SK) [Geophysical exploration of the Haragistya–Szilice–Borzova karst area (HU-SK)]. *Karsztfejlődés*, **20**, 81–99 [in Hungarian].

Gyuricza, G. 2004. A környezetföldtani térkép szerkesztésének módszertani kérdései az Aggtelek–rudabányai mintaterület példáján. [Methods of the environmental map edition in the case of the northern part of the Aggtelek–Rudabánya region]. *A Magyar Állami Földtani Intézet Évi Jelentése*, **2002**, 271–282 [in Hungarian].

Haviarová, D. & Gruber, P. 2006. Stopovacia skúška v jaskyni Milada. *Aragonit*, **11**, 43–45 [in Hungarian].

Iván, V. & Mádl-Szőnyi, J. 2017. State of the art of karst vulnerability assessment: overview, evaluation and outlook. *Environmental Earth Sciences*, **76**, 112.

Jimenez-Madrid, A., Carrasco-Cantos, F. & Martinez-Navarrete, C. 2012. Protection of groundwater intended for human consumption: a proposed methodology for defining safeguard zones. *Environmental Earth Sciences*, **65**, 2391–2406.

Karst Survey Konzorcium 2014. *Részjelentés "Az Aggteleki-karszt és a Szlovák karszt világörökség barlangjainak kezelése" keretében: Hidrodinamikai modellezés és forráshozamok hosszúemlékezetű modellezése. [Report in the frames of "Handling the UNESCO caves of Aggtelek karst and Slovak karst": Hydrodynamic modelling and long memory spring discharge modelling].* Report HUSK/1101/221/0180 [in Hungarian].

Kiss, K. 2012. Vörösagyag-talajok vizsgálata az Aggteleki-karszton (a Béke-barlang vízgyűjtőjén) [Analysis of red clay soils on the Aggtelek Karst (recharge area of Béke Cave)]. *Karsztfejlődés*, **18**, 89–103 [in Hungarian].

Kresic, N. 2012. *Water in Karst: Management, Vulnerability, and Restoration: Management, Vulnerability, and Restoration*. McGraw-Hill, New York.

Kresic, N. & Stevanovic, Z. 2009. *Groundwater Hydrology of Springs: Engineering, Theory, Management and Sustainability*. Butterworth-Heinemann, Oxford.

Less, G. 1997. *Az Aggtelek-Rudabányai-hegység földtani térképe*, 1:100 000 [Geological Map of the Aggtelek-Rudabánya-Mts., 1:100 000]. Geological Institute of Hungary, Budapest [in Hungarian].

Less, G., Mello, J. *et al.* 2004. *Geological Map of the Gemer–Bükk Area 1:100 000*. Geological Institute of Hungary, Budapest.

Less, S., Grill, J., Róth, L., Szentpétery, I. & Gyuricza, G. 1988. Geological Map of the Aggtelek-Rudabánya-Mts., 1: 25,000. Geological Institute of Hungary, Budapest.

Mádl-Szőnyi, J. & Tóth, Á. 2015. Basin-scale conceptual groundwater flow model for an unconfined and confined thick carbonate region. *Hydrogeology Journal*, **23**, 1359–1380.

Maillet, E. 1905. *Essais d'hydraulique souterraine et fluviale*. Hermann, Paris.

Marín, A., Ravbar, N., Kovačič, G., Andreo, B. & Petrič, M. 2014. Application of methods for resource and source vulnerability mapping in the Orehek Karst Aquifer, SW Slovenia. *In*: Mudry, J., Zwahlen, F., Bertrand, C. & LaMoreaux, J.W. (eds) *H2Karst Research in Limestone Hydrogeology*. Springer International, Berlin, 139–150.

Maucha, L. 1998. Az Aggteleki-hegység karszthidrológiai kutatási eredményei és zavartalan hidrológiai adatsorai. 1958–1993 [Results and undisturbed data of karst hydrological researches on Aggtelek Hills]. VITUKI Rt. *Hidrológiai kiadványa*, 414 [in Hungarian].

Neukum, C. 2013. Overview on methods and applications for the validation of vulnerability assessments. *Grundwasser*, **18**, 15–24.

Ravbar, N. 2007. *The Protection of Karst Waters: A Comprehensive Slovene Approach to Vulnerability and Contamination Risk Mapping*. Inštitut za raziskovanje krasa ZRC SAZU, Ljubljana.

Ravbar, N. & Goldscheider, N. 2007. Proposed methodology of vulnerability and contamination risk mapping for the protection of karst aquifers in Slovenia. *Acta Carsologica*, **36**, 397–411.

Ravbar, N. & Goldscheider, N. 2009. Comparative application of four methods of groundwater vulnerability mapping in a Slovene karst catchment. *Hydrogeology Journal*, **17**, 725–733.

Sásdi, L. 2006. Az *Aggtelek-Rudabányai hegység vízföldtani térképe* [Hydrogeological map of the Aggtelek-Rudabánya Mountains (1:100000)][in Hungarian].

Szentpétery, I. & Less, G. 2006. *Az Aggtelek–Rudabányai-hegység földtana. Magyarázó az Aggtelek–Rudabányai-hegység 1988-ban megjelent 1: 25,000 méretarányú fedetlen földtani térképéhez* [Geology of the Aggtelek–Rudabánya Hills. Explanatory Book to the Pre-Quaternary Geological Map of the Aggtelek–Rudabánya Hills, 1988, 1: 25,000]. Magyarország tájegységi térképsorozata, Magyar Állami Földtani Intézet, Budapest [in Hungarian with English abstract].

Tanács, E. & Barta, K. 2006. Talajvizsgálatok a Haragistya-Lófej erdőrezervátum területé n [Soil studies in the Haragistya-Lófej forest reserve]. *Karsztfejlődés*, **11**, 235–251, Eötvös Loránd University [in Hungarian].

Vias, J.M., Andreo, B., Perles, M.J., Carrasco, F., Vadillo, I. & Jimenez, P. 2006. Proposed method for groundwater vulnerability mapping in carbonate (karstic) aquifers: the COP method. *Hydrogeology Journal*, **14**, 912–925.

Witkowski, A.J., Kowalczyk, A. & Vrba, J. 2007. *Groundwater Vulnerability Assessment and Mapping: IAH-Selected Papers*. Taylor & Francis, London.

Zámbó, L. 1986. *A talaj-hatás karsztmorfogenetikai jelentősége* [The karstmorphogenetical importance of the soil effect]. Candidate thesis [in Hungarian].

Zámbó, L. 1998. Talajtakaró [Soil cover]. *In*: Baross, G. (ed.) *Az Aggteleki Nemzeti Park* [The Aggtelek National Park]. Mezőgazda Kiadó, Budapest, 97–117 [in Hungarian].

Zámbó, L. 2003. *Kiemelten veszélyeztetett, védett és érzékeny természeti területek (Aggteleki Nemzeti Park) talajtakarójának kezelését megalapozó kutatások* [Base research on soil cover management on highly endangered, protected and sensitive areas (Aggtelek National Park)]. Részjelentés a K-36-02-0013H sz. projekt 2002–2003 évi eredményeiről [Report on the Results (2002–2003) of Project No. K-36-02-0013H] [in Hungarian].

Zhang, Q. 2014. An assessment of groundwater resource vulnerability to pollution in the Jiangjia spring basin, China. *Environmental Earth Sciences*, **74**, 1–11.

Zwahlen, F. 2004. *Vulnerability and Risk Mapping for the Protection of Carbonate (Karst) Aquifers*. Final Report (COST action 620). European Commission, Directorate-General XII Science, Brussels.

Hydrochemical and isotopic characterization of carbonate aquifers under natural flow conditions, Sierra Grazalema Natural Park, southern Spain

DAMIÁN SÁNCHEZ*, JUAN ANTONIO BARBERÁ, MATÍAS MUDARRA, BARTOLOMÉ ANDREO & JOSÉ FRANCISCO MARTÍN

Centre of Hydrogeology of the University of Malaga (CEHIUMA) 29071, Malaga, Spain

**Correspondence: dsanchez@uma.es*

Abstract: We used hydrogeochemical techniques and environmental isotopes ($\delta^{18}O$ and $\delta^{2}H$) to characterize karst aquifers without a well-known hydrogeological conceptual model. The selected study area corresponds to carbonate outcrops located in the Sierra Grazalema Natural Park (SGNP) in southern Spain. The combination of high rainfall, large areas of carbonate outcrops and the high permeability of the outcropping rocks means that the SGNP is a strategic water reserve in a region that periodically suffers from a scarcity of water. There are still many uncertainties regarding the hydrogeological functioning of these aquifers, such as the groundwater flow paths, the identification of recharge areas and a possible hydrogeological connection with adjacent systems. The objectives of this work were: (1) to gain an insight into the hydrogeochemistry of groundwater in the SGNP aquifers; (2) to identify the recharge areas of the main springs of the karst system; and (3) to contribute to our understanding and conceptualization of the SGNP aquifers. Our results show marked differences in the chemical composition of the groundwater, which can be classified into three main groups: $CaHCO_3$ waters (southern sector); Ca–$MgHCO_3$ waters (northern sector); and those with higher SO_4^{2-}, Na^+ and/or Cl^- concentrations. These hydrochemical differences are principally dependent on the mineral composition of the rocks the groundwater flows through and comes into contact with. These results also shed light on the recharge areas associated with the main springs and seem to corroborate a possible net groundwater flow from the eastern part of the SGNP (Sierra del Endrinal) towards carbonate outcrops situated several kilometres to the SE.

Karst aquifers present particular hydrogeological characteristics, given their inherent heterogeneity and anisotropy, which differentiate them from other aquifers developed in porous or fractured media (White 1988; Bakalowicz 1995, 2005; Ford & Williams 2007; Abusaada & Sauter 2013). Karst aquifers can be characterized and studied by a wide range of approaches (Goldscheider & Drew 2007), including hydrochemical methods (e.g. chemographs and ionic relationships), hydrological methods (e.g. water balances and spring hydrographs), hydraulic approaches (aquifer pumping tests), isotopic methods (stable and radioactive isotopes), tracer techniques (artificial and natural dyes) or speleological investigations. The combined application of these methodologies leads to an understanding of the hydrogeological functioning of the studied system, such as the groundwater flow paths and velocities, the identification of recharge areas, the groundwater residence times and the available water resources.

Our knowledge of the intrinsic complexity of karst systems can be enhanced by studying the chemical composition of the water drained by springs, constituting a reliable set of tools for obtaining information about the structure, storage capacity and dynamics of carbonate aquifers (Mudry 1987; Lastennet & Mudry 1997; Mudarra & Andreo 2011; Sánchez *et al.* 2015). Hydrochemical techniques can contribute to locating and quantifying the natural acquisition of mineralization by water (considering the chemical compounds as natural tracers) or to identifying the extent and origin of water contamination if there are problems with water quality (Hunkeler & Mudry 2007). For instance, the relationship between the different chemical components dissolved in water can be useful in identifying and characterizing hydrogeochemical processes (Emblanch *et al.* 1998; Batiot *et al.* 2003*a*, *b*; Barberá & Andreo 2015).

Environmental isotopes are a powerful tool in characterizing karst aquifers (Collins & Gordon 1981). Hydrogen and oxygen isotopes are intrinsically associated with the water molecule and provide researchers with distinctive dual capacities in hydrogeological studies, especially in studies focused on tracking the origin and subsequent movement of water. Hydrogen and oxygen isotopic data can be used as a tracer of infiltration, recharge modalities and the regional distribution of groundwater flows whenever fractioning processes involving water

From: Parise, M., Gabrovsek, F., Kaufmann, G. & Ravbar, N. (eds) 2018. *Advances in Karst Research: Theory, Fieldwork and Applications*. Geological Society, London, Special Publications, **466**, 275–293.
First published online November 28, 2017, https://doi.org/10.1144/SP466.16

molecules occur in the system, resulting in characteristic δ^2H or $\delta^{18}O$ signatures (Araguás-Araguás *et al.* 2000; Gonfiantini *et al.* 2001; Andreo *et al.* 2004; Lambán *et al.* 2015). Favourable circumstances for isotope fractioning require certain conditions, such as altitude differences between the recharge and discharge points, the existence of long groundwater flow paths or when unusual sources, such as rivers or evaporated water bodies, contribute significantly to the recharge of groundwater reservoirs (Criss *et al.* 2007).

We present a practical and critical overview of the most common hydrochemical and isotopic approaches used to study carbonate karst aquifers. The combination of both classical hydrogeochemical techniques and the stable isotopes of the water molecule (δ^2H and $\delta^{18}O$) has allowed significant advances in our knowledge of many karst systems, from which the recharge area of the main drainage points has been inferred. The case study selected to jointly apply these methodologies is the Sierra Grazalema Natural Park (SGNP), a calcareous massif in southern Spain. The combination of large carbonate outcrops, high permeability and annual rainfall values as high as 2000–3000 mm means that the SGNP is a strategic groundwater reserve located in a region (southern Spain) that periodically suffers from droughts and a scarcity of water. In spite of this, the hydrogeological functioning of these aquifers is still uncertain (IGME 1984; ITGE 1992; Sánchez *et al.* 2014; Martín-Rodríguez *et al.* 2016) and needs to be resolved with new research. This work will contribute to our understanding of the hydrogeology of this karst system via the hydrochemical and isotopic characterization of its waters and a first attempt to map the main recharge areas.

Study area

The SGNP is located in the NE of Cádiz Province in southern Spain (Fig. 1). It covers an area of 260 km^2, of which *c.* 150 km^2 consists of permeable carbonate rocks. It was declared a Natural Park in 1985 as a result of its important environmental, floral and faunal values.

The relief is fairly rough, with altitudes ranging from 300 to *c.* 1650 m a.s.l. in the Sierra del Pinar (Fig. 1). The total population of *c.* 25 000 inhabitants is homogeneously distributed throughout several small towns. The declaration of Sierra Grazalema as a Natural Park has allowed the conservation of natural conditions, where livestock farming (primarily cattle), leather production and tourism represent the only significant economic activities in this area. Vegetation is mostly Mediterranean shrub and pasture, except for the highest areas, where there is neither soil development nor vegetation.

The prevailing climate is semi-continental Mediterranean. Rainfall mainly occurs in autumn, winter and, to a lesser extent, spring, and is associated with wet winds coming from the SW. Rainfall is scarce or non-existent during the rest of the year, particularly in summer. The historical mean annual precipitation is *c.* 1350 mm, although there is a high variability in the distribution of rainfall, ranging from 900 mm in the north–NW to 1800 mm in the central part (Sánchez *et al.* 2015). The rain gauge station located in Grazalema town (Fig. 1) is the wettest in southern Spain and an annual rainfall >3000 mm has been recorded there (Andreo *et al.* 2014). The mean annual air temperature calculated from the historical series is *c.* 16°C (Gálvez-Maestre 2005). The monitoring period considered in this work consists of the hydrological years 2012–13 and 2013–14. Compared with the mean historical annual precipitation, these years are classified as wet (2012–13, 1700 mm) and regular (2013–14, 1480 mm).

From a geological standpoint, the study area is located within the Betic Cordillera. The rocks that constitute the Sierra Grazalema belong to two different geological domains (Martín-Algarra 1987): the Penibetic and Middle Subbetic. The limit between these domains corresponds to the Boyar Corridor, a major tectonic disturbance with a NE–SW trend, which separates Penibetic outcrops (to the south) from Middle Subbetic outcrops (to the north; Fig. 1). Both domains consist of Jurassic carbonate rocks bounded at the base by Triassic clays and evaporites and are overlain at the top by lower Cretaceous–Oligocene marly limestones and marls. The Jurassic carbonate rocks in the Penibetic domain are mainly limestones, whereas in the Middle Subbetic domain there is a more significant representation of dolostones, which are stratigraphically below the limestones in both domains. The geological structure is made up by NE–SW-lying folds, which generate anticlines where limestones and dolostones appear, and synclines constituted by Cretaceous marls (Fig. 1).

The aquifers of the SGNP are made up of fractured and karstified Jurassic limestones and dolostones. Recharge takes place by the direct infiltration of rainwater (and sometimes snowmelt) and, to a lesser extent, by the infiltration of surface water into swallow holes; discharge is produced through springs located at the borders of the carbonate outcrops (Fig. 1). The impervious limits to groundwater flow correspond to Cretaceous marls, flysch-type clays and Triassic rocks. The southern (Penibetic) sector is drained by five springs located at the western edge (S1–S5, Fig. 1), whereas the natural discharge of the northern (Middle Subbetic) sector occurs through a greater number of outflow points (S6–S18). All 18 springs constitute the hydrochemical monitoring network used in this study.

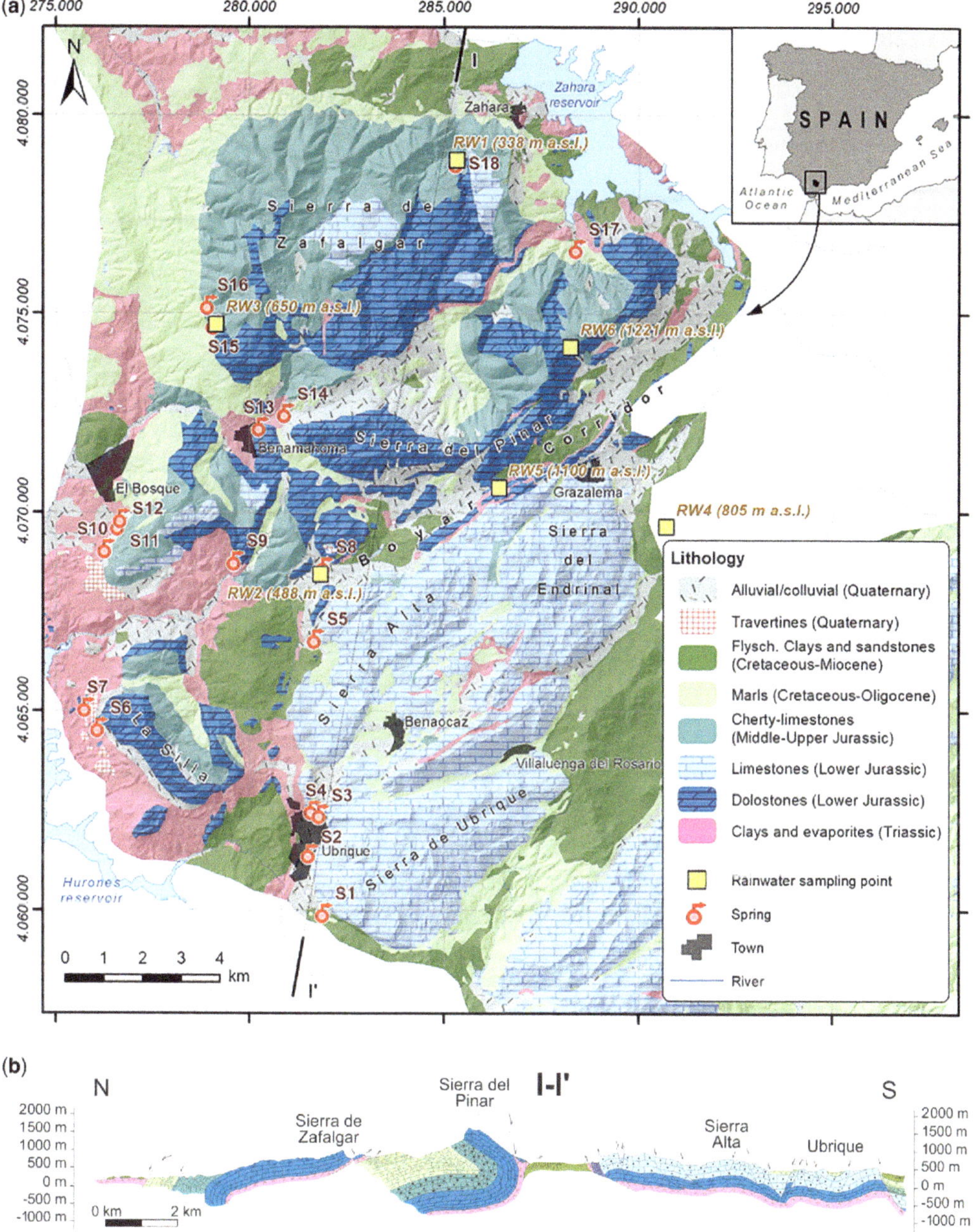

Fig. 1. (**a**) Location and geological setting of the Sierra Grazalema Natural Park showing the position of springs and rainwater sampling stations. (**b**) Geological cross-section.

The geometry of the SGNP aquifers is principally controlled by thrust faults and other tectonic disturbances bringing the clayey–gypsiferous Triassic substratum or other impervious units, such as the Flysch Complex, to the surface (for instance, between Sierra del Pinar and Sierra de Zafalgar or at the Boyar Corridor; Fig. 1). The absence of deep boreholes reaching the substratum of the aquifers in this area and the limited resolution of the only geophysical study carried out in the SGNP (IGME 1986) do not allow clarification of the three-dimensional geometry of this complex system. No reliable estimate has been made of the groundwater reserves, apart from an estimate of around 15 500 m^3 of reserves per 100 m of saturated zone (IGME 1984).

Exokarstic forms are well developed in the Penibetic sector, where dolines, swallow holes, caves and karren fields are common. By contrast, these forms do not exist, or are rare, on the Middle Subbetic outcrops. No known source of pollution exists on the recharge areas of the aquifers, apart from some extensive livestock farming and minor spills of urban wastewater during intense precipitation.

Regarding the temporal chemical and hydrodynamic variation of the SGNP springs, the base flow, which typically occurs during the summer, is characterized by the drainage of more mineralized waters and large decreases in the flow discharge, especially in the springs located in the Penibetic sector. The recharge produced by the first autumn rains has a more rapid and intense effect in the Penibetic sector than in the Middle Subbetic springs. Aquifer recharge usually generates dilution of the water drained by the springs as a result of the effect of recently infiltrated waters, which are typically less mineralized that those constituting the aquifer reserves. Nevertheless, some springs respond inversely, draining more mineralized waters during peak flow conditions (Andreo *et al.* 2014; Sánchez *et al.* 2014, 2017). Although the temporal evolution of stable isotopes ($\delta^{18}O$ and δ^2H) in the waters are not as clear as the chemographs, in general the range of isotopic data in Penibetic springs is larger than those found in Middle Subbetic springs, which might be related to the more developed karstic network of the Penibetic springs.

Methodology

Field measurements and spring water samples were taken from November 2012 to June 2014 with an average frequency of about two weeks. The number of water samples analysed per spring ranged from 20 to 68 (Table 1). The electrical conductivity, temperature and pH of the waters were measured *in situ* using portable WTW Cond 3310 and HACH HQ40d devices.

Spring and rainwater samples for isotopic analyses were stored in amber glass bottles and analysed in the laboratory a few days after the sampling campaigns. Rainwater was collected at five rain sampling stations distributed over the whole study area at altitudes ranging from 340 to 1220 m a.s.l. (Fig. 1). The collecting bottles were placed under the ground and filled with a thin layer of paraffin to avoid the evaporation of water.

Chemical analyses were carried out in the laboratory of the Centre of Hydrogeology, University of Malaga, Spain. The alkalinity (Alk, as HCO_3^-) was measured by volumetric titration using 0.02 N H_2SO_4 to pH 4.45. The major components (Ca^{2+}, Mg^{2+}, Na^+, K^+, Cl^-, SO_4^{2-} and NO_3^-) were determined by high-pressure liquid chromatography (Metrohm Model 791 Basic IC) with ±2% accuracy. Samples were filtered before being introduced into the system (in-line and pre-column filters).

The stable isotopes ^{18}O and ^2H in the water samples were determined at the same laboratory using a PICARRO Model L2120i-CRS cavity ring-down spectroscopy water isotope analyser. The raw data were processed with ChemCorrect software. The precision guaranteed by the manufacturer is <0.1‰ for $\delta^{18}O$ and <1.0‰ for δ^2H. The results are expressed as per mille (‰) deviations from the internationally accepted standard Vienna Standard Mean Ocean Water (V-SMOW; Craig 1961).

Results

Hydrogeochemical characterization

The water drained by the springs of the SGNP shows low to intermediate mineralization (mean electrical conductivity 260–860 μS cm^{-1}), except for spring S7, in which the flowing water has a considerably higher dissolved salt content (2273 μS cm^{-1}; Table 1). The mean water temperature varies from 13.0 to 17.5°C. As the mean annual air temperature in the area is *c.* 16°C, this indicates that the water temperatures of the springs are both colder (−3°C) and warmer (+1.5°C) than the average air temperature. When the mean water temperature of the springs is plotted against discharge altitude (Fig. 2), most of the points display a negative relationship between these two variables, showing a thermal gradient close to −1°C $(100\ m)^{-1}$. Springs S15 and S16 do not fit on the line because their waters are not as cold as might be expected based on their drainage altitudes (600–650 m a.s.l.). Springs S7 and S9–12 make up a fairly homogeneous group characterized by the drainage of warmer waters (17.0–17.5°C). All the waters are neutral or slightly alkaline based on their mean pH value (Table 1).

The discharge of springs in the aquifers of the SGNP shows large variations throughout the hydrological year. During base flow conditions (June–September), the discharge rate in the springs varies between <1 and 150 l s^{-1}, whereas under high flow conditions (around November–March), it can reach several cubic metres per second (Table 2). The hydrodynamic response of the outlets to recharge (precipitation) is more or less time-lagged depending on the spring and the intensity of karstification (Andreo *et al.* 2014; Sánchez *et al.* 2014, 2017). The difference between the maximum and minimum values in spring flow during the monitoring period

Table 1. *Mean and coefficients of variation (CV, %) of the main physical and chemical parameters of groundwater in the SGNP*

Spring	n	Electrical conductivity (µS cm^{-1})		Temperature (°C)		pH		Total alkalinity (mg l^{-1})		Cl^- (mg l^{-1})		SO_4^{2-} (mg l^{-1})		NO_3^- (mg l^{-1})		Ca^{2+} (mg l^{-1})		Mg^{2+} (mg l^{-1})		Na^+ (mg l^{-1})		K^+ (mg l^{-1})	
		Mean	CV	Mean	CV	Mean	CV	Mean	CV	Mean	CV	Mean	CV	Mean	CV	Mean	CV	Mean	CV	Mean	CV	Mean	CV
Penibetic																							
S1	24	412	33	14.5	11	7.85	7	266	34	11.6	66	11.0	34	1.2	43	87.8	28	3.9	63	9.2	79	1.0	22
S2	38	285	8	14.7	2	7.67	4	170	8	9.8	33	14.8	44	5.4	67	63.1	9	3.4	42	6.4	32	1.1	33
S3	68	260	6	14.4	1	7.69	4	170	5	4.8	20	9.2	24	2.4	42	62.9	5	2.2	26	2.9	15	0.6	24
S4	47	307	2	14.8	1	7.62	2	188	5	8.1	11	16.6	13	4.1	23	61	6	8.9	5	5.2	8	1.1	15
S5	44	287	5	14.4	1	7.51	3	189	5	5.2	15	9.6	29	4.3	34	64.6	4	5.4	40	3.6	18	0.7	16
Middle Subbetic																							
S6	28	402	5	15.5	2	7.66	3	236	18	14.3	20	25.0	20	3.5	26	61.6	18	22.5	18	7.4	16	1.0	18
S7	31	2273	6	17.5	1	7.09	3	280	10	335.3	12	842.7	14	5.5	11	326.7	9	68.1	10	197.2	7	3.1	16
S8	35	417	12	14.1	1	7.59	3	273	9	13.2	33	18.1	23	1.4	14	59.6	13	29.0	8	7.8	32	0.7	21
S9	30	566	3	17.4	2	7.34	2	310	2	11.8	8	90.5	6	1.9	12	94.4	5	31.8	2	7.7	3	1.2	22
S10	39	557	4	17.4	1	7.33	3	322	3	17.1	9	59.2	7	1.7	12	93.0	8	26.0	3	10.4	5	1.2	14
S11	38	637	7	17.1	1	7.54	2	327	3	42.9	33	64.2	9	1.4	13	93.2	8	25.3	3	29.1	33	1.3	15
S12	20	860	4	17.1	1	7.48	3	325	5	119.2	11	73.7	5	1.3	15	100.1	8	24.1	3	74.7	9	1.5	9
S13	40	391	8	13.5	2	7.46	3	235	7	9.6	38	35.4	38	2.1	25	67.2	10	19.7	13	5.9	33	0.8	19
S14	32	359	8	13.0	3	7.56	5	235	8	7.5	28	20.2	35	1.9	33	63.1	11	17.5	13	4.7	26	0.7	18
S15	33	442	5	13.9	2	7.66	6	282	5	7.3	8	35.3	27	1.5	16	78.3	12	21.1	12	3.8	8	0.9	16
S16	33	517	6	14.7	1	7.21	4	343	10	9.4	13	28.2	31	1.4	20	103.8	14	15.9	22	4.6	8	1.0	16
S17	37	452	6	14.5	2	7.52	5	254	9	20.9	33	38.1	24	2.0	31	75.1	10	18.9	12	13.3	29	1.1	17
S18	36	380	5	14.7	1	7.45	4	257	4	5.4	9	15.6	20	2.7	21	69.2	6	18.9	14	3.1	8	0.7	21

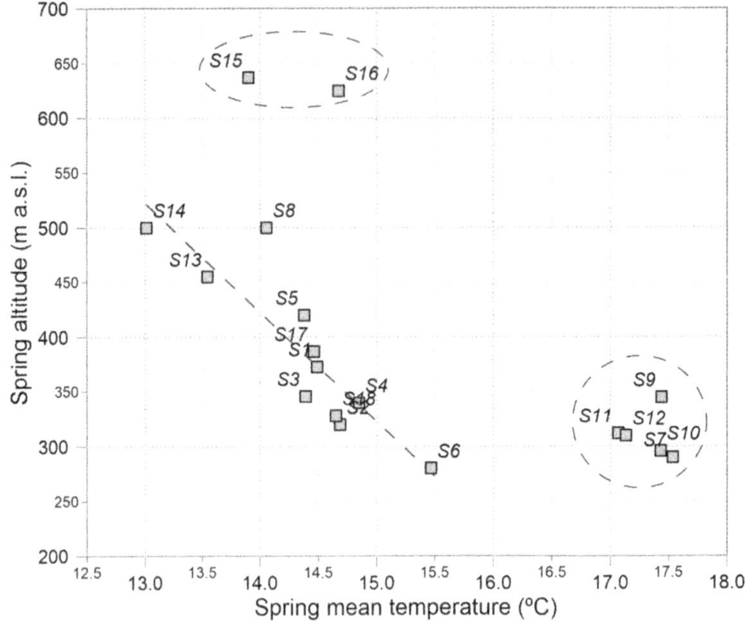

Fig. 2. Relationship between mean spring water temperature and discharge altitude.

ranged from 90 to >9 × 10^3 l s^{-1} and the coefficients of variation were as high as 212% (Table 2). Most of the waters drained by the springs located in the SGNP are of a Ca–Mg$(HCO_3^-)_2$ facies (Fig. 3). In general terms, the water samples present greater homogeneity with respect to the anions (Fig. 3, right-hand diagram) than the cations (Fig. 3, left-hand diagram); the latter displays a large range of Ca to Mg values. Water samples collected at spring S7 have a $CaSO_4^{2-}$ facies, with a significant contribution of

Table 2. *Main statistical parameters of the discharge rate (l s^{-1}) from the springs*

	No. of measurements	Minimum	Maximum	Mean	Range	Coefficient of variation (%)
Penibetic						
S1	35	0	6059	311	6059	212
S2	33	5	2625	157	2620	175
S3	32	5	2460	406	2459	106
S4	32	2	1610	172	1610	130
S5	35	4	2910	203	2906	162
Middle Subbetic						
S6	28	24	447	81	424	104
S7						
S8	26	10	147	34	137	85
S9	32	5	192	16	187	155
S10	34	17	212	33	195	97
S11	34	5	105	24	101	84
S12	25	0	90	9	90	118
S13	34	153	9338	606	9185	153
S14						
S15	29	6	1240	112	1234	149
S16						
S17	34	44	1780	156	1736	139
S18	34	152	5457	524	5305	137

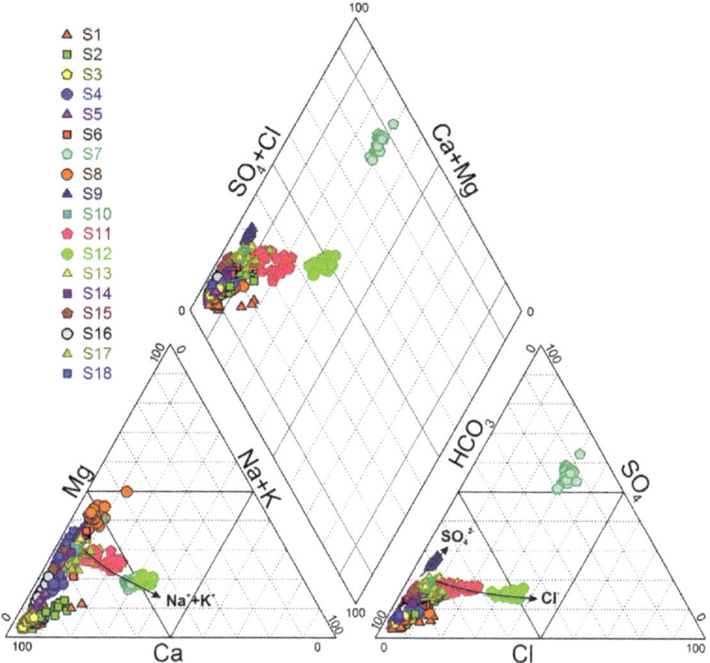

Fig. 3. Piper diagram showing the results for all water samples analysed at the Sierra Grazalema Natural Park.

Na^+ and Cl^-. Other springs draining waters with hydrochemical facies different from the general Ca–Mg/HCO_3^- pattern are S9, S11 and S12, where the proportions of SO_4^{2-}, Na^+ and Cl^- are higher than in the other springs.

Positive correlations between electrical conductivity and Alk and between electrical conductivity and the Ca^{2+} and Mg^{2+} concentrations were found in the spring waters (Fig. 4). All the water samples, except those collected at spring S7, plot close to

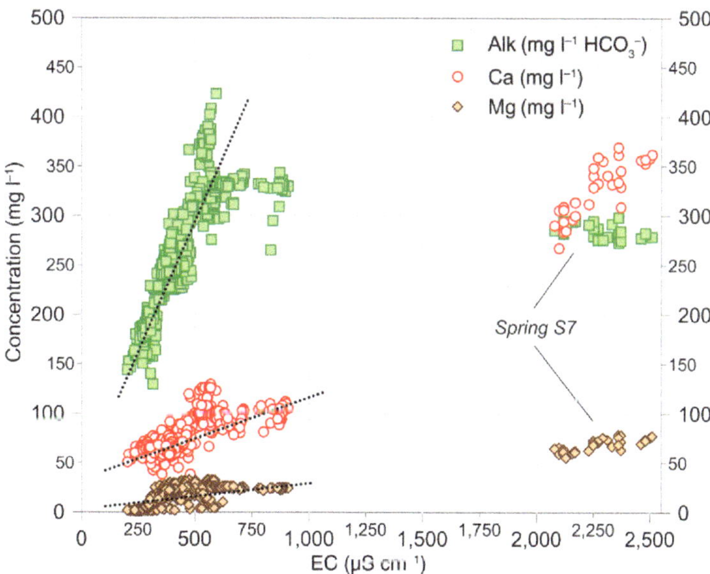

Fig. 4. Plot of electrical conductivity (EC) v. alkalinity, Ca^{2+} and Mg^{2+}.

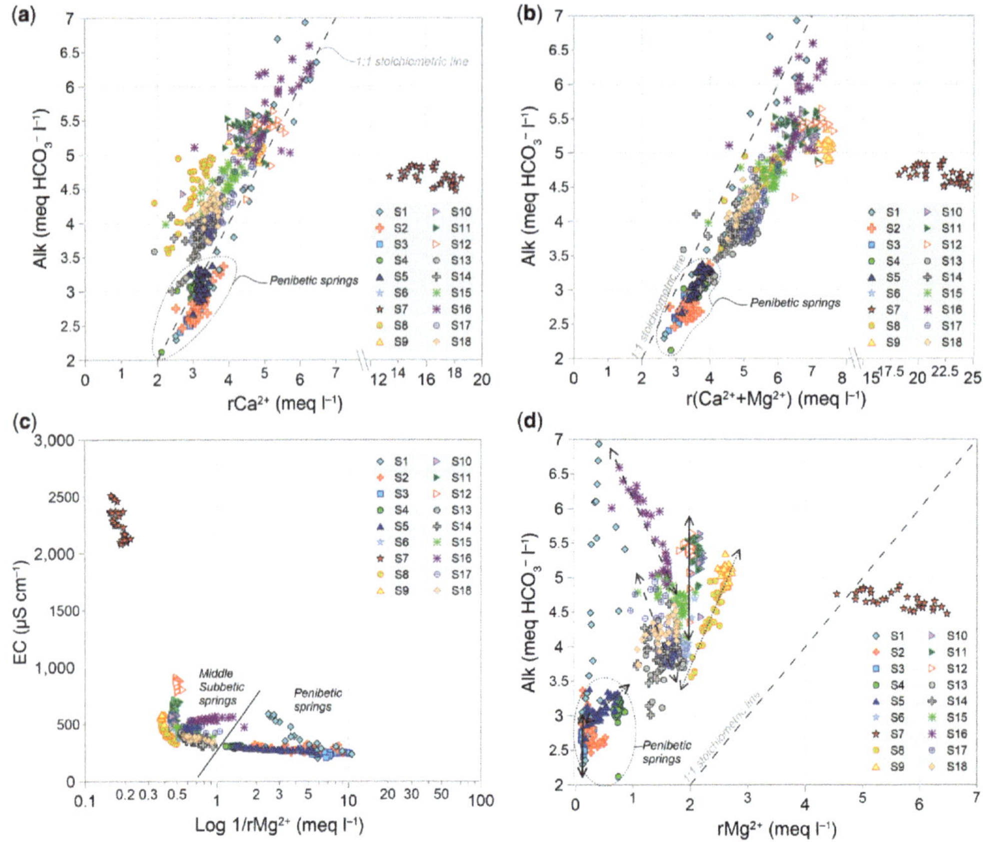

Fig. 5. Relationships between (**a**) Ca^{2+} and alkalinity, (**b**) (Ca+Mg) and alkalinity, (**c**) electrical conductivity (EC) and log (1/Mg), (**d**) alkalinity and Mg^{2+}, (**e**) SO_4^{2-} and Ca^{2+}, (**f**) Ca^{2+}/SO_4^{2-} and Mg^{2+} and (**g**) Cl^- and Na^+.

the 1:1 Ca–HCO_3^- stoichiometric line (Fig. 5a), suggesting a similar origin for both ions. Waters from spring S7 show excess $Ca^{2+}$with respect to HCO_3^- (around 3.5 times), which indicates another source of Ca^{2+} apart from the dissolution of carbonate minerals. The springs located in the Penibetic (southern) domain of the study area have lower Alk and Ca^{2+} values. In addition, these outlets plot either on or slightly under the 1:1 stoichiometric line because most of them present little excess Ca^{2+}. By contrast, the Middle Subbetic (northern) springs are characterized by higher values for both parameters and by a clear excess of HCO_3^- with respect to Ca^{2+}. When Alk is plotted against Ca+Mg (Fig. 5b), the water samples from these springs modify their position to the right-hand side of the stoichiometric line due to the contribution of Mg^{2+}, which is greater than in the Penibetic springs. Sampling point S7 again shows a particular chemical composition.

From a general point of view, springs in the Penibetic sector show lower Mg^{2+} concentrations than those draining the Middle Subbetic outcrops (Fig. 5c). The only water samples showing values of Mg^{2+} higher than those of Alk are those collected at spring S7 (Fig. 5d). These waters are also characterized by fairly stable Alk values, together with greater variations in Mg^{2+}. Some springs underwent very small changes in Mg^{2+} concentrations during the monitoring period (for instance, S3 and S12; vertical solid lines), whereas others show larger Mg^{2+} variations and a positive correlation with alkalinity (e.g. S5 and S8; dotted lines) or even a negative correlation (S16 and S17; dashed lines).

The only water samples displaying a stoichiometric ratio between SO_4^{2-} and Ca^{2+} concentrations are those flowing out of spring S7; all the other samples present greater contents of Ca^{2+} than SO_4^{2-} (Fig. 5e). Springs S9, S10, S11 and S12 – all draining the western sector of the Middle Subbetic domain (Fig. 1) – have greater SO_4^{2-} concentrations than the other outlets (excluding S7).

The Penibetic springs show narrow variations in Mg^{2+} concentrations (<1 meq l^{-1}), but large changes in the Ca^{2+}/SO_4^{2-} ratio (Fig. 5f). The Middle

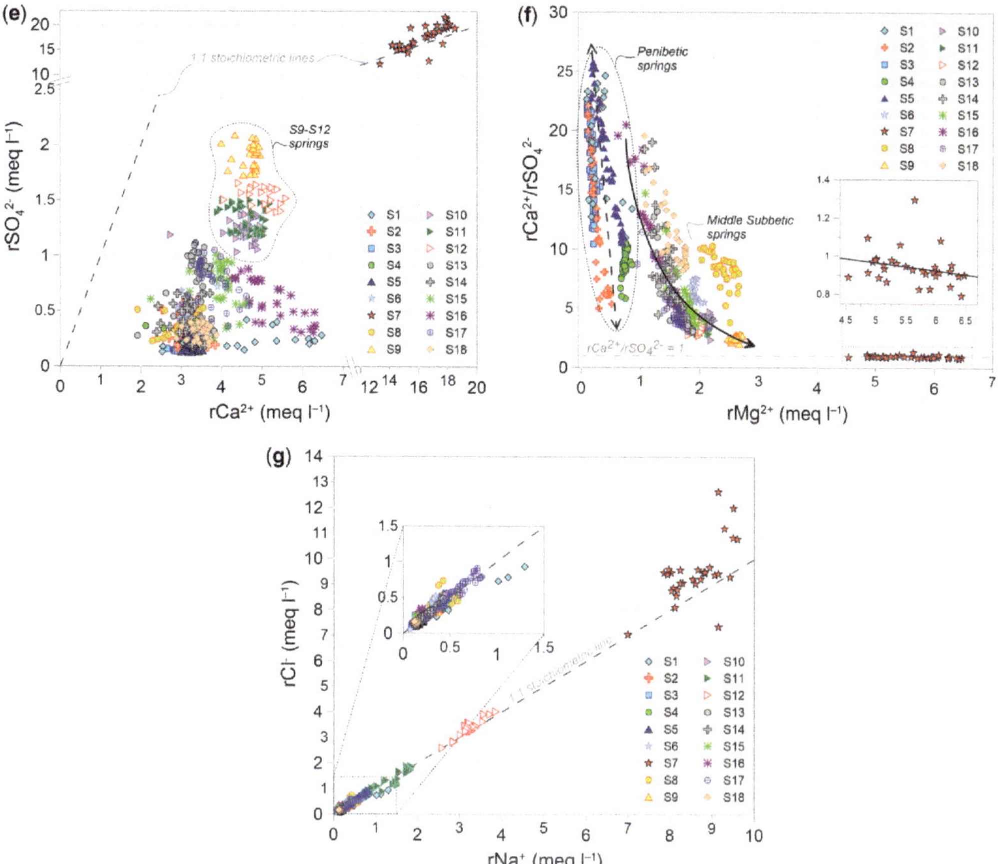

Fig. 5. *Continued.*

Subbetic springs show a general trend in which lower ratios of Ca^{2+}/SO_4^{2-} are found at higher Mg^{2+} concentrations, i.e. higher Mg^{2+} concentrations correlate with smaller Ca^{2+} and/or larger SO_4^{2-} concentrations. Spring S7 shows a constant Ca^{2+}/SO_4^{2-} ratio of about unity.

All the water samples plot on the 1:1 rCl/rNa line (Fig. 5g), which suggests a similar origin for both ions related to halite dissolution. The exception for this general pattern is again symbolized by spring S7 because the majority of its samples are located above the line and show an excess of Cl^- with respect to Na^+. This is more evident in the most mineralized water samples. Springs S12 and, to a lesser extent, S11 have higher Na^+ and Cl^- contents than the other springs.

Isotopic characterization

A sampling network consisting of six monitoring stations was set out in the study area to gather rainwater samples (yellow squares in Fig. 1). Selection of the sampling sites was made using geographical, altitude and accessibility criteria. They are located at altitudes ranging from 338 m (rain sampling station RW1) to >1220 m a.s.l. (RW6), covering a total range of *c.* 880 m (Table 3). The number of rainwater samples collected at each station varied between six and 20, with a total of 67. The $\delta^{18}O$ values range from −12.20 to −1.68‰ and the δ^2H values range from −83.91 to −4.75‰. The δ^2H values show a greater dispersion than $\delta^{18}O$ (mean coefficients of variation 58 and 42%, respectively).

The deuterium excess ($d = \delta^2H - 8\delta^{18}O$, after Dansgaard 1964) is acquired during evaporation and does not vary significantly during the later history of the cloud front. It is thus a valuable indicator of the source area of the water vapour (Rindsberger *et al.* 1983; Julián *et al.* 1992; Celle-Jeanton *et al.* 2001; Andreo *et al.* 2004). Values of deuterium excess in the rainwater sampled during the monitoring period ranged between 6.18 and 23.60‰ (Table 3).

Table 3. *Isotopic data obtained from the rainwater sampling stations of the SGNP*

Sampling point	Altitude (m a.s.l.)	$\delta^{18}O$ (‰)(SMOW)					δ^2H (‰)(SMOW)				Deuterium excess* (‰)			
		n	Mean	Minimum	Maximum	Coefficient of variation (%)	Mean	Minimum	Maximum	Coefficient of variation (%)	Mean	Minimum	Maximum	Coefficient of variation (%)
RW1	338	16	−4.91	−8.57	−1.68	39	−27.25	−62.37	−4.75	53	12.04	6.18	20.01	38
RW2	488	10	−4.45	−8.89	−2.28	56	−22.08	−58.68	−7.86	84	13.53	7.12	17.82	28
RW3	650	8	−5.29	−8.57	−3.13	37	−27.23	−60.50	−10.55	62	15.12	8.10	21.65	27
RW4	805	20	−5.97	−12.20	−2.60	44	−34.60	−83.91	−11.13	56	13.20	4.37	18.91	28
RW5	1100	7	−5.65	−10.06	−2.72	46	−28.45	−57.17	−9.57	58	16.74	11.02	23.31	28
RW6	1221	6	−6.74	−10.55	−4.71	32	−39.01	−60.81	−25.48	34	14.89	12.20	23.60	30
Mean		67	−5.50	−9.81	−2.85	42	−29.77	−63.91	−11.56	58	14.25	8.17	20.88	30

*Deuterium excess, $d = \delta^2H - 8\delta^{18}O$.

The isotopic signature of rainwater shows more dispersion than that for the spring waters (grey rectangle in Fig. 6a) for both $\delta^{18}O$ and δ^2H. Most rainwater samples plot between the global meteoric water line (GMWL, $d = +10$‰; Craig 1961) and the Mediterranean meteoric water line (MMWL, $d = +22$‰; Gat & Carmi 1970). The average deuterium excess of all rainwater samples analysed was an intermediate value (+14.25‰; Table 3). The local meteoric water line defined for the Sierra Grazalema is $\delta^2H = 8\delta^{18}O + 13.6$ ($R^2 = 0.94$), which is closer to the GMWL than to the MMWL. The most isotopically depleted rainwater was collected at sampling station RW4 ($\delta^{18}O = -12.20$‰ and $\delta^2H = -83.91$‰) and the heaviest rainwater was collected at station RW1 ($\delta^{18}O = -1.68$‰ and $\delta^2H = -4.75$‰). However, when the values weighted by the amount of rainfall are plotted (inset graph in Fig. 6a), the results show that the heaviest rainwater samples were gathered at station RW2 and the lightest samples at stations RW6 and, to a lesser extent, RW4. This pattern is congruent with the position of the sampling stations with respect to the path of the Atlantic wet air masses responsible for rainfall in the study area, as well as with their different altitudes. These air masses face the Sierra Grazalema from the west and move essentially to the east, in particular through the Boyar Corridor (Fig. 1). Station RW2, which is located at the westernmost edge of the corridor and at a low altitude (488 m a.s.l.), receives the first and isotopically heaviest rainwater. On the opposite side of the corridor and at considerably higher altitude (1221 m a.s.l.), station RW6 collects the last and lightest rainwater. $\delta^{18}O$ and δ^2H values registered at station RW4, the second most negative after those of RW6, are probably related to the position of this monitoring point on the leeward side of the >1500 m altitude Sierra del Endrinal peaks (Fig. 1).

The $\delta^{18}O$ composition of rainwater shows a linear dependence with altitude ($R^2 = 0.86$; Fig. 6b). A vertical isotopic gradient of −0.304‰ $\delta^{18}O$ (100 m)$^{-1}$ is obtained. This value is comparable with the vertical gradients reported by other researchers in the Mediterranean area (Bortolami *et al.* 1979; Julián *et al.* 1992; Andreo *et al.* 1997, 2004; Vallejos *et al.* 1997; Lambán *et al.* 2015).

Table 4 summarizes the main statistical parameters (maximum, minimum and mean values) of the spring water isotopic composition. As in Tables 1 and 2, the springs have been classified by the geological domain that they drain (Penibetic and Middle Subbetic). Isotopic data are available for all sampling points, except spring S17. The values range from −6.87 to −4.98‰ for $\delta^{18}O$ and from −39.34 to −26.92‰ for δ^2H. No significant difference was encountered between the isotopic compositions of the Penibetic and Subbetic springs ($\delta^{18}O$ and δ^2H

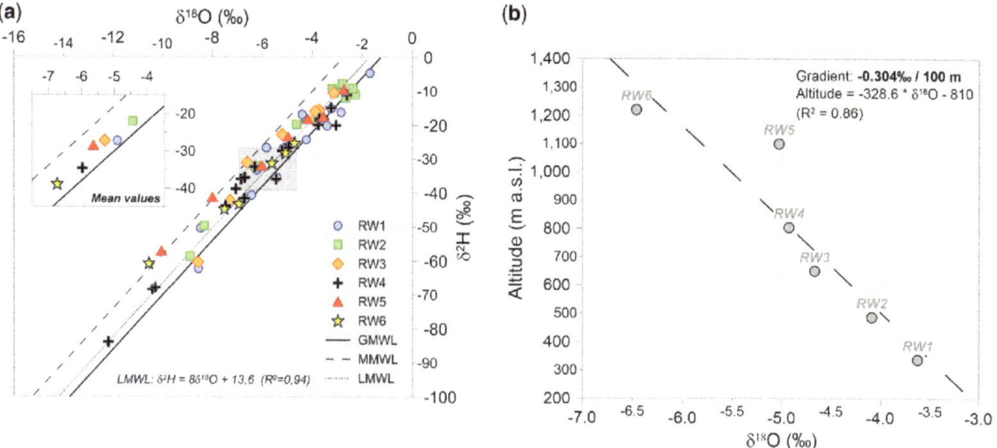

Fig. 6. (**a**) Isotopic composition of rainwater. GMWL, global meteoric water line; MMWL, Mediterranean meteoric water line; LMWL, local meteoric water line. The grey rectangle represents the limits of the isotopic data of all spring water samples. (**b**) Relationship between the average $\delta^{18}O$ composition and the altitude of monitoring stations taking into consideration the $\delta^{18}O$ values weighted by rainfall.

mean values of −5.61 to −30.34‰ and −5.61 to −30.27‰, respectively), although the latter springs show lower coefficients of variation (2% v. 4–5%; Table 4). The calculated deuterium excess varies between 10.59 and 16.82‰, with an average of 14.55‰ for the Penibetic and 14.45‰ for the Middle Subbetic springs.

On a $\delta^{18}O$–$\delta^{2}H$ diagram, the isotopic composition of all Sierra Grazalema spring water samples plots between the GMWL and the MMWL (Fig. 7). Together they define the local groundwater line (LGWL), the slope of which (5.8) is lower than that of the local meteoric water line (LMWL) (8).

Spring S13 displays the most negative $\delta^{18}O$ and $\delta^{2}H$ values of all the monitored points, whereas spring S6 and some isolated samples from spring S1 show the least negative values (Fig. 7). All the water samples plot around the LGWL and, when the average values are shown (inset plot in Fig. 7), all the springs are also placed next to this line.

Springs S13 and S14 share a close location (Fig. 1) and they have almost the same average isotopic signatures (plot with mean isotopic composition in Fig. 7), suggesting a similar recharge area for both discharge points. By contrast, other closely located pairs of springs, such as S3–S4, S6–S7 or S15–S16, show larger isotopic differences, which may be related to dissimilarities between their respective recharge areas.

The representation of the average $\delta^{18}O$ values of monitored springs and topographic elevations (Fig. 8) allows deductions about the recharge areas related to the main springs draining the SGNP aquifers. There is a general pattern of more negative $\delta^{18}O$ values from west to east and therefore the springs with the most positive values (mean $\delta^{18}O$ greater or equal to −5.40‰) are located at the westernmost part of the SGNP (springs S6–S7, S10 and S12). However, the springs showing the most negative values (mean $\delta^{18}O$ less than or equal to −5.90‰) are placed in the northern sector of the studied area (S13 and S14), in particular between the Sierra del Pinar and Sierra de Zafalgar ranges (Fig. 8).

Discussion

Groundwater from the SGNP aquifers in general shows low to intermediate mineralization and its chemical composition is mainly controlled by the mineral composition of the rocks through which the groundwater flows and the residence time of water in the aquifer. Most spring water samples analysed have a Ca–Mg$(HCO_3^-)_2$ facies (Fig. 3) as a consequence of the composition of the rocks constituting the unsaturated and saturated zones of the aquifers, principally limestones and dolostones. The fact that the ion HCO_3^- (Alk) is present in the mineral composition of both calcite ($CaCO_3$) and dolomite ($CaMg(CO_3)_2$) is reflected in the chemical facies of the groundwater, which is more homogeneous with respect to the anions than to the cations. With respect to the latter, the waters display different Ca/Mg ratios (Fig. 3) depending on the relative contribution of limestone and dolostone dissolution in their final mineralization and on the time the groundwater has been in contact with the rocks. This is a result of the different kinetics of the dissolution of calcite and dolomite (Palmer & Cherry 1984; Drever 1997; Appelo & Postma 2005). In this sense, if the

Table 4. *$\delta^{18}O$, $\delta^{2}H$ and deuterium excess values obtained in the spring water samples analysed*

Spring	Altitude (m a.s.l.)	$\delta^{18}O$ (‰)$_{(SMOW)}$					$\delta^{2}H$ (‰)$_{(SMOW)}$				Deuterium excess* (‰)			
		n	Mean	Minimum	Maximum	Coefficient of variation (%)	Mean	Minimum	Maximum	Coefficient of variation (%)	Mean	Minimum	Maximum	Coefficient of variation (%)
Penibetic														
S1	373	10	−5.39	−6.20	−4.98	7	−29.57	−37.22	−26.92	11	13.52	12.40	15.10	6
S2	320	17	−5.58	−6.08	−5.42	3	−30.08	−34.74	−28.60	5	14.54	12.70	15.61	5
S3	346	18	−5.91	−6.24	−5.55	4	−31.90	−33.47	−28.92	4	15.36	13.61	16.82	6
S4	340	21	−5.52	−5.69	−5.31	2	−29.70	−30.44	−28.93	1	14.47	13.35	15.92	5
S5	420	17	−5.66	−5.97	−5.27	4	−30.44	−31.75	−28.52	3	14.85	13.34	16.19	6
Mean	360	83	−5.61	−6.04	−5.30	4	−30.34	−33.52	−28.38	5	14.55	13.08	15.93	6
Middle Subbetic														
S6	280	15	−5.28	−5.59	−5.12	2	−28.17	−28.68	−27.43	1	14.06	12.68	16.19	7
S7	290	17	−5.39	−5.61	−5.24	2	−28.87	−29.43	−28.22	1	14.24	12.71	15.57	5
S8	500	16	−5.51	−5.76	−5.33	2	−28.88	−29.80	−27.71	2	15.22	14.24	16.38	4
S9	345	16	−5.46	−5.69	−5.28	2	−29.15	−29.74	−28.31	1	14.54	13.37	15.91	6
S10	296	17	−5.38	−5.63	−5.21	2	−29.22	−29.71	−28.56	1	13.85	12.49	15.30	5
S11	312	17	−5.48	−5.65	−5.34	2	−29.52	−29.79	−29.05	1	14.34	13.15	15.44	5
S12	310	11	−5.45	−5.52	−5.39	1	−29.39	−30.10	−27.29	2	13.71	10.59	14.66	8
S13	455	17	−6.11	−6.43	−5.52	4	−33.67	−34.99	−31.42	3	15.24	12.73	16.76	7
S14	500	11	−6.07	−6.32	−5.91	2	−33.40	−34.41	−31.92	2	15.14	14.17	16.19	4
S15	637	16	−5.70	−5.94	−5.53	2	−30.47	−31.20	−29.94	1	15.15	14.10	16.52	6
S16	625	16	−5.59	−5.86	−5.36	3	−29.90	−31.11	−28.86	2	14.79	13.30	16.13	6
S18	328	25	−5.84	−6.15	−5.56	3	−32.56	−33.79	−31.36	2	14.16	12.31	16.08	7
Mean	407	194	−5.61	−5.84	−5.40	2	−30.27	−31.06	−29.17	2	14.54	12.99	15.93	6

*Deuterium excess, $d = \delta^{2}H - 8\delta^{18}O$.

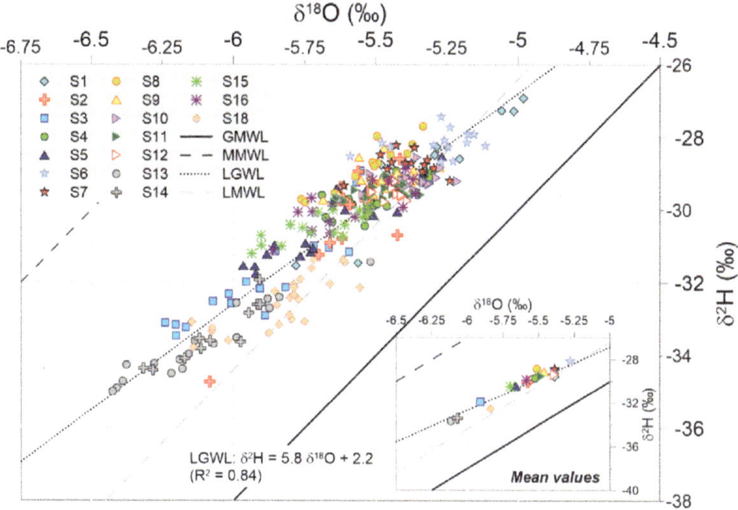

Fig. 7. Plot of $\delta^{18}O$ v. $\delta^{2}H$ of water molecules for the groundwater samples analysed.

residence time is sufficiently high, the waters may have relatively large concentrations of Mg^{2+}, even in the case of limestones (Edmunds *et al.* 1987). This hydrochemical pattern is coherent with those described in other carbonate systems in the presence of limestones and dolostones located in the same region (Mudarra & Andreo 2011; Barberá & Andreo 2015).

Groundwater mineralization is mainly dependent on the concentrations of carbonate-related ions: HCO_3^- (Alk), Ca^{2+} and Mg^{2+} (Fig. 4). Given that they have a similar source (calcite and dolomite dissolution), the chemical relationship between these components is close to the 1:1 stoichiometric line (Fig. 4). By contrast, the SO_4^{2-}, Cl^- and Na^+ concentrations are low in most springs because they do not constitute the principal components of carbonate rocks (Fig. 5e & g).

Although most waters sampled in the SGNP show some common hydrochemical characteristics, there are differences depending on the sector drained. Springs draining the southern sector of the study area (Penibetic), which is essentially made up of Jurassic limestones (Fig. 1), display the least mineralized waters because of their lower Alk and Ca^{2+} and Mg^{2+} contents (springs S1–S5; Figs 4 & 5). The role of HCO_3^- and Ca^{2+} in the chemical composition of the water flowing away from these springs is more relevant than that of Mg^{2+} (Fig. 5d). This is consistent with the almost complete absence of dolostone outcrops in the southern sector of the study area and suggests limited groundwater flow through these rocks in the saturated zone of the system. In addition, only a small contribution of SO_4^{2-} to the chemical composition of these waters has been observed (Fig. 5e), which indicates very little contact with evaporitic rocks (located principally under the carbonate aquifer).

The mountains located north of the Boyar Corridor and the La Silla range (SW sector, Fig. 1) consist of rocks belonging to the Middle Subbetic domain, in which lower Jurassic dolostones represent the main outcrop (Fig. 1). This lithological difference with respect to the southern sector has implications for the hydrochemistry of the groundwater. For instance, waters flowing out of the springs draining this sector show larger mineralization contents, mainly due to Alk and the concentrations of Ca^{2+} and Mg^{2+} (Fig. 5b). Mg^{2+} has a more significant role in the global chemical composition of these waters (Fig. 5c), reflected in the $Ca–Mg(HCO_3^-)_2$ chemical facies of most of the water samples (Fig. 3). Waters enriched in Mg^{2+} also have lower Ca/SO_4 ratios (Fig. 5f). Mg^{2+} has been suggested to be an indicator of the residence time in the saturated zone of the aquifer (Batiot *et al.* 2003*a*), although subsequent studies have questioned this statement – for instance, in carbonate systems where the unsaturated zone is formed by dolomitic rocks (Batiot *et al.* 2003*b*). Considering that the increase in Mg^{2+} concentration is accompanied by relative increases in SO_4^{2-} with respect to Ca^{2+}, a possible source of Mg^{2+} apart from the dissolution of dolomite might be the contact with evaporitic rocks (Mudarra & Andreo 2011). Nevertheless, the relationship between Mg^{2+}, SO_4^{2-} and Ca^{2+} seems to be more likely to be related to the dedolomitization process. This phenomenon has been reported by various researchers in other aquifers of the Betic Cordillera (Cardenal *et al.* 1994; Hidalgo &

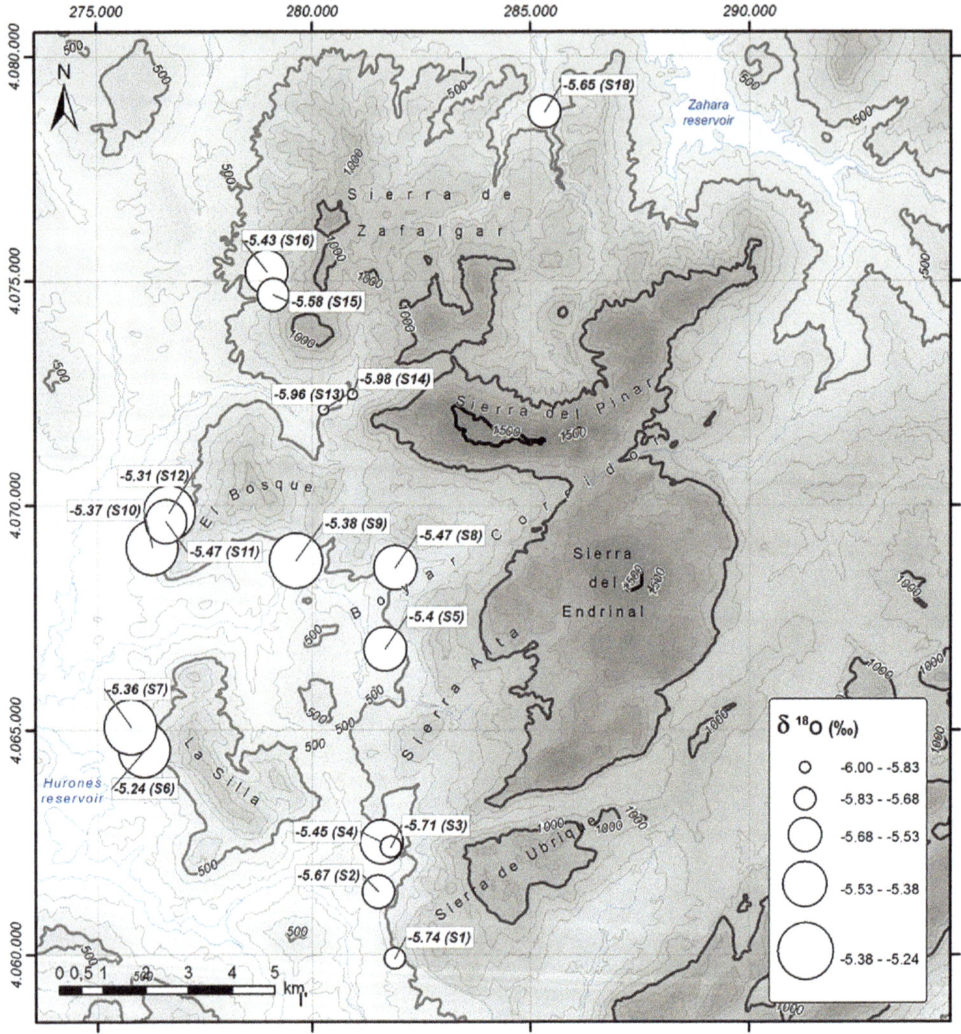

Fig. 8. Spatial distribution of spring water $\delta^{18}O$ composition in the Sierra Grazalema Natural Park ($\delta^{18}O$ values weighted by flow discharge).

Cruz-Sanjulián 2001; López-Chicano *et al.* 2001; Moral *et al.* 2008) as well as in other regions (Plummer 1977; Deike 1990; Plummer *et al.* 1990; Capaccioni *et al.* 2001; McMahon *et al.* 2004). It consists of the simultaneous dissolution of gypsum (eventually anhydrite) and dolomite, which produces an increase in the SO_4^{2-}, Mg^{2+} and Ca^{2+} concentrations in water, and the oversaturation and subsequent precipitation of calcite because of the common ion effect produced by the additional supply of Ca^{2+}. The result is an increase in SO_4^{2-}, Mg^{2+} and the SO_4^{2-}/Ca^{2+} ratio and a decrease in Alk and pH.

The chemical composition of waters drained by spring S7 strongly differs from all the other springs. They are of a Ca^{2+}–Na^{+} sulphate facies (Fig. 3), their total mineralization is 2.5–8.5 times higher than the rest (mean electrical conductivity 2273 µS cm^{-1}; Fig. 4) and the water temperature is the highest of all monitored springs with an average value of 17.5°C (Table 1). Water samples taken at this spring also show the greatest concentrations of Ca^{2+}, SO_4^{2-}, Na^{+} and Cl^{-} (Figs 4 & 5) and a Ca/SO_4 ratio close to unity (Fig. 5f), which indicates a common origin for both ions related to the dissolution of the gypsum, eventually anhydrite, present in the impervious Triassic bedrock of the La Silla aquifer (at the most southwestern part of the SGNP, Fig. 1). This process is also apparent in other carbonate aquifers located in similar geological contexts involving evaporite rocks at the bottom of the aquifer (López-Chicano *et al.* 2001; Martos-Rosillo *et al.* 2013). According to these results, groundwater drained by the spring S7 flows through deeper parts of the saturated zone of the carbonate aquifer, increasing the water temperature and coming into contact with Triassic evaporitic rocks, where not only gypsum, but also halite, dissolution takes place. The latter statement is based on the large Cl^- and Na^+ concentrations registered in this spring as well as on the ratio between them, which is close to unity (Fig. 5g). High concentrations of Mg^{2+} and Alk values of the same order of magnitude as in the other springs (Fig. 5d) suggest an evaporitic origin of most of the dissolved Mg^{2+} in the waters drained by this spring. Nevertheless, we cannot reject an influence of the dedolomitization process on the final Mg^{2+} content of this spring because its hypothermal nature indicates deeper flow paths where the groundwater would be simultaneously in contact with lower Jurassic dolostones and gypsum from the Triassic bedrock. In fact, water samples from spring S7 show a slight decrease in the Ca/SO_4 ratio with increasing Mg^{2+} concentrations (inset plot in Fig. 5f).

From a statistical standpoint, the water drained by the Penibetic springs has shown greater temporal variability than the Middle Subbetic springs, as indicated by the higher coefficients of variation with respect to the datasets for the physico-chemical composition (Table 1), flow discharge (Table 2) and isotopic signature (Table 4). This pattern, together with the existence of a more developed exokarstic landscape on the Penibetic sector, is consistent with a more evolved conduit network in the latter. Large conduits promote higher air ventilation of the unsaturated zone of the aquifer, leading to a decrease in the partial pressure of CO_2 and, consequently, the ability of groundwater to dissolve calcite and dolomite. This explains the lower alkalinity of the water drained by the springs located in the Penibetic sector of the SGNP (Table 1).

The group formed by springs S9, S10, S11 and S12 also shows particular hydrochemical characteristics that differentiates it from the other springs. In general, their waters show a higher mineralization (Table 1) and average temperatures (Fig. 2), larger concentrations of SO_4^{2-} (Fig. 5e), Mg^{2+} (especially S9; Fig. 5f) and both Cl^- and Na^+ (in particular S12 and, to a lesser extent, S11; Fig. 5g). The four springs drain the same carbonate outcrop, located near El Bosque town, which is surrounded by Triassic outcrops (Fig. 1). In agreement with these results, the general hydrochemical behaviour for the system drained by these springs is similar to that described for spring S7, resulting in the higher SO_4^{2-}, Cl^-, Na^+, and probably also Mg^{2+}, concentrations observed at these points. The difference between the S9–S12 group and spring S7 should be related to the groundwater–evaporite rocks interaction in terms of the time and/or volume of groundwater involved.

The deuterium excess can be used to characterize precipitation because it is a valuable indicator of the source area of water vapour (Dansgaard 1964; Rindsberger *et al.* 1983; Julián *et al.* 1992; Celle-Jeanton *et al.* 2001). On a global basis, the deuterium excess is, on average, *c.* 10‰ (GMWL), but regionally it varies due to variations in humidity, wind speed and sea surface temperature during primary evaporation (Clark & Fritz 1997). The deuterium excess in rainfall with a west Mediterranean origin generally has a relatively higher value (*c.* +15‰; Gat & Carmi 1970; Julián *et al.* 1992; Vallejos *et al.* 1997; Vandenschrick *et al.* 2002). This is due to a strong kinetic isotopic effect during evaporation in the summer period above the Mediterranean Sea, caused by the low relative humidity in the atmosphere. Conversely, Atlantic precipitation has a deuterium excess of *c.* +10‰ (Fritz *et al.* 1987; Clark & Fritz 1997).

The mean deuterium excess obtained in the SGNP rainwater ranges from 12.04 to 16.74‰ (average 14.25‰). The LMWL calculated for the study area from all rainwater samples analysed has a deuterium excess of 13.60‰ (Fig. 6a). These results indicate that air masses generating precipitation in the SGNP have a mixed Atlantic–Mediterranean origin, although with a more intense signature from the Atlantic.

The isotopic signature of groundwater samples shows a range of values much narrower than that of rainwater (Fig. 6a). This shows that the isotopic signature of groundwater in a single system is caused by the mixing of all the rainfall that has occurred over the years, reflecting an average value of all recharge events, the isotopic composition of which depends on the season, temperature, altitude and the amount of rain, among other factors.

The slope of the LGWL is lower than that of the LMWL (5.8 v. 8; Fig. 7). This is due to the evaporation experienced by rainwater during its route from clouds to the saturated zone of the aquifer. Partial evaporation can occur in the lowest part of clouds, at the land surface (Plata 1994; Andreo *et al.* 2004) and might also take place in the most superficial parts of the aquifer (Criss *et al.* 2007).

The isotopic signature of rainwater is controlled by latitude, the degree of continentality, temperature, the amount of rain and the altitude (Clark & Fritz 1997). On a local scale, such as that of the SGNP, all these variables can be considered to be homogeneous for the whole area, except for the altitude and wet air masses pathway (considering the same vapour mass, the first rains are isotopically enriched in $\delta^{18}O$ and δ^2H). In the study area, the acceptable correlation ($R^2 = 0.86$) between the mean $\delta^{18}O$

Table 5. *Mean recharge altitudes calculated for the SGNP springs and corresponding standard deviations*

Spring	Altitude of mean recharge area (m a.s.l.)	Standard deviation $\delta^{18}O$ (‰)	Standard deviation altitude (m)
S1	762	0.38	126
S2	740	0.16	53
S3	750	0.22	73
S4	667	0.11	36
S5	649	0.22	72
S6	616	0.13	43
S7	616	0.09	29
S8	672	0.12	40
S9	644	0.11	36
S10	638	0.11	35
S11	671	0.09	31
S12	621	0.22	73
S13	832	0.24	80
S14	841	0.15	49
S15	709	0.14	46
S16	661	0.16	52
S18	730	0.17	56

values of rainfall stations v. their altitude (Fig. 6b) demonstrates the existence of a stronger dependence on altitude rather than on the geographical distribution of rainfall in the isotopic composition of the recharge water. As the direct infiltration of rainwater represents the main recharge component for the SGNP aquifers (Andreo *et al.* 2014; Martín-Rodríguez *et al.* 2016), it is possible to establish a correlation between the mean isotopic composition of spring waters and their corresponding average recharge altitudes using the vertical gradient calculated in Figure 6b ($-0.304‰\ (100\ m)^{-1}$). To do so, it is first necessary to calibrate the $\delta^{18}O$–altitude pair with a known recharge area. This has been carried out using the La Silla aquifer because the limits of its recharge area are easy to identify as they coincide with the superficial geological contact between the permeable carbonate outcrops and the Triassic clays (Fig. 1). The results show a mean recharge altitude of 616 m a.s.l. and mean $\delta^{18}O$ values weighted for flow rate of −5.24 and −5.36‰ for the two discharge points of La Silla aquifer, S6 and S7, respectively (Fig. 8). The $\delta^{18}O$ value selected to carry out subsequent calculations was the average of these two values (−5.30‰).

Table 5 and Figure 9 show the mean altitudes obtained for the recharge area of the main springs in the SGNP. Springs S13 and S14 have the highest mean recharge altitudes (800–850 m a.s.l.). They are located in the westernmost part of the Sierra del Pinar carbonate range (Fig. 8), where the maximum elevations of the SGNP are reached (large areas >1500 m a.s.l. as well as the highest peak in the region with an altitude of *c.* 1650 m a.s.l.). These results indicate that the Sierra del Pinar aquifer is drained, at least partially, by springs S13 and S14.

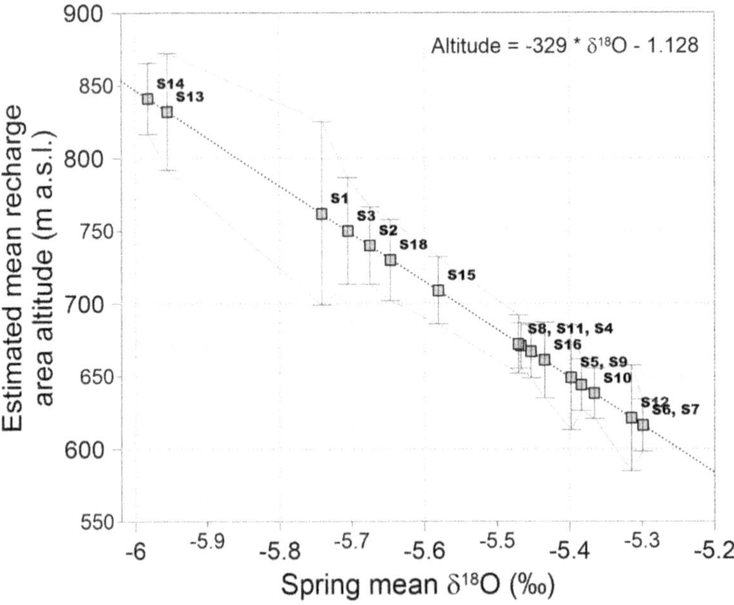

Fig. 9. Relationship between spring mean $\delta^{18}O$ values (weighted by spring flow) and estimated mean recharge altitude using the vertical gradient obtained in Figure 6b. Error bars have been calculated from $\delta^{18}O$ standard deviations using the same gradient.

S6 and S7 are the springs with the least negative mean $\delta^{18}O$ signatures and, consequently, with the lowest associated recharge areas (616 m a.s.l.; Fig. 9). Both discharge points drain the La Silla aquifer, in which the topography is also the smoothest out of the ranges constituting the SGNP (altitudes ranging from 280 to 850 m a.s.l.; Fig. 8). Outlets located near El Bosque town (S9–S12; Fig. 8) show recharge altitudes ranging from 621 to 671 m a.s.l.

Although situated near to each other, springs S3 and S4 show different recharge areas (Fig. 9). S3 drains waters infiltrated at higher altitudes (*c.* 750 m a.s.l.) than S4 (*c.* 665 m a.s.l.). Taking into account the position of both springs with respect to the carbonate outcrops, the Sierra de Ubrique (altitude >1000 m a.s.l.) might be the origin for the water drained by spring S3, whereas Sierra Alta (500–1000 m a.s.l.; Fig. 8) may be the main recharge area for spring S4. Nevertheless, this assumption should be confirmed by additional investigations, especially those based on tracer tests.

None of the springs draining the Penibetic sector (monitoring points S1–S5) seems to present an isotopic signature consistent with the high recharge altitudes (>1500 m a.s.l.) in the Sierra del Endrinal (eastern SGNP, Fig. 8). This may be considered evidence of a possible hydrogeological connection between the Sierra del Endrinal and adjacent carbonate outcrops located *c.* 4 km to the SE (Fig. 1), according to which rainwater infiltrated in the Sierra del Endrinal would be drained by springs located outside the SGNP. This hypothesis is not new and has already been suggested previously (Andreo *et al.* 2014). Recent studies based on water balance performed in the SGNP indicate a net deficit around 3000 $m^3\ a^{-1}$ between recharge and discharge in the Penibetic outcrops (Martín-Rodríguez *et al.* 2016).

Conclusions

We have presented a comprehensive view of the hydrogeological characteristics of heterogeneous and mountainous carbonate aquifers located in the SGNP, integrating results derived from commonly used hydrogeochemical and isotopic approaches. This characterization was carried out with information inferred from the major chemical components (Alk, SO_4^{2-}, Cl^-, Ca^{2+}, Mg^{2+}, Na^+, K^+ and NO_3^-) and the isotopic signature ($\delta^{18}O$ and δ^2H) of the waters drained by the principal springs in the study area. This information has enhanced our knowledge of the hydrogeological functioning of these karst aquifers.

The chemical composition of groundwaters in the SGNP is mostly dependent on the mineral composition of the rocks they flow through and come into contact with, especially those constituting the unsaturated and saturated zones of the aquifer and the impervious substratum. However, it also depends on the residence time and depth of the groundwater flows, which constrain the range of mineralization in the waters sampled, and the dedolomitization process. From a hydrochemical point of view, three groups of water have been identified according to their chemical composition: (1) waters from the Penibetic (southern) sector, which are characterized by a low mineralization content – they are of a $Ca(HCO_3)_2$ facies and show the lowest Alk as well as Ca^{2+} and Mg^{2+} concentrations; (2) waters from the Middle Subbetic (northern) sector, which present a higher mineralization content, primarily caused by greater Alk values and Ca^{2+} and Mg^{2+} concentrations – they are of a $Ca–Mg(HCO_3^-)_2$ facies; and (3) waters flowing out of specific springs (S7, S9, S10, S11 and S12), all located in the Middle Subbetic sector. These waters show a different physico-chemical composition characterized by higher temperatures and electrical conductivity, as well as SO_4^{2-}, Cl^- and Na^+ concentrations. This particular composition probably reflects deeper groundwater flows through the saturated zone of the aquifer and longer residence times in the system, favouring a marked water–rock interaction with mainly evaporitic rocks (sulphates and halite) in the Triassic clayey body. The latter constitutes the bedrock and lateral impervious limits of most of the aquifers in the study area. Nevertheless, the intensity of the evaporitic contribution to the chemical composition of groundwater is not homogeneous because it can be deduced from the large differences in electrical conductivity and ion concentrations between different outlets. The positive correlation between Mg^{2+} and the ratio SO_4^{2-}/Ca^{2+} in some springs suggests the existence of dedolomitization processes caused by the simultaneous dissolution of gypsum and dolomite.

Based on the isotopic signature of rainwater, the LWML of the SGNP has been calculated ($\delta^2H = 8\ \delta^{18}O + 13.6$), as well as the local vertical gradient ($-0.304‰\ \delta^{18}O\ (100\ m)^{-1}$). The combination of this gradient with the mean $\delta^{18}O$ signature of the water drained by the springs has allowed an estimate of the average recharge altitude for each spring. The results are congruent with the topographic distribution of permeable areas and the position of the springs. In addition, they allow assumptions regarding the flow paths of groundwater drained by different springs and a possible hydrogeological connection of the Sierra Grazalema aquifers with the adjacent permeable outcrops. The results obtained in this study have demonstrated that isotopic techniques based on stable isotopes of the water molecule are a valuable tool for use in hydrogeology to characterize aquifers and especially to clarify the recharge area related to each discharge point. The methodologies used in this work are applicable to any hydrological set-up and contain the potential to reveal deeper insights into any hydrogeological system.

This work is a contribution to the project CGL2015-65858-R and to the Research Group RNM-308 of the Junta de Andalucía, supported by the Environmental and Water Agency of Andalusia. We thank Dr Stephen Foster and an anonymous reviewer for their constructive comments and suggestions, which led to a substantial improvement of this paper.

References

ABUSAADA, M. & SAUTER, M. 2013. Studying the flow dynamics of a karst aquifer system with an equivalent porous medium model. *Groundwater*, **51**, 641–650.

ANDREO, B., MUDRY, J., CARRASCO, F. & VADILLO, I. 1997. Utilisation des traceurs météoriques (Cl^-, ^{18}O) à l'étude des aquifères carbonatés des Sierras Blanca et Mijas (Cordillère Bétique, Sud de l'Espagne) [in Spanish]. *In*: *VI Conference on Limestone Hydrology and Fissured Media and XII International Congress of Speleology*, 10–17 August 1997, La Chaux de Fonds, Switzerland, Vol. **2**, 251–254.

ANDREO, B., LINÁN, C., CARRASCO, F., JIMÉNEZ DE CISNEROS, C., CABALLERO, F. & MUDRY, J. 2004. Influence of rainfall quantity on the isotopic composition (^{18}O and ^{2}H) of water in mountainous areas. Application for groundwater research in the Yunquera-Nieves karst aquifers (S Spain). *Applied Geochemistry*, **19**, 561–574.

ANDREO, B., SÁNCHEZ, D. & MARTÍN-ALGARRA, A. 2014. *Hydrogeological Characterization and Water Resources Evaluation of Sierra de Grazalema Aquifers (Cádiz) for a Potential Implementation as Strategic Reserve in the Guadalete-Barbate River Water Basin* [in Spanish]. Andalusian Water Agency Technical Report.

APPELO, C.A.J. & POSTMA, D. 2005. *Geochemistry, Groundwater and Pollution*. 2nd edn. A.A. Balkema, Rotterdam.

ARAGUÁS-ARAGUÁS, L., FROEHLICH, K. & ROZANSKI, K. 2000. Deuterium and oxygen-18 isotope composition of precipitation and atmospheric moisture. *Hydrological Processes*, **14**, 341–355.

BAKALOWICZ, M. 1995. La zone d'infiltration des aquifères karstiques. Méthodes d'étude. Structure et fonctionnement. *Hydrogéologie*, **4**, 3–21.

BAKALOWICZ, M. 2005. Karst groundwater: a challenge for new resources. *Hydrogeology Journal*, **13**, 148–160.

BARBERÁ, J.A. & ANDREO, B. 2015. Hydrogeological processes in a fluviokarstic area inferred from the analysis of natural hydrogeochemical tracers. The case study of eastern Serranía de Ronda (S Spain). *Journal of Hydrology*, **523**, 500–514.

BATIOT, C., EMBLANCH, C. & BLAVOUX, B. 2003*a*. Carbone organique total (COT) et magnésium (Mg^{2+}): deux traceur complémentaires du temps de séjours dans l'aquifère karstique. *Comptes Rendus Geoscience*, **335**, 205–214.

BATIOT, C., LIÑÁN, C., ANDREO, B., EMBLANCH, C., CARRASCO, F. & BLAVOUX, B. 2003*b*. Use of total organic carbon (TOC) as tracer of diffuse infiltration in a dolomitic karst system: the Nerja Cave (Andalusia, southern Spain). *Geophysical Research Letters*, **30**, 2179.

BORTOLAMI, G.C., RICCI, B., SUSELLA, G.F. & ZUPPI, G.M. 1979. Isotope hydrology of the Val Corsaglia, Maritime Alps, Piedmont, Italy. *In*: *Isotope Hydrology 1978*, Vol. **I**. IAEA Symposium 228, 19–23 June 1978, Neuherberg, Germany, 327–350.

CAPACCIONI, B., DIDERO, M., PALETTA, C. & SALVADORI, P. 2001. Hydrogeochemistry of groundwaters from carbonate formations with basal gypsiferous layers: an example from the Mt Catria Mt Nerone ridge (Northern Appennnines, Italy). *Journal of Hydrology*, **253**, 14–26.

CARDENAL, J., BENAVENTE, J. & CRUZ-SANJULIÁN, J.J. 1994. Chemical evolution of groundwater in Triassic gypsum-bearing carbonate aquifers (Las Alpujarras, southern Spain). *Journal of Hydrology*, **161**, 3–30.

CELLE-JEANTON, H., TRAVY, Y. & BLAVOUX, B. 2001. Isotopic typology of the precipitation in the Western Mediterranean region at three different time scales. *Geophysical Research Letters*, **28**, 1215–1218.

CLARK, I. & FRITZ, P. 1997. *Environmental Isotopes in Hydrogeology*. Lewis Publishers, New York.

COLLINS, D.N. & GORDON, J.Y. 1981. Meltwater hydrology and hydrochemistry in snow- and ice-covered mountain catchments. *Nordic Hydrology*, **12**, 319–334.

CRAIG, H. 1961. Standard for reporting concentration of deuterium and oxygen-18 in natural waters. *Science*, **133**, 1702–1703.

CRISS, R., DAVISSON, L., SURBECK, H. & WINSTON, W. 2007. Isotopic methods. *In*: GOLDSCHEIDER, N. & DREW, D.P. (eds) *Methods in Karst Hydrogeology*. Taylor & Francis, London, 123–145.

DANSGAARD, W. 1964. Stable isotopes in precipitation. *Tellus*, **16**, 436–468.

DEIKE, R.G. 1990. Dolomite dissolution rates and possible Holocene dedolomitization of water-bearing units in the Edwards aquifer, south-central Texas. *Journal of Hydrology*, **112**, 335–373.

DREVER, J. 1997. *The Geochemistry of Natural Waters: Surface and Groundwater Environments*. 3rd edn. Prentice-Hall, Englewood Cliffs, NJ.

EDMUNDS, W.M., COOK, J.M., DARLING, W.G., KINNIBURGH, D.G. & MILES, D.L. 1987. Baseline geochemical conditions in the Chalk aquifer, Berkshire, U.K.: a basis for groundwater quality management. *Applied Geochemistry*, **2**, 251–274.

EMBLANCH, C., BLAVOUX, B., PUIG, J.M. & MUDRY, J. 1998. Dissolved organic carbon of infiltration within the autogenic karst hydrosystem. *Geophysical Research Letters*, **25**, 1459–1462.

FORD, D.C. & WILLIAMS, P.W. 2007. *Karst Hydrogeology and Geomorphology*. Wiley, Chichester.

FRITZ, P., DRINUNIE, R.J., FRAPE, S.K. & O'SHEA, O. 1987. The isotopic composition of precipitation and groundwater in Canada. *In*: *Isotope Techniques in Water Resources Development*. IAEA Symposium *299*, 30 March–3 April 1987, Vienna, Austria, 539–550.

GÁLVEZ-MAESTRE, M.J. 2005. Climatology [in Spanish]. *In*: LÓPEZ-GETA, J.A. (ed.) *Hydrogeologic Atlas of the Cádiz Province*. Spanish Geological Survey & Cádiz Council, Madrid, 53–58.

GAT, J.R. & CARMI, I. 1970. Evolution of the isotopic composition of atmospheric waters in the Mediterranean Sea area. *Journal of Geophysical Research*, **75**, 3039–3048.

GOLDSCHEIDER, N. & DREW, D.P. (eds) 2007. *Methods in Karst Hydrogeology*. Taylor & Francis, London.

GONFIANTINI, R., ROCHE, M.A., OLIVRY, J.C., FONTES, J.CH. & ZUPPI, G.M. 2001. The altitude effect on the isotopic composition of tropical rains. *Chemical Geology*, **181**, 147–167.

Hidalgo, M.C. & Cruz-Sanjulián, J.J. 2001. Groundwater composition, hydrochemical evolution and mass transfer in a regional detrital aquifer (Baza basin, southern Spain). *Applied Geochemistry*, **16**, 745–758.

Hunkeler, D. & Mudry, J. 2007. Hydrochemical methods. *In*: Goldscheider, N. & Drew, D.P. (eds) *Methods in Karst Hydrogeology*. Taylor & Francis, London, 93–121.

IGME 1984. *Hydrogeological Research Studies for the Regulation of Water Resources in the Guadalete-Guadiaro Watershed Border* [in Spanish]. Spanish Geological Survey, Technical Report.

IGME 1986. *Geo-electric Investigation in Sierra Grazalema Natural Park (Cádiz)* [in Spanish]. Spanish Geological Survey, Technical Report.

ITGE 1992. *Support Project for an Integrated Water Resource Management of the Carbonate Aquifers in Sierra de Grazalema* [in Spanish]. Spanish Geological Survey, Technical Report.

Julián, J.C.S., Araguás, L. *et al.* 1992. Sources of precipitation over south-eastern Spain and groundwater recharge. An isotopic study. *Tellus*, **44B**, 226–236.

Lambán, L.J., Jódar, J., Custodio, E., Soler, A., Sapriza, G. & Soto, R. 2015. Isotopic and hydrogeochemical characterization of high-altitude karst aquifers in complex geological settings. The Ordesa and Monte Perdido National Park (northern Spain) case study. *Science of the Total Environment*, **506–507**, 466–479.

Lastennet, R. & Mudry, J. 1997. Role of karstification and rainfall in the behavior of a heterogeneous karst system. *Environmental Geology*, **32**, 114–123.

López-Chicano, M., Bouamama, B., Vallejos, A. & Pulido, A. 2001. Factors which determine the hydrochemical behaviour of karstic springs: a case study from the Betic Cordilleras, Spain. *Applied Geochemistry*, **16**, 1179–1192.

Martín-Algarra, M. 1987. *Geological Alpine evolution of the contact between the Internal and External Zones of the Betic Cordillera* [in Spanish]. PhD thesis, University of Granada.

Martín-Rodríguez, J.F., Sánchez-García, D., Mudarra-Martínez, M., Andreo-Navarro, B., López-Rodríguez, M. & Navas-Gutiérrez, M.R. 2016. Water resources assessment and hydrogeologic water balance in mountainous karst aquifers. Case study of Sierra de Grazalema (Cádiz, Spain) [in Spanish]. *In*: Martínez-Cortina, L. & Martínez-Santos, P. (eds) *Hispanic–Lusitanian Conference on Groundwater under the Second Hydrological Planning Cycle*. International Association of Hydrogeologists, Reading, 163–170.

Martos-Rosillo, S., Rodríguez-Rodríguez, M., Pedrera, A., Cruz-SanJulián, J.J. & Rubio, J.C. 2013. Groundwater recharge in semi-arid carbonate aquifers under intensive use: the Estepa Range aquifers (Seville, southern Spain). *Environmental Earth Sciences*, **70**, 2453–2468.

McMahon, P.B., Bohlke, J.K. & Christenson, S.C. 2004. Geochemistry, radiocarbon ages, and paleorecharge conditions along a transect in the central High Plains aquifer, southwestern Kansas, USA. *Applied Geochemistry*, **19**, 1655–1686.

Moral, F., Cruz-Sanjulián, J.J. & Olías, M. 2008. Geochemical evolution of groundwater in the carbonate aquifers of Sierra de Segura (Betic Cordillera, southern Spain). *Journal of Hydrology*, **360**, 281–296.

Mudarra, M. & Andreo, B. 2011. Relative importance of the saturated and the unsaturated zones in the hydrogeological functioning of karst aquifers: the case of Alta Cadena (southern Spain). *Journal of Hydrology*, **397**, 263–280.

Mudry, J. 1987. *Apport du traçage physico–chimique naturel à la connaissance hydrocinématique des aquifères carbonatés*. PhD thesis, University of Franche-Comté.

Palmer, C.D. & Cherry, J.A. 1984. Geochemical reactions associated with low-temperature thermal-energy storage in aquifers. *Canadian Geotechnical Journal*, **21**, 475–488.

Plata, A. 1994. *Composición isotópica de las precipitaciones y aguas subterráneas de la Península Ibérica* [in Spanish]. Centro de Estudios y Experimentación de Obras Públicas, Madrid.

Plummer, L.N. 1977. Defining reactions and mass transfer in part of the Floridan aquifer. *Water Resources Research*, **13**, 801–812.

Plummer, L.N., Busby, J.F., Lee, R.W. & Hanshaw, B.B. 1990. Geochemical modelling of the Madison aquifer in parts of Montana, Wyoming, and South-Dakota. *Water Resources Research*, **26**, 1981–2014.

Rindsberger, M., Magaritz, M., Carmi, I. & Gilad, D. 1983. The relation between air mass trajectories and the water isotope composition in the Mediterranean Sea area. *Geophysical Research Letters*, **10**, 43–46.

Sánchez, D., Andreo, B., López, M., González, M.J. & Mudarra, M. 2014. Characterization of carbonate aquifers (Sierra de Grazalema, S Spain) by means of hydrodynamic and hydrochemical tools. *In*: Andreo, B., Carrasco, F., Durán, J.J., Jiménez, P. & LaMoreaux, J.W. (eds) *Hydrogeological and Environmental Investigations in Karst Systems*. Environmental Earth Sciences, **1**. Springer, Berlin, 171–180.

Sánchez, D., Barberá, J.A., Mudarra, M. & Andreo, B. 2015. Hydrogeochemical tools applied to the study of carbonate aquifers: examples from some karst systems of southern Spain. *Environmental Earth Sciences*, **74**, 199–215.

Sánchez, D., Martín-Rodríguez, J.F., Mudarra, M., Andreo, B., López-Rodríguez, M. & Navas, M.R. 2017. Time lag analysis of natural responses during unitary recharge events to assess the functioning of carbonate aquifers in Sierra de Grazalema Natural Park (southern Spain). *In*: Renard, P. & Bertrand, C. (eds) *EuroKarst 2016. Advances in Karst Science*. Springer, Berlin, 157–167.

Vallejos, A., Pulido-Bosch, A., Martín-Rosales, W. & Calvache, M.L. 1997. Contribution of environmental isotopes to the understanding of complex hydrologic systems. A case study: Sierra de Gádor, SE Spain. *Earth Surface Processes and Landforms*, **22**, 1157–1168.

Vandenschrick, G., van Wesemael, B., Frot, E., Pulido-Bosch, A., Molina, L., Stiévenard, M. & Souchez, R. 2002. Using stable isotope analysis (δD– $\delta^{18}O$) to characterise the regional hydrology of the Sierra de Gádor, south east Spain. *Journal of Hydrology*, **265**, 43–55.

White, W.B. 1988. *Geomorphology and Hydrology of Karst Terrains*. Oxford University Press, New York.

Climatological trends and anticipated karst spring quantity and quality: case study of the Slovene Istria

NATAŠA RAVBAR[1,2]*, GREGOR KOVAČIČ[3], METKA PETRIČ[1,2], JANJA KOGOVŠEK[1], CLARISSA BRUN[4] & ALENKA KOŽELJ[4]

[1]*Karst Research Institute ZRC SAZU, Titov trg 2, SI-6230 Postojna, Slovenia*

[2]*UNESCO Chair on Karst Education, University of Nova Gorica, Glavni trg 8, SI-5271 Vipava, Slovenia*

[3]*Faculty of Humanities, University of Primorska, Titov trg 5, SI-6000 Koper, Slovenia*

[4]*National Laboratory of Health, Environment and Food, Verdijeva 11, SI-6000 Koper, Slovenia*

**Correspondence: natasa.ravbar@zrc-sazu.si*

Abstract: The behaviour of aquifers with karst porosity is dependent on hydrological conditions. This is due to the peculiar characteristics of the groundwater flow and dynamics of hydrological processes in karst. As a result, karst aquifers are especially vulnerable to the effects of environmental change. We assessed the long-term climatological and hydrological trends and the short-term effects of increasingly frequent extreme hydrological events (droughts) for the Mediterranean karst spring Rižana in SW Slovenia. The findings predict higher mean annual air temperatures of 0.34°C decade^{-1}, lower annual precipitation of *c.* 60 mm decade^{-1} and higher annual actual evapotranspiration (especially during spring and summer) of 32–49 mm decade^{-1}. As a consequence, we can expect a decrease in the mean annual discharge of the spring of *c.* 480 l s^{-1} decade^{-1} with prolonged dry summer periods. Detailed monitoring of the physical, chemical and microbiological parameters showed that the flood pulses caused by precipitation events after a long dry period cause a significant deterioration in water quality. In such situations, contaminants stored in the unsaturated zone are flushed out and counts of coliform bacteria can reach >1400 cfu (100 ml)$^{-1}$, with total Al and Fe concentrations up to 206 and 474 μg l^{-1}, respectively. These results suggest that there should be urgent adherence to water quality standards to protect karst water sources in view of the anticipated climatological stresses. Management strategies should promote monitoring and the rational use of karst water supplies.

Both long-term measurements of meteorological and hydrological data and simulations indicate changed patterns in temperature and precipitation regimes around the world and their impacts on the water cycle (Bates *et al.* 2008; Kundzewicz *et al.* 2008; Gosling & Arnell 2016). Observed records and climate simulations are consistent in predicting the incidence of changes in the amount, intensity and variability of precipitation, and increases in temperature and evapotranspiration. These changes increase the deviations in hydrological regimes and have implications for both the quantity and quality of water in many areas (Jiménez Cisneros *et al.* 2014; Döll *et al.* 2015). Hagemann *et al.* (2013) reported that a considerable decrease in the availability of water resources is expected in central and southern Europe, the Middle East, southern Africa, southern China, the Mississippi River basin and southeastern Australia. The identified climatological and hydrological stresses may severely affect ecosystems that are dependent on freshwater and a number of socio-economic activities. The availability, stability of access and utilization of groundwater resources may, for example, provoke difficulties for many services, such as the supply of drinking water, agriculture, industry and hydropower.

The importance of aquifers with karst porosity for water supply is growing globally (Chen *et al.* 2017). However, karst aquifers are highly vulnerable to changes in the hydrological cycle as a result of their specific nature (i.e. the spatial and temporal heterogeneity of water flow and storage) (Siemers & Dreybrodt 1998; White 2002; Worthington 2007). Changes in hydrological conditions can result in either a shortage of high-quality stocks of drinking water (droughts) or the occurrence of more intensive floods. Hence there is a need for a better understanding of how and to what extent changes in climate are affecting karst aquifers and to what extent karst aquifers are vulnerable to global changes in climate (Di Matteo *et al.* 2012; Hartmann *et al.* 2014*a*, *b*; Jia *et al.* 2017).

Despite the important role of karst water resources in supplying populations with drinking

From: Parise, M., Gabrovsek, F., Kaufmann, G. & Ravbar, N. (eds) 2018. *Advances in Karst Research: Theory, Fieldwork and Applications*. Geological Society, London, Special Publications, **466**, 295–305.
First published online December 6, 2017, https://doi.org/10.1144/SP466.19

water, few studies have adequately evaluated the impact of predicted changes in climate on the amount of available water (e.g. Hartmann *et al.* 2012; Finger *et al.* 2013; Long & Mahler 2013). Research to date (e.g. Kogovšek 2001; Vesper *et al.* 2001; Williams *et al.* 2006; Hunkeler & Mudry 2007; Ravbar *et al.* 2012) has shown that changes in the quality of karst water are most pronounced in periods of low flow or during the flood pulses caused by intense precipitation events after dry periods lasting for several months. Flow dynamics and solute transport are even more pronounced in larger binary karst aquifers, where sinking rivers may represent a concentrated input of often polluted water into the highly permeable conduits of a karst aquifer and onwards towards karst springs (e.g. Kogovšek 2002; Bailly-Comte *et al.* 2007; Pronk *et al.* 2007).

The goal of this study was to gain more information about how, and to what extent, the quantity of karst groundwater sources and, as a consequence, the water quality of karst springs may be affected by anticipated climatological trends. A combined approach of statistical and specific spring monitoring methods was applied, which could be further utilized to model and simulate site-specific hydraulic responses to different climate scenarios. The focus is on the analysis of climatological and hydrological trends and monitoring the hydraulic behaviour of springs, as well as changes in water quality at short intervals during specific identified hydrological conditions. The selected case study is the Rižana spring, which has a complex karst and non-karst catchment and is the most important water source for the region of Slovene Istria, which is important for both agriculture and tourism. The spring supplies water for between 86 000 and 120 000 people (Ravbar & Kovačič 2015).

Study area

The Rižana River spring is located in northern Istria in the Mediterranean region of Slovenia. The region contains a transboundary karst aquifer between Slovenia and Croatia (Fig. 1). The region is characterized by the alternation of permeable Cretaceous and Paleocene limestone and dolomite and impermeable Eocene flysch rocks. The Rižana spring is fed by a complex aquifer structure that is recharged both through the direct infiltration of precipitation and through the sinking streams of the Brkini hills, which enter the highly permeable conduits of the karst aquifer. The spring discharges at an elevation of 70 m a.s.l. and the recharge area ranges from 500 to 1100 m a.s.l. Based on fundamental hydrogeological studies and numerous tracer experiments (Krivic *et al.* 1987, 1989; Biondić *et al.* 2015), the catchment of the Rižana spring is estimated to be 247 km^2. The region has a sub-Mediterranean climate characterized by hot summers and relatively cold winters (the average air temperature ranges from 20 to 22°C in July and from 0 to 4°C in January). The mean annual precipitation of the recharge area in the period 1961–2013 (Podgorje pod Slavnikom precipitation station) was 1492 mm with a mean annual runoff coefficient of 0.49. The greatest amount of precipitation typically occurs during November and December, with a secondary maximum in June.

The discharge of Rižana spring in the period 1966–2013 ranged from 10 l s^{-1} (absolute minimum) to 155 m^3 s^{-1} (absolute maximum) with an average flow rate of 3.8 m^3 s^{-1} (Janža 2010; Kovačič & Ravbar 2016). The lowest average monthly discharge generally occurs in July and August and the highest monthly discharge in November and December. The spring is a regionally important water source that has been utilized since 1935. The average amount of water abstracted for water supply is 200 l s^{-1}, but water abstraction is temporarily abandoned when the spring's discharge drops below the ecological minimum outflow value of 110 l s^{-1}. This usually happens during the dry summer months. Drinking water is then provided by numerous substitute water sources in a wider region and from neighbouring Croatia.

Methodology

Background

Karst aquifers are generally characterized by high permeability and, as a consequence, the rapid infiltration of recharge waters (Ford & Williams 2007). Heterogeneous groundwater flow along karst conduits towards springs prevails underground and may reach velocities of up to several hundred metres per hour. Subsequently, flow and transport are characterized by rapid recharge and discharge mechanisms, and by extreme dynamics. Therefore they can reflect the effects of climate change on annual or even seasonal timescales (Long & Mahler 2013). From this perspective, karst aquifers are suitable for researching these effects on a human timescale.

To meet the aims of this study, a cascade approach of region-scale to basin-scale to spring-scale was undertaken. The study was therefore divided into three successive steps.

(1) The statistical trends of historical climate series of basic regional climatological parameters (precipitation, air temperature and actual evapotranspiration) were combined with the trends of the basin-scale groundwater discharge dynamics.
(2) The study then focused on the responses of the observed water source to precipitation

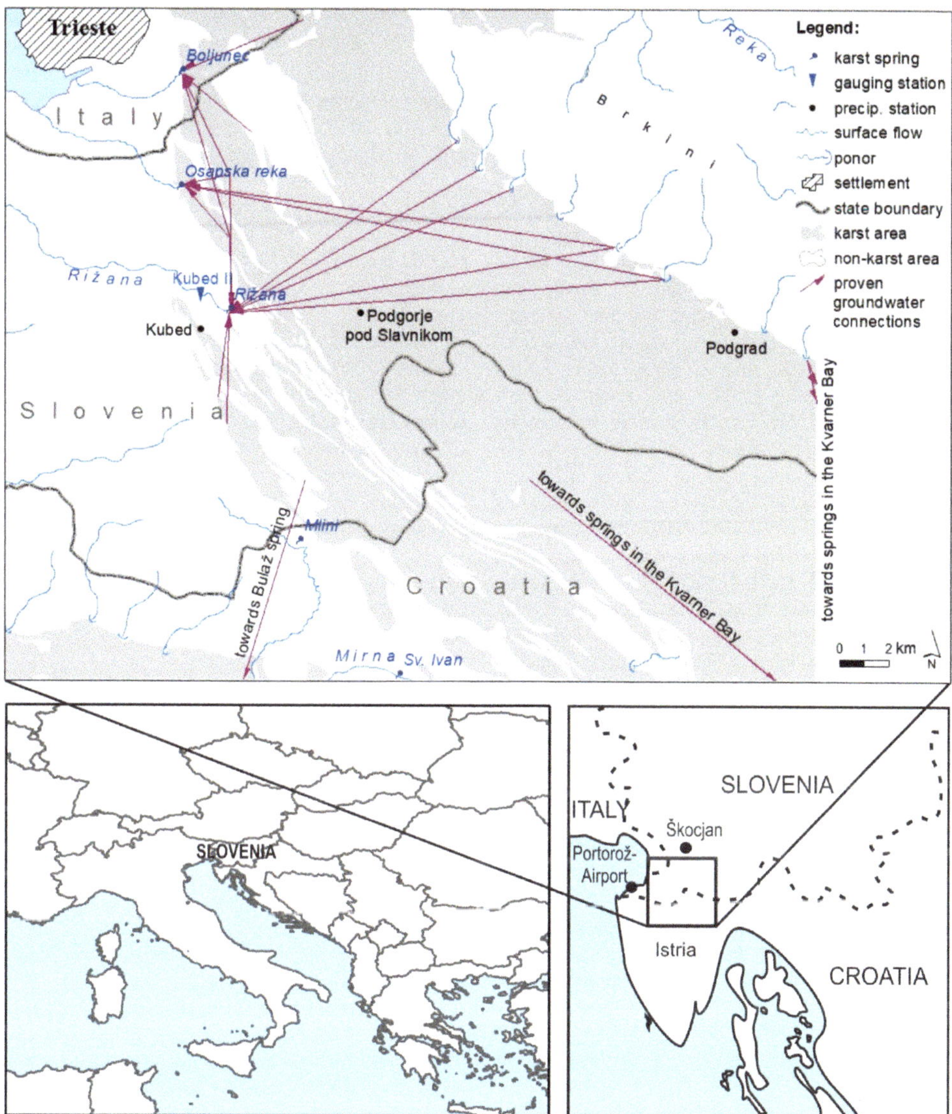

Fig. 1. Hydrogeological setting of the study area, the proved main groundwater flow connections, and location of the precipitation and gauging stations.

extremes and the accompanying changes in the physico-chemical characteristics of the spring. High-frequency monitoring of the spring, including measurements of the water temperature and electrical conductivity, was conducted for nine months. The observations were combined with the available precipitation, discharge and turbidity data. Analysis of the behaviour of the spring after individual recharge events indicated that emphasis should be given to monitoring the water quality during particular hydrological conditions.

(3) A typical flood pulse caused by precipitation events after a long dry period was extensively sampled and the water quality analysed.

A detailed description of the individual research methods and techniques is provided in the following sections.

Climatological and hydrological trends

The regional climatological and hydrological data were analysed statistically to gain information about the actual trends in the selected variables.

The trends of the changes in climate and hydrological variables were determined using the non-parametric Sen's slope test (the Theil–Sen estimator) (Theil 1950; Kraner Šumenjak & Šuštar 2011; Tilgenkamp 2011; Vannest *et al.* 2011; GraphPad Software 2016). Sen's slope analysis was performed using the Microsoft Excel template MAKESENS (Salmi *et al.* 2002). The statistical significance of the trends was verified at a confidence interval of 95% ($\alpha = 0.05$). Pearson's r coefficients were calculated among the hydrological and meteorological data series using Statistica software. The gauging station Kubed II and the meteorological stations Portorož-Airport (see inset map of Istria, Fig. 1), Podgorje pod Slavnikom, Kubed and Podgrad were included in the analysis. The trend analysis was conducted for the longest available monthly and yearly data series of the characteristic discharges of Rižana spring (1966–2013, 48 years) and for the seasonal and yearly precipitation (1961–2013; 53 years), air temperature (1961–2013, 53 years) and actual evapotranspiration (1971–2013, 43 years) data series from meteorological stations in the investigated area. All the data were obtained from the Slovenian Environmental Agency (2017). The evapotranspiration data series were calculated at the Slovenian Environmental Agency using the modified Hargreaves method based on the minimum and maximum air temperature and the precise geographical position of the stations (for further details, see Frantar 2008). The seasonal precipitation, actual evapotranspiration and air temperature were calculated for each station using the standard definition of the seasons (e.g. winter, December to February). Variations between the initial and final values of the variables during the study period and Pearson's coefficients between the meteorological and hydrological variables were also calculated.

Monitoring of spring behaviour

The water electrical conductivity and temperature were measured at Rižana spring from 20 September 2012 to 31 July 2013 using an Onset HOBO conductivity data logger. The electrical conductivity logger ranges between 0 and 1000 $\mu S\ cm^{-1}$ with an accuracy of 5 $\mu S\ cm^{-1}$ and a resolution of 1 $\mu S\ cm^{-1}$. The temperature logger ranges between 5 and 35°C with an accuracy of 0.1°C and a resolution of 0.01°C. The data logger was calibrated manually using a WTW Cond 330i conductivity meter. The data on turbidity were obtained from the Rižanski vodovod d.o.o. water supply company. All parameters were measured at 30 minute intervals.

The data for the total daily precipitation at Škocjan station (see insert map of Istria, Fig. 1) and the mean daily discharge of the Rižana spring at Kubed II gauging station from 15 May 2012 to 31 July 2013 were obtained from the Slovenian Environmental Agency (2017).

Detailed monitoring of the flood pulses after an almost three-month dry period was carried out in late June 2015. The recession period started at the end of March 2015 and the discharge of Rižana was <1 $m^3\ s^{-1}$ from 8 April to 23 June 2015. *In situ* monitoring of water electrical conductivity and temperature was carried out using the Onset HOBO conductivity data logger at 30 minute intervals. Data on precipitation for Škocjan station and the discharge of Rižana spring at 30 minute intervals were obtained from the Slovenian Environmental Agency (2017). A total of 36 water samples were taken for chemical and microbiological analysis, some as frequently as every two hours. All the sample analyses were performed in compliance with the standard procedures and accreditation document established by the National Laboratory of Health, Environment and Food. This paper presents the selected parameters of turbidity and the concentrations of nitrates, chlorides, total Fe and Al, and total organic carbon. Data on coliform bacteria were obtained from Pretnar (2015).

Results

The study was progressive. Each step in the study was made based on the results of the previous step. The results of each step are presented in a separate sub-section.

Climatological and hydrological trends

The results of the statistical analysis of the regional climatological and hydrological data show statistically significant trends in the decreasing annual precipitation of 57–61 mm decade^{-1} (4%). There are also increasing trends in the mean annual air temperature of 0.34°C decade^{-1} (2.6%) and in the annual actual evapotranspiration of 32–49 mm decade^{-1} (4–5%). The most noticeable result is a statistically significant decreasing trend in precipitation during the summer (Podgrad; 25 mm decade^{-1}) and the spring (Podgorje pod Slavnikom; 20 mm decade^{-1}), when the most evident and statistically significant increase in actual evapotranspiration was recorded at Portorož-Airport and Kubed. The increase in actual evapotranspiration is statistically significant for all seasons, with the exception of the winter period at Kubed station (Table 1).

The results show lower annual runoff in the region and, consequently, a lower mean monthly discharge of Rižana spring, especially in summer (June, 330 $l\ s^{-1}$ decade^{-1}; July, 110 $l\ s^{-1}$ decade^{-1}; August, 130 $l\ s^{-1}$ decade^{-1}; and September, 310 $l\ s^{-1}$ decade^{-1}) when the need for water also increases significantly

Table 1. *Selected results of the trend analysis for precipitation, actual evapotranspiration and air temperature data series*

Station Parameter	Temperature at Portorož-Airport	Actual evapotranspiration at Portorož–Airport	Precipitation at Podgorje pod Slavnikom	Precipitation at Podgrad	Actual evapotranspiration at Kubed
Annual mean (°C or mm)	12.94	980.0	1491.6	1532.3	762.5
Sen's slope for yearly data	0.034*	4.892*	-5.763*	-6.130*	3.183*
Change between the initial and end values of time series according to the Sens's linear trend – yearly data (°C or mm (%))	1.78 (14.7)	195.7 (22.2)	-299.6 (-18.7)	-318.7 (-19.0)	127.3 (18.1)
Sen's slope (winter)	0.027*	0.625*	-0.872	-0.554	-0.058
Sen's slope (spring)	0.031*	1.080*	-1.950*	-1.489	1.120*
Sen's slope (summer)	0.039*	2.308*	-1.459	-2.484*	1.653*
Sen's slope (autumn)	0.022*	0.875*	-1.367	-1.682	0.429*

For location of precipitation stations, see Figure 1.
*Statistically significant trend.

(crop irrigation and tourist season) (Kovačič 2016). The association between meteorological and hydrological variables was tested using Pearson's r coefficient. The mean annual discharge of Rižana spring shows a statistically significant strong positive correlation with the annual precipitation at Podgrad (r = 0.92) and Podgorje pod Slavnikom (r = 0.96) and a moderate negative correlation with the annual evapotranspiration at Kubed (r = −0.52). The weak negative correlation between the mean annual discharge of the Rižana spring and annual evapotranspiration at Portorož-Airport (r = −0.37) is not statistically significant.

A statistically significant decreasing trend in the mean annual discharge of Rižana spring of 480 l s^{-1} $decade^{-1}$ (12%) was confirmed. The decreasing trends of the mean annual minimum (110 l s^{-1} $decade^{-1}$) and maximum (1.92 m^3 s^{-1} $decade^{-1}$) discharges of the spring are also statistically significant. In the observed period, the mean annual discharge of Rižana spring decreased by 47% and the mean minimum annual discharge decreased by 77% (Table 2). Statistically significant negative trends in the absolute minimum discharge of Rižana spring from April to September (40–200 l s^{-1} $decade^{-1}$) and the mean minimum monthly discharge for the same period (50–180 l s^{-1} $decade^{-1}$) were identified.

Spring behaviour monitoring

High-frequency monitoring of physico-chemical parameters between September 2012 and July 2013. Given that an increasing trend in evapotranspiration and decreasing trends in precipitation and discharge can be anticipated in the future, the next phase of the study focused on the behavioural

Table 2. *Results of the trend analysis for the Rižana spring*

Parameter	Rižana–Kubed II
Q_{mean} (m^3 s^{-1})	3.83
Q_{min} (m^3 s^{-1})	0.01
Q_{max} (m^3 s^{-1})	153.39
Q_{mean}:Q_{max}	1:40
Sen's slope for Q_{mean} (year)	−0.048*
Sen's slope for mean Q_{min} (year)	−0.011*
Sen's slope for mean Q_{max} (year)	−0.192*
Change between the initial and end values of time series according to the Sens's linear trend – yearly data (m^3 s^{-1} (%))	−2.24 (−46.7)

*Statistically significant trend.

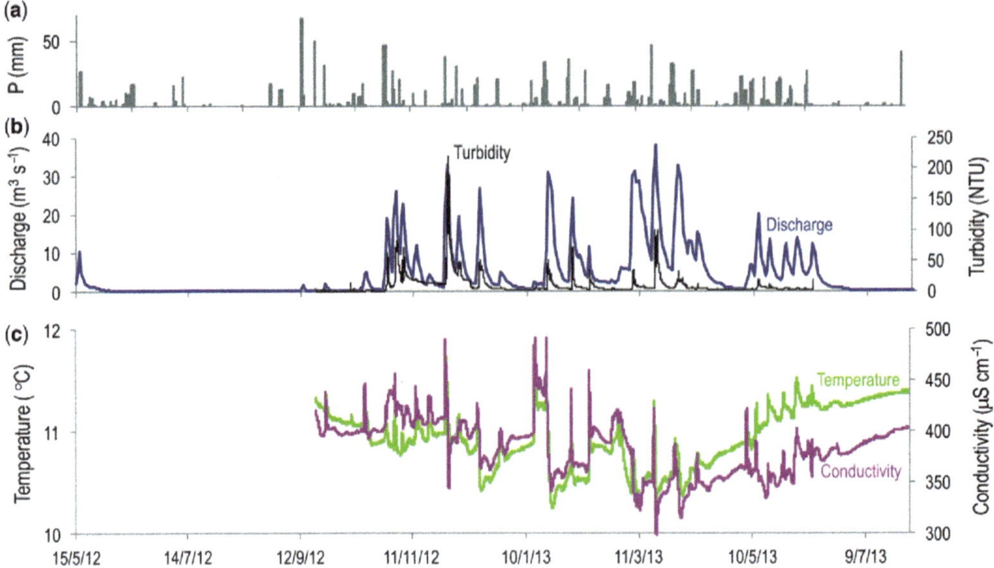

Fig. 2. Temporal evolution of (**a**) the total daily precipitation (P) at Škocjan station, (**b**) the mean daily discharge and turbidity of Rižana spring and (**c**) the temperature and conductivity of the water drained by Rižana spring measured at 30 minute intervals (for location of the precipitation station, see Fig. 1).

characteristics of the spring with an emphasis on low water conditions. The discharge, turbidity, temperature and electrical conductivity were monitored at high frequency to gain a better understanding of the hydrogeological functioning of the spring. Figure 2 presents the measured values combined with the available precipitation data.

The Rižana spring displays a typical karst hydrological regime, with short peak-flow events and prolonged periods of medium to low water levels. The spring is characterized by a fast reaction to hydrological events; maximum discharge often occurs a few hours after abundant rainfall. The turbidity can increase to as high as 70 or even >200 NTU during rainy periods. The mean (range) electrical conductivity of the water is *c.* 400 ± 70 µS cm^{-1}. Although detailed monitoring of the electrical conductivity and temperature started on 20 September 2012, the characteristics of the preceding low-flow period were defined based on precipitation and discharge data.

During the nine-month monitoring period, 27 high-flow events were observed with diverse responses of the electrical conductivity parameter. The spring usually exhibited behaviour typical of karst springs with an explicit piston-dilution effect, i.e. an increase in electrical conductivity during the rising limb of the spring hydrograph was followed by a minimum, indicating flushing of the water stored in the aquifer and the arrival of freshly infiltrated water (Ford & Williams 2007). The Rižana spring displayed this behaviour between the end of November 2012 and June 2013.

In the three-month period from 15 May to the middle of June 2012, only individual precipitation events (<20 mm daily) were observed, with no reaction of the discharge. When the first more intensive periods of rain (>40 mm daily) with high actual evapotranspiration occurred in late September after a dry period several months long, they resulted in small flood pulses at the spring with an insignificant increase in discharge and turbidity (Fig. 3). The discharge increased to *c.* 2 m^3 s^{-1} and the turbidity values were ≤5 NTU. However, there was a significant increase in the electrical conductivity, with a peak in increase of *c.* 40 µS cm^{-1}, and this did not subsequently, or only very slightly, fall below the pre-event values. This phenomenon has been observed in other springs and can be explained by the washing out of the highly mineralized water previously stored in the system (Vesper & White 2003; Doctor *et al.* 2006), with dilution having only a very limited effect.

As a result of the washing out effect and negligible dilution, such circumstances can lead to a deterioration in water quality. Detailed chemical and microbiological monitoring of a selected flood pulse after a period of low water several months long was carried out in the next phase of the research. This gave a better assessment of the particular behavioural characteristics of the spring and its dependence on the changes in hydrological conditions.

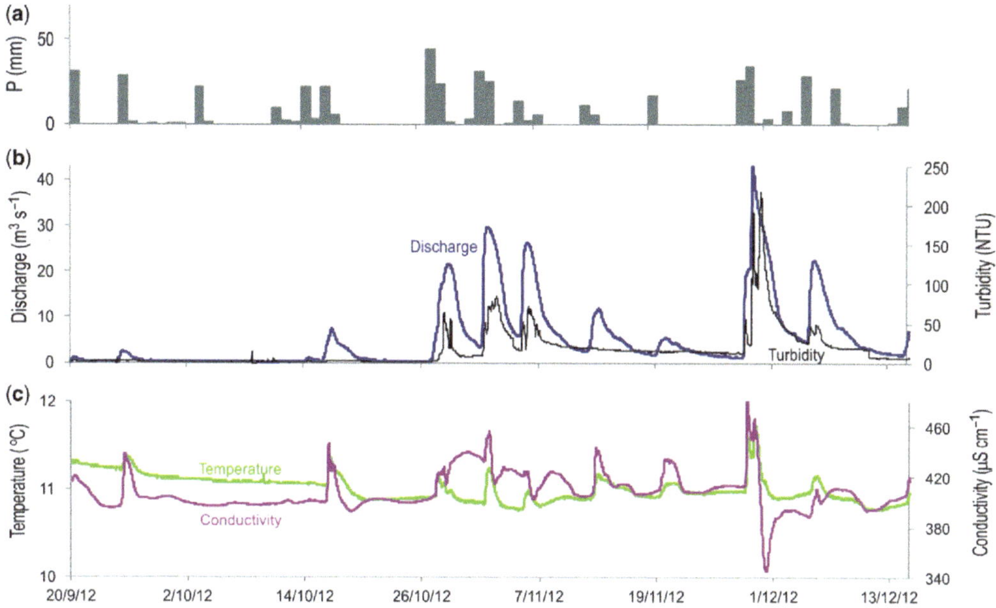

Fig. 3. Detailed view of the period from 20 September to 15 December 2012. Temporal evolution of (**a**) the total daily precipitation (P) at Škocjan station, (**b**) the mean daily discharge and turbidity of Rižana spring and (**c**) the temperature and conductivity of the water drained by Rižana spring measured at 30 minute intervals (for location of the precipitation station, see Fig. 1).

Detailed monitoring of physical, chemical and microbiological parameters during a flood pulse in late June 2015. The very wet hydrological year of 2014 was followed by an extremely arid semi-annual period. The first more abundant precipitation event occurred in the three-day period between 14 and 16 June 2015. In total, 25 mm of rain was recorded at the Škocjan precipitation station. After a few days, between 23 and 24 June, an additional 71 mm of rain fell, with a maximum hourly precipitation of 22 mm. This intensity of rainfall occurred with a one-year return period. The two recharge events provoked two different flood pulses of different magnitude at the observed spring (Fig. 4).

The first flood pulse induced a brief increase in the discharge of Rižana spring from 0.5 to 1.2 $m^3 s^{-1}$. This was mainly characterized by the washing out of highly mineralized water (an increase in electrical conductivity from an initial value of 386 to 468 $\mu S\ cm^{-1}$). The solute-enriched water is a result of storage in the conduits and fissures of the vadose zone. The slightly increased values of turbidity (up to 2.2 NTU) were due to the mobilization of sediments in the conduits. The influence of the allogenic recharge was accompanied by a decrease in electrical conductivity. Although the electrical conductivity of the sinking streams from the flysch was not measured simultaneously within the frame of this study, values of electrical conductivity of *c.* 200 $\mu S\ cm^{-1}$ are characteristic of these types of surface stream. The high number of coliform bacteria (up to 1400 cfu $(100\ ml)^{-1}$) was alarming, although these decreased in number with the decrease in flood pulse.

After the second precipitation event, the discharge of Rižana spring increased from 0.5 to 10.6 $m^3 s^{-1}$. The water level at the spring began to increase about five hours after the heavy rainfall and the pattern of supplanting the stored water was repeated. An increase in electrical conductivity by up to 17 $\mu S\ cm^{-1}$ in the next phase was a consequence of the washing out of water from less permeable areas of the vadose zone. The increase in flow dynamics also moved the sediments deposited in conduits and fissures, which was reflected by a small increase in turbidity. The beginning of a decrease in electrical conductivity and turbidity indicated a diffuse influx of newly infiltrated rainwater from the karst surface. The sharp decrease in electrical conductivity to values lower than before the flood pulse (*c.* 364 $\mu S\ cm^{-1}$) and the subsequent significant increase in turbidity of 13.9 NTU reflect the recharge of allogenic streams from the flysch margin. Towards the end of the flood pulse, the decreasing flow was accompanied by a gradual increase in electrical conductivity and lower turbidity. This pattern of changes in the observed parameters during a flood pulse has been recorded at numerous other

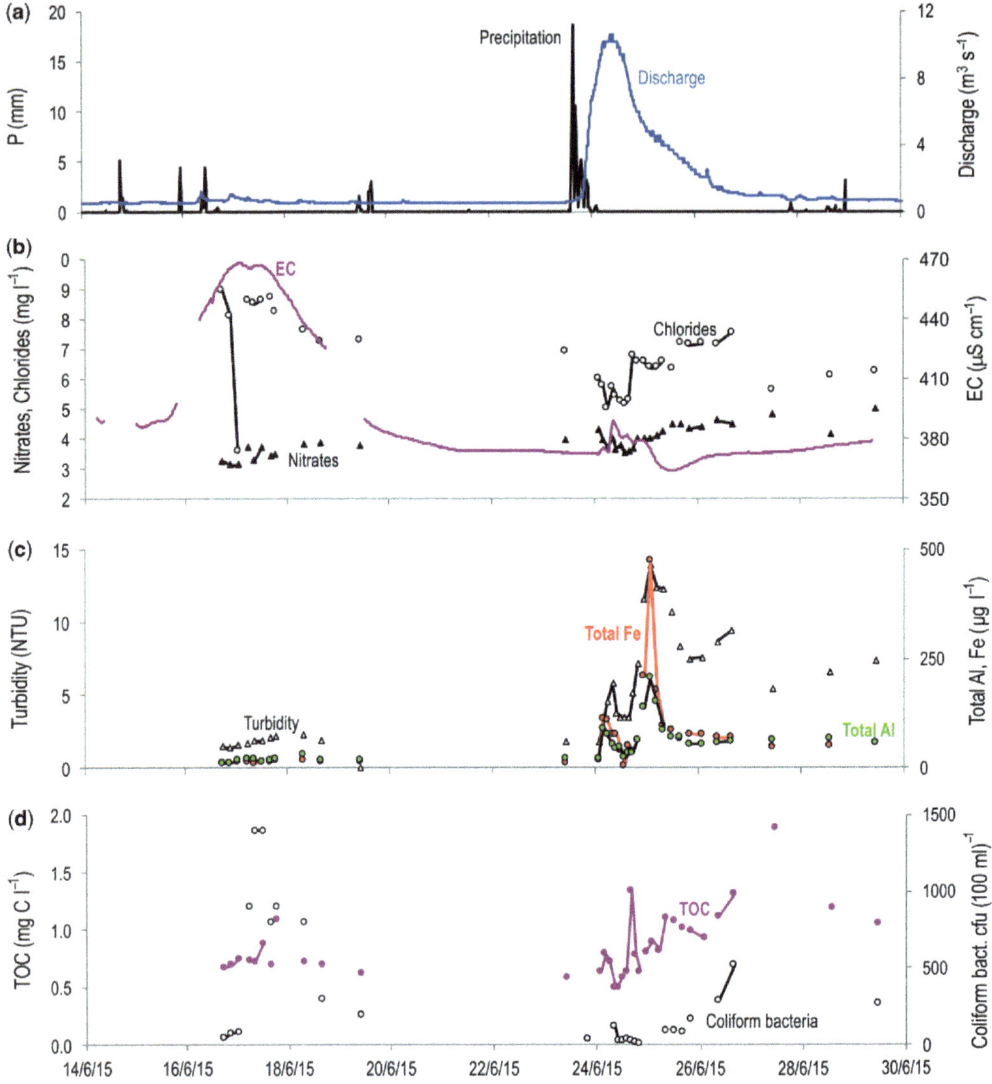

Fig. 4. (**a**) Precipitation (P) at Škocjan precipitation station and discharge of the Rižana spring measured at 30 minute intervals and (**b–d**) the physical, chemical and microbiological conditions at Rižana spring. A total of 36 water samples were taken for analysis, some as frequently as every two hours. Electrical conductivity (EC) was measured at 30 minute intervals. TOC, total organic carbon.

karst springs (e.g. Mahler *et al.* 2000; Pronk *et al.* 2006). Along with the increase in turbidity, the concentrations of total Fe and Al also increased (total Al up to 206 µg l^{-1}, total Fe up to 474 µg l^{-1}), confirming the transport of metals through particles suspended in the water. In the recession period of the second water pulse, the concentrations of total organic carbon, chloride and nitrate increased during the influx of new water, indicating the washing out of organic carbon and other contaminants from the surface.

Discussion

Linking the available historical meteorological data with gauging time series records allowed the projection of regional climatological and basin-scale hydrological trends for near-future predictions. As anticipated, the results of this analysis showed an increasing trend in the mean annual air temperatures (0.34°C $decade^{-1}$) and a decreasing trend in annual precipitation (*c.* 60 mm $decade^{-1}$). These results are consistent with the available climate simulations for the Mediterranean region (Hoerling *et al.* 2012; Flaounas *et al.* 2016). The results also prove that changes in the climatology have a direct effect on the decreasing annual actual evapotranspiration trend of 32–49 mm decade^{-1}. Consequently, a downward trend in groundwater outflow values is expected. This is of great concern because the decrease in the mean discharge of Rižana spring is currently 480 l s^{-1} decade^{-1}.

The projection of the annual and monthly outflow was a key variable in identifying the most vulnerable periods for the availability of groundwater in terms of the anticipated hydrological extremes of low flow. Especially concerning are the monthly upward trends in actual evapotranspiration and the negative trends in the absolute and mean minimum discharge, which generally occur during spring and the hot summer months (40–200 and 50–180 l s^{-1} decade^{-1}, respectively, in the period between April and September) when the need for drinking water (peak tourist season) and irrigation water is highest. These figures indicate that an increase in hydrological droughts can be expected.

Given the high dependence of the behaviour of karst aquifers on the hydrological conditions, it is expected that noticeable decreases in water level may have irreversible consequences on the water quality of springs (Hartmann *et al.* 2014*a*, *b*). High-frequency physico-chemical monitoring of the spring was carried out to evaluate its hydrological response to recharge events of various magnitudes in relation to water quality. The high variability of individual parameters indicated that changes in water quality were most pronounced during the flood pulses. A nine-month observation period showed that the deterioration in water quality was most pronounced for flood pulses caused by precipitation events after a long dry period.

This study showed that the response of the electrical conductivity and turbidity in the first event after the three-month period of low water differed from the response in the following periods of higher waters. Insignificant changes in discharge and turbidity and significant increases in electrical conductivity were typical. Water stored within the unsaturated zone was pushed out of the system and the effect of dilution due to newly infiltrated water was very small, causing an increase in the number of coliform bacteria to 1400 cfu (100 ml)$^{-1}$. The dynamics of matter or contaminant transport were greatest during the heavier rainfall events that provoked the discharge and turbidity to increase, accompanied by the transport of metals (e.g. total Al up to 206 µg l^{-1}, total Fe up to 474 µg l^{-1}) through particles suspended in the water. The secondary increase in coliform bacteria accompanied by the increase in total organic carbon indicated the arrival of water from the land surface. This shows that we can expect lower water quality in the first flood pulse triggered by intense precipitation after a long dry period.

Conclusions

Climate change may severely affect the availability of water in the Mediterranean region. Despite karst aquifers often being the only exploitable reserves of water in the region, we have a limited knowledge about their sensitivity to climate change. This study aimed to identify how the amount and quality of spring water might be affected by changes in air temperature, evapotranspiration and precipitation, focusing on the Rižana karst spring in SW Slovenia, which is the main water source for the Slovene Istria.

Although the study is site-specific, the results show that a direct effect of the anticipated climatological trends is an increase in hydrological droughts, especially during the spring and summer months. A deterioration in water quality can be expected in the periods of intense rainfall that follow prolonged periods of low water.

These findings are important in understanding the impact of predicted climate stresses on karst water resources and raise crucial concerns about the sustainable management and utilization of groundwater. Water resource managers can use this information to understand how the character of karst systems is changing and how they can prioritize conservation activities accordingly. The results could further be used in modelling approaches to predict the hydrologic responses of karst aquifers and the effects on water quality.

The study was conducted within the frame of the Živo! (Life – Water!) Project of the European Fund for Regional Development Cross-Border Cooperation Programme Slovenia–Croatia 2007–2013. The high-frequency monitoring of physical parameters was co-funded by the Slovenian Infrastructure Agency, Railway Sector.

References

Bailly-Comte, V., Jourde, H., Roesch, A. & Pistre, S. 2007. Mediterranean flash flood transfer through karstic area. *Environmental Geology*, **54**, 605–614.

Bates, B.C., Kundzewicz, Z.W., Wu, S. & Palutikof, J. (eds) 2008. *Climate Change and Water.* Technical Paper of the Intergovernmental Panel on Climate Change. IPCC Secretariat, Geneva.

Biondić, R., Petrič, M. & Rubinić, J. 2015. Overview of the hydrogeology. *In*: Zupan Hajna, N., Ravbar, N., Rubinić, J. & Petrič, M. (eds) *Life and Water on Karst: Monitoring of Transboundary Water Resources of Northern Istria.* ZRC Publishing, Ljubljana, 60–73.

Chen, Z., Auler, A.S. *et al.* 2017. The World Karst Aquifer Mapping project: concept, mapping procedure and map of Europe. *Hydrogeology Journal*, **25**, 771–785.

Di Matteo, L., Valigi, D. & Cambi, C. 2012. Climatic characterization and response of water resources to climate change in limestone areas: considerations on the importance of geological setting. *Journal of Hydrologic Engineering*, **18**, 773–779.

Doctor, D.H., Alexander, E.C., Petric, M., Kogovsek, J., Urbanc, J., Lojen, S. & Stichler, W. 2006. Quantification of karst aquifer discharge components during storm events through end-member mixing analysis using natural chemistry and stable isotopes as tracers. *Hydrogeology Journal*, **14**, 1171–1191.

Döll, P., Jiménez-Cisneros, B. *et al.* 2015. Integrating risks of climate change into water management. *Hydrological Sciences Journal*, **60**, 4–13.

Finger, D., Hugentobler, A. *et al.* 2013. Identification of glacial meltwater runoff in a karstic environment and its implication for present and future water availability. *Hydrology and Earth System Sciences*, **17**, 3261–3277.

Flaounas, E., Kelemen, F.D. *et al.* 2016. Assessment of an ensemble of ocean–atmosphere coupled and uncoupled regional climate models to reproduce the climatology of Mediterranean cyclones. *Climate Dynamics*, first published online 4 November 2016, https://doi.org/10.1007/s00382-016-3398-7

Ford, D. & Williams, P. 2007. *Karst Hydrogeology and Geomorphology*. Wiley, Chichester.

Frantar, P. (ed.) 2008. *Water Balance of Slovenia 1971–2000*. Slovenian Environmental Agency, Ljubljana.

Gosling, S.N. & Arnell, N.W. 2016. A global assessment of the impact of climate change on water scarcity. *Climatic Change*, **134**, 371–385.

GraphPad Software 2016. http://graphpad.com/quickcalcs/statratio1/ [last accessed 10 January 2016].

Hagemann, S., Chen, C. *et al.* 2013. Climate change impact on available water resources obtained using multiple global climate and hydrology models. *Earth System Dynamics*, **4**, 129–144.

Hartmann, A., Lange, J., Aguado, À.V., Mizyed, N., Smiatek, G. & Kunstmann, H. 2012. A multi-model approach for improved simulations of future water availability at a large Eastern Mediterranean karst spring. *Journal of Hydrology*, **468–469**, 130–138.

Hartmann, A., Goldscheider, N., Wagener, T., Lange, J. & Weiler, M. 2014*a*. Karst water resources in a changing world: review of hydrological modeling approaches. *Reviews of Geophysics*, **52**, 218–242.

Hartmann, A., Mudarra, M., Andreo, B., Marín, A., Wagener, T. & Lange, J. 2014*b*. Modeling spatiotemporal impacts of hydroclimatic extremes on groundwater recharge at a Mediterranean karst aquifer. *Water Resources Research*, **50/8**, 6507–6521.

Hoerling, M., Eischeid, J., Perlwitz, J., Quan, X., Zhang, T. & Pegion, P. 2012. On the increased frequency of Mediterranean drought. *Journal of Climate*, **25**, 2146–2161.

Hunkeler, D. & Mudry, J. 2007. Hydrogeochemical methods. *In*: Goldscheider, N. & Drew, D. (eds) *Methods in Karst Hydrogeology*. Taylor & Francis, London, 93–121.

Janža, M. 2010. Hydrological modeling in the karst area, Rižana spring catchment, Slovenia. *Environmental Earth Sciences*, **61/5**, 909–920.

Jiménez Cisneros, B.E., Oki, T. *et al.* 2014. *Freshwater Resources. Impacts, Adaptation, and Vulnerability. Part A: Global and Sectoral Aspects.* Contribution of Working Group II to the Fifth Assessment Report of the Intergovernmental Panel on Climate Change. Cambridge University Press, Cambridge, 229–269.

Jia, Z., Zang, H., Zheng, X. & Xu, Y. 2017. Climate change and its influence on the karst groundwater recharge in the Jinci Spring Region, northern China. *Water*, **9**, 267.

Kogovšek, J. 2001. Monitoring the Malenščica water pulse by several parameters in November 1997. *Acta Carsologica*, **30**, 39–53.

Kogovšek, J. 2002. Multiparameter observations of the Reka flood pulse in March 2000. *Acta Carsologica*, **31**, 61–73.

Kovačič, G. 2016. Discharge trends of the Adriatic Sea basin rivers in Slovenia, excluding the Soča river basin. *Geografski vestnik*, **88**, https://doi.org/10.3986/GV88201

Kovačič, G. & Ravbar, N. 2016. Characterisation of selected karst springs in Slovenia by means of a time series analysis. *In*: Stevanović, Z., Krešić, N. & Kukurić, N. (eds) *Karst without Boundaries*. CRC Press, Boca Raton, FL, 131–134.

Kraner Šumenjak, T. & Šuštar, V. 2011. Parametrični in neparametrični pristopi za odkrivanje trenda v časovnih vrstah. *Acta agriculturae Slovenica*, **97**, 305–312.

Krivic, P., Bricelj, M., Trišič, N. & Zupan, M. 1987. Sledenje podzemnih vod v zaledju Rižane. *Acta Carsologica*, **16**, 83–104.

Krivic, P., Bricelj, M. & Zupan, M. 1989. Podzemne vodne zveze na področju Čičarije in osrednjega dela Istre (Slovenija, Hrvatska, NW Jugoslavija). *Acta Carsologica*, **18**, 265–295.

Kundzewicz, Z.W., Mata, L.J. *et al.* 2008. The implications of projected climate change for freshwater resources and their management. *Hydrological Sciences Journal*, **53**, 3–10.

Long, A.J. & Mahler, B.J. 2013. Prediction, time variance, and classification of hydraulic response to recharge in two karst aquifers. *Hydrology and Earth System Sciences*, **17**, 281–294.

Mahler, B.J., Personne, J.C., Lods, G.F. & Drogue, C. 2000. Transport of free and particulate-associated bacteria in karst. *Journal of Hydrology*, **238**, 179–193.

Pretnar, G. 2015. Microbiological characteristics of selected karst springs. *In*: Zupan Hajna, N., Ravbar, N., Rubinić, J. & Petrič, M. (eds) *Life and Water on Karst: Monitoring of Transboundary Water Resources of Northern Istria*. ZRC Publishing, Ljubljana, 135–141.

Pronk, M., Goldscheider, N. & Zopfi, J. 2006. Dynamics and interaction of organic carbon, turbidity and bacteria in a karst aquifer system. *Hydrogeology Journal*, **14**, 473–484.

Pronk, M., Goldscheider, N. & Zopfi, J. 2007. Particle-size distribution as indicator for fecal bacteria contamination of drinking water from karst springs. *Environmental Science & Technology*, **41**, 8400–8405.

Ravbar, N. & Kovačič, G. 2015. Vulnerability and protection aspects of some Dinaric karst aquifers: a synthesis. *Environmental Earth Sciences*, **74**, 129–141.

RAVBAR, N., BARBERÁ, J.A., PETRIČ, M., KOGOVŠEK, J. & ANDREO, B. 2012. The study of hydrodynamic behaviour of a complex karst system under low-flow conditions using natural and artificial tracers (the catchment of the Unica River, SW Slovenia). *Environmental Earth Sciences*, **65**, 2259–2272.

SALMI, T., MÄÄTTÄ, A., ANTTILA, P., RUOHO-AIROLA, T. & AMNELL, T. 2002. *Detecting Trends of Annual Values of Atmospheric Pollutants by the Mann–Kendall Test and Sen's Slope Estimates –the Excel Template Application MAKESENS*. Finnish Meteorological Institute, Helsinki.

SIEMERS, J. & DREYBRODT, W. 1998. Early development of karst aquifers on percolation networks of fractures in limestone. *Water Resources Research*, **34**, 409–419.

SLOVENIAN ENVIRONMENTAL AGENCY 2017. Hydrological data series of the Rižana river (gauging station Kubed II) and meteorological data series for stations Kubed, Podgorje pod Slavnikom, Podgrad, Portorož-Airport and Škocjan, www.arso.gov.si/

THEIL, H. 1950. A rank-invariant method of linear and polynomial regression analysis. *Proceedings of the Royal Netherlands Academy of Sciences*, **53**, 386–392.

TILGENKAMP, A. 2011. Theil-Sen estimator, www.mathworks.com/matlabcentral/fileexchange/34308-theil-sen-estimator [last accessed 22 December 2015].

VANNEST, K.J., PARKER, R.I. & GONEN, O. 2011. Single Case Research: web based calculators for SCR analysis. (Version 1.0), www.singlecaseresearch.org/calculators/theil-sen [last accessed 10 January 2016].

VESPER, D.J. & WHITE, W.B. 2003. Metal transport to karst springs during storm flow: an example from Fort Campbell, Kentucky/Tennessee, USA. *Journal of Hydrology*, **276**, 20–36.

VESPER, D.J., LOOP, C.M. & WHITE, W.B. 2001. Contaminant transport in karst aquifers. *Theoretical and Applied Karstology*, **13**, 101–111.

WHITE, W.B. 2002. Karst hydrology: recent developments and open questions. *Engineering Geology*, **65**, 85–105.

WILLIAMS, S.D., WOLFE, W.J. & FARMER, J.J. 2006. Sampling strategies for volatile organic compounds at three karst springs in Tennessee. *Groundwater Monitoring & Remediation*, **26**, 53–62.

WORTHINGTON, S.R. 2007. Groundwater residence times in unconfined carbonate aquifers. *Journal of Cave and Karst Studies*, **69**, 94–102.

Preliminary analysis of the decrease in water level of Vrana Lake on the small carbonate island of Cres (Dinaric karst, Croatia)

OGNJEN BONACCI

Faculty of Civil Engineering, Architecture and Geodesy, Split University, 21000 Split, Matice hrvatske 15, Croatia
obonacci@gradst.hr

Abstract: A strong and potentially dangerous decreasing trend in the level of water in Vrana Lake over the last three decades was analysed. This freshwater lake is a unique karst hydrology feature located on the small Adriatic island of Cres (405.71 km^2), which is entirely composed of carbonate rocks. The lake is situated in a large cryptodepression and its base reaches a depth of 61.3 m below mean sea-level. The lake is a complex hydrological–hydrogeological system with an average water volume of *c.* 220 × 10^6 m^3. The larger geographical region has been affected by an increase in air temperature over the last *c.* 40 years. This exceptionally clean freshwater lake is the only source of potable water for the whole Cres archipelago. A dangerous drop in the water level of the lake started in 1983. This decreasing trend is driven by both global climate change and anthropogenic (the overexploitation of water) factors.

A great number of important problems related to water (drought, water scarcity, flash floods, groundwater pollution and landslides) are currently affecting the Mediterranean area and its karst regions in particular (Parise 2003; Wainwright & Thornes 2004; Myronidis *et al.* 2012; Alexakis *et al.* 2013; Pascuala *et al.* 2015; Martinotti *et al.* 2017). The numerous small karst islands of the Mediterranean Sea are now considered to be endangered (Terzić *et al.* 2010; Bonacci *et al.* 2012) and are found in an area with a serious deficit of freshwater (Zacharias & Koussouris 2000; Gikas & Tchobanoglous 2009). Small islands around the world, especially those composed of carbonate rocks, are highly vulnerable to the impacts of global climate change and rising sea-levels.

Water shortages and poor harvests during the droughts of the last 35 years have exposed the acute vulnerability of the Mediterranean region to climatic extremes (Karas 1998; Di Matteo *et al.* 2011; Dai 2013). Against this backdrop, the prospect of major global climate change brought about by human activities is a source of growing concern, raising serious questions over the sustainability of the region. The case study discussed here is a good example of this issue.

The study reported here analysed the causes of a dangerously decreasing trend in the mean annual water levels of Vrana Lake on the island of Cres, Croatia. The island is located along the northeastern Croatian coast of the Adriatic Sea (Fig. 1) and is part of the large area of the Outer Dinaric ranges. The island is largely composed of Cretaceous carbonate rocks, with limestones and dolomites as the prevalent lithology. It is an elongated island extending in a NW–SE direction and located between 44° 51′ 54″ N and 14° 26′ 41″ E.

With an area of 405.71 km^2, Cres is the largest Croatian island and has a population of 2879 (2011 census). Cres and the neighbouring island of Lošinj were once a single island, but were later divided by a channel and connected with a bridge. Lošinj is part of the Cres–Lošinj archipelago, which includes five smaller populated islands and a number of small, uninhabited islands. The highest elevation on Cres is 650 m above sea-level (m a.s.l.). The length of the coastline of Cres digitized from topographic maps at a scale of 1:25 000 is 268.205 km (Duplančić-Leder *et al.* 2004). It is 62 km long, with a width varying from 2 to 11 km. The index of indentedness is very high at 3.756. This index represents the relation between the coastline and the circumference of a circle with the same area as the island. If the index of indentedness in the narrow islands is much greater than 1, then the conditions for the formation of a deep and stable fresh groundwater lens are unfavourable (Diaz Arenas & Febrillet Huertas 1986; Falkenmark & Chapman 1989; Terzić *et al.* 2010). Cres and its archipelago are therefore crucially dependent on the freshwater resources stored in Vrana Lake.

The island has a mild climate and evergreen Mediterranean vegetation. More than 1300 plant species have been found on this small island, many of which are rare, endemic and endangered. A highly developed tourist industry during the summer months results in a high seasonal requirement for water. The economy of the island is mainly focused on tourism,

From: Parise, M., Gabrovsek, F., Kaufmann, G. & Ravbar, N. (eds) 2018. *Advances in Karst Research: Theory, Fieldwork and Applications*. Geological Society, London, Special Publications, **466**, 307–317.
First published online November 6, 2017, https://doi.org/10.1144/SP466.6

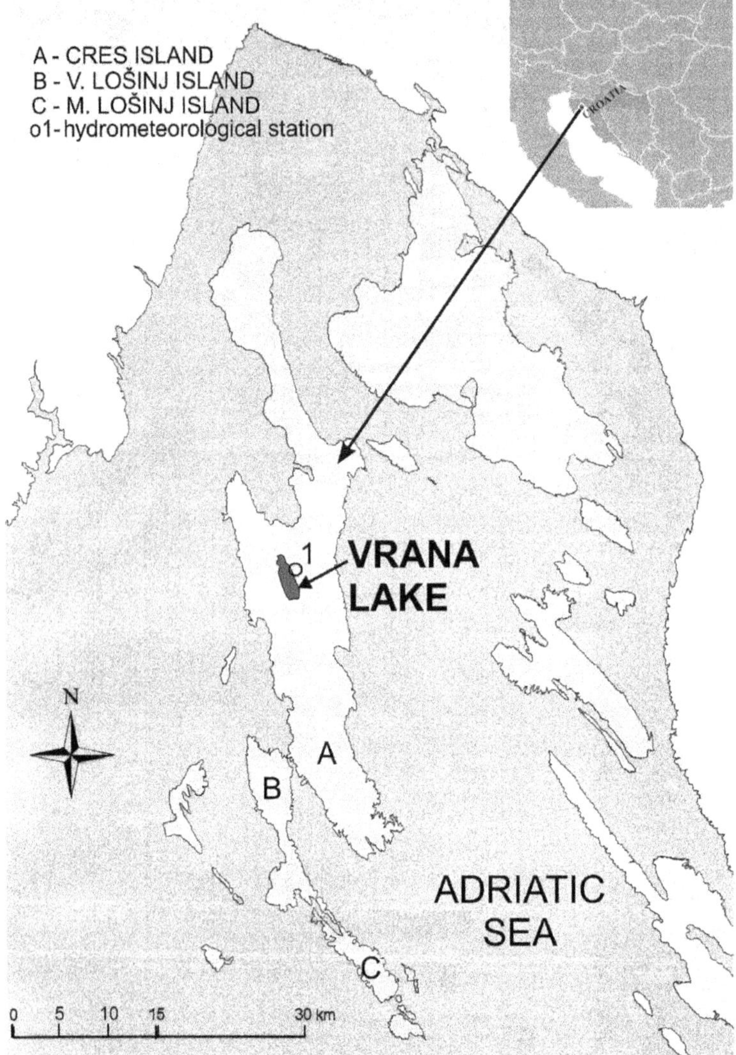

Fig. 1. Map of Cres island and Vrana Lake.

with some agriculture and fishing. More than 250 000 tourists visit the archipelago during the three summer months. The requirement for water for agricultural purposes is also high in this time period.

This paper addresses the possible influence of climatic and anthropogenic changes on the vulnerable karst water resources of the whole ecosystem of the island of Cres. The aim of the paper is to stimulate interdisciplinary monitoring and co-operation between hydrologists, hydrogeologists, ecologists and water resource managers to provide better protection and management of the vulnerable and exceptionally valuable Vrana Lake and the aquatic and terrestrial ecosystems of Cres.

Lake characteristics

Figure 2 shows the relationship between (1) the water depth (H) and the volume of water (V) and (2) the water depth (H) and the water surface area (A) for Vrana Lake. The minimum water level of 8.56 m a.s.l. was measured in 2012 and the maximum water level of 16.86 m a.s.l. was measured in 1936. The surface area (A) of water for the minimum and maximum water levels was 5.28 and 5.98 km^2, respectively. The volume of water (V) for the minimum and maximum water levels was 197×10^6 and 241×10^6 m^3, respectively. Vrana Lake is 5 km long with a maximum width of 1.45 km.

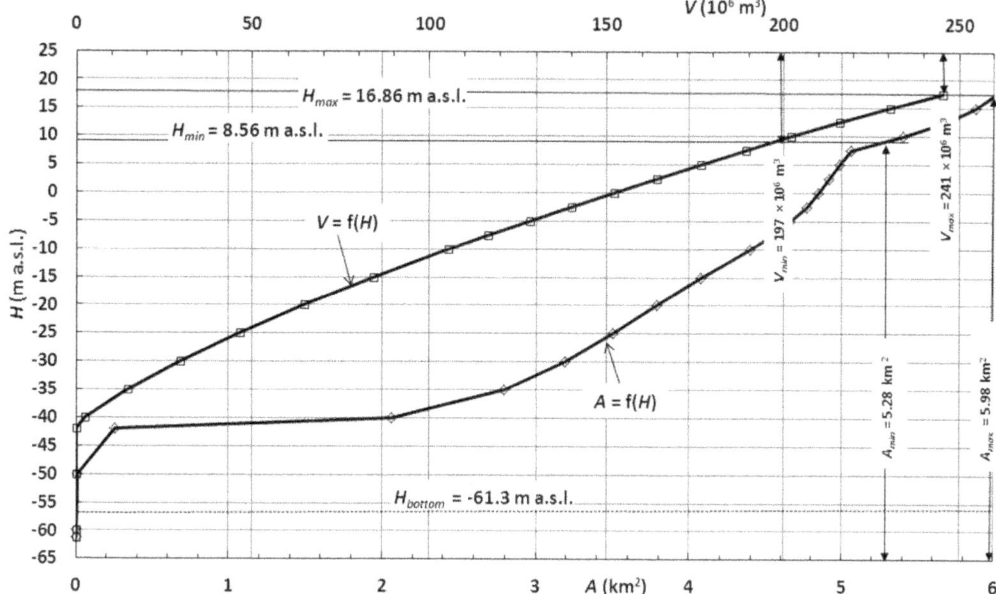

Fig. 2. Relationships between Vrana Lake and (1) the water depth (H) and the volume of water (V) and (2) the water depth (H) and the water surface area (A).

The salinity of the lake water varies in a narrow range between 62 and 92 mg l^{-1}, with an average of 72 mg l^{-1}. Many water bodies in karst islands are astatic and their water is brackish. Vrana Lake is the world's largest clean freshwater body located on a small completely karstified island. The lake is only 3–6 km away from the Adriatic Sea.

The lake was formed in the early Pleistocene. Quaternary deposits are of limited extent and are insignificant in terms of the hydrogeology. Geological investigations have indicated strong tectonic disturbance of the lake area (Terzić *et al.* 2010; Kuhta & Brkić 2013).

Direct measurement of the lake's inflow and outflow is not possible, which makes the definition and control of its water budget difficult. The water budget method was used to define the lake's catchment area as *c.* 25 km^2 (Bonacci 1993, 1995). This area was confirmed by the measured tritium values, which established the mean residence time of the lake water as 30–40 years. The same investigations showed that the atmosphere is the main source of water in the lake.

Time data series of the average annual water levels of Vrana Lake for the period 1929–2015 are presented in Figure 3. The average mean annual water level during the analysed period was 12.62 m a.s.l. The linear regression line shows a decreasing trend of 4.02 cm a^{-1} during this period. The linear correlation coefficient ($r = -0.678$) is statistically significant ($p < 0.01$). Using the rescaled adjusted partial sums method (Garbrecht & Fernandez 1994), the oscillations of the water level during the entire period analysed are clearly divided into two sub-periods: (1) 1929–85 and (2) 1986–2015. The average water level in the first sub-period is 2.36 m higher than that during the last 30 years. The decrease in the average water level between the two sub-periods (1929–83 and 1984–2015) is statistically significant ($p < 0.01$). The decrease in the mean annual water level is 4.37 m in the period 1983–90.

Precipitation, air temperature and evaporation

Figure 4 shows a time series of annual precipitation measured at the Cres meteorological station from 1926 to 2015. The average annual precipitation is 1067 mm, whereas the minimum and the maximum annual precipitation is 686 mm (in 2011) and 1743 mm (in 1960). There is no statistically significant decreasing trend in annual precipitation or in the redistribution of precipitation during the year over the whole period analysed (Bonacci 2015).

Using the rescaled adjusted partial sums method (Garbrecht & Fernandez 1994), the annual precipitation in the entire analysed period can be divided into four statistically significant ($p < 0.01$) sub-periods: (1) 1926–75, (2) 1976–81; (3) 1982–91; and (4) 1992–2015. Figure 5 shows these four time sub-series of annual precipitation measured at the Cres

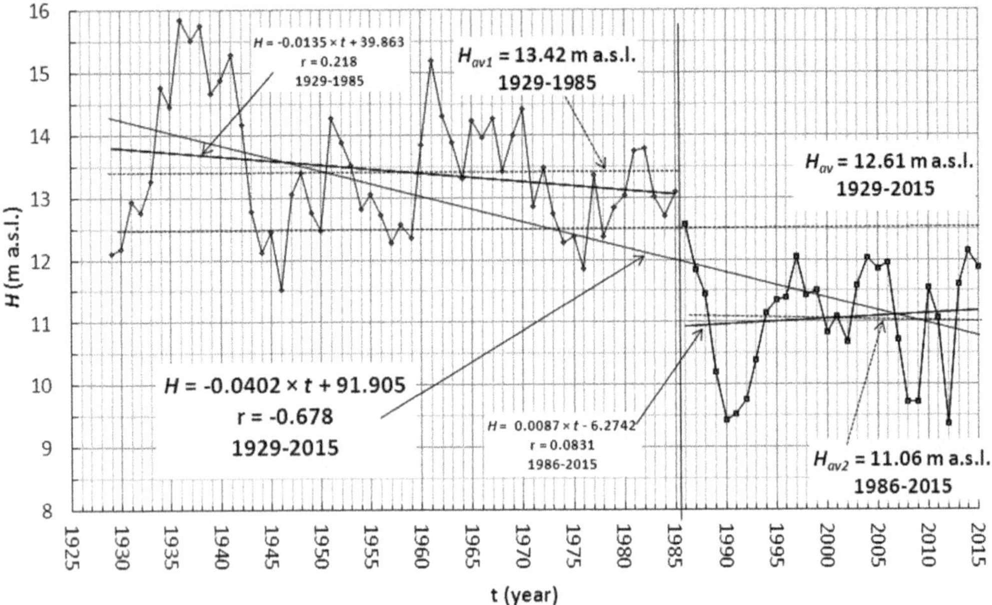

Fig. 3. Time data series for the average annual water levels for Vrana Lake in the time period 1929–2015.

meteorological station. The average annual precipitation in the third sub-period (1982–91) was 122.0 mm lower than over the whole analysed period (1926–2015) and was 356.8 mm lower than in the previous (second) sub-period (1976–81). The decrease in the mean annual water level may have been provoked by this sub-period of drought. The low amount of precipitation during the 1982–91

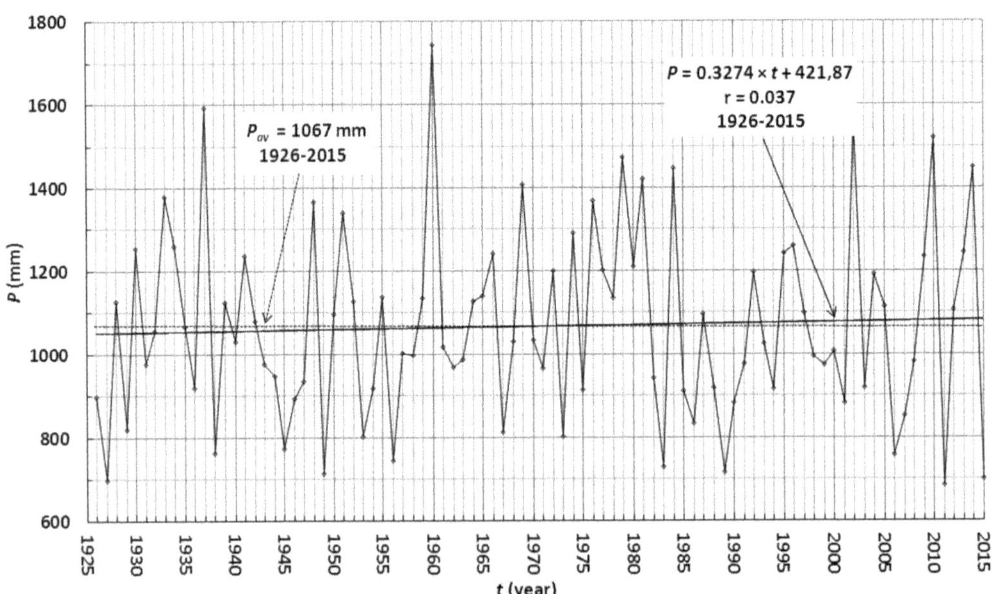

Fig. 4. Time series of annual precipitation measured at the meteorological station on Cres for the time period 1926–2015.

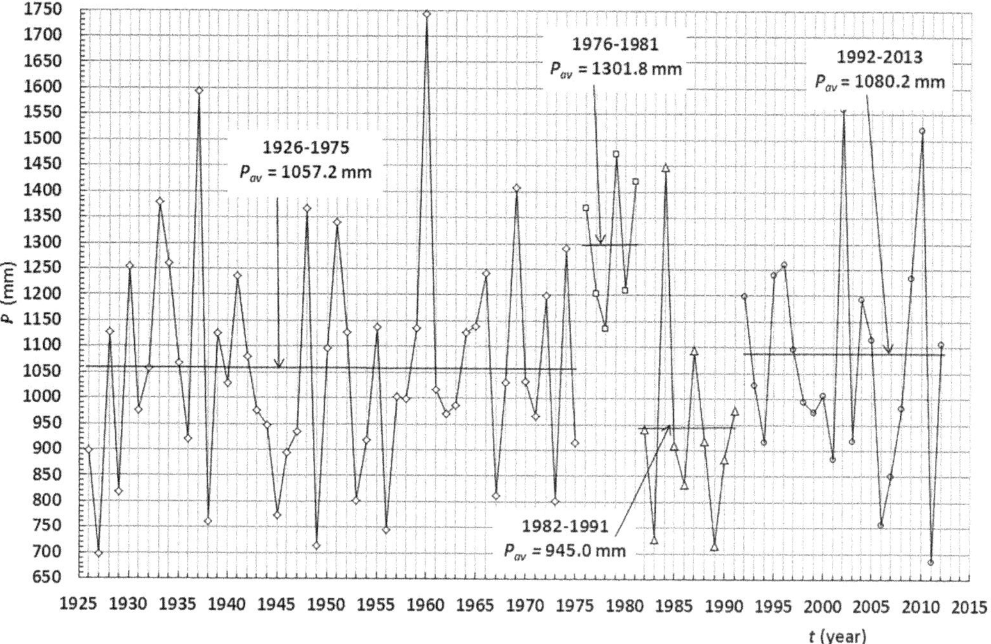

Fig. 5. Four time sub-series of the annual precipitation measured at the meteorological station on Cres for the four sub-periods 1926–75, 1976–81, 1982–91 and 1992–2015.

sub-period probably had a cumulative effect on the behaviour of the water level in the lake over the last 35 years.

Figure 6 shows a time series of the mean annual temperature recorded at the Cres meteorological station for the period 1948–2015. It includes linear and

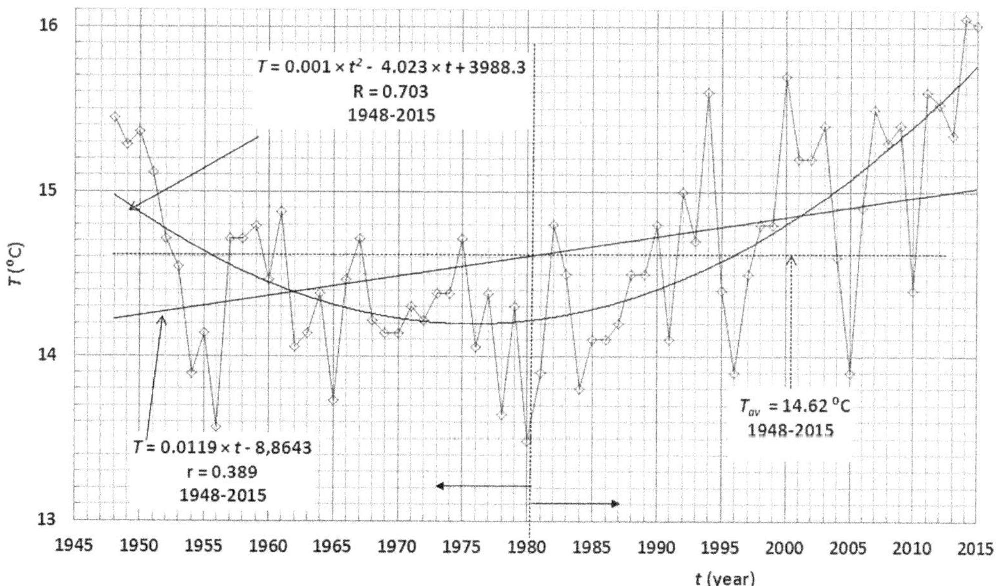

Fig. 6. Time series of the mean annual temperature for the meteorological station on Cres for the time period 1948–2015.

second-order polynomial trend lines. The average mean annual air temperature is 14.6°C. The minimum mean annual air temperature of 13.5°C was recorded in 1980 and the maximum of 16.05°C in 2014. The linear trend is statistically significant ($p < 0.01$) with a linear correlation coefficient of $r = 0.389$. The second-order polynomial trend with a coefficient of non-linear correlation of $R = 0.703$ is much higher. It provides a better fit to the analysed data than a linear trend. Using the rescaled adjusted partial sums method, the data series was divided into two subsets: (1) 1948–81 and (2) 1982–2015. There is a decreasing trend in the first sub-period, whereas the trend in the second sub-period is increasing and could be explained by global warming in this region. Analyses showed a strong jump in the regional annual air temperature over the last *c.* 25 years, which may have had an influence on the lake's hydrological regime (Bonacci 2012).

Evaporation was measured by a class A pan during 1991–95. The surface of Vrana Lake evaporates between 1.0 and 1.2 m a^{-1} (i.e. between 5×10^6 and 6×10^6 m^3 of lake water).

Relationships between water level and air temperature, precipitation and water pumping

Figure 7 shows the relationship between the mean annual water level and the mean annual air temperature for the period 1948–2015. The coefficient of linear correlation has a negative value ($r = -0.385$). Higher air temperatures increase the evaporation of lake water and therefore decrease the level of water in the lake.

Figure 8 shows a time series of the amount of water pumped annually from the lake for the period 1967–2015; water pumping started in 1965. The amount of pumping decreased in 1991 due to the war in Croatia and the subsequent decrease in tourism in the region. Pumping reached a maximum value of $C = 74$ l s^{-1} in 2010, when 2.33×10^6 m^3 of water were extracted from the lake.

Figure 9 shows the relationship between the mean annual water level and water pumping for the time period 1967–2015. The coefficient of linear correlation has a statistically significant negative value ($r = -0.629$), which confirms the important role of water extraction in the decrease in the water level of the lake.

The equation for the multiple linear correlations between the mean annual water level as a dependent variable and the three independent variables of annual precipitation, annual lake water pumping and mean annual air temperature defined for the period 1948–2015 is:

$$H = 19.70 + 0.3403 \times P - 0.03075 \times C - 0.4518 \times T \quad (1)$$

where H is the mean annual water level of Vrana Lake (m a.s.l.), P is the annual precipitation

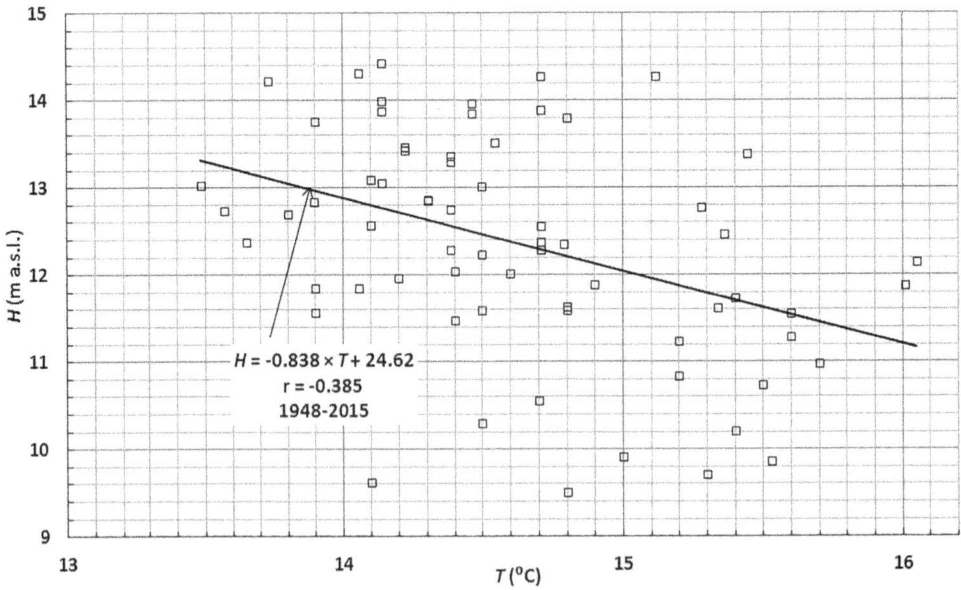

Fig. 7. Relationship between the mean annual water levels and the mean annual air temperature for the time period 1948–2015.

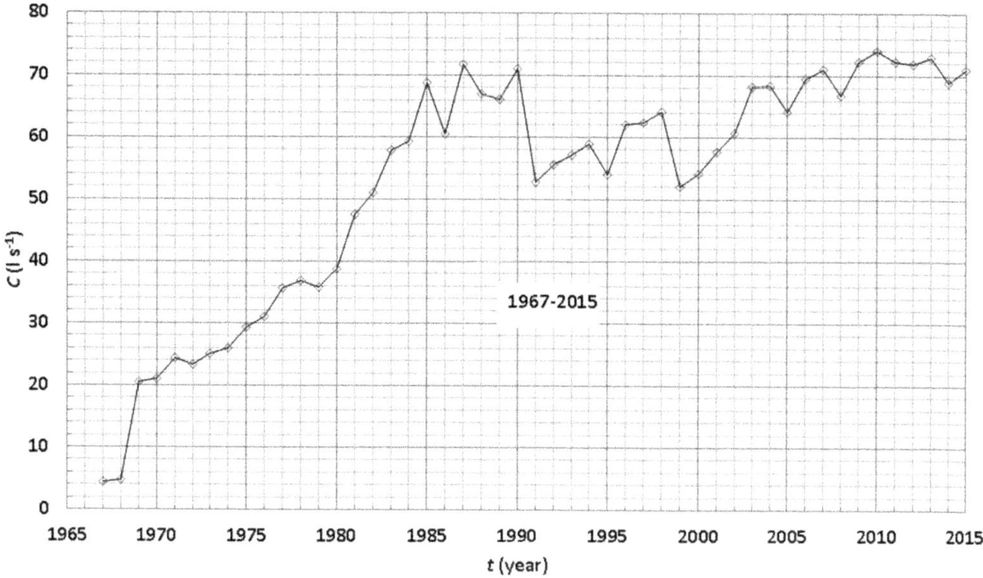

Fig. 8. Time series of annual water pumping from the lake for the time period 1967–2015.

measured at the meteorological station on Cres (m), C is the mean annual water pumping from the lake (l s^{-1}) and T is the mean annual air temperature measured at the Cres meteorological station (°C).

The coefficient of multiple linear correlation has a value of $r = 0.734$ and is statistically significant. Equation (1) shows that the most important influence on the decreasing trend for the mean annual lake water level is the mean annual air temperature. Higher air temperatures strongly influence the increase in evaporation from the lake's wide surface area (Bonacci *et al.* 2015; van der Schriek &

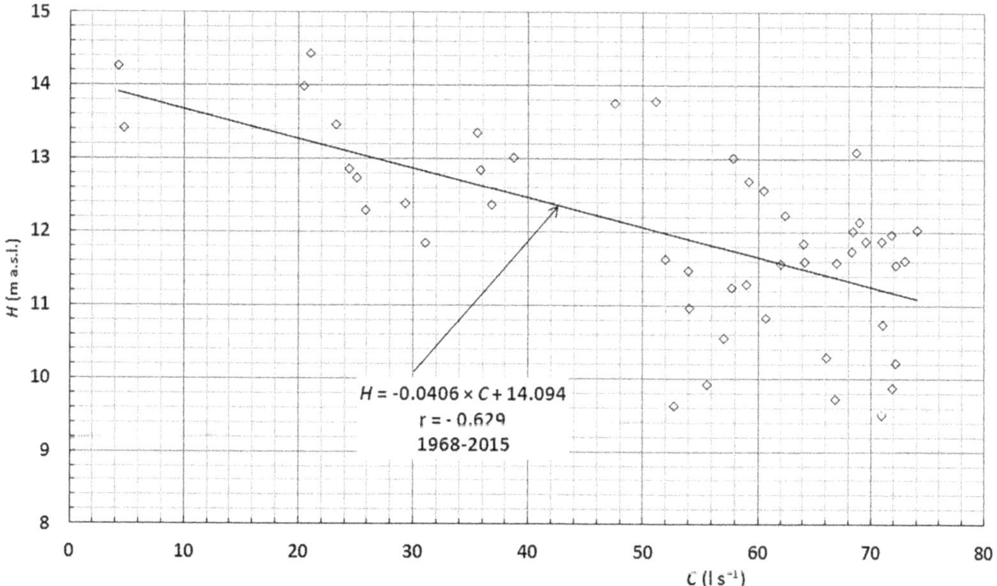

Fig. 9. Relationship between the mean annual water level and water pumping for the time period 1967–2015.

Giannakopoulos 2017). It should be stressed, however, that the complex water budget of Vrana Lake cannot be clearly explained by the multiple linear correlation model alone.

Figure 10 shows the autocorrelation function of the 87 year time series (1929–2015) for the mean annual water level of Vrana Lake. As each mean annual water level $H(i)$ in year i is significantly autocorrelated with the mean water level $H(i-1)$ of the previous year $(i-1)$, it is logical to conclude that the autocorrelation effect probably has a stronger influence on the mean lake water level than the effect of the mean annual air temperature. Figure 11 shows the relationship between the mean annual water level in year i, $H(i)$, and in the previous year $(i-1)$, $H(i-1)$. The coefficient of linear correlation is statistically significant ($r = -0.857$). The long-lasting memory effect of the large water body of Vrana Lake thus plays an important part in the behaviour of its water level.

Discussion and conclusions

Vrana Lake is an extraordinary and unique freshwater karst feature located in a cryptodepression on a small, narrow and completely karstifed island. The exceptionally clean freshwater is used as the water supply for the islands of Cres and Lošinj and is the only source of potable water in the whole archipelago. The lake environment is very important and is related to the present and future living conditions of the humans and other species on the archipelago. The Cres archipelago contains specific, valuable and vulnerable ecosystems and the water stored in Vrana Lake plays an important part in their protection and sustainable development.

Most climate change projections show important decreases in the availability of water in the Mediterranean region by the end of this century (Pascuala *et al.* 2015). Climate change poses particularly significant risks to all the small Mediterranean islands. The IPCC (2007) report warned that an increase of more than about 2°C could result in the expansion of desert and grassland at the expense of shrubland, and the expansion of mixed deciduous forest at the expense of evergreen conifer forest. Vulnerabilities vary depending on the geography, landscape characteristics and human activities on a particular island. Small islands have characteristics that make them especially vulnerable to the effects of climate change, rises in sea-level and extreme weather events (IPCC 2007). The principal effects of climate change will probably include the further erosion of coastlines and beaches, the salinization of groundwater, the loss of hilltop vegetation and flora, the loss of soil humus and increased erosion. The impacts of climate change on the biodiversity of islands are of

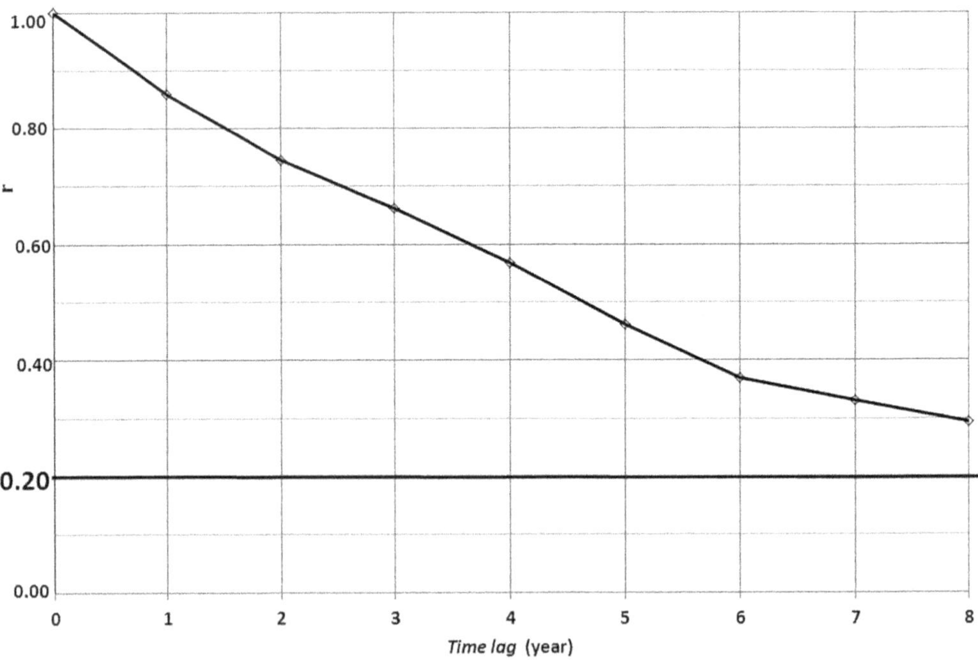

Fig. 10. Autocorrelation function of the 87-year time series (1929–2015) for the mean annual water level in Vrana Lake.

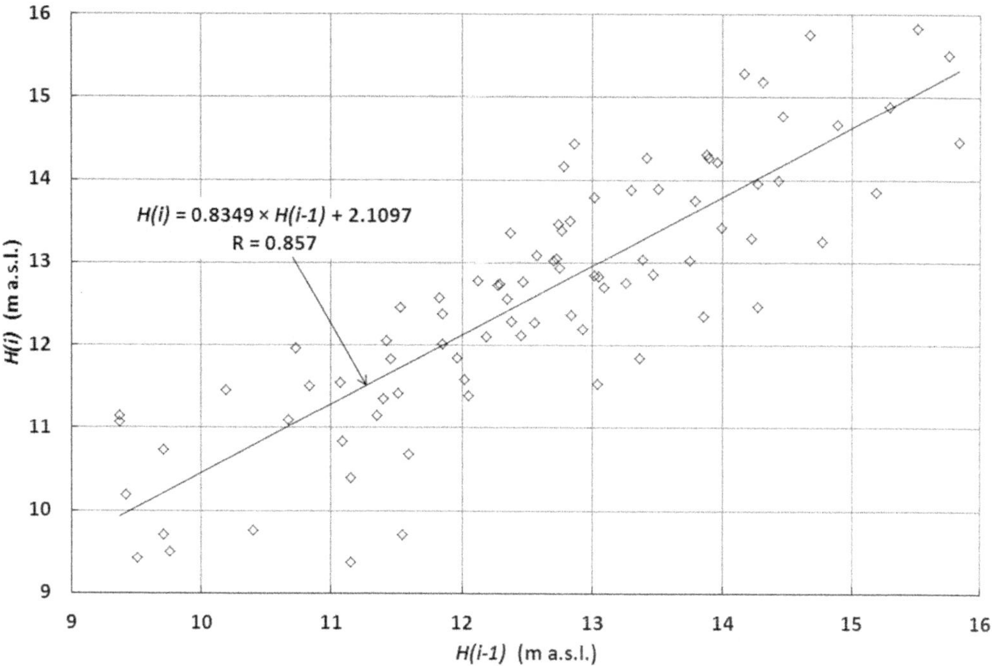

Fig. 11. Relationship between the mean annual water level in year i, $H(i)$, and the mean water level, $H(i-1)$, of the previous year, $(i-1)$.

particular concern. Erol *et al.* (2012) stressed that the potential impacts of climate change on the eco-hydrological processes of Mediterranean watersheds are significant and will require quick action towards improving the adaptation and management of fragile systems.

Global warming, a decrease in precipitation and a rise in sea-level may cause problems such as desertification, water scarcity, water pollution (salinization) and a decrease in food production, as well as threats to human health, ecosystems and economies. In the last few decades the number of heatwaves affecting the Mediterranean region has increased (Di Matteo *et al.* 2011; Dai 2013). On the Croatian islands, severe and long-lasting droughts are often interspersed with intense rainfall events causing dangerous flash floods; this may be particularly dangerous in karst areas (Parise 2003; Martinotti *et al.* 2017). The risk significantly increases with repeated dry years. Therefore it is particularly important to reduce and improve control of the use of lake water.

Climate change is a particular challenge to the island of Cres, the sustainable development of which strongly depends on the water stored in Vrana Lake. Both the land-based and marine biodiversity will be affected by climate change. The possible intrusion of seawater into the lake's freshwater body is to be expected if the strongly decreasing trend of water levels in the lake continues and if the rise in sea-level is significant.

The following two dangerous consequences due to natural processes (especially global warming and a rise in sea-level) and human intervention (over-pumping and pressure from tourism) can be expected for Vrana Lake: (1) the intrusion of seawater into the freshwater body of the lake and (2) pollution of its surface and underground catchment area. The natural inflows and outflows of the lake are too small for self-cleaning once pollution has occurred (Ožanić & Rubinić 2003). As a result of the extreme permeability of the coastal karst terrain, there is a serious risk of the intrusion of seawater into the freshwater aquifers of karst islands, especially during long, hot, dry summer periods typical of the Mediterranean climate.

The economy of Cres depends strongly on tourism and it is well known that this branch of the economy consumes large amounts of water. Vrana Lake is the only large source of freshwater on the Cres archipelago. Understanding the interaction between Vrana Lake and the groundwater in its catchment area is therefore essential for the development of the whole archipelago and to the preservation of its valuable ecosystem.

Our present knowledge of the functioning of Vrana Lake and its influence on the natural and social processes of the entire island is insufficient to determine the potential impacts likely to result from local and global natural and human-driven changes. Therefore it is of crucial importance and the highest priority to initiate the development of an assessment system to determine the vulnerability of the island of Cres and Vrana Lake.

It is not possible to determine an accurate value for the evaporation of water from the large surface area of Vrana Lake and for the rate of evapotranspiration from its catchment areas using the data available from the single meteorological station on Cres. Several deep piezometers are required around the lake to determine the directions of the complex and currently unknown local circulations of groundwater. The existing measurements do not provide a sufficiently detailed understanding of the hydrogeological–hydrological processes occurring in the lake and its watershed. There is a need for additional and detailed measurements of many different hydrological, hydrogeological and chemical parameters to ensure the sustainable development of the lake.

The successful management of the water resources stored in Vrana Lake is one of the key prerequisites for the sustainable ecological, economic and social development of the island. Successful management will give effective protection to the environment and safeguard the vulnerable and valuable biological diversity of Cres and the surrounding islands, which are all important regional centres for tourism. To fulfil these goals, it will be necessary to establish a truly interdisciplinary co-operation between different scientific disciplines (e.g. hydrology, climatology, hydrogeology, biology and ecology) and different governmental, regional and local participants, such as politicians, stakeholders and economists.

References

Alexakis, D., Kagalou, I. & Tsakiris, G. 2013. Assessment of pressures and impacts on surface water bodies of the Mediterranean. Case study: Pamvotis Lake, Greece. *Environmental Earth Sciences*, **70**, 687–698.

Bonacci, O. 1993. The Vrana Lake hydrology (Island of Cres-Croatia). *Water Resources Bulletin*, **29**, 407–414.

Bonacci, O. 1995. Investigations in karst hydrology of Croatia: the Vrana Lake on the Island of Cres. *Acta Geologica*, **25**, 1–15.

Bonacci, O. 2012. Increase of mean annual surface air temperature in the Western Balkans during last 30 years. *Vodoprivreda*, **44**, 75–89.

Bonacci, O. 2015. Karst hydrogeology/hydrology of Dinaric chain and isles. *Environmental Earth Sciences*, **74**, 37–55.

Bonacci, O., Ljubenkov, I. & Knezić, S. 2012. The water on a small karst island: the island of Korčula (Croatia) as an example. *Environmental Earth Sciences*, **66**, 1345–1357.

Bonacci, O., Popovska, C. & Geshovska, V. 2015. Analysis of transboundary Dojran Lake mean annual water level changes. *Environmental Earth Sciences*, **73**, 3177–3185.

Dai, A. 2013. Increasing drought under global warming in observations and models. *Nature Climate Change*, **3**, 52–58.

Diaz Arenas, A.A. & Febrillet Huertas, J. 1986. *Hydrology and Water Balance of Small Islands. A Review of Existing Knowledge.* Technical Documents in Hydrology. UNESCO, Paris.

Di Matteo, L., Valigi, D. & Cambi, C. 2011. Climatic characterization and response of water resources to climate change in limestone areas: some considerations on the importance of geological setting. Paper 41 presented at the 2011 Symposium on Data-Driven Approaches to Droughts, 21–22 June, Purdue University, West Lafayette, Indiana, http://docs.lib.purdue.edu/ddad2011/ [last accessed 10 October 2017].

Duplančić-Leder, T., Ujević, T. & Čala, M. 2004. Coastline lengths and areas of islands in the Croatian part of the Adriatic Sea determined from the topographic maps at the scale of 1:25,000. *Geoadria*, **9**, 5–32.

Erol, A., Timothy, O. & Randhir, T.O. 2012. Climatic change impacts on the ecohydrology of Mediterranean watersheds. *Climatic Change*, **114**, 319–341.

Falkenmark, M. & Chapman, T. 1989. *Comparative Hydrology. An Ecological Approach to Land and Water Resources.* UNESCO, Paris.

Garbrecht, J. & Fernandez, G.P. 1994. Visualization of trends and fluctuations in climatic records. *Water Resources Bulletin*, **30**, 297–306.

Gikas, P. & Tchobanoglous, G. 2009. Sustainable use of water in the Aegean Islands. *Journal of Environmental Management*, **90**, 2601–2611.

IPCC 2007. *Climate Change 2007: Impacts, Adaptation and Vulnerability.* Cambridge University Press, Cambridge.

Karas, J. 1998. Climate change and the Mediterranean region, http://www.greenpeace.org/international/Global/international/planet-2/report/2006/3/climate-change-and-the-mediter.pdf [last accessed 10 October 2017].

Kuhta, M. & Brkić, Ž. 2013. Seasonal temperature variations of Lake Vrana on the Island of Cres and possible influence of global climate changes. *Journal of Earth Science and Engineering*, **3**, 225–237.

Martinotti, M.E., Pisano, L. *et al.* 2017. Landslides, floods and sinkholes in a karst environment: the 1–6 September 2014 Gargano event, southern Italy. *Natural Hazards and Earth System Sciences*, **17**, 467–480.

Myronidis, D., Stathis, D., Ioannou, K. & Fotakis, D. 2012. An integration of statistics temporal methods to track the effect of drought in shallow Mediterranean lake. *Water Resources Management*, **26**, 4587–4605.

Ožanić, N. & Rubinić, J. 2003. The regime of inflow and runoff from Vrana Lake and the risk of permanent water pollution. *RMZ-Materials and Geoenvironment*, **50**, 281–284.

Parise, M. 2003. Flood history in the karst environment of Castellana-Grotte (Apulia, southern Italy). *Natural Hazards and Earth System Sciences*, **3**, 593–604.

Pascuala, D., Plaa, E., Lopez-Bustinsb, J.A., Retanaac, J. & Terradasac, J. 2015. Impacts of climate change on water resources in the Mediterranean Basin: a case study in Catalonia, Spain. *Hydrological Sciences Journal*, **60**, 2132–2147.

TERZIĆ, J., PEH, Z. & MARKOVIĆ, T. 2010. Hydrochemical properties of transition zone between fresh groundwater and seawater in karst environment of the Adriatic islands, Croatia. *Environmental Earth Sciences*, **59**, 1629–1642.

VAN DER SCHRIEK, T. & GIANNAKOPOULOS, C. 2017. Determining the causes for the dramatic recent fall of Lake Prespa (southwest Balkans). *Hydrological Sciences Journal*, **62**, 1131–1148.

WAINWRIGHT, J. & THORNES, J.B. 2004. *Environmental Issues in the Mediterranean – Processes and Perspectives from the Past and Present.* Routledge, London.

ZACHARIAS, I. & KOUSSOURIS, T. 2000. Sustainable water management in the European islands. *Physics and Chemistry of the Earth Part B Hydrology Oceans and Atmosphere*, **25**, 233–236.

Numerical groundwater modelling in karst

NEVEN KRESIC[1]* & SORAB PANDAY[2]

[1]*Amec Foster Wheeler, 1075 Big Shanty Road NW, Suite 100 Kennesaw, GA 30144, USA*

[2]*GSI Environmental inc., 626 Grant Street, Suite C, Herndon, VA 20170, USA*

**Correspondence: neven.kresic@rcn.com*

Abstract: The success of translating a conceptual site model for a karst site into a numerical groundwater model will depend on both the experience of the user and the capabilities and limitations of the selected computer program. Despite its numerous advantages, even MODFLOW – probably the most widely used, tested and verified modelling program currently available – has conceptual limitations that many karst hydrogeologists have to deal with on a routine basis while searching for an equivalent porous medium approach that may work. This includes assigning very high values of hydraulic conductivity to those model cells known, or suspected, to contain highly transmissive conduits, or assigning an unreasonable, very low effective porosity to the model cells with virtual 'conduits' to simulate high groundwater velocities. A new version of MODFLOW called MODFLOW-USG (UnStructured Grid) has been developed and released to the public domain. This new version retains full compatibility with previous versions of MODFLOW while taking advantage of unstructured grids and finite volume numerical solutions. It enables hydrogeologists to accurately translate even the most complex conceptual site models in karst into a numerical environment, thus eliminating the need for various surrogate modelling solutions based on an equivalent porous medium approach.

Karst groundwater systems present ongoing challenges for the application of physically based numerical models. The main explanation for these challenges is straightforward: in aquifers with intergranular porosity, the equations of groundwater flow are mostly based on the relatively simple Darcy's law. In karst aquifers, however, the nature of the porosity requires the application of different sets of flow equations for three distinct types of porous media: (1) the rock matrix; (2) rock discontinuities, such as faults, fractures and bedding planes; and (3) voids enlarged by solution, such as the channels and conduits developed from the initial discontinuities. Any meaningful quantitative integration of various equations describing these distinct flow regimes is further complicated by the uncertainties associated with the field distribution and identification of different porosity types. Unfortunately, the results of the related numerical modelling efforts have generally lagged behind the advances in the hydrogeology of intergranular (non-fractured, non-karstic) aquifers. As a consequence, the majority of modelling approaches in karst aquifers are still based on various applications of time series analyses, as well as general statistical and probabilistic methods developed in surface water hydrology. Such methods have one common thread: the need for a relatively long time series of data on aquifer recharge and spring discharge and the various input–output relationships. This, however, is also the key limiting factor for many practical engineering projects with short execution times, including some common tasks such as predicting the effects of new pumping activities or artificial aquifer recharge (Kresic 2009).

Because they are easy to design and understand – and require less mathematical involvement – finite difference numerical models have prevailed in hydrogeological practice. Several excellent finite difference modelling programs have been developed by the US Geological Survey and are in the public domain, which ensures their widest possible use. One of these is MODFLOW, which has become the industry standard as a result of its versatility and open structure. Independent subroutines called modules are grouped into packages that simulate specific hydrological features. New modules and packages can be easily added to the program without modifying the existing packages or the main code.

The main advantage of the classic MODFLOW program and the companion fate and transport models, such as MT3D and RT3D, is their broad user base and continuous updates, including the frequent introduction of new modules and numerical solvers. Most modelling concepts currently used across various computer programs are therefore indirectly or directly based on concepts first implemented in MODFLOW, so an experienced MODFLOW user usually has little trouble when transitioning to another program. Groundwater modelling is now more efficient than ever before because modern

From: PARISE, M., GABROVSEK, F., KAUFMANN, G. & RAVBAR, N. (eds) 2018. *Advances in Karst Research: Theory, Fieldwork and Applications*. Geological Society, London, Special Publications, **466**, 319–330.
First published online December 14, 2017, https://doi.org/10.1144/SP466.12

computers and computer operating systems do not place limitations on the model size for most practical applications. Models can have hundreds of layers and millions of cells and still be solved relatively quickly using a desktop PC.

Despite its numerous advantages, however, the classic MODFLOW program has conceptual limitations that many karst hydrogeologists have to deal with on a routine basis. Six limitations that immediately come to mind are: (1) the inadequacy of the equivalent porous medium (EPM) approach to quantifying flow through more discrete features such as karst conduits and channels; (2) the inadequacy of Darcy's equation to simulate turbulent flow; (3) the requirement that all model layers have to be continuous throughout the model domain; (4) the instability of the model when trying to simulate large vertical displacements caused by faults or artificial structures; (5) the condition that all the cells in the model must be rectangular, with rows and columns extending from one edge of the model to the other; and (6) the instability of the model when simulating the contacts between porous media with highly contrasting hydraulic conductivities. These limitations seriously affect attempts to simulate the complex three-dimensional geological relationships and discontinuities characteristic of karst, as well as the water fluxes between model cells.

Because of these limitations, hydrogeologists and groundwater modellers either avoid numerical modelling of karst aquifers altogether or are engaged in searching for an EPM approach that may work. This includes assigning very high values of the hydraulic conductivity to those model cell known, or suspected, to contain highly transmissive conduits, or assigning a very low effective porosity to simulate the high groundwater velocities. These solutions are inadequate and cannot then mimic the dynamics of flow and solute migration in karst environments. None of the EPM models can simulate an important hydraulic interaction between the conduits and the surrounding matrix following some rapid recharge episodes – namely, as the hydraulic head quickly rises in the conduits as a result of a recharge event, there may be a significant transfer of water from the conduits into the surrounding matrix (Kresic 2013). An EPM model will always keep the heads in the high hydraulic conductivity virtual conduit cells lower than in the surrounding model cells.

A new version of MODFLOW, called MODFLOW-USG (UnStructured Grid), that addresses these limitations has now been developed and released to the public domain, retaining full compatibility with previous versions of MODFLOW while taking advantage of unstructured grids and finite volume numerical solutions (Panday *et al.* 2013). It enables hydrogeologists to accurately translate even the most complex conceptual site models in karst into a numerical environment, thus eliminating the need for various surrogate modelling solutions based on an EPM approach. MODFLOW-USG is now part of several commercial graphical user interface (GUI) programs, including Groundwater Vistas.

There are other codes available that can solve for conduit flow coupled with porous medium flow. A popular code is MODFLOW-CFP (Shoemaker *et al.* 2008), which simulates the conduit flow process within the MODFLOW framework. The connected linear network (CLN) package of MODFLOW-USG is fashioned after the CFP package, with further generalizations that include different flow formulations such as the Darcy–Weisbach equation with the Colebrook and White formula for the friction factor (as in MODFLOW-CFP), Manning's equation, the Hazen–Williams equation and a laminar flow formulation (the Hagen–Poiseuille equation). MODFLOW-USG may further benefit from the flexibility in grid design, the adaptability of conduit connectivity with multiple nodes or layers, a greater robustness and efficiency in solution due to a fully coupled solution between groundwater and conduit domains, and a robust methodology of handling the wetting and drying of groundwater cells or conduits.

The MODFLOW-NLFP (Mayaud *et al.* 2015) code does not include conduit flow, but simulates the Darcy–Forchheimer equation in MODFLOW. The Darcy–Forchheimer equation has advantages compared with the equations solved in the CLN package in that it automatically accommodates the transition from laminar to turbulent flow, whereas the CLN package requires the selection of one of the options.

Other proprietary codes are available that solve for flow through both the matrix and fractures, including FEFLOW (DHI 2016) and FRAC3DVS, a precursor to the HydroGeoSphere code (Therrien & Sudicky 1996). These simulators are frequently developed and enhanced, but, because they are privately owned, their current capabilities in simulating conduit flow through karst conduit networks are unclear.

Features of MODFLOW-USG

MODFLOW-USG was developed to support a wide variety of structured and unstructured grid types, including nested grids and grids based on prismatic triangles, rectangles, hexagons and other cell shapes. Unstructured grids provide a high level of flexibility of discretization by implementing differently shaped grid block geometries and by using nested grid structures. The grid may be created as a combination of nested grids and polygons of different geometries. Pinching of layers, sub-layering and vertical displacements along faults may be directly accommodated without excessive discretization. Flexibility in grid

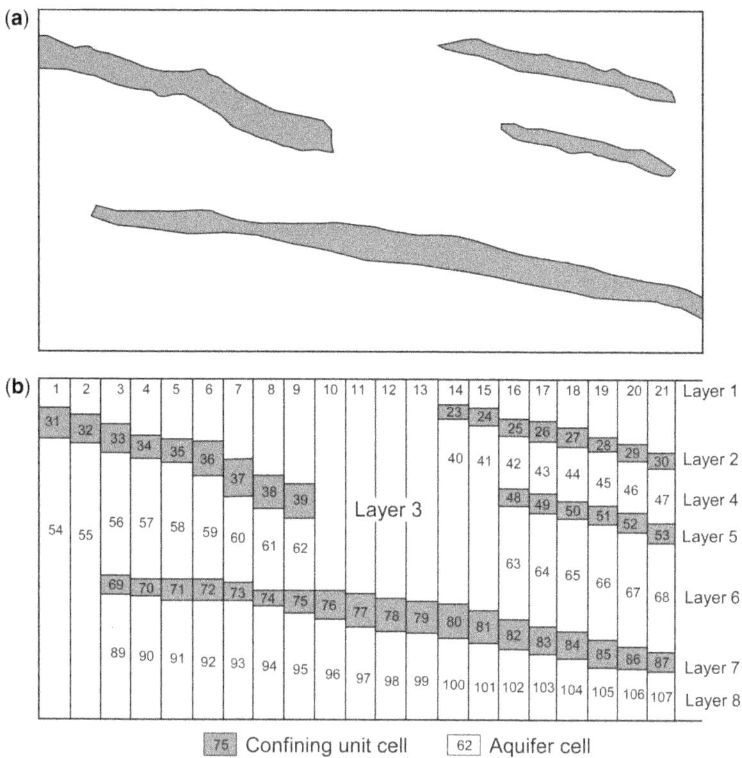

Fig. 1. Cross-sectional diagrams showing (**a**) pinching hydrostratigraphic layers and (**b**) an unstructured grid that could be used to represent the hydrostratigraphic layers (Panday *et al.* 2013).

design can be used to focus the resolution along karst conduits, channels, rivers and around wells, for example, or to sub-discretize individual layers to better represent hydrostratigraphic units (Figs 1 & 2).

As described in detail in the published MODFLOW-USG documentation (Panday *et al.* 2013), the program is based on an underlying control volume finite difference formulation in which a cell can be connected to an arbitrary number of adjacent cells. To improve the accuracy of the control volume finite difference formulation for irregular grid cell geometries or nested grids, a generalized ghost node correction package was developed, which uses interpolated heads in the flow calculation between adjacent connected cells.

The program includes a groundwater flow (GWF) process based on the GWF process in MODFLOW-2005, as well as a new CLN process to simulate the effects of karst conduits, multi-node wells and tile drains. The CLN process is tightly coupled with the GWF process in that the equations from both processes are formulated into one matrix equation and solved simultaneously. This robustness results from using an unstructured grid with unstructured matrix storage and solution schemes.

MODFLOW-USG also contains an optional Newton–Raphson formulation, based on the formulation in MODFLOW-NWT, for improving solution convergence and avoiding problems with the drying and rewetting of cells. Because the existing MODFLOW solvers were developed for structured and symmetrical matrices, they were replaced with a new sparse matrix solver package developed specifically for MODFLOW-USG. The sparse matrix solver package provides several methods for resolving non-linearities and multiple symmetrical and asymmetrical linear solution schemes to solve the matrix arising from the flow equations and the Newton–Raphson formulation, respectively (Panday *et al.* 2013).

Connected linear networks in karst

The CLN process was developed for MODFLOW-USG to provide the framework for incorporating one-dimensional connected features into a structured or unstructured three dimensional GWF process grid. A one-dimensional CLN feature is any hydrogeological or hydrological water conveyance feature

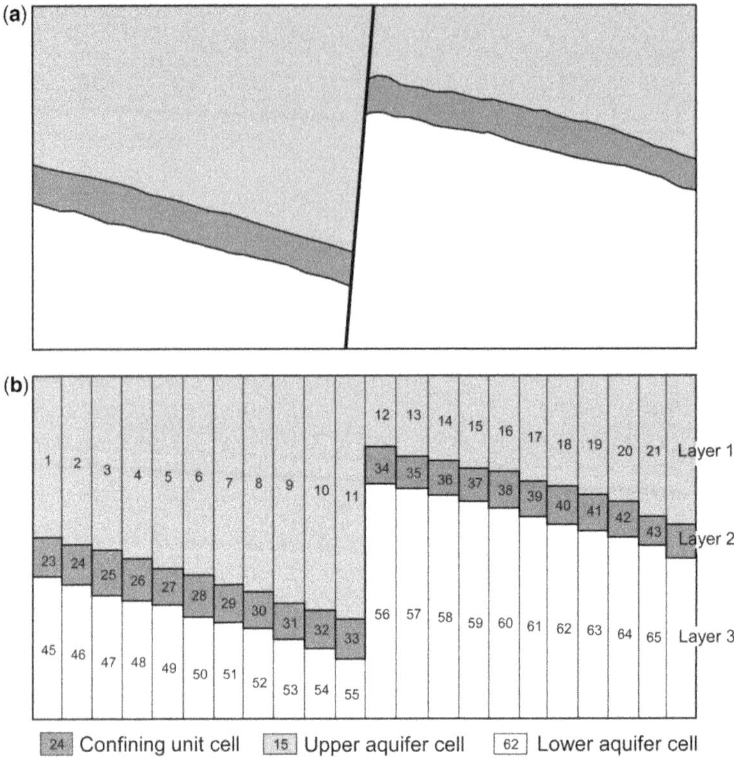

Fig. 2. Cross-sectional diagrams showing (**a**) the offset of hydrostratigraphic layers from a fault and (**b**) an unstructured grid that could be used to represent the faulted system (Panday *et al.* 2013).

that has a cross-sectional dimension much smaller than the longitudinal flow dimension and the size of the encompassing GWF cell. Flow is computed in the longitudinal direction of the network of connected one-dimensional features using specified cross-sectional properties; flow between the CLN and GWF cells is computed across the upstream wetted perimeter of the one-dimensional CLN feature. The CLN process thus provides a mechanism for including features with small cross-sectional areas relative to the GWF cell sizes without having to build this level of detail into the grid used for the GWF domain. An example problem is provided to further demonstrate the use of the CLN process.

As described in detail in the published MODFLOW-USG documentation (Panday *et al.* 2013), the CLN flow process is solved simultaneously with the GWF process. This means that the total number of cells in a MODFLOW-USG simulation is equal to the combined total of GWF and CLN cells. The CLN flow process solves for the flow of water within a network of linear features and for the interaction of the features with the porous medium. There are two types of flow calculation that occur with the addition of the CLN domain: flow within the CLN domain and flow between CLN and GWF cells. The CLN process does not inactivate dry cells as in MODFLOW-2005 for the GWF cells. Instead, a head value is always calculated for active CLN cells as in the upstream weighted formulation. As the saturated thickness approaches zero, the conductance values with connected cells also approach zero and the cell responds as if it were dry.

For flow within the CLN domain, the CLN process implements the solution of one non-linear equation per CLN cell. The CLN cells may represent wells, pipes, fractures, canals, rivers, streams or other linear one-dimensional features within a simulation domain that need to be represented by flow connections that are separate from those of the aquifer. The formulation is general enough that different types of feature can flow into one another. For instance, pipes can flow from or to open channel features. Extension of the code to include other features or geometries is a straightforward exercise in implementing functional forms for cross-sectional areas and volumes as a function of the flow depth at the CLN cells within the appropriate subroutines.

The current version of MODFLOW-USG incorporates cylindrical conduit geometry types. Flow within each segment can be laminar or turbulent and the formulations are derived in a general manner. Laminar flow is computed using the Hagen–Poiseuille equation. Three optional formulations are provided to simulate turbulent flow: the Darcy–Weisbach equation, the Hazen–Williams equation and Manning's equation. The turbulent flow formulations neglect the inertial term in the St Venant equations and therefore solve the diffusion wave equation. The dynamic wave equation for flow in open channels requires a solution to two equations per channel segment (one for continuity and a separate one for momentum) and therefore is not readily accommodated by the present form of the CLN process (Panday *et al.* 2013).

For flow between CLN cells and connected GWF cells, the CLN process implements the solution of one non-linear flow equation for each CLN to GWF connection. Various options are provided to compute the effective leakance between them, including input of the effective leakance value, computation from skin conductance and thickness or the use of a Thiem solution to provide radial influences and efficiency considerations for flow between CLN and GWF cells. The formulation is general enough that extension of the code to include other connection geometries or special considerations is straightforward as long as functional forms for the respective wetted surface areas with flow depths for the CLN to GWF connection can be provided. The CLN to GWF transfer equations are generalized to accommodate unstructured grid cells.

The CLN process uses the concept of a CLN segment, which is one or more CLN cells connected end to end. A CLN segment, of any length, may be used to represent a well, for example, or a karst spring represented by a drain cell. CLN segments can also be connected to one another to form a network (Fig. 3). A CLN network may be manually defined within the GUI by the user and is useful for representing karst conduits, radial collector wells or other connected linear features. Three-dimensional linear networks can also be easily imported from external programs such as geographical information systems (Table 1).

A CLN cell can be vertical, horizontal or tilted. Cell lengths may be dimensioned such that several CLN cells (from the same or different CLN segments) may be connected to one GWF cell, or one CLN cell may be connected to several GWF cells depending on the scale of the conceptualization of flow within the CLN and GWF domains. These connections can also be defined or re-defined after being imported from external geographical information system programs where the karst conduits are first conceptualized. In addition, there is no need for specific upstream to downstream sequential ordering of cells as may be necessary in other packages or codes.

The CLN process, when applied with cylindrical conduits and the Thiem solution for CLN to GWF connection, provides some of the functionalities of the multi-node well (MNW and MNW2) (Halford & Hanson 2002; Konikow *et al.* 2009) packages of MODFLOW-2005. A single cylindrical CLN cell

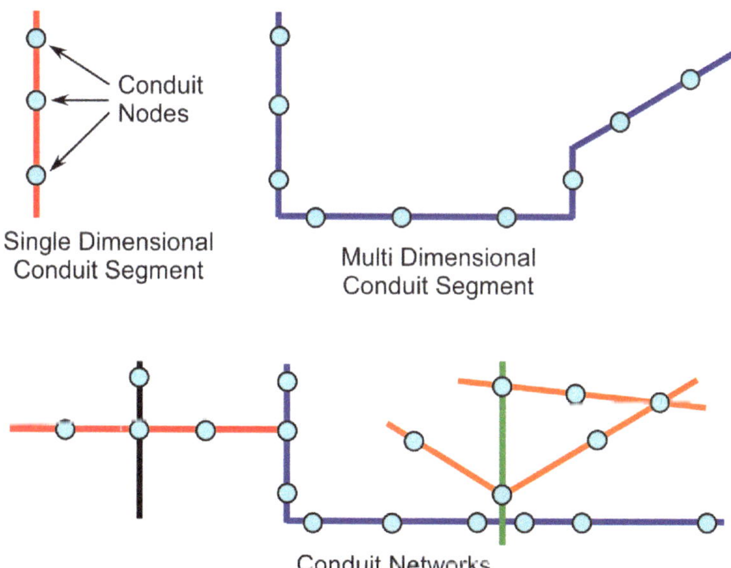

Fig. 3. Examples of conduit domain geometries supported by MODFLOW-USG (Panday *et al.* 2013).

Table 1. *Format of a three-dimensional shapefile table of a karst conduit suitable for import to MODFLOW-USG South Conduit in Figures 6 and 7*

ID	Xstart	Ystart	Zstart	Xend	Yend	Zend
1	2979.60	353.85	120.00	3025.70	368.15	93.00
2	3025.70	368.15	93.00	3065.45	371.33	90.00
3	3065.45	371.33	90.00	3152.89	369.74	89.00
4	3152.89	369.74	89.00	3284.85	369.75	85.00
5	3284.85	369.75	85.00	3412.03	361.79	80.00
6	3412.03	361.79	80.00	3593.27	363.38	70.00
7	3593.27	363.38	70.00	3729.99	366.56	62.00
8	3729.99	366.56	62.00	3947.80	369.74	57.00
9	3947.80	369.74	57.00	4062.27	368.15	56.00
10	4062.27	368.15	56.00	4345.26	404.72	55.00
11	4345.26	404.72	55.00	4513.78	434.93	54.00
12	4513.78	434.93	54.00	4644.15	477.85	52.00
13	4644.15	477.85	52.00	4760.20	503.29	51.00
14	4760.20	503.29	51.00	4831.74	531.91	45.00

Xstart and Ystart are, respectively, the X and Y coordinates of the beginning of the conduit segment; Xend and Yend are, respectively, the X and Y coordinates of the end of the conduit segment; Zstart and Zend are, respectively, the top and bottom elevations of conduit segment. All units are metres.

connected to multiple GWF cells may be pumped to simulate multi-node well conditions, in which the CLN cell extracts water from the GWF cells connected to it as part of the solution to the coupled CLN and GWF flow equations. A CLN cell can be pumped by use of the well (WEL) package by assigning a source or sink to that CLN cell. If flow through a narrow conduit is important to the production of a multi-layer well (for instance, the well bore resistance governs how much water is produced from each of the multiple aquifers), the well may also be represented by a segment consisting of multiple CLN cells. In this case, a CLN cell can be added to each GWF cell or layer and flow would be simulated within the conduit. This may be especially important when the conduit radius is small and resistance in the well bore affects the flow through the conduit. This approach is also useful for simulating borehole flow within wells that are screened to multiple aquifers.

The continuity equation for flow through a CLN cell is a function of flows from connected CLN cells and flows from connected GWF cells and may be written in difference form as

$$\frac{\Delta V_n}{\Delta t} = \sum_{m \in \eta_n} Q_{nm} + \sum_{p \in \eta_n} \Gamma_{cpn} \tag{1}$$

where ΔV_n is the change in the volume of water in CLN cell n at a given time, Δt is the time step size, Q_{nm} is the volumetric flow between connected CLN cells n *and* m and Γ_{cpn} is the volumetric flow from connected GWF cells p to CLN cell n.

The left-hand side of equation (1) is the storage term of the CLN cell, the first term on the right-hand side is the flow between cell n and the connected CLN cells, m, and the second term on the right-hand side is the interaction flow between cell n and all connected GWF cells, p. The summation in the first flow term on the right-hand side is over all m CLN cells connected to CLN cell n, whereas the summation in the second term on the right-hand side (the interaction term) is over all p GWF cells connected to CLN cell n.

Treatment of the term Γ_{cpn} in equation (1) is detailed in Panday *et al.* (2013). A flux coupling is provided between the GWF and CLN domains with three options to couple the domains. The first two options follow the MODFLOW-CFP (Shoemaker *et al.* 2008) formulation and include the user input of leakance or a skin hydraulic conductivity and thickness. The flow then occurs across this skin through the wetted perimeter between the CLN and GWF domains as a result of the hydraulic gradient. A third option uses a modified Thiem equation, which is particularly useful for vertical conduits or wells, to account for a radially logarithmic drawdown between large GWF cells and the comparatively small conduit radius.

Flow within the CLN domain may be computed by three alternative equations. For laminar flow between any two CLN nodes the volumetric flux is expressed by the Hagen–Poiseuille equation, whereas for turbulent flow it may be expressed either by the Darcy–Weisbach equation or the Hazen–Williams equation. Detailed formulations of the flow equations are provided in Panday *et al.* (2013).

Example simulation

Figures 4 and 5 illustrate the MODFLOW-USG model set-up for a hypothetical site with two karst

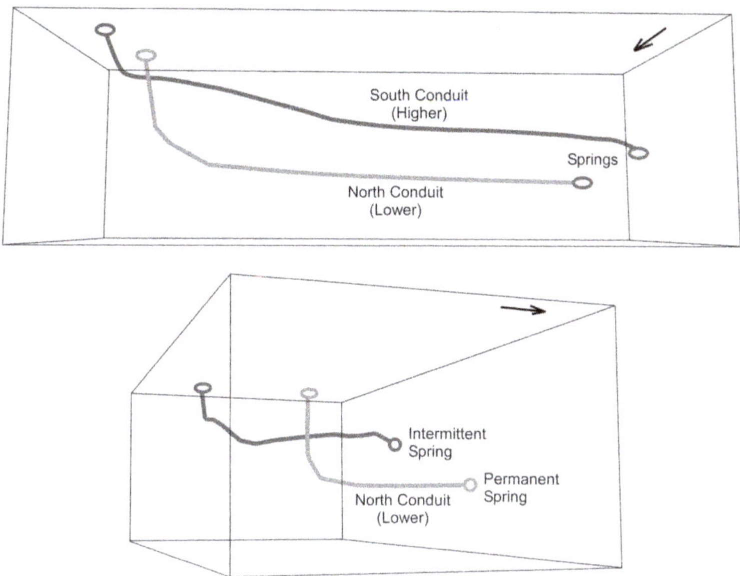

Fig. 4. Two views showing spatial position of the karst conduits at the modelled site.

conduits at different elevations, each connecting a sinking stream and a karst spring. The site is representative of typical karst conditions in the southeastern USA, developed in Mesozoic limestones and dolomites and with a subtropical humid climate. The northern conduit is lower and the flow in it is permanent, as is the flow of the sinking stream and the karst spring connected by the conduit. The southern conduit has intermittent function, connecting an intermittent sinking stream and an intermittent spring. The model has 12 layers, each 10 m thick, to accommodate the more precise vertical placement of various sections of the two conduits.

The model simulates three time periods: the first period is one year long and without recharge from the land surface; the second time period lasts for

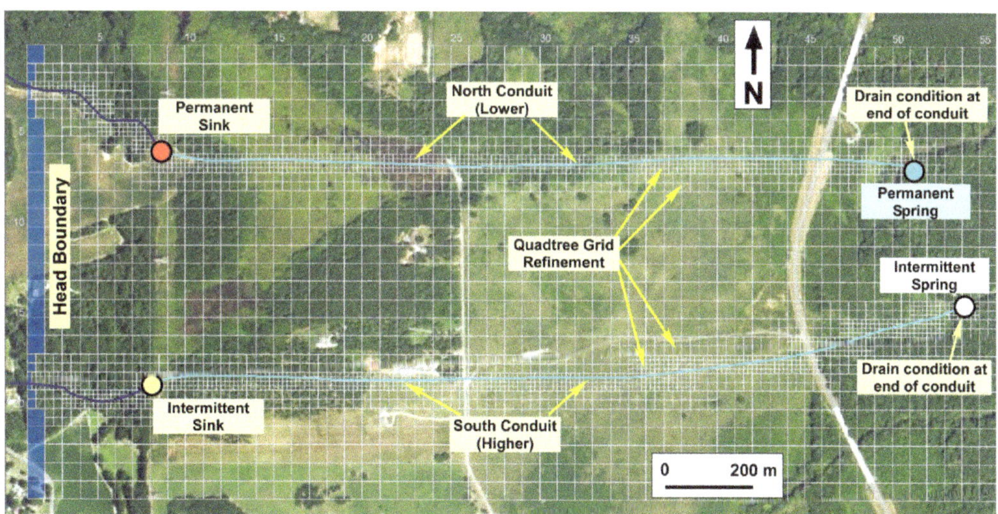

Fig. 5. General model set-up and boundary conditions.

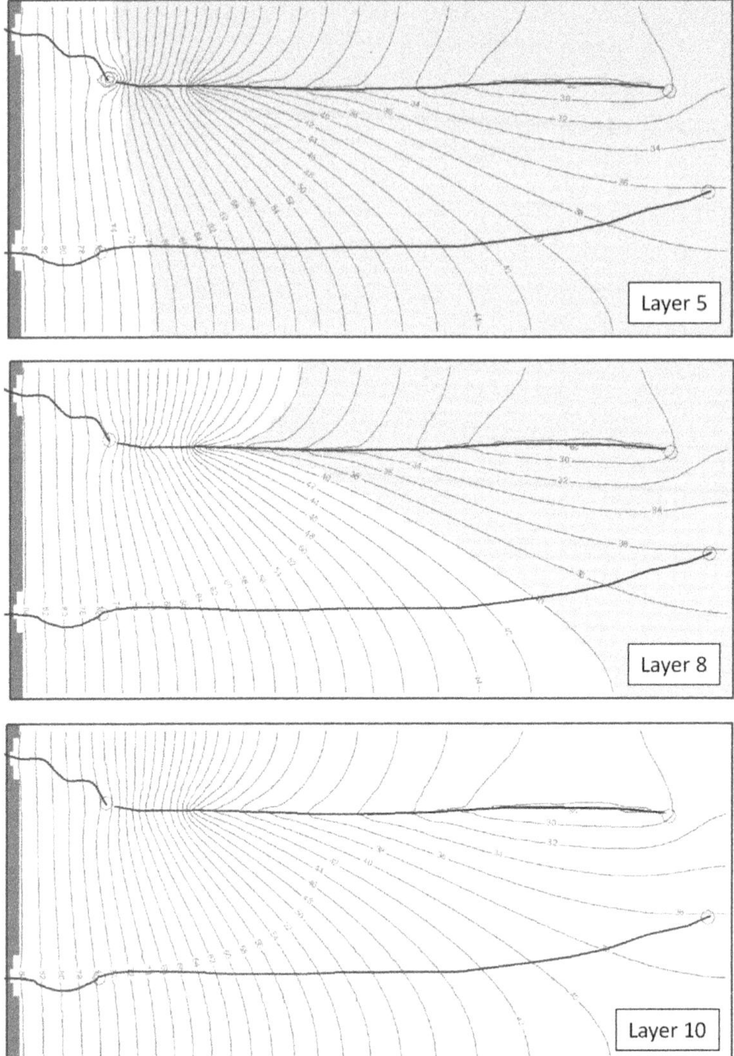

Fig. 6. Simulated potentiometric surface map at the end of time period 1 in layers 5, 8 and 10. The dry cells in a layer are shaded.

two days and simulates an areal recharge event from rainfall (0.15 m day^{-1}); the third period, 30 days long, simulates recession after the recharge event. Figures 6–8 show the modelled potentiometric surface contour lines at the end of the three time periods for select model layers. The simulated hydrographs at the two springs are shown in Figure 9.

Discussion

The CLN process in MODFLOW-USG provides for the physically based simulation of groundwater flow within preferential flow paths of karst aquifers, such as fully or partially water-filled conduits, which can have both permanent and intermittent flows and a varying three-dimensional geometry of the conduit networks. Conduits can be effortlessly brought into the model from external three-dimensional geographical information system shapefiles via user-friendly input screens, or can be added manually and connected with any of the existing CLNs in all three dimensions as needed. For example, Figure 10 illustrates the effects of adding and connecting one 'manual' conduit to the permanent North Conduit imported from a

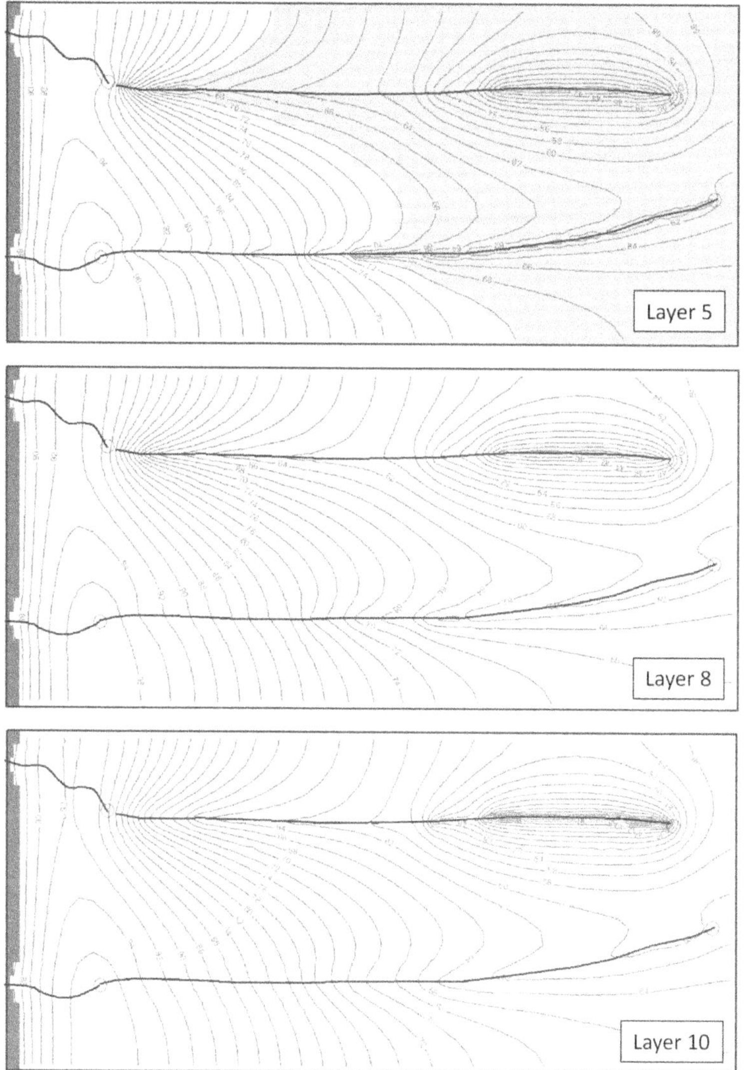

Fig. 7. Simulated potentiometric surface map at the end of time period 2 (end of high recharge episode) in layers 5, 8 and 10. The dry cells in a layer are shaded.

geographical information system file to the example model.

The CLN conduits can extend through both the saturated and unsaturated zones, can have vertical sections and can become fully submerged and/or dewatered depending on the assigned transient boundary conditions. Any section of a conduit can have a different radius from the adjacent sections and the groundwater flow within it can be described with any of the four available equations. In addition, the interaction between any individual conduit section and the surrounding aquifer matrix can be described with a set of unique parameters. This flexibility allows for a powerful sensitivity analysis of various CLN input parameters and assumptions. Notably, a section of a conduit can act as a linear drain for the surrounding aquifer matrix for a given set of boundary conditions and can also lose water to the matrix for another set of boundary conditions, such as during high recharge episodes, including recharge from distributed precipitation and from the focused sinking of surface water features, such as intermittent/activated sinks (Fig. 11).

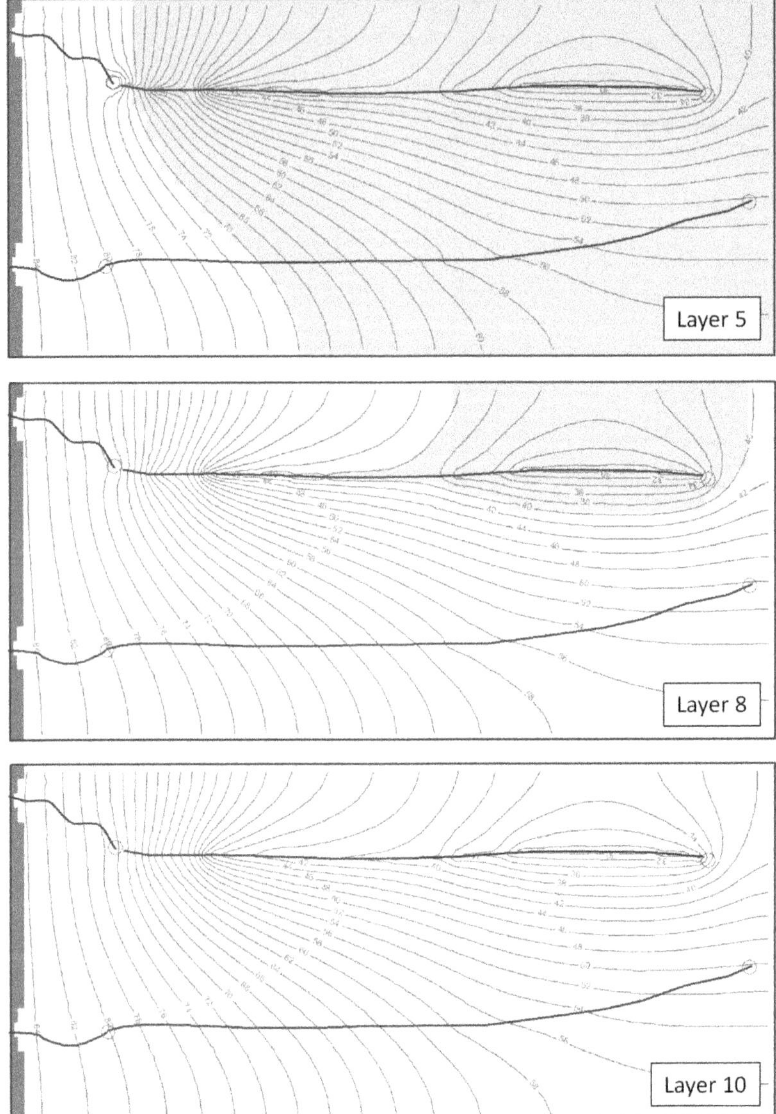

Fig. 8. Simulated potentiometric surface map for time period 3, 15 days after the end of the recharge episode, in layers 5, 8 and 10. The dry cells in a layer are shaded.

MODFLOW-USG includes simulations of the contaminant fate and transport (F&T) in both unsaturated and saturated zones and in both the aquifer matrix (Darcian porous media, i.e. GWF process model cells) and CLNs. Simulation options in the current beta version include linear and Freundlich types of sorption, diffusion, isotropic and anisotropic dispersion, zero- and first-order decay on soil, in water, and on soil and water, and dual domain transport. Contaminants can both enter and leave CLNs following the groundwater flow gradients.

Figure 12 is an example F&T simulation showing the development of a dissolved phase plume from a constant-strength source at the ground surface. The results in layer 10, which is always saturated and contains the longest stretch of the North Conduit, show the contaminant plume entering the conduit and flowing to the spring.

MODFLOW-USG is a powerful new groundwater flow and contaminant F&T numerical modelling program with a number of tools and solutions well suited for the simulation of karst aquifers and their

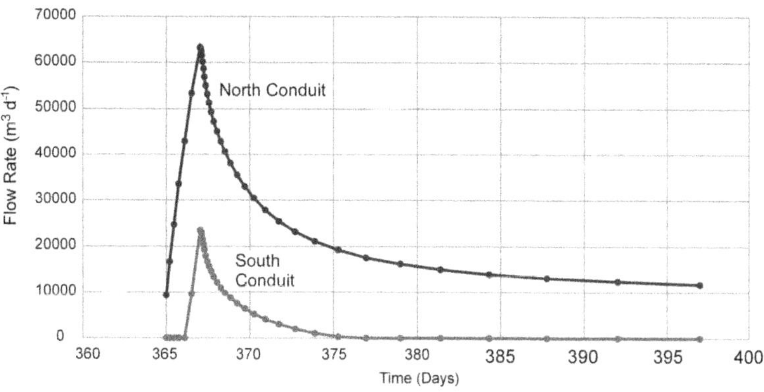

Fig. 9. Simulated flow rate at the two springs (drain cells at the end of conduits).

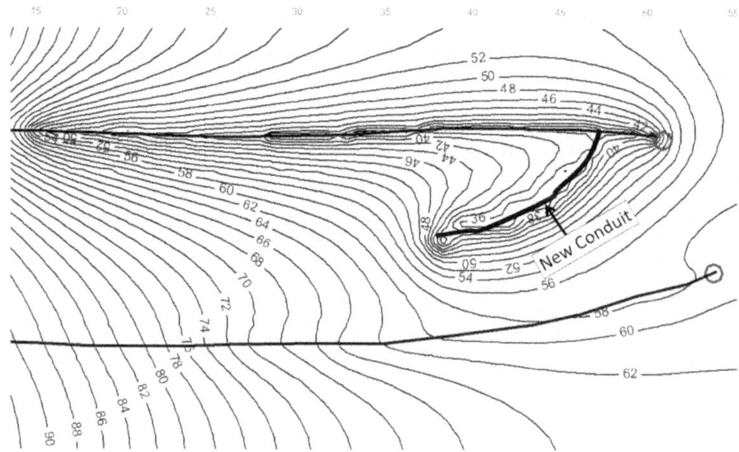

Fig. 10. Effect of an additional conduit in layer 10 on groundwater flow conditions. End of time period 2 (end of recharge episode).

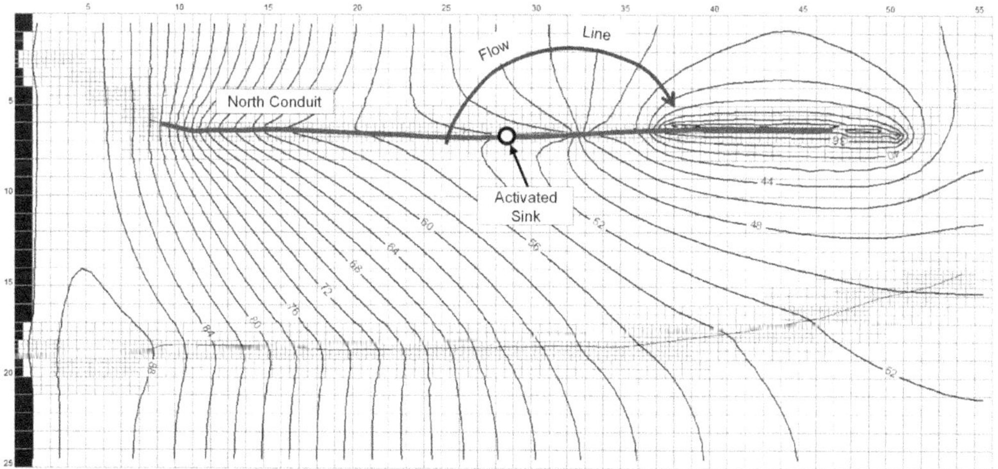

Fig. 11. Effect of a simulated intermittent sink connected to the North Conduit on the groundwater flow conditions. The North Conduit backs up and loses water to the surrounding matrix.

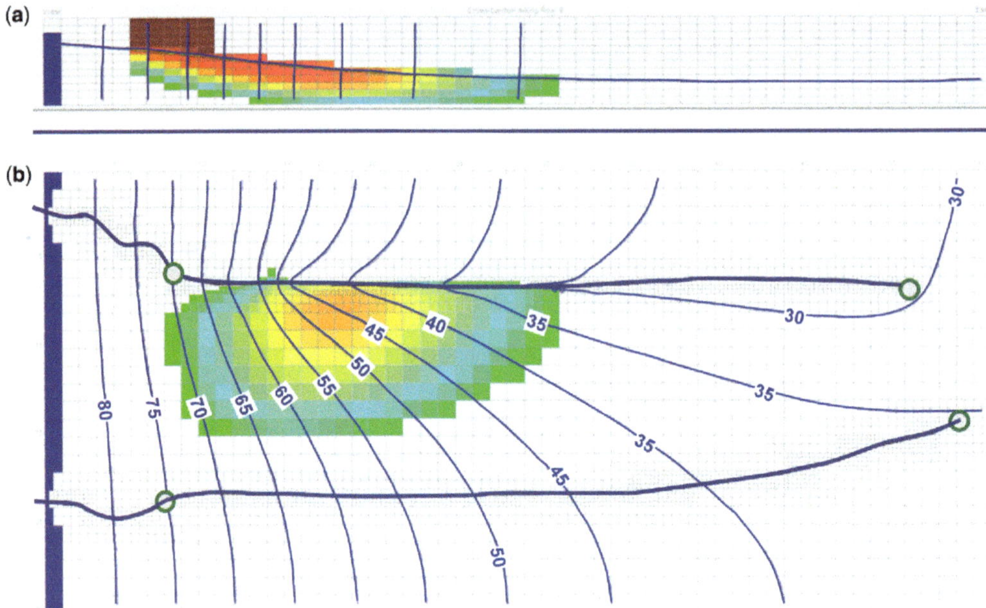

Fig. 12. Example simulation of contaminant fate and transport (F&T) with MODFLOW-USG. (**a**) Cross-sectional view along row 9. (**b**) Contaminant distribution in layer 10. The constant-strength source of contamination is at the land surface. The model solves flow and F&T in both unsaturated and saturated zones.

interactions with surface water features. It is in public domain, supported by a number of GUIs, and therefore easily accessible to various users.

References

DHI 2016. FEFLOW Release Note 2016, http://releasenotes.dhigroup.com/2016/FEFLOWrelinf.htm [last accessed 21 July 2017].

HALFORD, K.J. & HANSON, R.T. 2002. *User Guide for the Drawdown-limited, Multi-node Well (MNW) Package for the U.S. Geological Survey's Modular Three-dimensional Finite-difference Ground-water Flow Model, Versions MO-FLOW–96 and MODFLOW–2000.* US Geological Survey Open-File Report **02-293**.

KONIKOW, L.F., HORNBERGER, G.Z., HALFORD, K.J. & HANSON, R.T. 2009. *Revised Multi-Node Well (MNW2) Package for MODFLOW Ground-water Flow Model.* US Geological Survey, Techniques and Methods, **6-A30**.

KRESIC, N. 2009. *Groundwater Resources: Sustainability, Management, and Restoration.* McGraw-Hill, New York.

KRESIC, N. 2013. *Water in Karst. Management, Vulnerability, and Restoration.* McGraw-Hill, New York.

MAYAUD, C., WALKER, P., HERGARTEN, S. & BIRK, S. 2015. Nonlinear flow process: a new package to compute nonlinear flow in MODFLOW. *Groundwater*, **53**, 645–650.

PANDAY, S., LANGEVIN, C.D., NISWONGER, R.G., IBARAKI, M. & HUGHES, J.D. 2013. *MODFLOW-USG Version 1: an Unstructured Grid Version of MODFLOW for Simulating Groundwater Flow and Tightly Coupled Processes Using a Control Volume Finite-difference Formulation.* US Geological Survey, Techniques and Methods, **6-A45**.

SHOEMAKER, W.B., KUNIANSKY, E.L., BIRK, S., BAUER, S. & SWAIN, E.D. 2008. *Documentation of a Conduit Flow Process (CFP) for MODFLOW-2005.* US Geological Survey Techniques and Methods, **6-A24**.

THERRIEN, R. & SUDICKY, E.A. 1996. Three-dimensional analysis of variably-saturated flow and solute transport in discretely-fractured porous media. *Journal of Contaminant Hydrology*, **23**, 1–44.

Experiences in calibrating and evaluating lumped karst hydrological models

ANDREAS HARTMANN[1,2]

[1]*Institute of Earth and Environmental Sciences, Freiburg University, Germany*

[2]*Department of Civil Engineering, University of Bristol, Bristol, UK*

andreas.hartmann@hydrology.uni-freiburg.de

Abstract: Karst systems play an important part in providing clean drinking water for many countries worldwide. Their sustainable management often requires the application of karst hydrological models to quantify the water volumes available and to assess their sensitivity to changes in climate or land use. The available literature provides a broad overview of different modelling approaches to simulate karst hydrology, but guidance for the application of these approaches is scarce. This paper provides personal insights into the application, calibration and evaluation of lumped karst hydrological models. An introduction to model calibration and sensitivity analysis is given, with links to further reading and ready-to-use toolboxes. It is followed by three case studies of karst model applications at three different scales (the plot scale, aquifer scale and continental scale), which apply model calibration and sensitivity analysis to obtain realistic simulations. These three case studies elaborate how the necessary trade-off between the available data and process representation in the model structures was achieved. General recommendations and a workflow for future karst model applications are given.

Karst systems develop through the weathering of water-soluble rocks, a process referred to as karstification (Kiraly 2003). The consequent feedbacks between the atmosphere, biosphere, hydrosphere and geosphere (Goldscheider 2012) result in distinct surface and subsurface landform features, such as karren fields, dolines, sinking streams, cave systems and large karst springs (Ford & Williams 2007). Although beautiful, these features pose a challenge for hydrological and hydrogeological characterization (Goldscheider & Drew 2007). Karst systems are known for their pronounced heterogeneity and anisotropy of hydraulic properties (Bakalowicz 2005; Worthington *et al.* 2016), but also for their high groundwater storage capacities, which provide favourable conditions for groundwater extraction and water supply (Andreo *et al.* 2006). Some European countries – for instance, Austria and Slovenia – obtain around half of their drinking water from karst aquifers (European Commission 1995). Karst hydrological models are commonly applied to sustainably manage karst water resources and to understand their hydrological behaviour (Hartmann *et al.* 2014*a*).

The currently available karst hydrological models are adapted to the particular processes that are typically found in karst systems (White 2003). Some model structures consider the flow dynamics of the epikarst (Tritz *et al.* 2011; Hartmann *et al.* 2012), whereas others consider the transfer of groundwater between the cave, conduit or fracture systems and the fissured carbonate matrix (Liedl *et al.* 2003; Butscher & Huggenberger 2008; Reimann *et al.* 2011). Many of the reviews summarizing the range of different karst modelling techniques (Sauter *et al.* 2006; Ford & Williams 2007; Kovacs & Sauter 2007; White 2007; Ghasemizadeh *et al.* 2012; Hartmann *et al.* 2014*a*; Hartmann & Baker 2017) distinguish two main groups of karst modelling approaches by the manner in which the real karst system is considered within the model. Spatially distributed models represent the karst system as spatially discretized and apply the model equations at each point, whereas lumped approaches neglect the spatial distribution of the system parameters (e.g. the hydraulic conductivities or porosities) and focus on the representation of the dominant processes. Some models defined as lumped models sometimes include spatial information – for instance, subsections of the aquifer or spatial units of the karst system (Rimmer & Salingar 2006; Ladouche *et al.* 2014). Some recent approaches can be classified as semi-distributed because they represent spatial variability by distribution functions (Hartmann *et al.* 2014*b*, 2016). The selection of the final approach usually depends on the purpose, requirements, data availability and cost of the modelling project.

However, another factor that influences the reliability of the model simulation, with either a distributed or a lumped karst modelling approach, is the availability of observations from which to choose the most appropriate set of model parameters,

From: Parise, M., Gabrovsek, F., Kaufmann, G. & Ravbar, N. (eds) 2018. *Advances in Karst Research: Theory, Fieldwork and Applications*. Geological Society, London, Special Publications, **466**, 331–340.
First published online December 6, 2017, https://doi.org/10.1144/SP466.18

e.g. the soil thickness and storage capacity or the hydraulic conductivity of the matrix or conduits. As a result of the strong heterogeneity of karst systems, such information is difficult to obtain and is seldom available at the scale of the model application (Geyer *et al.* 2013). Although some of the necessary information can be obtained by experimental methods (e.g. soil cores or pumping tests), incommensurability prohibits the direct application of these measurements within the models (Beven 2006). As a consequence, many karst modellers use an inverse modelling approach to assess the most representative values of their model parameters by modifying them systematically until an acceptable fit between the observed and simulated model output – for instance, groundwater levels or karst spring discharge – is obtained. If the available observations contain enough information, such an inverse modelling approach allows an estimation of the most realistic set of effective model parameters and the model can be used for prediction. However, in many cases the available information is not sufficient to find a unique parameter set and many different parameter combinations result in similar model outputs (Perrin *et al.* 2003). In this case, the model predictions would have high uncertainties and the model's application for prediction would be strongly limited.

This paper provides some practical insights into my personal experience of identifying the most appropriate model structures and model parameters. It begins with an introduction into karst model calibration and the methods used to assess the information content of the available observations. This is followed by a set of three case studies performed at three scales (the plot, aquifer and continental scale) with strongly variable availability of data. An elaboration of the decisions and parameter estimation tools that lead to the final model structure and a discussion of the reliability of the simulations is presented. Although this paper represents an incomplete and personal view of karst model development and evaluation, it gives general recommendations for present and future karst modelling studies.

Calibration and parameter identifiability in karst hydrological models

Model calibration can be explained as a process of adjusting the model parameters to obtain an acceptable representation of the hydrological processes of interest that satisfies pre-defined criteria (Fig. 1). It is necessary because models are always simplifications of a real system. Naturally variable properties, such as hydraulic conductivities, have to be expressed by average or representative values within the chosen spatial model discretization (distributed or lumped). As pre-defined criteria, modellers commonly use error functions that express the deviation of simulations and observations by a single number. The coefficient of determination (r^2), the root-mean-square error, the Nash–Sutcliffe efficiency (Nash & Sutcliffe 1970) and the Kling–Gupta efficiency (Gupta *et al.* 2009) are popular examples. Prior knowledge about the real values of model parameters is included by choosing parameter ranges that envelope their observed values – for instance, a field capacity of 200–300 mm if previous field studies found values of *c.* 250 mm.

In early hydrological models, model calibration was carried out manually and required expert knowledge about the hydrological system and the chosen model (Boyle *et al.* 2000; Anderson 2002). As computational capacities increased, automatic calibration themes evolved that systematically explored the entire space of the model parameters within a predefined range to minimize the selected error function (Beven & Binley 1992). One of the most popular of these is the shuffled complex evolution algorithm (Duan *et al.* 1994), but many more have been developed in the last three decades (Gupta *et al.* 1999; Vrugt *et al.* 2009).

When the model structures became increasingly complex to incorporate more hydrological processes, it was found that an acceptable error function was often associated with a large number of different parameter combinations. This problem, referred to as equifinality (Beven 1989, 2006) arose because of uncertainties in the model input (e.g. precipitation),

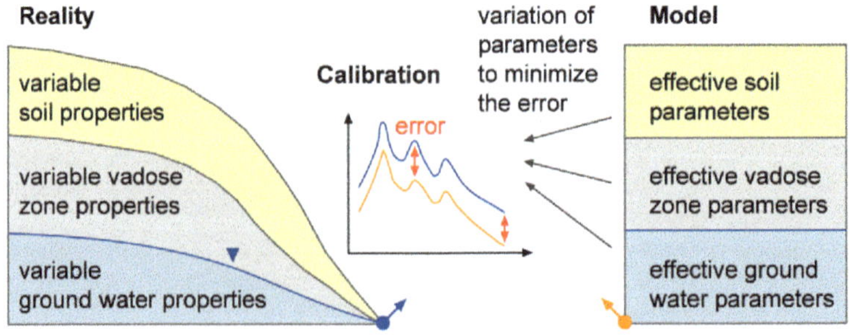

Fig. 1. Schematic diagram of the simplification of reality by models and the calibration procedure.

in the model structure (due to simplifications or the incorrect representation of processes) and in the observations (e.g. the uncertainties in the rating curve used to translate water level measurements in streams into discharge measurements) allowed a variety of acceptable parameter combinations. In addition, more complex model structures allowed the model internal processes to compensate for each other. For instance, simulated soil percolation rates that were too rapid could be compensated by simulated groundwater dynamics that were less dynamic than in the real system.

Because of these newly recognized problems, methods were developed that allowed an assessment of the sensitivity of model simulations to changes in the model parameters. Sensitivity analysis methods systematically vary the model parameters, most preferably within the same pre-defined ranges as the calibration, and analyse the variations in the model output (the error function or other metrics derived from the output, e.g. the mean discharge) according to those changes. Beginning with one-at-a-time sensitivity analysis, which varies only one model parameter while the remaining parameters are fixed, a wide range of approaches based on Monte Carlo sampling are classified as regional (Hornberger & Spear 1981; Young 1983) or global sensitivity analysis (Sarrazin *et al.* 2016; Chen *et al.* 2017). Apart from providing tables that list the percentage change in output according to a change in an individual parameter or the sensitivity indices, cumulative distribution functions of the normalized error functions of a large number of randomly sampled parameter sets (Monte Carlo sampling or similar) are a popular way to visualize the sensitivity of model parameters (Fig. 2). A parameter distribution differing from a uniform distribution is often regarded as an indicator of a sensitive parameter (Wagener *et al.* 2004). Similar to the automatic calibration schemes, guides and readily programmed toolboxes are available that provide a range of different procedures for sensitivity analysis (Pianosi *et al.* 2015, 2016).

If sensitivity analysis shows that many model parameters are insensitive, the information to identify all the model parameters by model calibration is probably insufficient, i.e. there is a strong risk of over-parametrization and equifinality (Fig. 3). Consequently, the model structure has to be simplified, insensitive parameters have to be excluded from the parameter estimation, or the information content of the available observations has to be increased. This can be achieved by considering separate time periods of observations instead of the entire time series (Gupta *et al.* 1998; Dunne 1999; Madsen *et al.* 2002; Wagener *et al.* 2003) or by the definition of multiple signatures that can be derived from the observed data (Wagener *et al.* 2007; Gupta *et al.* 2008; Yilmaz *et al.* 2008). Often, this increase in parameter identifiability results in small reductions in the overall model performance (Fig. 3; Kuczera & Mroczkowski 1998; Seibert & McDonnell 2002; Son & Sivapalan 2007). In particular, the inclusion of hydrochemical information in the calibration and sensitivity analysis provides a promising direction in karst systems where water quality data are often used to characterize the flow and storage behaviour (Birk *et al.* 2005; Hartmann *et al.* 2013*a*; Hartmann 2016).

Combined with model calibration, sensitivity analysis provides a powerful tool to assess the applicability of a chosen hydrological model for prediction and water resources management. It has to be acknowledged, however, that traditional model evaluation tools, such as spatial or temporal split-sample tests with independent observations (Klemeš 1986) and, if possible, the evaluation of the realism of model internal processes and parameters in discussion with field experts and their conceptual understanding of the systems, should not be omitted in the final model application.

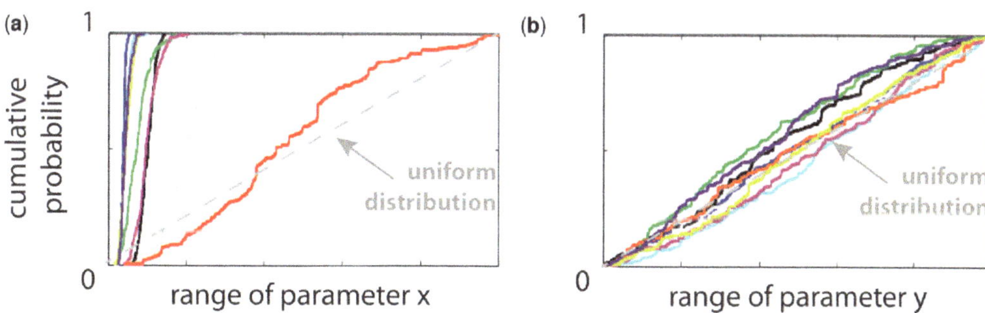

Fig. 2. Example results of regional sensitivity analysis for two model parameters (x, y) and different model applications (different colours) (modified from Schwerdtfeger *et al.* 2016). (**a**) The cumulative parameter distribution functions indicate that parameter x is sensitive because they strongly differ from the uniform distribution (except for the red model application), whereas in (**b**) parameter y remains insensitive for all model applications.

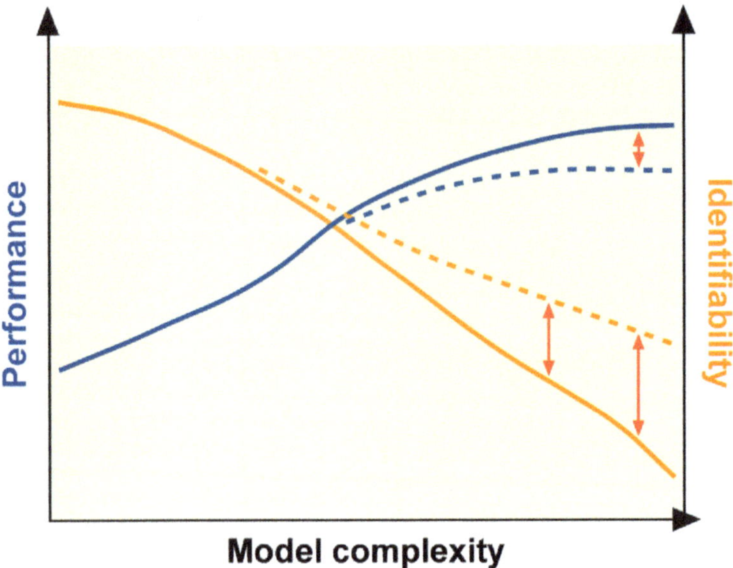

Fig. 3. Schematic diagram of the interplay between model complexity, model performance and parameter identifiability (modified from Wagener *et al.* 2004). Model complexity can be expressed by the number of degrees of freedom (or number of model parameters) given by a chosen model structure. Model performance is the ability of a model structure to obtain a good fit between observations and simulations. Identifiability can be expressed by the overall sensitivity of all model parameters. The red arrows indicate how additional information can increase the identifiability of model parameters by additional information. See Beven (2001) for more advice about how to approach this problem.

Case studies

The following case studies, with very different spatial scales and different data availability, elaborate the application of model calibration, sensitivity analysis and model evaluation procedures to obtain the most reliable simulation (Table 1).

Case study 1 addresses the simulation of drip rates at a karstic cave in northern Israel with just few square metres of areal extent (Hartmann *et al.* 2012). This study aimed to simulate the spatial and temporal variability of drip rates within a karstic cave in a semi-arid environment. The input and output of the system were measured at a six-hour resolution because of the strong dynamics of the semi-arid climate and the rapid flow dynamics of the karst system. In addition to climate data and the drip rates at several stalagmites, the input and output of an artificial tracer experiment were monitored. As there was no spatially distributed information on the structural properties of the karst system above the cave, a semi-distributed model, which considers the spatial distribution of karst properties by distribution functions, was chosen and set up to simulate water and tracer fluxes at the six-hour resolution of the available data. The structure of the model was based on the general conceptual model of the epikarst (Williams 2008). A multivariate calibration (mean drip rate, mean tracer concentration and coefficient of variation of drip rates and tracer concentration among all stalagmites) using the shuffled complex evolution metropolis algorithm (Vrugt *et al.* 2003) was performed with the Nash–Sutcliffe efficiency as the error function. Regional sensitivity analysis (similar to the example in Fig. 2) showed that the 13 model parameters were only identifiable by using all the available information. Using only the average drip rates provided a superior model performance, but, because of the reduced information, many model parameters remained unidentifiable. Hence the incorporation of tracer observations and information about the spatial variability of drip rates and tracer concentrations reduced the prediction uncertainty, which was also supported by the findings of a split-sample test (Klemeš 1986) and a discussion of the realism of the calibrated model parameters.

Case study 2 dealt with the identification and application of a karst model to a larger karst system in the Middle East (800 km^2, Hartmann *et al.* 2013*b*). Interpolated climate data at a daily time step was derived from various stations across the study area. Observations of daily discharge and weekly to bi-weekly observations of water quality parameters were available at the two major springs draining the system. Compared with case 1, the

Table 1. *Meta-information about the three case studies*

		Case study 1 ($150\ m^2$)	Case study 2 ($800\ km^2$)	Case study 3 ($7 \times 10^6\ km^2$)
Model input (climate data):	Type	Weather station	Weather stations	Globally gridded product
	Temporal resolution	6 hours	1 day	1 day
	Spatial resolution	Point information	Point information	0.25°
Model:	Type	Semi-distributed	Lumped (multiple model structures)	Distributed
	Temporal resolution	6 hours	1 day	1 day
	Spatial resolution	Entire plot	Entire catchment	0.25°
	Number parameters	13	7–12	4
Observations:	Type and spatial resolution	Discharge and tracer concentration at various observation points	Discharge and water quality at the gauging station	Actual evaporation and soil moisture at various point locations
	Temporal resolution	6 hours	1 day	1 month
Calibration:	Type	Multivariate	Individually on signatures	Estimation via 'soft rules'
	Error function	Nash–Sutcliffe	Eight hydrodynamic and hydrochemical signatures	Three 'soft rules' based on evaporation and soil moisture observations
	Algorithm	SCEM	SCEM	Based on GLUE
Sensitivity analysis:	Type	Regional	Global	Regional
	Algorithm	Derived from SCEM*	Sobol's method*	Based on HSY
Evaluation through:		Split-sample test and plausibility test	Model performance after combining independent calibrations and plausibility test	Independent point- and catchment-scale observations of recharge
Remaining uncertainty quantified through:		Implicitly by results of sensitivity analysis and by decrease in performance during split-sample test	Implicitly by results of sensitivity analysis and model performance with combined parameter set	Explicitly by large sample of remaining parameter sets the fulfil all 'soft rules'

Case study 1 from Hartmann *et al.* (2012), case study 2 from Hartmann *et al.* (2013*b*) and case study 3 from Hartmann *et al.* (2015).
GLUE, generalized likelihood uncertainty estimation (Beven & Binley 1992); HSY, Hornberger–Spear–Young method, which is a regional method for sensitivity analysis (Hornberger & Spear 1981; Young 1983); SCEM, shuffled complex evolution metropolis algorithm (Vrugt *et al.* 2003).
*Sobol's method is a global variance-based method for sensitivity analysis (Saltelli *et al.* 2008; Sarrazin *et al.* 2016).

larger extent of this system aggravated the identification of a unique conceptual model and hence the identification of a unique and adequate model structure. For that reason, four different spatially lumped karst model structures were prepared to run on a daily time step. Using the shuffled complex evolution metropolis algorithm, the models were calibrated individually on eight hydrodynamic and hydrochemical signatures. Sobol's sensitivity analysis (Saltelli *et al.* 2008; Sarrazin *et al.* 2016) was used to assess the information brought by each signature. The combination of the calibration and sensitivity analysis using the eight signatures allowed three of the four model structures to be iteratively discarded, hereby reducing the structural uncertainty of the model. The remaining model structure showed an acceptable performance for all eight signatures simultaneously and the sensitivity of all the model parameters was regarded as realistic in terms of the simulated fluxes and parameter values.

Case study 3 (Hartmann *et al.* 2015) was the first application of a karst recharge model on a continental scale (all carbonate rock regions in Europe, the Middle East and northern Africa, 7×10^6 km^2). Other than the previous case studies, which used data from rain gauges, the input of the model consisted of a gridded climate data set at a 0.25° × 0.25° resolution (Rodell *et al.* 2004; Rui & Beaudoing 2013). Observations for the calibration consisted of evaporative flux (Baldocchi *et al.* 2001) and soil moisture (Dorigo *et al.* 2011) observations at a point scale. The large scale of the application and the spatial variability of the landscape and climate characteristics required the application of a spatially distributed karst recharge model with a relatively simple structure (four model parameters) accounting for the limited availability of calibration data. For calibration, the entire model domain was split into four regions using cluster analysis and climatic and topographic descriptors according the concept of hydrological landscapes (Winter 2001). At each of these regions, the four model parameters were estimated using a newly developed parameter estimation approach accounting for uncertainties due to the differences in observation and simulation scales, data limitations and model simplifications (Hartmann *et al.* 2015). Using such an approach, groundwater recharge over the entire continental modelling domain could be estimated and the remaining uncertainty of the simulations was assessed explicitly. A comparison with independent observations of karstic groundwater recharge derived from the literature indicated that the model provided realistic recharge values and spatial patterns of groundwater recharge across the entire model domain.

To avoid the problem of equifinality, the complexity of the model structures had to be adapted to these data limitations because they were lumped structures, the parameters of which had to be found by some sort of calibration (Fig. 3). Although continuous input data was relatively easy to obtain, the availability of observations varied strongly for the three case studies, resulting in model structures that included varying degrees of complexity. This was mostly due to the scale of their application: at the plot scale (case study 1), dense measurement networks were capable of capturing the spatial and temporal variability of water and solute fluxes, allowing the application of a complex, process-based model and identification of its model parameters.

At the scale of an entire aquifer (case study 2), measurements were mostly available at the karst spring, providing integrated information on the flow, solute transport and storage behaviour of the system. For that reason, model structures with a lower number of model parameters provided acceptable process representation without too many insensitive parameters. In addition, in case studies 1 and 2, auxiliary information (artificial tracer experiment and water quality observations) was available to increase the available information (Fig. 3).

At the scale of an entire continent (case study 3), observations were only available at locations that were considered representative for a much larger area. Hence a rather simple model structure (four model parameters) was chosen, which only considered the most relevant processes (actual evapotranspiration, diffuse and concentrated recharge generation), relying on a rather general understanding of karst recharge processes. Sensitivity analysis revealed that the parameters of this simple model structure were identifiable at those regions where enough observations of evaporation and soil moisture were available. The estimates of karstic groundwater recharge at other regions with no observation of evaporation and soil moisture were less precise, as revealed by the uncertainty analysis.

Synthesis

The increase in computational capacities experienced over the last three decades has allowed the development of increasingly complex and process-based hydrological model structures. Likewise, improved tools for automatic model calibration and model evaluation have been developed showing that model parameters are often not uniquely identifiable, therefore limiting the prediction performance of hydrological simulation models. This is due to limited data availability, uncertainties in model input data and observation data, as well as model structure uncertainties. However, these model evaluation tools also allow us to adapt the complexity of hydrological models to the available data and to estimate the value of additional information to increase the process representation within these models and to obtain more robust and reliable predictions.

This paper has provided an overview of the methods available for automatic model calibration and model evaluation for karst system modelling, as well as some experience in applying these methods to three case studies at variable scales and with varying data availability. Even though this review is based on subjective and personal experience and is far from complete, some general recommendations and a workflow for future karst model applications are provided (similar to Beven 2001).

(1) Review all prior knowledge about the system to be modelled and collect all available data. For successful calibration, the selected model structure should be based on a realistic conceptual model of the system and should include all the relevant processes that previous research has found to be important. The input data (e.g. precipitation and potential evaporation) has to be complete and continuous for the entire period of observation. Observations for parameter estimations do not require continuous records, but they should be dense enough to capture the hydrological dynamics of the processes to be modelled. In addition, some of the input data should be reserved for model warm-up and some of the observations should be reserved for model evaluation.

(2) Choose the calibration procedure. Some possible algorithms for automatic calibration have been mentioned here, but a much wider range of methods is now available. All will require the definition of model parameter ranges that should reflect prior information about the system. Alternatively, parameter ranges can be derived from previous applications of the chosen model. In some cases, it may be possible to confine the model parameters by pre-analysis, e.g. recession analysis.

(3) Choose the error function. The calibration procedure will require an error function to be minimized. Some examples are provided in this paper, but any number that expresses the fit between the observations and simulations is possible. If multiple types of data are used for calibration (e.g. discharge and water quality data), chose a procedure to stepwise or simultaneously include the different types of observations in the calibration (e.g. a weighting scheme or a multi-criteria calibration approach).

(4) Test the performance of the model. First, it is necessary to test whether the model is generally able to obtain an acceptable fit of the observations and simulations. If the model fails, the selection of the chosen model structure has to be revised (back to step 1).

(5) Choose an approach to sensitivity analysis. As some of the methods are computationally demanding, the selected sensitivity analysis approach should depend on the calculation times of the model. Some review papers that provide recommendations for choosing the most appropriate approach are given in this paper and in the sensitivity analysis literature.

(6) Apply a sensitivity analysis. For consistency, choose the same parameter ranges as for the calibration. Evaluate whether the available observations contain enough information to identify all the model parameters. If a large number of insensitive parameters indicate that the information content of the observations is insufficient, then reduce the complexity of the model by structural simplifications or by excluding insensitive parameters from the parameter estimation and repeat the calibration (back to step 4) or include more observations – for instance, water quality data (back to step 3).

(7) Evaluate the plausibility and prediction performance of the model. Use previous studies and discussions with field experts to evaluate whether the calibrated and sensitive parameters reflect the conceptual model and prior understanding of the system. Use observations that were excluded from the calibration (see step 2) to test the prediction performance of the model (e.g. by a split-sample test). If the model shows a poor prediction performance or its parameters are inconsistent with the prior understanding of the system, go back to step 1.

(8) Apply the model for prediction. If the model passed all the previous steps, then there is a strong indication that it provides realistic and robust simulation, making it applicable for prediction.

It is necessary to acknowledge that this workflow is limited to the application of models that require calibration, i.e. it is meant to provide guidance for an inverse modelling problem. In some cases, the purpose of the model application may already define the model structure. For instance, if the spatial impact of pumping on the distribution of groundwater heads has to be assessed, then a distributed model has to be applied no matter how much information is available for parameter estimation. In such cases, the problem changes from an inverse problem to a forward modelling problem, for which model parameters are derived solely from prior information and expert knowledge. In such cases, computational limitations often limit the application of detailed uncertainty assessment schemes. However, a low number of manual variations of the model parameters that reflect the uncertainty of the prior information of the system (as, for instance, in Gunkel *et al.* 2015) may allow some preliminary estimate of the uncertainty of the simulation.

Many thanks to Jürgen Strub of the Chair of Hydrology, University of Freiburg, for designing Figures 1 and 3. Also thanks to Javier Martín Arias, Centre of Hydrogeology of the University of Málaga (ES), for his valuable recommendations to improve this paper.

References

ANDERSON, E.A. 2002. *Calibration of Conceptual Hydrologic Models for Use in River Forecasting.* Office of Hydrologic Development, US National Weather Service, Silver Springs, MD.

ANDREO, B., GOLDSCHEIDER, N. *ET AL.* 2006. Karst groundwater protection: first application of a pan-European approach to vulnerability, hazard and risk mapping in the Sierra de Líbar (southern Spain). *Science of the Total Environment*, **357**, 54–73.

BAKALOWICZ, M. 2005. Karst groundwater: a challenge for new resources. *Hydrogeology Journal*, **13**, 148–160.

BALDOCCHI, D., FALGE, E. *ET AL.* 2001. FLUXNET: a new tool to study the temporal and spatial variability of ecosystem-scale carbon dioxide, water vapor, and energy flux densities. *Bulletin of the American Meteorological Society*, **82**, 2415–2434.

BEVEN, K.J. 1989. Changing ideas in hydrology – the case of physically-based models. *Journal of Hydrology*, **105**, 157–172.

BEVEN, K.J. 2001. *Rainfall-Runoff Modelling: the Primer.* 1st edn. Wiley, Chichester.

BEVEN, K.J. 2006. A manifesto for the equifinality thesis. *Journal of Hydrology*, **320**, 18–36.

BEVEN, K.J. & BINLEY, A. 1992. The future of distributed models: model calibration and uncertainty prediction. *Hydrological Processes*, **6**, 279–298.

BIRK, S., GEYER, T., LIEDL, R. & SAUTER, M. 2005. Process-based interpretation of tracer tests in carbonate aquifers. *Ground Water*, **43**, 381–388.

BOYLE, D.P., GUPTA, H.V. & SOROOSHIAN, S. 2000. Toward improved calibration of hydrologic models: combining the strengths of manual and automatic methods. *Water Resources Research*, **36**, 3663–3674.

BUTSCHER, C. & HUGGENBERGER, P. 2008. Intrinsic vulnerability assessment in karst areas: a numerical modeling approach. *Water Resources Research*, **44**, W03408.

CHEN, Z., HARTMANN, A. & GOLDSCHEIDER, N. 2017. A new approach to evaluate spatiotemporal dynamics of controlling parameters in distributed environmental models. *Environmental Modelling & Software*, **87**, 1–16.

EUROPEAN COMMISSION 1995. *COST 65: Hydrogeological Aspects of Groundwater Protection in Karstic Areas. Final Report.* EUR 16547 EN (1995). European Commission, Brussels.

DORIGO, W.A., WAGNER, W. *ET AL.* 2011. The International Soil Moisture Network: a data hosting facility for global in situ soil moisture measurements. *Hydrology and Earth System Sciences*, **15**, 1675–1698.

DUAN, Q.Y., SOROOSHIAN, S. & GUPTA, H.V. 1994. Optimal use of the SCE-UA global optimization method for calibrating watershed models. *Journal of Hydrology*, **158**, 265–284.

DUNNE, S.M. 1999. Imposing constraints on parameter values of a conceptual hydrological model using baseflow response. *Hydrology and Earth System Sciences*, **3**, 271–284.

FORD, D.C. & WILLIAMS, P.W. 2007. *Karst Hydrogeology and Geomorphology.* Wiley, Chichester.

GEYER, T., BIRK, S., REIMANN, T., DÖRFLIGER, N. & SAUTER, M. 2013. Differentiated characterization of karst aquifers: some contributions. *Carbonates and Evaporites*, **28**, 41–46.

GHASEMIZADEH, R., HELLWEGER, F., BUTSCHER, C., PADILLA, I., VESPER, D., FIELD, M. & ALSHAWABKEH, A. 2012. Review: groundwater flow and transport modeling of karst aquifers, with particular reference to the North Coast Limestone aquifer system of Puerto Rico. *Hydrogeology Journal*, **20**, 1441–1461.

GOLDSCHEIDER, N. 2012. A holistic approach to groundwater protection and ecosystem services in karst terrains. *AQUA Mundi*, Am06046, 117–124.

GOLDSCHEIDER, N. & DREW, D. (eds) 2007. *Methods in Karst Hydrogeology.* I.A.H.: International Contributions to Hydrogeology, **26**. CRC Press, Boca Raton, FL.

GUNKEL, A., SHADEED, S., HARTMANN, A., WAGENER, T. & LANGE, J. 2015. Model signatures and aridity indices enhance the accuracy of water balance estimations in a data-scarce Eastern Mediterranean catchment. *Journal of Hydrology: Regional Studies*, **4**, 487–501.

GUPTA, H.V., SOROOSHIAN, S. & YAPO, P. 1998. Toward improved calibration of hydrologic models: multiple and noncommensurable measures of information. *Water Resources Research*, **34**, 751–763.

GUPTA, H.V., SOROOSHIAN, S. & YAPO, P.O. 1999. Status of automatic calibration for hydrologic models: comparison with multilevel expert calibration. *Journal of Hydrologic Engineering*, **4**, 135–143.

GUPTA, H.V., WAGENER, T. & LIU, Y. 2008. Reconciling theory with observations: elements of a diagnostic approach to model evaluation. *Hydrological Processes*, **22**, 3802–3813.

GUPTA, H.V., KLING, H., YILMAZ, K.K. & MARTINEZ, G.F. 2009. Decomposition of the mean squared error and NSE performance criteria: implications for improving hydrological modelling. *Journal of Hydrology*, **377**, 80–91.

HARTMANN, A. 2016. Putting the cat in the box: why our models should consider subsurface heterogeneity at all scales. *Wiley Interdisciplinary Reviews: Water*, **3**, 478–486.

HARTMANN, A. & BAKER, A. 2017. Modelling karst vadose zone hydrology and its relevance for paleoclimate reconstruction. *Earth-Science Reviews*, **1**–54.

HARTMANN, A., LANGE, J., WEILER, M., ARBEL, Y. & GREENBAUM, N. 2012. A new approach to model the spatial and temporal variability of recharge to karst aquifers. *Hydrology and Earth System Sciences*, **16**, 2219–2231.

HARTMANN, A., BARBERÁ, J.A., LANGE, J., ANDREO, B. & WEILER, M. 2013*a*. Progress in the hydrologic simulation of time variant recharge areas of karst systems – exemplified at a karst spring in southern Spain. *Advances in Water Resources*, **54**, 149–160.

Hartmann, A., Wagener, T., Rimmer, A., Lange, J., Brielmann, H. & Weiler, M. 2013*b*. Testing the realism of model structures to identify karst system processes using water quality and quantity signatures. *Water Resources Research*, **49**, 3345–3358.

Hartmann, A., Goldscheider, N., Wagener, T., Lange, J. & Weiler, M. 2014*a*. Karst water resources in a changing world: review of hydrological modeling approaches. *Reviews of Geophysics*, **52**, 218–242.

Hartmann, A., Mudarra, M., Andreo, B., Marin, A., Wagener, T. & Lange, J. 2014*b*. Modeling spatio-temporal impacts of hydro-climatic extremes on groundwater recharge at a Mediterranean karst aquifer. *Water Resources Research*, **50**, 6507–6521.

Hartmann, A., Gleeson, T., Rosolem, R., Pianosi, F., Wada, Y. & Wagener, T. 2015. A large-scale simulation model to assess karstic groundwater recharge over Europe and the Mediterranean. *Geoscientific Model Development*, **8**, 1729–1746.

Hartmann, A., Kobler, J., Kralik, M., Dirnböck, T., Humer, F. & Weiler, M. 2016. Model-aided quantification of dissolved carbon and nitrogen release after windthrow disturbance in an Austrian karst system. *Biogeosciences*, **13**, 159–174.

Hornberger, G.M. & Spear, R.C. 1981. An approach to the preliminary analysis of environmental systems. *Journal of Environmental Management*, **12**, 7–12.

Kiraly, L. 2003. Karstification and groundwater flow. *Speleogenesis and Evolution of Karst Aquifers*, **1**, 1–24.

Klemeš, V. 1986. Dilettantism in hydrology: transition or destiny. *Water Resources Research*, **22**, 177S–188S.

Kovacs, A. & Sauter, M. 2007. Modelling karst hydrodynamics. *In*: Goldscheider, N. & Drew, D. (eds) *Methods in Karst Hydrogeology*. CRC Press, Boca Raton, FL, 65–91.

Kuczera, G. & Mroczkowski, M. 1998. Assessment of hydrologic parameter uncertainty and the worth of multiresponse data. *Water Resources Research*, **34**, 1481–1489.

Ladouche, B., Marechal, J.-C. & Dorfliger, N. 2014. Semi-distributed lumped model of a karst system under active management. *Journal of Hydrology*, **509**, 215–230.

Liedl, R., Sauter, M., Hückenhaus, D., Clemens, T. & Teutsch, G. 2003. Simulation of the development of karst aquifers using a coupled continuum pipe flow model. *Water Resources Research*, **39**, 1057.

Madsen, H., Wilson, G. & Ammentorp, H.C. 2002. Comparison of different automated strategies for calibration of rainfall-runoff models. *Journal of Hydrology*, **261**, 48–59.

Nash, J.E. & Sutcliffe, J.V. 1970. River flow forecasting through conceptual models. Part I: a discussion of principles. *Journal of Hydrology*, **10**, 282–290.

Perrin, C., Michel, C. & Andréassian, V. 2003. Improvement of a parsimonious model for streamflow simulation. *Journal of Hydrology*, **279**, 275–289.

Pianosi, F., Sarrazin, F. & Wagener, T. 2015. A Matlab toolbox for global sensitivity analysis. *Environmental Modelling and Software*, **70**, 80–85.

Pianosi, F., Beven, K., Freer, J., Hall, J.W., Rougier, J., Stephenson, D.B. & Wagener, T. 2016. Sensitivity analysis of environmental models: a systematic review with practical workflow. *Environmental Modelling & Software*, **79**, 214–232.

Reimann, T., Geyer, T., Shoemaker, W.B., Liedl, R. & Sauter, M. 2011. Effects of dynamically variable saturation and matrix-conduit coupling of flow in karst aquifers. *Water Resources Research*, **47**, W11503.

Rimmer, A. & Salingar, Y. 2006. Modelling precipitation-streamflow processes in karst basin: the case of the Jordan River sources, Israel. *Journal of Hydrology*, **331**, 524–542.

Rodell, M., Houser, P.R. *et al.* 2004. The global land data assimilation system. *Bulletin of the American Meteorological Society*, **85**, 381–394.

Rui, H. & Beaudoing, H. 2013. *README Document for Global Land Data Assimilation System Version 2 (GLDAS-2) Products*. GES DISC/HSL, NASA Goddard Space Center, Greenbelt, MD.

Saltelli, A., Ratto, M. *et al.* 2008. *Global Sensitivity Analysis: The Primer*. Wiley, Chichester.

Sarrazin, F., Pianosi, F. & Wagener, T. 2016. Global sensitivity analysis of environmental models: convergence and validation. *Environmental Modelling & Software*, **79**, 135–152.

Sauter, M., Kovács, A., Geyer, T. & Teutsch, G. 2006. Modellierung der Hydraulik von Karstgrundwasserleitern – eine Übersicht. *Grundwasser*, **3**, 143–156.

Schwerdtfeger, J., Hartmann, A. & Weiler, M. 2016. A tracer-based simulation approach to quantify seasonal dynamics of surface-groundwater interactions in the Pantanal wetland. *Hydrological Processes*, **30**, 2590–2602, https://doi.org/10.1002/hyp.10904

Seibert, J. & McDonnell, J.J. 2002. On the dialog between experimentalist and modeler in catchment hydrology: use of soft data for multicriteria model calibration. *Water Resources Research*, **38**, 1241.

Son, K. & Sivapalan, M. 2007. Improving model structure and reducing parameter uncertainty in conceptual water balance models through the use of auxiliary data. *Water Resources Research*, **43**, W01415.

Tritz, S., Guinot, V & Jourde, H. 2011. Modelling the behaviour of a karst system catchment using non-linear hysteretic conceptual model. *Journal of Hydrology*, **397**, 250–262.

Vrugt, J.A., Gupta, H.V., Bouten, W. & Sorooshian, S. 2003. A shuffled complex evolution metropolis algorithm for optimization and uncertainty assessment of hydrologic model parameters. *Water Resources Research*, **39**, 1201.

Vrugt, J.A., Stauffer, P.H., Wöhling, T., Robinson, B.A. & Vesselinov, V.V. 2009. Inverse modeling of subsurface flow and transport properties: a review with new developments. *Vadose Zone Journal*, **7**, 843–864.

Wagener, T., McIntyre, N., Lees, M.J., Wheater, H.S. & Gupta, H.V. 2003. Towards reduced uncertainty in conceptual rainfall-runoff modelling: dynamic identifiability analysis. *Hydrological Processes*, **17**, 455–476.

Wagener, T., Wheater, H.S. & Gupta, H.V. 2004. *Rainfall–Runoff Modelling in Gauged and Ungauged Catchments*. Imperial College Press, London.

Wagener, T., Sivapalan, M., Troch, P. & Woods, R. 2007. Catchment classification and hydrologic similarity. *Geography Compass*, **1**, 901–931.

White, W.B. 2003. Conceptual models for karstic aquifers. *Speleogenesis*, **1**, 1–6.

White, W.B. 2007. A brief history of karst hydrogeology: contributions of the NSS. *Journal of Cave and Karst Studies*, **69**, 13–26.

Williams, P.W. 2008. The role of the epikarst in karst and cave hydrogeology: a review. *International Journal of Speleology*, **37**, 1–10.

Winter, T.C. 2001. The concept of hydrologic landscapes. *Journal of the American Water Resources Association*, **37**, 335–349.

Worthington, S.R.H., Davies, G.J. & Alexander, E.C. 2016. Enhancement of bedrock permeability by weathering. *Earth-Science Reviews*, **160**, 188–202.

Yilmaz, K.K., Gupta, H.V. & Wagener, T. 2008. A process-based diagnostic approach to model evaluation: application to the NWS distributed hydrologic model. *Water Resources Research*, **44**, W09417.

Young, P. 1983. The validity and credibility of models for badly defined systems BT. *In*: Beck, M.B. & van Straten, G. (eds) *Uncertainty and Forecasting of Water Quality*. Springer, Berlin, 69–98.

Geophysical observations and structural models of two shallow caves in gypsum/anhydrite-bearing rocks in Germany

GEORG KAUFMANN* & DOUCHKO ROMANOV

Institute of Geological Sciences, Geophysics Section, Freie Universität Berlin, Malteserstrasse 74-100, Haus D, 12249 Berlin, Germany

**Correspondence: georg.kaufmann@fu-berlin.de*

Abstract: The development of subsurface voids and cavities in soluble rocks is controlled by the hydrological and chemical processes in the host rock. Water (enriched with carbon dioxide) percolates through fractures and bedding partings of the host rock and removes material from the rock surface. As this enlargement is a highly heterogeneous process, only some fractures and bedding partings become significantly enlarged, evolving towards larger voids and caves. The size of the enlarged voids, often reaching the metre scale, can result in mechanically unstable structures, which, when close to the surface, are prone to collapse and thus are a hazard to infrastructure. We explored two caves in the anhydrite host rock of the Permian Zechstein sequences in northern Germany using geophysical measurements: the Kalkberghöhle close to Bad Segeberg (Hamburg region) and the Jettenhöhle close to Osterode (Harz region). Based on the results of gravity and electrical measurements, we were able to identify the cave voids and to characterize the local geological setting. Using these indirect geophysical observations, we deduced a structural model for both cave sites by numerical modelling. Our structural models were successfully calibrated against the Bouguer gravity data.

A wide variety of soluble host rocks can be found in different parts of Germany (Fig. 1). The soluble host rocks range from Devonian limestones in the west to Jurassic limestones in the south. The evaporitic sequences of the Permian Zechstein period characterize the landscape in central and northern Germany, either as outcrops exposed through the tectonic uplift of the southern Harz Mountains, or buried at *c.* 4–6 km depth beneath the entire northern German basin. In the latter area, localized salt diapirism has caused the uplift of soluble rocks, bringing them closer to the present day surface.

The soluble host rocks can be dissolved by water (enriched with carbon dioxide) percolating through the rock, removing material from fractures and bedding planes and later enlarging these flow paths (Ford & Williams 2007). With time, the permeability in the host rock increases substantially at the local scale and creates the larger voids and caves typical of subsurface karst features in the soluble host rock (Palmer 2007; Gutiérrez 2010; De Waele *et al.* 2011; Gutiérrez *et al.* 2014).

Geophysical methods can be used to detect these subsurface voids using indirect measurements. Most of these geophysical methods distinguish between the material properties (e.g. density, electrical resistivity and electrical permittivity) of the void and its infill and the significantly different material properties of the surrounding host rock. This contrast in material properties can be detected using specific geophysical techniques (e.g. Butler 1984; El-Qady *et al.* 2005; Dobecki & Upchurch 2006; Bechtel *et al.* 2007; Mochales *et al.* 2008; Parise & Lollino 2011; Margiotta *et al.* 2012, 2016; Carbonel *et al.* 2013, 2014; Kaufmann 2014; Kaufmann & Romanov 2016).

Different geophysical methods provide complementary information about the subsurface voids.

(1) Air- and water-filled voids have a much lower density (0 and 1000 kg m^{-3} for air and water) than the host rock (2200 and 2900 kg m^{-3} for gypsum and anhydrite, respectively). The sediment infill of a cavity also often has a lower density than the host rock. Gravity measurements can be used to detect the density differences between the low densities of the voids and the higher densities of the host rock and thus the likely location of subsurface voids.
(2) The electrical resistivity of the host rock is mainly determined by the fluids circulating in the void spaces of the rock. Thus the bulk electrical resistivity of the rock is mainly determined by the amount of fractures and their interconnections and infill. Although limestone, dolomite and anhydrite have a high resistivity (*c.* 1000–2000 Ω m; Telford *et al.* 2012), gypsum is often highly fractured close to the surface and provides interconnected pathways for circulating water and

From: Parise, M., Gabrovsek, F., Kaufmann, G. & Ravbar, N. (eds) 2018. *Advances in Karst Research: Theory, Fieldwork and Applications*. Geological Society, London, Special Publications, **466**, 341–357.
First published online December 6, 2017, https://doi.org/10.1144/SP466.13

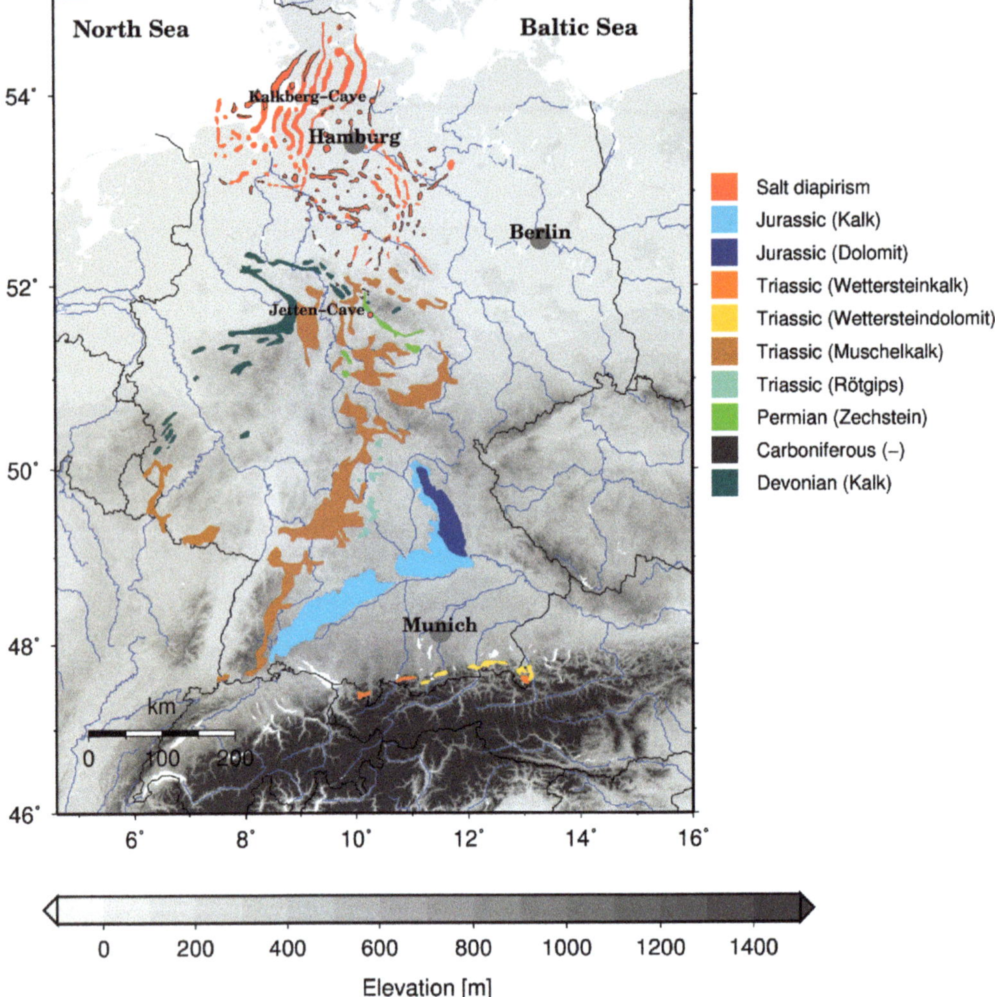

Fig. 1. Map of Germany showing outcrops of soluble rocks by geological epoch and the occurrence of salt diapirs in the northern German basin. The locations of major cities and the two studied caves are shown.

thus shows a lower resistivity. Cavities in the host rock represent either highly resistive areas if they are air-filled, or a lower resistivity if they are water-filled. A sediment infill can be highly resistive if it is dry or have a low resistivity if it is wet. Thus electrical resistivity imaging (ERI) mainly provides information about the amount of water in the subsurface.

(3) Locations where groundwater is accessible – such as sinks, resurgences, streams and lakes – in a cave can be sampled directly with electrical conductivity measurements. These measurements only probe the water component of the host rock and thus provide a different picture of the distribution of electrical resistivity from ERI. When combined with ERI, electrical conductivity measurements can distinguish the host rock matrix from water-filled fractures and bedding partings.

(4) Groundwater percolating through the unsaturated zone can drag excess electrical charges with it. These charges will induce an electrical potential difference, which can be mapped using the self-potential method. The groundwater-induced self-potential is often in the range of a few tens of millivolts.

(5) Ground-penetrating radar can reveal the subsurface structure from the reflection of electromagnetic waves.

The application of different geophysical methods provides a broad indirect picture of the voids in soluble host rocks and can help to unravel the strong preferential distribution of voids and cavities enlarged by dissolution in the subsurface. These indirect observations can then be used to infer the void geometry by forward and inverse modelling of the geophysical measurements.

We present here the results from geophysical surveys above two caves in Permian Zechstein rocks in Germany (Fig. 1). The Kalkberghöhle cave is located in the Hauptanhydrit caprock of a salt diapir in northern Germany, whereas the Jettenhöhle cave developed in the Hauptanhydrit formation exposed along the southern Harz Mountains in Germany. We aimed to identify both cave voids and, in some areas, the local stratigraphy from the geophysical signatures (gravity, ERI and self-potential measurements). We present structural models of the subsurface based on our geophysical results.

Geophysical methods

This section introduces the geophysical methods used to detect the subsurface structures in the areas of interest.

Gravity

We carried out the gravimetric survey using Lacoste-Romberg type D gravimeters with a precision estimated to be better than 0.03 mGal from repeated readings. The survey station coordinates were recorded with a hand-held global positioning system monitor to about 1 m accuracy, whereas the elevations were determined to 3 cm accuracy by levelling with a levelling rod. The raw data were processed with GRAViMAG software developed at the Geophysics Department, Freie Universität Berlin to derive the Bouguer gravity, Δg_b (mGal). We used four processing steps to derive the Bouguer gravity data from the raw measurements: (1) repeated base station measurements roughly every 2–3 h to monitor instrument drift; (2) correction of Earth tides based on the software package Eterna (Wenzel 1996); (3) linking the relative gravity measurements into the regional gravity network through a known station with absolute gravity; and (4) using latitude, free-air and Bouguer corrections, including a topographic Bouguer correction (where needed) derived from either the Shuttle Radar Topography Mission digital elevation model (Jarvis *et al.* 2008) or a local digital elevation model when a higher accuracy was needed. The densities ranged from 0 kg m^{-3} for the air-filled cave passages to *c.* 1500 kg m^{-3} for the cave sediments, and 2200 and 2900 kg m^{-3} for gypsum and anhydrite, respectively, as the host rock (e.g. Telford *et al.* 2012).

Electrical resistivity imaging

ERI was carried out along the profiles with a GeoTom MK8E1000 instrument and 25–75 steel electrodes, mostly in the Wenner and Schlumberger set-up. We also tested dipole–dipole configurations, often better suited to detecting lateral contrasts in electrical resistivity, but due to the poor signal-to-noise ratio of this set-up we decided not to use these measurements. The raw data were processed with the Res2DInv software package (Loke & Barker 1995, 1996), applying robust inversion methods, to derive the electrical resistivity, ρ_e (Ω m). Poor datum points were removed with the help of the software prior to inversion by identifying the gross outliers along each horizontal pseudo-depth section. Coordinates for the electrodes were taken with a hand-held global positioning system monitor and the elevation either from levelling or from the digital elevation map. Inversions of the profiles were carried out, including the topographic elevation. The electrical resistivity of the gypsum and anhydrite was *c.* 400–1000 Ω m, depending on the infill of fissures, whereas the air-filled cave voids had a resistivity >4000 Ω m (e.g. Telford *et al.* 2012).

Electrical conductivity

A hand-held electrical conductivity meter (ADWA Instruments) was used to measure the electrical conductivity, σ_e (S m^{-1}), in open water bodies (cave lakes, creeks and springs). The electrical conductivity values were calibrated to a temperature of 25°C.

Self-potential

Self-potential surveying was carried out with laboratory-made non-polarizable copper–copper sulphate electrodes and a Voltcraft multimeter. One electrode served as the base from which the potential difference, ΔU (mV), was mapped across the survey area. The base electrode was located outside the sampling profiles. A hole was dug for all the electrodes to achieve good coupling conditions between the electrodes and the soil. Measurements were carried out within a time frame of 2 h and therefore the diurnal variation was small. We expected streaming potentials in the range of tens of millivolts from subsurface water channelled into the wider fractures opened by dissolution, thus inducing preferential groundwater flow towards a base level.

Cave sites

This section presents the results of the geophysical surveys above the two cave sites (Fig. 1). Both sites are located in Hauptanhydrit, the anhydrite

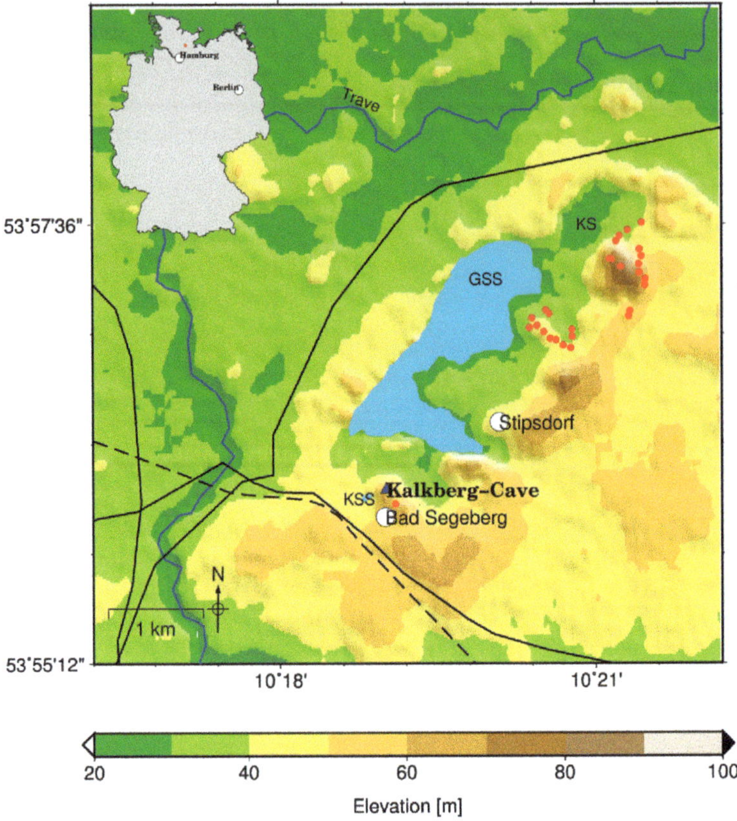

Fig. 2. Map of Bad Segeberg showing the study site (red square) and the Kalkberg cave (blue triangle), cities (white dots), sinkholes (red circles), rivers (light blue lines), roads (black solid lines) and railways (black dashed lines). GSS, Grosser Segeberger See; KSS, Kleiner Segeberger See; KS, Klüthsee. The inset shows the location (red square, north of Hamburg) of the figure within Germany.

part of the z3 (Leine) sequence of the Permian Zechstein deposits in Germany.

Kalkberghöhle

The Kalkberghöhle is located close to the city of Bad Segeberg (Fig. 2) in the northern German basin, with elevations *c.* 10–40 m above sea-level (a.s.l.). Salt diapirism since the Keuper period has caused uplift of the deeply buried soluble Zechstein rocks closer to the surface, with the Segeberg salt diapir responsible for exposing the Hauptanhydrit along the north-trending axis of the diapir (e.g. Kardel *et al.* 2009; Ipsen & Mucke 2011). On the surface, the lakes Grosser Segeberger See (28.9 m a.s.l.), Kleiner Segeberger See (37.9 m a.s.l.) and Klüthsee (*c.* 28 m a.s.l.) mark the axis of the salt diapir as subrosion depressions. Three major fault zones are mapped perpendicular to the long axis of the salt diapir and the two northern fault zones are lined by a series of sinkholes (Ross 1993). The Kalkberg, a hill with a former height of 114 m a.s.l., was topped by a castle in early Medieval times (e.g. Sparr 1997; Ipsen & Mucke 2011). Its name, including the German word Kalk (limestone), is misleading because of the exposed anhydrite, but stems from the Medieval mining use of Kalk as a term to describe soluble rocks.

Drinking water for the Medieval castle was recovered from a well, with its opening originally at *c.* 110 m a.s.l. After the castle had been destroyed, open-pit mining of the anhydrite and its gypsum shell substantially lowered height of the hill, with its highest peak today at 90 m a.s.l. The main quarry has been re-cultivated and rebuilt to host an open-air theatre (Fig. 3). In 1807, the former well was excavated again and a drillhole was lowered into the Hauptanhydrit from the bottom of the well at 26.4 m a.s.l. in hope of finding salt (Fig. 3d). The drillhole ended at −62 m a.s.l., still within the Hauptanhydrit. Two further drillholes along the eastern shores of the Grosser Segeberger See then reached

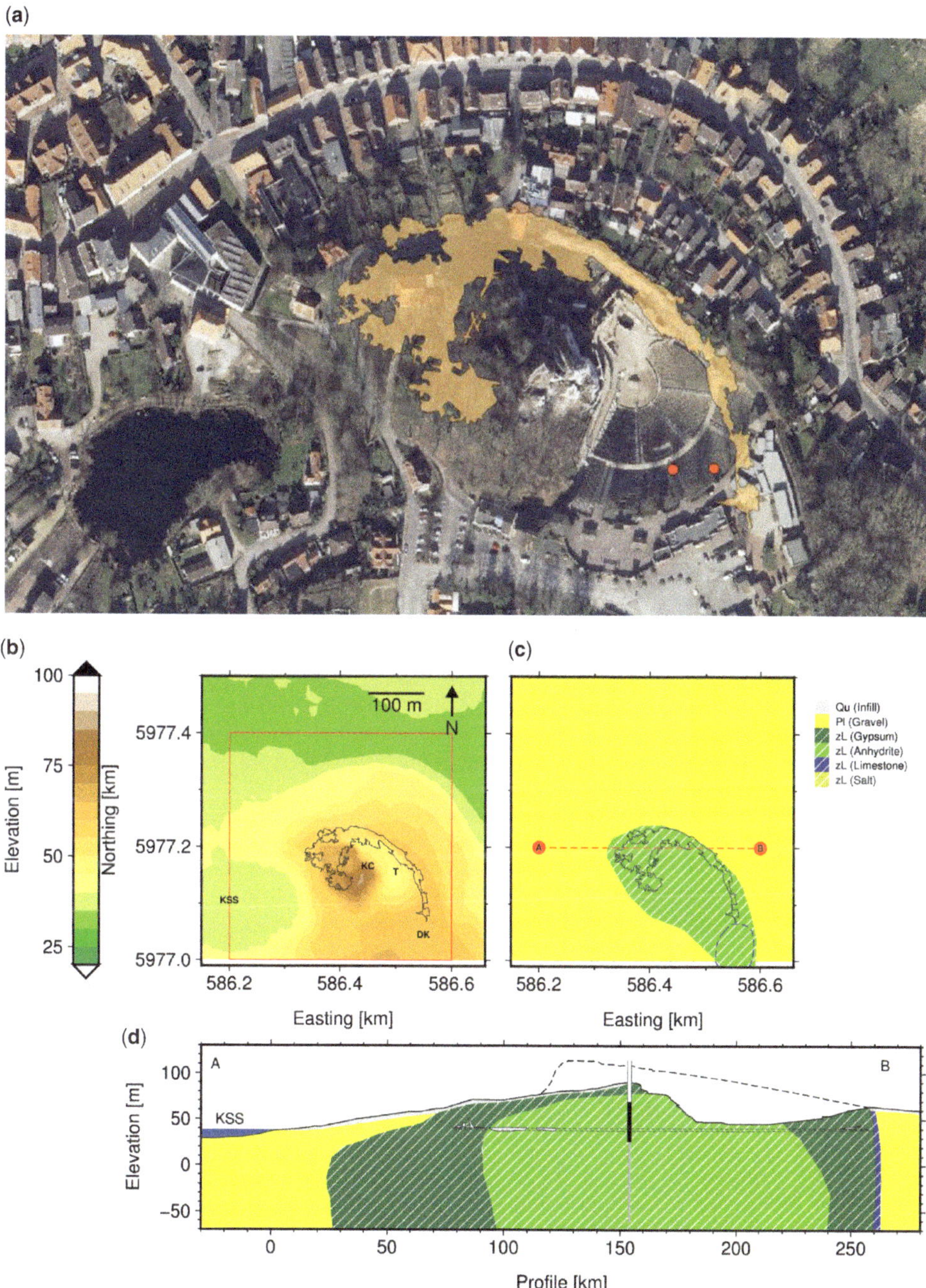

Fig. 3. Kalkberghöhle, Bad Segeberg, Schleswig-Holstein, Germany. (**a**) Map of the Kalkberghöhle cave superimposed onto the digital ortho-photograph of the area. The two red dots mark the former mining shafts. (**b**) Map view of the study area. KC, Kalkberghöhle cave; KSS, Kleiner Segeberger See; T, open-air theatre; DK, Damloser Kuhle sinkhole. The survey area is shown by the red rectangle. (**c**) Geological map showing the lithological units. The location of the geological cross-section in part (d) is marked by the red dashed line. (**d**) Geological cross-section along the profile shown in part (c). The former land surface is shown as a dashed line, the well removed in part by mining is shown in white and existing part of the well is shown in black. The drillholes lowered from the bottom of the well are shown in grey. The blue area is the KSS. The cave is shown as a white area, with the cave level indicated by dashed lines.

the Stassfurt salt at about 103 and 149 m depth, respectively. With the idea of starting salt mining, two shafts were lowered in the main quarry (now the open-air theatre), reaching depths of 88 and 116 m. However, both shafts flooded due to unmanageable water inflow from the surrounding rock (Sparr 1997).

A substantial maze cave, the Kalkberghöhle, with about 2 km of passages, has been explored beneath the remnants of the Kalkberg (Fricke 1990). Detailed geological mapping of the cave (Ross 1990) revealed that although the central parts of the Kalkberghöhle are located in the Hauptanhydrit and thus fairly stable, the outer parts of the cave (in the west and SE) are developed in the gypsum part and are fairly unstable and prone to roof collapse. The passage towards the SE, which extends underneath the open-air theatre, is prone to collapse and has been secured with concrete pillars (Meier 2003; Konietzky *et al.* 2007; Mucke *et al.* 2008). Along the northwestern part of the known cave, the overburden above the cave is very thin, sometimes only a few metres. Residential buildings in this area are thus potentially in danger of collapse. An old sinkhole called Damloser Kuhle in the direction of the southeastern cave passages indicates a probable former extension of the cave system.

The evolution of the cave has been discussed in Kupetz & Brust (2008) and Ipsen & Mucke (2011), with its early evolution dating back to the last interglacial (*c.* 125 ka BP) as an active water-table cave along the 34.0–37.5 m a.s.l. level with autogenic recharge. When the palaeoclimate became colder and the Fennoscandian ice sheet approached the region, the supply of water stopped and the maze of enlarged cave passages was filled with glaciofluvial sediments. The infill blocked most of the formerly active passages and, since the Last Glacial Maximum (*c.* 21 ka BP), the reactivated infiltration of surface water has been efficiently blocked by sediments. Cave enlargement progressed under almost stagnant conditions, with typical solution forms (flat ceilings and steeply dipping side walls) developed along the 38–50 m a.s.l. level.

Gravity survey. Figure 4a projects the cave map (white outline) onto the topography and shows the locations of the three gravity profiles (red dots). The summit of the Kalkberg and the open-air theatre are marked as topographic features. The old sinkhole Damloser Kuhle is located south of the mapped cave passages. Some of the larger rooms – Zentralhalle, Barbarossahalle and Säulenhalle – are located in the transition zone from mechanically stable anhydrite to mechanically unstable gypsum, whereas some of the other rooms, such as the Gonzohalle, are entirely in the unstable gypsum.

Figure 4b shows the Bouguer gravity map (Δg_b). A value of 2200 kg m^{-3} has been chosen as the reference density and represents the average rock density for gypsum. With this choice, Bouguer gravity values below zero represent mass deficits relative to the gypsum rock and are thus possible voids and cave rooms. Note that the signal from the less dense salt is constant over the size of our working area as a result of the large extent of the salt dome

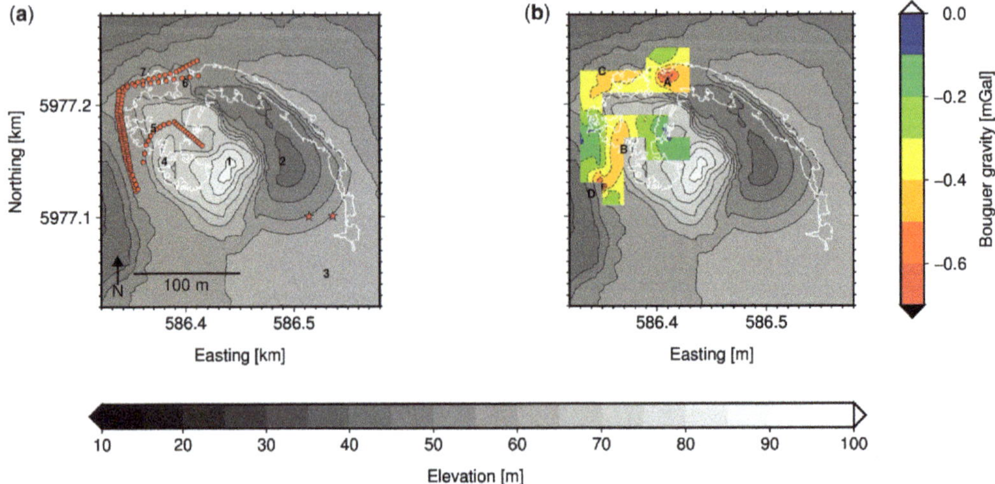

Fig. 4. Kalkberghöhle, Bad Segeberg, Schleswig-Holstein, Germany. (**a**) Location map showing gravity stations (red dots) and two former mining shafts (red stars). 1, Summit of the Kalkberg; 2, open-air theatre; 3, the old sinkhole Damloser Kuhl; 4, Zentralhalle; 5, Barbarossahalle; 6, Säulenhalle; 7, Gonzohalle. (**b**) Bouguer gravity map. A, Säulenhalle; B, Zentralhalle; C, Gonzohalle; D, artificial manhole on the road.

and is accounted for as a regional signal. The Bouguer gravity values are negative (−0.1 to −0.6 mGal) above the entire cave, indicating the karstified gypsum present underneath the gravity profiles. Three locations with a strong focused negative Bouguer gravity signal down to −0.6 mGal correspond to the rooms Säulenhalle, Zentralhalle and Gonzohalle. The Zentralhalle, however, has only been touched in its northern parts by the profile. Clearly the larger rooms dominate the Bouguer gravity signal, whereas the main cave level provides the broad negative background signal. The large gravity minimum in the far southwestern corner corresponds to the void of an artificial manhole on the road.

Jettenhöhle

The Jettenhöhle is located along the southern rim of the Harz Mountains (Fig. 5). The asymmetrical uplift of the Harz Mountains caused tilting of the soluble rocks of the Permian Zechstein period, which are exposed along the entire southern extension of the Harz Mountains and form a zone of intense karstification a few kilometres wide, with numerous caves, sinkholes and karst resurgences.

The area of interest, close to the city of Osterode and covering the karst regions Hainholz and Bollerkopf (Fig. 6), is characterized by hilly relief at elevations between 175 and 350 m a.s.l. Surface flow is only present in areas where the soluble Zechstein rocks are covered by insoluble marls and clays. Hydrologically, the region is subdivided into two catchments (Brandt *et al.* 1976). In the northwestern part, the Bollerkopfbach sinks into the active cave Marthahöhle. The Heiligentalbach, the remnant of an old valley coming from the Harz Mountains, with its headwaters lost by erosion, is a dry valley with an active surface creek still present in only a small portion, the continuation of which is unknown. The main drainage of the northwestern part is carried via the Schurfbach towards the Hackenbach creek and drains to the River Oder. In the southeastern

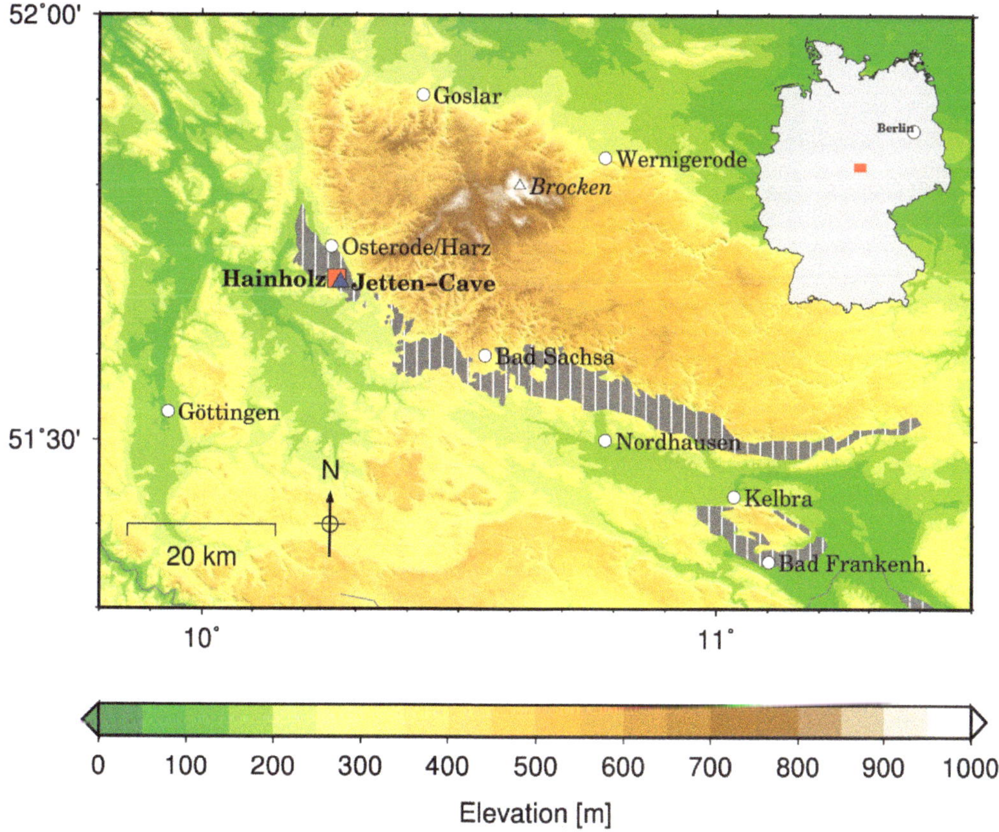

Fig. 5. Topographic map of the Harz Mountains showing the study site (red square) and cave (blue triangle), cities, geographical locations and outcrops of soluble Zechstein rocks in Lower Saxony, Thuringia and Saxony-Anhalt (grey hashed area). The inset shows the location (red square) of the figure within Germany.

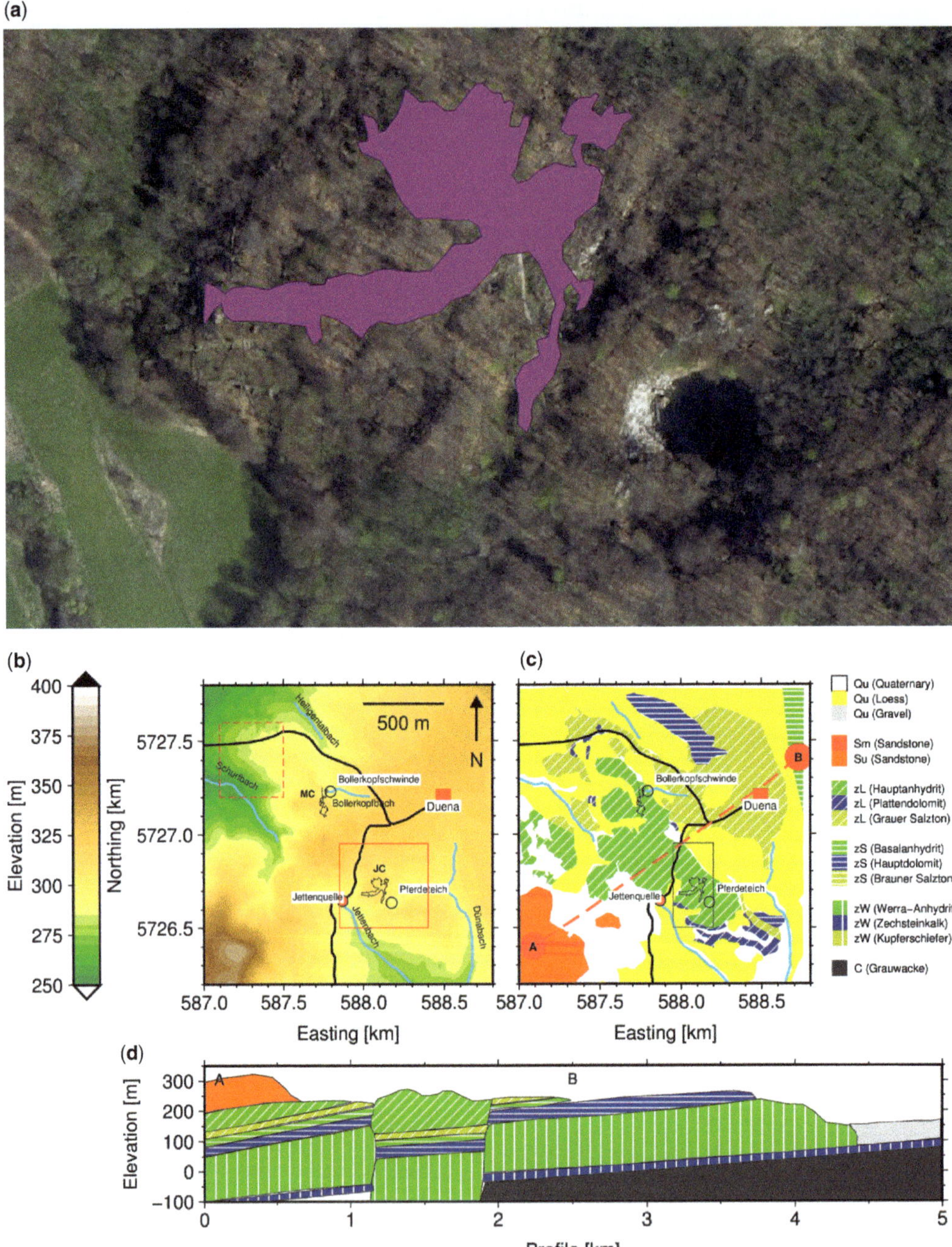

Fig. 6. Jettenhöhle, Osterode, Lower Saxony, Germany. (**a**) Map of the cave Jettenhöhle superimposed onto the digital ortho-photograph of the area. (**b**) Map view showing roads (solid black lines), rivers (solid light blue lines), sinkholes (open circles), springs (red closed circles), villages (red squares) and caves. JC, Jettencave; MC, Marthacave. The survey area is shown as a red solid rectangle. (**c**) Geological map showing lithological units. The location of the geological cross-section in part (d) is marked by the red dashed line. (**d**) Geological cross-section along profile shown in part (c).

part, the Düna Plateau is drained via the Dünabach creek, a subsurface drainage through the Jettenhöhle towards the Jettenbach creek and the Pferdeteich sinkhole, also draining towards the Jettenbach. The water from the Jettenbach flows to the River Sieber and also reaches the River Oder.

Three Zechstein sequences are exposed from north to south in the Hainholz/Bollerkopf area (Fig. 6) (e.g. Herrmann 1969, 1981*a*, *b*; Vladi 1972; Brandt *et al.* 1976; Jordan 1981; Kempe 1997, 2008). In the north, the sequence starts with the greywacke from the Paleozoic Harz Mountains. The Zechstein sequence follows, with z1/zW (Werra-Anhydrit) on top of the Paleozoic rocks, followed by z2/zS (Hauptdolomit, Basalanhydrit) and z3/zL (Grauer Salzton, Plattendolomit, Hauptanhydrit). Although the Werra-Anhydrit forms a prominent step, the zS sequence is barely visible. However, a tectonic graben structure has preserved large parts of the Hauptanhydrit of the zL sequence, which forms an extended and extremely karstified area. To the south, the Zechstein sequences are capped by Triassic Buntsandstein formations (mainly sandstone).

Our survey area was located in the graben structure with the zL sequence dominating the landscape. The Hauptanhydrit, with its shallow parts converted to gypsum from the surface towards the fluctuating water-table, is an extensively karstified area with deep karren, pits, numerous partially water-filled sinkholes and no significant surface runoff. The Plattendolomit exposed along the southwestern part of the Hainholz is characterized by a less rugged surface. Lithologically, the Plattendolomit is located below the Werra-Anhydrit, but the graben structure caused partial exposure due to bending of the rock layers.

The cave Jettenhöhle is located in the Hauptanhydrit (Fig. 7a). This cave, *c.* 748 m in total length (Kempe 1997, 2008), is developed in the gypsified

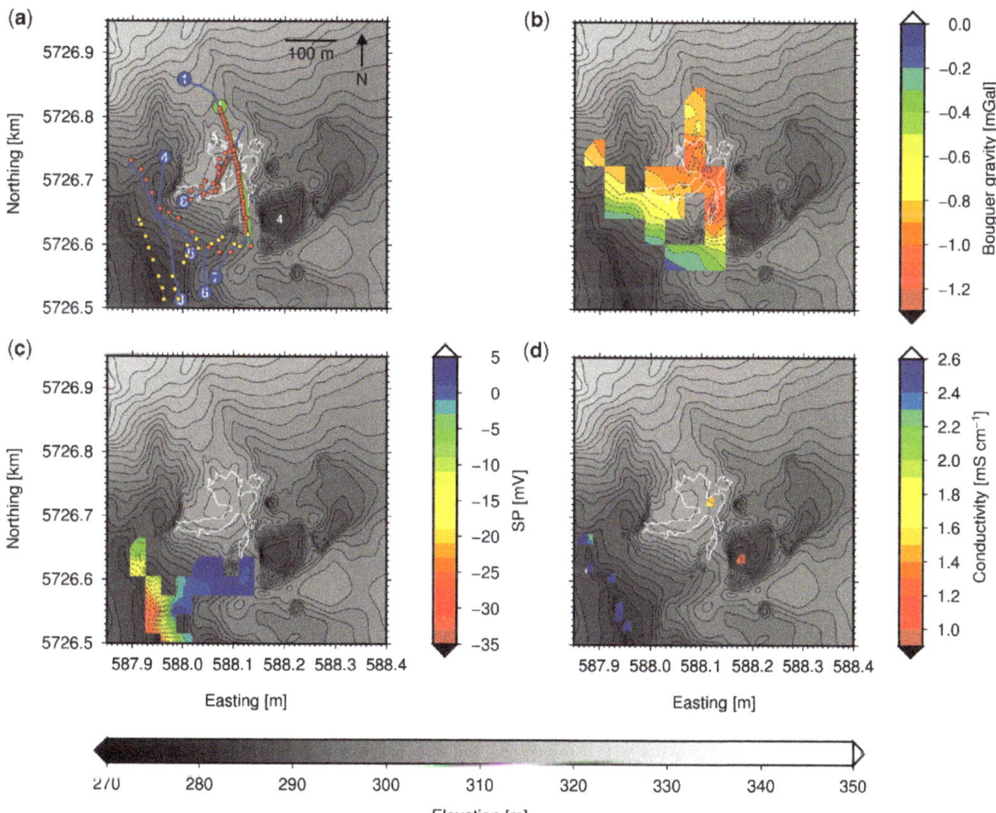

Fig. 7. Jettenhöhle, Osterode, Lower Saxony, Germany. (**a**) Location map with gravity stations (red dots), self-potential stations (yellow dots), electrical resistivity imaging profiles (blue lines with first electrode marked with profile number) and ground-penetration radar profile (green line). 1, Romarhalle; 2, Kreuzdom; 3, Jettenstube; 4, Pferdeteich collapse sinkhole; 5, Hirschzungenerdfall collapse sinkhole. (**b**) Bouguer gravity map. (**c**) Self-potential map. (**d**) Electrical conductivity map of open water.

part of the Hauptanhydrit. The cave consists of a large tunnel (Romarhalle), *c.* 10–20 m wide and several metres high, its deepest point being the large room Kreuzdom, 30 × 30 m wide and *c.* 13 m below the entrance. Two small lakes cover the Kreuzdom, about 1–2 m deep, but fluctuations in the lake level are small (10–20 cm on average). From the Kreuzdom, a large block pile leads to a further room, the Jettenstube, mostly filled with breakdown deposits and containing small lakes. All the cave lakes are a result of insoluble residuals from the dissolution of the gypsum (e.g. Kempe 2008), which cover the bottom parts of the cave and inhibit the water from totally disappearing. To the north of this room, a prominent collapse sinkhole (Hirschzungenerdfall) marks a former continuation of the cave. The Jettenhöhle has only a thin surface cover and collapse sinkholes mark older parts of the cave along the western and eastern sides. The large collapse sinkhole Pferdeteich is located to the east of the cave and receives water from the north, which disappears into a sink along the southern rim of the Pferdeteich (Hauptanhydrit). The lake in the Pferdeteich fluctuates substantially, with variations in the lake level of up to 7 m during the year.

We continue to obtain results from our geophysical surveys above and around the Jettenhöhle cave. We follow a strategy that picks up the signal from the cave and then use the knowledge gained from the surveys above the cave to extend our interpretation beyond the known cave passages into the surroundings.

The locations of all measurements are shown in Figure 7a. Gravity measurements were taken directly above the cave and in its vicinity. The station distances varied between 5 and 30 m. ERI measurements were made along seven profiles, mostly in the Wenner and Schlumberger set-up. Profiles 1 and 8 were located above the Jettenhöhle to track air-filled cave voids in the gypsum rock, whereas profiles 3–7 were located between the collapse sinkhole Pferdeteich and the creek Jettenbach. Depending on the length of the ERI profiles, 25 or 50 steel electrodes were used, with an electrode spacing between 4 and 5 m. ERI profile 1 was also explored with ground-penetrating radar measurements to obtain additional information on the cave voids. Self-potential measurements between the Pferdeteich and Jettenbach were carried out to shed light on the subsurface drainage of the Pferdeteich sinkhole. These measurements were complemented by electrical conductivity measurements in the sinkhole Pferdeteich, the lake in the Kreuzdom of the Jettenhöhle, the spring Jettenquelle and the Jettenbach.

Gravity survey. Figure 7b shows the Bouguer gravity map (Δg_b). A value of 2800 kg m^{-3} was chosen as the reference density, which represents the lower bound of the rock density for both the anhydrite and the dolomite. With this choice, Bouguer gravity values below zero represent mass deficits relative to the anhydrite/dolomite rock. Bouguer gravity values are negative (−0.3 to −0.5 mGal) above the entire cave, reflecting the less dense gypsum rock ($CaSO_4 xH_2O$, density *c.* 2300 kg m^{-3}), which is a result of the hydration of the anhydrite ($CaSO_4$) within the groundwater zone. Cave rooms such as the Kreuzdom and Jettenstube are clearly visible by large negative Bouguer gravity values (−0.8 to −1.2 mGal). The large, sharply confined negative anomaly below −1.0 mGal is related to small breakdown passages in the south of the Kreuzdom, which are close to the surface and end on the side of a small collapse sinkhole (Kleine Jettenhöhle). Thus the cave voids have been clearly identified by gravity measurements. A second feature beyond the known cave limits is the regional trend in the Bouguer gravity data towards slightly positive values in the SW. The trend has a sharp gradient along the slopes of the valley of the creek Jettenbach. Comparing the Bouguer gravity data with the geological map (Fig. 6), we find a correlation of the gravity trend with the mapped outcrop of the Plattendolomit in the south. Our gravity data suggest that the denser dolomite extents even further north and limits the Werra-Anhydrit along the southern graben shoulder.

Electrical resistivity imaging. Figure 8 shows five ERI profiles above the cave Jettenhöhle. All the profiles have been carried out in the Wenner set-up, with 64 electrodes for profile 1 and 34 electrodes for profile 8. The electrical resistivity images ($\rho_{e,b}$), characterizing the bulk electrical resistivity of the rock were obtained from robust inversion with Res2DInv. They are both characterized by small root-mean-square values. Profile 1 is a 300 m long transect above the main cave rooms. Both rooms, the Jettenstube and the Kreuzdom, can be identified by resistivities >4000 Ω m, a value characteristic of air-filled voids. Although the Kreuzdom has an overburden of *c.* 10 m, the Jettenstube has been traversed at its northern end, where the overburden is already *c.* 15 m. The third high-resistivity anomaly, which is very pronounced, reflects the small passages in the southern part of the cave close to the small sinkhole. The gypsum rock itself is characterized by resistivities of *c.* 200–1000 Ω m. The low-resistivity anomaly (E) in the deeper section along the southern end of the profile, with values down to 20 Ω m, indicates flowing groundwater. Profile 8 is located above the main cave passages and the high-resistivity anomaly (D) at 15 m depth reflects the large cave passages Romarhalle and Jettenstube.

Profiles 3, 6 and 7 are all located south of the cave. In contrast with profiles 1 and 8, all three of these profiles are characterized by low resistivities

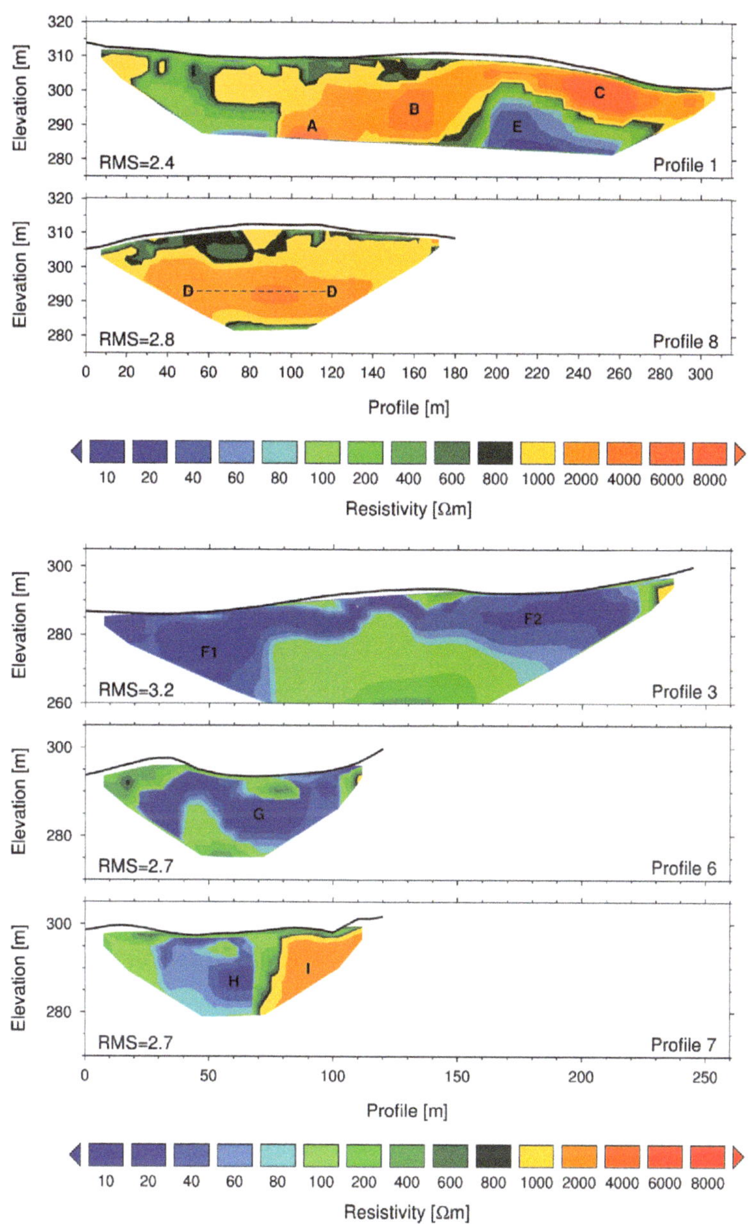

Fig. 8. Jettenhöhle, Osterode, Lower Saxony, Germany. Electrical resistivity imaging profiles 1, 8, 3, 6 and 7. A, Jettenstube; B, Kreuzdom; C, high-resistivity anomaly; D, large cave passages Romarhalle and Jettenstube; E, low-resistivity anomaly; F1, F2, G and H, low-resistivity anomalies; I, high-resistivity anomaly.

between 20 and 60 Ω m, indicating groundwater movement, and some patches with higher resistivities above 100 Ω m, possibly drier parts. The three low-resistivity parts (F1, G, H) are located along the same flow path and, together with anomaly E in profile 1, seem to map the subsurface drainage of the water from the collapse sinkhole Pferdeteich towards the creek Jettenbach. The sharp contrast between the low-resistivity (H) and the high-resistivity (I) parts in profile 7 marks the transition from gypsum to another rock type, which we interpret as Plattendolomit. We speculate that the Plattendolomit adjacent to the Werra-Anhydrit (converted to gypsum) is karstified, but to a lesser degree than the gypsum. Therefore the gypsum drains surface water efficiently towards a lower phreatic zone, whereas the Plattendolomit, with its lower hydraulic conductivity, keeps the water coming from the Pferdeteich and thus reflects phreatic conditions with low electrical conductivities.

Electrical conductivity. Electrical conductivity measurements were taken in several open water bodies using a hand-held conductivity meter. Conductivities are automatically referenced to a temperature of 25°C. The measured electrical conductivity (Fig. 7d) increases from the input point to the karst aquifer, from the Pferdeteich (1.0 mS cm^{-1}), the small lake inside the Jettenhöhle (1.5 mS cm^{-1}), towards the Jettenquelle spring (2.2 mS cm^{-1}). We can compare these electrical conductivity readings ($\sigma_{e,f}$) measured in open water locations with the electrical resistivities ($\rho_{e,b}$) obtained from the ERI measurements. The electrical resistivity for the water samples is around $\rho_{e,b} = 4.5–10\ \Omega$ m, which is comparable with the low-resistivity values found in ERI profiles 3, 6 and 7 (20–40 Ω m). This is already a clear hint that the gypsum aquifer is strongly karstified (well-developed secondary porosity).

Self-potential. Self-potential measurements were carried out with non-polarizable copper–copper sulphate electrodes and a multimeter along the small valley between the Pferdeteich collapse sinkhole and the local base level Jettenbach. The self-potential measurements were referenced to a base electrode located on the rim of the sinkhole in the vadose zone. The mapped self-potential measurements are shown in Figure 7c. The self-potential is around the zero reference value along the higher parts of the valley and decreases to negative values downstream, with a final strong gradient along the banks of the Jettenbach creek to values below −30 mV. Both the amplitude and the pattern are indicative of a self-potential induced by groundwater flow, thus it is likely that the water disappearing into the Pferdeteich sinkhole reappears along the Jettenbach creek.

Structural models

We assembled three-dimensional structural models of the subsurface for both cave sites based on our geophysical data and information about the local and regional geological setting. We built a three-dimensional structural model of the cave sites and their surroundings within the PREDICTOR software package (e.g. Kaufmann *et al.* 2012, 2015*a*, *b*). The PREDICTOR package uses a given digital topography, which is extended to depth with different lithological layers. Each layer can have different physical properties and can accommodate three-dimensional structures such as caves, voids or other objects. This numerical model is then used to predict different geophysical signals. The program basically performs the calculations in three parts: (1) the assembly of the model; (2) the solution of the governing equations; and (3) the prediction of geophysical signals. Details can be found in Kaufmann *et al.* (2015*a*).

The model is discretized in three spatial dimensions and then boundary conditions are set for the hydraulic problem (recharge as fixed flow, resurgences as fixed head). The groundwater equation is then solved as a transient, three-dimensional problem for both laminar and turbulent flow, thus appropriately describing the flow in porous, fractured rocks. To predict the gravity signal (used in this work), we used the discretized parallelepipedal elements to predict a Bouguer gravity signal for each block and then added all the block contributions to obtain a large-scale signal.

Kalkberghöhle

We used a digital elevation model with 5 m resolution for the Kalkberghöhle in Bad Segeberg in northern Germany. The subsurface structure was defined as gypsum. We neglected the anhydrite core of the structure, but as our gravity readings had only been taken above the gypsified part of the cave, our simplification was acceptable. The underlying salt diapir was a regional structure in our model, thus it affected the total gravity signal by the lower mass of the salt, but with no lateral variation on our scale. The structural model is shown in Figure 9a, with the cave Kalkberghöhle in dark red and the Bouguer gravity data as colour-coded squares on top.

The cave void has been discretized from the map of Fricke (1990). As the ceiling height of some of the cave rooms – such as the Zentralhalle, Barbarossahalle, Säulenhalle and Gonzohalle – has only been measured locally, we extended the room height to satisfy our gravity measurements.

The reason for the extension of the mapped cave passages towards the surface can be seen in Figure 9b, where we compare the observed and predicted Bouguer anomalies above the western part of the Kalkberghöhle. We can satisfactorily predict both the broad negative anomaly above the entire cave, representing the highly karstified gypsum, and the three local minima above the larger cave rooms, which seem to extend close to the surface, as our prediction suggests.

Jettenhöhle

We used a digital elevation model with 5 m resolution for the Hainholz area in the southern Harz Mountains and defined two lithological units, the Hauptanhydrit and the Plattendolomit below.

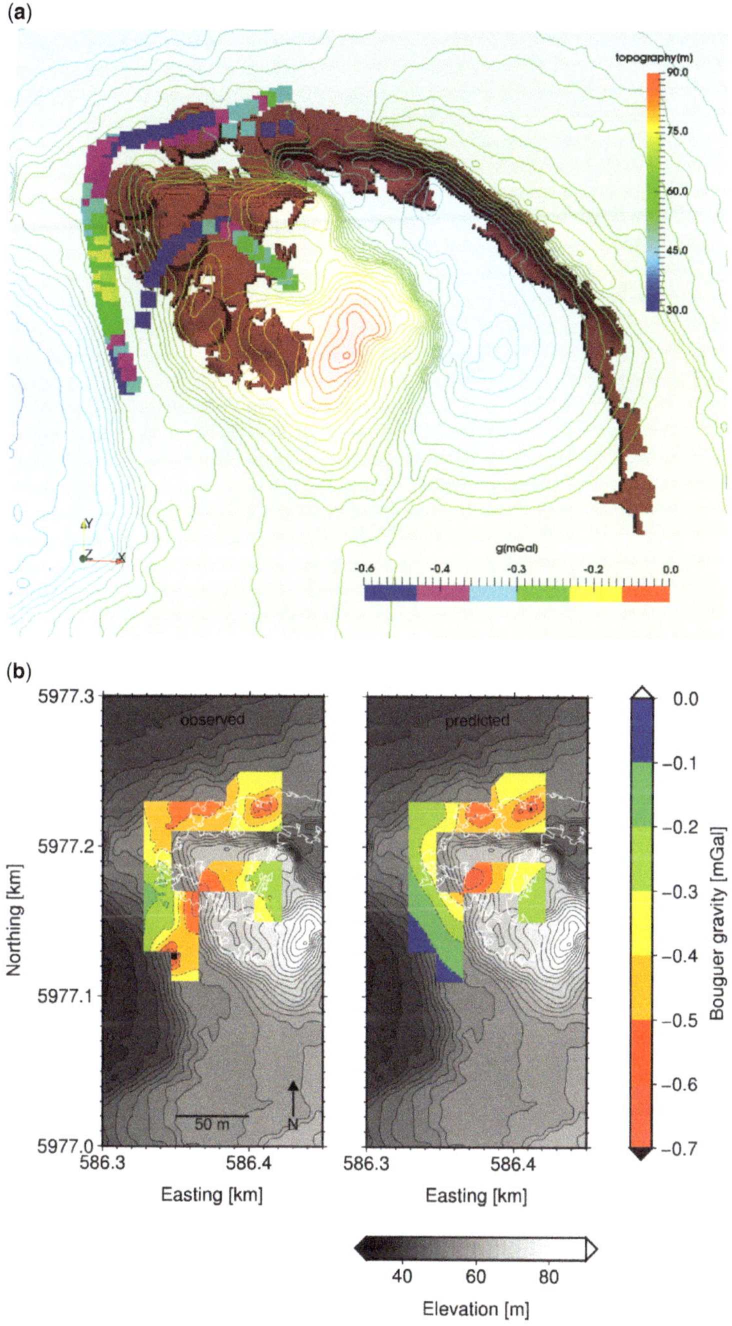

Fig. 9. Kalkberghöhle, Bad Segeberg, Schleswig-Holstein, Germany. (**a**) Topographic contour map showing the Kalkberghöhle cave as a dark red block, gravity profiles as map. (**b**) Observed and modelled Bouguer gravity anomaly over the area.

A structural model of this set-up is shown in Figure 10a. Here the Jettenhöhle is shown in dark red, the topography as colour-coded contours and the ERI profiles as cross-sections. The Pferdeteich sinkhole is the large depression at the front and the valley in the upper left is the Jettenbach creek.

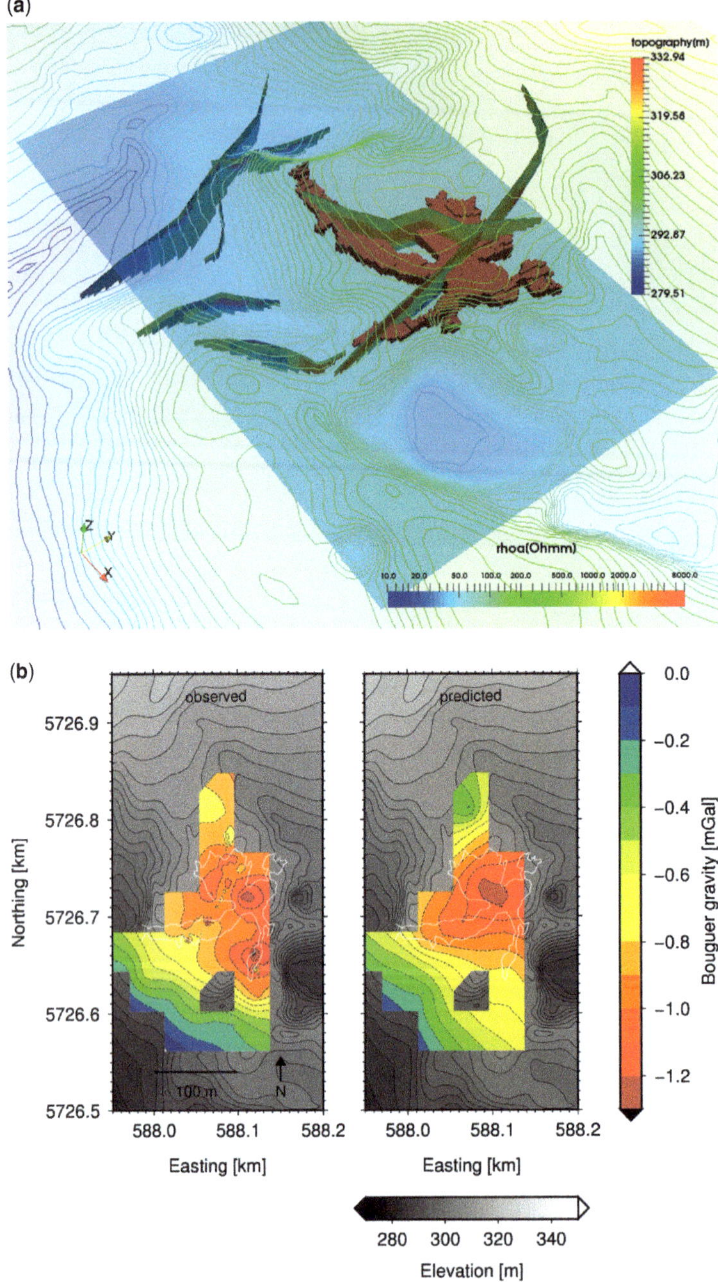

Fig. 10. Jettenhöhle, Osterode, Lower Saxony, Germany. (**a**) Topographic contour map showing the Jettenhöhle cave as a dark red block and electrical resistivity imaging profiles as cross-sections. (**b**) Observed and modelled Bouguer gravity anomaly over the area.

We used this structural model to predict the Bouguer anomaly of the area (Fig. 11). Our model prediction can explain the strong negative signal (down to −1.2 mGal) caused by the larger cave voids. The small passages to the south, which cause a strong local negative Bouguer signal, are not

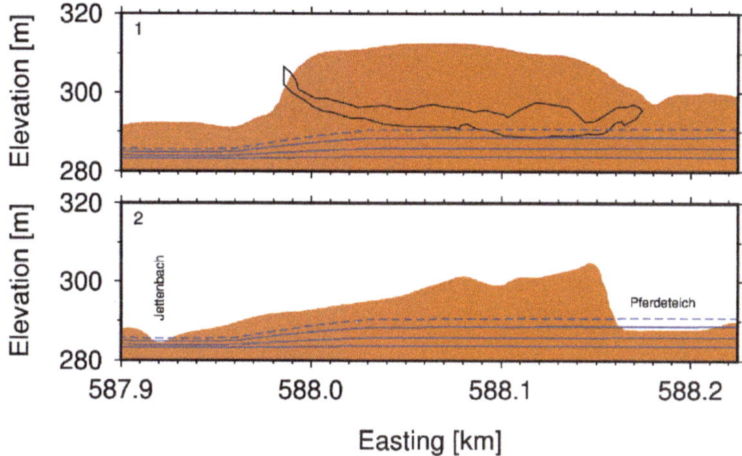

Fig. 11. Jettenhöhle, Osterode, Lower Saxony, Germany. Two cross-sections along the axis Jettenbach-Pferdeteich (2, bottom) and the Jettenhöhle (1, top). Topography in brown, cave cross-section as black line, and three water table models (blue solid lines) for 0, 5 and 10 mm/day recharge, and a higher water table (dashed blue line) for 15 mm/day recharge.

captured as a result of the limited resolution of the structural model. In addition to the local Bouguer signal, the regional trend towards larger values close to the Jettenbach creek is also modelled satisfactorily. The amplitude of this regional signal is slightly underestimated, which was probably caused by the limited information on the geological map (Fig. 6) about the outcrop of the Plattendolomit, which is partially covered by loess in this area.

We add a transient three-dimensional groundwater flow model, modelled on a daily basis with recharge by precipitation (up to 15 mm day^{-1}) and flow towards the local base level Jettenbach. The subsurface consists of gypsum (hydraulic conductivity $K_f = 10^{-4}$ m s^{-1}, specific storage $S_s = 5 \times 10^{-6}$ 1 m^{-1}), dolomite ($K_f = 10^{-7}$ m s^{-1}, $S_s = 5 \times 10^{-7}$ 1 m^{-1}) and the cave void. The resulting water-table gently dips towards the creek Jettenbach, with a distinct gradient along the transition from gypsum to dolomite. ERI profiles 3, 6 and 7, and, in part, profile 1, show low electrical resistivity in the phreatic zone below the water-table; ERI profiles 1 and 8 show high electrical resistivity around the cave voids in the vadose zone. Thus the groundwater flow explains the mapped ERI measurements. The low hydraulic conductivity of the dolomite is responsible for keeping the water-table high after substantial rain, with possible back-flooding of the Pferdeteich sinkhole. This is often observed after snowmelt in spring, with levels rising by 2–4 m, as can be seen in the two cross-sections below the structural model.

The water-table may rise even higher during stronger recharge events and flood the lower part of the cave, as can be seen in the two cross-sections between Pferdeteich–Jettenbach and along the Jettenhöhle (Fig. 11b, dashed blue line). This rare flooding of the cave has been documented (Vladi 1972).

Conclusions

We have compiled a large dataset of geophysical surveys in and above the Kalkberghöhle and Jettenhöhle caves, developed in the Hauptanhydrit formation of the z3 (Leine) sequence of Permian Zechstein rocks. We chose these two cave sites because the caves have a shallow overburden of 10–40 m and are characterized by small passages and large rooms. With these favourable settings, we expect the caves to be detectable by indirect geophysical measurements.

We have shown that the known and surveyed cave passages can be traced with both gravity and ERI measurements, which reflect the lower density and the different electrical resistivities of the cave voids and host rock. Gravity is a useful method here as we can identify the cave voids in the Bouguer gravity signal as local negative anomalies and the geological setting as a regional trend in the Bouguer gravity data. The ERI measurements, mostly carried out in the Wenner set-up, provide additional information about the subsurface geology and water content, which are useful in characterizing the hydrological situation. Other ERI set-ups, such as the Schlumberger or dipole dipole set-ups, essentially confirmed our results, with the latter set-up being less useful due to the poor signal-to-noise ratio.

For the Kalkberghöhle locality, we were able to explain large parts of the Bouguer gravity signal of the cave void using a numerical prediction of the digitized cave map. We were able to address the ceiling heights in the larger rooms, which have not been accurately mapped in the field.

For the locality Hainholz in the southern Harz Mountains, we have shown that, in addition to the strong negative Bouguer data above the cave, a regional trend towards larger Bouguer values is observed in the western direction. This increase in gravity indicates the denser Plattendolomit partly outcropping along the western edge of the graben structure. Observations of lake level fluctuations then enabled us to propose a simple hydrogeological model of the Hainholz area.

The geophysical surveys were part of the BSc theses of Alexandra Werner and Johann Diezel, supervised by GK. Field support from Ron Freibothe, Julio Galindo-Guerreros, Thomas Hiller, Grit Jahn, Lucinda Gürlich, Johannes Mayr, Douchko Romanov and Alexandra Werner is greatly acknowledged. This work was funded by the Deutsche Forschungsgemeinschaft under research grant KA1723/6.

References

Bechtel, T.D., Bosch, F.P. & Gurk, M. 2007. Geophysical methods. *In*: Goldscheider, N. & Drew, D. (eds) *Methods in Karst Hydrogeology*. Balkema, Rotterdam, 171–199.

Brandt, A., Kempe, S., Seeger, M., Vladi, F. 1976. Geochemie, Hydrographie und Morphogenese des Gipskarstgebietes von Düna/Südharz. *Geologisches Jahrbuch Reihe C*, **15**, 3–55.

Butler, D. 1984. Microgravimetric and gravity-gradient techniques for detection of subsurface cavities. *Geophysics*, **49**, 1084–1096.

Carbonel, D., Gutiérrez, F. et al. 2013. Differentiating between gravitational and tectonic faults by means of geomorphological mapping, trenching and geophysical surveys. The case of Zenzano Fault (Iberian Chain, N Spain). *Geomorphology*, **189**, 93–108.

Carbonel, D., Rodríguez, V. et al. 2014. Evaluation of trenching, ground penetrating radar (GPR) and electrical resistivity tomography (ERT) for sinkhole characterization. *Earth Surface Processes and Landforms*, **39**, 214–227.

De Waele, J., Gutierrez, F., Parise, M. & Plan, L. 2011. Geomorphology and natural hazards in karst areas: a review. *Geomorphology*, **134**, 1–8.

Dobecki, T.L. & Upchurch, S. 2006. Geophysical applications to detect sinkholes and ground subsidence. *The Leading Edge*, **25**, 336–341.

El-Qady, G., Hafez, M., Abdalla, M. & Ushijima, K. 2005. Imaging subsurface cavities using geoelectric tomography and ground-penetrating radar. *Journal of Cave & Karst Studies*, **67**, 174–181.

Ford, D.C. & Williams, P.W. 2007. *Karst Geomorphology and Hydrology*. Wiley, Chichester.

Fricke, U. 1990. Ein neuer Plan der Segeberger Kalkberghöhle. *Mitteilungen des Verbandes der Deutschen Höhlen- und Karstforscher*, **36**, 77–85.

Gutiérrez, F. 2010. Hazards associated with karst. *In*: Alcántara, I. & Goudie, A. (eds) *Geomorphological Hazards and Disaster Prevention*. Cambridge University Press, Cambridge, 161–175.

Gutiérrez, F., Parise, M., De Waele, J. & Jourde, H. 2014. A review on natural and human-induced geohazards and impacts in karst. *Earth Science Reviews*, **138**, 61–88.

Herrmann, A. 1969. Einführung in die Geologie, Morphologie und Hydrogeologie des Gipskarstgebietes am südwestlichen Harzrand. *In*: Herrmann, A. & Pfeiffer, D. (eds) *Der Südharz – seine Geologie, seine Höhlen und Karsterscheinungen*. Verband der deutschen Höhlen- und Karstforscher, **9**, 1–10.

Herrmann, A. 1981*a*. Eine neue geologische Karte des Hainholzes bei Düna/Osterode am Harz. *Berichte der Naturhistorischen Gesellschaft Hannover*, **124**, 17–33.

Herrmann, A. 1981*b*. Zum Gipskarst am südwestlichen und südlichen Harzrand. *Berichte der Naturhistorischen Gesellschaft Hannover*, **124**, 35–45.

Ipsen, A. & Mucke, D. 2011. *Die Segeberger Höhle – eine Welt im Verborgenen*. Fledermaus-Zentrum, Bad Segeberg.

Jarvis, A., Reuter, H., Nelson, A. & Guevara, E. 2008. Hole-filled Seamless SRTM Data V4. International Centre for Tropical Agriculture, http://srtm.csi.cgiar.org

Jordan, H. 1981. Karstmorphologische Kartierung des Hainholzes (Südharz). *Berichte der Naturhistorischen Gesellschaft Hannover*, **124**, 47–53.

Kardel, J., Mucke, D. & Frühwirt, T. 2009. Felssicherung in Anhydrit und Gips am Beispiel des Kalkbergstadions Bad Segberg (Rock stabilization of anhydrite and gypsum by the example of the Kalkbergstation Bad Segeberg). *In*: *VerÃ¶ffentlichungen 17*. Tagung für Ingenieurgeologie und Forum Junge Ingenieurgeologen, 6–9 May 2009, Horchschule Zittau/Görlitz, 149–154.

Kaufmann, G. 2014. Geophysical mapping of solution and collapse sinkholes. *Journal of Applied Geophysics*, **111**, 271–288.

Kaufmann, G. & Romanov, D. 2016. Structure and evolution of collapse sinkholes: combined interpretation from physico-chemical modelling and geophysical field work. *Journal of Hydrology*, **540**, 688–698.

Kaufmann, G., Jahn, G., Galindo-Guerreros, J. & Nielbock, R. 2012. Geophysical exploration of cave sites: the case of the Unicorn Cave, Scharzfeld/Harz, Germany. *Braunschweiger Naturkundliche Schriften*, **11**, 69–80.

Kaufmann, G., Ullrich, B. & Hoelzmann, P. 2015*a*. Two iron-age settlement sites in Germany: from field work via numerical modelling towards an improved interpretation. *Archaeological Discovery*, **3**, 1–14.

Kaufmann, G., Nielbock, R. & Romanov, D. 2015*b*. The Unicorn Cave, Southern Harz Mountains, Germany: from known passages to unknown extensions with the help of geophysical surveys. *Journal of Applied Geophysics*, **123**, 123–140.

Kempe, S. 1997. Gypsum karst of Germany. *International Journal of Speleology*, **25**, 209–224.

KEMPE, S. 2008. Gipskarst – ein Überblick. *Exkursionsführer und Veröffentlichung der Deutschen Gesellschaft für Geowissenschaften*, **235**, 30–41.

KONIETZKY, H., WALTER, K. & FRÜHWIRT, T. 2007. *Standsicherheitsgutachten Kalkberghöhle Bad Segeberg.* Unpublished report, Geomontan GmbH.

KUPETZ, M. & BRUST, M. 2008. Neue Beobachtungen in der Segeberge Kalkberghöhle – Gerinnehöhlen-Stockwerk, Eiskeil-Pseudomorphose und Fazetten. *Mitteilungen des Verbandes der deutschen Höhlen- und Karstforscher*, **54**, 76–84.

LOKE, M. & BARKER, R. 1995. Rapid least-squares inversion of apparent resistivity pseudosections using a quasi-Newton method. *Geophysical Prospecting*, **44**, 131–152.

LOKE, M.H. & BARKER, R.D. 1996. Practical techniques for 3D resistivity surveys and data inversion. *Geophysical Prospecting*, **44**, 499–523.

MARGIOTTA, S., NEGRI, S., PARISE, M. & VALLONI, R. 2012. Mapping the susceptibility to sinkholes in coastal areas, based on stratigraphy, geomorphology and geophysics. *Natural Hazards*, **62**, 657–676.

MARGIOTTA, S., NEGRI, S., PARISE, M. & QUARTA, T.A.M. 2016. Karst geosites at risk of collapse: the sinkholes at Nociglia (Apulia, SE Italy). *Environmental Earth Sciences*, **75**, 1–10.

MEIER, G. 2003. Ingenieurgeologische Ergebnisse bei der Standsicherungsanalyse der Kalkberghöhle in Bad Segeberg. *In*: *Veröffentichungen 14.* Tagung für Ingenieurgeologie, CAU Kiel, 353–358.

MOCHALES, T., CASAS, A.M., PUEYO, O., ROMAN, M.T., POCOVI, A., SORIANO, M.A. & ANSON, D. 2008. Detection of underground cavities by combining gravity, magnetic and ground penetrating radar surveys: a case study from the Zaragosa area, NE Spain. *Environmental Geology*, **53**, 1067–1077.

MUCKE, D., RAHMEL, U., DENSE, C., GLOZA-RAUSCH, F., IPSEN, A., VÖLKER, C. & VÖLKER, R. 2008. Vertretbare Eingriffe im Segeberger Karst in Ausführung. *Mitteilungen des Verbandes der deutschen Höhlen- und Karstforscher*, **54**, 68–75.

PALMER, A. 2007. *Cave Geology*. Cave Books, Dayton, OH.

PARISE, M. & LOLLINO, P. 2011. A preliminary analysis of failure mechanisms in karst and man-made underground caves in Southern Italy. *Geomorphology*, **134**, 132–143.

ROSS, P.-H. 1990. Die Tektonik der Segeberger Höhle. *Mitteilungen des Verbandes der deutschen Höhlen- und Karstforscher*, **36**, 29–32.

ROSS, P.-H. 1993. Die Segeberger Karstlandschaft. *Berichte aus dem Geologischen Landesamt Schleswig-Holstein*, **2**, 33–48.

SPARR, H.-P. 1997. *Der Kalkberg: Naturdenkmal und Wahrzeichen der Stadt Bad Segeberg*. Christinas, Hamburg.

TELFORD, W.M., GELDART, L.P. & SHERIFF, R.E. 2012. *Applied Geophysics*. 2nd edn. Cambridge University Press, Cambridge.

VLADI, F. 1972. Verkarstung und Höhlenentwicklung im Gips. *In*: KEMPE, S., MATTERN, E., REINBOTH, F., SEEGER, M. & VLADI, F. (eds) *Die Jettenhöhle bei Düna und ihre Umgebung Abh.* Karst- und Höhlenkunde A6, Verband der deutschen Höhlen- und Karstforscher, Munich, Germany.

WENZEL, F. 1996. The nanogal software: Earth tide data processing package eterna 3.30. *Bulletin d'Informations Marees Terrestres*, **124**, 9425–9439.

Models of temperature, entropy production and convective airflow in caves

GIOVANNI BADINO†

Dipartimento di Fisica, University of Torino, I-10125 Torino, Italy

Associazione Geografica La Venta, Treviso, Italy

m.parise@ba.irpi.cnr.it

Abstract: Three major problems in cave micrometeorology are analysed: the concept of the temperature of a cave and its phenomenology; the internal energy flows and consequent local entropy production; and a non-hydrostatic physical model of the underground convective air circulation. A cave's temperature is rich in information, but it is often difficult to obtain because it requires experimental accuracies to be pushed beyond the reach of common external meteorological instruments to detect a variety of factors, such as thermal sedimentation, seasonal variations and the effects of external morphology. Energy flow has an essential role in estimating the sensitivity of a cave to external inputs. A model for evaluating the local entropy production is developed. Entropy seems to be the most basic parameter, explaining both the sensitivity of the environment and its tendency to form complex structures. Analysis of the second-order terms of convective air circulation is more complex than expected; nevertheless, only such an analysis is able to explain the behaviours (e.g. continuity in the airflow) that cannot be predicted by the first-order models.

Giovanni Badino sadly died on 8 August 2017, shortly after receiving the reviews for this paper. Mario Parise and Franci Gabrovsek revised the manuscript in the light of the reviewers' reports as well as they could without the input of the author. Both reviewers were of the opinion that the manuscript has scientific interest and recommended publication. The Geological Society and the other volume editors are very grateful to Mario and Franci for their work, which has allowed this paper to be published. Mario Parise's email address is given for correspondence.

This paper differs from those that are usually expected in a book on karst phenomena. Mathematical modelling is, however, becoming increasingly important in karst studies, although perhaps more painfully than in other fields. In particular, during the last few decades, much work has been carried out with the aim of analysing the phenomenology of the subterranean microclimate and micrometeorology, part of the emerging field known as the physics of caves.

The traditional approach of this type of research has usually been strictly naturalistic, with qualitative discussions and models, typically avoiding any kind of mathematical model or quantitative approach. There are exceptions, such as the work of Lomonosov (1732), described in Badino (2017), and Bock (1913), described in Cigna (2017).

Hence there is a gap between theoretical work and reality, suggesting the need to carry out experiments in certain directions to confirm or deny a particular model. Many of the current interpretations of cave phenomena, if translated into computer programs, could produce results that are very different from the real world; the assumptions behind many interpretations often remain hidden.

In the past, it was difficult to carry out accurate experimental studies in subterranean environments, often as a result of the absence of accurate instruments, especially data-loggers. Measurements were usually occasional and often only at shallow depths. Suitable instruments have been available for the last few decades, but the data are often presented as raw numbers, without any attempt to confirm theoretical models. This situation is changing as new theoretical studies are improving our understanding of the physical phenomenology of caves (Cigna 1967; De Freitas *et al.* 1982; Gadoros 1989; Lewis 1991; Badino 1995; Jernigan & Swift 2001; Lismonde 2002; Pflisch *et al.* 2010; Luetscher 2013; Mammola *et al.* 2016).

This work seeks to define more clearly, and on a physical basis, some major problems related to energy and air fluxes and the impacts of humans in caves. The first part is an introductory section outlining the underground thermal regimes, distinguishing

†Deceased 8 August 2017.

From: Parise, M., Gabrovsek, F., Kaufmann, G. & Ravbar, N. (eds) 2018. *Advances in Karst Research: Theory, Fieldwork and Applications*. Geological Society, London, Special Publications, **466**, 359–379.
First published online January 31, 2018, https://doi.org/10.1144/SP466.24

between the layers nearest to the surface that are still affected by external temperature changes, a deeper layer crossed by air and water fluxes in thermal equilibrium with the rocks (usually the karstified strata) and a deep zone crossed by the geothermal flux. The paper then focuses on three topics.

- First, we discuss what establishes the cave temperature, which is a concept much more complex, variable and more rich in information than is often thought. It is easy to show that the underground temperature is closely related to the external climate and the local morphology; it changes with altitude, at a rate which is different from that typical of the external meteorology. However, it is difficult to sustain the usual assumption that it is possible to define the temperature of a certain conduit in an unambiguous way – this is because thermal sedimentation is commonly present. The variation in temperature between the floor and the roof of a gallery has a large impact on speleogenetic processes, on the formation of underground glaciers and, above all, on the creation of distinct ecological niches. This means that it is necessary to discuss its phenomenology.
- The second topic, closely related to the first, is an attempt to estimate the sensitivity of a cave, or a part of it, using the tools of ecological thermodynamics (Swenson 1988). The usual approach to assessing the sensitivity of a cave – for example, to the impact of tourism – is essentially qualitative and uses energy terms. Caves are ranked by estimating their energy in the sense of the maximum energy flow passing through them. The concept is ambiguous and not easy to express mathematically, particularly because it is easy to show that the average energy flowing in almost any cave equals zero. If this energy flow was not zero, then the internal energy of the cave, and thus its temperature, would vary. From an ecological standpoint, the external systems, always far from thermodynamic equilibrium, are more difficult to approach in terms of entropic production than the subterranean systems. For the latter, it is possible to estimate the local entropy production using radical simplifications. It is easy to deduce the energy degraded in the process (a thing which principally has an educational value, since energy appears, wrongly, to be an easier concept than entropy), which then becomes the key parameter in assessing the cave sensitivity and its carrying capacity. It is the deconstructive part of the total amount of energy flowing through a certain underground area.
- The third theme is convective air circulation inside caves. Although there are models and advanced experimental measures (Lewis 1991) for barometric air circulation (due to variations in atmospheric pressure), convective circulation – such as the airflow between entrances at different altitudes or the circulation related to external variations in air density – is usually explained using a hydrostatic approximation. During the last decade, low-cost sonic anemometers have enabled accurate experimental measurements to be made, which have shown that convective air circulation is far more complex than expected and that the reality is in conflict with the usual simplified models. In practice, air circulation in caves is a result of second-order effects, which are completely lost in coarse model approximations. The absence of a reliable model, as well as making some of the air circulation frequently encountered underground inexplicable, also means that it is impossible to make predictions about a topic that is currently attracting a great deal of interest: whether the conditions for the formation of ice water deposits exist in volcanic caves on Mars. We present an advanced model that includes external atmospheric parameters and show that there are two other reasons for air movement, one linked to variations in the atmospheric pressure–altitude profile due to changes in temperature and the other related to variations in the external lapse rate – that is, the variation of temperature with altitude.

Under the apparent stability of cave atmospheres, complex processes occur that play an important part in subterranean processes due to their long timescales.

Traditional measurements of the microclimate of caves were usually made relatively close to the surface and with external instruments that did not allow high accuracies and the acquisition of long time series. These results were often qualitative descriptions of the cave atmosphere and indicated an apparent stability everywhere. Nevertheless, a measure of the Earth's temperature by a thermometer with a resolution of ±50°C would indicate the same value everywhere, leaving unexplained why here there is a desert, there a forest and there an ice cap. This situation is now quickly changing, thanks to technological advances in the fields of data acquisition and storage, in the development of physical models and to some advances in measurement techniques and data processing. Studies of cave climate and meteorology are important for a number of reasons:

(1) to estimate the reliability of palaeoclimatic data. Although climate records essentially refer to hypogean deposits (McDermott 2004; Fairchild & Baker 2012), the direct interest is on epigean palaeoclimates. The correlation between the two is, however, not simple;

(2) to understand local thermal insulation (Badino 2004), connecting it to the cave sensitivity and the occurrence of complex structures;
(3) to understand the speleogenetic role of condensation (Dreybrodt *et al.* 2005; Gabrovšek *et al.* 2010);
(4) to characterize parts of caves as ecological islands;
(5) to protect caves, especially show caves, against anthropogenic impacts (Cigna 1993, 2002; North *et al.* 2009; De Freitas 2010; Parise 2011).

Temperature of caves

Underground thermal regimes

It is well known that the temperature underground increases with depth, but it is also a common experience that caves are fairly cold. This is because there are underground layers with different thermal regimes: the upper levels (the heterothermic and neutral levels), crossed by external fluids, are in thermal equilibrium with the atmosphere, whereas at lower levels, in absence of external fluid flows, the thermal contact is with deeper layers of the Earth's crust (the geothermic layer).

The specific thermal capacity of water is *c.* 4.2 kJ kg^{-1} K^{-1}, whereas that of rock is five times less. The thermal capacity of water per unit volume is therefore about double that of the thermal capacity of rock. In temperate climates, the usual infiltration rate of water in karst is *c.* 1 m^3 a^{-1} m^{-2} (1000 mm a^{-1}). This means that, at a first approximation, meteoric waters should be able to equilibrate a rock layer 2 m thick each year. It is then obvious that, whatever the thickness of rock crossed by infiltrating water, it will acquire the temperature of the water over a very short period of geological time. Thus mountains with deep karst are in equilibrium with water that has infiltrated the rock in the recent past.

It is possible to distinguish a first layer in which there are daily temperature fluctuations: the heterothermic daily layer. The Fourier law of conduction shows that, in compact rock, this layer has a typical thickness of 15–20 cm. Seasonal variations in temperature can be measured 20 times deeper because the conductive penetration length of sinusoidal alterations in temperature depends on the square root of the period (Isachenko *et al.* 1969). This depth defines the lower limit of the heterothermic seasonal layer. In permafrost studies, these two first layers are referred to together as the active layer because this is where ice can be formed and melted, if allowed by the weather conditions (Shiklomanov & Nelson 2013). The heterothermic seasonal layer is usually a few metres thick and the seasonal variation in temperature decreases to almost zero within this depth range.

Water infiltrating beneath the heterothermic seasonal layer is no longer affected by seasonal changes and therefore the rock and water both take on the average local temperature of the infiltrating waters (T_{IW}). The neutral layer is crossed by airflows and water fluxes in both the vadose and phreatic regimes and extends down to the water table to the level where the rock permeability is sufficiently high to allow a significant flux of external water. This permeability horizon is the lower boundary of the neutral layer.

Below this level, the rock (and interstitial water, here almost motionless) are affected by the geothermal flux and the temperature steadily increases in a conductive regime. In the absence of a general naming agreement, it is possible to call this horizon the geothermic layer. It extends downwards for several kilometres, far beyond the depths relevant to the present discussion.

The neutral layer is thus enclosed between two surfaces: an upper surface where the yearly variation in temperature is close to zero and the interface between the neutral and geothermal layers, where the rock temperature starts to increase with the local geothermal lapse rate. The thickness of this stratum depends on the local rock permeability. In non-karst environments it is usually *c.* 50–100 m, but it can be much more along major rock discontinuities that are able to drain water, as observed during the excavation of the Mt Blanc tunnel (Guichonnet 1967).

In karst environments, the neutral layer can extend throughout a whole mountain (Sendra & Reboleira 2012). The deepest cave in the world (Krubera cave, Abkhazia, Western Caucasus) crosses 2200 m of rock vertically (Klimchouk 2012), with a temperature increase of <5°C over the whole depth (Provalov, pers. comm.). The temperature of the neutral layer is not independent of depth because the crossing fluids increase their temperature with a typical karst lapse rate of 3–5°C km^{-1}. This explains why these mountains are fairly cold and, to a first approximation, unaffected by the geothermal flux.

Geothermal energy does not flow through these upper layers because it is intercepted by deep flowing water and carried away. The obvious consequence is that this water is heated along its travel path between the lowest parts of ponor caves and springs. This heating has a significant role in phreatic speleogenesis (Badino 2017).

Details on continental geothermal flow have been given by Davies & Davies (2010) and the world average is $F_{gt} = 0.07$ W m^{-2}. Therefore it is easy to show that, under stationary conditions, the temperature increase is given by $500/P_i$, where P_i is the infiltration intensity in units of mm a^{-1}. In alpine karst, water is expected to be heated by

c. 0.5°C, but in areas with low infiltration rates the increase in temperature can be much greater. The geothermal flux is roughly homogeneous, however, and the water flux is primarily limited to conduits. We would expect that conduits would intercept only a small part of the geothermal heat flow, but Badino (2005) has shown that, due to lensing effects, this is not the case.

Therefore the temperature of water flowing in contact with the geothermal layer invariably increases along its route, regardless of the type and shape of the drainage system, and this heating is able to increase the solubility of gypsum, quartzite and halite rocks.

Average external temperature

To a first approximation, we can assume that a cave has the temperature T_C of the surrounding rock T_R, which has the temperature of the infiltrating water T_{IW}, which is almost equal to the local yearly average temperature T_L. The estimation of the local average temperature T_L is therefore important; it depends on the local morphology, but, above all, on the latitude and altitude of the cave's location.

The dependence of temperature on latitude is far from regular, but it is possible to see in Figure 1 (Pinna 1977) that the temperature T_L at sea-level in southern Europe decreases by c. 0.7°C per degree of latitude. The dependence of temperature on altitude is fairly regular, with the temperature decreasing by c. 6.5°C per kilometre of increasing height, which is the lapse rate of the International Standard Atmosphere (Bohren & Albrecht 1998).

It is easy to estimate T_L anywhere in the world from the local temperature T_{L0} of a reference point:

$$T_L = T_{L0} - 0.7 \times \Delta\text{Lat} - 6.5 \times \Delta\text{Alt} \quad (1)$$

where the Δ values represent the difference in latitude (°) and altitude (km) between the particular location and the reference station.

Accuracy of temperature measurements

The previous sections have examined the general concepts of heat exchange between the external and subterranean environments, without entering into details of the physical terms. The concept of temperature is univocally defined only for systems at equilibrium; nevertheless, the concept can be also extended to systems in local quasi-equilibrium, although with intrinsically low accuracies. Measurements of the surface temperature have an intrinsically low precision because the external environment is far from equilibrium: an external high-resolution temperature measurement would therefore have no physical meaning. However, this does not apply to caves, which are in a near-equilibrium condition. As a consequence, it is, in principle, possible to reach accuracies of the order of 1 mK. At such a level of precision, cave atmospheres show thermal and meteorological exchanges (i.e. transient processes). To a first approximation, the cave

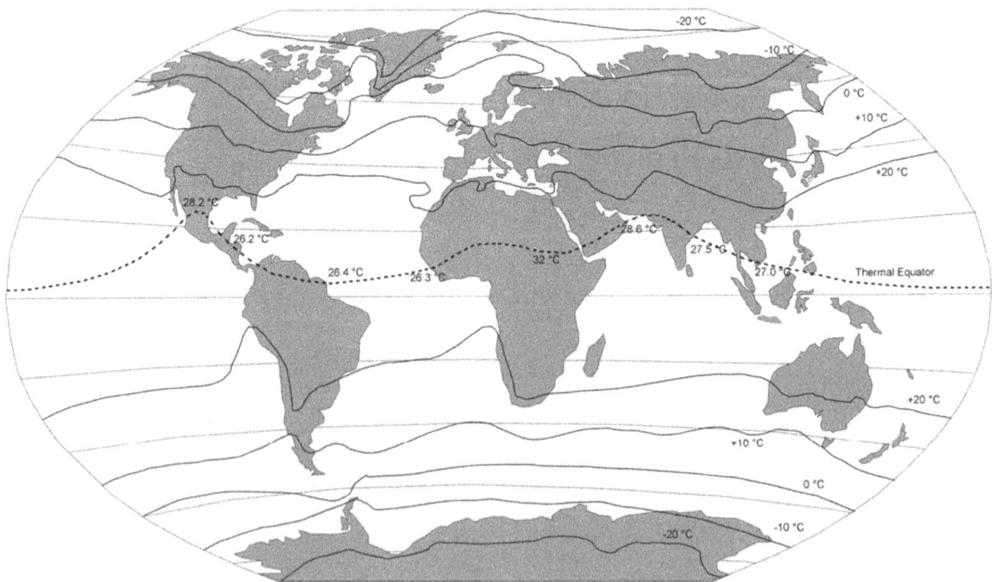

Fig. 1. Dependence of temperature on latitude.

temperature T_C is almost equal to the temperature T_{IW} of the infiltrating water, which is almost equal to T_L. We next discuss the processes that are able to introduce systematic differences between these temperatures.

Effect of seasonal rain on cave temperatures

The first approximation that we are going to consider is connected to the differences between T_{IW} and T_L, which are basically an effect of the local climate. Rain water is usually cooler than the average temperature of the atmosphere and the rain is formed at an altitude that is always higher than the local altitude. Therefore the resulting water is usually colder than the local air temperature. The value of ΔT depends on the climate and location and is assumed to be *c.* 2°C in temperate regions (Celico 1986).

The most important effect is that precipitation is concentrated in certain time periods, depending on the regional climate, to give a temperature effect that is dependent on the rainy season. Climatic data from many different regions (www.worldclimate.com) can be compared with the Köppen classification of climates.

It is easy to roughly estimate the temperature of infiltrating water as a weighted average value from the average monthly temperature, T_j, and the average monthly precipitation, P_j:

$$T_{IW} = \frac{1}{P_T} \sum_{j=1}^{12} T_j \times P_j \qquad (2)$$

where P_T is the yearly precipitation. This approach is inaccurate based on observations of the differences between the temperatures of air and rain and fluid selection, but it is the best possible estimation in the absence of specific measurements. It is interesting to discuss some extreme cases to estimate ΔT_u – that is, the difference between the underground temperature and T_L.

Singapore, located in the Intertropical Convergence Zone (Fig. 2), has an almost constant temperature and rainfall and therefore the temperature of rain is close to the average yearly temperature (Table 1). By contrast, in Turkmenistan, rainfall is concentrated during the cold season and thus the water infiltrating underground is much colder than the air and, consequently, the subsoil is colder than the atmosphere.

In steppe climates (e.g. Chihuahua, Mexico), rain is concentrated during the hot season and the large yearly temperature range (15°C) allows the subsoil to be heated relative to the external average temperature. This effect is important in continental type climates, where it is much stronger than in Chihuahua. In Ulan Bator (Mongolia), precipitation is concentrated during the hot season, but the enormous temperature range (30°C) increases the effect and therefore the subsoil temperature is expected to be *c.* 11°C warmer than T_L.

In temperate climates the differences between the temperature of infiltrating water and the average local temperature are always small. In the Mediterranean area (e.g. Pisa, Italy), rain is concentrated during the spring and autumn seasons, which is why the approximation $T_C \approx T_L$ is fairly precise for most caves.

Effects of snow

Moving northwards, the effect of a continental climate on the precipitation season holds, but an important new phenomenon is seen and is discussed here for the case of Perm (Russia). The average temperature in Perm is +1.7°C, but for almost six months of the year the average temperature is below zero. The average temperature of precipitation is 5.2°C, but this is not a good estimate of the temperature in the subsoil because water does not flow below 0°C. A negative temperature does not participate in establishing the temperature of infiltrating water, T_{IW}, and the negative parts of the yearly variation in temperature, which usually describe T_L, become unimportant. During melting, the water inflows to the subsurface are at temperatures close to 0°C. This creates a new, paradoxical effect: warm episodes during late winter. To provide an example, the inflow of snowmelt underground significantly cools the cave of Rio Martino in Italy (Fig. 3).

During melting, the air temperature is expected to be much higher than 0°C, which would compensate for such an effect, but it is easy to prove that it cannot actually balance the cooling effect. Therefore, in regions where snow is a significant part of the yearly precipitation, the rock temperature, T_R, may be significantly higher than T_L and always >0°C; the same can be stated for the cave temperature. Seeping water warms the rock (especially in karst areas) and extends the internal isothermal zero to altitudes much higher than those outside. The recalculated T_{IW} in Perm is *c.* 8°C.

The case of Perm is very common in mountains. As an example, the Ulugh Beg cave (Hodja Gur Gur Atà, Uzbekistan), at an altitude of 3750 m a.s.l., has an estimated T_L of −6°C. The altitude of the external isothermal 0°C is *c.* 2800 m a.s.l. The cave has important ice deposits in the first 300 m and the measured internal temperatures are −0.8°C at 3700 m a.s.l. and 0°C at 3550 m a.s.l. (Bernabei & De Vivo 1992).

This case is fairly extreme (ΔT_u = 5°C), but it is normal to measure smaller, though significant, differences in temperature. The isothermal zero extends underground much further than outside, which is the

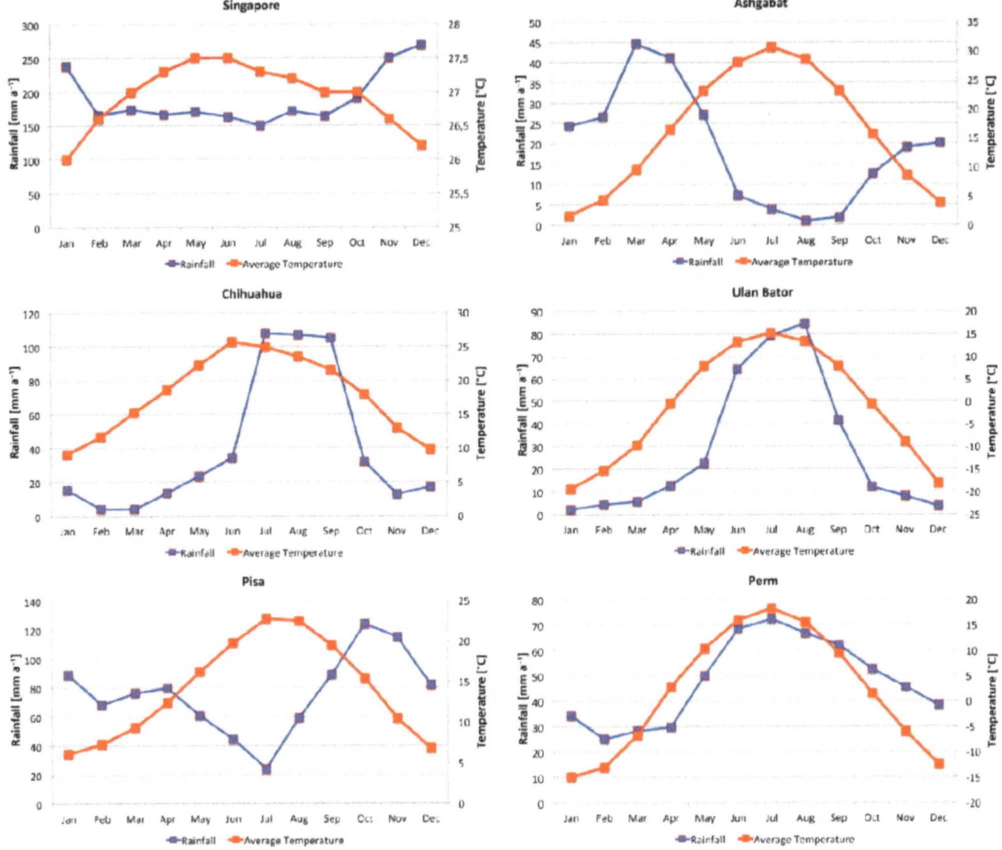

Fig. 2. Rainfall and temperature in different climatic areas.

main reason why cave temperatures below zero are extremely rare (Table 1).

Underground rocks appear to be in thermal equilibrium with the atmosphere in temperate environments because the highest precipitation occurs in spring and autumn, but also in extreme climates (tropical and polar) as a result of their very small temperature ranges.

The disequilibrium can be positive in arid regions (e.g. in Chihuahua) or negative (e.g. in Ashgabat, Turkmenistan). The extremes occur in continental climates where the subsoil is typically much warmer

Table 1. *Climate, rainfall and temperature data for some of the caves discussed in the text*

	Cave type	Climate	T_L	Rainfall	T_{IW}	ΔT_u
Singapore	*Af*	Tropical rainforest	26.9	2270	26.9	0
Ashgabat, Turkmenistan	*B*Wk	Desert, mid-temperature	16.1	229	11.8	−4.3
Khartoum, Sudan	*B*Wh	Desert, hot	29.2	155	31.2	+1.0
Chihuahua, Mexico	*B*Sh	Steppe, hot	17.8	476	21.6	+3.8
Pisa, Italy	*C*sa	Mediterranean	14.0	907	13.0	−1.0
Edinburgh, UK	*C*fb	Oceanic	8.5	665	8.8	+0.3
Pjatigorsk, Russia	*C*fa	Humid subtropical	8.7	603	12.4	+3.7
Ulan Bator, Mongolia	*D*wc	Continental boreal	−1.3	340	9.9	+11.2
Perm, Russia	*D*fb	Continental hemiboreal	1.7	573	8.2 (5.1)	+6.5
Ushuaia, Argentina	*ET*	Tundra	5.5	555	5.6	+0.1

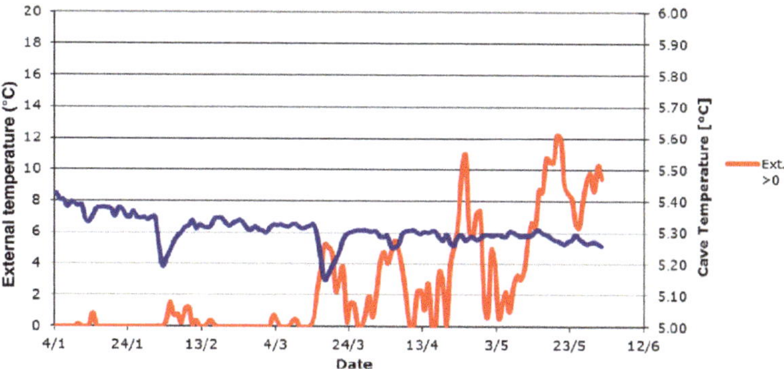

Fig. 3. External snow melting, Rio Martino cave, Italy. Cave temperature shown in blue; external temperature in red.

than the external atmosphere. The same occurs in high mountains, just below and above the altitude of isothermal zero. These estimations are based on external infiltration, therefore the temperature is significantly higher than these figures allow, if other fluid selection effects are excluded, because deep geothermal waters are involved in cave equilibrium (Badino 2010).

There are other effects resulting from the external morphology that can produce large differences between T_{IW} and T_L and cause anomalies in the temperature of water infiltration. A cave can be crossed by water coming from a lake, or slow fluxes of water that have been exposed to sunlight, which, during summer, could attain temperatures significantly higher than that of the atmosphere. A large cave entrance can accumulate thick snow deposits that cool the cave temperature during the hot season with melting water, always at 0°C, when T_L is much higher. An example is found in the Slovenian karst at Velika Ledenica cave in Paradani, an ice cave at an altitude of *c.* 1200 m a.s.l., where T_L is around 6–8°C.

Airflow selection and glaciers

The convective movement of air, discussed in detail in the following sections, plays an important part in establishing temperatures and thermal sedimentation in caves, especially near cave entrances. If the internal atmosphere is warmer than the external atmosphere (in terms of the virtual temperature, which includes the humidity; Badino 1995), the airflow moves upwards. Cold, external air enters the lower entrances and the internal air passes through the upper entrances.

By contrast, during summer, the internal atmosphere is denser and colder than the external atmosphere and therefore moves downwards, drawing down external air from the upper entrances. These entrances are now 'washed' by external air, which is warmer than the air in the cave. The internal air, which is relatively cold, descends down the mountain.

Caves therefore filter airflows based on their temperature, systematically choosing between the external and internal flows, with the hotter air passing through the upper entrances and the colder air through the lower entrances. The result is that the average temperature of an airflow that passes across the entrance is not T_L, but the average local temperature during the months when the internal temperature is lower (in the upper parts) or higher (in the lower parts).

An important example of this process is the classic cold air trap, which causes the formation of many internal glaciers (Luetscher 2013), especially in small caves. If the cave morphology tends to descend from the entrance, the air flow in winter moves downwards, cooling the system, in a similar process to the case of cave entrances that accumulate snow. This is seen, for example, in Askynskaya cave in the Ural mountains (270 m a.s.l., latitude 54° N, $T_L = 2.1$°C), which has a north-facing entrance that is small enough (18 m^2) to reduce the radiative energy input. Its main hall is completely occupied by an 8000 m^2 glacier, one of the largest in the world (Mavlyudov 2005). Its temperature is fairly stable at *c.* −4°C. The selection effect is active, with significant airflows in caves with many entrances creating thermally insulated branches and extreme thermal sedimentation.

Another interesting case is the Cucchiara cave (Mt Cronio, Sicily, Italy). Along the entrance route of the external air flux, the temperature is 36.6°C near the roof and 26.3°C at the floor, just 2 m below, whereas in the middle there is a thin mixing cloud at the contact layer between the two atmospheres (Badino 2013).

Internal thermal sedimentation

Temperature sedimentation can be generally observed in the upper parts of conduits, where the air is warmer than the air near the bottom of the conduit (Badino 2010). In a closed system, these differences are equalled out over a short time period, mainly by the exchange of radiative energy. It is easy to show that the Stefan–Boltzmann radiation between rock surfaces with temperatures different by a factor ΔT is given by:

$$I = 4\sigma T^3 \Delta T \approx 5\Delta T (\mathrm{W\ m^{-2}}) \quad (3)$$

where σ is Stefan's constant ($\sigma = 5.67 \times 10^{-8}$ W m^{-2} K^{-4}). The right-hand side of this approximation is valid for temperatures of *c.* 10°C (283 K). The sedimentation always implies an exchange of energy, instabilities and, in general, that the system is not thermodynamically closed, but in contact with some time-dependent external process.

Vertical temperature gradients also exist in ventilated conduits and are usually *c.* 0.1°C m^{-1}. There is evidence that these are actually seasonal effects related to the main underground wind flows. As air particles move upwards, they tend to cool with the moist adiabatic lapse rate, which is steeper than the karst lapse rate. The result is that air tends to flow along the cave floor, releasing water in the form of condensation. The opposite occurs during the downward movement of air in winter, when it tends to flow against the roof, causing evaporation. This effect can create seasonal condensation and a systematic difference in the temperature and vapour content between the floor and roof of the cave.

The seasonal flow lines differ depending on the morphology of the conduit, creating sedimentation and allowing the formation of glaciers. This process is active in the Kinderlinskaya cave near Ufa, Russia, which is traversed by a significant airflow that is unable to destroy the cold air trap near the entrance, where there is another large subterranean glacier (Kuzmina *et al.* 2012).

Internal temperature gradients

The main variation in underground temperature is a result of differences in altitude. The vertical movements of fluid cause variations in temperature of *c.* −4 to −6°C km^{-1} for moist air and −2.34°C km^{-1} for water.

A typical cave atmosphere is saturated with water vapour and therefore we can expect internal lapse rates near to the moist adiabatic rates. Surprisingly, this does not occur. The usual lapse rate is between −2.8 and −4°C km^{-1} (Badino 2000), a fact which has enormous consequences for the energy balance in caves. This karst lapse rate has an intermediate value between the adiabatic lapse rates of water and moist air. This result is reasonable because the average lapse rate derived from the thermal exchange between air and water is the average of the two lapse rates weighted with their thermal capacity flows.

The lowest lapse rates measured are related to very wet caves (e.g. the Ceki Due cave, Slovenia, and other alpine caves); the highest, near to the moist air lapse rate, are measured into the large shafts of Tepui cave in Venezuela, which is crossed by strong air flows and minimum water fluxes. Cave atmospheres show lapse rates significantly lower (in absolute values) than the external lapse rate and these lapse rates are not adiabatic. There is therefore a continuous energy flow between the air, water and rock created by the vertical movement of air, which has many different consequences.

Water flowing out from deep karst is significantly cooler than the local temperature T_L at springs (and the temperature of infiltrating water, T_{IW}), because, in crossing the mountain, the water has been heated by only 3.5°C km^{-1}, whereas the external heating is usually *c.* 6.5°C km^{-1}. Thus the usual impression at a spring that, 'this water is very cold because it comes from very far away' could be correct.

Water flows out from deep karst with more energy (thermal energy) than at infiltration (potential energy). This means that it subtracts energy from karst and, more precisely, looking at the energy balance, from the internal airflow. In caves, water looks to be the chisel and air the hammer. It is obvious that thermal exchanges between the two fluids are only possible if they have different temperatures and only if the system is not in the maximum entropy state.

Therefore the general water–air energy exchange along a vertical transfer is complex, seasonal and probably responsible for local condensation, thermal stratification and the difference in temperature between the two fluids (Trombe 1952) and among different cave regions. Much work still has to be done to understand these processes.

Energy and entropy fluxes underground

Stationary condition for energy flow

Cave systems are not isolated from the external environment. Convective heat transfer between the rock and air and heat conduction within the rock both release thermal energy to the heterothermic and neutral layers. Direct heat flow through the rock, either by pure conduction or by the convection of interstitial water, is low and is only important in the case of superficial systems, limited to a few metres in depth.

It is useful to make some preliminary observations. The first, fairly trivial, is that the internal energy of the cave depends on its average temperature. However, the average cave temperature is essentially constant, which indicates that the sum of energy flows into the system, averaged on a geological timescale, is zero:

$$\int_{\tau} \left(\sum_{j} F_{\mathrm{j}} \right) \mathrm{d}t = 0 \qquad (4)$$

The observation seems paradoxical, given that a useful parameter for assessing the cave sensitivity is its energy level (Heaton 1986).

Thermal capacity of caves

When considered as an underground system, the thermal capacity of a rock body is simply its mass multiplied by the specific heat capacity of the rock. However, for a cave, which is actually an absence of rock, the problem becomes more complex. The external fluids exchange energy with the cave air, water and rock. Unfortunately, it is not possible to uniquely define the energy exchange with the rock. The rock depth involved in the thermal exchange is proportional to the square root of its duration (Isachenko *et al.* 1969).

In addition, the incoming fluids enter in thermal contact with only certain parts of the cave. Water flows along the lower parts of conduits, but the airflows behave in a similar way, with different seasonal paths. A rising air particle tends to cool (with a moist adiabatic lapse rate) and then to flow like water in the lower parts of conduits. However, when it descends, its temperature increases and thus it tends to remain in contact with the ceiling. These processes are mainly seasonal (as is the global convective airflow underground) and therefore create local thermal sedimentation, which is common and easy to measure. They play an essential part in cave ecology and in the formation of underground glaciers, but also complicate estimations of the thermal capacity of the cave.

It is easy to solve the Fourier equation for conduction (Badino 2010), which shows that the depth of the heat exchange in compact rock is 15–20 cm for diurnal variations and a few metres for annual variations. This means that we can safely assume that, for annual thermal disturbances in a cave, the thermal capacity is essentially that of the first few metres of rock containing it and always much higher than the thermal capacity of the air and water present in the galleries. The presence of fluids in a cave can, however, dampen sudden fluctuations in the temperature of the external inputs.

A model

It is easy to make a simple black box model considering a cave with a heat capacity C_{c} and a temperature T_{c}. The inflowing energy comes from water and air, heat conduction through the rock, the enthalpies of solution of salts and possibly other characteristics of the cave, such as the oxidation of organic materials, geothermal energy and biological releases.

The most important terms are those of water and air. The air and water flows (kg s^{-1}), are $F_{\mathrm{a}}(t)$ and $F_{\mathrm{w}}(t)$, with specific heat capacities of C_{a} and C_{w} (J kg^{-1} K^{-1}) and temperatures $T_{\mathrm{a}}(t)$ and $T_{\mathrm{w}}(t)$, respectively. The thermal energy deposited in C_{c} in a time Δt causes a change in the temperature of the system given by:

$$\Delta T_{\mathrm{t}} = \frac{E}{C_{\mathrm{c}}} = \frac{1}{C_{\mathrm{c}}} \times [F_{\mathrm{w}}(t)C_{\mathrm{w}}(T_{\mathrm{w}} - T_{\mathrm{c}}) + F_{\mathrm{a}}(t)C_{\mathrm{a}}(T_{\mathrm{a}} - T_{\mathrm{c}})]\Delta t \qquad (5)$$

The temperature of the incoming flow varies throughout the year with a period $\tau = 1$ a. The temperature of the system is constant, T_{c}, which means that its value is such that the heat exchange between the outside and the cave is, on average, zero. Theoretically, T_{c} is defined by the condition:

$$\int_{0}^{\tau} \{F_{\mathrm{w}}(t)[T_{\mathrm{w}}(t) - T_{\mathrm{c}}] + F_{\mathrm{a}}(t)[T_{\mathrm{a}}(t) - T_{\mathrm{c}}]\}\mathrm{d}t = 0 \qquad (6)$$

We can also compare the relative weight of the flows of air and water, at least to a first approximation. The airflows in typical alpine caves are of the order of 10 m^{3} s^{-1} per square kilometre of karst (*c.* 10 kg s^{-1} km^{-2}) – that is, a thermal capacity flow of *c.* 10^{4} J s^{-1} K^{-1} km^{-2}.

The typical water infiltration intensity P_{i} is of the order of 1000 mm a^{-1} (30 kg s^{-1} km^{-2}), which means a thermal capacity flow of 10^{5} J s^{-1} K^{-1} km^{-2}, an order of magnitude greater than that due to airflow. Even including the presence of water vapour in the estimation, which, in many conditions, can dominate the enthalpy content of air, the water flow is the main term in the equation. It is possible to conclude that, at least in temperate climates, the incoming water establishes the cave temperature.

The energy intake due to the dissolution of calcium carbonate is negligible. The energy required to dissolve calcium carbonate is −28.8 ± 0.3 kJ mol^{-1} (288 kJ kg^{-1}), resulting a fraction of a watt of energy per square kilometre of karst.

Internal energy and fluxes

There are different ways to express the idea of cave energy. We could refer to the total energy U, in joules, or to its local density in J m^{-3}. As an alternative, we could consider the time-dependent energy flow through the cave, expressed as a power (in watts). This parameter, in turn, can be estimated for each subterranean region and thus expressed in W m^{-3}. However, given the constancy of subterranean environmental conditions over time, the most important term is not the total energy content U, which remains constant, but the exchange of energy with the external environment.

The energy flows infiltrating into the ground have roughly the same outside temperatures, so they behave like a sinusoidal wave, both during the day and throughout the year:

$$f_{in}(t) = F[T_{ave} + T_A \sin(\omega t)] \quad (7)$$

The meteorological data that are most significant for this discussion are the monthly average temperature, available online (www.worldclimate.com). The term T_A, the amplitude of the sinusoidal wave, is equal to half the total thermal excursion between the monthly average during hot and cold seasons (see Appendix A).

To simplify the calculations, we use a square wave instead of a sinusoidal wave, assuming that the temperature oscillates between $(T_L - T_A)$ during the cold season and $(T_L + T_A)$ during the hot season. In numerical estimations, to maintain the energy flow in the individual periods equal to the sinusoidal case, this should be considered as a square wave amplitude reduced by a factor $2/\pi \approx 0.64$. This approach allows us to quantify the power released to a cave or to some underground region.

The thermal capacity of air is *c.* 10^3 J kg^{-1} K^{-1} and therefore the average energy released during the hot season by 1 kg of air to a cave colder by ΔT is $10^3 \Delta T$. An important energy release may be the latent heat of evaporation if the cooling air causes condensation, but this is strongly variable and can be estimated as a correction.

It is then possible to estimate the typical power of an alpine cave. As shown here, the typical fluid fluxes in 1 km^2 are 10 kg s^{-1} of air and 30 kg s^{-1} of infiltrated water. Taking T_A = 8°C, an average temperature difference of 5°C can be considered, including the square wave correction. Therefore the power released by air is 50 kW km^{-2}, whereas that released by water is *c.* 600 kW km^{-2}. The air term is underestimated because the condensation term has been ignored; however, the water term always dominates and the order of magnitude of the power released to a similar cave is between 0.1 and 1 MW km^{-2}.

It is interesting to consider the extreme case of the Subterranean River of Sabang (Philippines). The region is in the Intertropical Convergence Zone and T_A = 1.6 (www.worldclimate.com). It is a cave *c.* 33 km in length, essentially horizontal, with 25 km^2 of surface with an average freshwater flow of 1000 kg s^{-1}, a seawater flow during tides of 800 kg s^{-1}, an airflow of *c.* 100 m^3 s^{-1} and a rich biological activity (Badino *et al.* 2018).

The number of swallows and bats can be estimated in tens of thousands, with an energy release (mainly outside, during hunting flights) of the order of *c.* GJ day^{-1} – that is, a negligible 10^4 W on average. In conclusion, the power released to the cave from air is 100 kW and 8 MW from water. The cave is one of the most powerful in the world, although the power density per square kilometre is similar to that estimated for alpine caves.

The scenario changes in the case of very small caves, with entrances that do not allow significant convective airflows, such as caves with prehistoric paintings. The water fluxes can be negligible and the energy flow is due to barometric airflows (Lewis 1991). Each time that the external pressure P_A changes, an excess or a deficiency (ΔV) of the cave air volume V_c, is observed, which is given by:

$$\Delta V = V_c \frac{\Delta P}{P_A} \quad (8)$$

The atmospheric pressure is constant over a long period of time, but the total entering flux depends on the sum of the decrease in atmospheric pressure (i.e. half of the total sum of variations):

$$\frac{1}{2}\sum_j \frac{|\Delta P_j|}{P_A} = 0.01[\text{days}^{-1}] \quad (9)$$

The external air completely replaces the internal air volume in *c.* 100 days if the inflow is uniformly mixed (which is, in fact, a very strong assumption).

The airflow intensity therefore depends on the cave volume and must be studied in each separate case. However, considering a typical volume of 10^4 m^3, a flux of 100 m^3 day^{-1} is expected and the power flow becomes a few watts or less. Even with this rough estimation, it is obvious that the entry of a person into a cave of this type, with a typical energy release of 100–200 W (neglecting water vapour and carbon dioxide) has an impact far from zero. It is not surprising that painted caves are very small and have usually been found without significant entrances.

It is therefore possible to state that the average maximum power released from the external environment to a cave system is *c.* 0.1–1 MW km^{-2}, but this can decrease to 1 kW km^{-2} in the case of isolated caves and probably increases to several tens of MW km^{-2} for ponor. There is therefore a factor of 10^4 between the maximum and minimum power caves. For comparison, the Sun deposits a maximum power on the Earth's surface (Barry & Chorley

2010) of *c.* 320 kW m^{-2} in the Sahara Desert and a minimum of 50 kW m^{-2} in the North Atlantic, a mere factor of six between the two. Therefore the energetic variability of the underground world is much larger than that of the outside world.

The fluids entering underground release these energies to the surface of caves, but a similar (and more important) discussion may be carried out for the inner part of caves, considering the local fluid flows and local thermal disequilibria, and then repeating the same calculations with the corresponding local terms.

The total average energy flow in a cave over a long period of time has to be zero. Therefore the previous calculations, concentrated on the average maximum energy deposition, have a doubtful physical meaning: do they correspond to a maximum impact of the external environment on the cave, or is there a more precise parameter to characterize the environment's sensitivity to the impact of its surroundings? To answer this, it is necessary to use the second law of thermodynamics.

Entropy fluxes

The production of entropy is a key concept in studies of ecological thermodynamics (Swenson 1988), but its estimation for external systems is extremely complex (Ozawa *et al.* 2003) and has received little quantitative verification (Meysman & Bruers 2010). Caves are much simpler and it is possible to simplify the approach.

Consider an ideal cave, C_I, with entering fluids as described in the preceding section, but remaining unchanged. In particular, it is assumed that the temperature is constant. The thermal exchanges with the outside world create a variation in the entropy of the system universe + cave. In particular, there is a release of thermal energy ΔQ_H from the universe at temperature $T + \Delta T$ to the cave at temperature T, and an energy subtraction $-\Delta Q_L$ from the universe at temperature $T - \Delta T$ to the cave at T.

If the energy transfers were reversible, as in the case of an ideal Carnot engine, the total entropy of the universe and C_I would be unchanged because some work $L = \Delta Q_H - \Delta Q_L$ from the universe at temperature $T + \Delta T$ (given by Carnot cycle efficiency) would be carried out.

In fact, no work is performed in this unchangeable cave, therefore $\Delta Q_H = \Delta Q_L$ (ΔQ from here on), the process is irreversible and the entropy increases by a factor L/T:

$$\Delta S_U = -\frac{\Delta Q}{T+\Delta T} + \frac{\Delta Q}{T-\Delta T} \approx \frac{\Delta Q}{T}\left(1\frac{\Delta T}{T} + 1 + \frac{\Delta T}{T}\right) = 2\Delta Q\frac{\Delta T}{T^2} \quad (10)$$

The state of C_I is unchanged and as the entropy is a property of C_I, the cave entropy remains constant. What changes is the entropy of the universe, which decreases during the hot season (irreversible fluid cooling) and increases by a smaller factor during the cold season (irreversible fluid heating). The total increase is as expressed in equation (10).

In equation (10), ΔT is the total seasonal temperature range, ΔQ is the amplitude of seasonal energy deposition and T is the absolute cave temperature. The increase in entropy is proportional to ΔQ, which shows that the approach to classifying caves in terms of the average maximum energy absorbed (Heaton 1986) is also reasonable when applying the second law of thermodynamics.

Until now, the discussion has involved an ideal cave C_I, which is unchangeable by definition, performs a closed thermodynamic cycle and moves the entropic impact of inflowing fluids to the universe. This is unsatisfactory. Real caves have small, but non-zero, temperature ranges, show large local heterogeneities, their thermodynamics cycles are barely open and they are able to evolve because they are crossed by fragments of the universe (e.g. air, water, cavers) that release energy. It is therefore necessary to add an additional step.

Ideal and real caves

It is possible to consider a real cave C_R, with many different ($j = 1, 2, \ldots n$) local temperature ranges, ΔT_{Lj}, and energy deposition ΔQ_{Lj}. The described thermal exchange process for the universe minus C_I can be divided into two steps:

(1) an irreversible energy exchange between the universe and C_R;
(2) an irreversible energy exchange between C_R and C_I (which is in a closed cycle, $dT = 0$, $dS = 0$), $C_R - C_I$.

In the first step, the entropy production of the universe is smaller than in the ideal case and then step 2 concludes the 'fall' to the ideal situation. This means that the second step allows us to compare the difference between the real and ideal caves or, in other words, that the entropy increase of C_R measures its degradation compared with the ideal situation.

We can identify C_R with the real cave and its entropy increase as a measure of its degradation. A most precise term should be deconstruction, to indicate irreversibility and, probably, the local suppression of complex formations such as speleothems, large crystals or, on Earth, macromolecules.

$$\Delta S_{Lj} = 2\Delta Q_{Lj}\frac{\Delta T_{Lj}}{T_{Lj}^2} \quad (11)$$

The various terms now refer to local parameters and then the equation gives the local entropy production for a ΔQ_{Lj} energy inflow. The index L_j can be omitted for T because the absolute temperature of caves is almost the same everywhere in the world and the other terms are much more uncertain. Equation (11), modified in terms of fluxes, using the inflowing power W_{Lj}, becomes

$$\frac{\Delta S_{Lj}}{dt} = 2W_{Lj}\frac{\Delta T_{Lj}}{T^2} \tag{12}$$

The entropy increase ΔS gives a measure of the irreversibility of the process, in this case quantifying the gap between the real and ideal conditions of a cave.

Deconstructive power

Another, more friendly, measure of irreversibility is the term $T\Delta S$, which has the meaning of energy degraded by the irreversibility of a process or work lost due to the irreversibility.

Returning to fluxes and using dQ_{Lj}/dt – that is, the power flow to the cave region L_j – it is possible to move from the local deposited energy to the applied power ΔW_{Lj}; the $T\Delta S$ flux (W). This results in a measure of irreversible energy released locally, or deconstructive power, D_w:

$$D_w = T\frac{\Delta S_{Lj}}{dt} = W_{Lj}\frac{2\Delta T_{Lj}}{T} \tag{13}$$

that is:

$$[\text{deconstructive power}] = 2[\text{power}]\frac{[\text{temperature range}]}{[\text{temperature}]} \tag{14}$$

This parameter is fundamental. It depends on the absolute temperature of the cave and, unfortunately, on the local power fluxes (usually unknown) and the local temperature ranges (seldom measured). Estimating the local value of D_w is therefore difficult, but possible, and appropriate measures will have to be taken in the future. Experienced cavers, however, know very well that speleothems and crystals are usually associated with the more stable parts of caves, whereas they are less common along the main streams of water and air.

A rough estimation from original data taken at the Rio Martino cave (Western Alps, Italy) is indeed possible. This is a sub-horizontal resurgence 2.9 km long, where the regional cavers' association installed an internal meteorological station in 2004. The average water flow in the cave is *c.* 60 kg s^{-1} and the airflow is *c.* 2 kg s^{-1}. The temperature range near the station is *c.* 1–2°C. It is possible to estimate the power flux into the cave as 3–500 kW. This results in a global D_w of *c.* 1.5 kW. The graph in Figure 4 shows the variation in temperature due to the visit of a large group of cavers with carbide lamps. The total energy released was 140 MJ over 2 h, an average release of 20 kW. ΔT_{Lj} was measured as 0.4°C. D_w is then 25 W, smaller than the previous value. Is it possible to conclude that the impact of visits is negligible? These data are too rough and not sufficiently oriented to the local situation, but this approach is interesting.

To find a meaning for the deconstructive power, it is possible to extend the approach to the external environment. It is obvious that the external energy budget is completely different from the internal budget, which has been estimated earlier with dramatic simplifications. Nevertheless, climatic data are easily accessible (www.worldclimate.com; insolation from www.geocoops.com) and it is intriguing to see what happens when we consider the Earth as a cave. Table 2 gives the external deconstructive power along the central parts of North and South America, including Arctic and Antarctic stations. As expected, D_w reaches a minimum in the most humid and equatorial climates and a maximum in the desert regions. Are tropical rainforests the Earth's speleothems?

Conclusions

The term D_w ($T\Delta S$), estimated for local transformations underground, has many interesting characteristics:

(1) it is proportional to the energy released by flowing fluids, consistent with more traditional approaches to cave sensitivity estimations;
(2) it is always positive, both for water and air fluxes, which have different behaviours underground;
(3) it includes the internal local temperature range, which is a fundamental parameter with which to estimate the local thermal insulation;

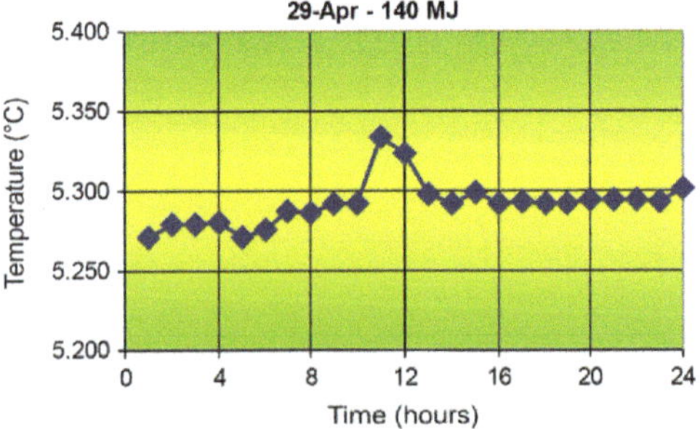

Fig. 4. Variation in temperature in the Rio Martino cave due to the visit of a large group of cavers with carbide lamps.

Table 2. *Environmental data and deconstructive power in North and South America*

	Latitude	T_L	$2\Delta T_L$	Insolation (W m^{-2})	D_w
Sachs Harbour	72	−14.2	36.1	100	14.0
Calgary	51	3.6	26.2	125	11.8
Denver	39	10.1	23.7	200	16.7
El Paso	32	17.3	21.9	250	18.9
Torreon	25	22	13.1	250	11.1
Mexico City	20	16	5	220	3.8
Managua	12	27.3	1	205	0.7
Bogota	5	13.9	0.9	200	0.6
Manaus	−3	26.6	1.5	175	0.9
Recife	−8	25.5	2.9	200	1.9
Brasilia	−16	20.6	3.3	190	2.1
São Paulo	−24	18.3	6.6	180	4.1
Cordoba	−32	17.1	13.1	225	10.2
Neuquen	−39	14.3	17.8	150	9.3
Comodoro Rivadavia	−46	12.6	12.3	130	5.6
Rio Gallegos	−52	7.1	12	100	4.3
Marambio	−64	−8.9	13	100	4.9

(4) it is proportional to the irreversibility of local thermal processes.

For these reasons, it looks to be useful in estimating the local speed of adaptation and sensitivity in caves. It will be interesting to measure it accurately, both in different caves and in different regions of the same cave, to study the morphological differences corresponding to different powers of adaptation.

Convective subterranean winds

Driving pressure of subterranean airflows

The main causes of airflows in caves are usually considered to be changes in air pressure (barometric circulation) and variations in the external temperature (convective circulation) (Bock 1913; Trombe 1952). It is shown here that, in the details of convective circulation, two further causes are hidden: barometric changes induced by variations in the outside temperature and airflow due to anomalous (non-neutral equilibrium) external atmospheric lapse rates.

The traditional treatments of convective flow use hydrostatic approximations (a constant air density with various corrections). The driving pressure is then dependent on the differences in density between the internal and external columns of air. This term is related to the differences in temperature and air humidity. It is common in meteorology to use the virtual temperature, which is the dry air temperature with the same density as that considered. Normally it is higher than the real temperature by several fractions of a degree. This convention is used here.

It is expected that the convective currents are proportional to $T_{ext} - T_{int}$. Therefore these currents stop when the outside temperature is very close to that of the cave. Original measurements made with sonic anemometers and external meteorological stations do not, in fact, match, if not in first approximation, the usual barometric plus convective model. The measurements show that there are other hidden terms in addition to the driving forces described here.

Lapse rates

The main parameter considered here is the change in temperature with altitude (lapse rate). An air particle moving vertically can become more or less dense than the surrounding air and then can be forced back, stay where it is (in neutral equilibrium) or be forced to continue rising. In dry air and in neutral equilibrium, the lapse rate is $G_{da} = -9.7°C\ km^{-1}$ (the dry adiabatic lapse rate).

This thermal gradient changes in a moist atmosphere. When an air particle rises and cools, part of its water vapour condenses, releasing the enthalpy of evaporation, which then reduces cooling. Its value depends on the pressure and temperature (Badino 2010), but in typical karst conditions it is *c.* −4 to −6°C km^{-1} (the moist adiabatic lapse rate).

Another important lapse rate for the caves is that which describes the heating of water in dissipative flow in a thermally insulated conduit, converting its potential energy into thermal energy. This is −2.34°C km^{-1} (Badino 1995) (the water adiabatic lapse rate).

In caves with a large vertical development, there is mixing of water and airflows and typical gradients in the great alpine caves are between −3 and −4°C km^{-1} (Badino 2000) (Fig. 5). On average, the atmosphere has a lapse rate of −6.5°C km^{-1} (G_{ISA}, International Standard Atmosphere).

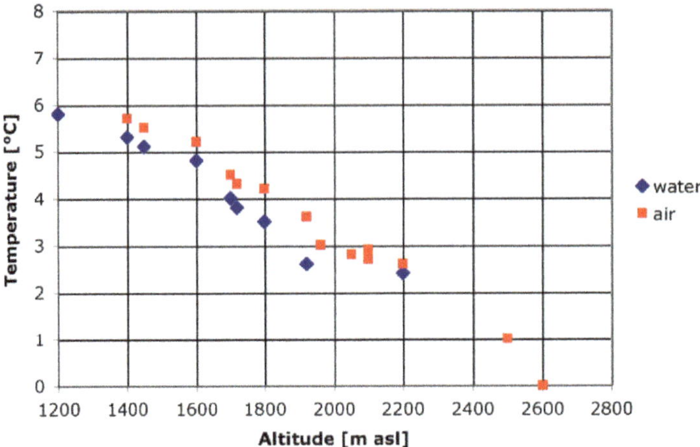

Fig. 5. Typical temperature gradients in the Complesso di Piaggia Bella cave.

The reality is more complex because air masses are not always in equilibrium. Cold air can immobilize under a hotter mass, creating thermal inversion, especially in winter. In other cases, the real lapse rate G_A is almost zero or it varies rapidly with altitude (Fig. 6). A knowledge of the actual lapse rate is always important for weather forecasting, therefore worldwide there are *c.* 800 sounding stations, which launch weather balloons every day at 00:00 and 12:00 h, rising up to *c.* 30 km altitude. The data sent to the ground are instantly available for weather forecasts (weather.uwyo.edu/upperair/sounding.html). These data show that in alpine environments (data from Levaldigi airport, Western Alps) during summer, the G_A lapse rate is between −9 and −4°C km^{-1}, whereas in winter it ranges from −10 to >5°C km^{-1}.

Driving pressure of convective flows

The calculation will be developed in this way:

(1) to determine how the external pressure varies with altitude, with different lapse rates;
(2) to determine the pressure profile in the underground atmosphere;
(3) to connect the two atmospheres by two extreme entrances;
(4) to calculate the driving pressure at the entrances.

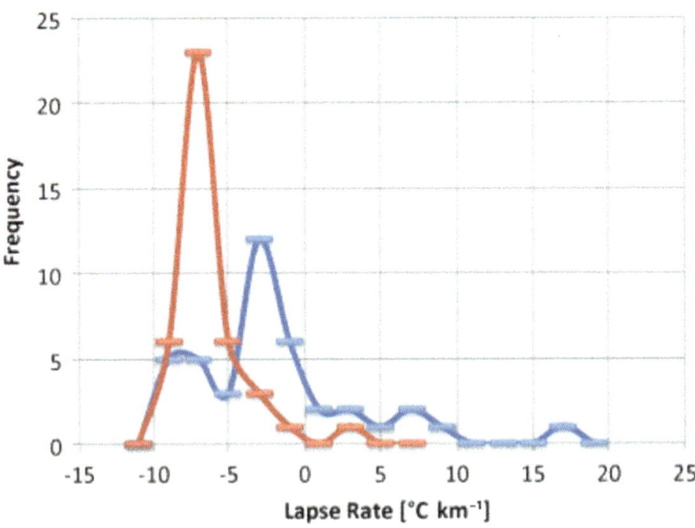

Fig. 6. Lapse rate distribution summer and winter, Levaldigi airport, Western Alps.

The term G_A is the external lapse rate, but we now use the convention to eliminate the negative sign, assuming that $T(z) = T_{A0} - G_A z$, with a negative mark.

It follows that the pressure change with altitude is:

$$\frac{dP}{P} = -\frac{M_{mol}g}{R(T_{A0} - G_A z)}dz \tag{15}$$

with the solution:

$$P(z) = P_{A0}\left[1 - \frac{G_A}{T_{A0}}z\right]^s \tag{16}$$

where P_{A0} and T_{A0} are the atmospheric pressure and temperature at $z = 0$. A dimensionless exponent s appears, which depends on the thermodynamic transformation performed by the air particle moving vertically:

$$\begin{cases} s_{da} = \dfrac{M_{mol}g}{RG} = \dfrac{c_p}{R} = 3.5 \rightarrow \text{dry air} \\ s_{ISA} = s_{da}\left(\dfrac{G_{da}}{G_{ISA}}\right) = 3.5 \times \left(\dfrac{9.7}{6.5}\right) \approx 5.22 \rightarrow \text{in ISA} \\ s_c = s_{da}\left(\dfrac{G_{da}}{G_c}\right) = 3.5 \times \left(\dfrac{9.7}{3.5}\right) \approx 9.70 \rightarrow \text{in caves} \end{cases} \tag{17}$$

In the following, equation (16) is approximated by:

$$(1 - x)^n \approx 1 - nx + \frac{n(n-1)}{2}x^2 \tag{18}$$

For $x \ll 1$, only linear and quadratic terms are used. The linear term gives:

$$P(z) = P_{A0}\left(1 - s_A \frac{G_A}{T_{A0}}z\right) \tag{19}$$

which is the classical hydrostatic approximation (constant air density) with an error <1%, but is already too large to discuss air currents in caves. The pressure differences that move even strong air currents (5 m s^{-1}) are of the order of 100 Pa, which is one-thousandth of the atmospheric pressure at ground level. Therefore it is necessary to move to the second-order approximation, which allows a qualitative analysis of the terms involved.

Figure 7 illustrates the parameters for the linear approximation in equation (19) on a graph $P(z)$. The cave is encased in the altitude range $h < z < h + H$.

The pressure gradient is constant and depends only on the temperature at $z = 0$:

$$\frac{dP(z)}{dz} = -B_{A0} = -\frac{M_{mol}g}{R}\left(\frac{1}{T_{A0}}\right) = 0.034\left(\frac{1}{T_{A0}}\right) \tag{20}$$

The pressure differences ΔP_1 and ΔP_2 at each entrance are different because their value depends on the local friction loss (Badino 1995), but the total algebraic sum of the two differences must be the total driving pressure of the system, ΔP_{ie}.

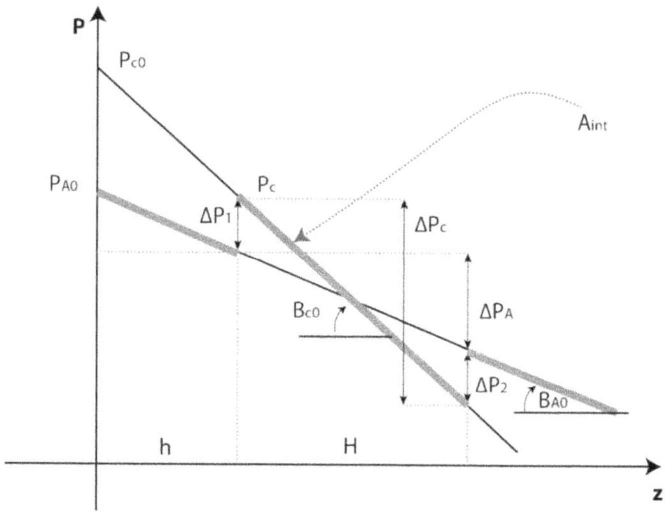

Fig. 7. Illustration of the parameters for the linear approximation in equation (19).

This is independent of friction losses, either at the entrances or along the internal conduits. In the following it is generally assumed that a large upper entrance is present (that is, the internal and external pressure are equal) and that the lower entrance acts as the bottleneck of the flow system (Badino 1995).

Figure 8 shows the situation in different seasons. In area A, there is a sudden drop in pressure at the lower entrance, causing an airflow to exit. The case C is winter and B is an intermediate condition with no airflow. It appears that the convective circulation depends on the gradient of the curve $P(z)$ and not directly on the differences among the external and internal temperatures.

The equation giving $P(z)$ shows that, even maintaining a constant ground pressure P_{A0}, if the temperature T_{A0} increases, the atmosphere swells and, if it drops, the atmosphere deflates, just like a balloon. Therefore at mountain altitudes, part of the atmosphere goes either over or below the cave entrances, changing the local pressure. As a consequence, when the temperature T_{A0} drops, the pressure at $z > 0$ also goes down.

Second-order solutions: new driving pressures

It is possible to develop the calculation for the two atmospheres, external and internal. Assuming that the regulating bottleneck is at the lower entrance, the exact solution is obtained by subtracting the change in pressure between outside and the cave atmosphere in the altitude range of the cave ($h < z < h + H$). Applying equation (16) and subtracting, we obtain:

$$\Delta P_1 = P_{A0}\left[\left(1-\frac{G_c}{T_{c0}}h\right)^{S_c}\left(1-\frac{G_A}{T_{A0}}(h+H)\right)^{S_A} - \left(1-\frac{G_A}{T_{A0}}h\right)^{S_A}\left(1-\frac{G_c}{T_{c0}}(h+H)\right)^{S_c}\right] \times \left(1-\frac{G_c}{T_{c0}}(h+H)\right)^{-S_c} \quad (21)$$

Applying a second-order approximation to each term gives:

$$\Delta P_{ieB} \approx \frac{\rho_{A0}gH}{T_{c0}}\left[(T_{A0}-T_{c0})\left(1-\frac{h}{L_{c0}}\right) + \frac{(H+2h)}{2}(G_c - G_A)\right] \quad (22)$$

where $L_{c0} = 1/B_{c0} \approx 8$ km. The first term in square brackets is the classic, hydrostatic, approximation, whereas the second term depends on the actual differences in lapse rate, which can cause a driving pressure if the first term is zero. There is also an airflow, with thermal equilibrium between inside and outside. This is called the gradient of convective circulation.

The dependence on G_A (and, of course, on G_c, which nevertheless is always close to 3.5°C km^{-1}) becomes increasingly important with the altitude of the lower entrance. Low values of G_A, as on days of high humidity with air particles moving upwards, create driving pressures that are higher during the

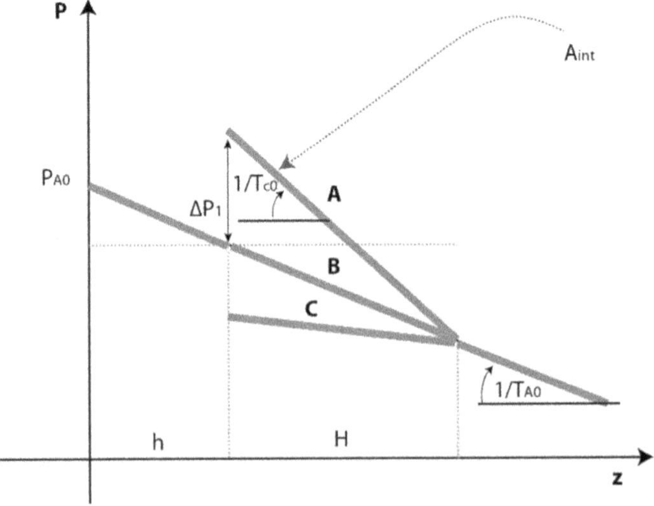

Fig. 8. Illustration of the parameters for the linear approximation in equation (19) in different seasons.

summer than in winter. In a similar way, high values of G_A, as occur on dry days with downwards moving air particles, create different driving pressures depending on the season. These results show that at high altitude the intensity of convective movement is highly dependent on the external atmospheric gradient and is more intense during summer if the external gradient is small and during winter if the external gradient is large. In the intermediate stages, even the direction of the movement depends on G_A, especially at high altitudes.

Equation (16) gives another interesting result: a temperature change at ground level ($z = 0$) provokes a general variation in the volume of the atmosphere. This, in turn, causes a pressure change with altitude, even at constant pressure P_{A0}. Such a temperature change causes barometric flows if the cave has a significant volume. In fact, a partial derivative of the function $P(z)$ gives, at constant pressure P_{A0}:

$$\frac{dP(z)}{dt} = P_{A0}s\frac{G_A z}{T_{A0}^2}\left(1 - \frac{G_A}{T_{A0}}z\right)^{s-1}\left(\frac{\partial T_{A0}}{\partial t}\right) \quad (23)$$

Therefore a change in the external temperature corresponds to a phase of barometric air movement into the cave, although $dP_{A0} = 0$ because the external atmosphere in front of the cave swells ($dT_{A0} > 0$) or contracts ($dT_{A0} < 0$) by sending air masses above or below the entrance. It is easy to show that this term is very important. Under real conditions, the variations in pressure have a typical order of magnitude of 100 Pa h^{-1}, whereas those of temperature have a typical order of magnitude of 1°C h^{-1}. This corresponds to:

$$\frac{1}{P_{A0}}\frac{dP_{A0}}{dt} = 10^{-3}[h^{-1}]$$
$$\frac{1}{T_{A0}}\frac{dT_{A0}}{dt} = 3 \times 10^{-3}[h^{-1}] \quad (24)$$

This shows that the barometric circulation due to variations in temperature could dominate those due to changes in pressure at sea-level for caves at altitudes of <1000 m. Each cave with a significant vertical development therefore shows barometric airflows induced by changes in temperature. The driving pressure depends on the altitude and is different at each entrance, which is probably a significant difference compared with the barometric circulation due to changes in P_{A0}. This implies, for example, that the caves tend to blow air out in the morning ($dT_{A0} > 0$) and to take air in during the evening. It is possible to call this a thermal barometric circulation, pointing out that the usual barometric circulation is the air pressure measured at the cave entrance and not at $z = 0$.

Neutral temperature of convection

The hydrostatic approach to subterranean convective winds predicts that the driving pressure, and then the airflow, reach zero when the external temperature is close to the internal temperature. Experimentally, this statement has been shown to be false. Previous equations allow the estimation to be improved.

The exact solution of $P(z)$ can be developed to obtain the temperature T_{A0} at which the convection stops, at $z = 0$, referred to as X_{A0}. The solution is:

$$X_{A0} = G_A\left[h + \frac{H}{1 - (1 - (G_c H/(T_{c0} - G_c h)))^{S_c/S_A}}\right] \quad (25)$$

For more intuitive results, we give a second-order approximation at $z = h$:

$$X_A(h) = T_c + \frac{1}{2}(G_A - G_c)H \quad (26)$$

If the external gradient is constant up to the upper cave entrance at $h + H$, the air circulation stops when the external temperature at the lower entrance is the cave temperature at $z = h$, plus a corrective term. This term is given by half the difference in lapse rate multiplied by the vertical extent of the cave.

It has been shown above that if the summer lapse rates are usually between 9 and 4°C km^{-1}, then

$$T_c + 0.25H < X_A(h) < T_c + 1.25H \quad (27)$$

During the winter this is between 10 and −5°C km^{-1}:

$$T_c - 8.25H < X_A(h) < T_c + 1.75H \quad (28)$$

This shows that in the case of caves with a large vertical development, the temperature correction, especially in winter and at night, could be several degrees centigrade. It is therefore not true that convective flows stop when $T_{ext} = T_{int}$.

Case study: Su Bentu cave

Su Bentu cave, located in NE Sardinia, is one of most important caves in Italy. About 15 km long, it is the final part of the enormous subterranean drainage system of Supramonte, extended over 170 km^2 (Cabras *et al.* 2008). A strong airflow passes through it, exiting during the summer months. The entrance is in the Lanaittu valley, a dry syncline basin at 200 m a.s.l., just 100 m above the resurgence of Su Cologone, surrounded by mountains up to 1000 m high. Near the entrance there is a meteorological station installed by the Servizio Agrometeorologico

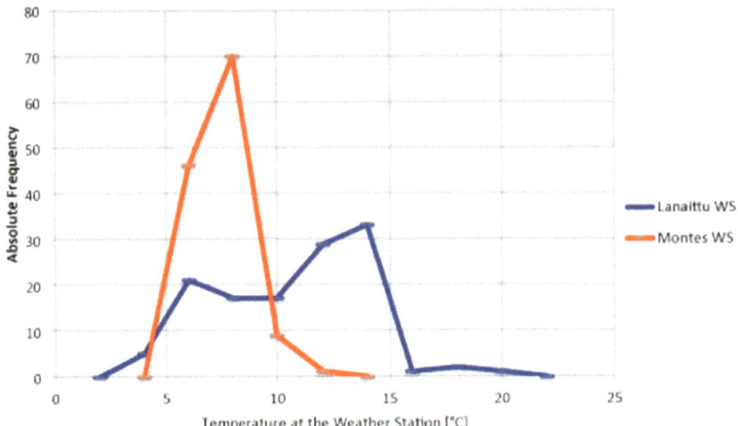

Fig. 9. Neutral temperature of convection with lapse rate ($v < 0.2$ m s^{-1}), Su Bentu Cave, Sardinia.

Regionale (Valle di Lanaittu, 135 m a.s.l.); another station is located nearby (Orgosolo Montes, 1211 m a.s.l.), in a more favourable place on the slope of a mountain 19 km SW of the entrance station.

An internal meteorological station was installed *c.* 500 m from the Su Bentu cave entrance in 2011 under the framework of ESA-Caves, a training course for astronauts organized by the European Space Agency (Bessone *et al.* 2013). The station also measures the airflow using a Gill sonic anemometer, giving unusually complete experimental data for micrometeorological cave studies.

The cave temperature at the lower altitude entrance is 14.9°C, with a relative humidity of 100% and a virtual temperature of 16.7°C. It is possible to estimate the experimental neutral temperature. The data relate to the period from the end of September to the end of December, the humid season in Sardinia. Only night-time data are considered to eliminate instrumental effects related to solar radiation. Figure 9 shows the two external temperatures at the time the airflow stops ($v < 0.2$ m s^{-1}). The neutral temperature given by Lanaittu station is *c.* 4°C lower than expected and has a wide variation ($X_A(200) = 10.6 \pm 3.2$°C), whereas the neutral temperature at Orgosolo Montes station is more regular, with $X_A(1200) = 7.5 \pm 1.2$°C. Using the International Standard Atmosphere lapse rate of 6.5° C km^{-1}, this corresponds to $X_A(200) = 14 \pm 1.2$°C, close to that of the cave. The two distributions are completely different. In particular, it is difficult to explain why the mass of the cave air, which is relatively warm, is unable to stay in balance with the outside atmosphere even when the air near the cave is considerably colder.

The presence of two stations at different altitudes have allowed data to be analysed by (1) calculating the experimental lapse rate G_A at the moment

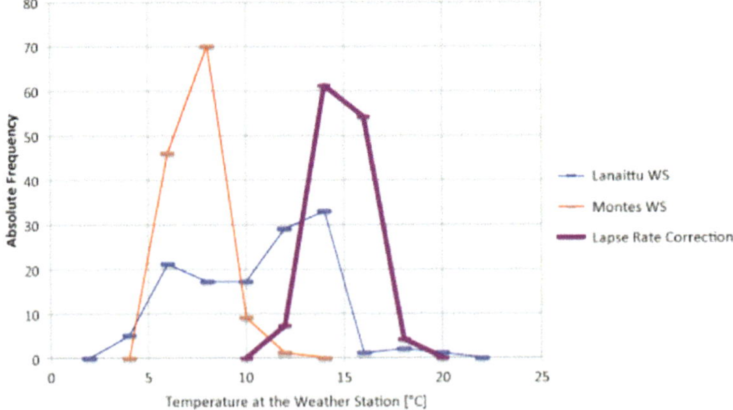

Fig. 10. Neutral temperature of convection with lapse rate ($v < 0.2$ m s^{-1}) at the moment when the airflow stops, Su Bentu Cave, Sardinia. The graph shows also the lapse rate correction; see text for explanation.

when the airflow stops; and (2) applying the equation with $H = 900$ m and $G_c = 3.5°C\ km^{-1}$ to obtain the expected value of X_A under these conditions. The results are shown in Figure 10. The distribution for Orgosolo Montes is completely superimposed on the experimental distribution, but is shifted by 6.5° C. The variability in lapse rate explains the distribution width of X_A. The Lanaittu data are not coherent, but are easily explained: the station is exposed to katabatic, night-time winds from the surrounding mountains. The local, external temperature at the cave entrance is not representative of the external air columns that drive the Su Bentu wind. The traditional description of convective circulations does not work here from a physical point of view, but is a reasonable approximation compared with the typical uncertainties involved in these problems.

A more accurate calculation showed that the convective circulations are linked to the pressure profiles of the internal and external atmospheres, with a key part played by their lapse rates. Variations in the outside temperature, which cause variations in the volume of the atmosphere, induce variations in pressure that cause barometric circulations with a constant sea-level pressure. The experimental verification of this more complete model is difficult, but will probably lead to an understanding of the airflow phenomenology of special caves, such as those on Mars.

Conclusions

A cave is commonly considered as a nearly invariable environment. Daily or seasonal climatic cycles cannot infiltrate the cave, the weather remains stable and meteorological fluctuations are absent. These stable conditions occur homogeneously anywhere inside a cave; the temperature does not change throughout the year and constant values are maintained throughout the cave system. Caves thus appear to be in the realm of invariability.

These beliefs are generated by the observation of only small variations in amplitude of the both internal temperature and humidity relative to external variations; however, this does not justify the common thinking that neither climatic nor meteorological processes take place in caves. In the real world, external cycles appear underground in the form of variations in the direction and intensity of airflows, almost imperceptible variations in the seasonal and daily temperatures, local disequilibria in air and water temperatures, local thermal air sedimentation or the condensation of water onto the cave walls. The best way to highlight these processes is to carefully analyse the energy exchanges, which always occur locally with small differences of temperature, condensation–evaporation processes and air flux.

The phenomenology of the two main flowing fluids in a cave (water and air) is completely different. Water can only flow downwards on the lowest part of conduits. As a result of its enormous thermal capacity, water generally dominates thermal exchanges. Water usually subtracts energy from the underground systems.

By contrast, air can flow everywhere and can reverse its direction of flow and, consequently, thermal exchanges. Air can make rock surfaces wet and can smooth variations in temperature and dominate energy exchanges via its water vapour content. More importantly, the thermodynamic transformations of air are not adiabatic. This means that a thermal exchange will take place continuously between air and water. The airflow generally releases energy to the cave as a whole.

Measurements suggest that the meteorological variability among caves, and also among different points of a cave, are much larger than that of the external environment, where the astonishing variability of the Earth's surface is created by relatively small differences in precipitation, temperature and absolute humidity. The characterization of these micrometeorological variations is probably the key to a better understanding of the underground world.

Appendix A

Symbol	Meaning	Dimensions
T_L	Local average yearly temperature	°C
T_{IW}	Infiltration water temperature	°C
T_j	Average monthly temperature	°C
P_j	Average monthly precipitation	mm or kg m^{-2}
T_R	Rock temperature	°C
T_c	Cave temperature	°C
ΔT_u	Difference between T_R and T_L	K or °C
F_j	Generic energy flow	W
C_c	Cave heat capacity	J K^{-1}
C_a	Air specific heat capacity at $dP = 0$	J K^{-1} kg^{-1}
C_w	Water specific heat capacity	J K^{-1} kg^{-1}
F_a	Air flow	kg s^{-1}
F_w	Water flow	kg s^{-1}
T_a	Inflowing air temperature	°C
T_w	Inflowing water temperature	°C

(*Continued*)

Symbol	Meaning	Dimensions
ΔT_c	Cave temperature variation	K or °C
T_A	Amplitude of local monthly T variation	K or °C
P_A	Atmospheric pressure	Pa
V_c	Cave volume	m^3
ΔP_j	Atmospheric pressure change	Pa
ΔQ	Thermal energy exchange	J
ΔS_U	Entropy change of the universe	$J\ K^{-1}$
ΔT_{Lj}	Local temperature range in cave	K or °C
ΔQ_{Lj}	Local thermal energy exchanged	J
ΔS_{Lj}	Local entropy variation	$J\ K^{-1}$
ΔW_{Lj}	Local inflowing power	W
D_w	Deconstructive power	W
z	Altitude	km
h	Lower entrance altitude	km
H	Cave vertical development	km
M_{mol}	Air molar mass	$g\ mol^{-1}$
R	Ideal gases constant	$J\ K^{-1}$
P_{A0}	External pressure at $z = 0$	Pa
G_{da}	Dry adiabatic lapse rate	$K\ km^{-1}$ or $°C\ km^{-1}$
G_{ISA}	International Standard Atmosphere lapse rate	$K\ km^{-1}$ or $°C\ km^{-1}$
G_A	Generic external lapse rate	$K\ km^{-1}$ or $°C\ km^{-1}$
G_c	Internal lapse rate	$K\ km^{-1}$
g	Gravitational acceleration	$m\ s^{-2}$
$P(z)$	Pressure at z	Pa
P_c	Cave pressure at $z = h$	Pa
B_{A0}	External pressure gradient	$Pa\ km^{-1}$
B_{c0}	Internal pressure gradient	$Pa\ km^{-1}$
$T(z)$	Temperature at z	K or °C
T_{A0}	External temperature at $z = 0$	K or °C
T_{c0}	Cave temperature at $z = 0$	K or °C
T_c	Cave temperature at $z = h$	K or °C
s	Politropic index	Non-dimensional
s_A	External politropic index	Non-dimensional
s_c	Internal politropic index	Non-dimensional
ΔP_1	$P_{ext} - P_{int}$ at $z = h$ (lower entrance)	Pa
ΔP_2	$P_{ext} - P_{int}$ at $z = h + H$ (higher entrance)	Pa
ΔP_{ie}	Total driving pressure	Pa

References

Badino, G. 1995. Fisica del clima sotterraneo. *Memorie Istituto Italiano di Speleologia, Bologna*, **7**, II.

Badino, G. 2000. I gradienti di temperatura nei monti, un indicatore esplorativo. *Talp*, **21**, 72–80.

Badino, G. 2004. Cave temperatures and global climatic changes. *International Journal of Speleology*, **33**, 103–114.

Badino, G. 2005. Underground drainage systems and geothermal flux. *Acta Carsologica*, **34**, 277–316.

Badino, G. 2010. Underground meteorology, what's the weather underground? *Acta Carsologica*, **39**, 427–448.

Badino, G. 2013. Micrometeorology of Mt Cronio Caves, Sicily. *In*: Moore, K. & White, S. (eds) *Proceedings of the 17th International Congress of Speleology*, 22–28 July 2017, Sydney, Australia, Vol. 2. Australian Speleological Federation, Sydney, 339.

Badino, G. 2017. Driving pressure of subterranean airflows: an analysis. *In*: Moore, K. & White, S. (eds) *Proceedings of the 17th International Congress of Speleology*, 22–28 July 2017, Sydney, Australia, Vol. 2. Australian Speleological Federation, Sydney, 205–208.

Badino, G., De Vivo, A., Forti, P. & Piccini, L. 2018. The Puerto Princesa underground river (Palawan, Philippines): some peculiar features of a tropical, high-energy coastal karst system. *In*: Parise, M., Gabrovsek, F., Kaufmann, G. & Ravbar, N. (eds) *Advances in Karst Research: Theory, Fieldwork and Applications*. Geological Society, London, Special Publications, **466**. First published online January 24, 2018, https://doi.org/10.1144/SP466.22

Barry, R. & Chorley, R. 2010. *Atmosphere, Weather and Climate*. Routledge, London.

Bernabei, T. & De Vivo, A. 1992. *Grotte e Storie dell'Asia Centrale*. CEV, La Venta, Treviso.

Bessone, L., Beblo-Vranesevic, K. *et al.* 2013. Esa Caves: training astronauts for SPACE exploration. *In*: Filippi, M. & Bosak, P. (eds) *Proceedings of the 16th International Congress of Speleology*, 19–27 July 2013, Brno, Czech Republic, Vol. 1. Czech Speleological Society, Brno, 321–327.

Bock, H. 1913. Mathematisch-physicalishe Untersuchung der Eishöhlen und Windröhren. *In*: Simonys, F. (ed.) *Höhlen im Dachstein*. Varlage des Vereines für Höhlenkunde in Österreich, Graz, 102–147.

Bohren, C. & Albrecht, B. 1998. *Atmospheric Thermodynamics*. Oxford University Press, Oxford.

Cabras, S., De Waele, J. & Sanna, L. 2008. Caves and karst aquifer drainage of supramonte (Sardinia, Italy): a review. *Acta Carsologica*, **37**, 227–240.

Celico, P. 1986. *Prospezioni Idrogeologiche*. Liguori, Naples.

Cigna, A. 1967. An analytical study of air circulation in caves. *International Journal of Speleology*, **3**, 41–54.

Cigna, A.A. 1993. Environmental management of tourist caves: the examples of Grotte di Castellana and Grotta Grande del Vento, *Italy*. *Environmental Geology*, **21**, 173–180.

Cigna, A.A. 2002. Modern trend in cave monitoring. *Acta Carsologica*, **31**, 35–54.

Cigna, A.A. 2017. Hermann bock, a forgotten precursor of cave meteorology. *In*: Moore, K. & White, S. (eds) *Proceedings of the 17th International Congress of Speleology*, 22–28 July 2017, Sydney, Australia, Vol. 1. Australian Speleological Federation, Sydney, **1**, 423–425.

Davies, J.H. & Davies, D.R. 2010. Earth's surface heat flux. *Solid Earth*, **1**, 5–24.

De Freitas, C.R. 2010. The role and importance of cave microclimate in the sustainable use and management of show caves. *Acta Carsologica*, **39**, 477–489.

De Freitas, C.R., Littlejohn, R.N., Clarkson, T.S. & Kristament, I.S. 1982. Cave climate: assessment of airflow and ventilation. *International Journal of Climatology*, **2**, 383–397.

Dreybrodt, W., Gabrovšek, F. & Perne, M. 2005. Condensation corrosion: a theoretical approach. *Acta Carsologica*, **34**, 317–348.

Fairchild, I.J. & Baker, A. 2012. *Speleothem Science*. Wiley-Blackwell, Chichester.

Gabrovšek, F., Dreybrodt, W. & Perne, M. 2010. Physics of condensation corrosion in caves. *In*: Andreo, B., Carrasco, F., Durán, J.J. & LaMoreaux, J.W. (eds) *Advances in Research in Karst Media*. Environmental Earth Sciences. Springer, Berlin, 491–496.

Gadoros, M. 1989. The physical system of speleoclimate. *In*: Filippi, M. & Bosak, P. (eds) *Proceedings of the 16th International Congress of Speleology*, 19–27 July 2013, Brno, Czech Republic, Vol. 3. Czech Speleological Society, Brno, 752–754.

Guichonnet, P. 1967. *Il Traforo del Monte Bianco*. Mondadori, Milan.

Heaton, T. 1986. A tremendous range in energy environments on earth. *NSS News*, **44**(8), 301–304.

Isachenko, V., Osipova, V. & Sukomel, A. 1969. *Heat Transfer*. Mir Publishers, Moscow.

Jernigan, J. & Swift, R. 2001. A mathematical model of air temperature in Mammoth Cave, Kentucky. *Journal of Cave and Karst Studies*, **63**, 3–8.

Klimchouk, A. 2012. Krubera (Voronja) cave. *In*: White, W.B. & Culver, D.C. (eds) *Encyclopedia of Caves*. Elsevier, New York, 443–450.

Kuzmina, L.Y., Galimzianova, N.F., Abdullin, S.R. & Ryabova, A.S. 2012. Microbiota of the kinderlinskaya cave (South Urals, Russia). *Microbiology*, **81**, 251–258.

Lewis, W. 1991. Atmospheric pressure changes and cave airflow. *NSS Bulletin*, **53**(1), 1–12.

Lismonde, B. 2002. *Climatologie du monde souterrain*. Vols 1 & 2. Comité Départemental de Spéléologie de Isère, Echirolles.

Lomonosov, M. 1732. *De motu aeris in fodinis observatock*. Novi Comm. Acad. Petropolitanae, **I**.

Luetscher, M. 2013. Glacial processes in caves. *In*: Frumkin, A. (ed.) *Karst Geomorphology*. Treatise on Geomorphology, **6**. Elsevier, Oxford, 258–266.

Mammola, S., Giachino, P.M., Piano, E., Jones, A., Barberis, M., Badino, G. & Isaia, M. 2016. Ecology and sampling techniques of an understudied subterranean habitat: the Milieu Souterrain Superficiel (MSS). *The Science of Nature*, **103**, 88.

Mavlyudov, B.R. 2005. Glacial karst as possible reason of quick degradation of Scandinavian glacier sheet. *In*: Mavlyudov, B.R. (ed.) *Glacier Caves and Glacial Karst in High Mountains and Polar Regions*. Institute of Geography of the Russian Academy of Sciences, Moscow, 68–73.

McDermott, F. 2004. Palaeo-climate reconstruction from stable isotope variations in speleothems: a review. *Quaternary Science Reviews*, **23**, 901–918.

Meysman, F. & Bruers, S. 2010. Ecosystem functioning and maximum entropy production: a quantitative test of hypotheses. *Philosophical Transactions of the Royal Society of London Series B: Biological Sciences*, **365**, 1405–1416.

North, L.A., van Beynen, P.E. & Parise, M. 2009. Interregional comparison of karst disturbance: west central Florida and southeast Italy. *Journal of Environmental Management*, **90**, 1770–1781.

Ozawa, H., Ohmura, A., Lorenz, R. & Pujol, T. 2003. The second law of thermodynamics and the global climate system: a review of the maximum entropy production principle. *Reviews in Geophysics*, **41**, 1018.

Parise, M. 2011. Some considerations on show cave management issues in southern Italy. *In*: Van Beynen, P.E. (ed.) *Karst Management*. Springer, Berlin, 159–167.

Pflisch, A., Wiles, M., Horroks, R., Piasecki, J. & Ringeis, J. 2010. Dynamic climatologic processes of barometric cave systems using the example of Jewel Cave and Wind Cave in South Dakota, USA. *Acta Carsologica*, **39**, 449–462.

Pinna, M. 1977. *Climatologia*. UTET, Torino.

Sendra, A. & Reboleira, A.S.P. 2012. The world's deepest subterranean community – Krubera-Voronja Cave (Western Caucasus). *International Journal of Speleology*, **41**, 221–230.

Shiklomanov, N. & Nelson, F. 2013. Thermokarst and civil infrastructure. *In*: Giardini, R. & Harbor, J. (eds) *Glacial, Periglacial Geomorphology*. Treatise on Geomorphology, **8**. Elsevier, Oxford, 354–373.

Swenson, R. 1988. Emergence and the principle of maximum entropy production. *In*: *Proceedings of the 32nd Annual Meeting of the International Society for General Systems Research*, **32**.

Trombe, F. 1952. *Traité de spéléologie*. Payot, Paris.

Evaluating the susceptibility to anthropogenic sinkholes in Apulian calcarenites, southern Italy

A. FIORE[1], N. L. FAZIO[2], P. LOLLINO[2], M. LUISI[1], M. N. MICCOLI[1], R. PAGLIARULO[2], M. PERROTTI[2], L. PISANO[2], L. SPALLUTO[1,3], C. VENNARI[2], G. VESSIA[2,4] & M. PARISE[2,3]*

[1]*Basin Authority of Apulia, c/o Innova Puglia S.P.A (ex tecnopolis CSATA), Str. Prov. per Casamassima km 3 – 70010 Valenzano, Bari, Italy*

[2]*CNR-IRPI, Via Amendola 122-I, 70126, Bari, Italy*

[3]*Department of Earth and Environmental Sciences, University Aldo Moro, Via Orabona 4, 70125, Bari, Italy*

[4]*University of Chieti – Pescara, Via dei Vestini 31, Chieti Scalo, Italy*

**Correspondence: mario.parise@uniba.it*

Abstract: Sinkholes are the main hazard related to underground voids of both natural and anthropogenic origin. Instabilities developing underground may propagate upwards in a dramatic manner and reach the surface in the form of a sinkhole. The Apulia region in southern Italy is an interesting case study due to the outcropping of soluble rocks throughout the region. These rocks are affected by karst processes and have a high number of anthropogenic cavities. The latter were excavated by humans at different times for a variety of purposes. The worrying recent increase in the number of sinkhole events registered in Apulia led us to collect information on natural and anthropogenic sinkholes in Apulia. We focused on anthropogenic cavities, mostly excavated in Plio-Pleistocene calcarenites, and characterized the rock masses before using two- and three-dimensional parametric numerical analyses to model the instability processes, with the aim of exploring the failure mechanisms that lead to the occurrence of sinkholes. The parametric studies allowed us to carry out a preliminary evaluation of the stability conditions through simple charts designed for use in the field.

The occurrence of sinkholes, which may be related to both natural caves and anthropogenic cavities, has received increasing attention from scientists and land use planners in recent years due to a number of collapses affecting many urban areas and infrastructures worldwide (Yan *et al.* 2001; Closson *et al.* 2005, 2010; Canakci 2007; Parise & Gunn 2007; Waltham & Lu 2007; Gutiérrez 2010; De Waele *et al.* 2011; Nof *et al.* 2013). Collapse events at Guatemala City in February 2007 and May 2010 (Hermosilla 2012) and in Florida in 2013 (which resulted in one casualty and was reactivated in 2014) also attracted the attention of the mass media. These sinkhole events triggered a number of monitoring studies with the aim of assessing the possibility of identifying precursors to the final stage of collapse (Swedzicki 2001; Baer *et al.* 2002; Abelson *et al.* 2003; Aydan *et al.* 2005; Castañeda *et al.* 2009; Filin & Baruch 2010; Gutiérrez *et al.* 2011; Klein *et al.* 2011).

Sinkholes often occur in Italy and many regions consider sinkholes to be among the many geological hazards active in their territory. Typically triggered by meteoric events, sinkholes have also been produced (or reactivated) by some of the seismic shocks that have occurred in central Italy in recent years (Kawashima *et al.* 2010; Parise *et al.* 2010, in the aftermath of the 2009 L'Aquila earthquake; Borgatti *et al.* 2013, following the Emilia Romagna earthquake in May–June 2012; and several online newspapers reporting sinkholes after the 2016 seismic sequence in Latium, Abruzzo and Marche).

If we focus our attention on those sectors of the country characterized by soluble rocks, and which are therefore affected by karst processes, the number of hazardous events is high and sinkholes are the main geological hazard (Parise & Vennari 2013; Parise *et al.* 2013*a*). This is definitely the case for Apulia in southeastern Italy, a region where karst is well developed with >2000 natural caves and the presence of a high number of anthropogenic cavities. The increase in the frequency of sinkholes observed during the last few decades led to the Basin Authority of Apulia (AdB Puglia) issuing specific rules addressing the assessment of risk. The main mission of AdB Puglia is land planning and management

From: Parise, M., Gabrovsek, F., Kaufmann, G. & Ravbar, N. (eds) 2018. *Advances in Karst Research: Theory, Fieldwork and Applications*. Geological Society, London, Special Publications, **466**, 381–396.
First published online January 4, 2018, https://doi.org/10.1144/SP466.20

and the control of existing rules and procedures in construction practices with the aim of safeguarding the natural landscape and private and public properties.

This paper focuses on anthropogenic cavities carved within the calcarenites of Apulia; cavities of natural origin and cavities in different lithologies are excluded from the analysis. We give a brief introduction to the geology of Apulia and its sinkhole hazard and describe the collection of information on sinkholes in Apulia, characterization of the physical and mechanical properties of the calcarenite rocks, the implementation of numerical analyses for modelling the instability processes and the design of monitoring activities in selected natural and anthropogenic cavities.

Sinkhole hazards in Apulia

The Apulia region in southern Italy represents the foreland of the southern Italian Apennines (Selli 1962; D'Argenio *et al.* 1973; Ricchetti *et al.* 1988). It is mostly made up of a 6–7 km thick succession of Mesozoic shallow water limestones and dolostones, locally covered by thin and discontinuous Tertiary and Quaternary carbonate and clastic deposits (Ciaranfi *et al.* 1988; Ricchetti *et al.* 1988). Mesozoic rocks are well-bedded and form a hard bedrock affected by gentle folds and by several sets of faults and joints (Doglioni *et al.* 1996; Pieri *et al.* 1997; Festa 2003). Larger brittle deformation zones segment the foreland into several blocks with different degrees of uplift (Doglioni *et al.* 1994; Gambini & Tozzi 1996; Pieri *et al.* 1997). Poorly deformed Tertiary and Quaternary deposits fill the tectonic depressions among the uplifted blocks and unconformably overlie the Mesozoic bedrock (Fig. 1).

As a result of the long subaerial exposure of the Mesozoic succession, the carbonate bedrock records the development of a dense network of karst cavities in the subsurface, at least partly controlled by tectonic discontinuities (e.g. Festa *et al.* 2014; Pepe & Parise 2014). As a result, the Apulia region has a strong susceptibility to natural sinkholes. These are initiated by the frequent breakdown processes occurring in karst caves as part of their natural evolution (White & White 1969; Andrejchuk & Klimchouk

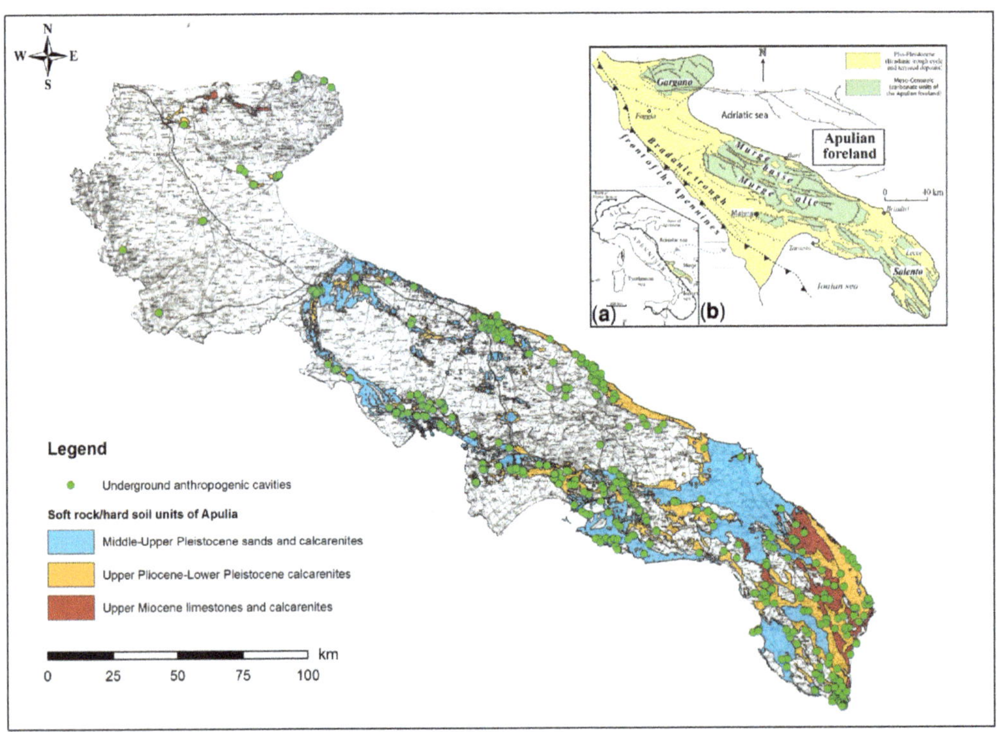

Fig. 1. Geological map showing only soft rock/hard soil units cropping out in the Apulia region (modified after Lepore *et al.* 2014). Map location of underground anthropogenic cavities obtained from the census of the Apulian Speleological Federation (www.catasto.fspuglia.it/). Inset: (**a**) location of the Apulian foreland in a synthetic structural map of Italy; (**b**) simplified geological map of the Apulia region (modified after Pieri *et al.* 1997).

2002; Klimchouk & Andrejchuk 2002; Palmer 2007; Parise 2008). To this hazard of natural origin, the high number of anthropogenic cavities contributes to the occurrence of additional sinkholes. There is a great variety of different types of anthropogenic cavity, including underground quarries, worship sites, oil mills and settlements. Most of these cavities are excavated in soft calcarenites of Plio-Pleistocene age. It is worth noting that the Federazione Speleologica Pugliese (the association that covers all caving grottos in the region) has surveyed and mapped 2311 natural caves and 1059 anthropogenic cavities in Apulia (both registers can be consulted at www.fspuglia.it/catastogrotte.htm).

Almost all the categories of anthropogenic cavities ranked in the classification defined by the Commission of the Italian Speleological Society and later approved at an international level by the Commission of the International Union of Speleology (Fig. 2; Galeazzi 2013; Parise *et al.* 2013*b*, *c*) are present in Apulia, creating an extremely varied and complex setting (Fig. 3). The interactions among the geological features of the region, well suited to the excavation of cavities, and historical vicissitudes have often led to the underground environment being considered of high strategic value to the local population. For instance, civilian settlements (category B, Fig. 2) are widely distributed in Apulia and nearby regions due to the development of the rupestrian culture (Fonseca 1970; Fonseca *et al.* 1979).

Extensive and complex settlements consisting of houses, shelters and cavities with different typologies and functions characterize the land between Basilicata and Apulia, corresponding to the outcrop areas of the Plio-Pleistocene calcarenites (Laureano 1993, 1995; Cotecchia & Grassi 1997). In particular, many cavities were used as worship sites or churches; the hypogean churches and crypts of Apulia are well known worldwide as a result of the great abundance of frescos on the walls of the sites (Fonseca 1980, 1991; Falla Castelfranchi 1991; Dell'Aquila & Messina 1998). However, much of this heritage has been lost due to a lack of maintenance and safeguarding and many frescos have been illegally removed.

As a consequence of the wide urban expansion recorded in Apulia in the last century, several cavities now lie below densely populated neighbourhoods or roads with heavy traffic. These conditions represent the main geomorphological hazard for human society in Apulia and require careful evaluation to protect and safeguard human life and to provide the necessary information for correct land use planning and management.

A worrying increase in the number of sinkhole events (Fig. 4) has been registered in recent years

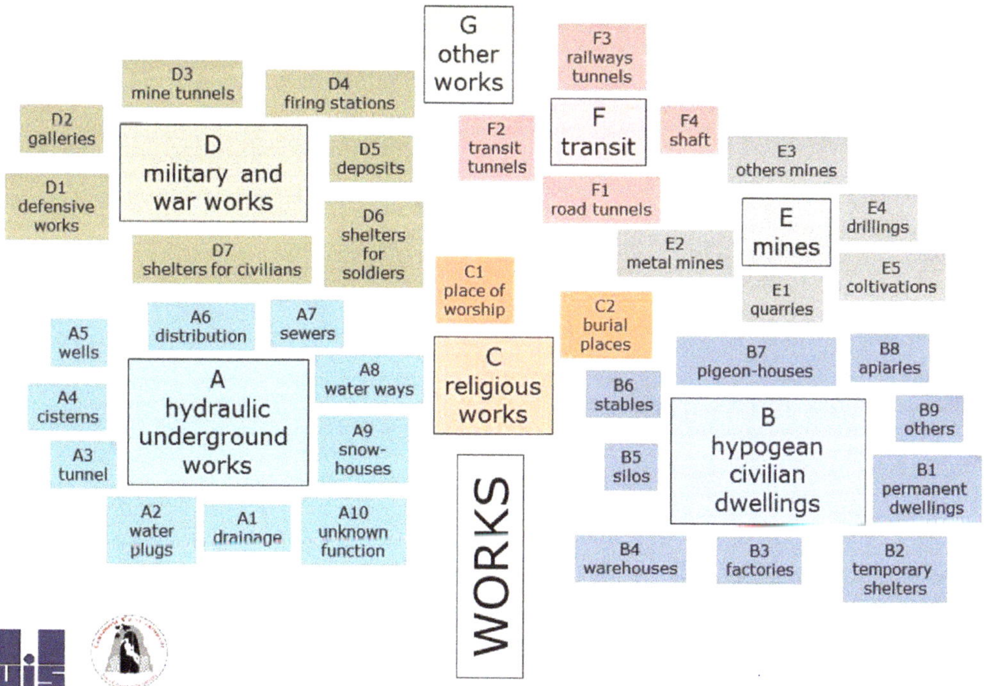

Fig. 2. Classification of anthropogenic cavities (modified after Parise *et al.* 2013*c*).

Fig. 3. Types of anthropogenic cavity in Apulia (from AdB Puglia photo archive). (**a**) Fallen material obstructs a passage in an underground calcarenite quarry, Altamura (Upper Pliocene–Lower Pleistocene calcarenites); (**b**) simple cavity, Bari (Upper Pliocene–Lower Pleistocene calcarenites); (**c**) underground calcarenite quarry, Canosa di Puglia (Upper Pliocene–Lower Pleistocene calcarenites); (**d**) sinkhole that appeared in 2007 at Gallipoli after the partial collapse of an underground calcarenite quarry (Middle–Upper Pleistocene calcarenites); (**e**) underground oil mill, Melpignano (Miocene calcarenites); and (**f**) underground calcarenite quarry, Mottola (Upper Pliocene–Lower Pleistocene calcarenites).

(Delle Rose *et al.* 2004; Festa *et al.* 2012); this reached a peak in 2009 and 2010 (Barnaba *et al.* 2010; Parise 2012; Fiore & Parise 2013; Parise & Vennari 2017). The sinkholes are of both natural and anthropogenic origin, although those related to anthropogenic cavities are more common. Sinkholes of natural origin include the sinkhole clusters registered in Gargano after a September 2014 rainstorm (Martinotti *et al.* 2017; Parise *et al.* 2017). Anthropogenic sinkholes affected many different sectors of Apulia, with the highest frequency at sites such as Altamura (Pepe *et al.* 2013), Cutrofiano (Toni & Quartulli 1986; Negri *et al.* 2015) and Casalabate (Margiotta *et al.* 2012).

The increasing number of sinkholes in Apulia is reflected in coverage by the local and national

Fig. 4. Two of the most recent sinkholes occurred at Altamura as a result of failures within the complex network of underground quarries (see Pepe *et al.* 2013).

press. Newspaper clippings are an important source of information on geohazards, especially local daily newspapers or magazines. As proved by several previous studies of natural hazards (e.g. Calcaterra & Parise 2001, and references cited therein), this source often provides data that would otherwise be lost. Newspaper clippings supply much more detail in terms of the temporal occurrence of geohazards than other sources of documentation and are generally more reliable. However, in many cases the records do not highlight the nature of the underground cavity at the origin of a sinkhole, leaving some degree of uncertainty. Several sources were therefore scrutinized with the aim of collecting the highest number of well-documented events and to cross-check information whenever possible. The sources included the scientific literature, bachelor theses, newspaper clippings (at both the regional and local level), historical books and technical reports. Direct investigation and field surveys were also added to these sources.

A chronological reference was found for *c.* 150 sinkhole events in Apulia, with the oldest records dating back to 1925 (Parise & Vennari 2017). These, of course, represent only some of the events that occurred, but clearly indicate the relevance of the hazard in the region. The documented sinkholes cover the whole regional territory and are mostly related to antropogenic cavities (89 events, *c.* 64%), with about one-third originating from natural karst caves (48 events, 34%). The origin of the documented sinkholes is unknown for only three events (2%). Looking at their temporal distribution, a clear increase in number is seen from the year 2000, with the frequency at least doubling with respect to the years pre-2000.

The type of anthropogenic cavity that was most commonly at the origin of sinkholes in Apulia, with frequent and repeated events, was category E (extraction works), especially in relation to underground mining of the local calcarenites while trying to preserve and use the ground for agricultural practices (Parise 2010, 2016). The strong link between underground quarries and sinkholes is not surprising because it has been already recorded and has become the object of studies in many other countries and regions (Bekendam 1998; Hutchinson *et al.* 2002; Ferrero *et al.* 2010; Sunwoo *et al.* 2010; Vattano *et al.* 2013). Long and complex systems of underground quarries have been realized in many sectors in Apulia. Once quarrying activity has stopped, some of these sites are re-used for other purposes (e.g. mushroom farming). Most of them, however, are abandoned, often with their access closed. This has resulted in a loss of memory of the underground sites, which, combined with urban expansion towards the outskirts of the settlement areas (i.e. where the original quarries were located), has resulted in new buildings and communication roads being built just above the old quarries. Following many recent sinkhole events, numerical modelling techniques were used to assess the stability of these underground quarries through an integrated geological–speleological–geotechnical approach (Parise & Lollino 2011; Lollino *et al.* 2013, 2014; Parise 2013, 2015; Vattano *et al.* 2013).

Although underground quarries are important in terms of the occurrence of sinkholes, the other types of anthropogenic cavity in Apulia cannot be excluded when studying the possibility of underground failures. We therefore took into account the most typical cross-sections of the main anthropogenic cavities in the region, not limiting our analysis to only category E. Figure 5 shows sketches of the main cavity cross-sections, including different types of quarries, from those with a trapezoidal shape at Canosa di Puglia (Società Italiana per Condotte d'Acqua S.P.A. 1989) to the more regular quarries at Cutrofiano (Negri *et al.* 2015), aqueducts (Parise *et al.* 2015), olive oil mills (Monte 1995; Stendardo 1995) and some civilian settlements.

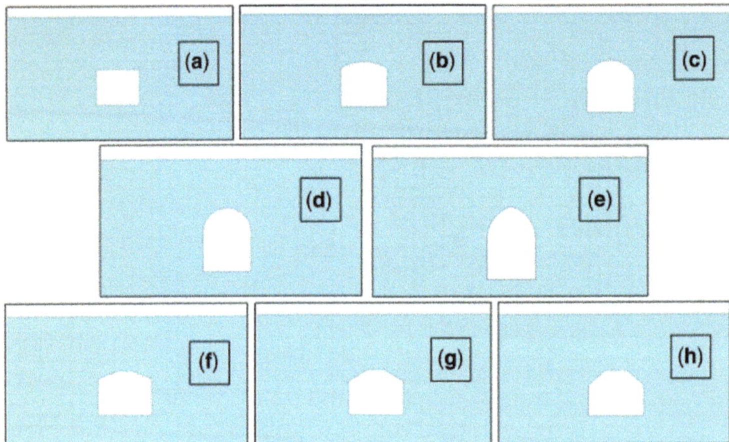

Fig. 5. Sketches of the main types of anthropogenic cavities in Apulia. (**a**) Cavity with square section, e.g. the underground quarries at Cutrofiano; (**b**, **c**) cavities with low arch (olive mill, civilian settlement); (**d**, **e**) cavity with (d) semi-circular arch and (e) with lancet arch, with height greater than width (underground aqueducts); and (**f**, **g**, **h**) cavity with trapezoid section (underground quarry at Canosa di Puglia).

Technical properties of calcarenite rocks

The different types of anthropogenic cavities in Apulia were excavated in the calcarenite deposits that crop out in many sectors of the region (Fig. 1). Calcarenite are soft, or very soft, rocks, making them easy to work, at the same time guaranteeing sufficient strength to support excavations. This has led to the calcarenite rocks becoming the main lithotype used for building purposes in Apulia. The calcarenite deposits cropping out in the study area fall into three groups (Fig. 1): Miocene calcarenites; Upper Pliocene–Lower Pleistocene calcarenites; and Middle–Upper Pleistocene sands and calcarenites.

All the Miocene calcarenites, belonging to different geological units, are merged here into a single group containing shallow and deep marine bioclastic and glauconite-rich carbonate deposits cropping out at the margin of Gargano Promontory and in Salento. Plio-Pleistocene calcarenites are mostly made up of shallow marine bioclastic and lithoclastic carbonate deposits belonging to the Calcarenite di Gravina Formation, which crops out widely in the region. The Middle–Upper Pleistocene sands and calcarenites consist of shallow marine terraced carbonate and mixed deposits cropping out at the margin of Murge and in Salento. The focus of this paper is the Upper Pliocene–Lower Pleistocene calcarenites because most of the Apulian anthropogenic cavities occur in this unit (Fig. 1).

The Miocene calcarenites generally have a higher uniaxial compression strength than the other two units, ranging from 10 to 23 MPa for dry rocks (Andriani & Walsh 2010; Fiore *et al.* 2015). The Upper Pliocene–Lower Pleistocene calcarenites have a dry uniaxial compression strength between 1 and 9 MPa (Coviello *et al.* 2005; Andriani & Walsh 2010; Ciantia *et al.* 2014; Fiore *et al.* 2015). The degree of saturation has a significant influence on the rock strength, so that the compression strength strongly decreases in saturated rocks, in some cases by >50%, resulting in a range of 0.3–6 MPa. Based on the International Society of Rock Mechanics classification, the rock can be therefore described as a very soft rock. The tensile strength shows a wide range, typically between 0.2 and 2.0 MPa (Coviello *et al.* 2005; Ciantia *et al.* 2014). The Middle–Upper Pleistocene sands and calcarenites have the same range of uniaxial strength as the previous unit (Fiore *et al.* 2015), although these were only measured in the calcarenitic stratigraphic intervals of the unit. All these values have to be considered as approximate because the formations are highly variable in terms of their petrographic and structural features, which influence the mechanical response of the rock and lead to a high degree of heterogeneity among the different sites.

Weathering processes of both chemical and physical origin also cause degradation of the rock mass and a reduction in mechanical strength (Fookes & Hawkins 1988; Zupan Hajna 2003; Ghabezloo & Pouya 2006; Andriani & Walsh 2010; Calcaterra & Parise 2010, and references cited therein; Ciantia & Castellanza 2016). Such processes, especially those related to an increase in the presence of water up to the saturation point, might bring the calcarenite rock mass to conditions close to those of general failure. The effects of a progressive decay in the mechanical properties of the rock mass were therefore analysed with the aim of evaluating the

Table 1. *Values of the Hoek–Brown parameters assigned in the analyses (see text for definitions)*

Parameter	γ (kN m^{-3})	E' (MPa)	ν	σ_c (kPa)	Geological strength index	ψ (°)	m_i	D
Adopted value	16	100	0.3	Variable	100	0	3, 8, 16	0

development of instability and its propagation towards the surface until general collapse in the form of a sinkhole.

To characterize the mechanical properties of the different types of calcarenites, we collected a variety of geotechnical data from previous studies. Given the difficulties in assessing the data quality of this large number of sources of different ages, these were integrated using new laboratory tests performed at the CNR IRPI Laboratory.

Numerical modelling and results

Numerical simulations were performed using the finite element software Plaxis-2D for two-dimensional analyses (PLAXIS 2D 2015) and Plaxis-3D for three-dimensional analyses (PLAXIS 3D 2015). Based on mesh discretization of the calculation domain, the finite element method (FEM) allows the integration of equilibrium and compatibility equations, the constitutive laws of the materials, and the initial and boundary conditions of the problem within the framework of continuum mechanics. In the specific FEM adopted here, deformation is considered to be small and therefore the calculations were carried out with reference to the original undeformed geometry.

The main goal of this numerical modelling was to define cave stability charts so that a trained technician could preliminarily evaluate the stability conditions based on the main geometric features of the cave and the geotechnical properties of the rock mass. A large set of two-dimensional sensitivity finite element analyses was carried out. In addition, taking into account the main failure mechanisms observed, we analysed the influence of the shape of the vault on the behaviour of a cavity and the effects of the interactions of two close cavities on the overall stability conditions of the whole system. Three-dimensional FEM analyses were performed to investigate the role of the third dimension in the general stability of the cavities and the results were compared with the two-dimensional analyses with the same assumptions.

The calculation domain for the two-dimensional-analyses consisted of a 100 m wide and 60 m high rectangle, where the cavity was defined by its width (L), height (h) and the overburden thickness or roof thickness (t). The cavity sizes were chosen in accordance with the values typically observed in Apulia for anthropogenic cavities, with L ranging from 1 to 30 m, h from 3.5 to 7 m and t from 2.5 to 10 m. The risk of boundary effects is low for these cavity sizes because the calculation domain is chosen to be much larger than the cavity. No horizontal displacement was imposed along the sides and no vertical or horizontal displacement was assigned along the mesh bottom, although free displacements were allowed at the ground surface.

The two-dimensional finite element mesh is formed of three-noded triangular elements with three degrees of freedom corresponding to the displacement components. The discretization mesh was defined by searching for a compromise between the need for calculation accuracy in the area of interest (i.e. around the cavity) and the need to reduce the computational costs. A highly refined mesh (average element size *c.* 0.1 m^2) was therefore chosen in the area around the cavity, whereas coarser elements were chosen in the areas far from the cavity.

Based on the non-linear strength envelope of the calcarenite rock in Mohr's plane, we adopted an elastic, perfectly plastic constitutive model with a Hoek–Brown failure criterion (Hoek & Brown 1997; Fraldi & Guarracino 2009; Suchowerksa *et al.* 2012). The Hoek–Brown parameters adopted are listed in Table 1. Based on the rock geotechnical characterization performed in the laboratory, a unit weight γ of 16 kN m^{-3}, an elasticity modulus E' of 100 MPa and a Poisson ratio ν of 0.3 were assumed as fixed in the numerical analyses. The variable chosen in the parametric analyses is the threshold value of uniaxial compressive strength, $\sigma_{c,min}$, corresponding to the onset of a failure mechanism within the examined domain. Many field surveys have indicated that the calcarenite rock masses are rarely jointed and therefore the value of the geological strength index (Hoek 2007) was considered to be equal to 100 in the analyses (i.e. a massive rock mass). As a consequence of this assumption, the numerical results presented here cannot be considered valid for rock masses characterized by single joints or by joint sets, which might play a significant part in the control of the stability of the rock mass.

Most of the anthropogenic cavities in Apulia have been excavated by hand and therefore the D value was assumed to be zero to simulate a very low disturbance of the rock mass or negligible stress release processes as a result of the excavation technique. Following Cai (2010), the parameter m_i, a constant of the rock material, was assumed to be the ratio between the uniaxial compressive and the tensile strength of the rock. Therefore the corresponding value assumed in the analyses ranges between 3 and 16 in accordance with the results of laboratory tests carried out on samples of the Calcarenite di Gravina Formation. As a consequence, the numerical analyses were carried out assuming values of m_i equal to 3, 8 and 16, respectively. To summarize, the threshold value, $\sigma_{c,min}$ – that is, the uniaxial compressive strength mobilized at the triggering of a local or global failure mechanism – was chosen as the variable dependent on two non-dimensional geometrical parameters, L/t and L/h, respectively, whereas the values of the geological strength index, D and m_i were prescribed as constant in the analyses.

The numerical analyses were carried out following three calculation stages:

- the initialization of the stress state within the rock mass domain (first stage) was pursued by assigning K_0, i.e. the 'at rest' coefficient of lateral earth pressure, equal to 1 to follow a conservative assumption – in fact, lower values of K_0 are unlikely to be measured for shallow rock masses (Brown & Hoek 1978; Gonzalez de Vallejo & Hijazo 2008);
- the elasto-plastic analysis (second stage) in accordance with the assumed mechanical properties of the calcarenite; and
- the simulation of the cavity excavation (third stage) – in this stage, the whole portion of the domain within the cavity was simultaneously deactivated (this represents a conservative solution) and a new plastic calculation performed.

A general failure mechanism was represented by the development of a proper sinkhole (Waltham *et al.* 2005; Gutiérrez *et al.* 2008, 2014), whereas local failure mechanisms were characterized by the propagation of closed-shape failures along the cavity roof or by plastic zones along the lateral walls. From a numerical viewpoint, the generation of a failure mechanism was detected through a combination of the following criteria: a lack of numerical convergence; the generation of a closed yielding mechanism involving a well-defined rock volume delimited by clear plastic zones or deviatoric strains; and vertical displacements of selected nodes belonging to the mesh region involved in the failure mechanism increasing with numerical time steps (Fig. 6).

The final goal of the analyses was the definition of a simple chart to be used for the preliminary assessment of the stability of the cavity (Goodings & Abdulla 2002). The numerical results were therefore represented as plots showing the $\sigma_{c,min}/\sigma_v$ ratio – that is, the ratio between the threshold value of the uniaxial compressive strength mobilized at failure and the vertical stress before excavation at the depth of the cavity roof along the y-axis and the non-dimensional ratio L/t, here defined as the width-to-depth ratio, along the x-axis, keeping the ratio L/H constant. A typical example of the numerical results in terms of a curve delimiting the 'stable' area from the 'unstable' area is shown in Figure 7.

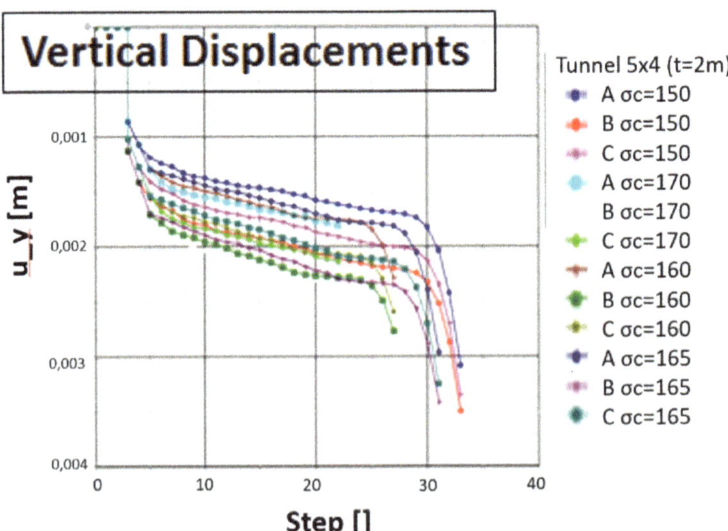

Fig. 6. Vertical displacements of selected nodes belonging to the mesh region in the failure mechanism. The different colours of the lines show the time steps during the computation.

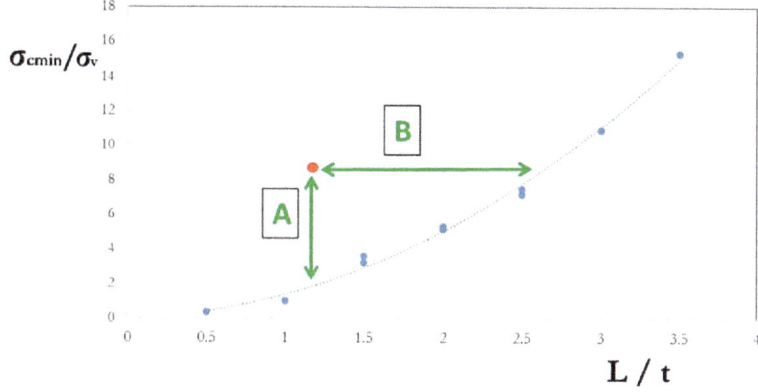

Fig. 7. Graph of the ratio between L/t and $\sigma_{c,lim}/\sigma_v$ for the assessment of the cavity stability: the position of the red dot shows a stable condition (after Perrotti *et al.* in press).

A point representative of the *in situ* strength and stress conditions of the rock mass and the geometrical features of the cavity that lies above the curve indicates that the cavity is stable, whereas a point that lies below the curve indicates that a possible instability is to be expected. The stability chart can be read in two alternative ways: (1) to calculate the safety margin with respect to failure (segment A in Fig. 7); or (2) to calculate the maximum value of the width-to-depth ratio L/t allowed for stability (segment B in Fig. 7) given the assigned value of the ratio between the *in situ* uniaxial compressive strength and the vertical stress.

The numerical results were summarized in charts such as Figure 8, which reports the threshold stability curves for different L/H ranges accounting for different values of m_i. Figure 8 highlights the effect of the m_i parameter: for a specific range of L/H, at the same value of L/t, the values of $\sigma_{c,min}/\sigma_v$ increase with m_i. However, this effect is still limited at L/h values <1, whereas it becomes more evident with higher L/h values. It also appears that, with increasing L/H values, the values of $\sigma_{c,min}/\sigma_v$ corresponding to the same value of L/t become higher.

The numerical results confirm that both the geometrical features of the cavity (the shape factor,

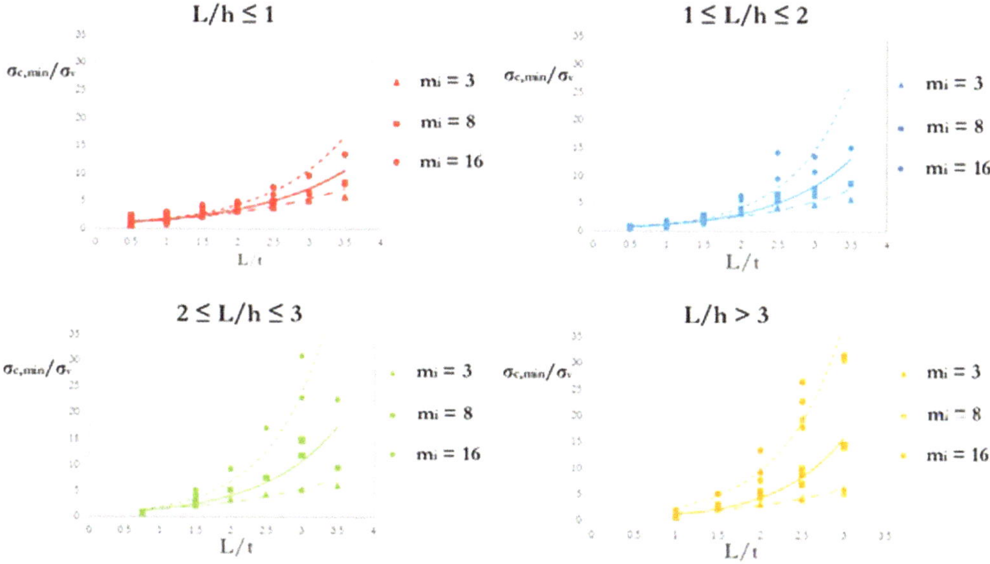

Fig. 8. Curves L/t–$\sigma_{c,lim}/\sigma_v$ for different values of the ratio L/h obtained for a rectangular cavity with different values of m_i (3, 8 and 16) (after Perrotti *et al.* in press).

L/h, and the width-to-depth ratio, L/t) have a strong influence on the failure mechanism (Tharp 1995; Diederichs & Kaiser 1999*a*, *b*; Liu *et al.* 2000; Hatzor *et al.* 2002). The failure mechanisms resulting from the numerical analyses can be classified according to one of the following schemes (Fig. 9): (1) arch-shaped general failure, formed by plastic shear zones following an arch scheme that does not reach the ground surface; (2) beam-shaped general failure, with plastic shear zones starting from the upper lateral edges of the cavity and propagating to the ground surface, thus giving rise to a sinkhole; and (3) arch-shaped local failure, corresponding to a plastic shear mechanism involving a small portion of rock in the middle of the cavity roof and generating fissures therein (Fig. 9).

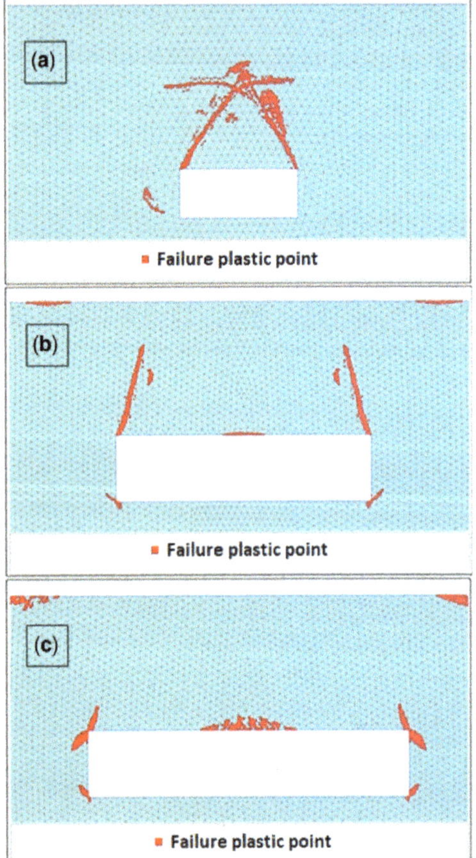

Fig. 9. Arch failure mechanisms. (**a**) Formation of an unstable wedge along the whole vault formed by plastic shear zones. (**b**) Mechanism of beam-shaped general failure of the vault, with possible evolution into a sinkhole. (**c**) Mechanism of local failure of the vault corresponding to a plastic shear mechanism involving a small portion in the middle vault.

To investigate the influence of the shape of the vault on the behaviour of a cavity, two-dimensional parametric analyses were performed with respect to a cavity with $L = 5$ m and $h = 4$ m, with variable shapes of the vault. For a rectangular cavity (Fig. 10a), plastic zones develop from the upper edges towards the ground at *c.* 45° inclination. Consequently, a failure arch involving about half of the roof thickness is developed, corresponding to a uniaxial compression strength of *c.* 140 kPa. For a semi-circular vault, collapse of the cavity occurs at a lower uniaxial compression strength (50 kPa) and a general failure mechanism, involving both the vault and the walls of the cavity (Fig. 10b), is obtained. For a cavity with a lancet vault, the threshold value of uniaxial compression is 60 kPa. In this case, the failure mechanism affects a limited portion in the middle vault and the cavity walls (Fig. 10c). This confirms that the two last vault geometries (Fig. 10b, c) represent more stable configurations than the rectangular geometry.

Further two-dimensional analyses were carried out to evaluate the effects of two close parallel

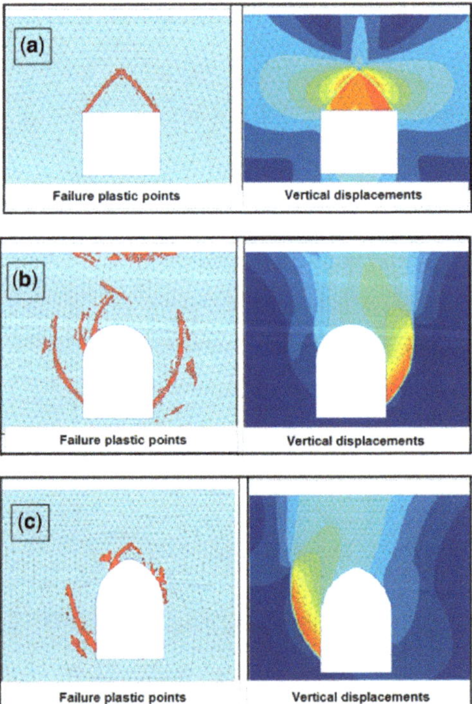

Fig. 10. Influence of vault shape on the evolution of the cavity. The left-hand panels show the plastic zones and the right-hand panels show the vertical displacements for (**a**) a rectangular cavity ($L = 5$ m, $h = 4$ m, $t = 5$ m), (**b**) a semi-circular arch cavity and (**c**) a lancet arch cavity.

cavities (twin cavities) and the effect of their interactions on the overall stability of the whole system. This represents a common situation in areas characterized by a high density of excavations, where cavities are numerous and often separated by thin walls. First, the influence of the distance d between the two cavities (located at the same depth t) was considered by taking into account variable geometric parameters, with an L/t ratio between 0.5 and 2 and a distance between the cavities ranging from a maximum value of $d = 4L$ and a minimum of $d = 0.5L$ (Fig. 11). The numerical results show that until the distance between the twin cavities is greater than $d = 2L$, no interaction between the cavities is observed. As a consequence, the corresponding threshold values of the compressive strength are comparable with those obtained for a single cavity of the same size ($\sigma_{c,lim} = 120$ kPa). When the distance d decreases, a clear increase in the interaction between the two nearby cavities appears. In detail, when $d = L$, a moderate interaction gives a uniaxial compressional strength $\sigma_{c,lim}$ of 150 kPa. When $d < L$, the numerical results show a clear increase in the threshold strength value; at $\sigma_{c,lim} = 220$ kPa, the wall between the cavities fails first and then a general failure mechanism involving the whole system of two cavities is developed, resembling the failure mechanism for a single cavity of width 2.5L.

The last stage of the parametric study consisted in three-dimensional FEM analyses to investigate the

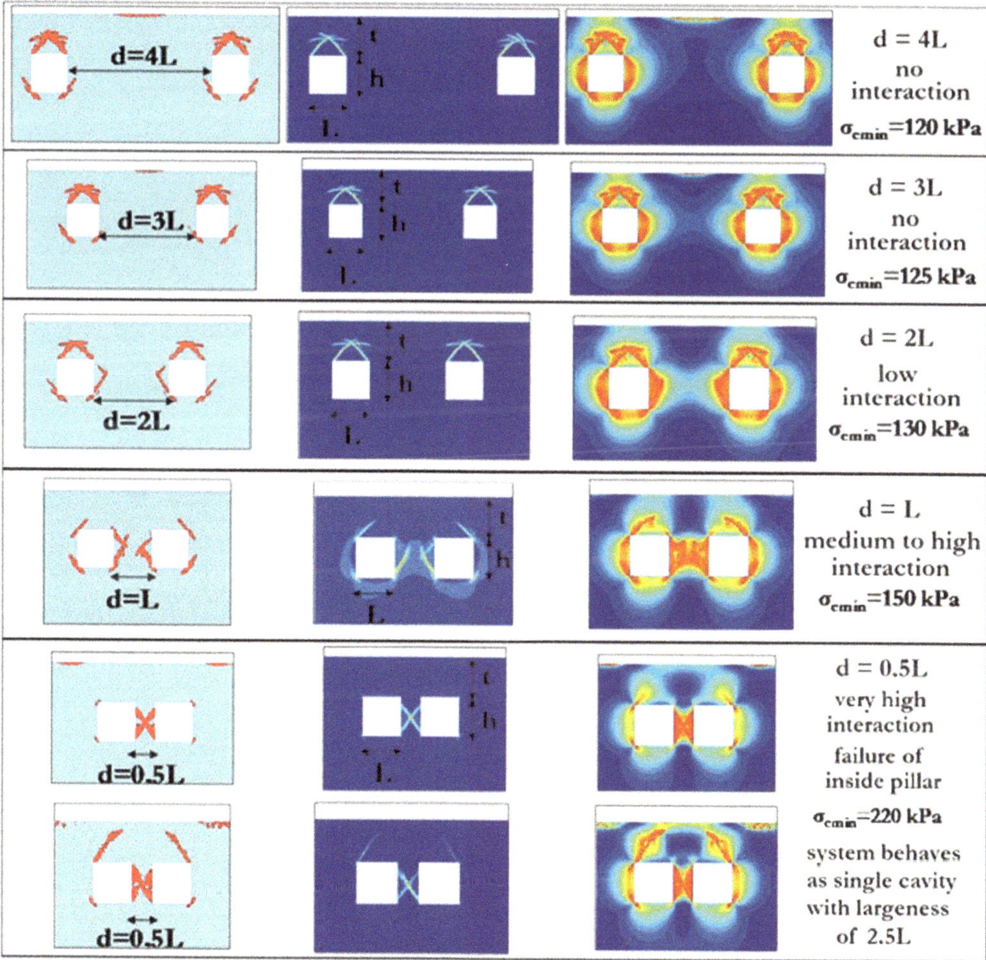

Fig. 11. System formed by two nearby cavities showing the influence of the distance d between them. From the top to the bottom, the distance between the two cavities decreases ($d = 4L$; $d = 3L$, $d = 2L$, $d = L$, $d = 0.5L$) corresponding to a greater interaction, as shown in the plastic zone images (left-hand panels), in the geometry of the cavity images (centre panels) and in the vertical displacements images (right-hand panels).

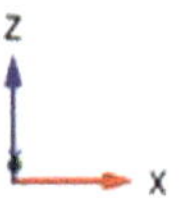

Fig. 12. Plastic points (red) obtained from the finite element model analysis of the Barletta case study after the strength reduction phase.

role of the third dimension in the general stability of the cavities (Fig. 12) compared with two-dimensional analyses with the same assumptions. The influence of the cavity length (z-direction) on cavity stability was investigated by simulating a progressive reduction in the cavity length from a maximum value of $d = 60$ m to a minimum of $d = 5$ m (Fig. 13). The analyses were performed by taking into account different types of cavity cross-section (one square and two rectangular cross-sections with different widths). The results of the three-dimensional FEM analyses were then compared with the corresponding two-dimensional analyses assuming the same geometrical features in the x–y plane. Simulation of the reduction in the cavity length, d, was carried out by reducing the rock volume involved in the deactivation process (Fig. 13). The three-dimensional numerical results obtained confirmed that the two-dimensional analyses are only representative of the actual geometry of the cavity when the cavity is significantly long, with d roughly equal to 50–60 m. By contrast, for lower cavity lengths ($d < 50$–60 m), the effect of the third dimension is not negligible and two-dimensional analyses seem to be more conservative.

Fig. 13. Parametric analyses of a cavity with variable depth carried out by a finite element model three-dimensional analysis.

Conclusions and future perspectives

Studies and research aimed at evaluating sinkhole hazards are important in civil defence and land safeguarding for both karst (natural) caves and cavities of anthropogenic origin (Brinkmann & Parise 2012; Parise 2012, 2015). We considered the instability problems and related sinkholes resulting from anthropogenic cavities in Apulia, southern Italy. With our preliminary evaluation of the stability conditions, the availability of charts for use by practitioners and technicians is an important step in the assessment of the possible public use of anthropogenic cavities covered by the Apulian regional law L.R. 33 2009 Safeguard and Promotion of the Geological and Speleological Heritage.

The outcomes of the parametric studies presented here are preliminary and are strongly dependent on our hypothesis and assumptions. They therefore need to be calibrated and validated through applications to real case studies. They cannot be considered as an applicative tool in their present form. Nevertheless, this approach can give a preliminary view of the stability conditions and provide a basis on which develop further analyses taking into account the local geometric, stratigraphic and geomechanical features.

Based on the two-dimensional numerical results, it appears that the threshold strength for the stability of the cavity increases with the ratio L/t: an arch-shaped general failure occurs for $L/t < 1.5$, whereas for $L/t > 1.5$ the most common situation is represented by the occurrence of beam-shaped general failure. Better stability conditions occur when the cavity width is comparable with the roof thickness. The absolute values of the ratio $\sigma_{c,min}/\sigma_v$ increase as m_i increases from 3 to 16. The results of the two-dimensional FEM simulations show that the L/t ratio affects the failure mechanism, whereas m_i affects the strength ratio $\sigma_{c,min}/\sigma_v$, which increases with m_i.

To avoid many of the problems related to the implementation of the two-dimensional numerical codes, which often give only partial results due to the lack of analysis in the third dimension, we compared the two- and three-dimensional numerical results. The third dimension is an additional source of strength if the length of the cavity is short. When the cavity length increases, the threshold strength for stability tends to the corresponding value of the two-dimensional model as a function of the shape of the cavity cross-section.

Most of the models discussed here can only be used for a preliminary assessment of the instability conditions related to underground voids due to the approximations used to detect failure within the numerical study. Therefore the charts proposed in the study need to be validated by comparison with real case studies before being used as an effective tool in engineering practice.

This paper summarizes part of the activities carried out during the project 'Stability of artificial cavities with relation to the influence by geometrical and geomechanical factors, and to weathering processes', which involved CNR IRPI and AdB Puglia in 2016. This was part of the larger project 'Monitoring geomorphological instabilities in the Apulia region' developed by AdB Puglia.

References

Abelson, M., Baer, G. et al. 2003. Collapse-sinkholes and radar interferometry reveal neotectonics concealed within the Dead Sea basin. *Geophysical Research Letters*, **30**, 1545.

Andrejchuk, V. & Klimchouk, A. 2002. Mechanisms of karst breakdown formation in the gypsum karst of the Fore-Ural regions, Russia (from observations in the Kungurskaja Cave). *International Journal of Speleology*, **31**, 89–114.

Andriani, G.F. & Walsh, N. 2010. Petrophysical and mechanical properties of soft and porous building rocks used in Apulian monuments (south Italy). *In*: Přikryl, R. & Török, Á. (eds) *Natural Stone Resources for Historical Monuments*. Geological Society, London, Special Publications, **333**, 129–141, https://doi.org/10.1144/SP333.13

Aydan, O., Tano, H., Sakamoto, A., Yamada, N. & Sugiura, K. 2005. A real-time monitoring system for the assessment of stability and performance of in abandoned room & pillar lignite mines. Paper presented at Post-Mining 2005, November 16–17, Nancy, France.

Baer, G., Schattner, U., Wachs, D., Sandwell, D., Wdowinski, S. & Frydman, S. 2002. The lowest place on Earth is subsiding; an InSAR Interferometric Synthetic Aperture Radar perspective. *Geological Society of America Bulletin*, **114**, 12–23.

Barnaba, F., Caggiano, T. et al. 2010. Sprofondamenti connessi a cavità antropiche nella regione Puglia. *In*: *Proceedings of the 2nd International Workshop Catastrophic Sinkholes in the Natural and Anthropogenic Environment*, 3–4 December 2009, Rome, Italy, 653–672.

Bekendam, R.F. 1998. *Pillar stability and large-scale collapse of abandoned room and pillar limestone mines in south-Limburg, the Netherlands*. PhD thesis, University of Utrecht.

Borgatti, L., Bianchi, E., Bonaga, G., Gottardi, G., Landuzzi, A., Vittuari, L. & Pellegrini, M. 2013. Sinkholes in the Po Plain as evidence of paleoliquefaction events. *Rendiconti Società Geologica Italiana Online*, **24**, 34–36.

Brinkmann, R. & Parise, M. 2012. Karst environments: problems, management, human impacts, and sustainability. An introduction to the special issue. *Journal of Cave and Karst Studies*, **74**, 135–136.

Brown, E.T. & Hoek, E. 1978. Trends in relationships between measured in-situ stresses and depth. *International Journal of Rock Mechanics and Mining Science & Geomechanical Abstracts*, **15**, 211–215.

Cai, M. 2010. Practical estimates of tensile strength and Hoek–Brown strength parameter m_i of brittle rocks. *Rock Mechanics and Rock Engineering*, **43**, 167–184.

Calcaterra, D. & Parise, M. 2001. The contribution of historical information in the assessment of landslide hazard. *In*: Glade, T., Albini, P. & Frances, F. (eds) *The Contribution of Historical Data in Natural Hazard Assessment*. Kluwer, Dordrecht, 201–217.

Calcaterra, D. & Parise, M. (eds) 2010. *Weathering as a Predisposing Factor to Slope Movements*. Geological Society, London, Engineering Geology Special Publications, **23**, http://egsp.lyellcollection.org/content/23/1

Canakci, H. 2007. Collapse of caves at shallow depth in Gaziantep city center, Turkey: a case study. *Engineering Geology*, **53**, 915–922.

Castañeda, C., Gutiérrez, F., Manunta, M. & Galve, J.P. 2009. DInSAR measurements of ground deformation by sinkholes, mining subsidence, and landslides, Ebro River, Spain. *Earth Surface Processes and Landforms*, **34**, 1562–1574.

Ciantia, M.O. & Castellanza, R. 2016. Modelling weathering effects on the mechanical behaviour of rocks. *European Journal of Environmental and Civil Engineering*, **20**, 1054–1082.

Ciantia, M.O., Castellanza, R. & Di Prisco, C. 2014. Experimental study on the water-induced weakening of calcarenites. *Rock Mechanics and Rock Engineering*, **48**, 441–461.

Ciaranfi, N., Pieri, P. & Ricchetti, G. 1988. Note alla carta geologica delle Murge e del Salento (Puglia Centro

Meridionale). *Memorie Società Geologica Italiana*, **41**, 449–460.

Closson, D., Karaki, N.A., Klinger, Y. & Hussein, M.J. 2005. Subsidence and sinkhole hazards assessment in the southern Dead Sea area, Jordan. *Pure and Applied Geophysics*, **162**, 221–248.

Closson, D., Karaki, N.A., Milisavljevic, N., Hallot, F. & Acheroy, M. 2010. Salt-dissolution-induced subsidence in the Dead Sea area detected by applying interferometric techniques to ALOS Palsar Synthetic Aperture Radar images. *Geodinamica Acta*, **23**, 65–78.

Cotecchia, V. & Grassi, D. 1997. Incidenze geologico-ambientali sull'ubicazione e lo stato di degrado degli insediamenti rupestri medioevali della Puglia e della Basilicata. *Geologia Applicata e Idrogeologia*, **32**, 1–10.

Coviello, A., Lagioia, R. & Nova, R. 2005. On the measurement of the tensile strength of soft rocks. *Rock Mechanics and Rock Engineering*, **38**, 251–273.

D'Argenio, B., Pescatore, T. & Scandone, P. 1973. Schema geologico dell'appennino meridionale (Campania e Lucania). *Accademia Nazionale dei Lincei Quaderni*, **183**, 49–72.

Dell'Aquila, F. & Messina, A. 1998. *Le chiese rupestri di Puglia e Basilicata*. Mario Adda editore, Bari.

Delle Rose, M., Federico, A. & Parise, M. 2004. Sinkhole genesis and evolution in Apulia, and their interrelations with the anthropogenic environment. *Natural Hazards and Earth System Sciences*, **4**, 747–755.

Diederichs, M.S. & Kaiser, P.K. 1999*a*. Stability of large excavations in laminated hard rock masses: the voussoir analogue revisited. *International Journal of Rock Mechanics and Mining Science*, **36**, 97–117.

Diederichs, M.S. & Kaiser, P.K. 1999*b*. Tensile strength and abutment relaxation as failure control mechanisms in underground excavations. *International Journal of Rock Mechanics and Mining Science*, **36**, 9–96.

De Waele, J., Gutiérrez, F., Parise, M. & Plan, L. 2011. Geomorphology and natural hazards in karst areas: a review. *Geomorphology*, **134**, 1–8.

Doglioni, C., Mongelli, F. & Pieri, P. 1994. The Puglia uplift (SE Italy): an anomaly in the foreland of the Apenninic subduction due to buckling of a thick continental lithosphere. *Tectonics*, **13**, 1309–1321.

Doglioni, C., Tropeano, M., Mongelli, F. & Pieri, P. 1996. Middle–Late Pleistocene uplift of Puglia: an 'anomaly' in the Appenninic foreland. *Memorie Società Geologica Italiana*, **51**, 101–117.

Falla Castelfranchi, M. 1991. *Pittura monumentale bizantina in Puglia*. Electa, Milan.

Ferrero, A.M., Segalini, A. & Giani, G.P. 2010. Stability analysis of historic underground quarries. *Computers and Geotechnics*, **37**, 476–486.

Festa, V. 2003. Cretaceous structural features of the Murge area (Apulian Foreland, southern Italy). *Eclogae Geologicae Helvetiae*, **96**, 11–22.

Festa, V., Fiore, A., Parise, M. & Siniscalchi, A. 2012. Sinkhole evolution in the Apulian karst of southern Italy: a case study, with some considerations on sinkhole hazards. *Journal of Cave and Karst Studies*, **74**, 137–147.

Festa, V., Fiore, A., Miccoli, M.N., Parise, M. & Spalluto, L. 2014. Tectonics v. karst relationships in the Salento Peninsula (Apulia, southern Italy): implications for a comprehensive land-use planning. *In*: Lollino, G., Manconi, A., Guzzetti, F., Culshaw, M., Bobrowsky, P. & Luino, F. (eds) *Engineering Geology for Society and Territory – Volume 5: Urban Geology, Sustainable Planning and Landscape Exploitation*. Springer, Berlin, 493–496.

Filin, S. & Baruch, A. 2010. Detection of sinkhole hazards using airborne laser scanning data. *Photogrammetric Engineering & Remote Sensing*, **76**, 577–587.

Fiore, A. & Parise, M. 2013. Cronologia degli eventi di sprofondamento in Puglia, con particolare riferimento alle interazioni con l'ambiente antropizzato. *Memorie Descrittive della Carta Geologica d'Italia*, **93**, 239–252.

Fiore, A., Luisi, M., Miccoli, M.N., Muzzicato, P. & Spalluto, L. 2015. *Monitoraggio dei dissesti di carattere geomorfologico del territorio pugliese. Relazione finale*, www.adb.puglia.it/public/page.php?112

Fonseca, C.D. 1970. *Civiltà rupestre in terra ionica*. Bestetti, Rome.

Fonseca, C.D. 1980. La civiltà rupestre in Puglia. *In*: AA.VV., *La Puglia tra Bisanzio e l'Occidente*. Milano, 36–116.

Fonseca, C.D. 1991. Le grotte della civiltà rupestre. *Itinerari Speleologici*, **5**, 13–25.

Fonseca, C.D., Bruno, A.R., Ingrosso, V. & Marotta, A. 1979. *Gli insediamenti rupestri medioevali nel Basso Salento*. Congedo, Galatina.

Fookes, P.G. & Hawkins, A.B. 1988. Limestone weathering: its engineering significance and a proposed classification scheme. *Quarterly Journal of Engineering Geology*, **21**, 7–31, https://doi.org/10.1144/GSL.QJEG.1988.021.01.02

Fraldi, M. & Guarracino, F. 2009. Limit analysis of collapse mechanisms in cavities and tunnels according to the Hoek–Brown failure criterion. *International Journal of Rock Mechanics and Mining Science*, **46**, 665–673.

Galeazzi, C. 2013. The typological tree of artificial cavities: a contribution by the Commission of the Italian Speleological Society. *Opera Ipogea*, **1**, 9–18.

Gambini, R. & Tozzi, M. 1996. Tertiary geodynamic evolution of the Southern Adria microplate. *Terra Nova*, **8**, 593–602.

Ghabezloo, S. & Pouya, A. 2006. Numerical modeling of the effect of weathering on the progressive failure of underground limestone mines. *In*: Van Cotthem, A., Charlier, R., Thimus, J.F. & Tshibangu, J.P. (eds) *Eurock 2006 – Multyphysics Coupling and Long Term Behavior in Rock Mechanics*. Taylor and Francis, London, 233–240.

Gonzalez de Vallejo, L.I. & Hijazo, T. 2008. A new method of estimating the ratio between in situ rock stresses and tectonics based on empirical and probabilistic analyses. *Engineering Geology*, **101**, 185–194.

Goodings, D.J. & Abdulla, W.A. 2002. Stability charts for predicting sinkholes in weakly cemented sand over karst limestone. *Engineering Geology*, **65**, 179–184.

Gutiérrez, F. 2010. Hazards associated with karst. *In*: Alcantara, I. & Goudie, A. (eds) *Geomorphological Hazards and Disaster Prevention*. Cambridge University Press, Cambridge, 161–175.

Gutiérrez, F., Guerrero, J. & Lucha, P. 2008. A genetic classification of sinkholes illustrated from evaporite paleokarst exposures in Spain. *Environmental Geology*, **53**, 993–1006.

Gutiérrez, F., Galve, J.P., Lucha, P., Castañeda, C., Bonachea, J. & Guerrero, J. 2011. Integrating geomorphological mapping, trenching, InSAR and GPR for the identification and characterization of sinkholes in the mantled evaporite karst of the Ebro Valley (NE Spain). *Geomorphology*, **134**, 144–156.

Gutiérrez, F., Parise, M., De Waele, J. & Jourde, H. 2014. A review on natural and human-induced geohazards and impacts in karst. *Earth Science Reviews*, **138**, 61–88.

Hatzor, Y.H., Talesnick, M. & Tsesarksky, M. 2002. Continuous and discontinuous stability analysis of the bell-shaped caverns at Bet Guvrin, Israel. *International Journal of Rock Mechanics and Mining Science*, **39**, 867–886.

Hermosilla, R. 2012. The Guatemala city sinkhole collapses. *Carbonates and Evaporites*, **27**, 103–107.

Hoek, E. 2007. *Practical Rock Engineering*, https://www.rocscience.com/documents/hoek/corner/Practical-Rock-Engineering-Full-Text.pdf

Hoek, E. & Brown, E.T. 1997. Practical estimates of rock mass strength. *International Journal of Rock Mechanics and Mining Science*, **34**, 1165–1186.

Hutchinson, D.J., Phillips, C. & Cascante, G. 2002. Risk considerations for crown pillar stability assessment for mine closure planning. *Geotechnical and Geological Engineering*, **20**, 41–63.

Kawashima, K., Aydan, O. et al. 2010. Reconnaissance investigation on the damage of the 2009 L'Aquila, Central Italy earthquake. *Journal of Earthquake Engineering*, **14**, 817–841.

Klein, E., Contrucci, I., Ngoc-Tuyen, C. & Bigarre, P. 2011. Mining induced seismicity: monitoring of a large scale salt cavern collapse. Paper presented at the EAGE Conference, 23–26 May, Vienna, Austria.

Klimchouk, A. & Andrejchuk, V. 2002. Karst breakdown mechanisms from observation in the gypsum caves of the Western Ukraine: implications for subsidence hazard assessment. *International Journal of Speleology*, **31**, 55–88.

Laureano, P. 1993. *Giardini di pietra*. Bollati Boringhieri, Torino.

Laureano, P. 1995. *La piramide rovesciata*. Bollati Boringhieri, Torino.

Lepore, D., Spalluto, L., Fiore, A., Luisi, M. & Miccoli, M.N. 2014. Distribuzione delle cavità antropiche nel territorio pugliese in relazione alle caratteristiche litostratigrafiche delle calcareniti neogeniche. Paper presented at the 3rd International Workshop Voragini in Italia, ISPRA, 8 May 2014, Rome, Italy.

Liu, D., Wang, S. & Li, L. 2000. Investigation of fracture behavior during rock mass failure. *International Journal of Rock Mechanics and Mining Science*, **37**, 489–497.

Lollino, P., Martimucci, V. & Parise, M. 2013. Geological survey and numerical modeling of the potential failure mechanisms of underground caves. *Geosystem Engineering*, **16**, 100–112.

Lollino, P., Parise, M. & Vattano, M. 2014. Some considerations on 3-D and 2-D numerical models for the assessment of the stability of underground caves. *In*: Lollino, G., Manconi, A., Guzzetti, F., Culshaw, M., Bobrowsky, P. & Luino, F. (eds) *Engineering Geology for Society and Territory. Volume 5 – Urban Geology, Sustainable Planning and Landscape Exploitation*. Springer, Berlin, 585–588.

Margiotta, S., Negri, S., Parise, M. & Valloni, R. 2012. Mapping the susceptibility to sinkholes in coastal areas, based on stratigraphy, geomorphology and geophysics. *Natural Hazards*, **62**, 657–676.

Martinotti, M.E., Pisano, L. et al. 2017. Landslides, floods and sinkholes in a karst environment: the 1–6 September 2014 Gargano event, southern Italy. *Natural Hazards and Earth System Sciences*, **17**, 467–480.

Monte, A. 1995. *Frantoi ipogei del Salento*. Edizioni del Grifo, Lecce.

Negri, S., Margiotta, S., Quarta, T.A.M., Castiello, G., Fedi, M. & Florio, G. 2015. Integrated analysis of geological and geophysical data for the detection of underground man-made caves in an area in southern Italy. *Journal of Cave and Karst Studies*, **77**, 52–62.

Nof, R.N., Baer, G., Ziv, A., Raz, E., Atzori, S. & Salvi, S. 2013. Sinkhole precursors along the Dead Sea, Israel, revealed by SAR interferometry. *Geology*, **49**, 1019–1022.

Palmer, A.N. 2007. *Cave Geology*. Cave Books, Dayton, OH.

Parise, M. 2008. Rock failures in karst. *In*: Cheng, Z., Zhang, J., Li, Z., Wu, F. & Ho, K. (eds) *Landslides and Engineered Slopes. Proceedings of the 10th International Symposium on Landslides*, 30 June–4 July 2008, Xi'an, China. Vol. **1**. CRC Press, Boca Raton, FL, 275–280.

Parise, M. 2010. The impacts of quarrying in the Apulian karst. *In*: Carrasco, F., La Moreaux, J.W., Duran Valsero, J.J. & Andreo, B. (eds) *Advances in Research in Karst Media*. Springer, Berlin, 441–447.

Parise, M. 2012. A present risk from past activities: sinkhole occurrence above underground quarries. *Carbonates and Evaporites*, **27**, 109–118.

Parise, M. 2013. Recognition of instability features in artificial cavities. *In*: Filippi, M. & Bosak, P. (eds) *Proceedings of the 16th International Congress of Speleology*, 21–28 July 2013, Brno, Czech Republic, Vol. **2**, 224–229.

Parise, M. 2015. A procedure for evaluating the susceptibility to natural and anthropogenic sinkholes. *Georisk*, **9**, 272–285.

Parise, M. 2016. Modern resource use and its impact in karst areas – mining and quarrying. *Zeitschrift für Geomorphologie*, **60**, 199–216.

Parise, M. & Gunn, J. (eds) 2007. *Natural and Anthropogenic Hazards in Karst Areas: Recognition, Analysis and Mitigation*. Geological Society, London, Special Publications, **279**, http://sp.lyellcollection.org/content/279/1

Parise, M. & Lollino, P. 2011. A preliminary analysis of failure mechanisms in karst and man-made underground caves in southern Italy. *Geomorphology*, **134**, 132–143.

Parise, M. & Vennari, C. 2013. A chronological catalogue of sinkholes in Italy: the first step toward a real evaluation of the sinkhole hazard. *In*: Land, L., Doctor, D.H. & Stephenson, B. (eds) *Proceedings of the 13th Multidisciplinary Conference on Sinkholes and the Engineering and Environmental Impacts of Karst*, Carlsbad, New Mexico, USA. National Cave Karst Research Institute, Carlsbad, NM, 383–392.

Parise, M. & Vennari, C. 2017. Distribution and features of natural and anthropogenic sinkholes in Apulia. *In*: Renard, P. & Bertrand, C. (eds) *EuroKarst 2016, Neuchatel. Advances in the Hydrogeology of Karst and Carbonate Reservoirs*. Springer, Berlin, 27–34.

Parise, M., Perrone, A., Violante, C., Stewart, J.P., Simonelli, A. & Guzzetti, F. 2010. Activity of the Italian National Research Council in the aftermath of the 6 April 2009 Abruzzo earthquake: the Sinizzo Lake case study. *In*: *Proceedings of the 2nd International Workshop on Catastrophic Sinkholes in the Natural and Anthropogenic Environment*, 3–4 December 2009, Rome, Italy, 623–641.

Parise, M., Vennari, C., Guzzetti, F., Marchesini, I. & Salvati, P. 2013*a*. Preliminary outcomes from a catalogue of natural and anthropogenic sinkholes in Italy, and analysis of the related damage. *Rendiconti Online della Società Geologica Italiana*, **24**, 225–227.

Parise, M., De Giovanni, A. & Martimucci, V. 2013*b*. Sinkholes caused by underground quarries: the case of the 2–3 May 2010, event at Barletta (southern Italy). *In*: *Speleology and Spelestology, Proceedings of the IV International Scientific Conference*, November 2013, Nabereznye Chelny, Russia, 158–162.

Parise, M., Galeazzi, C., Bixio, R. & Dixon, M. 2013*c*. Classification of artificial cavities: a first contribution by the UIS Commission. *In*: Filippi, M. & Bosak, P. (eds) *Proceedings of the 16th International Congress of Speleology*, 21–28 July 2013, Brno, Czech Republic, Vol. **2**, 230–235.

Parise, M., Galeazzi, C., Germani, C., Bixio, R., Del Prete, S. & Sammarco, M. 2015. The map of ancient underground aqueducts in Italy: updating of the project, and future perspectives. *In*: *Proceedings of the International Congress in Artificial Cavities Hypogea 2015*, 11–17 March 2015, Rome, Italy, 235–243.

Parise, M., Pisano, L. & Vennari, C. 2017. Sinkhole clusters after heavy rainstorms. *Journal of Cave and Karst Studies*, **79**.

Pepe, M. & Parise, M. 2014. Structural control on development of karst landscape in the Salento Peninsula (Apulia, SE Italy). *Acta Carsologica*, **43**, 101–114.

Pepe, P., Pentimone, N., Garziano, G., Martimucci, V. & Parise, M. 2013. Lessons learned from occurrence of sinkholes related to man-made cavities in a town of southern Italy. *In*: Land, L., Doctor, D.H. & Stephenson, B. (eds) *Proceedings of the 13th Multidisciplinary Conference on Sinkholes and the Engineering and Environmental Impacts of Karst, Carlsbad, New Mexico, USA*. National Cave Karst Research Institute, Carlsbad, NM, 393–401.

Perrotti, M., Lollino, P. *et al.* In press. Finite element-based stability charts for underground cavities in soft calcarenites. *International Journal of Geomechanics*.

Pieri, P., Festa, V., Moretti, M. & Tropeano, M. 1997. Quaternary tectonic of the Murge area (Apulian foreland – southern Italy). *Annali di Geofisica*, **40**, 1395–1404.

PLAXIS 2D 2015. Tutorial Manual, Reference Manual, Material Models Manual, Scientific Manual. Plaxis Academy, Delft.

PLAXIS 3D 2015. Tutorial Manual, Reference Manual, Material Models Manual, Scientific Manual. Plaxis Academy, Delft.

Ricchetti, G., Ciaranfi, N., Luperto Sinni, E., Mongelli, F. & Pieri, P. 1988. Geodinamica ed evoluzione sedimentaria e tettonica dell'avampaese apulo. *Memorie Società Geologica Italiana*, **41**, 57–82.

Selli, R. 1962. Il Paleogene nel quadro della geologia dell'Italia meridionale. *Memorie Società Geologica Italiana*, **3**, 737–789.

Società Italiana per Condotte d'Acqua S.P.A. 1989. *Interventi urgenti a salvaguardia della pubblica e privata incolumità. Rilevamento cavità sotterranee della città, studi e indagini geognostiche del territorio. Relazione generale*. Comune di Canosa di Puglia, Canosa.

Stendardo, A. (ed.) 1995. *Presicce sotterranea*. Congedo Editore, Galatina.

Suchowerksa, A.M., Merifield, R.S., Carter, J.P. & Clausen, J. 2012. Prediction of underground cavity roof collapse using the Hoek–Brown failure criterion. *Computers and Geotechnics*, **44**, 93–103.

Sunwoo, C., Song, W.K. & Ryu, D.W. 2010. A case study of subsidence over an abandoned underground limestone mine. *Geosystem Engineering*, **13**, 147–152.

Swedzicki, T. 2001. Geotechnical precursors to large-scale ground collapse in mines. *International Journal of Rock Mechanics and Mining Science*, **38**, 957–965.

Tharp, T.M. 1995. Mechanics of upward propagation of cover-collapse sinkholes. *Engineering Geology*, **52**, 23–33.

Toni, L. & Quartulli, S. 1986. Coltivazione di calcareniti in sotterraneo nel comune di Cutrofiano (Lecce). *Quarry and Construction*, **feb**, 23–26.

Vattano, M., Di Maggio, C., Madonia, G., Parise, M., Lollino, P. & Bonamini, M. 2013. Examples of anthropogenic sinkholes in Sicily and comparison with similar phenomena in southern Italy. *In*: Land, L., Doctor, D.H. & Stephenson, B. (eds) *Proceedings of the 13th Multidisciplinary Conference on Sinkholes and the Engineering and Environmental Impacts of Karst, Carlsbad, New Mexico, USA*. National Cave Karst Research Institute, Carlsbad, NM 263–271.

Waltham, T. & Lu, Z. 2007. Natural and anthropogenic rock collapse over open caves. *In*: Parise, M. & Gunn, J. (eds) *Natural and Anthropogenic Hazards in Karst Areas: Recognition, Analysis and Mitigation*. Geological Society, London, Special Publications, **279**, 13–21, https://doi.org/10.1144/SP279.3

Waltham, T., Bell, F. & Culshaw, M. 2005. *Sinkholes and Subsidence*. Springer, Berlin.

White, E.L. & White, W.B. 1969. Processes of cavern breakdown. *Bulletin of the National Speleological Society*, **31**, 83–96.

Yan, H., Zhiyi, L., Zhifa, Y. & Jianfeng, W. 2001. Engineering geology study of Lingquansi Cave Temple, People's Republic of China. *Bulletin of Engineering Geology and the Environment*, **60**, 43–57.

Zupan Hajna, N. 2003. *Incomplete Solution: Weathering of Cave Walls and the Production, Transport and Deposition of Carbonate Fines*. Carsologica, Postojna-Ljubljana.

Groundwater flood hazards and mechanisms in lowland karst terrains

OWEN NAUGHTON[1,2]*, TED McCORMACK[2], LAURENCE GILL[1] & PAUL JOHNSTON[1]

[1]*Department of Civil, Structural and Environmental Engineering, University of Dublin Trinity College, Ireland*

[2]*Geological Survey Ireland, Beggars Bush, Haddington Road, Dublin, Ireland*

**Correspondence: naughto@tcd.ie*

Abstract: The spatial and temporal complexities of flooding in karst terrains pose unique challenges in flood risk management. Lowland karst landscapes can be particularly susceptible to groundwater flooding due to a combination of low aquifer storage, high diffusivity and limited or absent surface drainage. Numerous notable groundwater flood events have been recorded in the Republic of Ireland throughout the twentieth century, but flooding during the winters of 2009 and 2015 was the most severe on record, causing widespread and prolonged disruption and damage to property and infrastructure. Effective flood risk management requires an understanding of the recharge, storage and transport mechanisms governing water movement across the landscape during flood conditions. Using information gathered from recent events, the main hydrological and geomorphological factors influencing flooding in these complex lowland karst groundwater systems are elucidated. Observed flood mechanisms included backwater flooding of sinks, high water levels in ephemerally flooded basins (turloughs), overtopping of depressions, and discharges from springs and resurgences. This paper addresses the need to improve our understanding of groundwater flooding in karst terrains to ensure efficient flood prevention and mitigation in the future, and thus helps to achieve the aims of the European Union Floods Directive.

Karst landscapes present a unique set of environmental and engineering challenges to planners, stakeholders and the scientific community. Geohazards such as subsidence, sinkholes, landslides, flooding and water contamination are common in karst environments (Santo *et al.* 2007; Zhou 2007; Maréchal *et al.* 2008; Worthington *et al.* 2012; Gutiérrez *et al.* 2014; Martinotti *et al.* 2017). Flooding and flood risk management is a major challenge facing society in the coming decades, especially in the light of the increased frequency of extreme weather events as a result of climate change (Intergovernmental Panel on Climate Change 2014). Effective flood risk management requires an understanding of the recharge, storage and transport mechanisms in operation during flood conditions. In the context of groundwater flooding within karst systems, the heterogeneous and anisotropic nature of water carrying fractures and conduits beneath the surface lead to obvious problems in developing such an understanding (Field 1993).

Karst terrains are uniquely susceptible to flooding from groundwater sources due to a combination of the low storage and high diffusivity characteristics of these aquifers (Parise *et al.* 2015). Water levels within karst groundwater systems can rise dramatically during periods of intense or prolonged rainfall. As the subsurface storage and drainage reaches capacity, the rising water table can reach the topographic surface and produce floods (Gutiérrez *et al.* 2014; Naughton *et al.* 2017). However, unlike linear flood features such as river channels or coastlines, the manifestation of groundwater flooding may be discontinuous and determined by the spatially variable hydrodynamic properties and responses within the karst system. The surface expression of groundwater flooding may only occur during extreme weather events and at relatively long recurrence intervals. Thus the flood frequencies traditionally used in flood risk assessment (such as 10 or 1% annual exceedance probability) may be undefinable. These inherent difficulties are explicitly acknowledged within the European Union Floods Directive 2007/60/EC, whereby Member States are permitted to limit groundwater flood hazard maps to extreme event scenarios only.

Groundwater flooding has not traditionally been recognized as posing a significant risk and so remains relatively less well understood than other forms of flooding (Bonacci *et al.* 2006; Morris

From: Parise, M., Gabrovsek, F., Kaufmann, G. & Ravbar, N. (eds) 2018. *Advances in Karst Research: Theory, Fieldwork and Applications*. Geological Society, London, Special Publications, **466**, 397–410.
First published online December 11, 2017, https://doi.org/10.1144/SP466.9

et al. 2007). Consequently, investigations into the contribution of karst hydrology to surface flooding are still in their infancy (Gutiérrez *et al.* 2014). This has begun to change in the last decade or so, driven, in part, by the introduction of the European Union Floods Directive and its requirement to consider all forms of flooding, including groundwater, but primarily due to a series of groundwater-related flood events across Europe – in France (Pinault *et al.* 2005; Maréchal *et al.* 2008), Spain (Lopez-Chicano *et al.* 2002), the UK (Hughes *et al.* 2011; Morris *et al.* 2015), Croatia (Bonacci *et al.* 2006) and Italy (Parise 2003). Studies of polje hydrology and flooding are perhaps the best described in the literature.

In the Republic of Ireland, the last decade has seen the worst groundwater flooding in living memory. The dramatic floods during the winters of 2009 and 2015 caused widespread damage and disruption to communities across the country, particularly in the extensive karstic limestone lowlands on the western seaboard (Naughton *et al.* 2017). This paper presents a detailed example of the phenomenon of groundwater flooding in the lowland karst terrains of western Ireland. Using examples and insights gained during the recent unprecedented flood events, we describe the main hydrological and geomorphological characteristics influencing flooding in these complex lowland karst groundwater systems.

Background and geological context

Limestone accounts for >40% (30 000 km^2) of the surface or near-surface outcrop in the Republic of Ireland, making it the most prevalent bedrock type and primary regionally important aquifer lithology in the country (Simms 2004; Drew 2008). The main Irish limestones were formed during the Early Carboniferous or Dinantian (Drew *et al.* 1996) when a marine transgression in the Tournaisian period inundated much of the Old Red Sandstone continent and provided the depositional environment necessary for limestone formation (Guion *et al.* 2000; Sevastopulo & Wyse Jackson 2009). The telegenetic origin of Irish limestones has resulted in little primary (matrix) porosity; instead, modern groundwater circulation is dominated by secondary (joints, fractures and bedding planes) and tertiary (solutionally widened) porosity. Irish limestones have undergone karstification many times since their formation, with the most significant period being the Tertiary (65–2 Ma) (Williams 1970; Drew 1990). Karst features have been documented in >80% of the limestone outcrop, indicating that karstification has occurred across most, if not all, of the limestone formations in the country (Drew *et al.* 1996). More recent dissolution processes during the Holocene (10 ka to present) has also resulted in the development of a weathered epikarst zone near the bedrock surface, as well as active karst features, such as stream caves in the Burren (White 1988; Drew & Jones 2000; Drew 2008).

The degree of karstification varies significantly across the country due to variabilities in the bedrock purity, fracturing and landscape history. Karst features are sparse or absent across much of the central and eastern limestones, where normal fluvial drainage systems have developed with little interaction with the underlying karst aquifer (Coxon 1986). By contrast, karst groundwater flow systems dominate on the relatively pure, well-bedded lowland limestones in the west of the country (Fig. 1). These lowlands experience a western maritime climate with a long-term average annual rainfall (1981–2010) of *c.* 1100 mm (Walsh 2012). Recharge is principally autogenic in the form of direct (diffuse) and local point recharge; significant allogenic recharge is relatively uncommon (Drew 2008). One notable exception to this is the Gort Lowland catchment in south Galway, which receives the majority of catchment recharge from the adjoining sandstone uplands (Gill *et al.* 2013; McCormack *et al.* 2014).

Over 90% of Irish karst occurs in a low-lying or lowland setting, typically <100 m above sea-level (Drew 2008). A distinguishing feature of Irish lowland karst terrains is the relatively shallow depth of the vadose zone, wherein the phreatic surface remains close to or at the surface throughout the year. There is widespread water circulation between the surface and groundwater flow systems in this hydrogeological setting, with frequent reversals of the hydraulic gradients. The shallow depth to groundwater limits the buffering effect of aquifer storage during recharge events. The lack of vadose zone storage is mitigated by the presence of a well-developed epikarst, where significant weathering, fracturing and dissolution of the near-surface bedrock provides additional storage. Nonetheless, ephemeral surface flooding from groundwater sources is a prevalent feature of lowland karst terrains in Ireland.

During periods of high rainfall, excess recharge that cannot be accommodated by the subterranean network of water-bearing fractures and conduits is temporarily stored in ephemeral, geographically isolated water bodies known as turloughs. Turloughs vary in size from ≤1 ha to >250 ha and more than 400 active turloughs have been documented (Sheehy Skeffington & Gormally 2007). Turloughs are usually located along lines of concentrated flow within an aquifer and thus play a key part in lowland karst hydrology (Sheehy Skeffington *et al.* 2006); they act as temporary storage for local and regional recharge in a role akin to that of temporary bank and floodplain storage in fluvial systems (Naughton

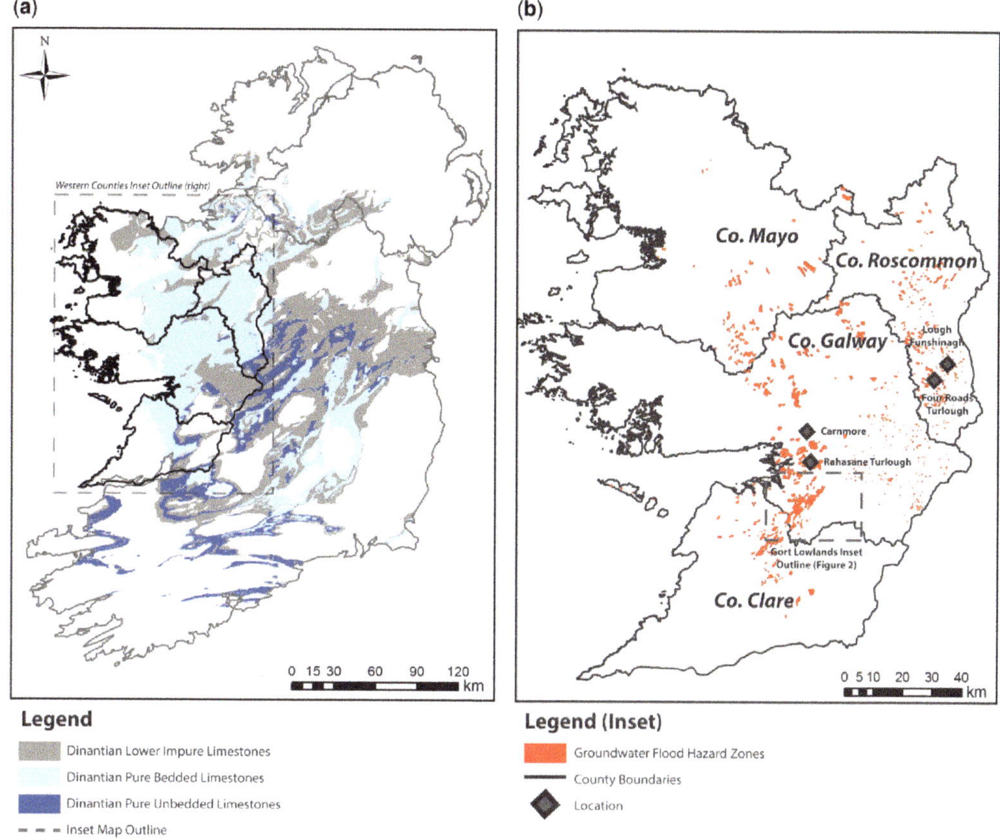

Fig. 1. (**a**) Areas of Carboniferous limestone and distribution of turloughs in the Republic of Ireland (geological data from the Geological Survey Ireland). (**b**) Location of groundwater flood hazard zones and site location map.

et al. 2017). Turlough flooding thus bears many similarities to that which occurs within poljes in karst, in that both act as a subsystem of surface and groundwater flow through the karst groundwater flow system (Bonacci 2013). In fact, turloughs have been considered to be a subtype of the polje landform. Both karst depressions display complex hydrological and hydrogeological characteristics, such as periodic inundation, temporary springs, lacustrine sediment deposition, swallow holes and estevelles. As with turloughs, the flooding within poljes can be severe and pose a significant flood hazard to surrounding properties and infrastructure (Mijatovic 1988; Kovacic & Ravbar 2010).

During typical winter rainfall levels, flooding is confined within the basin and acts as an environmental supporting condition for the wetland floral and faunal species that have colonized the turlough habitat (Sheehy Skeffington *et al.* 2006; Moran *et al.* 2008*a*; Porst *et al.* 2012). During extreme and/or prolonged rainfall, floodwaters within the basins can reach extreme levels and cause widespread damage and disruption to the surrounding areas. It is in this context that turloughs represent the principal form of recurrent, extensive groundwater flooding in Ireland. Historically, groundwater flooding has been centred on the karst limestone plains of the western lowlands, principally in counties Roscommon, Mayo, Galway and Clare (Fig. 1). The last decade has seen the worst groundwater flooding that the western limestone lowlands of Ireland have experienced in living memory.

Groundwater flood events of 2009–10 and 2015–16

The winters of 2009–10 and 2015–16 were exceptionally wet seasons across the Republic of Ireland. Although both winters represent extreme meteorological events, they differed in the intensity and duration over which rainfall persisted. The heavy rainfall events of 2009 were caused by a series of deep, fast-moving Atlantic depressions crossing the country during November. Over twice the long-term average amounts of rainfall were recorded at stations across Ireland, making November 2009 the wettest on record (Walsh 2010; McCarthy *et al.* 2016). By contrast, the winter of 2015–16 saw more persistent wet weather. A succession of storm fronts moved across Ireland from November to February, bringing with them exceptional rainfall accumulations across much of the country (McCarthy *et al.* 2016). Between December and February, a total of >600 mm (189% of the long-term average) fell across the island of Ireland, making it the wettest winter on record in a rainfall time series stretching back to 1850 (McCarthy *et al.* 2016; Noone *et al.* 2016). December 2015 was also the wettest of any month on record in Ireland, with five stations exceeding the previous Irish record for the highest monthly rainfall total (McCarthy *et al.* 2016).

The unprecedented rainfall events during the winters of 2009–10 and 2015–16 caused widespread damage and disruption to residential houses, businesses, infrastructure and agriculture across the region. The sustained nature of the flooding, lasting for more than three months in some cases, caused prolonged hardship to rural communities, who struggled to prevent the inundation of homes and workplaces amid unparalleled disruption to transport networks. The flooding of large tracts of agricultural land severely affected agricultural activity and posed a serious welfare risk to livestock, while anoxic soil conditions due to prolonged submergence damaged hundreds of hectares of valuable pasture land. An idea of the scale of flooding is given in Figure 2, which shows the extent of floods for south Co. Galway during the winter of 2015–16 derived from field and satellite synthetic aperture radar measurements. This region is effectively devoid of permanent surface water. The flooded extents shown, encompassing an area >38 km^2, are primarily associated with flooding of the karstic groundwater system. Further widespread flooding was also reported in counties Roscommon and Mayo, with more localized events in Clare, Longford and Westmeath.

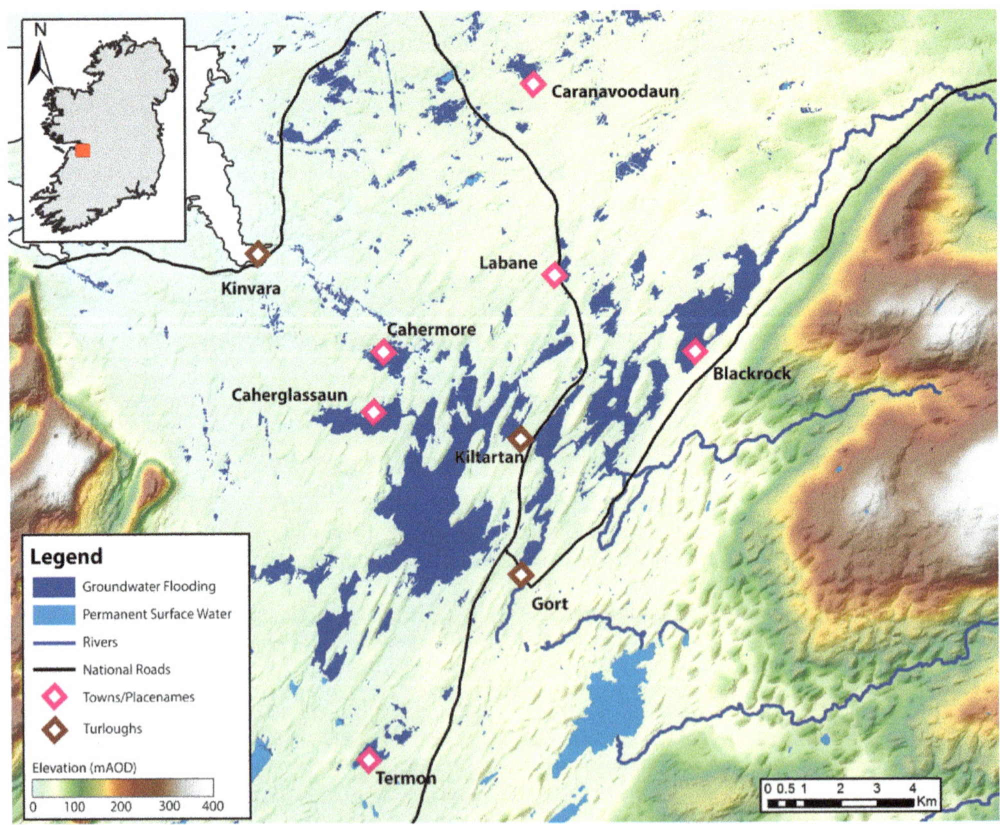

Fig. 2. Groundwater flood extent map for 2015–16 flooding in the Gort Lowlands, Co. Galway, Ireland (derived from field measurements and SAR imagery courtesy of Copernicus Emergency Management Service).

Groundwater flood response

Of these two exceptional winters, it was 2015–16 which generally saw the highest groundwater levels and most widespread flooding. Although some of the difference can be accounted for by regional variations in rainfall, the duration (or persistence) of heavy rainfall was the primary cause. The crucial durations governing groundwater flooding can vary dramatically depending on the hydraulic properties and structure of the aquifer system, with response times ranging from minutes and hours up to multi-annual timescales (Maréchal *et al.* 2008; De Waele *et al.* 2010; Hughes *et al.* 2011). In the case of lowland karst groundwater flow systems in Ireland, surface flooding is strongly related to the cumulative rainfall typically measured in weeks to months (Moran *et al.* 2008*b*; Naughton *et al.* 2012). Figure 3 shows the maximum rainfall depths for a range of durations from the Irish Meteorological Service (Met Eireann) rainfall station in Gort, south Galway. Both winters showed rainfall totals significantly above the median (1981–2010) for the area. For shorter durations (<40 days), the rainfall totals in 2009 exceeded those experienced in 2015, whereas for higher durations this reversed, with the rainfall totals in 2015 significantly exceeding those of 2009. This greater flooding experienced during 2015–16 is an indication that the crucial rainfall durations influencing the magnitude of groundwater floods in Irish karst systems lies in this range.

The relationship between antecedent rainfall and flood magnitude varies significantly within and between systems, reflecting heterogeneities in the extent of karstification, hydraulic connectivity and aquifer storage. This variability gives rise to a spectrum of flooding regimes within turloughs, ranging from short duration flooding in basins with a rapid response to rainfall events, to long duration flooding in response to longer term precipitation patterns (Naughton *et al.* 2012).

An example of the variability in water level response during flood conditions is demonstrated in Figure 4, which shows normalized water level hydrographs (relative to their peak level) for three turloughs in south Co. Galway during the 2009 flood event. Substantial differences are evident in hydrograph shape and the timing of peak flood levels, despite comparable inputs of rainfall. The flood maximum in Blackrock turlough occurred on 26 November 2009, about three weeks after the onset of flooding within the basin. Over this period, the floodwaters reached depths of up to 18 m, representing a flood volume of >15.6×10^6 m^3, which included a 2.9×10^6 m^3 increase over a single day. Blackrock turlough has an extensive allogenic upland catchment with the capacity to rapidly deliver large amounts of recharge to the turlough, which is, in turn, drained by a large conduit network (McCormack *et al.* 2016). By contrast, the nearby Caranavoodaun and Termon turloughs have more localized autogenic catchments. Caranavoodaun turlough showed a damped response compared with Blackrock, with peak levels lagging by about eight days and showing a relatively prolonged recession. Termon South turlough did not peak until much

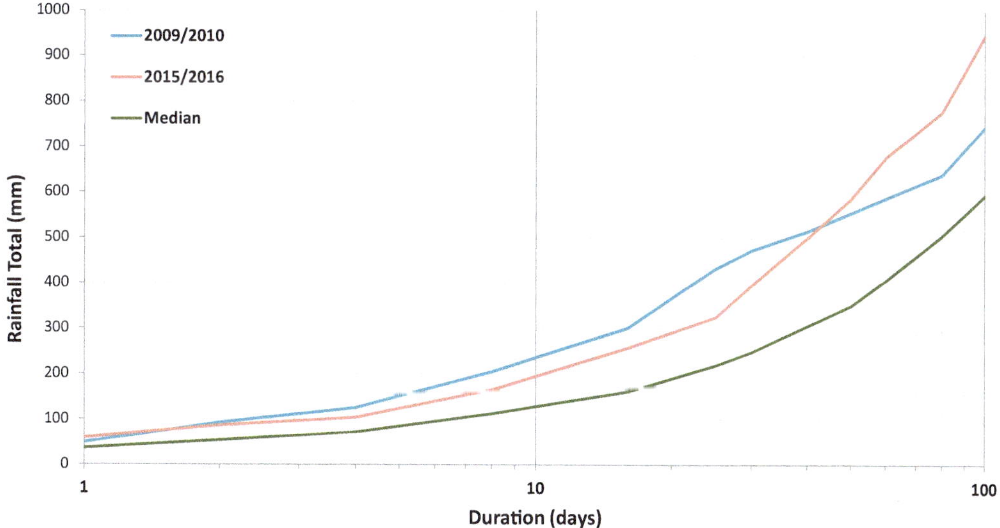

Fig. 3. Maximum rainfall depths recorded at the Gort (Derrybrien) rainfall station, Co. Galway for winter 2009–10 and 2015–16.

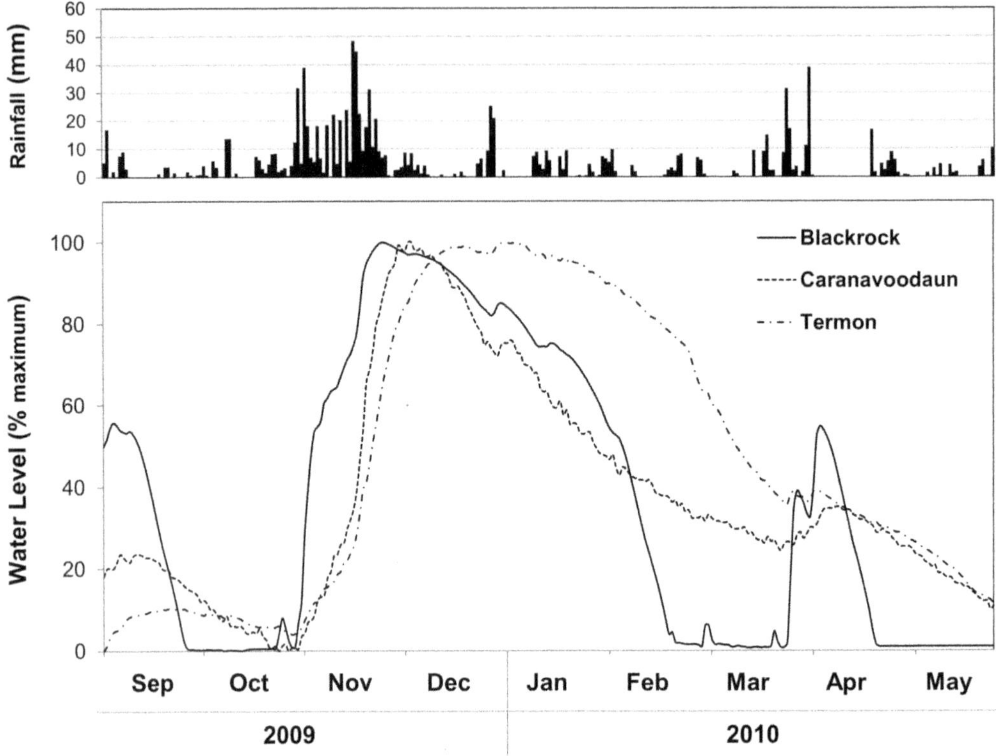

Fig. 4. Rainfall record from Gort rainfall station and normalized water level hydrographs for Blackrock, Caranavoodaun and Termon South turloughs, Co. Galway.

later in the flooding season, on 6 January 2010. The heavy rainfall of November 2009 contributed significantly to the stored floodwaters within Termon, but the slow drainage characteristics of the underlying groundwater flow system meant that peak levels occurred much later in the season after further rainfall. The flood maximum in Termon turlough is thus a function of a rainfall duration measured in months rather than the weeks of Blackrock turlough.

Groundwater flooding mechanisms

In response to the flooding in 2009 and 2015, we carried out a series of studies for key locations identified by the Office of Public Works and local authorities as potentially affected by groundwater flooding. The objective was to assess the extent, nature and mechanisms of flooding, whether a significant flood risk existed and whether groundwater was the key contributor to that risk. Consistent long-term data on groundwater flooding in Ireland do not exist and information was therefore derived from diverse sources, including field measurements, hydrometric data, aerial photography, satellite imagery, historical land maps, technical reports, local authority road closure notices, local accounts and media sources. What became apparent during the study was that although the primary form of extensive, recurrent groundwater flooding in Ireland originates in turloughs, a range of mechanisms beyond simple turlough flooding play a key part during extreme groundwater flood events.

From experience in the Chalk aquifers of southern England, Robins & Finch (2012) proposed two distinct types of groundwater flood event: groundwater flooding and groundwater-induced flooding. The former is considered as a true groundwater flood in which the water table rises above the ground elevation, whereas a groundwater-induced flood occurs when intense groundwater discharge via springs and highly permeable shallow horizons discharges to the surface water, causing overbank flooding (Robins & Finch 2012). A similar division is proposed here for lowland karst groundwater systems, wherein flood mechanisms can be broadly divided into those where the damage mechanism is primarily due to either hydrostatic action or hydrodynamic action. The principal mechanisms identified

Table 1. *Groundwater flooding mechanisms in lowland karst groundwater systems in Ireland*

Type	Damage	Description
Turlough flooding	Hydrostatic	Turlough floodwaters rise to extreme levels and pose a flood risk to the surrounding area
Backwater flooding	Hydrostatic	Point recharge (sinking streams/rivers) exceeds the groundwater drainage capacity, causing inundation of the sink itself and backwater flooding upstream
Overtopping of sinks/ basins	Hydrodynamic	Ephemeral overland flow due to overtopping of flooded depressions
Discharge from springs and resurgences	Hydrodynamic	(a) Groundwater springs and risings at the periphery of upland areas exceed normal discharge levels, causing flooding around and downstream of the resurgence (b) Shallow lateral flow paths are activated within the epikarst by high groundwater levels, triggering ephemeral springs and flooding of adjacent depressions

are given in Table 1 and Figure 5 and represent the type examples of how groundwater flooding manifests in Irish lowland karst catchments. The main damage mechanism in turlough and backwater flooding of sinks is by hydrostatic action, whereby elevated water levels with low or negligible water velocities pose a risk to surrounding receptors. Mechanisms where hydrodynamic action (flowing water) posed a risk included ephemeral overland flow due to the overtopping of flooded depressions, and excess discharge from permanent/transient springs and resurgences.

Turlough flooding

Turloughs represent the principal form of recurrent, extensive groundwater flooding in Ireland. Dozens of examples of flooding around turlough basins were identified across the western lowlands (e.g. Fig. 2). Although the numbers of receptors affected at any one site were relatively low, cumulatively turlough flooding caused extensive damage and disruption to communities across the region. For example, Rahasane turlough, in the Dunkellin River catchment, Co. Galway, flooded 12 houses along its

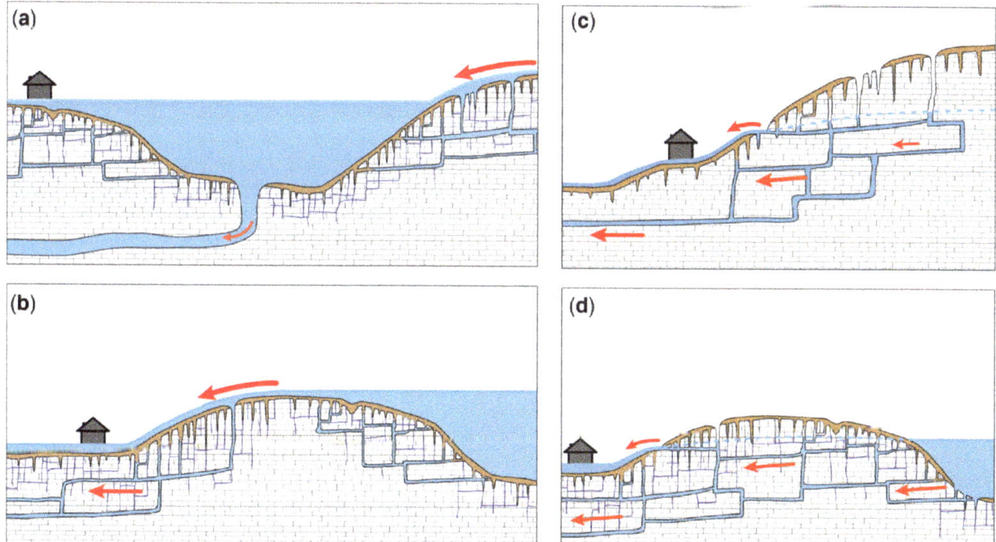

Fig. 5. Groundwater flood mechanisms in lowland karst groundwater flow systems: (**a**) turlough/backwater flooding of sinks; (**b**) overtopping of basins and sinks; (**c**) discharge from spring and resurgences at the periphery of upland areas; and (**d**) lateral flow through shallow epikarst pathways.

banks during the November 2009 event. Flooding at Labane turlough, Co. Galway, forced the closure of the N18 road between Galway City and Limerick City for more than two months. Further north at Lough Funshinagh, Co. Roscommon, floodwaters in 2015–16 were the highest in living memory, covering an area of 4.6 km^2 with a peak volume of >16 × 10^6 m^3. Lough Funshinagh was notable due to the length of time the floodwaters persisted, with water levels falling at a rate of only a few centimetres per week. Flood levels remained high throughout 2016 and were still above the previous record flood (from 2009 to 2010) a full six months after the peak.

The nature of flooding, in terms of timing, extent and duration, varied substantially, both locally and regionally, in line with the spectrum of flooding regimes and modus operandi characteristic of turloughs (Naughton *et al.* 2012). This may in some part be due to their polygenetic origins and the complex landscape history of Irish limestones. Turloughs were originally considered as hollows in glacial drift with underlying karst drainage systems (Williams 1964). However, Drew (1973) asserted that turloughs invariably lie in bedrock hollows and were solutional features requiring a far longer period to develop than has passed since the last glaciation. Coxon & Coxon (1994) suggested that turloughs are polygenetic, with glacial deposition influencing their morphology, but solutional rather than glacial processes being the determining factor in turlough formation. The lines of high permeability associated with turloughs may thus represent the re-use of remnants of karst drainage systems created during Tertiary dissolution, but subsequently partially blocked by glacial drift, rather than post-glacial dissolution pathways. Coxon & Drew (1986) suggested three models to explain turlough origin and the presence of the high permeability zones required for turlough formation: (1) glacial hollows with flow paths developed post-glacially; (2) glacial hollows that developed along the line of existing pre-glacial flow routes; and (3) pre-glacial karst features with associated flow paths modified by glaciation.

The hydrological budget of turloughs can be described using two general conceptual models: through-flow systems and surcharged tank systems (Naughton *et al.* 2012). Rainfall onto, and evaporation from, the water body is common across all models, as well as surface runoff from the surrounding slopes. Surface evaporation is generally of minor importance to the water budget due to the seasonality of turlough flooding because it typically occurs during the winter months. Direct precipitation and runoff can be significant, particularly in shallow, flat basins, but under flood conditions groundwater flow is the dominant hydrological process.

In through-flow systems the recharge and discharge processes work in partial isolation, with no direct transfer of water between them without first passing through the main water body. The turlough basin effectively acts as a sink, receiving recharge from the surrounding vadose zone, shallow groundwater systems and/or point recharge. In the case of a purely distributed through-flow system, drainage occurs via a distributed network of shallow fractures and conduits (Fig. 6a). Through-flow systems can also consist of point recharge and discharge (Fig. 6b), but groundwater flow within the system elements is unidirectional. In a surcharged tank system (Fig. 6c), the main recharge and discharge processes do not occur simultaneously. Instead, the water budget is controlled by a bidirectional flow system located at or near the turlough base, with the turlough acting as overflow storage for the underlying conduit network. Under this scenario there is no significant discharge from the turlough during filling periods. During recession periods recharge is still derived from the local (proximal) shallow groundwater systems, but not from the distal catchment (Naughton *et al.* 2012).

Understanding the nature of a turlough's hydrological budget is crucial if active flood management measures are to deliver the intended outcomes. For example, the construction of surface drainage is often the first alleviation option considered after a flood event. A key element of drainage design is an estimate of the required channel conveyance capacity. In the case of a through-flow system, a reasonable basis for such a calculation would be the net volume changes within the basin during a representative flood season. However, this is not the case for a surcharged tank system. Artificially lowering the hydraulic head would increase the gradient into the basin, as the water budget is controlled by the head difference between the turlough and the underlying groundwater system. The extra conveyance capacity provided by the channel would thus be at least partially offset against increased recharge from the distal catchment. Although the effective catchment area required to provide sufficient recharge to flood a turlough basin may be relatively modest if operating as a through-flow system, of the order of a few square kilometres, the catchment from which floodwaters can potentially be derived can be orders of magnitude greater in surcharge tank systems. In this case, the capacity of the drain/culvert may need to be significantly greater than that in a through-flow turlough of comparable size.

The excavation and clearance of swallow holes is often cited as a potential solution to turlough flooding. Although this may improve drainage in some circumstances, the drainage rate is often not limited by localized constrictions at the inlet, but by the capacity of the underlying groundwater flow system. In the case of surcharged tank systems, any surface modification of the estevelle is unlikely to reduce

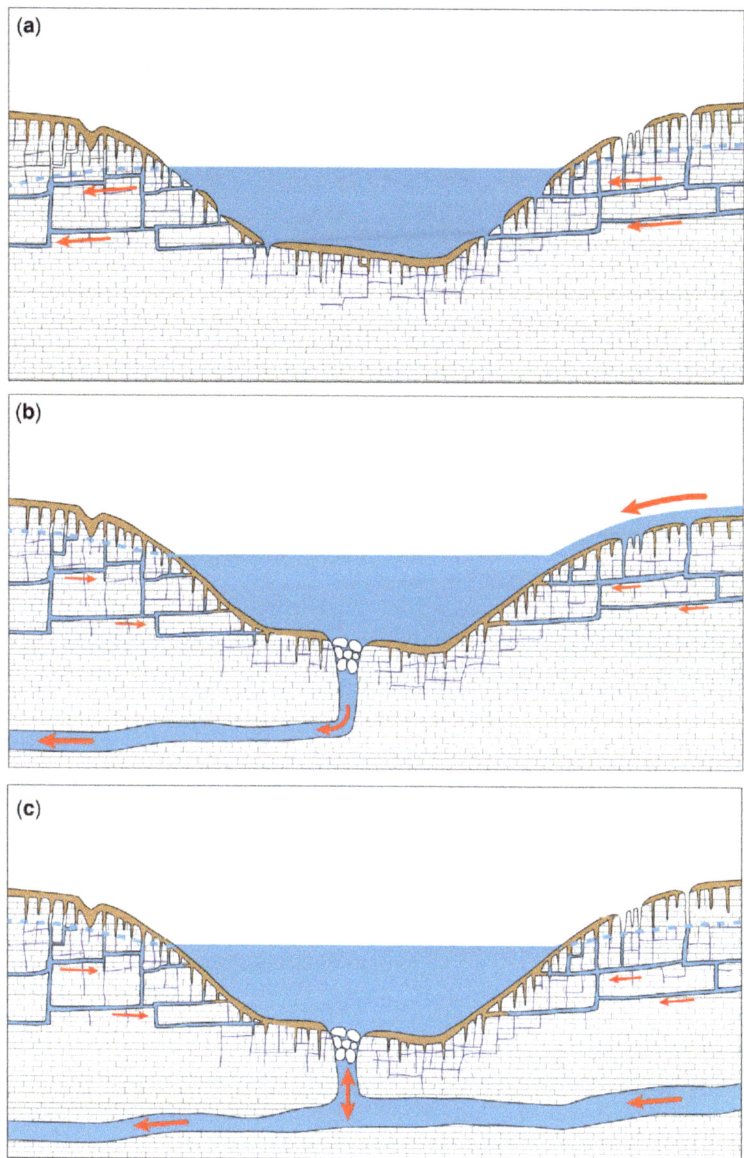

Fig. 6. Conceptual diagrams representing possible turlough water budgets: (**a**) distributed through-flow; (**b**) through-flow with distributed/point recharge and point discharge; and (**c**) surcharged tank.

flooding because it serves as both the entry and exit points for floodwaters. In through-flow systems the rate of drainage is dependent on the flow capacity and the relative hydraulic head within the turlough and receiving groundwater system. If this gradient is sufficiently small, as is often the case during flood conditions, outflow may cease altogether and so any perceived improvement in drainage due to swallow hole enlargement is unlikely to improve the situation. Moreover, the role of turloughs as flood attenuation devices within lowland karst systems means that the reduction of flood risk in one turlough is likely to be at the expense of raising it in another, so a solid understanding of both site and catchment hydrodynamics is key.

Backwater flooding of sinks

Backwater flooding occurs when excess point recharge (sinking streams or rivers) causes the inundation of dolines or sinks capable of accommodating recharge under normal conditions (Fig. 5a). This mode is analogous to the recharge-related sinkhole flooding described by Zhou (2007), whereby flooding occurs when the capacity of the sinkhole is not sufficient to transfer storm water runoff into the subsurface. The damage mechanism in backwater flooding is principally hydrostatic, but such cases clearly have a strong fluvial component given their dependence on the discharge of the influent surface water. Backwater flooding is common across the Irish karst lowlands, but in the clear majority of cases it is related to small autogenic streams with low baseflow discharges and so does not pose a significant flood risk. Historically, large-scale backwater flooding in karst areas would have been relatively common. However, many areas formerly characterized by internal drainage have been systematically modified by arterial drainage schemes built during the late nineteenth and early twentieth centuries (Drew & Coxon 1988). For example, the 1000 km^2 Clare River catchment in north Co. Galway originally discharged underground via a series of large sinks, turloughs and springs. Subsequent construction and channelization of the Clare River altered the natural karstic groundwater system and it is surface water, rather than groundwater, that is now the controlling factor in present day flooding.

One catchment where the karst system has remained effectively unmodified is in the Gort Lowlands, south Co. Galway, and here backwater flooding persists as a significant flood risk. The 500 km^2 catchment is divided into sandstone uplands to the east and a lowland limestone plain to the west. Backwater flooding occurs where point allogenic recharge from the sandstone uplands, in the form of three rivers, discharges onto the limestone lowlands and sinks underground. The mean flows in the rivers range between 1.2 and 3 $m^3 s^{-1}$, but discharges can reach >40 $m^3 s^{-1}$ during flood conditions, causing widespread flooding upstream and inundating hundreds of hectares in the process (McCormack & Naughton 2016). Backwater flooding also occurs on the limestone plain due to the intermittent rising and sinking of discharges from the well-developed conduit network. Backwater flooding at one such sink in Kiltartan in 2009 incurred an estimated cost of €540 000 to the local communities (Jennings O'Donovan & Partners 2011) and comparable damage was caused again during the winter of 2015–16.

Overtopping of basins and sinks

This flood mechanism is intrinsically linked to flooding within turlough and sink depressions because it occurs when floodwaters build up in surface depressions to such an extent that the level exceeds and overtops the surrounding topographic divide (Fig. 5b). When overtopping occurs, ephemeral overland flow routes develop, bypassing the groundwater flow systems normally governing water movement through the catchment. It thus differs from turlough and sink flooding in that the damage mechanism is hydrodynamic and relates to floodwaters moving across the landscape. This flood mechanism bears some similarity to the karst flash floods described by Bonacci *et al.* (2006), whereby overland flow plays the dominant part in flood formation and inter-basin overflow and/or redistribution of the catchment areas occurs due to rising groundwater. However, where flash flooding is typically in response to short (minutes to hours), high-intensity storms, the crucial recharge duration for the equivalent in lowland karst can be measured in weeks to months. This is due to the significant surface storage present within the gently undulating topography and low relief characteristic of the lowland landscape.

The delayed build-up of waters makes this mechanism easier to foresee than flash flooding, but that is not to say that it is easily preventable or managed. For example, a build-up of floodwaters around Kiltartan, south Co. Galway in 2009 caused overtopping of the N18 National Road with transient flow rates of >30 $m^3 s^{-1}$ (Fig. 7), forcing the closure of the highway and a nearby railway line for more than two weeks. Another example of overtopping occurred during the floods of 2015–16 further west in the Gort catchment at Caherglassaun turlough. Flooding within the turlough reached record depths of 14.6 m, causing overland discharge of >5 $m^3 s^{-1}$ northwards towards Cahermore, damaging properties along the flow route and around Cahermore turlough.

Discharge from springs and resurgences

Flooding in lowland karst aquifers can also be caused by high discharges of groundwater, via springs and resurgences, during which time the hydrodynamic force of the floodwater is the main cause of damage. This flood mechanism can be considered as groundwater-induced, in that flooding occurs due to intense groundwater discharge via springs or shallow, highly permeable horizons within the epikarst (Robins & Finch 2012). In lowland karstic systems, a distinction can be made between two discharge scenarios: (1) groundwater springs and risings on the periphery of upland areas exceeding normal discharge levels and causing flooding around and downstream of the resurgence (Fig. 5c); and (2) shallow flow paths within the epikarst zone are activated by high groundwater levels, triggering ephemeral springs and flooding of adjacent depressions (Fig. 5d).

The first scenario occurs where the lowland karst landscape is characterized by flat and undulating plains separated by isolated areas of higher ground,

Fig. 7. Overflow across the N18 National Road, Kiltartan, Co. Galway (image provided by Galway County Council).

such as can be found in Co. Roscommon and north Co. Galway. Here the recharge zones are located on topographic plateaus, which typically have thin or absent subsoil, have a high density of recharge landforms and a well-developed epikarst zone (Hickey 2010). Infiltration is transmitted through a well-developed epikarst system to springs located at the periphery of the upland areas, where groundwater is discharged via a combination of perennial and/or overflow springs. During periods of intense recharge, discharge from these peripheral springs can pose a significant flood risk. Excess spring discharge was the primary cause of groundwater flooding in Four Roads, Co. Galway, during November 2009. Springs discharging from the base of an adjacent karst plateau caused localized flooding around and downstream of the resurgence, inundating six houses and a community centre, as well as causing the prolonged closure of roads and limiting access to the local school.

The second scenario arises when elevated groundwater levels cause significant lateral flow through the uppermost weathered zone of the bedrock, the epikarst. Karst aquifers can have substantially enhanced and homogeneously distributed porosity and permeability in the epikarst (Klimchouk 2004). Under normal hydrological conditions this enhanced permeability plays an important part in regulating recharge to the phreatic zone by concentrating diffuse recharge towards areas of high vertical permeability. However, when phreatic groundwater levels rise sufficiently high, these pathways can transfer substantial lateral flows, giving rise to ephemeral springs and seeps in adjacent topographic depressions previously unaffected by flooding. This mechanism contributed to the flooding in Carnmore, Co. Galway in November 2009, when a series of temporary springs activated in response to high water levels in an adjacent turlough. Discharge from epikarst springs caused the flooding of four houses, with a further seven houses and two business premises at high risk. There was also a significant hydrostatic element due to the ponding of spring discharge, further highlighting that flooding in lowland karst groundwater systems is often the result of multiple mechanisms acting in combination.

Conclusions

Lowland karst groundwater systems represent a challenging environment from a flood risk management perspective. The diversity of flood mechanisms identified in lowland karst terrains reinforces the need to develop a greater understanding of the complex hydrological and hydrogeological processes in operation during flood conditions. Although an important evidence base has been collated on groundwater flooding from recent extreme events, significant gaps remain in our knowledge. The first and most pressing is the lack of hydrological data, which could be addressed through the establishment of a permanent monitoring network to provide long-term quantitative data at flood-prone locations. Methodologies for improving groundwater flood hazard maps and real-time flood monitoring are also required. A prerequisite to effective flood risk management and mapping is the ability to monitor spatial and temporal changes in flood conditions and extent at a catchment scale. Traditional field-based methods are heavily reliant on labour-intensive, site-specific visits and instrumentation. The increasing ability and availability of remote sensing data, such as synthetic aperture radar, offers the potential to describe flood conditions quickly, accurately and at a large spatial scale, even over remote and rugged terrain.

The floods of 2009–10 and 2015–16 have brought into focus society's close, and often turbulent, relationship with the water cycle in karst areas. Internationally, there has been increasing recognition of the flood mitigation benefits provided by functioning wetlands, nowhere more so than in the lowland karst landscapes of Ireland and the turloughs therein. However, the often-competing priorities of flood management and ecological conservation mean that inevitable conflicts lie ahead. An interdisciplinary approach is thus crucial to enable communities in lowland karst regions to develop the adaptation and mitigation strategies needed in the face of future climate uncertainty.

This work was carried out as part of the scientific project GWFlood: Groundwater Flood Monitoring, Modelling and Mapping, funded by the Geological Survey Ireland and also represents outputs from research funded by the Office of Public Works and the Irish Research Council. The authors thank the Irish Meteorological Service (Met Eireann) for the provision of rainfall data, Galway County Council for the provision of aerial photography and geographical information system data, and the Office of Public Works for the provision of LIDAR, hydrometric and aerial photography data.

References

Bonacci, O. 2013. Poljes, ponors and their catchments. *In*: Shroder, J. (Editor-in-Chief) & Frumkin, A. (eds) *Karst Geomorphology*, Treatise on Geomorphology, **6**. Academic Press, San Diego, CA, 112–120.

Bonacci, O., Ljubenkov, I. & Roje-Bonacci, T. 2006. Karst flash floods: an example from the Dinaric karst (Croatia). *Natural Hazards and Earth System Sciences*, **6**, 195–203.

Coxon, C.E. 1986. *A study of the hydrology and geomorphology of turloughs*. PhD thesis, Trinity College Dublin.

Coxon, C.E. & Coxon, P. 1994. Carbonate deposition in turloughs (seasonal lakes) on the western limestone lowlands of Ireland. II: The sedimentary record. *Irish Geography*, **27**, 28–35.

Coxon, C.E. & Drew, D.P. 1986. Groundwater flow in the lowland limestone aquifer of eastern Co. Galway and eastern Co. Mayo western Ireland. *In*: Paterson, K. & Sweeting, M.M. (eds) *New Directions in Karst*. Geo Books, Norwich, 259–279.

De Waele, J., Martina, M.L., Sanna, L., Cabras, S. & Cossu, Q.A. 2010. Flash flood hydrology in karstic terrain: Flumineddu Canyon, central-east Sardinia. *Geomorphology*, **120**, 162–173.

Directive 2007/60/EC. 2007. Directive 2007/60/EC of the European Parliament and of the Council of 23 October 2007 on the assessment and management of flood risks. *Official Journal of the European Union*, **L288/27**.

Drew, D.P. 1973. Hydrogeology of the north Co. Galway & south Co. Mayo lowland karst area, western Ireland. *In*: Panos, V. (ed.) *Proceedings of the 6th International Congress of Speleology*, Academic Press, Oloumec, C57–C61.

Drew, D. 1990. The hydrology of the Burren, Co. Clare. *Irish Geography*, **23**, 69–89.

Drew, D. 2008. Hydrogeology of lowland karst in Ireland. *Quarterly Journal of Engineering Geology and Hydrogeology*, **41**, 61–72, https://doi.org/10.1144/1470-9236/07-027

Drew, D.P. & Coxon, C.E. 1988. The effects of land drainage on groundwater resources in karstic areas of Ireland. *In*: Yuan, D. (ed.) *Proceedings of the International Association of Hydrogeologists 21st Congress of Karst Hydrogeology and Karst Environment Protection*, **4**. Geological Publishing House, Beijing, 204–209.

Drew, D. & Jones, G.L. 2000. Post-Carboniferous pre-Quaternary karstification in Ireland. *Proceedings of the Geologists' Association*, **111**, 345–353.

Drew, D., Burke, A.M. & Daly, D. 1996. Assessing the extent and degree of karstification in Ireland. *In*: Rozkowski, A., Kowalczyk, A., Motyka, J. & Rubin, K. (eds) *International Conference on Karst Fractured Aquifers – Vulnerability and Sustainability*, June 10–13 1996, Katowice-Ustron, Poland. Silesia University Press, Katowice, 37–47.

Field, M.S. 1993. Karst hydrology and chemical contamination. *Journal of Environmental Systems*, **22**, 1–26.

Gill, L.W., Naughton, O., Johnston, P.M., Basu, B. & Ghosh, B. 2013. Characterisation of hydrogeological connections in a lowland karst network using time series analysis of water levels in ephemeral groundwater-fed lakes (turloughs). *Journal of Hydrology*, **499**, 289–302, https://doi.org/10.1016/j.jhydrol.2013.07.002

Guion, P.D., Gutteridge, P. & Davies, S.J. 2000. Carboniferous sedimentation and volcanism on the Laurussian margin. *In*: Woodcock, N. & Strachan, R. (eds)

Geological History of Britain and Ireland. Blackwell Science, Oxford, 227–270.

GUTIÉRREZ, F., PARISE, M., DE WAELE, J. & JOURDE, H. 2014. A review on natural and human-induced geohazards and impacts in karst. *Earth-Science Reviews*, **138**, 61–88.

HICKEY, C. 2010. The use of multiple techniques for conceptualisation of lowland karst, a case study from County Roscommon, Ireland. *Acta Carsologica*, **39**, 331–346.

HUGHES, A.G., VOUNAKI, T. *ET AL*. 2011. Flood risk from groundwater: examples from a Chalk catchment in southern England. *Journal of Flood Risk Management*, **4**, 143–155.

INTERGOVERNMENTAL PANEL ON CLIMATE CHANGE 2014. Summary for policymakers. *In*: *Climate Change 2014 – Impacts, Adaptation and Vulnerability: Part A: Global and Sectoral Aspects: Working Group II Contribution to the IPCC Fifth Assessment Report*. Cambridge University Press, Cambridge, xv–xvi.

JENNINGS O'DONOVAN & PARTNERS 2011. *Engineering Proposals for the Reinstatement of Culverts on the N18 and the Provision of New Culverts on Minor Roads at Kiltartan/Feasibility of an Overland Channel from Coole to Kinvarra*. Office of Public Works, Galway.

KLIMCHOUK, A. 2004. Towards defining, delimiting and classifying epikarst: its origin, processes and variants of geomorphic evolution. *Speleogenesis and Evolution of Karst Aquifers*, **2**, 1–13.

KOVACIC, G. & RAVBAR, N. 2010. Extreme hydrological events in karst areas of Slovenia, the case of the Unica River basin. *Geodinamica Acta*, **23**, 89–100.

LOPEZ-CHICANO, M., CALVACHE, M.L., MARTIN-ROSALES, W. & GISBERT, J. 2002. Conditioning factors in flooding of karstic poljes – the case of the Zafarraya polje (south Spain). *Catena*, **49**, 331–352.

MARÉCHAL, J.-C., LADOUCHE, B. & DÖRFLIGER, N. 2008. Karst flash flooding in a Mediterranean karst, the example of Fontaine de Nîmes. *Engineering Geology*, **99**, 138–146.

MARTINOTTI, M.E., PISANO, L. *ET AL*. 2017. Landslides, floods and sinkholes in a karst environment: the 1–6 September 2014 Gargano event, southern Italy. *Natural Hazards and Earth System Sciences*, **17**, 467–480.

MCCARTHY, M., SPILLANE, S., WALSH, S. & KENDON, M. 2016. The meteorology of the exceptional winter of 2015/2016 across the UK and Ireland. *Weather*, **71**, 305–313.

MCCORMACK, T. & NAUGHTON, O. 2016. Groundwater flooding in the Gort Lowlands. *Groundwater Newsletter*, **53**, 7–11.

MCCORMACK, T., GILL, L.W., NAUGHTON, O. & JOHNSTON, P.M. 2014. Quantification of submarine/intertidal groundwater discharge and nutrient loading from a lowland karst catchment. *Journal of Hydrology*, **519** (Part B), 2318–2330.

MCCORMACK, T., NAUGHTON, O., JOHNSTON, P.M. & GILL, L.W. 2016. Quantifying the influence of surface water–groundwater interaction on nutrient flux in a lowland karst catchment. *Hydrology and Earth System Sciences*, **20**, 2119–2133.

MIJATOVIC, B.F. 1988. Catastrophic flood in the polje of Cetinje in February 1986, a typical example of the environmental impact of karst. *Environmental Geology*, **12**, 117–121.

MORAN, J., SHEEHY SKEFFINGTON, M. & GORMALLY, M. 2008*a*. The influence of hydrological regime and grazing management on the plant communities of a karst wetland (Skealoghan turlough) in Ireland. *Applied Vegetation Science*, **11**, 13–24.

MORAN, J., KELLY, S., SHEEHY SKEFFINGTON, M. & GORMALLY, M. 2008*b*. The use of GIS techniques to quantify the hydrological regime of a karst wetland (Skealoghan turlough) in Ireland. *Applied Vegetation Science*, **11**, 25–36.

MORRIS, S.E., COBBY, D. & PARKES, A. 2007. Towards groundwater flood risk mapping. *Quarterly Journal of Engineering Geology and Hydrogeology*, **40**, 203–211, https://doi.org/10.1144/1470-9236/05-035

MORRIS, S., COBBY, D., ZAIDMAN, M. & FISHER, K. 2015. Modelling and mapping groundwater flooding at the ground surface in Chalk catchments. *Journal of Flood Risk Management*, https://doi.org/10.1111/jfr3.12201

NAUGHTON, O., JOHNSTON, P.M. & GILL, L. 2012. Groundwater flooding in Irish karst: the hydrological characterisation of ephemeral lakes (turloughs). *Journal of Hydrology*, **470–471**, 82–97.

NAUGHTON, O., JOHNSTON, P.M., MCCORMACK, T. & GILL, L.W. 2017. Groundwater flood risk mapping and management: examples from a lowland karst catchment in Ireland. *Journal of Flood Risk Management*, **10**, 53–64.

NOONE, S., MURPHY, C., COLL, J., MATTHEWS, T., MULLAN, D., WILBY, R.L. & WALSH, S. 2016. Homogenization and analysis of an expanded long-term monthly rainfall network for the island of Ireland (1850–2010). *International Journal of Climatology*, **36**, 2837–2853.

PARISE, M. 2003. Flood history in the karst environment of Castellana-Grotte (Apulia, southern Italy). *Natural Hazards and Earth System Science*, **3**, 593–604.

PARISE, M., RAVBAR, N., ŽIVANOVIĆ, V., MIKSZEWSKI, A., KRESIC, N., MÁDL-SZŐNYI, J. & KUKURIĆ, N. 2015. Hazards in karst and managing water resources quality. *In*: STEVANOVIĆ, Z. (ed.) *Karst Aquifers – Characterization and Engineering*. Professional Practice in Earth Sciences. Springer, Berlin, 601–687.

PINAULT, J.L., AMRAOUI, N. & GOLAZ, C. 2005. Groundwater-induced flooding in macropore-dominated hydrological system in the context of climate changes. *Water Resources Research*, **41**, W05001.

PORST, G., NAUGHTON, O., GILL, L., JOHNSTON, P. & IRVINE, K. 2012. Adaptation, phenology and disturbance of macroinvertebrates in temporary water bodies. *Hydrobiologia*, **696**, 47–62.

ROBINS, N. & FINCH, J. 2012. Groundwater flood or groundwater-induced flood? *Quarterly Journal of Engineering Geology and Hydrogeology*, **45**, 119–122, https://doi.org/10.1144/1470-9236/10-040

SANTO, A., DEL PRETE, S., DI CRESCENZO, G. & ROTELLA, M. 2007. Karst processes and slope instability: some investigations in the carbonate Apennine of Campania (southern Italy). *In*: PARISE, M. & GUNN, J. (eds) *Natural and Anthropogenic Hazards in Karst Areas: Recognition, Analysis and Mitigation*. Geological Society, London, Special Publications, **279**, 59–72, https://doi.org/10.1144/SP279.6

SEVASTOPULO, G.D. & WYSE JACKSON, P.N. 2009. Carboniferous: Mississippian (Tournasian and Visean). *In*: HEPWORTH HOLLAND, C. & SANDERS, I.S. (eds) *The Geology of Ireland*. Dunedin Academic Press, Edinburgh.

SHEEHY SKEFFINGTON, M. & GORMALLY, M. 2007. Turloughs: a mosaic of biodiversity and management systems unique to Ireland. *Acta Carsologica*, **36**, 217–222.

SHEEHY SKEFFINGTON, M., MORAN, J., O CONNOR, A., REGAN, E., COXON, C.E., SCOTT, N.E. & GORMALLY, M. 2006. Turloughs – Ireland's unique wetland habitat. *Biological Conservation*, **133**, 265–290.

SIMMS, M.J. 2004. Tortoises and hares: dissolution, erosion and isotasy in landscape evolution. *Earth Surface Processes and Landforms*, **29**, 447–494.

WALSH, S. 2010. *Report on Rainfall of November 2009*. Met Eireann, Dublin.

WALSH, S. 2012. *A Summary of Climate Averages for Ireland 1981–2010*. Met Eireann, Dublin.

WHITE, W.B. 1988. *Geomorphology and Hydrology of Karst Terrains*. Oxford University Press, Oxford.

WILLIAMS, P.W. 1964. *Aspects of the limestone physiography of parts of counties Clare and Galway, Western Ireland*. PhD thesis, University of Cambridge.

WILLIAMS, P.W. 1970. Limestone morphology in Ireland. *In*: STEPHENS, N. & GLASSCOCK, R.E. (eds) *Irish Geographical Studies*. Queen's University, Belfast, 105–124.

WORTHINGTON, S.R., SMART, C.C. & RULAND, W. 2012. Effective porosity of a carbonate aquifer with bacterial contamination: Walkerton, Ontario, Canada. *Journal of Hydrology*, **464**, 517–527.

ZHOU, W. 2007. Drainage and flooding in karst terrains. *Environmental Geology*, **51**, 963–973.

Škocjan Caves, Slovenia: an integrative approach to the management of a World Heritage Site

VANJA DEBEVEC[1]*, BORUT PERIC[1], SAMO ŠTURM[1], TOMAŽ ZORMAN[1] & PETER JOVANOVIČ[2]

[1]*Škocjan Caves Park, Škocjan 2, SI-6215 Divača, Slovenia*

[2]*Institute of Occupational Safety, Chengdujska 25, SI-1000 Ljubljana, Slovenia*

**Correspondence: vanja.debevec@psj.gov.si*

Abstract: The Škocjan Caves are included in UNESCO's World Heritage List due to their outstanding natural features. The caves include a large underground canyon containing the Reka River, collapse dolines with vegetation in rock fissures and impressive archaeological sites with a rich history of speleological and scientific research. They are also included in the Ramsar Directory of Wetlands of International Importance. Together with their broader surface area, the site is known as the UNESCO Karst Biosphere Reserve. The aim of the management of the reserve is to protect the World Heritage Site and to preserve its outstanding universal value for future generations. The protection activities are regulated by the provisions of international documents, the Škocjan Caves Regional Park Act and the park's management plan. These activities include monitoring of the water quality in the Reka River and meteorological surveys on the surface. Monitoring of the microclimate of the caves focuses on measuring the effects of tourism and monitoring the levels of radon, with the aim of the ensuring the safety of the park's employees. Ensuring a favourable status for the underground habitats and species is laid down in the Natura 2000 management programme. Particular attention is paid to ensuring high-quality, safe visits to the caves and providing educational and awareness-raising activities on the surface of the park.

The Škocjan Caves in Slovenia were included in UNESCO's World Heritage List in 1986 due to their outstanding universal value. They are known to contain one of the largest underground canyons in the world, a rich biodiversity of cave fauna and chasmophytic vegetation in collapse dolines, impressive archaeological sites and pioneering research activities of karst explorers. The World Heritage Site of the Škocjan Caves (Fig. 1) meets the following criteria for the assessment of outstanding universal value from the Operational Guidelines for the Implementation of the World Heritage Convention (UNESCO World Heritage Centre 2012): Criterion vii (contain superlative natural phenomena or areas of exceptional natural beauty and aesthetic importance) and Criterion viii (be outstanding examples representing major stages of Earth's history, including the record of life, significant on-going geological processes in the development of landforms, or significant geomorphic or physiographic features).

With the aim of regulating the conservation and exploration of outstanding geomorphological, geological and hydrological features, rare and endangered animal and plant species, palaeontological and archaeological sites, ethnological and architectural characteristics and the cultivated landscape, and to ensure appropriate conditions for the development of the region, the Parliament of the Republic of Slovenia adopted the Škocjan Caves Regional Park Act in 1996 (Act of Škocjan Caves Regional Park 1996) and in the same year the Government of the Republic of Slovenia established the Škocjan Caves Public Service Agency as the park's managing authority. The Škocjan Caves became a Wetland of International Importance in 1999, as defined in the Ramsar Convention. They were the first location in Europe to be included in this list in line with the criteria for the identification of underground wetlands. In 2004, the Škocjan Caves Park was included in the UNESCO Man and the Biosphere Programme and became a member of the world network of biosphere reserves as the Karst Biosphere Reserve.

With the aim of ensuring the protection of the caves and promoting sustainable development in the park, three areas were established with different levels of protection in line with the Škocjan Caves Regional Park Act (Act of Škocjan Caves Regional Park 1996) (Fig. 2). The core region encompasses a surface area of 401 ha above the Škocjan Caves, where 70 inhabitants live in three villages: Škocjan, Betanja and Matavun. Activities that may endanger valuable natural features, surface and underground ecosystems, and cultural heritage, or that may undermine the integrity of the settlements, are prohibited

From: PARISE, M., GABROVSEK, F., KAUFMANN, G. & RAVBAR, N. (eds) 2018. *Advances in Karst Research: Theory, Fieldwork and Applications*. Geological Society, London, Special Publications, **466**, 411–429.
First published online December 18, 2017, https://doi.org/10.1144/SP466.14

Fig. 1. View towards Murmuring Cave (Šumeča jama) in the Škocjan Caves, Slovenia.

in this narrow protected area. The area of influence of the Škocjan Caves Regional Park encompasses the entire Reka River watershed and covers 450 km^2. Activities affecting the environment, the harmful effects of which may endanger the core area, or indirectly or directly aggravate the situation in the core area, are prohibited. The same applies to activities that may change the existing water regime of the Reka River and affect water quality, except in cases of flood control (Act of Škocjan Caves Regional Park 1996). With the designation of the Karst Biosphere Reserve, the park's wider transition area of 14 377 ha was established in the territory of the municipality of Divača, covering the surface above the underground Reka River stream. In this area, the park carries out awareness-raising and educational activities with an emphasis on understanding the importance of protecting the underground water and the unique and vulnerable karst land (Ščuka *et al.* 2014).

The inclusion of the Škocjan Caves in UNESCO's World Heritage List means that the state institutions and the park's managing authority have to fulfil two important tasks: (1) to protect the outstanding universal value and preserve the integrity of the site and (2) to enable high-quality public access to the sights (Hamilton-Smith 2002). Quality management of this UNESCO site is a significant challenge because it requires appropriate knowledge and tools to find and establish an ecological balance (Berce-Bratko 1996).

The Škocjan Caves Park Act (Act of Škocjan Caves Regional Park 1996) lays down the obligations of the managing authority and provides the basis for the integrated management and monitoring of the cave system and the surface area of the park. The provisions of Articles 3, 4, 5 and 16 of the Act regulate the constant monitoring and analysing of the natural features and cultural heritage, the preparation of additional technical proposals for protection and maintenance measures in cooperation with other competent institutions, the coordination of research tasks and the organization of research activities in the park (Act of Škocjan Caves Regional Park 1996). The Škocjan Caves Regional Park management plan for 2013–17 was approved by the Park's Agency Council and adopted by the National Assembly of the Republic of Slovenia (Programme for the Protection & Development of the Škocjan Caves Park from 2013–2017).

The guidelines provided in the relevant international documents have led to a comprehensive approach to the conservation of the World Heritage Site, supported by monitoring the influence of the wider environment and evaluating the benefits of the protected area beyond its borders with the key

Fig. 2. Map showing the location of the Škocjan Caves Park, Slovenia.

goal of establishing integrated monitoring. One of the aims of integrated monitoring is to achieve sustainable management by integrating all the existing monitoring activities to provide a comprehensive and broad insight into the current situation and emerging demands in the Škocjan Caves Regional Park. This approach enables the managing authority to anticipate the consequences of influences, to understand the effects of managing activities and to implement monitoring in a cost-efficient way (World Heritage Resource Manual 2012; Hedge *et al.* 2013). The aim is to develop a system that will encompass the results of surface and underground monitoring and introduce appropriate measures based on new scientific evidence. This system will reflect the karst disturbance index. This index has been successfully adopted elsewhere in the assessment of the hazards and conservation status of karst land and is a useful tool for the evaluation of protection measures in long-term surveys (Van Beynen & Townsend 2005; Calò & Parise 2006; North *et al.* 2009).

Further training of the park's personnel (whose tasks include the protection of cultural heritage in the core area, as well as education and awareness-raising activities) is required to enable them to apply the scientific results obtained in the implementation of the park's management plan.

Hydrological and geomorphological features of the Škocjan Caves

The Škocjan Caves were formed at the juncture of upper Cretaceous (Turonian and Senonian) thick-bedded limestone and Paleocene thin-layered limestone (Gospodarič 1984; Knez 1996). They are located on the southeastern edge of the central karst and belong tectonically to the Trieste–Komen anticline (Kranjc *et al.* 1999). Two large tectonic lines were cut in the direction of the Dinarides (NW–SE) in this area and these tectonic lines have affected the underground streams and the general direction of the main underground water bed of the Škocjan Caves system. In geological terms, this area is severely broken tectonically, resulting in the creation of underground halls and the main underground water bed, best described as an underground canyon.

After flowing on the impermeable flysch rock surface for 50 km, the Reka River passes onto permeable limestone bedrock, where it disappears below ground through a blind sinkhole, creating large caves. The Reka River has created the largest blind valley in Slovenia, with karst landforms typical of contact karst (Mihevc 2008). After disappearing into the first sinkhole under the village of Škocjan, the Reka River reappears twice at the surface in the Little Collapse Doline (Mala dolina) and the Big Collapse Doline (Velika dolina), where it flows on the bottom of >150 m deep collapse dolines, and then continues its underground journey for *c.* 35 km to the springs of the Timavo River on the other side of the karst plateau.

In the first part of the underground canyon, the Reka River flows through Murmuring Cave (Šumeča jama), which is open to tourists. High above the ceiling of this large hall, there is an entrance to the side passage called Silent Cave (Tiha jama). This is a *c.* 600 m long relict passage, thickly covered with alluvial sediments of mostly sand, clay and flowstone. It is in this part of the underground system that the largest dripstone formations have formed, with stalagmites up to 15 m high. The river turns NW under Cerkvenik Bridge and continues along the underground canyon through Hanke Channel (Hankejev kanal). Just before the first sump in the Škocjan Caves, 2 km from the last sinkhole, the river flows through Martel Hall, which is the largest underground chamber in Slovenia with a volume of 2.1×10^6 m^3.

The rest of the underground stream of the Reka River, between the Škocjan Caves and the Timavo River springs, is accessible through deep abysses in the karst, through which it is possible to descend to the underground river and follow its flow for a few hundred metres. The Reka River is a highly complex karst aquifer with a >300 m deep vadose zone, large underground halls and significant fluctuations of the water level. The epiphreatic zone, where the piezometric level fluctuates due to frequent flooding, is >100 m high. The floods are the result of the river's torrential nature: in the summer months, its flow is measured in litres per second, whereas after heavy rains, especially in its upper part, the streamflow increases up to 3000 times and may reach a few cubic metres per second. The highest streamflow measured during the last 50 years was 305 m^3 s^{-1}, while its medium value is *c.* 8 m^3 s^{-1} (Kranjc *et al.* 1999).

In co-operation with the Slovenian Environment Agency, the streamflow is regularly monitored at five stations located along the surface stream of the Reka River. The last hydrological station is located just before the sinkhole where the river disappears into the Škocjan Caves. The temperature and water levels are measured in addition to streamflow (Fig. 3).

Sampling is carried out periodically to monitor the water quality at both the surface and underground. The 2009–13 chemical status of the whole Reka River stream, the 2009–15 ecological status and the 2009–13 status of underground water and the volume of water in the aquifer were all assessed as good. The park's employees have, in co-operation with the Karst Research Institute and the University of Trieste, installed instruments for measuring temperature, streamflow and water level in some of the

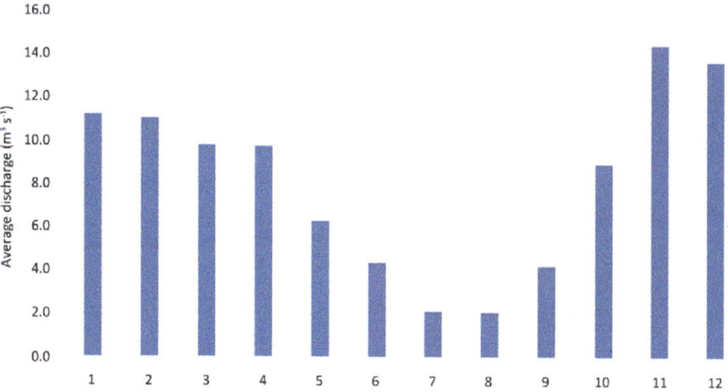

Fig. 3. Average discharge of the Reka River 1952–2014.

caves along the underground stream of the Reka River with the aim of monitoring the dynamics of the river's underground flow. The results obtained have shown fast movement of the flood wave along the complex system of karst channels (Gabrovšek & Peric 2006). This thesis is supported by tracer tests performed in cooperation with representatives of the Berchtesgaden National Park and the University of Trieste in 2006 and 2016 (Galli 2012). Uranin tracer was injected at the point where the Reka River passes from impermeable flysch rock onto the permeable limestone bedrock at very low water levels. The water that sinks underground at that point flows separately from the main passage of the underground Reka River stream through the Škocjan Caves and Labodnica Cave towards the springs of the Timavo River, Brojnice and Sardoč.

Natura 2000 and the Škocjan Caves Park

In its Resolution on National Environmental Action Plan 2005–12 (Bibič 2007), the National Assembly of the Republic of Slovenia laid down the operational programme for the management of Natura areas as one of the key natural protection programmes. Guidelines for protection measures were provided for various types of activities and applied to the habitat types and species present in the Škocjan Caves Park. The most important measures include direct control in the field, limits on tourist visits and sightseeing, and bans on any behaviour or activity that may pose a threat to protected species. Such limits and bans were introduced in the areas of the park where the presence of a large number of visitors jeopardizes the achievement of the protection goals. In the underground caves, the limitations apply to tourist visits that may disturb the animals that live or occasionally stay in caves, in line with the cave conservation goals that are crucial for the protection of Natura areas. In Slovenia, all underground animal species, including bats, are protected and therefore there is a strong emphasis on limiting activities that represent a threat to protected animal species. The Škocjan Caves Park is included in Natura 2000 as the Karst Special Protection Area SI5000023 (Council Directive 79/409/EEC on the Conservation of Wild Birds) and the Karst Special Area of Conservation SI3000276.

The karst is the third largest Special Protection Area for the protection of endangered bird species. The eagle owl (*Bubo bubo*), barred warbler (*Sylvia nisoria*), tawny pipit (*Anthus campestris*), rock partridge (*Alectoris graeca*) and short-toed snake eagle (*Circaetus gallicus*) have all been recorded in the karst grasslands, scrubs and cliff walls. The ornithological importance of the area derives from strong populations of species listed in Annex I to the Birds Directive with high shares of national nesting populations. The first two species sufficiently define the borders of the whole area. Some species and their habitats in the park are affected by factors such as light pollution at night and visitors. The area where some species look for food stretches beyond the park's borders. The diversity and abundance of selected qualifying species are low within the boundaries of the Škocjan Caves Park. In addition to the five bird species nesting in the park and the three species that are probably doing so, nightjar (*Caprimulgus europaeus*) is the most abundant qualifying species within the park's boundaries.

The Karst Special Conservation Area consists of a large limestone plateau in SW Slovenia and the NW part of the Dinaric Karst, with numerous surface and underground karst phenomena and habitat types of European importance, such as caves, dry grassland, juniper formations, evergreen oaks and cliff walls. The area provides a habitat for endangered European animal and plant species (bats, butterflies, beetles, amphibians and birds) and is a migratory corridor for birds of prey and large mammals.

According to the latest data, the following species from the Habitats Directive (Annex IV) specific to the Karst Special Conservation Area have been recorded in the park: olm (*Proteus anguinus*), crested newt (*Triturus cristatus*), yellow-bellied toad (*Bombina variegata*), narrow-mouthed whorl snail (*Vertigo angustior*), marsh fritillary (*Euphydryas aurinia*), stag beetle (*Lucanus cervus*), longhorn beetle (*Morinus funereus*), lesser horseshoe bat (*Rhinolophus hipposideros*), greater horseshoe bat (*Rhinolophus ferrumequinum*), long-fingered bat (*Myotis capaccinii*), common bentwing bat (*Miniopterus schreibersii*) and greater mouse-eared bat (*Myotis myotis*). Among these species, special importance is attributed to those that have large populations in the park, such as the common bentwing bat, known for its hibernation and maternity colonies of several thousand individuals hanging from the ceilings of the Reka River underground canyon. Colonies of the greater horseshoe bat, long-fingered bat and greater mouse-eared bat are also regarded as important, although they are not so numerous. The forests, hedges, meadows, water surfaces and bushes in the park are sufficiently large to provide habitats for bats. Regular monitoring of bats is carried out in the Škocjan Caves. Some species and their habitats are affected by various factors. The hibernation roosts of some bat species are located along tourist routes in the cave. The area where some species look for food (the food habitat) stretches beyond the park's borders. As a general rule, bats are not to be disturbed by noise, lighting or microclimate changes while they are hibernating. Female bats in maternity roosts are not to be disturbed because this may have fatal consequences for their offspring. When disturbed, bats usually leave their roosts, which increases their mortality rate.

Special attention should be given to the olm, which is a qualifying species. It is an endemic Dinaric species and the only cave vertebrate in Europe. It is acutely sensitive to any disturbance and pollution of the underground water. The factors endangering this species include organic and chemical pollution of the Reka River. Dry grassland is the home of the marsh fritillary butterfly. There are about 40 ha of dry grassland areas in the park, so the species is regularly recorded. However, it is endangered due to the loss of pasture land and the growth of grasslands. The stag beetle and the longhorn beetle are dependent on wood biomass, dead trees and forest land. They are regularly recorded in the park, although they are affected by traffic and the stag beetle is also targeted by collectors and is affected by the excessive use of chemicals and fertilizers. The crested newt and the yellow-bellied toad are dependent on wetlands and water bodies. Amphibians are endangered due to the introduction of non-native species into the water bodies, water pollution (chemical pollutants, eutrophication) and the loss of habitat. The narrow-mouthed whorl snail is dependent on marshy meadows and can be found between high-stemmed plants, on riverside willows and on alder. It is endangered by water management measures, such as cleaning of the riverside and the removal of fallen trees.

The part of the Karst Special Conservation Area (SCI Kras SI3000276) in the Škocjan Caves Park encompasses the following habitat types laid down in the Habitats Directive (Directive 92/43/EEC): caves not open to the public, siliceous rocky slopes with chasmophytic vegetation, Medio-European calcareous screes of hill and montane levels, eastern sub-Mediterranean dry grasslands and *Juniperus communis* formations. The last two habitat types cover *c.* 14% of the park's area. They have a good level of representativeness and a favourable conservation status, except for the grasslands, which are gradually being overgrown. There is lush vegetation with prevailing karst pastures and thermophilic forests of mixed deciduous trees on the less steep parts of walls. Scree slopes are also overgrown by vegetation in areas without intensive movement of material. The moist cliff walls with thermophilic and glacial remnants in the lower part of cliff walls above the Reka River bed and its sinkhole represent one of the special features of the park. The species from the Habitats and Birds Directives related to these habitat types include all the previously mentioned bat species, the olm, the marsh fritillary and the southern festoon (*Zerynthia polyxena*) butterfly. Species from the Birds Directive include the eagle owl, nightjar, woodlark (*Lullula arborea*), peregrine falcon (*Falco peregrinus*), barred warbler, hoopoe (*Upupa epops*) and Scops owl (*Otus scops*). The factors disturbing and endangering these habitat types include the excessive use of fertilizer, excessive pasture on one site or in a small area, the discharge of municipal wastewater, the use of pesticides, the introduction of non-native species, sports and leisure activities, climbing, speleology, cycling, hiking, changes in the ecological conditions, the removal of edges and the overgrowth of vegetation. The park's management plan lays down measures for the protection of Natura 2000 species and habitats and the guidelines for research and development aimed at achieving the sustainable use of resources.

Microclimate in the Škocjan Caves

Monitoring of the cave microclimate focuses on measuring the air temperature, relative humidity, carbon dioxide levels, air flow, and the concentrations of radon and its decay products. The microclimate parameters change due to changes in the external conditions of the air or water masses outside the cave and as a result of the presence of visitors, who change the dynamics of the microclimate parameters by introducing expiratory air, warmth, dust and other particles (Cigna 2004). Many changes remain unnoticed or marginal as a result of the extremely large underground passages of the Škocjan Caves.

Visitors can follow two types of guided tour of the Škocjan Caves: the first tour takes them through Silent Cave, Murmuring Cave and the Big Collapse Doline and the second tour shows them the Mahorčič and Marinič caves with the Little Collapse Doline. Microclimate parameters are measured in Silent Cave and Murmuring Cave and at certain places along the 3.5 km long trail, where guided tours are organized throughout the year (Fig. 4).

Silent Cave is entered from the Globočak Collapse Doline through an artificial tunnel, which is closed with doors at both ends. The doors are opened during visiting hours. The visitors first enter the part of Silent Cave called Paradise (Paradiž), which is covered with rich flowstone deposits. They then go to Calvary (Kalvarija) and descend to the Tent (Šotor), which is the deepest point of the cave. The path continues past Škavna, through the Great Hall (Velika dolina) and past the Organs Hall to Murmuring Cave. The river enters the cave through a sinkhole, which continues into an 80 m high canyon and flows along the Hanke Channel towards the sump.

Periodic measurements of the microclimate of the caves have been carried out since 2006, with occasional interruptions due to maintenance work or the disconnection of the electrical power supply. Three meteorological stations in the Škocjan Caves regularly keep track of the temperature, carbon dioxide levels and humidity at 15 min intervals in Calvary, Tent and Murmuring caves. Other measuring points are only monitored by temperature and humidity sensors.

Silent Cave is a closed system with an entrance that is regularly closed, so there is very little air circulation. At the Tent location, the passage forms a pocket structure where air remains. In this part of the cave system, from Calvary and Tent up to the Great Hall, the microclimate parameters are relatively constant. The influence of external air is more noticeable in Murmuring Cave than in Silent Cave. Figure 5 shows the monthly average temperature when the influence of the outer air is observed at the entrance to Murmuring Cave, Bridge (Most) and Rimstone Pools (Ponvice).

Radon has been measured in the Škocjan Caves for >15 years. Slovenian legislation lays down the scope of measurements as well as the supervision of radiation protection for people working in the caves. ^{222}Rn is a radioactive gas with a half-life of 3.8 days. It is formed as a decay product of ^{226}Ra and decays into the short-lived decay products (radon decay products) of ^{218}Po, ^{214}Pb, ^{214}Bi and ^{210}Po (Nazaroff & Nero 1988). All the stated radon decay products are heavy metals that stay in the air as atmospheric particulates before they stick to water molecules, aerosols and the cave walls. The radon isotope ^{222}Ra (hereinafter referred to simply as radon) enters the caves by diffusion from the surfaces of the walls and floors and by convection from deep within the Earth (Nazaroff & Nero 1988; Dueñas *et al.* 1999). When inhaled, radon decay products are deposited on the surfaces of the respiratory tract from the nose to the alveolus, where they decay and emit alpha, beta and gamma particles. The most hazardous are ^{218}Po and ^{214}Po, which emit alpha particles that can damage cells in the epithelial tissue of the bronchial walls in the lungs (Nazaroff & Nero 1988; Samet *et al.* 1991; National Research Council 2009).

Measurements in the Škocjan Caves are focused on the concentration of radon and its decay products, the concentration of positive and negative ions in the air, the concentration of $PM_{2.5}$ and PM_{10} particles and the size distribution of particles in the underground air. This paper presents the results of measurements of radon concentrations. Two methods are used for measuring radon concentrations: the passive method using detectors for traces of radioactive nuclei and the active method using measuring instruments. The nuclei tracer method is used to determine the three-month average radon concentration for a period of one year at six locations near the tourist trail. The measuring locations are in the following order from the entrance to the exit: Paradise, Tent, the entrance to the Great Hall, Murmuring Cave and Rimstone Pools. The measuring instruments method is used to determine the time course of changes in radon concentrations over a period of several days in different seasons. These measurements are taken at two locations: Calvary and Tent.

Concentrations of the radon decay products are determined with measuring instruments in a similar manner to the radon concentration. Again, the changes in the concentrations of radon decay products were measured over a period of several days, in different seasons, at the Calvary and Tent measuring locations.

The measurements of the concentrations of radon and radon decay products with measuring instruments are carried out in line with the DP-LMSAR-3.02 procedure. Measurements of radon concentrations with nucleus traces detectors are carried out in line with the DP-LMSAR-3.03 procedure laid down by the Slovenian Institute of Occupational Safety.

The highest radon concentrations measured with the trace detectors were recorded in the upper part of the cave from Calvary to the Great Hall during summer (April to the end of October), when the highest differences between the temperatures in the cave and on the surface are registered. The lowest values were

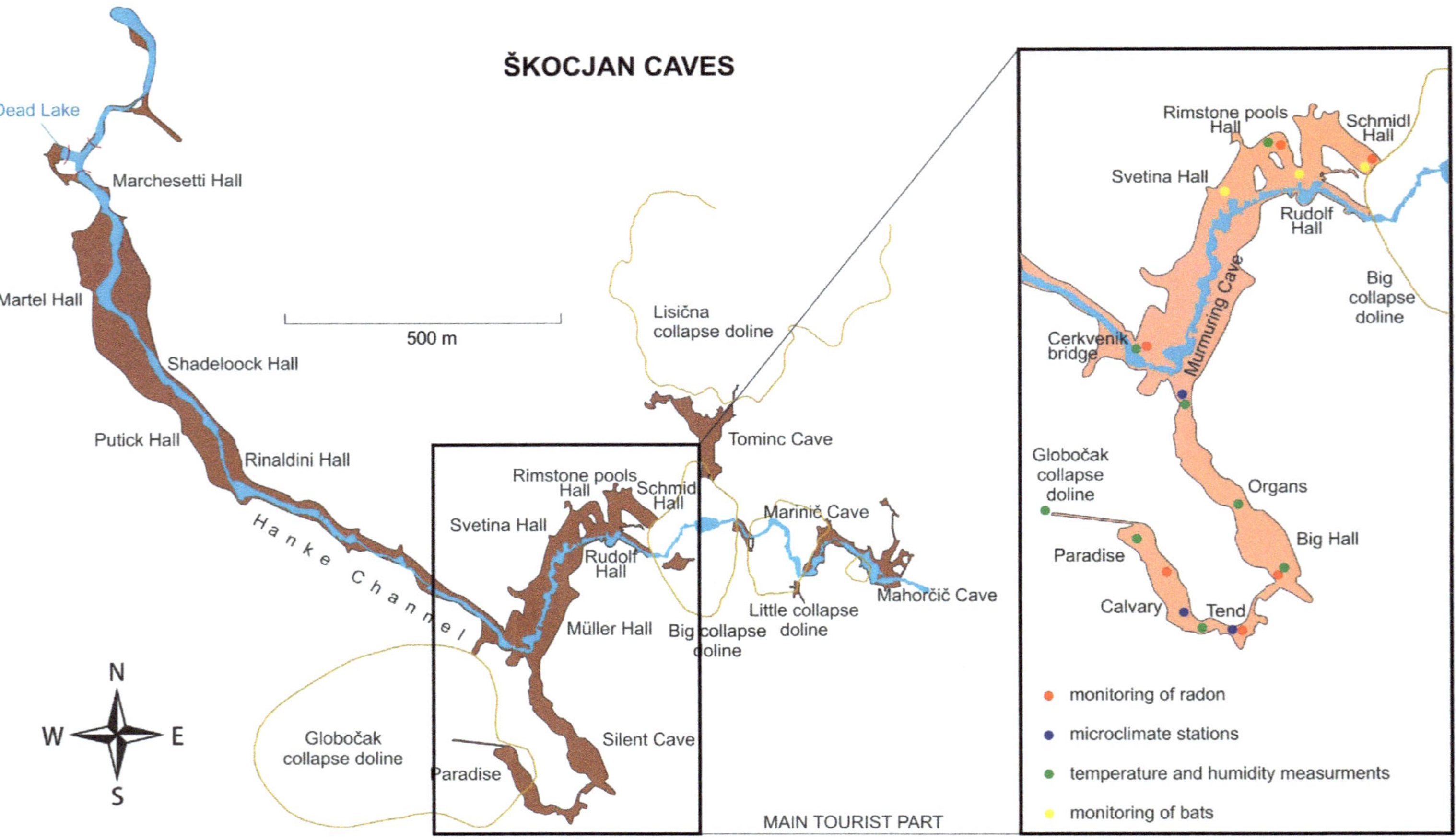

Fig. 4. Map of the tourist sector in the Škocjan Caves showing the locations of monitoring stations.

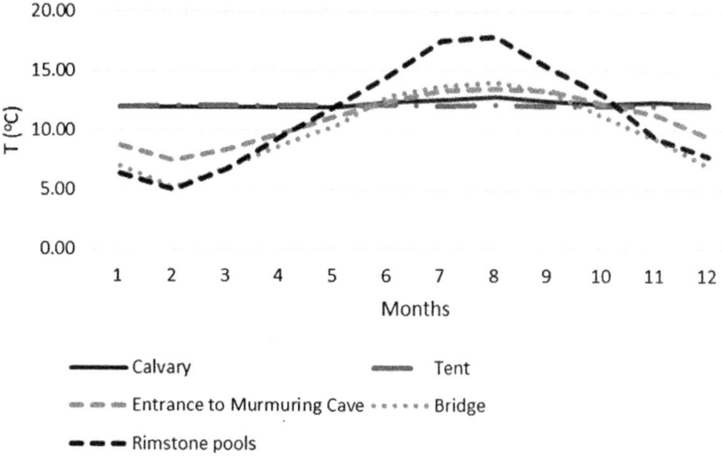

Fig. 5. Average monthly temperature in the Škocjan Caves.

measured from the entrance to Murmuring Cave and across Cerkvenik Bridge towards Rimstone Pools. Figure 6 gives the average radon concentrations measured with trace detectors at various locations along the tourist trail during the period 2000–15. It also shows the average radon concentrations during summer and winter for the same years. The radon concentration also depends on the measurement location as a result of natural air circulation in the cave system.

Anthropogenic aspects of cave use

Tourists in the Škocjan Caves

The Škocjan Caves and their collapse sinkholes were mentioned in Antiquity; they are also shown on maps dating back to the sixteenth century. Many travellers stopped at the caves in the eighteenth century. In 1823, a trail was constructed to the bottom of the Big Collapse Doline, which enabled visitors to access Tominc Cave. Local people, who helped

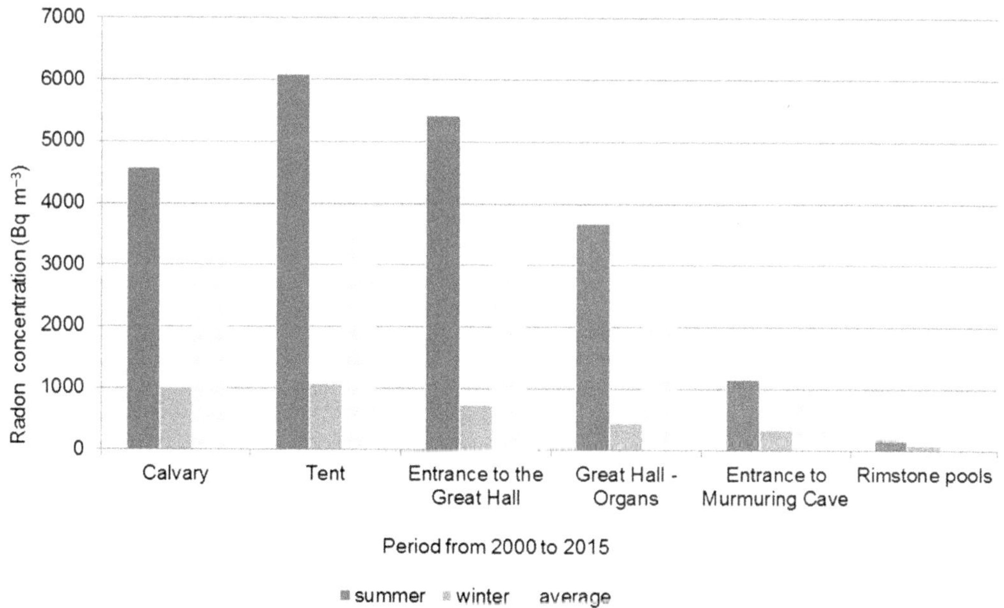

Fig. 6. Radon concentrations at various locations in the Škocjan Caves.

during explorations of the cave, also guided visitors through the cave. Proper development of tourism started only after 1885, however, when local people began systematically collecting entrance fees and prepared recommendations for the illumination of some trails (Shaw 1998) (Fig. 7).

Figure 8 shows that the number of visitors has been increasing over time. The total number of visitors exceeded 120 000 at the end of 2015. The increase is most notable during the summer months, with peak periods in April and May, and later in October. In recent years, the park's marketing service has prepared a number of action plans aimed at achieving a more even distribution of visitors outside the high season. On average, visitors spend an hour and a half in the Škocjan Caves. This amount of time is divided between the visit to Silent Cave, which is completely underground and where they spend *c.* 35 minutes on average, and the visit to Murmuring Cave and the bottom of the Big Collapse Doline, where they usually spend 60 minutes. Since 2011, visitors can also choose a 45-minute walk through Mahorčič and Marinič caves. The time spent in the caves influences the period during which trails, stalactites and other special features are illuminated, which causes the growth of lampenflora in the vicinity of the lighting sources.

In terms of the microclimate, the three main consequences of anthropogenic impacts are an increase in carbon dioxide levels and temperature, and changes in moisture. Each group of tourists in the cave causes an increase in temperature and carbon dioxide levels. Longer term changes have not been observed. At the Calvary and Globočak measuring locations, the parameters change when tourists enter the cave and a decrease in the relative humidity is especially notable when the doors are opened. When visitors leave the cave, the parameters return to their base values. During the summer months, when the number of visitors increases, this only occurs after the last group has left the cave. Figure 9 shows an example of temperature measurements in May 2012 at Tent and illustrates the dynamics of temperature in relation to the number of visitors and the time of day.

Visitors to the cave are exposed to harmless doses of radiation. The ratio between the time of exposure and the effective radiation dose received is higher and more important for the park rangers who work in the cave.

Along with the conservation of natural and cultural heritage, which enables the preservation of the outstanding universal value of the World Heritage Site, the manager of a show cave is also obliged to take due care for the safety of visitors and the quality of their experience. The movements of visitors are directed by the cave infrastructure, which also prevents them from accessing dangerous areas or the habitats of underground animals and enables employees to carry out maintenance work smoothly and safely.

The basic infrastructure (e.g. paths, platforms, viewpoints, bridges and railings) had mostly been erected and put in place by the beginning of the nineteenth century; electrical power and lighting with reflectors were introduced in 1959. A large proportion of the original infrastructure has been substantially renovated during the last ten years. Non-slip coatings were applied to trails and old railings and the main bridges were removed and replaced with new ones. Illumination along the whole length of the tourist trail was equipped with LED lights with a colour temperature of 3000 K, reducing the overall illumination of surfaces and features. A total of 160 lights with a power of 5–30 W and a colour temperature of 4000 K illuminate the space in the cave seen from the pathways. Further growth of lampenflora on speleothems was decreased, although more studies are needed to assess the efficiency of the new system. An optical cable in the cave allows park personnel to monitor the movements of tourist groups from a control centre. The new lightning system is managed by the programme DALI, which enables changes in lighting. The new infrastructure in the cave also enabled the creation of a communications network and the direct monitoring of basic physical and chemical parameters, which are automatically entered into a database. This project will start by 2018.

The majority of tourists wish to see the protected habitat types in the cave and steep walls as these offer the most interesting experience. This is an inevitable consequence of marketing activities and the cave manager has to create a balance between the use of natural assets by tourists and the protection of vulnerable ecosystems. The protection regime of the Škocjan Caves Park (Act of Škocjan Caves Regional Park 1996) ensures a supervised entrance into the cave, in combination with the constant presence of park personnel in the cave during visiting hours. Visitors only walk along marked tourist trails and are not allowed to leave them. Article 9 of the Škocjan Caves Park Act imposes a strict regime that prohibits the disturbance, hunting and killing of wild animals; all such acts may lead to fines imposed in Articles 28 to 34. The entire area of the park is supervised by a ranger service. Visitors are not allowed to use additional lights or to take photographs. The park personnel also ensure that the visitors do not make too much noise. The guides explain the rules of conduct before people enter the cave. The employees who work towards achieving the primary goals of the protection and conservation of habitats and protected species and interpret these to visitors are well acquainted with these issues. This is of utmost importance for the safety of visitors and for their understanding of the significance of

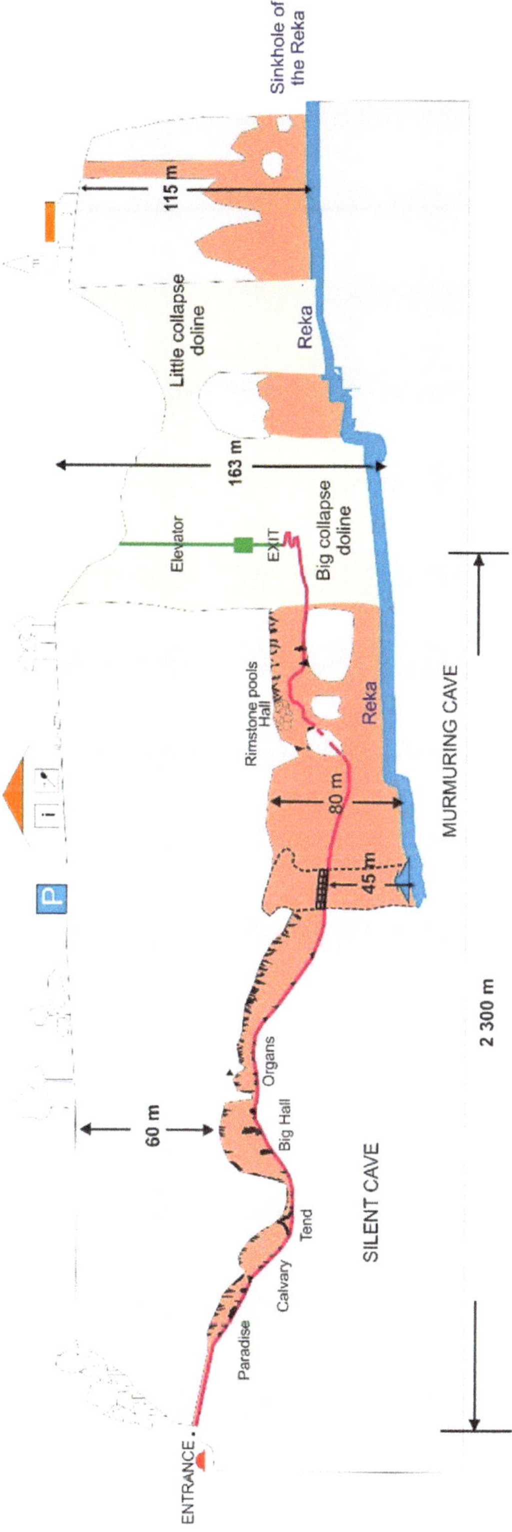

Fig. 7. Cross-section of the Škocjan Caves showing a marked tourist path

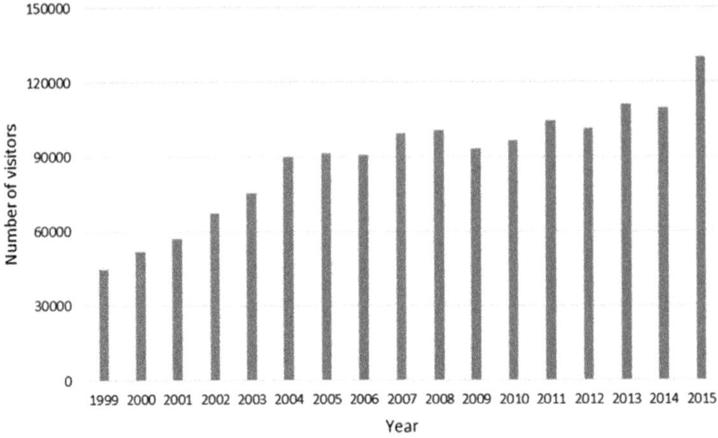

Fig. 8. Number of tourists visition the Škocjan Caves from 1999 to 2015.

nature conservation and protection measures, along with safeguarding the universal value of the World Heritage Site. This target is achieved by museum collections, information and education boards at the information centre and along the Škocjan Educational Trail, and other technical facilities that support interpretation and the education of the general public (Presetnik 2015).

People working in the Škocjan Caves

The Škocjan Caves Park employs seven cave guides, who guide tourists through the caves and act as rangers. An additional 38 seasonal workers are employed as guides during the high season from May to October. Rangers who work in the caves every day need to be in good physical shape. Their work is responsible and requires them to be focused and to manage different settings depending on the morphology of the cave. The characteristics of their working environment (low temperatures, a lack of natural light and high levels of moisture) have multiple effects on their bodies. People who work in the caves are exposed to radiation from the radon and its decay products that are emitted from the cave walls. Legislation obliges the manager to monitor concentration of radon and its decay products in the cave (Rules on the Requirements and Methodology of Dose Assessment for the Radiation Protection of the Population and Exposed Workers 2003; Rules on the Monitoring of Radioactivity 2007).

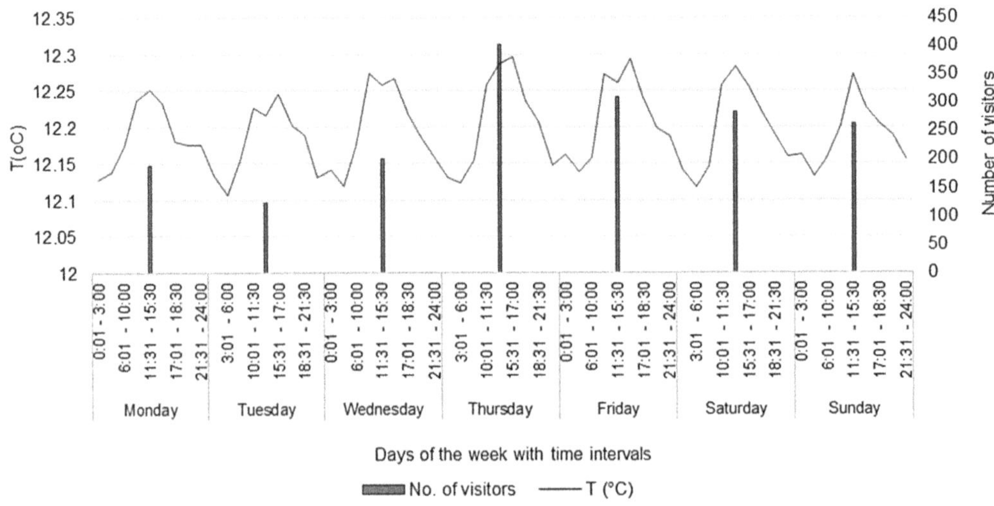

Fig. 9. Temperature in the Škocjan Caves in relation to the number of visitors at the location Tent in May 2012.

Measurements are made by a laboratory certified to the SIST EN ISO/IEC 17025 standard.

The Ionising Radiation Protection and Nuclear Safety Act (2002) and Rules on the Obligations of the Person Carrying Out a Radiation Practice and Person Possessing an Ionising Radiation Source (2004) oblige workers in tourism to act in accordance with the radiation protection provisions so that they do not exceed the legal dose, to have regular medical examinations and to receive training in protection from radiation.

The Škocjan Caves Park has prepared a risk assessment that sets out adjustments to working conditions and the environment and suitable protection measures. The time spent in the caves (guiding visitors, maintenance or exploration) is carefully recorded. The time spent in the cave is an important input for calculating the annual effective dose of radiation and is carried out in accordance with the ICRP32 model (Jovanovič 2016). All workers in the cave belong to a category for which the highest permitted annual dosage is 6 mSv, which helps to guarantee high-quality tourist management and the allocation of work schedules. The limit values are rarely exceeded.

A comparative analysis of lung function in cave guides working in the Škocjan and Postojna caves was conducted in 2010 to establish the effects of the time spent in the cave on the general well-being of cave guides. Members of the control group were employed in administration at the two sites. Thirty cave guides and 28 administrative workers were examined at Sežana Hospital. Lung function values were slightly higher in the group of guides, but the difference was not statistically significant. There was a positive relation between peak expiratory flow and the time spent in the cave, as well as the effective dose ($r = 0.39$, $p < 0.05$) (Debevec Gerjevič 2010). However, the nature of the work of the cave guides requires annual medical examinations assessing their lung function, eyes and larynx, and taking blood samples. The checks are carried out by the Occupational Medicine Department of the Slovenian Institute of Occupational Safety (Ljubljana), which specializes in radiation exposure.

Training of cave guides

There are two types of caves that can be visited in Slovenia: caves with visiting restrictions that are open to the public and caves from which the public are banned, but which can be accessed by experts. In the first category, anyone can enter the show cave as long as they are supervised, recorded and do not move beyond the regulated route. The caves allow different activities, ranging from cave tourism to cultural, scientific, educational and medical activities, and can be used as a source of drinking water. All these activities should be carried out with respect to protection regimes and in line with laws and regulations for the protection of nature and in a manner that does not harm or destroy the natural features of the caves. Entry into caves closed to the public is prohibited, with the exception of scientific or archaeological research, which can only be conducted with the permission of the Ministry of the Environment and Spatial Planning of the Republic of Slovenia.

Visits to show caves are led by appropriately trained cave guides; visits should be supervised and their condition monitored. A registry of cave guides is kept by the competent ministry. In line with the applicable legislation, the Škocjan Caves Park public agency is authorized to organize training and to test the competence of cave guides. Training is divided into theoretical and practical courses. Theoretical training takes place in the park's congress and promotional centre (Pr Nanetov'h), whereas the practical course includes outdoor and underground work in the Škocjan Caves Park. Candidates for the final exam are students and part-time employees, along with members of the different caving societies that manage other show caves. People who pass the exam have permission to work in all the show caves in Slovenia.

The main themes covered in training and testing are simulations of guiding through a cave, safety at work, the safety of visitors and emergency responses. Other important fields are didactics, pedagogics and psychology. The environment in which cave guides work calls for a knowledge of cave biology, as well as the knowledge and skills to ensure that the visits do not harm the cave environment. A knowledge is also required of karstology, history, the organization of cavers and knowledge about caves in Slovenia and other countries of the world. Other obligatory parts of the education of cave guides are field trips on the surface area and underground, where participants learn about karst geology, the development of karst phenomena, as well as cave protection and interpretation and where they can practise their presentation skills. As the caves are in a protected area and are considered to be natural assets, the themes of nature and environmental protection, parks in Slovenia and abroad, and the protection of natural features and cultural heritage are discussed. Slovenian legislation on protection from radiation obliges everybody who works in caves, as guides or otherwise, to take a course and pass an exam in protection from radiation. The exam is valid for five years and training is obligatory for full-time employees, students, part-time employees and everybody who occasionally works in caves. Training sessions have been organized regularly since 2002. The Škocjan Caves

Park has trained and certified 47 cave guides, who work in the park and in other caves in Slovenia. Such training sessions are essential to offer visitors an enjoyable experience and to monitor the environment in the caves and at the surface.

Adaptive management and management effectiveness

Adaptive management is a quality process of monitoring the state of conservation and protection of natural and cultural heritage in protected areas. Along with checking whether standards and indicators are complied with, it includes mechanisms for minimizing the effects of changes. Tourism is an important risk factor in preserving cave ecosystems. For the last few years, larger groups have been booked in advance and additional part-time rangers were employed in 2017 with the aim of ensuring a quality experience, better supervision of the World Heritage Site, the provision of information to visitors and a personal approach from the guide with smaller groups. Special attention is given to the employees so that they do not exceed the permitted dose of radiation. The effective doses have been successfully reduced by strict monitoring of the time spent in the cave and regular training in protection from radiation.

The safety of visitors has also been improved with renovated paths and new railings. The new lighting system has enabled a reduction in illumination time and the use of energy; the growth of lampenflora has also been reduced. In the past two years, new methods of the removal of lampenflora from surfaces along the tourist trail in the cave have been successfully explored. Some sections of the trail are regularly checked for the existence of fissures; all the trees and rocks above the paths in the collapse dolines are checked each year. Any objects that threaten the safety of visitors are removed. In 2009, the path through Mahorčič and Marinič caves was renovated with the aim of reducing the strain on the Silent and Murmuring caves. Mahorčič and Marinič caves (Fig. 10) are only open from May to October and may be closed due to high water levels in the Reka River during periods of heavy rainfall. In the framework of the Climaparks Project, a hydrological model was developed in 2013 that enables monitoring of the level of the Reka River (Vidmar & Čotar 2013). A useful warning tool produces estimates of the lapse of time required for swollen waters to travel from the measuring location to the sinkhole in the Škocjan Caves.

A significant reduction in the numbers of bats in their winter shelters was observed in 2013–14. A sizeable colony of common bentwing bats had moved to a new location *c.* 1 km deeper in the

Fig. 10. High-water level in Murmuring Cave (Šumeča jama) in the Škocjan Caves.

cave. It is assumed that the main reason was disturbance due to lengthy renovation works in the cave (Presetnik 2015). In winter 2015–16, the colony was again seen in its previous location in the Rimstone Pools Hall, but bat numbers had fallen from 5000 bats before 2012 to 3000 bats. The management authority imposed measures to limit visits to this part of the cave in the most crucial period for the bats. During their winter hibernation, visitors take an alternative trail, which ensures that the bat locations in the Rimstone Pools Hall remain in the dark and tourists do not approach them. Another measure was a change in the direction and intensity of lighting so that the reflectors did not illuminate the ceiling throughout the year. The alternative route remains in place, at least until the bat colonies return to their pre-2012 size. Table 1 gives an overview of the bat species and measures for improving their populations and habitats.

A task of every manager of a World Heritage Site is to guarantee conservation of the site's outstanding universal values. Given the awareness that species habitats are complex and karst terrain is vulnerable, planning and participation in activities are carried out with good care, so all measures for adjustments and conservation are carried out at the proper time, in a wider area and sometimes even outside the protected area. An assessment of management effectiveness gives information on how well a protected area is managed, how the management protects values and how it achieves set goals and results. In 2014, the International Union for Conservation of Nature assessed the Škocjan Caves in the framework of the programme World Heritage Outlook. The assessment of the state of conservation was positive, which makes the Škocjan Caves one of 21% sites fulfilling this responsible task (World Heritage Outlook 2014).

Challenges for the future

World Heritage Sites throughout the world are facing increased numbers of visitors, which brings challenges and opportunities for their management. The impact of tourism depends on the interactions of a location with the places surrounding it. Good and effective management is directed towards a destination as a whole, which is widely accepted by local inhabitants and visitors. Good tourism management calls for good connections between tourist activities, the World Heritage Site and local communities (Tourtellot 2010). Planned socioeconomic analyses will enable us to gain insights into tourist activities in the wider park area and to strengthen the local economies and sustainable development. Repeating the survey conducted among visitors and local people in 2010 would enable the use of an up-to-date evaluation of the quality of visits and the acceptability of tourism to local people. The latter is an important element in making sound assessments of the carrying capacity of the cave and surface areas, which includes all aspects and stakeholders, and sets up a system of adaptation measures. Cave monitoring will be complemented with new geological and speleological research and a series of new parameters for measuring cave microclimates, including aerosols. The renovated cave infrastructure with its communication networks will make monitoring much easier.

One of the main tasks for managers is to ensure favourable conditions for life and a favourable status of the habitats of the Natura 2000 species in the park. A major project on the management of Natura 2000 sites is being prepared. It focuses on conservation measures to ensure a favourable status of species included in Bird and Habitats Directive and a favourable status of the habitats of species in the Natura 2000 area of Kras, which also covers the Škocjan Caves Park. The main focus of the project will be to protect the habitats of birds from the Birds Directive (Directive 2009/147/EC), i.e. the Eurasian eagle owl, peregrine falcon and Eurasian skylark. Collapse doline walls are nesting sites for the Eurasian eagle owl and peregrine falcon; they are a habitat type in which bats look for food and Protected Habitat Type 8210 calcareous rocky slopes with chasmophytic vegetation. One of the main activities will be the establishment of safe zones for the Eurasian eagle owl; some purchases of land in the park are also envisaged. Designed areas of strict conservation will help us to prevent improper use, visits that could cause disturbance and other negative impacts arising from the presence of humans. Another important measure that will help to fight a major threat to the owl in the Natura area Kras SI5000023 will be the insulation of the medium voltage transmission lines in the vicinity of known nesting sites of the species. This will help to reduce mortality from electric shocks (Šturm 2015).

Conservation activities and ensuring a favourable habitat for species from Annex II to the Habitats Directive, which are included in SCI Kras SI3000276 and are present in the Škocjan Caves Park, include bat species with hibernating wintering colonies and breeding summer colonies in the Škocjan Caves system, and the amphibians Italian crested newt and yellow-bellied toad. Renovations of karst ponds, which are their basic habitat, will help to preserve and improve the condition of several ponds in the area. It will also help to conserve the population of amphibians, which are among the most threatened species in Europe. They still thrive in the karst only because of a network of water biotopes (karst ponds). The Italian crested newt and yellow-bellied toad are among the most threatened

Table 1. *Bat species in the Škocjan Caves Park and measures for the improvement of their populations or habitats*

Species of bat	State of conservation for bats	Estimated number of individuals in the Škocjan Caves Park 1999–2015	Reference values (Škocjan Caves Park)	Endangering factors
Rhinolophus hipposideros	Assessment not possible	5–20	Wintering	The bats shelter in house attics during the day, only occasionally in caves. Breeding sites in buildings are threatened by humans (from vandalism to improper house renovations). Main threats for the species in caves are vandalism and disturbance by visitors
Rhinolophus ferrumequinum	Unfavourable state – trend unknown	20–90	40 breeding, 60 wintering	Main causes of endangerment are competing against insects in farming and forestry, and disturbances in shelters and breeding sites (from vandalism, disturbance, improper house renovations or building interventions)
Myotis capaciinii	Assessment not possible	1–50	Constant presence, wintering	Vulnerable to disturbance in caves and destruction of cave shelters
Miniopterus schreibersii	Unfavourable state – trend unknown	300–8000	6000 constant presence	Species is vulnerable due to its strict ecological demands, which mean that an important part of the local population gathers in a very limited space. Main threats are a lack of shelters and disturbance in shelters and breeding sites
Myotis myotis	Unfavourable state – trend unknown	0–100	Presence	Main threats are a loss of habitat and improper renovations of buildings, as well as restricted access to shelters (cave and other underground habitats, church attics, bell towers, attics of other buildings)
Myotis blythii	Unfavourable state – trend unknown	0–100	Breeding	Threat factors are: renovations of buildings or other infrastructure during the wrong season; use of chemical substances for protection of wood in attics; removal of trees with cavities; disturbance in breeding and wintering sites (vandalism, tourist use of caves and buildings, e.g. castles; speleological explorations); isolation of populations; lack of suitable environments where food can be found and use of insecticides

Bat species are from the Habitats Directive (Annex II, species for SCI Karst; SCI Kras SI5000023 present in the Škocjan Caves Park). The estimate of species abundance in the Škocjan Caves is for 1999–2014. Reference values of populations and potential measures to alleviate the unfavourable state of the population or habitat in the Škocjan Caves from Šturm (2014) and Presetnik (2015).

animal species. The project envisages the renovation of karst ponds, which will help to improve the state of the environment and increase the population of amphibians. Another major goal of the project is to prevent the overgrowing of karst grasslands, which will help to improve the condition of this habitat type.

Tourist visits to the cave are not the only important factor that influences the cave environment, although their impact can be direct (e.g. physical damage on flowstone, litter in the cave or the introduction of spores). We should not forget the water that flows into the cave. The Reka River, which is the main allogeneous inflow into the karst aquifer, can bring, with substantial speed, pollution that occurs far from the point where it sinks underground (Gutierrez *et al.* 2014). The Reka watershed is relatively large (*c.* 450 km^2), with a considerable number of potential polluters. Sources of pollution include the collection and treatment of urban wastewater, industrial plants, illegal discharge, traffic and, to a lesser extent, farming. This is the reason why efforts in the management of a relatively large area are directed predominantly into ranger services, awareness-raising activities for the general public, the organization and coordination of cleaning campaigns, the co-financing of equipment for rescue teams (who would be called on in the event of spills of dangerous substances into the Reka River) and other measures aimed at the protection of water sources.

All these factors should be taken into consideration when devising management measures. The conservation and protection of the cave, which is a natural monument and World Heritage Site, should always be considered first because its use as a tourist destination is of only secondary importance. With this clear hierarchy of management measures in mind, a five-year management and conservation plan for the cave has been prepared. Its primary goals are the conservation of the caves, natural assets, plant and animal species, and cultural heritage. The organization of quality guiding through the cave, focused on first-hand experiences, is secondary to this aim. Workshops and training session are organized for local people in the park. The 59 trained volunteer rangers are actively involved in monitoring activities and encourage visitors and local inhabitants to behave responsibly.

Conclusions

In 2016, the Škocjan Caves celebrated the 30th anniversary of their inclusion on the UNESCO World Heritage List and the 20th anniversary of the establishment of the Škocjan Caves Park Public Agency, which manages the cave. In 2014, the International Union for Conservation of Nature assessed the conservation of the site (World Heritage Outlook 2014). In 2015, the International Coordinating Council of the UNESCO Man and Biosphere Programme confirmed that all the criteria for biosphere reserves are met and suggested that the location should be a model in the fields of education and participation of local communities for biosphere reserves worldwide (Periodic Review Recommendations for Karst Biosphere Reserve 2015).

Services for the organization of tourist visits, rangers, and research and development work in close cooperation. Creating the ideal balance between visits and the protection of natural assets and cultural heritage calls for adaptations and measures for improving the state. The Škocjan Caves Park runs a hydrological station and a warning system for swollen waters. Regular training sessions and awareness-raising activities take place in the area of influence. The monitoring of cave animals is established and regular summer and winter inspections of bats take place. The monitoring of the cave microclimate includes the monitoring of radon and its decay products. Regular medical examinations are organized to ensure proper health and safety at work, along with training sessions in protection from radiation. The Škocjan Caves Park is also a certified provider of training sessions for cave guides. Cave visits are supervised. In 2016, the cave infrastructure was renovated, new illumination was introduced, and trails and railings were repaired. A system for monitoring lampenflora is in place. Visitors can choose the trail through Mahorčič and Marinič caves, which helps to redirect the flow of tourists from the Silent and Murmuring caves.

Interdisciplinary studies and activities from the management programme of Natura 2000 are planned for the future. The socioeconomic impact of visitors and its acceptability for the local community will be assessed. The karst disturbance index will be included in further research studies in the core and transition areas to evaluate anthropogenic impacts on the underground ecosystems. A communications network between individual measuring locations in the cave and on the surface will be established. Our integrated approach to the management of the Škocjan Caves aims to emphasize the connections between research studies that deal with the implementation of conservation measures, the education of personnel and the general public, and the presentation of a location that teaches people how to take care of our natural world.

Special gratitude is paid to all employees of the Škocjan Caves Park, Slovenia, for their efforts in carrying out the work, measurements and monitoring in the cave. The same holds true for high-quality guiding through the cave and on the surface, as well as for the contribution of all the park services towards the enforcement of integrative management in the protected area. Special acknowledgements go to the institutions for their long years of cooperation with Škocjan Caves Park: the Karst Research Institute, Postojna; the Centre for Cartography of Fauna and Flora, Ljubljana; the Institute of Occupational Safety, Ljubljana; the Institute of the Republic of Slovenia for Nature Conservation, Nova Gorica Regional Unit; and the Slovenian Environment Agency, Ljubljana. Their input has made a valuable contribution towards the exploration and monitoring of the Škocjan Caves.

Note

Full details of the Directives and Annexes to the Directives mentioned in this paper are all available at http://ec.europa.eu/environment/nature and http://natura2000.eea.europa.eu/Natura2000/SDF.aspx?site=SI3000276

References

Act of Škocjan Caves Regional Park 1996. *Official Gazette of the Republic of Slovenia*, **57/96**.

Berce-Bratko, B. 1996. Človek in Kras. *Acta Carsologica*, **XXV**, 25–39.

Bibič, A. 2007. *Natura 2000 site management programme: 2007â 2013: operational programme*. Ministry of the Environment and Spatial Planning, Ljubljana.

Calò, F. & Parise, M. 2006. Evaluating the human disturbance to karst environment in southern Italy. *Acta Carsologica*, **35**, 47–56.

Cigna, A.A. 2004. Climate of caves. *In*: Gunn, J. (ed.) *Encyclopedia of Caves & Karst Sciences*. Routledge, London, 228–230.

Debevec Gerjevič, V. 2010. *Functional aspects of caves environments impact on man*. Master's thesis, University of Ljubljana.

Dueñas, C., Fernández, M.C., Cañete, S., Carretero, J. & Liger, E. 1999. ^{222}Rn concentrations, natural flow rate & the radiation exposure levels in the Nerja Cave. *Atmospheric Environment*, **33**, 501–510.

Gabrovšek, F. & Peric, B. 2006. Monitoring the flood pulses in the Epiphreatic Zone of Karst Aqifers: the case of Reka River System, Karst Plateau, SW Slovenia. *Acta Carsologica*, **35/1**, 35–45.

Galli, M. 2012. *I traccianti nelle ricerche sul Timavo*, Edizioni di Univerita' di Trieste, Trieste.

Gospodarič, R. 1984. Cave sediments and Škocjanske jame speleogenesis. *Acta Carsologica*, **12**, 27–48.

Gutierrez, F., Parise, M., De Waele, J. & Jourde, H. 2014. A review on natural and human-induced geohazards and impacts in karst. *Earth Science Reviews*, **138**, 61–88.

Hamilton-Smith, E. 2002. Management assessment in karst areas. *Acta Carsologica*, **31**, 13–20.

Hedge, P., Molloy, F. *et al.* 2013. *An Integrated Monitoring Framework for the Great Barrier Reef World Heritage Area*. Department of the Environment, Canberra.

Ionising Radiation Protection and Nuclear Safety Act 2002. *Official Gazette of the Republic of Slovenia*, **50/2003**.

Jovanovič, P. 2016. *Poročilo o sevealni obremenjenosti zaposlenih v Škocjanskih jamah za leto 2015*. Institute of Occupational Safety, Ljubljana.

Knez, M. 1996. Development of phreatic channels in Škocjanske jame Caves Slovenia [abstract]. Paper presented at the 30th International Geological Congress. Beijing, China.

Kranjc, A., Aljančič, M., Fišer, P., Klemen Krek, Z., Lah, L., Oražm Adamič, M. & Šumrada, J. 1999. Slovene classical karst. *In*: Kranjc, A. (ed.) *Znanstveno raziskovalni center SAZU, Založba ZRC*. Inštitut za raziskovanje krasa ZRC SAZU, Ljubljana.

Laboratory procedure DP-LMSAR-3.02. n.d. Zavod za varstvo pri delu, d.d.-Merjenje koncentracije radona v radonovih potomcev-rev9.

Laboratory procedure, Institute of Occupational Safety DP-LMSAR-3.03. n.d. Določevanje koncentracije radona z detektorji Sledi-Rev6.

Mihevc, A. 2008. Kraški relief in reliefne enote na obravnavanem območju. *In*: Luthar, O., Dobrovoljc, H., Fridl, J., Mulec, J. & Pavšek, M. (eds) *Kras, trajnostni razvoj kraške pokrajine*. Založba ZRC, Ljubljana, 29–38.

National Research Council 2009. *Comparative Dosimetry of Radon in Mines and Homes*. National Academy Press, Washington, DC.

Nazaroff, W.W. & Nero, A.V. 1988. *Radon and its Decay Products in Indoor Air*. Wiley, Chichester.

North, L.A., van Beynen, P.E. & Parise, M. 2009. Interregional comparison of karst disturbance: west-central Florida and southeast Italy. *Journal of Environmental Management*, **9**, 1770–1781.

Presetnik, P. 2015. *Raziskave netopirjev (Chiroptera) v Parku Škocjanske jame v letu 2015*. Center za kartografijo favne in flore, Ljubljana.

Programme for the Protection & Development of the Škocjan Caves Park from 2013–2017. *Official Gazette of the Republic of Slovenia*, **11/2014**.

Periodic Review Recommendations for Karst Biosphere Reserve 2015. Paper presented at the 27th Session of the MAB International Coordination Council, Paris, France, 8–12 June 2015, SC-15/CONF.227/10, http://www.unesco.org/new/fileadmin/MULTIMEDIA/HQ/SC/pdf/SC-15-CONF-227-10_PR_and_follow-up_of_recommendations_en.pdf

Rules on the Monitoring of Radioactivity 2007. *Official Gazette of the Republic of Slovenia*, **20/07, 97/09**.

Rules on the Requirements and Methodology of Dose Assessment for the Radiation Protection of the Population and Exposed Workers 2003. *Official Gazette of the Republic of Slovenia*, **115/2003**.

Rules on the Obligations of the Person Carrying Out a Radiation Practice and Person Possessing an Ionising Radiation Source 2004. *Official Gazette of the Republic of Slovenia*, **13/2004**.

Samet, M.J., Alber, E.R. *et al.* 1991. *Comparative Dosimetry of Radon in Mines and Homes*. National Academy Press, Washington.

Ščuka, S., Peric, B., Cerkvenik, R., Šturm, S., Zorman, T., Kranjc, D. & Debevec, V. 2014. *The Karst Biosphere Reserve Periodic Review*. Škocjan Caves Park, Škocjan.

Shaw, R.T. 1998. Early tourist at Škocjanske jame – 18th century to 1914. *Acta Carsologica*, **27**, 235–264.

SIST En ISO/IEC 17025. 2005. *General requirements for the competence of testing and calibration laboratories (ISO/IEC 17025:2005), SIST En ISO/IEC 17025*, Slovenian institute for Standardisation.

Šturm, S. 2014. *Predlog conacije jamskega sistema škocjanskih jam iz vidika varstva in ohranjanja populacij netopirjev v jamah*. Škocjan Caves Park, internal report.

Šturm, S. 2015. *Strokovne usmeritve za upravljalca pri oblikovanje načrta vodenja obiskovalcev v jami v ključnih obdobjih prezimovanja in razmnoževanja netopirjev, vrste Natura 2000, v Parku Škocjanske jame*. Škocjan Caves Park, internal report.

Tourtellot, J.B. 2010. Part threat, part hope – the challenge of tourism. *World Heritage*, **58**, 6–17.

UNESCO World Heritage Centre 2012. *Operational Guidelines for the Implementation of the World Heritage Convention*, **WHC 05/02**, http://whc.unesco.org/archive/opguide12-en.pdf

Van Beynen, P.E. & Townsend, K.M. 2005. A disturbance index for karst environments. *Environmental Management*, **36**, 101–116.

Vidmar, A. & Čotar, A. 2013. *Izdelava hidrološkega modela CLIMAPARKS, KSH-80*. Škocjan Caves Park, internal report.

World Heritage Outlook 2014, www.worldheritageout look.iucn.org/

World Heritage Resource Manual 2012. *Managing Natural World Heritage*. UNESCO, Paris.

Karst-specific composite model for informed resource management decisions on the Biosfera de la Reserva Selva el Ocote, Chiapas, Mexico

JOHANNA L. KOVARIK & PHILIP E. VAN BEYNEN*

School of Geosciences, University of South Florida, Tampa, FL 33620, USA

**Correspondence: vanbeyne@usf.edu*

Abstract: High permeability and rapid recharge in karst aquifers make them susceptible to contamination. We combined a groundwater vulnerability map with an environmental disturbance index to give an adaptable spatial tool for developing management strategies for a karst environment in the Reserva de la Biosfera Selva el Ocote (el Ocote), Chiapas, Mexico. Seventy-two per cent of the study area is classified as an area of least concern for management, with 60% falling within el Ocote. Consequently, although there are concerns regarding the vulnerability of the karst ecosystem, the lack of development and the natural protection of the ecosystem, the immediate need for remedial action by the area's managers is currently minimal. About 27% of the study area is classified by the composite model as of moderate concern, with 34% within el Ocote. This reflects a balance between areas of moderate and high vulnerability, but little disturbance. Based on the management zones created by this study, much of the sub-catchment is zoned as of least or moderate concern, where disturbance has not occurred. As such, the opportunity exists to prevent major human impacts on vulnerable areas and the entire ecosystem, but only if local stakeholders are incorporated into this process of limiting development.

Karst areas cover *c.* 12% of the Earth's surface and serve as important sources of water and carbon sinks (Ford & Williams 2007; Liu *et al.* 2008; Zhang 2011). Karst is formed by the dissolution of soluble rocks that promote the rapid subsurface hydrological flow and characteristic features such as springs, sinking streams, caves and sinkholes. The distinctive nature of the subsurface hydrological system of European karst was recognized during the early part of the twentieth century (Grund 1903; Katzer 1909; Cvijić 1918). High permeability and rapid recharge of karst aquifers lead to a high potential for groundwater contamination and impaired springs and wells. This groundwater pollution is fostered by poor agricultural practices, deforestation and other detrimental human land use (Cigna 1993; Hamilton Smith 2002; Parise 2011).

Public exposure to karst is usually in the form of caves; millions of people visit show caves each year globally. In addition to providing spectacular natural features, caves also preserve archaeological and palaeontological artefacts, create habitats for rare and endangered biota and protect records of past climate conditions. Humans have used caves over the millennia for recreational and religious purposes, which can have a negative impact on these subsurface environments. Caves are fragile environments, despite being surrounded by bedrock. They have long been sites of vandalism and garbage disposal, and even tourism has many negative impacts, such as changing cave microclimates, the removal of decorations, the installation of infrastructure and disturbance of the endemic biota (Cigna & Forti 1988; Huppert *et al.* 1993; Aley 2004; Hamilton Smith 2004).

An increase in our understanding of the natural processes operating within karst landscapes has led to the investigation of environmental issues, such as the overexploitation and pollution of water resources and caves, sinkhole collapse and desertification due to the removal of vegetation (Stringfield & LeGrand 1969; Parise & Pascali 2003). Various studies have examined these fragile landscapes and human impacts on them, and possible approaches to their adaptive and holistic management to reduce negative human activities (Drew & Hötzl 1999; van Beynen & Townsend 2005; Parise & Gunn 2007; van Beynen 2011; Gutiérrez *et al.* 2014). Advances in technology, such as computer-based geographical information systems, have aided the development of management tools to assist the prevention and mitigation of human impacts through the development of best management practices. Geographical information systems have been used to measure human disturbance in karst (van Aken *et al.* 2014; Kovarik & van Beynen 2015; Porter *et al.* 2016) and have allowed the downscaling of measurements to the catchment scale

From: Parise, M., Gabrovsek, F., Kaufmann, G. & Ravbar, N. (eds) 2018. *Advances in Karst Research: Theory, Fieldwork and Applications*. Geological Society, London, Special Publications, **466**, 431–442.
First published online November 6, 2017, https://doi.org/10.1144/SP466.1

(van Beynen & Bialkowska-Jelinska 2012). Groundwater vulnerability maps (GVMs) based on geographical information systems have been used to designate zones of vulnerability for aquifers and to delineate protection areas for particular water sources, such as springs and wells (Daly *et al.* 2002; Zwahlen 2004; Goldscheider 2005; Vías *et al.* 2006; Kattaa *et al.* 2010; Kavouri *et al.* 2011; van Beynen *et al.* 2012; Jiménez-Madrid *et al.* 2013). Risk and hazard maps measure the potential vulnerability of a particular area to contamination, but, as with the development of GVMs, these mapping techniques do not address the impacts on the karst ecosystem as a whole (Ravbar & Kranjc 2007; Bonacci *et al.* 2009; Goldscheider 2012).

Resource managers have used the environmental impact statement process in the USA to evaluate potential impacts on karst systems; however, as with the GVM approach, this approach often does not fully account for the complexity of karst ecosystems (Veni 1999). Investigators have applied environmental indices to assist managers with identifying environmental problems, deciding which problems require immediate attention and to provide a framework for effectiveness monitoring (Smeets & Weterings 1999). Van Beynen & Townsend (2005) developed the karst disturbance index (KDI), although this index scores impacts to the environment without consideration of the intrinsic susceptibility of the particular karst system (Calò & Parise 2006; van Beynen *et al.* 2007; De Waele 2009; North *et al.* 2009; Bauer & Kellerer-Pirklbauer 2010; Day *et al.* 2011). Harley *et al.* (2011) and Van Aken *et al.* (2014) created a method for modelling and predicting cave disturbance; however, their approach assumes that only vadose cave entrances (as opposed to flooded entrances) are associated with disturbance and ignored the impacts caused by disturbance above the cave.

This study addresses these problems by incorporating a GVM with a holistic environmental disturbance index (KDI) as an adaptable spatial tool for management planning and protection for all components of the karst environment. Within this new composite model, the combination of the KDI and the GVM recognizes both human disturbance and how the physical nature of the karst will affect this impact. The composite model creates the most comprehensive approach currently available, allowing resource managers to quickly identify the areas of highest concern and the implementation of adaptive holistic management strategies to mitigate these problems. The location for applying this model is a sub-catchment of the Rio la Venta watershed, a portion of which is incorporated by the Reserva de la Biosfera Selva el Ocote, Chiapas, Mexico. Our first goal was to calculate the factors of the GVM and the KDI and then to combine them in a geographical information system. We then used this composite model to delineate zones of concern requiring protective measures by evaluating the final scores of the composite model. Management strategies are discussed based on the results obtained.

Study area

The Reserva de la Biosfera Selva el Ocote (el Ocote) is located within the Highlands of Chiapas, Mexico (Fig. 1). El Ocote is part of the Grijalva–Usumacinta hydrological region, which consists of three main sub-catchments, including the Rio la Venta sub-catchment. The Rio la Venta watershed consists mainly of Middle to Upper Cretaceous Cintalapa limestone and Cantelha dolostone of the Sierra Madre Formation, where karstification has produced cockpit karst, large sinkholes, caves, sinking streams and springs. The longest mapped cave system in the area is the Cueva del Rio la Venta at 13 km in length (Concha 2009). Within el Ocote, the Rio la Venta flows through a 400 m deep canyon incised within the Sierra Madre Formation and travels *c.* 97 km from Aguacero at the southeastern corner of el Ocote to el Encajonado in the west.

The study area is a 368 km^2 (36 800 ha) portion of the larger Rio la Venta watershed, of which 60% is Biosfera Selva el Ocote land and the remaining 40% *ejidos* (communal lands designated by the state) or private land. The NE–SW-trending Sierra Monterrey and the Sierra Veinte Casas form the boundary to the north and the Uxpanapa fault forms the southern boundary (Fig. 1). Annual precipitation averages 1500 mm and there is a pronounced wet season from May to September. Temperatures reflect the humid–warm climate, reaching annual maxima of 30–34.5°C. Vegetation consists of >286 species of flora and is 82% primary and second-growth forest of tall perennial, sub-perennial, or intermediate sub-perennial and sub-deciduous forest depending on elevation (Breedlove 1981). The remaining areas are slash-and-burn agriculture. El Ocote contains a broad diversity of fauna (Medellin 1994), including large mammals such as jaguar (*Panthera onca*) and howler monkeys (*Alouatta palliata mexicana*). There are 16 communities, smaller settlements and ranches, including six main villages. The major economic activities within and around Rabasa, Emiliano Zapata, Venustiano Carranza, General Cardenas, Unidad Modelo and Adolfo Lopez Mateo are agriculture and animal husbandry. More information on the caves in the Rio la Venta watershed can be found in the text *Rio La Venta Treasures of Chiapas* (Bernabei & De Vivo 1999).

Methodology

KDI and GVM development

Two previous studies using geographical information systems calculated a spatially explicit map of the KDI (Kovarik & van Beynen 2015) and a validated GVM (Kovarik 2015) for this study area. The KDI was based on the methodology of van Beynen & Townsend (2005) (Fig. 2a), whereas the GVM used the European approach, specifically the COP method (Vías *et al.* 2006), with modifications from the Slovene approach (Ravbar & Kranjc 2007) (Fig. 2b). To summarize the procedures used in Kovarik & van Beynen (2015) and Kovarik (2015), vector layers were converted into raster (grid) format using ESRI's ArcGIS software. Each cell within the raster was assigned a value based on the original data of the study area according to the rating scheme of the disturbance index or vulnerability assessment. The rasters for each indicator or subfactor were then combined for each map to produce the final KDI and GVM in the geographical information system. Twelve rasters (one for each indicator with a score greater than zero) were created and combined to score the final KDI map (values used from North *et al.* 2009) and 12 rasters were combined to create three intermediate rasters and one final map to score resource vulnerability (Fig. 2a, b). All rasters and the final maps for both have a resolution of 30 m × 30 m in UTM NAD 83 Zone 15 North. The final values for the KDI ranged from 0 to 0.22 (pristine to little disturbance) and the GVM values ranged from 0 to 14.28 (very high vulnerability to low vulnerability). No area was found to be disturbed, highly disturbed or severely disturbed for the KDI and areas of very low vulnerability were not present in the study area for the GVM.

Composite model

Incorporating caves. Caves are an important part of karst environments and are particularly sensitive to disturbance; once disturbed, caves cannot be restored to their pristine state (Elliott 2004). For this particular project, if disturbance occurs within a cave, then the KDI incorporates the cave locations and subsurface development in the spatial model for disturbance. Conduit development is considered through the O factor in the GVM, which includes subsurface cave passage development. However, the GVM does not include cave entrances unless they are also insurgences. To ensure that this major component of the karst environment is incorporated into the composite model, the final GVM scores are altered to incorporate cave locations. A raster is created based on cave locations, where each cell containing the location of a cave is given a value of zero and all other cells are given a value of one. The value of zero is considered to be appropriate in this study due to the sensitivity of the caves to disturbance. This raster is then multiplied by the final GVM so that each cell containing a cave entrance is given the highest vulnerability rating in the final raster.

Aggregating values. To create the composite model, the final GVM, including cave values and the KDI created for the study area, are reclassified based on the scores and values of disturbance and vulnerability, respectively, using ArcGIS (Table 1). These two rasters are added together in ArcGIS to obtain the final values for the study area (Table 2). Values in the final raster between seven and ten indicate the areas of highest concern. Seven was chosen as the cut-off value because any disturbance on very high vulnerability areas could result in a serious impact on the karst ecosystem. Values between five and six indicate areas of moderate concern, as areas of high and very high vulnerability karst should be monitored even when lacking disturbance, and values between one and four indicate areas of least concern (Table 2).

Results and discussion

Creation of management zones based on the composite model

With the pristine or little disturbed nature of the study area, the final map reflects the GVM more than the KDI. The majority of the cells score three and four, capturing the low to moderate vulnerability, pristine state and little disturbance of the study area (Fig. 3a). Seventy-two per cent of the study area (265 km^2) is classified as an area of least concern in terms of management, with 60% of this area (146 km^2) within el Ocote. Consequently, although there are concerns regarding the vulnerability of the karst ecosystem, the lack of development and the natural protection of the ecosystem, the immediate need for remedial action by the area's managers is currently minimal. Seventy-nine per cent of the area of the zone classified as of least concern falls within natural areas, such as forest or savanna. These are areas on non-carbonate rock, or with slope values <31% on carbonate bedrock. Twenty per cent of this zone is cleared land, again in areas on predominantly non carbonate rock, or with slopes <31% on carbonate. The remaining 1% is developed, containing either settlements or roads on areas of non-carbonate or less steep slopes on carbonate. Where the zone of least concern falls within el Ocote, further development or land clearing should be discouraged or prevented where possible. However, the area outside el Ocote is suitable for development and agriculture as

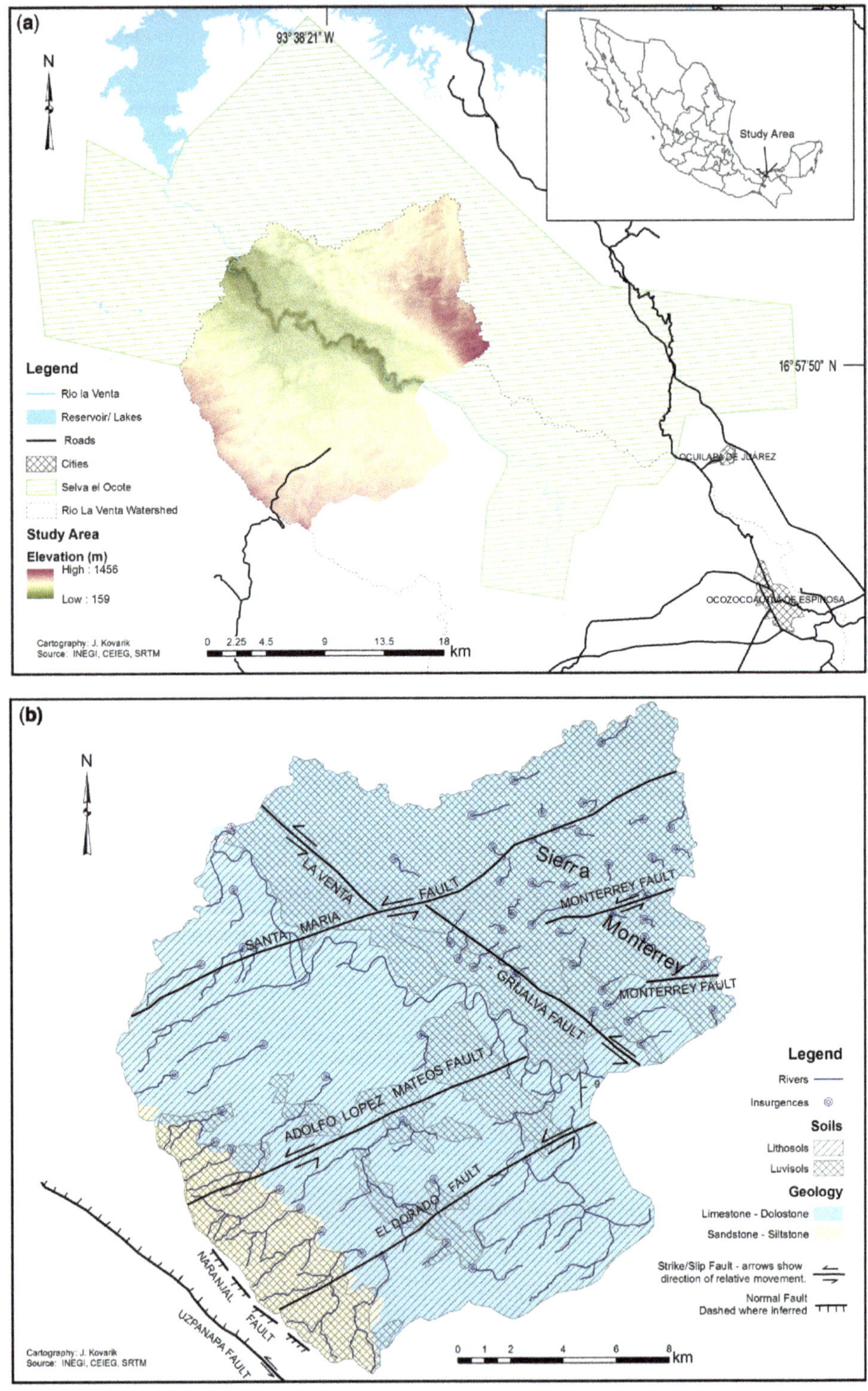

Fig. 1. (**a**) Location map of the Biosfera Selva el Ocote, Chiapas, Mexico and the study area. (**b**) Geological structure of study area.

Fig. 1. (*Continued*) (**c**) Google Earth satellite image of the study area.

current disturbance is minimal and the vulnerability of the ecosystem is lower in these areas.

About 27% of the study area (102 km^2) is of moderate concern, of which 34% (75 km^2) is within el Ocote. This reflects the moderate and high vulnerability areas, where a little disturbance does occur, and the high- and very high vulnerability areas still classified as pristine. Eighty-nine per cent (91 km^2) of the area of moderate concern is natural area, 10% (10 km^2) is cleared land and *c.* 1% (1 km^2) is developed land. This classification covers sites within the outer buffer zones for insurgences (500–1000 m) and streams (100 m) that sink directly into the subsurface, cave entrance areas and slopes >31% in areas of carbonate rock. Areas of moderate concern indicate zones where some development is possible, but requires guidance. Deforestation and agricultural practices on slopes >31% and within stream buffers should be avoided as a result of thin soils and an increased risk of the movement of contaminants into the subsurface. Quarrying should not occur and any future road and settlement construction should be limited to the areas of least concern, or should utilize mitigation strategies to avoid impacting karst features in areas that are already developed.

Areas of highest concern exist in the study area where settlements and roads correspond to areas of very high vulnerability, which includes <1% of the sub-catchment (Fig. 3b–d). No area of this zone occurs within the reserve. Only values of seven were scored, which fall within the lowest end of the highest concern category for the study area, indicating that disturbance may not be permanent because scores of ten are possible for this category through this quantitative measure. Managers can concentrate their limited resources on the small discrete locales with scores between 6 and 7, as opposed to areas which are so highly damaged (>8) that no remediation is possible or areas that currently lack any development (<3). Another benefit of a quantitative approach is that it provides a snapshot of the environmental state of an area so that future studies can determine whether remediation efforts have been successful or require more attention. Consequently, to ensure that these areas will improve, it is imperative that local residents and natural resource managers work together to create a plan for mitigation.

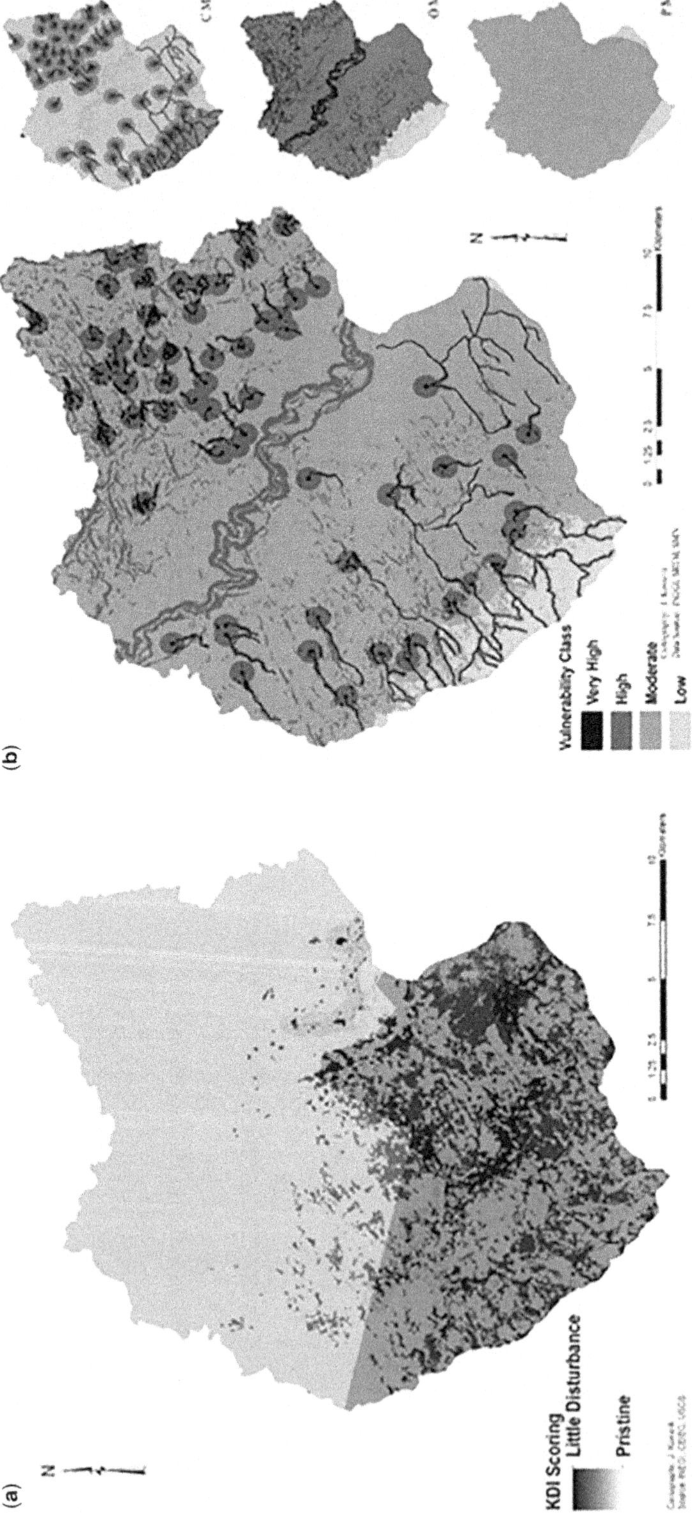

Fig. 2. **(a)** Karst disturbance index and **(b)** groundwater vulnerability model for the study area.

Table 1. *Reclassified KDI and GVM scores*

KDI			COP GVM		
Disturbance	Original class values	Adjusted class values	Vulnerability	Original class values	Adjusted class values
Severe	0.8–10	5	Very high	0–0.5	5
High	0.6–0.79	4	High	0.5–1	4
Disturbed	0.4–0.59	3	Moderate	1.0–2.0	3
Little	0.2–0.39	2	Low	2.0–4.0	2
Pristine	0–0.19	1	Very low	4.0–15.0	1

Angulo *et al.* (2013) is one of the few published studies to combine environmental indices investigating human–environmental interactions for karst settings. They combined the KDI (van Beynen & Townsend 2005) with a karst sustainability index to produce a priority management index. The priority management index can be used to '... determine the different management needs of the territory' (Angulo *et al.* 2013). They suggest that their index helps better manage protected karst areas and provide a benchmark for the state of the area, thereby allowing temporal comparisons. The main difference between our composite index and that of Angulo *et al.* (2013) is that they did not specifically consider the unique properties of karst groundwater parameters incorporated in the COP GVM. Consequently, we feel that our approach provides a more comprehensive strategy for managing karst environments.

Management strategies

Policy-based strategies for the management and protection of karst landscapes in developed countries are often inadequate and have only been implemented within the last 30 years (Fleury 2009). Federal and state agencies have created best management practices – standards and guidelines for management – ordinances and comprehensive plans to mitigate or prevent impacts from human actions to the natural environment. Best management practices can be defined as methods or practices that are either structural or non-structural and/or land use planning (Evans & Corradini 2001). Such management strategies for karst environments need to incorporate specific regulations and best management practices for: (1) roads built over well-developed karst; (2) cattle exclusions from sinkholes, stream ways and springs; (3) buffer zones around both urban and rural karst, including the installation of buffer strips; and (4) the transition of land from agriculture to forest through conservation reserves. Best management practices and strategies for caves and karst areas are well-documented worldwide (Clarke 1997; Watson *et al.* 1997; Eberhard 1998; Stokes & Griffiths 2000; Jones *et al.* 2003; Zhou & Beck 2005; Hildreth-Werker & Werker 2006; USDA 2008; Williams 2008; van Beynen 2011; Burt *et al.* 2013).

Management strategies are most successful on federal or state government lands where there is a structured process for addressing resistance to

Table 2. *Aggregate values*

Vulnerability		Disturbance				
		Pristine	Little	Disturbed	High	Severe
		1	2	3	4	5
Very low	1	2	3	4	5	6
Low	2	3	4	5	6	7
Moderate	3	4	5	6	7	8
High	4	5	6	7	8	9
Very high	5	6	7	8	9	10

(light shading)	Least
(medium shading)	Moderate
(dark shading)	Highest concern

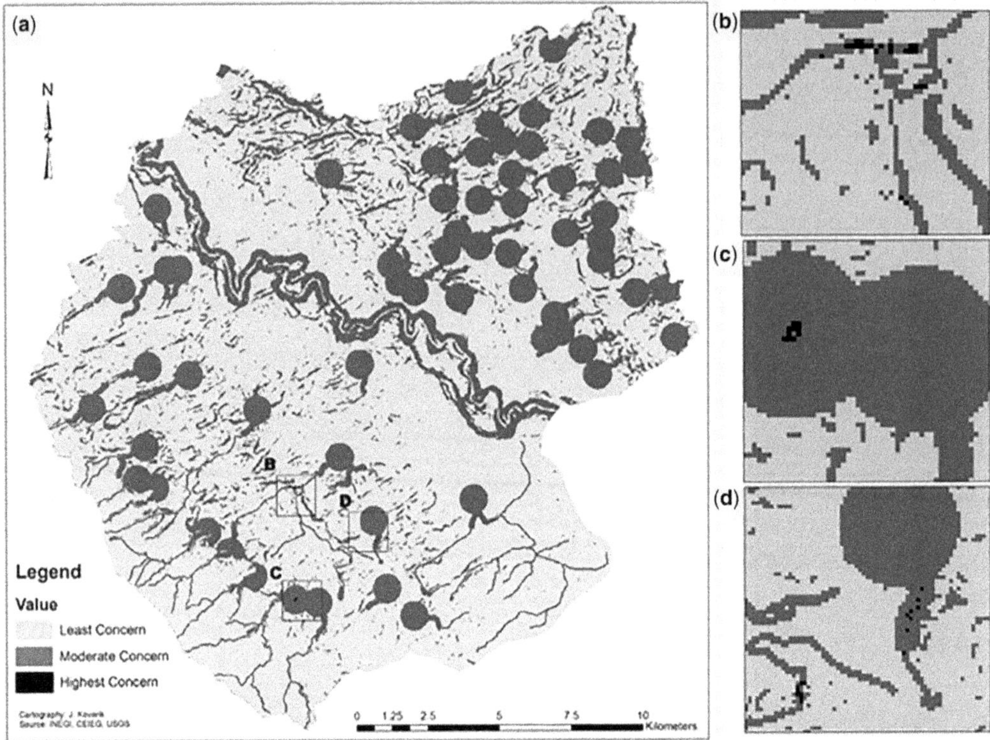

Fig. 3. Management zone mapping within the study area (**a**) based on the KDI and GVM, with the areas of highest concern as insets (**b–d**).

implementation and the associated costs. Conversely, on private land, participation and the implementation of such management strategies are uncommon. Federal and state lands do not always encompass the entire hydrological or ecological unit and so practices from non-managed lands will impact the protected areas. In developing countries, policy and practice are forced on local populations through a top-down approach as the land status on which they subsist changes to that of protected area. Many best management practices represent a significant change to the current practices of local farmers and can increase the costs of farming or result in the loss of workable land (Urich *et al.* 2001). Local residents often have a limited understanding of natural karst-related processes or the larger picture of what best management practices are meant to accomplish.

The health of the overall karst ecosystem is reliant on the participation of all concerned parties within a watershed. Case studies have demonstrated that environmental education and resource monitoring involving local residents from the beginning of the projects greatly improve the understanding of environmental problems as well as the relations between environmental issues and everyday life (Elke *et al.* 2007; Bocchino *et al.* 2014). Innovative ideas and solutions that work for the local community and environment originate through processes such as this. Environmental education can also provide information on simple measures that individual property owners can use to help alleviate or prevent negative impacts on karst ecosystems, thereby improving community health without the onerous pressure of restrictive government policy. Federal, state and non-governmental organizations in the USA have created informational booklets and websites to share with private landowners, containing tips and information regarding the susceptibility of karst landscapes. Best management practices are outlined that inform landowners about how they can be applied to ensure good water quality and maintain soil productivity (Zokaites 1997; Veni *et al.* 2001).

For our area, a top-down approach was attempted to manage natural resources with the implementation of policy and best management practices. In the study area, 60% of the land is currently part of el Ocote, however, the remaining portion within the sub-catchment is private land and *ejidos*. With the lack of enforcement on reserve property and the incorporation of private land within the greater watershed, a top-down approach will probably be ineffective. Through working to educate local people about the natural processes within karst environments and the impact of local land use on ecosystem health, a co-operative environment will hopefully be created where local residents will begin to work with local specialists to create solutions mitigating impacts. Based on the management zones created in this study, much of the sub-catchment is within the zones of least or moderate concern, where disturbance has not occurred. This presents an opportunity to prevent any major human impact on vulnerable areas and the ecosystem as a whole.

Within the study area a water quality monitoring project and an environmental education programme were started to target the areas of highest concern based on the results of the composite model. Water quality sampling was initially conducted by the US Forest Service International Program and sampling was then continued by a local university. Fish inventories were carried out in the Rio la Venta to provide environmental data, hopefully leading to new conservation strategies in the watershed (Gonzalez-Diaz *et al.* 2008). The reserve is currently working with local community members to conduct water quality sampling at sites around the larger watershed. Karst education materials were adapted to the local region and translated into Spanish.

Settlements within the study area in an area of highest concern, as well as larger communities with easy access to the reserve, were selected for the pilot project. Workshops introducing a karst curriculum to teachers and reserve workers were developed to evaluate the usefulness of the materials and to gather information to alter the curricula specifically for the study area. Best management practices for local landowners regarding cave visitation and tours were translated into Spanish and provided to el Ocote to share with local landowners. Best management practices such as sinkhole management, environmentally considerate septic tank practices and pesticide management targeted at landowners for agriculture, forestry and development in rural karst areas have been translated into Spanish. In cooperation with el Ocote, they are currently being adapted into a booklet that can be shared with local residents and industry, incorporating the composite model along with information about the natural processes at work within the Rio la Venta watershed.

Some development is occurring within the watershed, in the buffer zone as well as directly outside the reserve, and more development will be planned for the future. The careful management of such development will reduce any resulting negative impacts. One management strategy that could be adopted is to integrate the composite model into a broader system using the US Environmental Protection Agency's three-tiered monitoring and assessment programme (US Environmental Protection Agency 2006). Landscape-scale assessment indicators scored using land use maps and geographical information systems constitute tier one. The second tier consists of rapid assessment protocols using a combination of data analysis and fieldwork of the physical landscape. Tier three involves intensive site monitoring. The composite model can be incorporated into the first two levels. Because of the model's assumptions and generalizations of the physical landscape, intensive site assessment and monitoring must be conducted before any actual development occurs. This three-pronged approach will promote the most informed management decisions possible. Such a method is imperative for non-resilient natural systems such as karst.

Conclusions

This study successfully calculated and applied a composite model based on the KDI and a GVM creating zones of least, moderate and highest concern in the study area. Seventy-two per cent of the study area (265 km^2) is classified as an area of least concern in terms of management, with 60% of this area (146 km^2) within el Ocote. Consequently, although there are concerns regarding the vulnerability of the karst ecosystem, the lack of development and the natural protection of the ecosystem, the immediate need for remedial action by the area's managers is currently minimal. About 27% of the study area is classified by the composite models as moderate concern, of which 34% is within el Ocote. The various zones of the model can be used to evaluate which locations within the sub-catchment of the Rio la Venta are of highest concern to land managers and local residents in relation to karst ecosystem health; these are where management strategies should be focused. The composite model can also guide development and provide a simple and easy to understand representation for inclusion in educational materials for local residents and industry.

Top-down approaches to natural resource management, such as requiring best management practices through policy and regulatory control, are likely be ineffective in the study area due to a lack of enforcement and understanding of the environment. As a result of these issues, karst-specific environmental education and water quality monitoring were the first management tools applied within the study area and were focused on the areas of highest concern from the composite model. Although the composite model was applied in a remote rural area of a developing country, it is anticipated that it is also applicable to other locations and would benefit planners and managers in developed countries.

Future work within the study area should focus on the continued characterization of the natural hydrological processes of the watershed and an inventory of the biota and karst features to incorporate into the model. This information can be utilized to further refine the composite model and to adapt site-specific management strategies where needed within the zones of highest concern. It is hoped that education of local residents about their karst environment and the impacts of their actions on various facets of the ecosystem will result in a reduced need for future mitigation. An essential component of adaptive management, of which the composite model can be included, is to have the most current, accurate data regarding the human/environmental changes within the application area. An appropriate timeframe for the collection of such data to continue the effectiveness of the model would be every five years.

Support for environmental education materials, translation and workshops was provided by Project Underground, the National Speleological Society, the Tharp Fellowship from the University of South Florida, the US Forest Service International Programs and through the generous volunteer work of Andrea Kuhlman and Miriam Rosario-Toro.

References

ALEY, T. 2004. Tourist caves: algae and lampenflora. *In*: GUNN, J. (ed.) *The Encyclopedia of Caves and Karst Science*. Routledge, New York, 733–734.

ANGULO, B., MORALES, T., URIARTE, J.A. & ANTIGÜEDAD, I. 2013. Implementing a comprehensive approach for evaluating significance and disturbance in protected karst areas to guide management strategies. *Journal of Environmental Management*, **130**, 386–396.

BAUER, C. & KELLERER-PIRKLBAUER, A. 2010. Human impacts on karst environment: a case study from Central Styria. *Zeitschrift für Geomorphologie, Supplementary Issues*, **54**, 1–26.

BERNABEI, T. & DE VIVO, A. 1999. *Rio La Venta Treasures of Chiapas*. Association La Venta, Coneculta, Turra, Padua.

BOCCHINO, B., DEL VECCHIO, U. *ET AL.* 2014. Increasing people's awareness about the importance of karst landscapes and aquifers: an experience from southern Italy. *In*: KUKURIC, N., STEVANOVIC, Z. & KRESIC, N. (eds) *Proceedings of the International Conference and Field Seminar Karst Without Boundaries*, 11–15 June 2014, Trebinje–Dubrovnik (Bosnia and Herzegovina–Croatia), 398–405.

BONACCI, O., PIPAN, T. & CULVER, D.C. 2009. A framework for karst ecohydrology. *Environmental Geology*, **56**, 891–900.

BREEDLOVE, D.E. 1981. *Flora of Chiapas. Part 1. Introduction to the Flora of Chiapas*. California Academy of Sciences, San Francisco, CA.

BURT, C., BACHOON, D.S., MANOYLOV, K. & SMITH, M. 2013. The impact of cattle farming best management practices on surface water nutrient concentrations, faecal bacteria and algal dominance in the Lake Oconee watershed. *Water and Environment Journal*, **27**, 207–215.

CALÒ, F. & PARISE, M. 2006. Evaluating the human disturbance to karst environments in southern Italy. *Acta Carsologica*, **35**, 47–56.

CIGNA, A.A. 1993. Environmental management of tourist caves: the examples of Grotte di Castellana and Grotta Grande del Vento, Italy. *Environmental Geology*, **21**, 173–180.

CIGNA, A.A. & FORTI, P. 1988. The environmental impact assessment of a tourist cave. *Proceedings of the International Symposium, 170th Anniversary Postojnska, Jama*, 10–12 November. Center for Scientific Research SAZU and Postojnska Jama Tourist and Hotel Organization, 29–38.

CLARKE, A. 1997. *Management Prescriptions for Tasmania's Cave Fauna. Tasmanian Regional Forest Agreement*. Environment and Heritage Technical Committee, Hobart.

CONCHA, C. 2009. *Tra deserti e foreste: Viaggio nelle grotte del messico*. La Venta esplorazioni geografiche, Treviso.

CVIJIĆ, J. 1918. *Hydrographie souterraine et évolution morphologique du karst*. Recueil des Travaux de l'Institut Géographie Alpine, Grenoble.

DALY, D., DASSARGUES, A. *ET AL.* 2002. Main concepts of the 'European approach' to karst-groundwater-vulnerability assessment and mapping. *Hydrogeology Journal*, **10**, 340–345.

DAY, M., HALFEN, A. & CHENOWETH, S. 2011. The Cockpit Country, Jamaica: boundary issues in assessing disturbance and using a karst disturbance index in protected areas planning. *In*: VAN BEYNEN, P.E. (ed.) *Karst Management*. Springer, Dordrecht, 399–414.

DE WAELE, J. 2009. Evaluating disturbance on Mediterranean karst areas: the example of Sardinia (Italy). *Environmental Geology*, **58**, 239–255.

DREW, D. & HÖTZL, H. 1999. *Karst Hydrogeology and Human Activities: Impacts, Consequences and Implications*. A.A. Balkema, Rotterdam.

EBERHARD, R. 1998. *Planning for Karst Management in Multiple-use Forest: the Junee-Florentine Karst Study*. Vol. 10. Tasforests, Hobart, 33–48.

ELKE, M., GLASER, D., SETTY, K., SUSSMAN, D. & YOCUM, D. 2007. Design and Implementation of Sustainable Water Resources Programs in San Cristobal de las Casas, Mexico, http://www2.bren.ucsb.edu/~keller/courses/GP_reports/FinalReport_Chiapas2.pdf

ELLIOTT, W.R. 2004. Protecting caves and cave life. *In*: CULVER, D.C. & WHITE, W.B. (eds) *Encyclopedia of Caves*. Elsevier, Amsterdam, 428–467.

EVANS, B.M. & CORRADINI, K.J. 2001. *BMP Pollution Reduction Guidance Document*. Environmental Resources Research Institute, Pennsylvania State University, State College, PA.

FLEURY, S. 2009. *Land Use Policy and Practice on Karst Terrains: Living on Limestone*. Springer Science and Business Media, Dordrecht.

FORD, D. & WILLIAMS, P.W. 2007. *Karst Hydrogeology and Geomorphology*. Revised edition. Wiley, Chichester.

GOLDSCHEIDER, N. 2005. Karst groundwater vulnerability mapping: application of a new method in the Swabian Alb, Germany. *Hydrogeology Journal*, **13**, 555–564.

GOLDSCHEIDER, N. 2012. A holistic approach to groundwater protection and ecosystem services in karst terrains. *AQUA Mundi*, **3**, 117–124.

GONZALEZ-DIAZ, A.A., QUINONES, R.M., VELAZQUEZ-MARTINEZ, J. & RODILES-HERNANDEZ, R. 2008. Fishes of La Venta River in Chiapas, México. *Zootaxa*, **1685**, 47–54.

GRUND, A. 1903. *Die karsthydrographie studien aus west-bosnien*. Leipzig.

GUTIÉRREZ, F., PARISE, M., DE WAELE, J. & JOURDE, H. 2014. A review on natural and human-induced geohazards and impacts in karst. *Earth-Science Reviews*, **138**, 61–88.

HAMILTON SMITH, E. 2002. Management assessment in karst areas. *Acta Carsologica*, **31**, 13–20.

HAMILTON SMITH, E. 2004. Tourist caves. *In*: GUNN, J. (ed.) *The Encyclopedia of Caves and Karst Science*. Routledge, New York.

HARLEY, G.L., POLK, J.S., NORTH, L.A. & REEDER, P.P. 2011. Application of a cave inventory system to stimulate development of management strategies: the case of west-central Florida, USA. *Journal of Environmental Management*, **92**, 2547–2557.

HILDRETH-WERKER, V. & WERKER, J.C. (eds) 2006. *Cave Conservation and Restoration*. National Speleological Society of America, Huntsville, AL.

HUPPERT, G., BURRI, E., FORTI, P. & CIGNA, A. 1993. Effects of tourist development on caves and karst. *Catena Supplement*, **25**, 251–268.

JIMÉNEZ-MADRID, A., CARRASCO, F., MARTÍNEZ, C. & GOGU, R.C. 2013. DRISTPI, a new groundwater vulnerability mapping method for use in karstic and non-karstic aquifers. *Quarterly Journal of Engineering Geology and Hydrogeology*, **46**, 245–255, https://doi.org/10.1144/qjegh2012-038

JONES, W., HOBBS, H., III *ET AL*. 2003. *Recommendations and Guidelines for Managing Caves on Protected Lands*. Special Publications, **8**. Karst Waters Institute, Leesburg, PA.

KATTAA, B., AL-FARES, W. & AL CHARIDEH, A.R. 2010. Groundwater vulnerability assessment for the Banyas Catchment of the Syrian coastal area using GIS and the RISKE method. *Journal of Environmental Management*, **91**, 1103–1110.

KATZER, F. 1909. *Karst und Karsthydrographie, von dr. Friedrich Katzer*. D.A. Kajon, Sarajevo.

KAVOURI, K., PLAGNES, V., TREMOULET, J., DÖRFLIGER, N., REJIBA, F. & MARCHET, P. 2011. PaPRIKa: a method for estimating karst resource and source vulnerability – application to the Ouysse karst system (southwest France). *Hydrogeology Journal*, **19**, 339–353.

KOVARIK, J.L. 2015. *A composite spatial model incorporating groundwater vulnerability and environmental disturbance to guide land management*. PhD thesis, University of South Florida.

KOVARIK, J.L. & VAN BEYNEN, P.E. 2015. Application of the karst disturbance index as a raster-based model in a developing country. *Applied Geography*, **63**, 396–407.

LIU, Z., DREYBRODT, W. & WANG, H. 2008. A possible important CO_2 sink by the global water cycle. *Chinese Science Bulletin*, **53**, 402–407.

MEDELLIN, R.A. 1994. Mammal diversity and conservation in the Selva Lacandona, Chiapas, Mexico. *Conservation Biology*, **8**, 780–799.

NORTH, L.A., VAN BEYNEN, P.E. & PARISE, M. 2009. Interregional comparison of karst disturbance: west-central Florida and southeast Italy. *Journal of Environmental Management*, **90**, 1770–1781.

PARISE, M. 2011. Some considerations on show cave management issues in southern Italy. *In*: VAN BEYNEN, P.E. (ed.) *Karst Management*. Springer, Dordrecht, 159–167.

PARISE, M. & GUNN, J. (eds) 2007. *Natural and Anthropogenic Hazards in Karst Areas: Recognition, Analysis and Mitigation*. Geological Society, London, Special Publications, **279**, http://sp.lyellcollection.org/content/279/1

PARISE, M. & PASCALI, V. 2003. Surface and subsurface environmental degradation in the karst of Apulia (southern Italy). *Environmental Geology*, **44**, 247–256.

PORTER, B.L., NORTH, L.A. & POLK, J.S. 2016. Comparing and refining karst disturbance index methods through application in an island karst setting. *Environmental Management*, **58**, 1027–1045.

RAVBAR, N. & KRANJC, A. 2007. *The Protection of Karst Waters: a Comprehensive Slovene Approach to Vulnerability and Contamination Risk Mapping*. Inštitut za raziskovanje krasa ZRC SAZU, Postojna, Slovenia.

SMEETS, E. & WETERINGS, R. 1999. *Environmental Indicators Typology and Overview*. European Environment Agency, Copenhagen.

STOKES, T. & GRIFFITHS, P. 2000. *A Preliminary Discussion of Karst Inventory Systems and Principles (KISP) for British Columbia*. B.C. Ministry of Forests, Research Branch, Victoria, BC.

STRINGFIELD, V.T. & LEGRAND, H.E. 1969. Hydrology of carbonate rock terranes – a review: with special reference to the United States. *Journal of Hydrology*, **8**, 377–417.

URICH, P.B., DAY, M.J. & LYNAGH, F. 2001. Policy and practice in karst landscape protection: Bohol, the Philippines. *The Geographical Journal*, **167**, 305–323.

US DEPARTMENT OF AGRICULTURE FOREST SERVICE 2008. *Tongass Land and Resource Management Plan*. US Department of Agriculture Forest Service Report, R10-MB-603b.

US ENVIRONMENTAL PROTECTION AGENCY 2006. *Application of Elements of a State Water Monitoring and Assessment Program for Wetlands*. Wetlands Division, Office of Wetlands, Oceans and Watersheds, US Environmental Protection Agency, Washington DC.

VAN AKEN, M., HARLEY, G.L., DICKENS, J.F., POLK, J.S., & NORTH, L. & A. 2014. A GIS-based modeling approach to predicting cave disturbance in karst landscapes: a case study from west-central Florida. *Physical Geography*, **35**, 123–133.

VAN BEYNEN, P., FELICIANO, N., NORTH, L. & TOWNSEND, K. 2007. Application of a karst disturbance index in Hillsborough county, Florida. *Environmental Management*, **39**, 261–277.

VAN BEYNEN, P.E. 2011. *Karst Management*. Springer, Dordrecht.

VAN BEYNEN, P.E. & TOWNSEND, K. 2005. A disturbance index for karst environments. *Environmental Management*, **36**, 101–116.

van Beynen, P.E. & Bialkowska-Jelinska, E. 2012. Human disturbance of the Waitomo catchment, New Zealand. *Journal of Environmental Management*, **108**, 130–140.

Van Beynen, P.E., Niedzielski, M.A., Bialkowska-Jelinska, E., Alsharif, K. & Matusick, J. 2012. Comparative study of specific groundwater vulnerability of a karst aquifer in central Florida. *Applied Geography*, **32**, 868–877.

Veni, G. 1999. A geomorphological strategy for conducting environmental impact assessments in karst areas. *Geomorphology*, **31**, 151–180.

Veni, G., DuChene, H. et al. 2001. *Living with Karst: A Fragile Foundation*. American Geological Institute, Alexandria.

Vías, J.M., Andreo, B., Perles, M.J., Carrasco, F., Vadillo, I. & Jiménez, P. 2006. Proposed method for groundwater vulnerability mapping in carbonate (karstic) aquifers: the COP method. *Hydrogeology Journal*, **14**, 912–925.

Watson, J., Hamilton-Smith, E., Gillieson, D. & Kiernan, K. (eds) 1997. *Guidelines for Cave and Karst Protection*. IUCN, Gland.

Williams, P. 2008. *World Heritage Caves and Karst*. IUCN, Gland.

Zhang, C. 2011. Carbonate rock dissolution rates in different land uses and their carbon sink effect. *Chinese Science Bulletin*, **56**, 3759–3765.

Zhou, W. & Beck, B.F. 2005. Roadway construction in karst areas: management of stormwater runoff and sinkhole risk assessment. *Environmental Geology*, **47**, 1138–1149.

Zokaites, C. (ed.) 1997. *Living on Karst: a Reference Guide for Landowners in Limestone Regions*. Cave Conservancy of the Virginias, Glen Allen, VA.

Zwahlen, F. 2004. *Vulnerability and Risk Mapping for the Protection of Carbonate (Karst) Aquifers*. Final Report (COST Action 620). European Commission, Directorate-General XII Science, Research and Development, Brussels.

A review and statistical assessment of the criteria for determining cave significance

AUGUSTO S. AULER[1,2*], TATIANA A. R. SOUZA[1,2], DANIELA C. SÉ[2] & GUSTAVO A. SOARES[2]

[1]*Instituto do Carste, Rua Barcelona 240/302, Belo Horizonte, MG, 30360-260, Brazil*

[2]*Carste Ciência e Meio Ambiente, Rua Aquiles Lobo 297, Belo Horizonte, MG, 30150-160, Brazil*

**Correspondence: aauler@gmail.com*

Abstract: Determining the significance of caves is challenging due to a lack of consensus on which parameters to consider and their relative importance, in addition to difficulties in applying the parameters in a repeatable way. However, classifying caves by levels of significance is unequivocally important because it allows the prioritization of caves for future environmental protection. In countries where the subsurface belongs to the government, such as Brazil, the decision process is coordinated at the government level and a comprehensive approach balancing environmental protection and economic/social interests should be applied. Brazil has the most comprehensive set of criteria for assessing the significance of caves, encompassing parameters applied in other countries. A sample of 401 Brazilian caves in limestone and iron-rich rocks was analysed statistically to infer the relative frequency of each criterion in assigning significance to caves. The analysis included 70 parameters; 30 were present in caves, but on average less than five of these parameters occurred together at each cave. Subjective parameters tended to be less represented. Biotic parameters were dominant and both abiotic and biotic parameters displayed a correlation with length and area, suggesting that these parameters could be good indicators of cave significance. Applying the set of criteria proposed by the Brazilian government to our sample, there was no marked difference between caves in different rock types. A better approach to defining cave relevance is required, with an emphasis on science-based parameters.

Caves are common landforms in most rock types and environments, although they tend to be more characteristic of karst terrains. Caves have been important features during all stages of human evolution and cultural development, providing shelter and substrates for cultural and religious practices (Kempe 1988; Lewis-Williams 2002; Goldberg & Bar-Yosef 2012). As a result of the protected nature of caves, common surface-level weathering agents are largely absent and therefore archaeological and palaeontological remains tend to be especially well preserved; these remains have been essential in understanding the fossil record (Andrews 1990; Auler *et al.* 2006; Schubert & Mead 2012). The mechanisms involved in cave formation can provide clues to unravelling the genesis of economic mineral deposits or provide reservoirs for late mineral or oil emplacement (Hill 1990, 1995). Caves also represent a little-known habitat for several species, harbouring a diverse fauna and fragile ecosystems (Culver & Pipan 2009; Romero 2009); they have the potential for novel geomicrobiological interactions, with applications in areas as diverse as drug development and mineral generation (Cheeptham 2013; Parker *et al.* 2013). Caves may be a valuable sporting and tourism asset in many areas; an estimated 650 caves worldwide are open to tourism, generating a gross annual revenue of *c.* $US2.5 billion (Ford & Williams 2007). Thus it is natural that caves are considered to be important landforms worldwide; an emphasis on cave conservation commonly results in the creation of conservation areas centred around caves.

The dimensional concept of a cave is subject to debate. The most commonly accepted definition has an anthropocentric bias and considers a cave as a natural void beneath the land surface that is large enough to admit humans (Palmer 2007). However, many caves lack a natural opening; statistical studies by Curl (1958) suggested that caves without an entrance are far more abundant than accessible caves. As a habitat for subterranean fauna, caves include small, inaccessible channels in which cave organisms can thrive. Caves are usually analysed through a human perspective, but a broader and more scientific definition should take into consideration a portion of the rock and soil space surrounding them. The lower dimensional limit of a cave is open to debate, but a literal interpretation has been adopted by the Brazilian government, taking into consideration the accessibility to humans (an

From: Parise, M., Gabrovsek, F., Kaufmann, G. & Ravbar, N. (eds) 2018. *Advances in Karst Research: Theory, Fieldwork and Applications*. Geological Society, London, Special Publications, **466**, 443–459.
First published online November 28, 2017, https://doi.org/10.1144/SP466.8

underground space that a human can enter and fit) (Brasil 2008), which includes some voids <2 m long. Along with the cave itself, Brazilian law (Brasil 1990) also requires the protection of an area 250 m wide perpendicular to the surface projection of the cave walls. This 250 m buffer zone has a preliminary status and can be modified by specific studies (Auler 2016).

Environmental impacts and the protection of karst terrains have been considered by many researchers (Parise & Gunn 2007; Gutiérrez *et al.* 2014) and methods for determining karst and cave disturbance and sustainability have been devised (van Beynen & Townsend 2005; van Beynen *et al.* 2012; van Aken *et al.* 2014) and applied (Calò & Parise 2006; van Beynen *et al.* 2007; De Waele 2009; North *et al.* 2009; van Beynen & Bialkowska-Jelinska 2012; Kovarik & van Beynen 2015). Although the significance of karst areas has been assessed (Angulo *et al.* 2013), limited amounts of data are available regarding the protocols for determining the significance of caves and thus electing priorities in cave conservation. Choosing the caves worth protecting carries an intrinsically complex dimensional difficulty and deals with several parameters relevant to the cave environment. Ranking these parameters to achieve an objective and repeatable classification of the importance of caves and thus enabling the ranking of caves by conservation priority is a challenging proposition; nevertheless, a few countries, including the USA and Brazil (Kovarik 2013; Donato *et al.* 2014; Auler & Piló 2015), have attempted to do this. In many situations, caves may interfere with economically and socially significant projects, such as hydroelectric reservoirs, mining and urbanization (Auler & Piló 2015; Jaffé *et al.* 2016). In these scenarios, establishing levels of cave significance and protection is mandatory.

This paper reviews the existing protocols for establishing the importance of caves and provides a statistical assessment of the criteria applied in Brazil to determine the level of importance. The relative weight of each criterion and their interrelationships are analysed, with the aim of providing a baseline to determine the most influential parameters in Brazilian law. This review highlights the difficulties in dealing with both quantitative and subjective criteria and providing a balance in which truly significant caves receive a proper level of protection. This analysis can be useful to decision-makers in clarifying the classification schemes and the selection of speleological sites worthy of environmental protection.

Review of protocols for determining the significance of caves

Middleton (2016*a*, *b*) has reviewed the legal instruments for protecting caves in some Western countries, while Huppert (1995) and Lera (2002) reviewed cave protection laws in the USA.

In Brazil, caves are protected by the 1988 Constitution (Brasil 1988) and are owned by the federal government. Federal and state level decrees, directives and recommendations follow the constitutional protection – for a review, see Auler & Piló (2015) and Auler (2016). Federal Decree 6640 aims to 'protect the caves in the Brazilian territory' (Brasil 2008) and, together with the Normative Instruction 2 (MMA 2009), provides detailed information about how to classify caves according to levels of importance. The increase in the price of mineral commodities between 2008 and 2014 led to the classification of thousands of caves according to the 6640 Federal Decree as part of environmental assessments for mining projects. The Decree allows for four levels of cave significance, with differing degrees of protection (Table 1).

According to Brazilian law, 70 parameters must be considered when assessing the significance of caves (Tables 2 & 3) (Brasil 2008; MMA 2009). These criteria consist of mostly biological parameters, with a lesser emphasis on geological and cultural parameters. Some parameters (Table 3) need to be combined with others to attain a given significance level. The analysis must be performed at both the local and regional scale, which denote a geomorphological landform (e.g. ridge, valley and basin)

Table 1. *Levels of cave significance according to the Federal Decree 6640 (Brasil 2008) and associated environmental compensation*

Level of significance	Degree of protection	Environmental compensation
Maximum	Permanently protected	Not applicable
High	May be impacted, but subject to compensation	Two similar caves of equal significance that become permanently protected
Medium	May be impacted, but subject to compensation	Monetary compensation or programmes/actions aimed to protect caves
Low	May be impacted, no compensation required	Not applicable

Table 2. *Parameters that must be evaluated during the assessment of cave significance. If one criterion is met, then the cave will have either a maximum or high significance*

Parameter	Type
Maximum significance	
I: Peculiar or rare genesis	RL/SB
II: Peculiar morphology	RL/SB
III: Notable dimensions in length, area or volume	AB
IV: Peculiar speleothems	RL/SB
V: Geographical isolation	AB
VI: Essential shelter for threatened species	RL/SB
VII: Essential habitat for endemic or relict troglobites	RL/SB
VIII: Habitat for a rare troglobite	RL
IX: Peculiar ecological interactions	RL/SB
X: Significant environmental/palaeoenvironmental importance or cave used for compensation	RL/SB
XI: Significant historical, cultural or religious value	RL/SB
High significance	
I: Type locality	AB
II: Established populations of species with important ecological role	RL/SB
III: New taxa	AB
IV: High species richness	AB
V: High relative abundance of species	AB
VI: Peculiar faunal composition	RL/SB
VII: Troglobites that are not rare, endemic or relict	RL
VIII: Troglomorphic species	RL
IX: Obligatory trogloxenes	RL
X: Population of exceptional size	RL/SB
XI: Rare species	RL/SB
XII: High length when compared with other caves at the regional scale	AB
XIII: High area when compared with other caves at the regional scale	AB
XIV: High volume when compared with other caves at the regional scale	AB
XV: Significant presence of rare speleogenetic features	RL/SB
XVI: Perennial lake or underground drainage, with marked influence over any of the criteria in this table	RL/SB
XVII: Diversity of chemical deposition reference types of speleothem and processes of deposition	AB
XVIII: Notable configuration of speleothems	RL/SB
XIX: High influence of the cave over the dynamics of the karst system	RL/SB
XX: Interrelation of the cave with a maximum relevance cave	AB
XXI: Aesthetic and scenic values of national and international significance	SB
XXII: Systematic public visitation with a regional or national scope	RL/SB

AB, absolute; RL, relative; SB, subjective. See text for details.

and a karst area (usually with limits related to a geological formation or group), respectively. Caves must be compared according to the rock in which they develop. Calculation of the level of significance involves sequential steps, as summarized in Figure 1.

In the USA, the Federal Cave Resources Protection Act (FCRPA) is a federal law approved in 1988 that aims 'to secure, protect, and preserve significant caves on Federal lands for the perpetual use, enjoyment, and benefit of all people' (United States Congress 1988). Because ownership of land in the USA also includes what lies below (e.g. caves and mineral deposits), the FCRPA applies only to federal lands, where regulations already infer some degree of protection to caves. The FCRPA provided a starting point to assess cave significance; a set of criteria was established by Regulation 36 CFR 290.3 (United States Government 2016*a*) and Regulation 43 CFR 37.11 (United States Government 2016*b*). If one or more of these criteria (Table 4) are met, then the cave is granted the status of significant.

In the Republic of the Philippines, the National Caves and Cave Resources Management and Protection Act was passed as the Republic Act 9072 in 2001 (Republic of the Philippines 2001); it is further detailed in an internal memorandum (Republic of the Philippines 2007). It is based mostly on the US FCRPA, but the procedure for classifying caves is different. In the Philippines, three classes of caves are recognized (Table 5). Class I caves allow restricted access, mostly for documentation (mapping and photography), scientific and educational purposes. Class II caves may have some sections at

Table 3. *Additional parameters that, if present and combined, may result in high or medium significance caves*

Parameter	Type
I: Resident population of Chiroptera	AB
II: Cave used for nesting by wild birds	AB
III: High diversity of organic substrates	AB
IV: Medium species richness	AB
V: Medium relative abundance of species	AB
VI: Use of cave by migratory species	RL/SB
VII: Presence of faunistic singularities at the local level	RL/SB
VIII: Presence of geological structure of scientific interest	RL/SB
IX: Presence of palaeontological remains	AB
X: Aesthetic and scenic value at the local scale	SB
XI: Systematic visitation at the local scale	RL/SB
XII: Presence of condensation or percolation water with important influence on other criteria	RL/SB
XIII: Lake or underground drainage with very important influence on other criteria	RL/SB
XIV: Presence of faunistic singularities at the regional scale	RL/SB
XV: Medium length compared with other caves within the same karst area	AB
XVI: Medium area compared with other caves within the same karst area	AB
XVII: Medium depth compared with other caves within the same karst area	AB
XVIII: Medium volume compared with other caves within the same karst area	AB
XIX: Presence of rare speleogenetic features	RL/SB
XX: Perennial lake or underground drainage with important influence on other criteria	RL/SB
XXI: Diversity of chemical deposition with many types of speleothems and genetic processes	AB
XXII: Presence of chemical or clastic sedimentation of scientific interest	RL/SB
XXIII: Aesthetic or scenic value at the regional scale	SB
XXIV: Constant use (periodical or systematic) for educational, recreational or sporting purposes	RL/SB
XXV: Low diversity of organic substrates	AB
XXVI: Low species richness	AB
XXVII: Low relative abundance of species	AB
XXVIII: Presence of faunistic singularities at the local scale	RL/SB
XXIX: Medium length compared with other caves within the same geomorphological landform	AB
XXX: Medium area compared with other caves within the same geomorphological landform	AB
XXXI: High depth compared with other caves within the same geomorphological landform	AB
XXXII: Medium volume compared with other caves within the same geomorphological landform	AB
XXXIII: Few types of speleothems/depositional processes in terms of diversity of chemical sedimentation	AB
XXXIV: Sporadic or infrequent use for educational, recreational or sporting purposes	RL/SB
XXXV: Public visitation sporadic or infrequent	RL/SB
XXXVI: Presence of condensation or percolation water with some influence on other criteria	AB
XXXVII: Intermittent lake or underground drainage with some influence on other criteria	AB

AB, absolute; RL, relative; SB, subjective. See text for details.

the restriction levels of Class I caves, but will mostly be open to experienced cavers, guided educational tours or visits. Class I and II caves are thus granted some degree of protection. Class III caves are mostly devoid of value; they have free access and may be used for economic purposes such as guano mining (Republic of the Philippines 2007).

In the Canadian province of British Columbia, another set of suggested criteria was established to assess not only the significance of caves, but also of karst landforms in general (British Columbia 2003). Regarding caves, it is unclear how many parameters are needed to grant a cave significant status (Table 6).

There is considerable variation in the parameters taken into account when determining the importance of caves. Of the parameters listed in the protocols discussed here, only five parameters (threatened and rare species, archaeology and palaeontology, and cultural resources) stand out as occurring in all protocols.

Both the Brazilian and US regulations have seen numerous applications. The former has been routinely used to classify caves during environmental licensing for mining, hydroelectric power plants, roads, railways, electrical transmission lines and other projects (Auler & Piló 2015). More than 2000 caves had been analysed using the 6640 Federal Decree criteria by December 2016. The FCRPA has also been applied extensively, notably for management plans on US federal lands (Kovarik 2013). The large number of criteria adopted by Brazilian

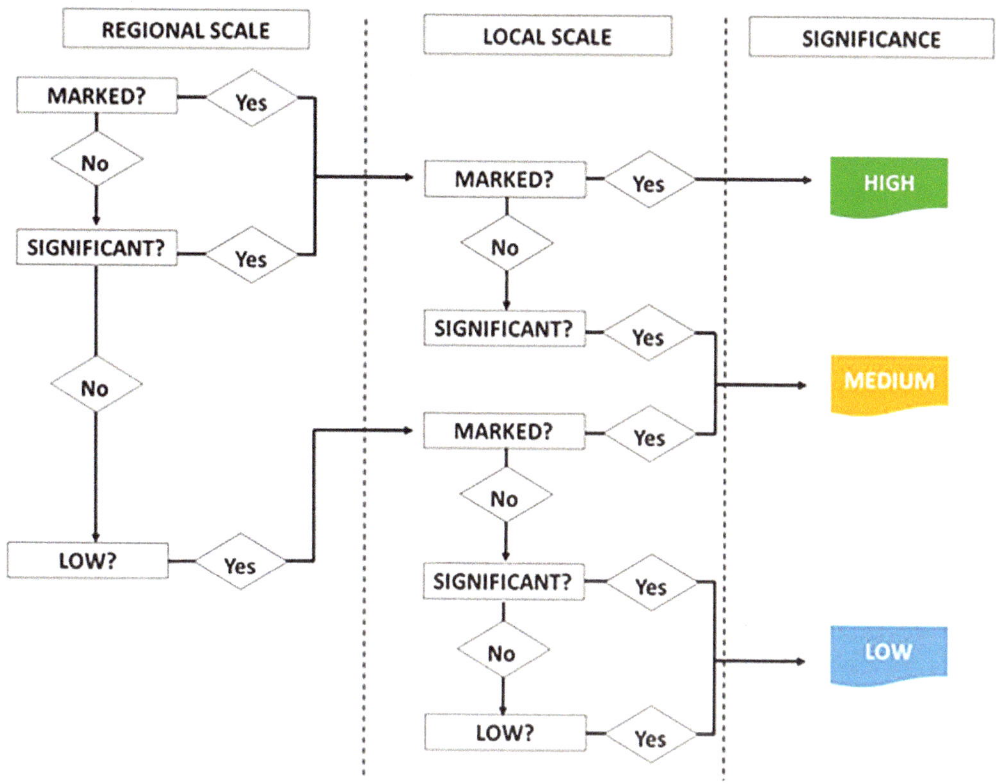

Fig. 1. Steps involved in analysing the parameters for determining cave relevance at the high, medium and low levels, based on the 6640 Decree (Brasil 2008) and associated Normative Instruction 2 (MMA 2009).

Table 4. *Criteria for determining cave significance according to the FCRPA. The cave is classified as significant if at least one of the criteria is met*

I: Provides seasonal or year-long habitat for organisms and animals
II: Contains species or subspecies of flora or fauna native to caves
III: Contains species or subspecies of flora or fauna sensitive to disturbance
IV: Contains species or subspecies of flora or fauna found on state or federal sensitive, threatened or endangered species lists
V: Contains historical properties or archaeological resources or other features included or eligible for inclusion on the National Register of Historic Places because of their research importance for history or prehistory, historical associations, or other historical or traditional significance
VI: Possesses geological or mineralogical features that are fragile, or that exhibit interesting formation processes of scientific interest or that are otherwise useful for study
VII: Possesses deposits of sediments or features useful for evaluating past events
VIII: Possesses palaeontological resources with the potential to contribute useful educational or scientific information
IX: Is part of a hydrological system or contains water important to humans, biota or the development of cave resources
X: Provides or could provide recreational opportunities or scenic values
XI: Offers opportunities for educational or scientific use
XII: Is virtually in a pristine state, lacking evidence of contemporary human disturbance or impact
XIII: Length, volume, total depth, height or similar measurements are notable
XIV: Located on lands administered by the National Park Service
XV: Located within special management areas designated wholly or in part due to the cave resources found therein

United States Government (2016*a*, *b*).

Table 5. *Criteria used for classifying caves in the Philippines*

Cave type	Criteria
Class I	Contains delicate and fragile geological formations, threatened species, and archaeological and palaeontological value
Class II	Contains areas or portions with sensitive geological, archaeological, cultural, historical and biological value or a high-quality ecosystem
Class III	No known threatened species and archaeological, geological, natural history, cultural and historical value

Republic of the Philippines (2007).

law encompasses nearly all the parameters listed by the other countries. As such, the Brazilian parameters will be further analysed to gain knowledge of the most frequent criteria and how they relate to each other in providing importance and protection to caves.

Methods and sample

A sample of 401 Brazilian caves was analysed. The caves in the sample are located in three geographical areas: the Iron Quadrangle area, the Espinhaço mountain range and the Lagoa Santa area in south-central Brazil (Fig. 2). The sample contains two different rock types. A total of 322 caves are in Proterozoic limestone of the Sete Lagoas Formation, Bambuí Group in the karst area of Lagoa Santa (Auler 2004). Seventy-nine caves are in iron-rich rocks in two different areas: the 2.5 Ga rocks of the Cauê Formation (Babinski *et al.* 1995) in the Iron Quadrangle area in south-central Brazil and the southern extension of the Espinhaço mountain range, within 1.99 Ga rocks of the Serra da Serpentina Formation (Rolim 2016). Many of these caves are of small dimensions as a result of the definition of a cave as an underground space a human can access (Brasil 2008).

The cave significance studies were performed by the environmental consultancy Carste Ciência e Meio Ambiente from 2010 to 2016 and followed the complete procedure for cave relevance assessment. All the studies adopted similar methods, following the steps dictated by Brazilian law, and were coordinated by the same team of professionals, minimizing any possible variations caused by personal and technical disparities.

Although there are four levels of cave significance and only the maximum level provides permanent protection, caves with a high significance status are also considered to have some degree of protection. This is due to inherent difficulties in destroying a high relevance cave, which normally requires two other caves of similar characteristics located in the same geological context to be permanently preserved (Auler & Piló 2015; Auler 2016). The vast majority of high relevance caves remain protected at present due to a lack of proper compensation. Therefore our sample of 401 caves contained maximum and high relevance caves. The analysis aimed to determine which parameters are crucial in providing protection to speleological sites at the maximum or high level. Parameters that occurred but did not play a part in classifying a given cave at these two levels were not included in the analysis. Some parameters can occur at the high, medium or low level (without a relationship to similar terms in the final relevance status), such as length, area, volume, species richness and relative species abundance. The high and medium levels in these parameters were combined as meaning 'presence'. The low level was assumed to mean 'absence' because it does not provide any protection to caves.

Statistical analysis was performed using R software (R Core Team 2016) via the packages psych and Itm. Correlation analyses were performed to evaluate the relationship between parameters. Relevance parameters were analysed separately according to rock type to allow for the manifestation of lithology-specific correlations. Tetrachoric

Table 6. *In British Columbia, significant caves might include the listed criteria*

Well-developed decorations
Significant hydrological, archaeological, palaeontological or cultural value
Bat hibernacula or rare cave-dwelling organisms
Significant recreational opportunities
Unique intrinsic values, such as large dimensions, unusual configurations and rare/uncommon location

British Columbia (2003).

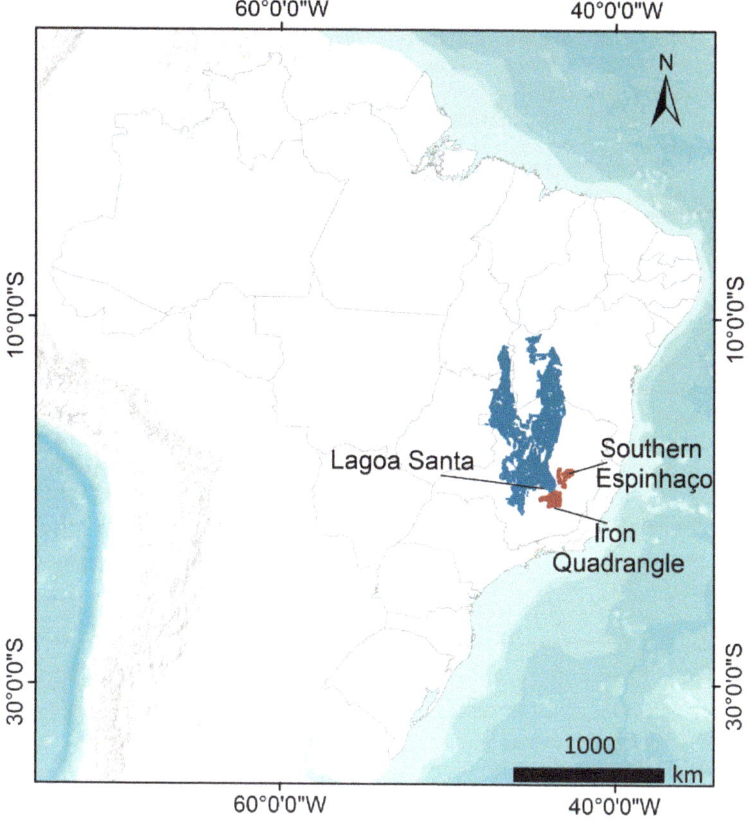

Fig. 2. Location of studied caves. Blue, limestone; red, iron-rich formations.

correlation was applied, i.e. an inferred Pearson correlation with the assumption of bivariate normality for dichotomous data. Marginal frequencies were converted to normal theory thresholds and the resulting table was converted to the (inferred) latent Pearson correlation, resulting in the observed cell frequencies with the observed marginals.

Critical analysis of the parameters

The large number of parameters was classified into four major groups: (1) absolute parameters that require measurement (AB); (2) relative parameters that require observation and/or documentation and comparison with a larger dataset (RL); (3) relative parameters that require observation and/or documentation and may be biased by subjective interpretation (RL/SB); and (4) subjective parameters (SB). The classification of parameters is shown in Tables 2 and 3.

Absolute parameters are usually related to dimensional criteria – such as cave length, depth, area and volume – and are sorted into frequency classes of high/medium/low and then compared with a larger dataset or evaluated through a simple presence/absence criterion. Thirty-two parameters were considered to be absolute parameters (*c.* 46% of the total). Dimensional parameters can be obtained through cave mapping and have some variation depending on the level of precision and the expertise of the team. In particular, the maximum criteria 'notable dimensions in length, area and volume' is currently assumed to be present when a given cave attains a value five times larger than the average of the comparative sample (at the regional and local scales). Other absolute parameters include the species richness, the relative abundance of species and the diversity of organic substrates (whether guano, vegetal debris, miscellaneous detritus, roots, carcasses, faeces of non-flying vertebrates or regurgitation pellets). These parameters require a quantitative approach.

Other absolute criteria can be assessed by a simpler presence/absence approach, such as the presence of palaeontological remains, a resident population of Chiroptera (a species living in the cave uninterruptedly for a minimum of 30 days), the presence of new taxa, the type locality, use of the cave for nesting by wild birds, the interrelation of the cave with one of maximum relevance (assessed through overlapping protection buffers) and geographical isolation (the absence of other caves in the surroundings). Absolute criteria are subject to uncertainties during sampling (e.g. the experience of the team and the duration of sampling), but tend to provide less disputable results overall.

The relative criteria (RL) are represented by 35 parameters (50% of the total). They consist of biological, geological and cultural attributes and do not require quantification. They are assessed through documentation and comparison with a larger sample at the local and regional scales. Most of these parameters carry some degree of subjectiveness caused by the need to take into consideration debatable adjectives such as notable, essential, peculiar, significant, rare and exceptional. Furthermore, particularly in the biotic attributes, uncertainties related to taxonomic limitations (Jaffé *et al.* 2016) can prevent a more precise classification. Relative criteria also benefit from a broader knowledge of cave characteristics at the regional and local scales because these are the caves used for comparison. Experience and expertise are of foremost importance; variable team skills can significantly influence the final outcome. These attributes are labelled RL and RL/SB in Tables 2 and 3. The last group of parameters is mostly subjective (SB) and therefore tends to admit variation during interpretation. This is true for attributes related to scenic and aesthetic values.

Although 70 parameters must be taken into consideration for each cave evaluation, a scrutiny of our sample of 401 caves shows that only 30 actually played a part in providing relevance (Table 7); the remaining parameters have either not been observed or do not contribute to the calculation of the significance level. We grouped similar parameters to simplify the analysis. Dimensional parameters related

Table 7. *Parameters occurring in the study sample*

Parameter	Type	Abiotic (A)/ biotic (B)
I: Notable dimensions in length, area or volume	AB	A
II: Essential shelter for threatened species	RL/SB	B
III: Essential habitat for endemic or relict troglobites	RL/SB	B
IV: Habitat for a rare troglobite	RL	B
V: Relative abundance of species	AB	B
VI: Peculiar fauna composition	RL/SB	B
VII: Troglobites that are not rare, endemic or relict	RL	B
VIII: Troglomorphic species	RL	B
IX: Population of exceptional size	RL/SB	B
X: Rare species	RL/SB	B
XI: Significant length compared with other caves within the same karst area*	AB	A
XII: Significant area compared with other caves within the same karst area*	AB	A
XIII: Significant volume compared with other caves within the same karst area*	AB	A
XIV: Presence of faunistic singularities	RL/SB	B
XV: Diversity of chemical deposition	AB	A
XVI: Notable configuration of speleothems	RL/SB	A
XVII: Influence of the cave over the karst system	RL/SB	A
XVIII: Interrelation of the cave with a maximum relevance cave	RL/SB	A
XIX: Cave used for nesting by wild birds	AB	A
XX: Diversity of organic substrates	AB	B
XXI: Presence of geological structure of scientific interest	RL/SB	A
XXII: Established populations of species with important ecological role	RL/SB	B
XXIII: Presence of chemical or clastic sedimentation of scientific interest	RL/SB	A
XXIV: Lake or underground drainage with important influence on other criteria	RL/SB	A
XXV: Significant depth compared with other caves within the same karst area*	AB	A
XXVI: New taxa	AB	B
XXVII: Species richness	AB	B
XXVIII: Resident population of Chiroptera	AB	B
XXIX: Presence of palaeontological remains	AB	A
XXX: Presence of condensation or percolation water	AB	A

Renumbered from parameters in Tables 2 and 3.
Parameters in bold are related to maximum relevance status.
*Denotes parameters removed from the statistical analyses (see text).

to both high and medium length, area, volume and depth were combined into four parameters (XI, XII, XIII and XXV;– named as 'significant' in Table 7). However, due to statistical mistakes in the Normative Instruction 2 (MMA 2009) in which high (caves with dimensions larger than the average plus the standard deviation) and medium (caves with values between the average plus the standard deviation and the average minus the standard deviation) these dimensional parameters are overrepresented, and in some cases reach 100% of the caves in the sample. As a result of these inconsistencies, they were removed from our statistical analyses. Parameters that have to be analysed at both a local and regional scale were also combined as a single parameter, with the emphasis on the regional scale.

The list of criteria is evenly distributed between abiotic (14) and biotic (16) parameters. None of the purely subjective parameters (SB in Tables 2 & 3) were present. In addition to the difficulties in objectively assessing whether a cave is aesthetically or scenically important, the sample does not include caves known for their beauty or that are comparable with other scenic caves in the same karst area. Overall, the parameters that present a subjective component were more likely to be absent, perhaps reflecting uncertainty and hesitation by the technicians. Table 8 lists the missing parameters and the possible reasons why they were not represented in our sample. The 30 parameters that played a part in assigning maximum or high levels of relevance will be the focus of further statistical analysis.

Statistical assessment

The relative frequency of the parameters varied considerably. For limestone caves (Fig. 3) biotic parameters are more prevalent. The most recurrent biotic parameters are species richness (XXVII, 78.3%), diversity of organic substrates (XX, 69.9%), relative abundance of species (V, 62.4%), troglobites that are not rare, endemic or relict (VII, 37.9%) and troglomorphic species (VIII, 36%). Among the parameters that assign maximum relevance, the most frequent is a habitat for a rare troglobite (IV, 6.5%) and an essential habitat for endemic or relict troglobites (III, 6.2%). Regarding the abiotic parameters, none

Table 8. *Parameters that are not present or do not influence cave relevance in the studied population of caves, with possible reasons for absence*

Parameter	Type	Abiotic (A)/ biotic (B)	Reason
Peculiar or rare genesis	RL/SB	A	It is unlikely that a cave in the studied sample will present a genesis that is rare or distinct
Peculiar morphology	RL/SB	A	It is unlikely that a cave in the studied sample will have a morphology not seen elsewhere
Peculiar speleothems	RL/SB	A	It is unlikely that a cave in the studied sample will have speleothems not seen elsewhere
Geographical isolation	AB	A	It is highly improbable that no other cave will exist in the surroundings of the analysed cave
Peculiar ecological interactions	RL/SB	B	Difficult to assess; depends on the experience of the collector, time taken for observation and some knowledge about the structure of the community
Significant environmental/ palaeoenvironmental importance or cave used for compensation	RL/SB	A	Few caves contain such values and no cave already used for compensation is likely to be included in the sample
Significant historical, cultural or religious value	RL/SB	A	Few caves contain such values. Usually evaluated through archaeological studies
Type locality	AB	B	No specimen was first described in those caves because the caves were unknown to science
Presence of rare speleogenetic features	RL/SB	A	It is unlikely that a cave in the studied sample will have distinct speleogenetic features
Obligatory trogloxenes	RL	B	Lack of studies on habits of trogloxenes
Aesthetic and scenic value	SB	A	Difficult to assess, caves lack parameters
Public visitation	RL/SB	A	Few caves are subject to frequent visits; difficult to assess
Use of cave by migratory species	AB	B	No migratory species has been observed in caves

Parameters shaded in light grey result in maximum relevance. Note that similar parameters were combined for simplification.

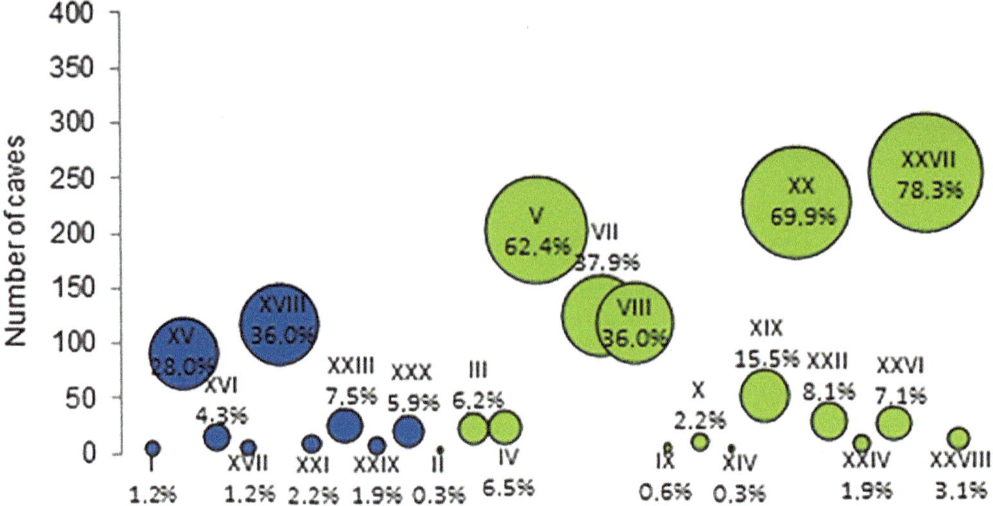

Fig. 3. Parameters (as for Table 7) occurring in 322 limestone caves. Blue, abiotic parameters; green, biotic parameters.

is present in >40% of the sample. The most frequent are the interrelation of the cave with a maximum relevance cave (XVIII, 36%) and diversity of chemical deposition (XV, 28%). The only abiotic maximum relevance parameter, notable dimensions, was present in only 1.2% of the sample. Some attributes that carry intrinsic subjectiveness (peculiar faunal composition or notable configuration of speleothems) are absent or among the least frequent in the sample of limestone caves.

To assess whether rock type plays a part in parameter manifestation, we reanalysed the parameters in the sample of caves in iron-rich rocks (Fig. 4). Biotic parameters are again dominant,

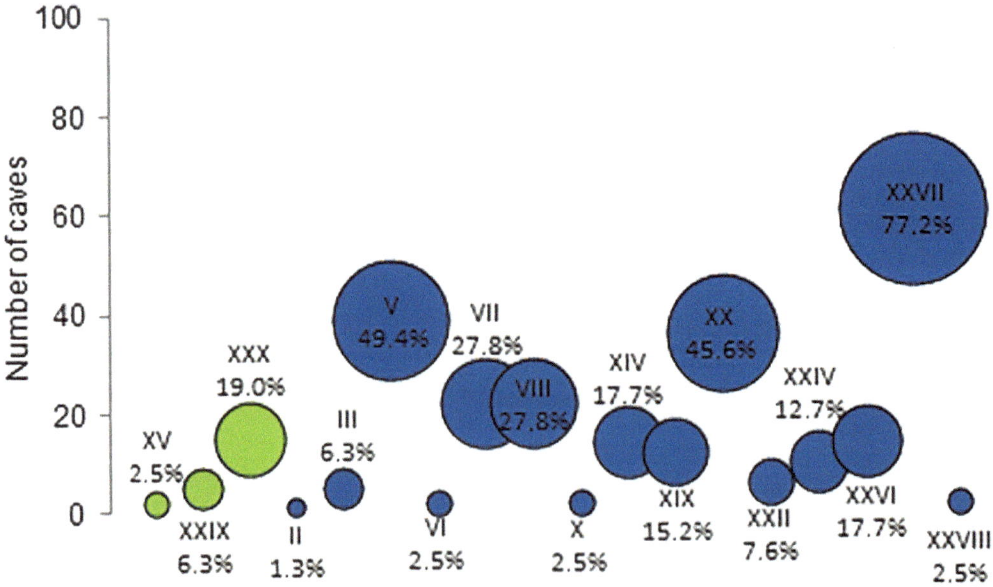

Fig. 4. Parameters (as for Table 7) occurring in 79 caves in iron-rich formations. Blue, abiotic parameters; green, biotic parameters.

although with a lower frequency. The same parameters predominate: species richness (XXVII, 77.2%), relative abundance of species (V, 49.4%) and diversity of organic substrates (XX, 45.6%). For maximum relevance, an essential habitat for endemic or relict troglobites (III, 6.3%) predominates. Abiotic parameters are much less frequent in caves in iron-rich rocks. Only three parameters occur, all at a much smaller frequency. The most common is the presence of condensation or percolation water (XXX, 19%). The lack of notable clastic and chemical deposition and the absence of hydrological features make these caves less suitable for the manifestation of abiotic parameters commonly observed in limestone caves.

We tested whether the dimensional parameters length and area correlate with other parameters. For limestone caves (Fig. 5a) the correlation between these two dimensional parameters and most parameters is positive and is stronger for biotic parameters. A resident population of Chiroptera (XXVIII) shows the highest correlation with both area and length because long and spacious caves tend to favour the existence of bat colonies. Correlation is also very strong for the presence of a lake or underground drainage. A positive correlation is also found for the habitat for a rare troglobite, as pointed out by Jaffé *et al.* (2016) for Amazonian iron ore caves. The parameters troglobites that are not rare, endemic and relict (VII), population of exceptional size (VII), presence of faunistic singularities (XIV) and established populations of species with an important ecological role (XXII) also show a marked correlation with length and area. A negative correlation

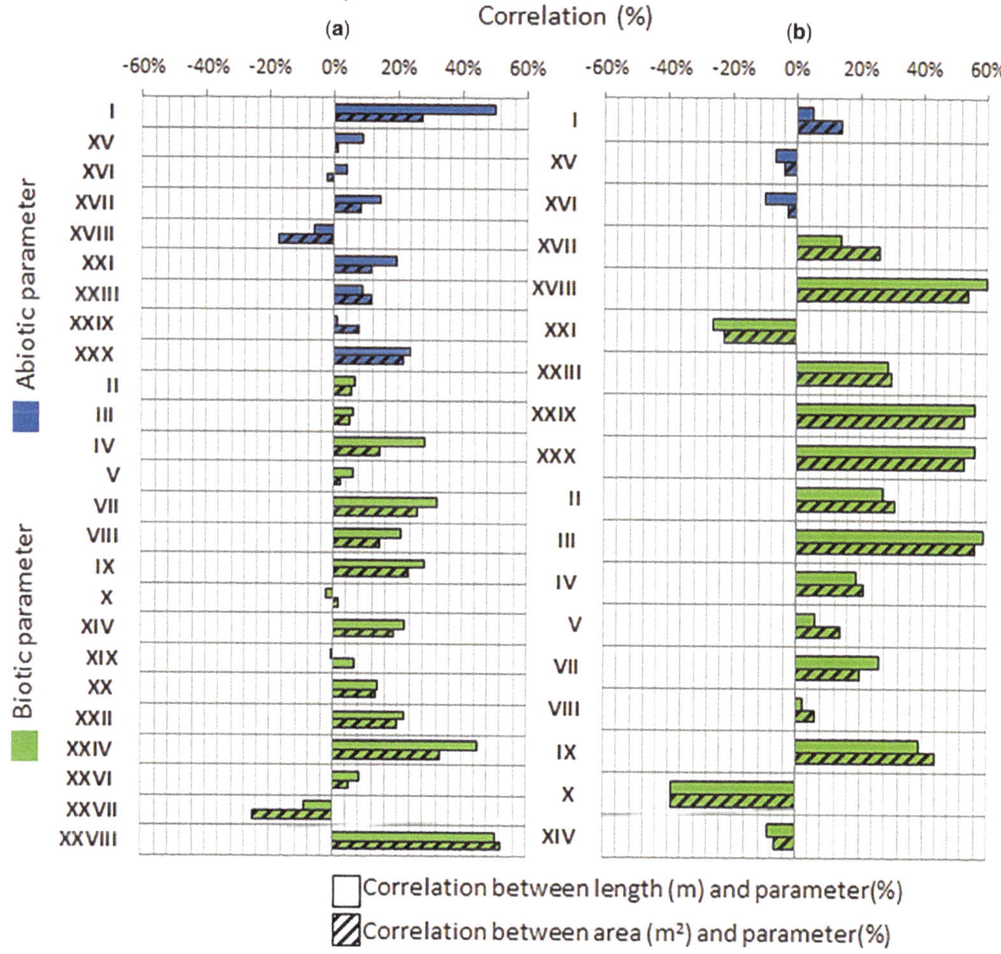

Fig. 5. Correlation between length/area and biotic (green) and abiotic (blue) parameters for caves in (**a**) limestone and (**b**) iron-rich formations.

(especially for area) occurs only for species richness (XXVII). This could be related to sampling differences (sampling smaller caves may be more efficient because they are easier to access for assessing microhabitats and larger caves require more biologists, increasing the chance of collector interference in the results).

Most abiotic parameters also correlate positively with length and area, although in a more discrete way. Among the stronger correlations, in addition to the obvious maximum relevance parameter notable dimensions (I), there is the presence of a geological structure of scientific interest (XXI) and the presence of condensation or percolation water (XXX). Only one parameter shows a clear negative correlation with length and area: the interrelation of the cave with a maximum relevance cave (XVIII). Because many caves occur in clusters, it is fairly common that the protection buffer zone of a cave will encompass other caves. This parameter does not depend on the characteristics of the cave, but on its location.

A similar pattern emerges for the smaller sample of caves in iron-rich formations, with some noteworthy differences (Fig. 5b). A higher positive correlation than for limestone caves is obtained for most biotic parameters and this especially strong for essential habitat for endemic or relict troglobites (III), as pointed out by Jaffé *et al.* (2016), the presence of faunistic similarities (XIV), troglobites that are not rare, endemic or relict (VII), troglomorphic species (VIII) and new taxa (XXVI). Our knowledge of cave fauna in iron-rich formations is still limited, increasing the chance of sampling new species. A negative correlation occurs for species richness (XXVII), as discussed for limestone caves. Unlike limestone caves, however, there is a negative correlation with the relative abundance of species (V) and a discrete negative correlation with a resident population of Chiroptera (XXVIII), contrary to the findings of Jaffé *et al.* (2016). We attribute this unexpected find to our limited sample of caves in iron-rich formations.

Tetrachoric analysis demonstrates correlations between some of the parameters in limestone caves (Table 9). There are expected high correlations between endemic and relict troglobites (III) with both rare troglobites (IV) and new taxa (XXVI). A very strong and expected correlation is also found between troglobites that are not rare, endemic or relict (VII) and troglomorphic species (VIII). Notable dimensions (I) also show a marked correlation with the presence of faunistic singularities (XIV). Tetrachoric correlation for iron formation caves are more marked for biotic parameters (Table 10). Essential shelter for threatened species (II) correlated well with habitat for a rare troglobite (IV). Essential habitat for endemic or relict troglobites (III) shows a marked correlation with rare troglobite (IV), the presence of faunistic singularities (XIV) and new taxa (XXVI), all expected correlations. The same occurs in the strong correlations between the presence of faunistic singularities (XIV) with both troglobites that are not rare, endemic and relict (VII) and troglomorphic species (VIII). A marked correlation was observed between essential shelter for threatened species (II) and diversity of chemical deposition (XV). Both parameters could benefit from larger caves. Unfortunately, this possibility cannot be tested because our sample of iron-rich caves does not contain caves that present the parameter notable dimensions in length, area or volume (I).

The relationship between cave area and the number of parameters (Fig. 6) shows that most caves in limestone and iron-rich formations display only a few parameters, on average four and two, respectively; very large caves tend to present a larger number of parameters, especially in iron-rich formations.

Discussion and final remarks

Environmental studies to determine cave importance in Brazil must consider a large set of criteria. These parameters show considerable variations in the type and mode of interpretation; some contain an important subjective component, resulting in difficulties regarding their application and replicability. Not all parameters are present and, out of 70 parameters, only 30 were found to occur in our sample of 401 caves of maximum and high relevance in limestone and iron-rich formations. A much smaller number of parameters occurs in a single cave, four predominate in limestone caves and only two in iron-rich caves. A maximum of 13 parameters occurs in a single cave. Parameters with a subjective component have a higher tendency of being absent.

Regardless the rock type, biotic parameters are the most recurrent, with species richness, diversity of organic substrates and relative abundance of species being the most frequent for both limestone and iron-rich caves. Among the abiotic parameters, the interrelation of the cave with a maximum relevance cave, diversity of chemical deposition and the presence of condensation and percolation water are the most common, although with a lower frequency.

Most parameters, biotic and abiotic, show a positive correlation with length and area, suggesting that these two parameters could provide a simpler alternative to the current complex Brazilian protocol of cave relevance, as suggested by Jaffé *et al.* (2016). This relationship holds true for both limestone and iron-rich caves, despite minor differences in a few parameters, indicating that the manifestation of parameters is largely independent of lithology.

Table 9. *Tetrachoric correlation between parameters in limestone caves*

	I (%)	II (%)	III (%)	IV (%)	VI (%)	VII (%)	VIII (%)	IX (%)	X (%)	XIV (%)	XV (%)	XVI (%)	XVII (%)	XVIII (%)	XIX (%)	XX (%)	XXI (%)	XXII (%)	XXIII (%)	XXIV (%)	XXVI (%)	XXVII (%)	XXVIII (%)	XXIX (%)	XXX (%)
I	100																								
II	−1	100																							
III	9	22	100																						
IV	20	−1	56	100																					
VI	9	4	17	18	100																				
VII	9	7	20	18	17	100																			
VIII	9	7	34	19	15	88	100																		
IX	−1	0	−2	−2	−2	10	2	100																	
X	−2	−1	14	13	3	6	11	−1	100																
XIV	50	0	22	21	4	7	7	0	−1	100															
XV	−1	−3	−13	6	3	−2	−12	−5	−5	−3	100														
XVI	−2	−1	−5	1	7	2	3	−2	−3	−1	14	100													
XVII	24	−1	−3	−3	9	9	9	−1	−2	−1	−1	−2	100												
XVIII	3	−4	−14	6	11	7	0	2	−11	−4	31	13	9	100											
XIX	−5	−2	3	−8	3	−12	−14	−3	−6	−2	2	−5	−5	−5	100										
XX	7	4	6	4	5	1	0	5	−4	4	9	11	1	20	15	100									
XXI	18	−1	−4	−4	3	6	7	26	−2	−1	0	7	37	7	−6	5	100								
XXII	17	−2	7	−3	14	14	11	27	−4	19	7	5	−3	13	6	14	3	100							
XXIII	7	−2	7	2	−27	5	6	−2	4	20	24	−6	−3	−19	−6	−7	4	−4	100						
XXIV	−2	−1	−4	6	−4	13	4	28	−2	−1	12	−3	−2	−1	−6	4	−2	13	5	100					
XXVI	−3	20	48	37	19	28	17	−2	4	−2	4	−6	−3	−3	5	8	−4	1	1	14	100				
XXVII	6	3	7	8	28	4	7	−5	−8	3	14	4	6	19	2	13	3	7	−5	−9	12	100			
XXVIII	14	31	18	10	6	16	5	21	−3	31	−3	−4	−2	−6	12	12	10	28	2	24	16	−4	100		
XXIX	−2	−1	6	6	6	13	14	−1	−2	−1	−3	−3	−2	−6	−6	4	−2	−4	14	−2	14	−4	−2	100	
XXX	−3	−1	−1	15	6	13	6	−2	−4	−1	2	−5	−3	0	0	2	−4	12	8	26	19	0	11	16	100

Parameter numbering as in Table 7.

Table 10. *Tetrachoric correlations between maximum relevance parameters in iron-rich formations*

	II (%)	III (%)	V (%)	VI (%)	VII (%)	VIII (%)	X (%)	XIV (%)	XV (%)	XIX (%)	XX (%)	XXII (%)	XXIV (%)	XXVI (%)	XXVII (%)	XXVIII (%)	XXIX (%)	XXX (%)
II	100																	
III	44	100																
V	−11	−26	100															
VI	70	62	−16	100														
VII	18	42	−5	26	100													
VIII	18	42	−5	26	100	100												
X	−2	29	0	−3	26	26	100											
XIV	24	56	−33	35	75	75	35	100										
XV	70	29	0	49	8	8	−3	14	100									
XIX	−5	3	−21	−7	29	29	−7	27	−7	100								
XX	−10	−13	37	−15	17	17	18	4	1	−10	100							
XXII	−3	12	10	−5	35	35	−5	−1	−5	28	3	100						
XXIV	−4	6	16	−6	2	2	18	2	18	−16	34	100	100					
XXVI	24	56	−6	35	30	30	35	48	14	8	−9	−1	2	100				
XXVII	6	−23	29	−10	−13	−13	−10	−30	9	−2	−17	4	12	−14	100			
XXVIII	−2	−4	0	−3	8	8	−3	−7	−3	16	1	56	−6	−7	9	100		
XXIX	−3	−7	16	−4	−16	−16	−4	−12	−4	−11	28	−7	−10	−12	−23	−4	100	
XXX	−5	−13	30	−8	−16	−16	−8	−14	−8	−11	−5	−14	1	20	19	−8	−13	100

Parameter numbering as in Table 7.

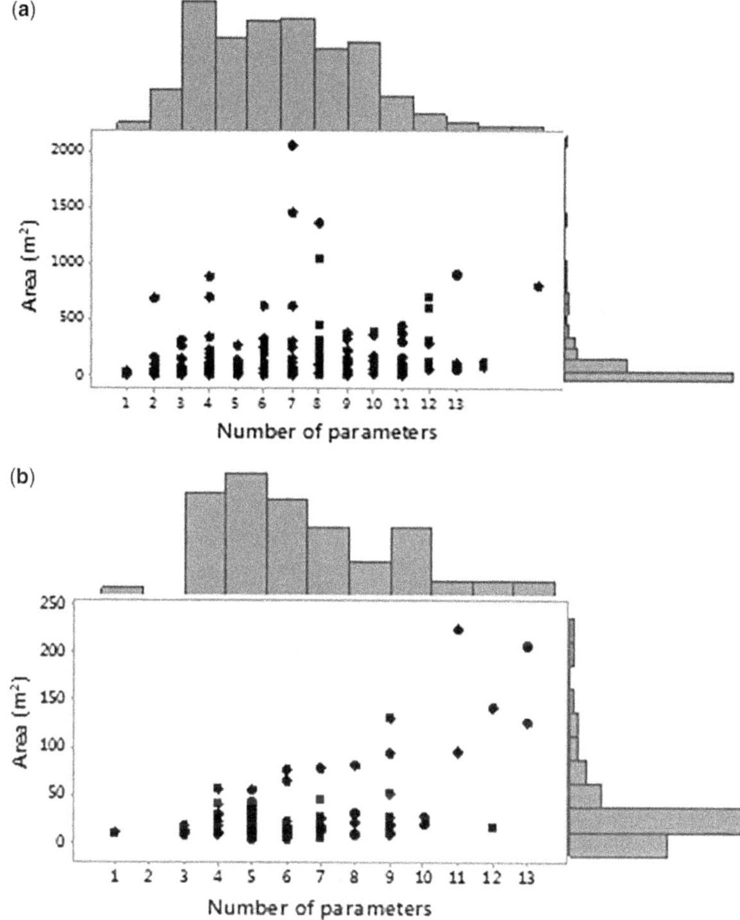

Fig. 6. Marginal plot of area v. number of parameters for (**a**) limestone caves and (**b**) caves in iron-rich formations.

There are various protocols for assessing cave relevance and those applied in Brazil and in the USA have seen numerous applications. The Brazilian legislation, with its large number of criteria, encompasses nearly all the parameters applied in other countries and thus serves as a useful basis for assessing the relative significance of the parameters. Our analysis demonstrates that only a few parameters usually occur, suggesting that a complex, subjective and excessively detailed approach may not be warranted. The present scheme, in addition to being costly and time consuming, requires the consideration of many parameters that have no practical role in defining cave significance. Debatable and subjective parameters tend to be less useful because of the bias and uncertainty in their application, casting doubts on their practicality. More well-defined and quantifiable parameters are recommended.

This review shows that the conservation of caves in Brazil is biased towards biological parameters and changes should be made to better contemplate the abiotic value of caves. The protection of caves is part of a complex political and economic scenario, of paramount importance when the cave location interferes with planned or existing spatially restrictive projects, such as mines and hydroelectric power plants. In such situations, cave relevance can be the key issue in determining the viability of a project. A well-balanced approach for cave classification is required, with a stronger emphasis on measurable and scientifically accurate parameters that can provide a robust scheme of cave importance.

We thank the mining companies for allowing us to use data from their environmental assessment reports. We acknowledge Jim Goodbar for providing information on the USA,

Lígia Saback for providing a critical review of the biological parameters, and Rafael Cruz and Lorenzza França for helping with figures. Comments by Greg Middleton, Paolo Forti and an anonymous reviewer improved the manuscript.

References

ANDREWS, P. 1990. *Owls, Caves and Fossils*. Natural History Museum, London.

ANGULO, B., MORALES, T., URIARTE, J.A. & ANTIGÜEDAD, I. 2013. Implementing a comprehensive approach for evaluating significance and disturbance in protected karst areas to guide management strategies. *Journal of Environmental Management*, **130**, 386–396.

AULER, A.S. 2004. Bambuí karst. *In*: GUNN, J (ed.) *Encyclopedia of Cave and Karst Science*. Taylor & Francis, New York, 133–135.

AULER, A.S. 2016. Cave protection as a karst conservation tool in the environmentally sensitive Lagoa Santa Karst, southeastern Brazil. *Acta Carsologica*, **45**, 131–145.

AULER, A.S. & PILÓ, L.B. 2015. Caves and mining in Brazil: the dilemma of cave preservation within a mining context. *In*: ANDREO, B., CARRASCO, F., DURÁN, J.J., JIMÉNEZ, P. & LAMOREAUX, J.W. (eds) *Hydrogeological and Environmental Investigations in Karst Systems*. Springer, Berlin, 487–496.

AULER, A.S., PILÓ, L.B. *ET AL.* 2006. U-series dating and taphonomy of Quaternary vertebrates from Brazilian caves. *Palaeogeography, Palaeoclimatology, Palaeoecology*, **240**, 508–522.

BABINSKI, M., CHEMALE, F., JR & VAN SCHMUS, W.R. 1995. The Pb/Pb ages of the Minas Supergroup carbonate rocks, Quadrilátero Ferrífero, Brazil. *Precambrian Research*, **72**, 235–245.

BRASIL 1988. Constituição da República Federativa do Brasil de 1988. Presidência da República, www.planalto.gov.br/ccivil_03/constituicao/constituicao.htm [last accessed 27 January 2014].

BRASIL 1990. Decreto Federal 99.556. Presidência da República, www.planalto.gov.br/ccivil_03/decreto/1990-1994/D99556.htm [last accessed 27 January 2014].

BRASIL 2008. Decreto Federal 6.640. Presidência da República, www.planalto.gov.br/ccivil_03/_Ato2007-2010/2008/Decreto/D6640.htm [last accessed 27 January 2014].

BRITISH COLUMBIA 2003. *Karst Management Handbook for British Columbia*. Ministry of Forests, Victoria, BC.

CALÒ, F. & PARISE, M. 2006. Evaluating the human disturbance to karst environments in southern Italy. *Acta Carsologica*, **35**, 47–56.

CHEEPTHAM, N. 2013. *Cave Microbiomes: a Novel Resource for Drug Discovery*. Springer, New York.

CULVER, D.C. & PIPAN, T. 2009. *The Biology of Caves and Other Subterranean Habitats*. Oxford University Press, New York.

CURL, R.L. 1958. A statistical theory of cave entrance evolution. *National Speleological Society Bulletin*, **20**, 9–22.

DE WAELE, J. 2009. Evaluating disturbance on Mediterranean karst areas: the example of Sardinia (Italy). *Environmental Geology*, **58**, 239–255.

DONATO, C.R., RIBEIRO, A.S. & SOUTO, L.S. 2014. A conservation status index as an auxiliary tool for the management of cave environments. *International Journal of Speleology*, **43**, 315–322.

FORD, D. & WILLIAMS, P. 2007. *Karst Hydrogeology and Geomorphology*. Wiley, Chichester.

GOLDBERG, P. & BAR-YOSEF, O. 2012. Cave dwellers in the Middle East. *In*: CULVER, D.C. & WHITE, W.B. (eds) *Encyclopedia of Caves*. Academic Press, Waltham, MA, 94–99.

GUTIÉRREZ, F., PARISE, M., DE WAELE, J. & JOURDE, H. 2014. A review on natural and human-induced geohazards and impacts in karst. *Earth-Science Reviews*, **138**, 61–88.

HILL, C.A. 1990. Sulfuric acid speleogenesis of Carlsbad Caverns and its relationship to hydrocarbons, Delaware Basin, New Mexico and Texas. *American Association of Petroleum Geologists Bulletin*, **74**, 1685–1694.

HILL, C.A. 1995. Sulfur redox reactions, native sulfur, Mississippi-Valley-type deposits and sulfuric acid karst, Delaware Basin, New Mexico and Texas. *Environmental Geology*, **25**, 16–23.

HUPPERT, G.N. 1995. Legal protection of caves in the United States. *Environmental Geology*, **26**, 121–123.

JAFFÉ, R., PROUS, X. *ET AL.* 2016. Reconciling mining with the conservation of cave biodiversity: a quantitative baseline to help establish conservation priorities. *PLoS One*, **11**, e0168348, https://doi.org/10.1371/journal.pone.0168348

KEMPE, D. 1988. *Living Underground*. Herbert Press, London.

KOVARIK, J. 2013. Geologic management of cave and karst resources on national forest system lands. *In*: LAND, L. & JOOP, M. (eds) *Proceedings of the 20th National Cave and Karst Management Symposium NCKRI Symposium*. National Cave and Karst Research Institute, Carlsbad, **3**, 135–142.

KOVARIK, J.L. & VAN BEYNEN, P.E. 2015. Application of the karst disturbance index as a raster-based model in a developing country. *Applied Geography*, **63**, 396–407.

LERA, T. 2002. Legal protection of caves and bats at the turn of the millennium, www.espeleoastur.as/Espeleolex/pdf/LEXUSA1.PDF [last accessed 14 April 2017].

LEWIS-WILLIAMS, D. 2002. *The Mind in the Cave*. Thames & Hudson, New York.

MIDDLETON, G.J. 2016*a*. Legal status of caves and karst areas. *Zeitschrift für Geomorphologie*, **60** (Suppl. 2), 319–335.

MIDDLETON, G.J. 2016*b*. Necessary qualities in legislation to protect karst areas. *Zeitschrift für Geomorphologie*, **60** (Suppl. 2), 337–352.

MMA 2009. Instrução Normativa N. 2. Ministério do Meio Ambiente, www.icmbio.gov.br/cecav/images/download/IN%2002_MMA_criterios_210809.pdf [last accessed 27 January 2014].

NORTH, L.A., VAN BEYNEN, P.E. & PARISE, M. 2009. Interregional comparison of karst disturbance: west-central Florida and southeast Italy. *Journal of Environmental Management*, **90**, 1770–1781.

PALMER, A.N. 2007. *Cave Geology*. Cave Books, Dayton.

PARISE, M. & GUNN, J. (eds) 2007. *Natural and Anthropogenic Hazards in Karst Areas: Recognition, Analysis and Mitigation*. Geological Society, London, Special Publications, **279**, http://sp.lyellcollection.org/content/279/1

Parker, C.W., Wolf, J.A., Auler, A.S., Barton, H.A. & Senko, J.M. 2013. Microbial reducibility of Fe (III) phases associated with the genesis of iron ore caves in the Iron Quadrangle, Minas Gerais, Brazil. *Minerals*, **3**, 395–411, https://doi.org/10.3390/min3040395

R Core Team 2016. *R: A Language and Environment for Statistical Computing*. R Foundation for Statistical Computing, Vienna, www.R-project.org

Republic of the Philippines 2001. National Caves and Cave Resources Management and Protection Act. Republic Act No. 9072, www.chm.ph/index.php?option=com_docman&task=doc_details&gid=17&Itemid=101 [last accessed 05 February 2017].

Republic of the Philippines 2007. Memorandum 04. Department of Environment and Natural Resources, http://bmb.gov.ph/downloads/DMC/DMC%202007-04.pdf [last accessed 12 November 2017].

Rolim, V.K. 2016. *As formações ferríferas da região de Conceição do Mato Dentro – MG: Posicionamento estratigráfico, evolução tectônica, geocronologia, características geoquímicas e gênese dos minérios*. PhD thesis, Universidade Federal de Minas Gerais.

Romero, A. 2009. *Cave Biology. Life in Darkness*. Cambridge University Press, Cambridge.

Schubert, B.W. & Mead, J.I. 2012. Paleontology of caves. *In*: Culver, D.C. & White, W.B. (eds). *Encyclopedia of Caves*. Academic Press, Waltham, MA, 590–598.

United States Congress 1988. *Federal Cave Resources Protection Act 1988 (FCRPA)* **16 U.S.C. 4301–4310**.

United States Government 2016*a*. Code of Federal Regulations 290. Cave Resources Management. *In*: *Code of Federal Regulations Title 36*. US Government Publishing Office, Washington, DC, www.ecfr.gov/cgi-bin/text-idx?SID=5a9055cbca86bd3eae706b0306150501&mc=true&node=pt36.2.290&rgn=div5 [last accessed 5 February 2017].

United States Government. 2016*b*. Code of Federal Regulations 37. Cave Resources Management. *In*: *Code of Federal Regulations Title 43*. US Government Publishing Office, Washington, DC, www.ecfr.gov/cgi-bin/text-idx?SID=5a9055cbca86bd3eae706b0306150501&mc=true&node=pt43.1.37&rgn=div5 [last accessed 5 February 2017].

van Aken, M., Harley, G.L., Dickens, J.F., Polk, J.S. & North, L.A. 2014. A GIS-based modelling approach to predicting cave disturbance in karst landscapes: a case study from west-central Florida. *Physical Geography*, **35**, 123–133.

van Beynen, P. & Bialkowska-Jelinska, E. 2012. Human disturbance of the Waitomo catchment, New Zealand. *Journal of Environmental Management*, **108**, 130–140.

van Beynen, P. & Townsend, K. 2005. A disturbance index for karst environments. *Environmental Management*, **36**, 101–116.

van Beynen, P., Feliciano, N., North, L. & Townsend, K. 2007. Application of a karst disturbance index in Hillsborough County, Florida. *Environmental Management*, **39**, 261–277.

van Beynen, P., Brinkmann, R. & van Beynen, K. 2012. A sustainability index for karst environments. *Journal of Caves and Karst Studies*, **74**, 221–234.

What will be the future of the giant gypsum crystals of Naica mine?

PAOLO FORTI

La Venta Esplorazioni Geografiche & Italian Institute of Speleology, via Zamboni 67, 40 126 Bologna, Italy

paolo.forti@unibo.it

Abstract: The mine caves of Naica (Chihuahua, Mexico) are famous because they host large gypsum crystals. Mine works intersected new caves hosting the largest crystals in the world in the year 2000. From 2006 these caves became the object of a multidisciplinary research project with the goal of inferring their ages, the boundary conditions for their formation and the mechanisms inducing their development. Several other scientific aspects were also considered, including palynology, mineralogy, microbiology, physiology, hydrogeology and astrobiology. From 2006 to 2009, scientists and explorers tried to ensure the complete documentation of these natural wonders because they were expected to be accessible for only a few years. As a result of their location *c.* 160 m below the natural groundwater level, they were predicted to be flooded with thermal water as soon as dewatering of the mine ceased. This occurred at the end of 2015, so that the lower part of the mine is already submerged and in the near future the giant crystal caves will also disappear. Theoretically, it is still possible to maintain these incredible wonders for future generations, but this seems highly unlikely. Soon the crystals will be submerged below *c.* 150 m of hot water, restarting their incredible slow growth.

The caves of Naica (Chihuahua, Mexico) are perhaps the most famous mine caves in the world due to the presence of gigantic gypsum crystals. The first reports of these cavities date back to the early twentieth century (Degoutin 1912; Foshag 1927) and related to the first cave (Cueva de las Espadas, Sword Cave) intersected by mine works *c.* 120 m below the Naica mine entrance. The largest crystals in this cave reached 2 m in length and were the largest in the world at that time (Fig. 1). Unfortunately, most of the crystals were extracted and donated to the principal natural museums of the world.

The mine galleries at the −290 level intercepted several natural cavities in the year 2000, the most important of which were Cueva de los Cristales (Crystal Cave) (London 2003), Ojo de la Reina (Queen's Cave) and Cueva de las Velas (Sails Cave). A few years later the Cueva del Tiburon (Shark Cave) was also intercepted (Badino *et al.* 2009). All these caves host gypsum crystals much larger that those in the Cueva de las Espadas. The largest of these crystals, *c.* 13 m in length, is hosted in Cueva de los Cristales (London 2003). The +50 cave was explored in 2006 (Badino *et al.* 2009) and a borehole for mine ventilation crossed the Palacios Cave in 2009 (Beverly & Forti 2010).

Little research was carried out on the Naica caves until 2005, mainly because it was difficult to obtain permission to perform scientific research inside an active mine owned by a private company (the Peñoles Company, Mexico City). Speleoresearch & Films of Mexico City and La Venta Esplorazioni Geografiche (Italy) (Bernabei *et al.* 2009) were allowed to organize a multidisciplinary international research team at the end of 2005 (Fig. 2) to infer not only the genesis and age of the Naica gypsum crystals, but also to explore other important scientific aspects, from palynology (Garofalo *et al.* 2010) to mineralogy (Forti *et al.* 2009), micrometeorology (Badino 2009) to physiology (Giovine *et al.* 2009) and hydrogeology (Gázquez *et al.* 2012) to astrobiology (Gázquez *et al.* 2013*c*). The aim of the research was to ensure the complete documentation of the crystals and to gain a better knowledge of these natural wonders, which will not be accessible at the end of exploitation of the mine. The Naica Project continued to the end of 2009 (Forti & Sanna 2010; Gázquez *et al.* 2016) when it was suddenly interrupted before its planned end date by the decision of the mine owner.

A mine gallery intercepted a huge hydrothermal vein in 2015. This flooded the lower part of the mine, which was later abandoned. Over the next few months the caves with the giant crystals were lost under *c.* 160 m of hot water.

This paper presents a short geological outline of the area, followed by the main results obtained by the multidisciplinary studies performed within the framework of the Naica Project. A possible solution to allow future generations to see the giant crystals of Naica is presented.

Geological setting

The entrance to Naica mine lies at 1385 m a.s.l. on the southern side of the Sierra de Naica, an anticline structure consisting of carbonate rocks of

From: Parise, M., Gabrovsek, F., Kaufmann, G. & Ravbar, N. (eds) 2018. *Advances in Karst Research: Theory, Fieldwork and Applications*. Geological Society, London, Special Publications, **466**, 461–475.
First published online November 6, 2017, https://doi.org/10.1144/SP466.4

Fig. 1. Cueva de las Espadas: a general view of the *espadas*, the large prismatic gypsum crystals along the main cave corridor (after Degoutin 1912).

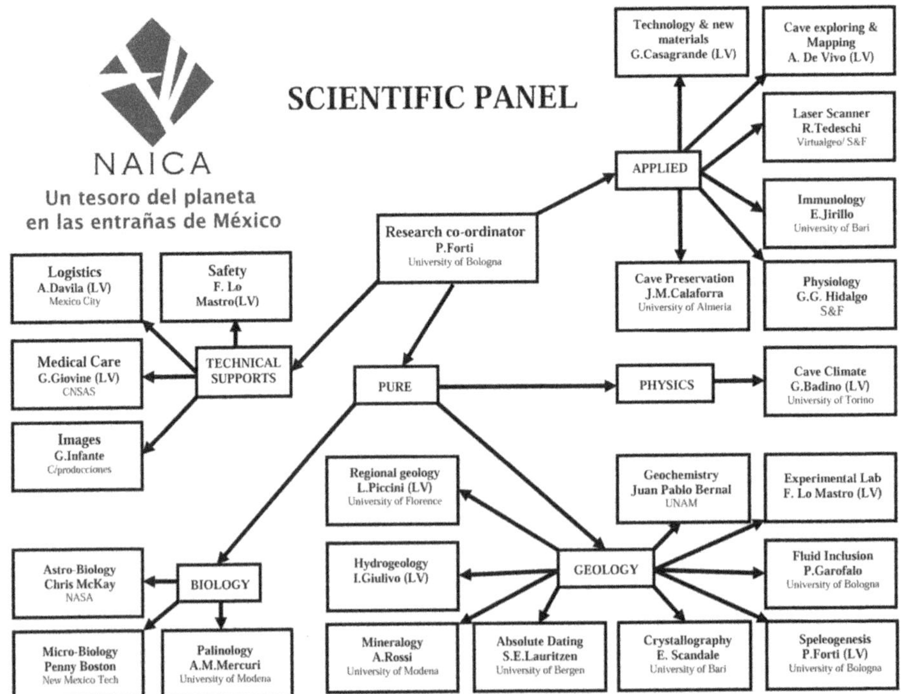

Fig. 2. Structure of Project Naica (after Bernabei *et al.* 2009).

Albian–Cenomanian age extending NE to SW (Fig. 3). The geological sequence consists of limestone and dolostone with interbedded clays and silts (Franco-Rubio 1978). The polysulphide (Pb, Zn and Ag) ore bodies were deposited by hydrothermal flows (Erwood *et al.* 1979). The mineral deposits (consisting mainly of pyrite, pirrotine, sphalerite, galena and chalcopyrite) occur in chimney and interbedded bodies developed within dykes and hosting carbonate formations. The latter are consequently strongly altered and partially transformed into metasomatic calc-silicates. Intrusive magmatic activity during the Tertiary is evidenced by felsic dykes emplaced in the carbonate series, recently dated to 30.2 myr BP (Alva-Valdivia *et al.* 2003).

The contact between the groundwater and these igneous bodies created a hydrothermal system containing brines, which flowed along the lines of weakness, following the alignment of the dykes and faults. In this hypogenic system, the brines interacted with felsic dykes and limestone, giving rise to new minerals (Megaw *et al.* 1988).

The gradual cooling of the aquifer water resulted in the precipitation of low-temperature hydrothermal minerals. The mineral paragenesis consists of pyrite, pyrrhotite, sphalerite, galena and chalcopyrite, all formed in the hypersaline high temperature brines (Erwood *et al.* 1979). During colder stages, minerals such as quartz, calcite, aragonite, anhydrite and, eventually, gypsum precipitated (Erwood *et al.* 1979; Forti 2010).

Speleogenesis

The Naica caves are probably one of the best examples of hypogenic speleogenesis in the world (Klimchouk 2009; Palmer 2011). Their genesis and evolution were almost completely controlled by the ascending flow of thermal water, which was, in turn, closely related to the main regional tectonic structures: the Naica Fault and the Montaña Fault (Fig. 3) (Forti 2010). About 1–2 myr ago, tectonic stresses partially displaced the ore bodies, giving rise to open joints and fractures, which still control the ascending flow of thermal fluids. The high temperature thermal fluids responsible for the evolution of the ore bodies were always characterized by net deposition and therefore the permeability of the host formation was reduced and no karst voids developed. A short time later, a progressive decrease in temperature made the thermal waters aggressive with respect to carbonate formation, so that cavities developed at different levels inside the aquifer (Fig. 3). This process, characterized by net corrosion, lasted for a relatively short time interval, as testified by the clear tectonic shape maintained by all the known Naica caves (Forti 2010).

The evolutionary stages of the caves were complex and related to different speleogenetic mechanisms and processes (corrosion, double exchange, acid aggression, CO_2 diffusion and condensation corrosion), most of which led to the incongruent dissolution of the carbonate formations and the deposition of gypsum and other minerals. Although they were always controlled by the presence of the thermal aquifer, the resulting evolution was different from cave to cave and was related to deep phreatic, epiphreatic and vadose environments (Table 1).

The scarcity of corrosional features did not allow a detailed reconstruction of cave evolution, which was only partially achieved by a detailed analysis of the thick chemical deposits hosted inside each cave (Fig. 4). From this point of view the most interesting cave is Cueva de las Espadas, the evolution of which was characterized by several changes between these three hydrogeological environments through time (Gázquez *et al.* 2012), whereas the deeper caves (Cristales, Ojo de la Reina, Velas and Tiburon) suddenly changed from deep phreatic to vadose *c.* 30 years ago due to mine dewatering (Bernabei *et al.* 2007). The Palacios and the +50 caves are the only caves in which the deposition of phreatic gypsum never occurred. This is because the thermal water left them before they cooled down below the gypsum–anhydrite equilibrium temperature.

The complete dewatering of the caves did not represent the end of cave development, which was characterized by a later active stage in all the cavities. The dewatering of the caves and subsequent exposure to air greatly affected their development, giving rise to the evolution of several new diagenetic minerals, but also resulting in condensation, corrosion and dissolution processes. In a few years these mechanisms were responsible for damage (due to partial redissolution) to the giant gypsum crystals, which were also partially transformed into calcite speleothems (Fig. 5). The +50 cave is the only cave in which a clear epigenic evolution occurred, giving rise to common gravitative speleothems (Forti *et al.* 2009).

Genesis and evolution of the giant gypsum crystals

The first research target was to explain the development of the giant gypsum crystals. It was proved that their evolution was possible because of the extreme temperature stability within the thermal aquifer; heat transfer was induced only by convective movement due to the absence of any natural thermal spring around Naica. However, although necessary, thermal stability is not sufficient to allow giant crystals to develop. It was demonstrated that their genesis was controlled by a completely new genetic

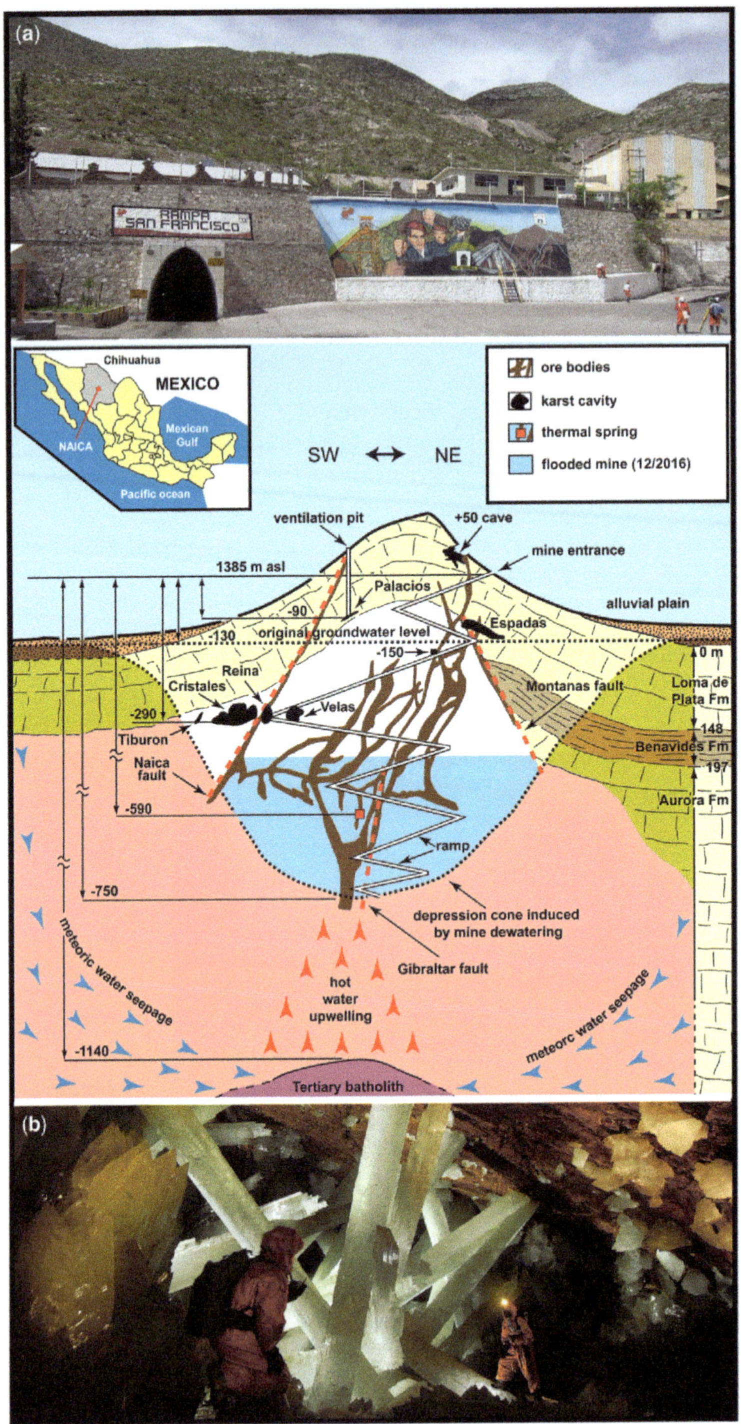

Fig. 3. Index map and sketch of the Naica mine. The largest natural cavities are closely related to the main faults. On the right is the carbonate sequence of the Naica ridge: darker colours relate to flooded formations (after Forti 2010, modified). (**a**) The mine entrance (from the La Venta Esplorazioni Geografiche Archive); (**b**) general view of the giant crystals in the Cueva de los Cristales (from the La Venta Esplorazioni Geografiche Archive).

Table 1. *Environments, rate of temperature decrease and operational range, active speleogenetic processes and related cave deposits*

Environment	Temperature (°C) decrease range		Processes	Chemical deposits	Reference
Vadose	Fast	20–25	Mineralization of organic matter	Organic mineral and phosphate deposition	Forti *et al.* (2009)
		30–20	CO_2 diffusion	Calcite speleothem evolution	Gázquez *et al.* (2012)
		54–25	Acid aggression	Gypsum, metallic sulphate and silicate deposition	Forti *et al.* (2009)
		54–25	Inorganic oxidation	Oxide–hydroxide deposition	Badino *et al.* (2011)
		54–40	Aerosol/vapour deposition	Gypsum needles and flowers	Beverly & Forti (2010)
		54	Capillary uplift and evaporation	Development of gypsum 'sails', sulphate and halide deposition	Bernabei *et al.* (2007)
Epiphreatic	Slow	54–40	CO_2 diffusion	Aragonite deposition	Forti *et al.* (2008)
Deep phreatic	Very slow	57–54	Gypsum–anhydrite disequilibrium	Giant gypsum crystals and celestine development	García-Ruiz *et al.* (2007)
		60–57	Organic oxidation	Oxide–hydroxide deposition	Forti *et al.* (2007)

mechanism: the anhydrite–gypsum solubility disequilibrium, which is still restricted to the Naica environment (García-Ruiz *et al.* 2007).

Anhydrite is less soluble than gypsum at temperatures >58°C and more soluble at lower temperatures. Therefore gypsum is forced to precipitate from a solution saturated with respect to anhydrite at temperatures <58°C, whereas anhydrite lenses, dispersed throughout the carbonate sequence, are dissolved. The combination of these two opposite processes causes a slight supersaturation to be maintained with respect to gypsum (Fig. 6). The extremely low supersaturation induced by this process, avoiding new nucleation, explains why only a few giant gypsum crystals grew rather than a great number of smaller crystals (García-Ruiz *et al.* 2007; Van Driessche *et al.* 2011).

The anhydrite–gypsum solubility control is indirectly corroborated by the presence of small celestine crystals just below, or even inside, the gypsum crystals because the structure of the dissolving anhydrite is able to host more strontium than the growing gypsum (Gázquez *et al.* 2013*a*). This mechanism explains why gypsum deposition started earlier in Cueva de las Espadas, located 160 m higher and close to the water table; the decrease in temperature was faster inside this cave than in caves at −290 m, which allowed earlier deposition. No phreatic gypsum was deposited inside Cueva Palacios, <30 m higher than Cueva de las Espadas, except for thin, clearly vadose needles and helictites, because the groundwater level was below that of Palacios, but above Espadas, when phreatic gypsum deposition started.

Age of the crystals

To evaluate the time interval over which the giant crystals developed, a few samples of already broken crystals were taken from Cueva de los Cristales, Ojo de La Reina and Cueva de las Espadas to be dated by the $^{230}Th/^{234}U$ method (Sanna *et al.* 2010, 2011). A few samples were analysed using thermal ionization mass spectrometry on a Finnigan 262RPQ instrument (Lauritzen *et al.* 2008), but this dating methodology proved to be inefficient due to the low amount of uranium trapped within the gypsum. Only one sample, taken *c.* 50 mm below the outer surface of a prismatic crystal from Cueva de los Cristales, gave a good result (34.544 ± 0.819 kyr).

To improve the U/Th dating, multi-collector inductively coupled plasma mass spectrometry and a new extraction chromatographic method were used (Sanna *et al.* 2010). The results achieved were satisfactory (Table 2), although the analysed samples displayed a low uranium concentration and high background thorium level. The ages obtained were

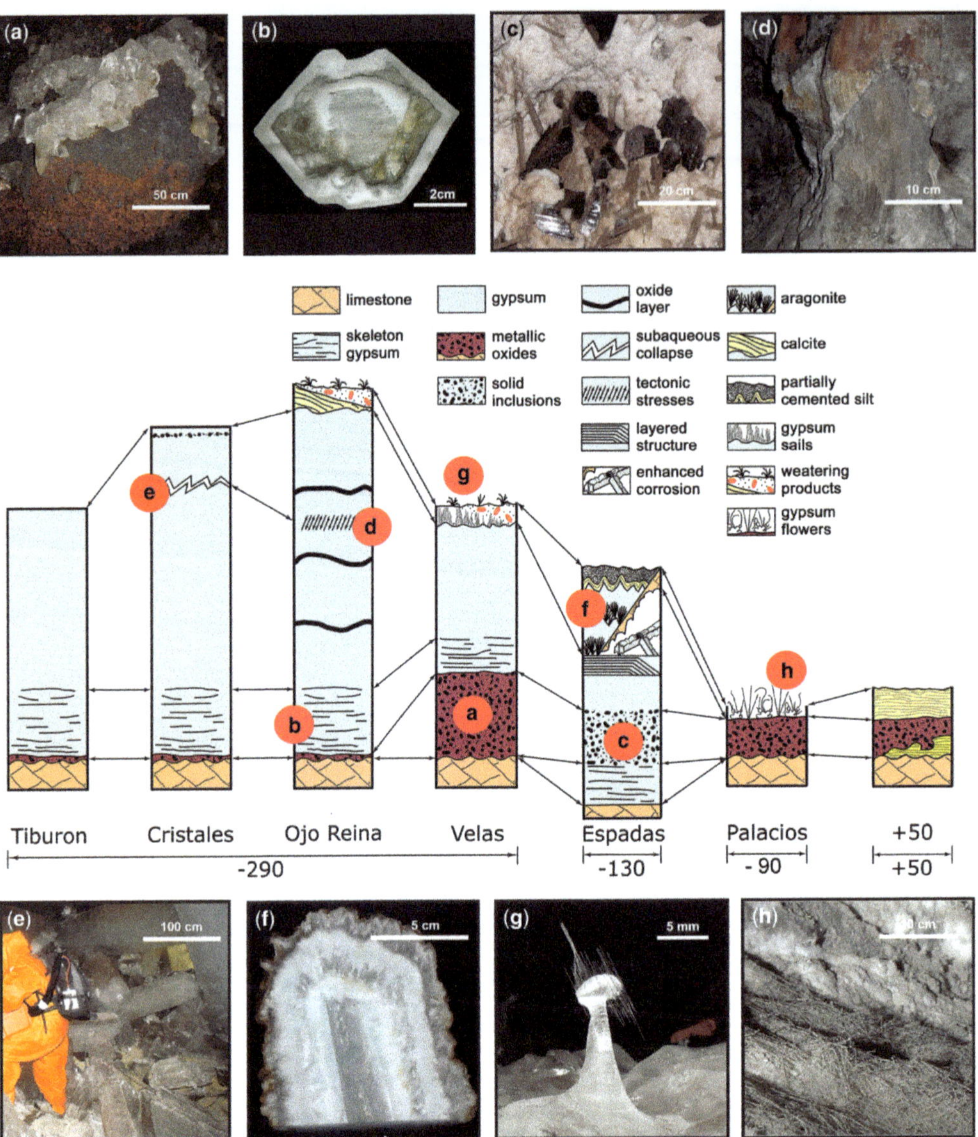

Fig. 4. Stratigraphy of the chemical deposits of the Naica caves. Depositional sequences vary from cave to cave, but many of the processes are similar, so it was possible to correlate most of them (modified from Forti 2010). (**a**) Cueva de las Velas: thick metallic oxide–hydroxide deposits over which the gypsum started to grow. (**b**) Cueva de las Espadas: polished section of a gypsum crystal showing skeleton structures in its core area. (**c**) Cueva de las Espadas: black gypsum crystals due to the presence of many solid inclusions. (**d**) Ojo de la Reina, within the red circle: micro-grooves induced by tectonic stresses along the main exfoliation plane of a giant crystal. (**e**) Cueva de los Cristales: earthquake-induced collapse of large crystals. (**f**) Cueva de las Espadas: section of a stalagmite showing alternation of gypsum and aragonite and the final calcite crust. (**g**) Cueva de las Velas: a 'sail' developed during the short interval of cave dewatering. (**h**) Cueva Palacios: gypsum needles deposited by the upwelling of hot moist air when the groundwater was just in between this cave and Cueva de las Espadas (images from La Venta Esplorazioni Geografiche Archive).

within reliable ranges. The first sample was taken from the inner core (*c.* 13 cm from the outer surface) of a broken prismatic crystal in Cueva de los Cristales and gave an age of 158 ± 75 kyr. The second sample was taken at the entrance of the Ojo de la Reina cave, which is only a few tens of metres

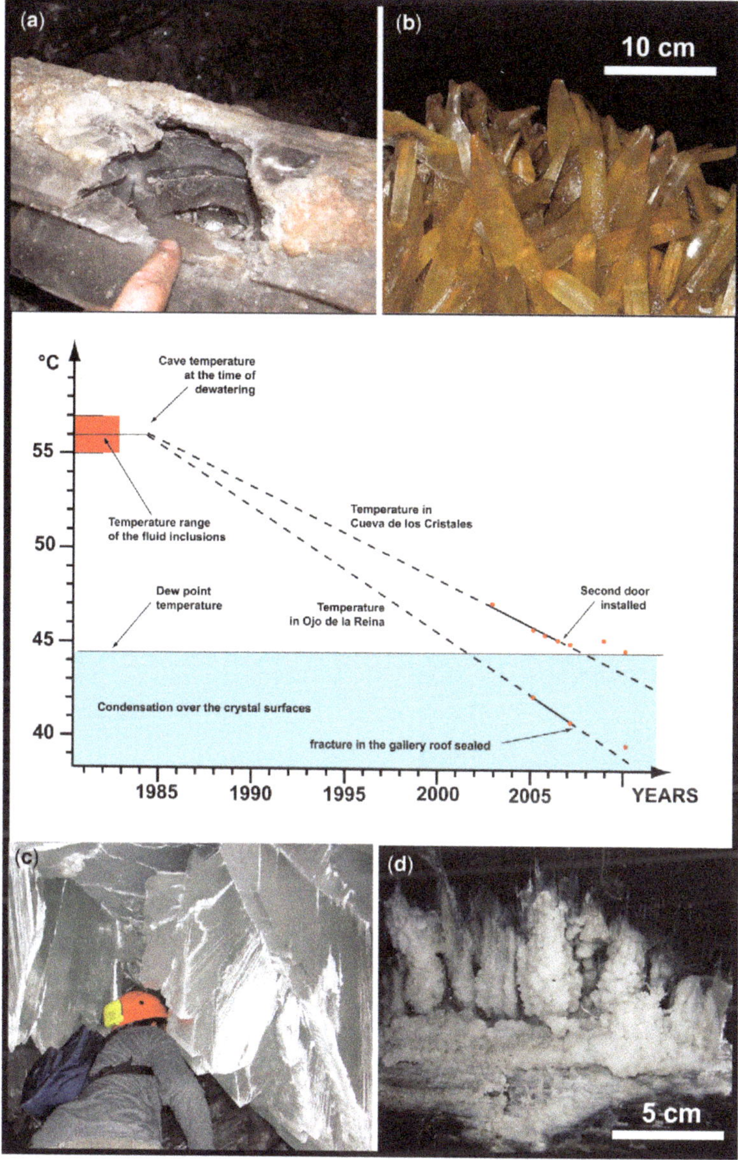

Fig. 5. Temperature evolution in Cueva de los Cristales and Ojo de la Reina after dewatering. Condensation over the gypsum crystals occurred in Ojo before 2005 (Badino 2009), while in Cristales the dew point was not reached until the end of the Naica Project in 2009 (modified after Badino *et al.* 2011). (**a**) Cueva de las Espadas: large dissolution cup over a large gypsum crystal. (**b**) Cueva de las Espadas: hook-shaped gypsum crystals after strong condensation processes. (**c**) In the Ojo de la Reina cave, condensation partially transformed the gypsum megacrystals into white calcite crusts, but where the condensation rate was fast enough, calcite coralloids (**d**) were formed over the gypsum crystal faces (photos from La Venta Esplorazioni Geografiche Archive).

away from Cueva de los Cristales. The sample was extracted from the base of a crystal a few centimetres from the contact with the limestone rock of the cave floor and the resulting age was 191 ± 13 kyr (Sanna *et al.* 2011). Based on its location with respect to the host rock, the data from this sample should represent the beginning of gypsum deposition within the caves at the −290 level of the Naica mine.

To confirm the validity of this assumption, an experimental unit was built and placed in one of

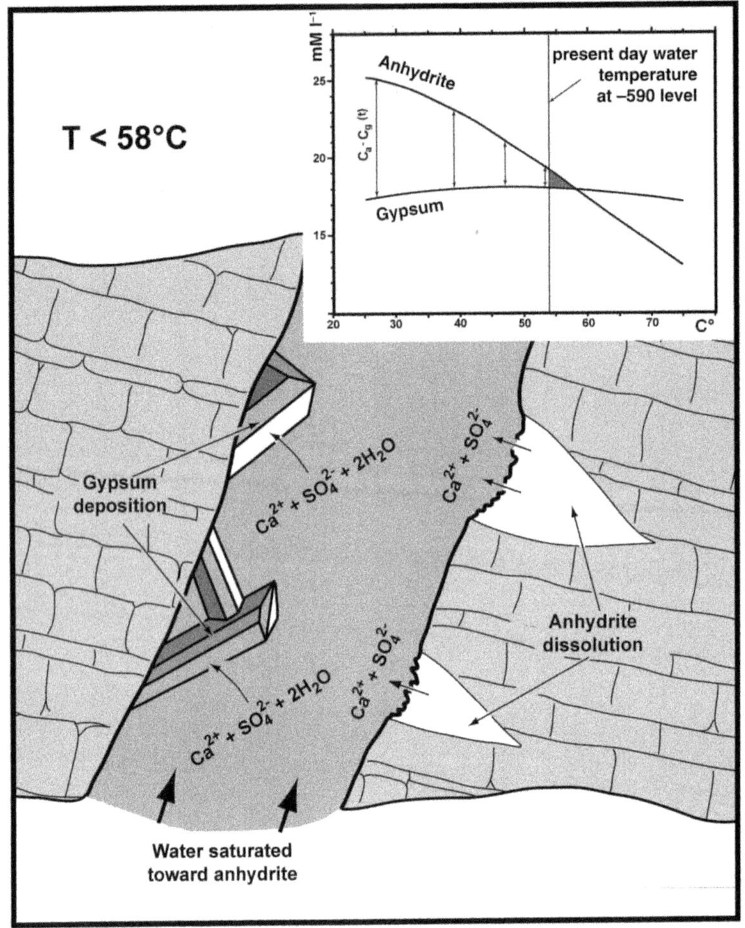

Fig. 6. Chemical reaction inducing the development of the giant crystals controlled by the gypsum–anhydrite solubility disequilibrium (after Forti *et al.* 2009, modified) and gypsum–anhydrite solubility diagram (inset), with the temperature range in the caves of Naica at the time when gypsum crystals were growing indicated by shading (modified from García-Ruiz *et al.* 2007).

the thermal springs (Fig. 3) at the −590 level, where the thermal water, not yet contaminated, spills out from the ceiling of a mine gallery (Forti & Lo Mastro 2008). The single unavoidable difference between the present day water and the water feeding the crystals in the caves until 20 years ago was its temperature, which decreased to 51 °C. The device reproducing the cave conditions for gypsum deposition

Table 2. *Ages of Naica crystals*

Sample ID	Mineral	Cave	Age (ka)	2σ+	2σ−
N01-1	Gypsum	Ojo	191.018	13.75	12.5
N1	Gypsum	Cristales	34.544	0.82	0.81
N07-10	Gypsum	Cristales	158.526	101.64	51.96
ESP1-1	Gypsum	Espadas	57.01	1.77	1.77
ESP1-2	Aragonite	Espadas	14.491	4.15	4.03
ESP-surf	Calcite	Espadas	7.863	0.04	0.04

Modified from Sanna *et al.* (2011).

was installed at the end of 2006 and the experiment was successful. Euhedral gypsum monocrystals started to grow and the first available data covering an interval of 480 days showed a good correlation between crystal growth and time (Forti *et al.* 2009). The resulting average growth (corrected for the temperature factor) was 0.004 ± 0.0002 mm a^{-1}. The experimentally measured growth rate gives an extrapolated age for the largest crystals of 250 ka, a value extremely close to that obtained with the U/Th method for the crystals of the Ojo de la Reina. Considering the experimental errors characterizing the independent methods of dating, the agreement between the two values can be considered to be very good.

Using recent data on the dissolution kinetics of crystalline gypsum and assuming that the dissolution and precipitation kinetics are symmetrical around the saturation point (i.e. no nucleation threshold), the growth rate of 0.004 mm a^{-1} converts to a supersaturation of about 1.005. This value corresponds to a nucleation probability (García-Ruiz *et al.* 2007; Van Driessche *et al.* 2011) of less than one in a million, which is reasonable for the growth of 100 giant crystals (the total number of giant crystals in the caves of Naica is 120–130) over a period of 200–300 ka.

Ten years of research

Much other scientific research was carried out in addition to the study of the genesis and evolution of the giant gypsum crystals, mainly in the framework of the Naica Project (Forti & Sanna 2010; Gázquez *et al.* 2016), the most important of which are reported in the following sections.

Mineralogy

Despite the fact that most of their walls, roofs and floors are covered by gypsum, the Naica caves proved to be among the most interesting minerogenetic environments in the world (Forti *et al.* 2007, 2009). Over 40 different minerals (Table 3), ten of which are novel to cave environments, have been found and several different minerogenetic mechanisms (Table 1) were simultaneously active inside the caves (Forti *et al.* 2009).

Apart from the +50 cave, which was drained hundreds of thousands of years ago, epiphreatic and/or aerobic conditions were reached only recently in the other caves of Naica (a few hundred years for Cueva de las Espadas and *c.* 30 years for the caves of the −290 level). Nevertheless, it is evident that the presence of an atmosphere has been crucial for the development of the majority of cave minerals in Naica. Only four of a total of 40 minerals (coronadite, opal, quartz and hectorite) are restricted to the deep phreatic environment, whereas five other minerals (gypsum, goethite, celestine, calcite and dolomite), although in different quantities, are present in all the environments, including the deep phreatic environment. Four minerals (aragonite, apatite, fraipontite and a still-undetermined silicate) developed in both the epiphreatic and vadose environments, whereas more than two-thirds of the cave minerals of Naica (28 of 40 minerals) developed only in the vadose zone. The abundance of active minerogenetic processes within this vadose environment (six processes, while in the other two environments there were only three processes; see Table 1) is confirmed by the presence of ten new cave minerals, among which only one (hectorite) grew under phreatic conditions while the other nine (antlerite, copper pentahydrite, orientite, plumbojarosite, starkeyite, szmikite, szmolnokite, woodruffite and a still undetermined Mg–Zn silicate) developed in direct contact with the cave atmosphere.

Palaeoclimatology

Fluid inclusions trapped in the gypsum crystals (Fig. 7a) allowed the definition of the salinity and temperature of the feeding water at the time of their deposition, thus giving an insight into the palaeoenvironment and palaeoclimate of the Naica region (Garofalo *et al.* 2010).

To evaluate the climatic evolution of the Naica region at the time of crystal growth, the pollen composition of cave crystals and sediments was compared with pollen found in the desert soils of Naica. The best data were provided by the analyses of the pollens (Fig. 7b) trapped within the crystal structures (Garofalo *et al.* 2007). This was the first time that well-preserved pollen grains had been found inside euhedral crystals (Holden 2008).

A total of 52 pollen grains were extracted from two gypsum samples of Las Espadas and Los Cristales caves. Despite the low abundance, many pollen taxa of woody plants (*Pinus*, Pinaceae undiff., *Quercus*, *Q.* cf. *garryana*, *Lithocarpus densiflora* cf., *Juniperus* type, *Taxus*), herbs (*Artemisia*, *Plantago*, Poaceaerated) and spores (*Lycopodium*) were present within the crystals. Interestingly, *Quercus* and *Juniperus* were found in both cave crystals, which suggests similar climatic conditions in the outside environment at the time of crystal growth. Even considering the limitation of the dataset, this floral assemblage indicates a homogenous catchment basin, which is in contrast with the actual desert–scrub communities and desert grassland of the Chihuahua Desert. This assemblage is similar to the mixed broadleaf wet forests present in the southwestern USA, characterized by pine–oak woodland and pine forest. Significant amounts of *Quercus*

Table 3. *Minerals observed in the Naica caves*

Mineral No.	Cave	Mineral	Chemical formula	Environment
1	7	Anglesite	$PbSO_4$	Vadose
2	6	Anhydrite	$CaSO_4$	Vadose
3	4	Antlerite*	$Cu_3(SO_4)(OH)_4$	Vadose
4	3	Apatite	$Ca_5(PO_4)_3(C,F,OH,Cl,O)$	Vadose–epiphreatic
5	3	Aragonite	$CaCO_3$	Vadose–epiphreatic
6	4	Azurite	$Cu_3(CO_3)_2(OH)_2$	Vadose
7	6	Bassanite	$CaSO_4 \cdot 0.5H_2O$	Vadose
8	3, 6	Blödite	$Na_2Mg(SO_4)_2 \cdot 4H_2O$	Vadose
9	4	Calcantite	$CuSO_4 \cdot 5H_2O$	Vadose
10	1, 3, 5, 6, 7	Calcite	$CaCO_3$	Vadose–epiphreatic–phreatic
11	3, 4, 6, 7	Celestine	$SrSO_4$	Vadose–epiphreatic–phreatic
12	1, 6, 7	Coronadite	$Pb(Mn^{4+}, Mn^{2+})_8O_{16}$	Phreatic
13	4	Copper pentahydrite*	$Mg_{0.45}Cu_{0.55}(SO_4) \cdot 5H_2O$	Vadose
14	3, 5	Dolomite	$CaMg(CO_3)_2$	Vadose–epiphreatic–phreatic
15	3	Hematite	α-Fe_2O_3	Vadose
16	3, 6	Epsomite	$MgSO_4 \cdot 7H_2O$	Vadose
17	6	Hexaidrite	$MgSO_4 \cdot 6H_2O$	Vadose
18	3, 6, 7	Fluorite	CaF_2	Vadose–epiphreatic–phreatic
19	1	Phraipontite	$(Zn,Al)_3(Si,Al)_2O_5(OH)_4$	Vadose–epiphreatic
20	2, 3, 4, 5, 6, 7, 8	Gypsum	$CaSO_4 \cdot 2H_2O$	Vadose–epiphreatic–phreatic
21	1, 5, 7	Goethite	α-$Fe^{3+}O(OH)$	Vadose–epiphreatic–phreatic
22	1	Guanine	$C_5H_3(NH_2)N_4O$	Vadose
23	3, 6	Halite	$NaCl$	Vadose
24	3	Hectorite*	$Na_{0.3}(Mg,Li)_3Si_4O_{10}(F,OH)_2$	Phreatic
25	1, 7	Jarosite	$KFe_3^{3+}(SO_4)_2(OH)_6$	Vadose
26	7	Kieserite	$MgSO_4 \cdot H_2O$	Vadose
27	1	Magnetite	Fe_3O_4	Vadose
28	1, 4	Malachite	$Cu_2(CO_3)(OH)_2$	Vadose
29	2	Norstrandite	$Al(OH)_3$	Vadose
30	7	Opal	$SiO_2 \cdot nH_2O$	Phreatic
31	7	Orientite*	$Ca_2Mn^{2+}Mn_2^{3+}Si_3O_{10}(OH)_4$	Vadose
32	1	Pirolusite	MnO_2	Vadose
33	1	Plumbojarosite*	$PbFe_6^{3+}(SO_4)_4(OH)_{12}$	Vadose
34	6	Quartz	SiO_2	Phreatic
35	7	Rozenite	$FeSO_4\ 4H_2O$	Vadose
36	6, 7	Starkeyite*	$MgSO_4 \cdot 4H_2O$	Vadose
37	7	Szmikite*	$MnSO_4 \cdot H_2O$	Vadose
38	7	Szmolnokite*	$FeSO_4 \cdot H_2O$	Vadose
39	7	Woodruffite*	$(Zn,Mn^{2+})Mn_3^{4+}O_7 \cdot 1–2H_2O$	Vadose
40	1, 3	Not determined	Si, Al, Mg, Cu, Zn	Vadose–epiphreatic

*New cave mineral.
Cave numbers: 1, +50; 2, Cueva Palacios; 3, Cueva de las Espadas; 4, −150; 5, Cueva de los Cristales; 6, Ojo de la Reina; 7, Cueva de las Velas; 8, Tiburon.
Modified after Forti *et al.* (2009).

pollen were recorded within lake sediments from the Alta Babicora basin of northern Mexico (NW of Naica) from 38 ka to *c.* 29 ka. Therefore isotopic fluid inclusion (Garofalo *et al.* 2010; Gázquez *et al.* 2013*a*, *b*) and pollen data (Garofalo *et al.* 2007) from Naica indicate that the climatic changes occurring in the region at least over the last 35 ka substantially controlled the composition of the source fluid present within the caves and hence the growth of gypsum.

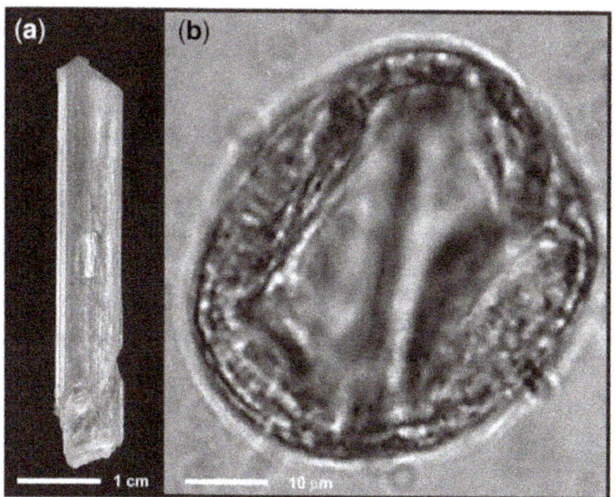

Fig. 7. (**a**) Huge biphasic fluid inclusion in a gypsum crystal of Cueva de las Espadas. (**b**) The first pollen grain (*Quercus garryana* cf.) extracted from a fragment of gypsum crystal of Cueva de los Cristales (photos from La Venta Esplorazioni Geografiche Archive).

Other palaeoenvironmental studies were performed utilizing different methodologies, such as those based on the progressive variation of the crystal size at the bottom of Cueva de las Espadas or the relationships between gypsum and calcite–aragonite deposits inside the same cave (Gázquez *et al.* 2011; Calaforra *et al.* 2013). From these studies it was possible to define the variations in the groundwater level within the Naica ridge in the last 60 ka (Gázquez *et al.* 2013*a*).

Astrobiology

Many fossilized (by the precipitation of Fe/Mn oxides/hydroxides) biogenic structures were found among the solid inclusions inside the gypsum crystals, which must therefore be coeval with the crystals (Panieri *et al.* 2008). The presence of widespread fluid inclusions suggested that still-living organisms may have survived trapped inside them. Therefore several large fluid inclusions trapped inside the giant crystals of Cueva de los Cristales were sampled (Fig. 8a) and the possible organisms inside them fed under controlled (aerobic and anaerobic) atmospheres. This experiment started in 2007, lasted for more than four years and was successful: organic matter developed from which it was possible to extract DNA to recognize the type of bacteria (acidobacteria, bitrospira and others) that were able to survive the extreme environmental conditions of the caves (Boston *et al.* 2012).

The bacteria living within these crystals are new organisms with some similarities to the extremophiles living in thermal springs or remote mines (Boston *et al.* 2012). These findings strengthened

Fig. 8. Naica mine. (**a**) Drilling a core over a broken piece of a giant gypsum crystal from Cueva de los Cristales for fluid inclusion analyses. (**b**) Testing a NASA laser-induced miniaturized Raman spectrometer for future studies of the Mars surface.

Fig. 9. Three-dimensional rendering of Cueva de los Cristales (quartered drawing). The data utilized for the construction of the three-dimensional model came from (**a**) detailed traditional mapping, (**b**) high-definition photographs and (**c**, **d**) laser scanner acquisition (photos from La Venta Esplorazioni Geografiche Archive).

the idea that the Naica caves might be similar to extra-terrestrial sites, such as Mars (Gázquez *et al.* 2012, 2013*c*). The National Aeronautics and Space Administration (NASA) therefore joined the project and tested new instruments inside the Naica caves (Fig. 8b) to investigate the organic matter trapped within the rocks. It is planned to use these instruments on the surface of Mars in the near future.

Flooding of the mine

As soon as the caves at the −290 m level were discovered, it was clear that they would remain accessible for only a short time. The Peñoles Company clearly said that the maintenance of mine dewatering would become economically non-viable after the exploitation of the ore bodies. In 2002, they predicted that this would occur in about 10 years (Peñoles Company, pers. comm.).

Thus one of the main tasks of the Naica Project was to decide the best way to preserve at least the memory and the records or, even better, a significant part of this incredible underground world for future generations. La Venta Esplorazioni Geografiche realized detailed maps of all the Naica caves. Speleoresearch & Films of Mexico City shot a full-length film documenting the explorative and

Fig. 10. Naica mine: the huge and dry −120 gallery giving access to the Cueva de las Espadas. The owners of Naica mine could realize this project very easily, without spending huge amounts of money … but time is against the project and therefore it seems unlikely that future generations will have the chance to see the giant crystals directly (photos from La Venta Esplorazioni Geografiche Archive).

scientific aspects of the Naica caves, with special reference to Cueva de los Cristales; several shooting sessions produced a spectacular array of thousands of photographs from all the caves. In 2007 a three-dimensional topography of Cueva de las Espadas and Cueva de los Cristales was carried out by Virtualgeo in co-operation with La Venta Esplorazioni Geografiche using laser scanning technology (Canevese *et al.* 2009) to produce exact high-resolution three-dimensional maps and virtual reality trips (Fig. 9).

What was more worrying, however, was the material conservation of the crystals. It would be good to preserve at least a portion of these caves, avoiding what happened in Cueva de las Espadas at the beginning of the last century, when most of the large crystals were extracted by mine workers and visitors and sold to museums and mineral collectors. The same occurred immediately after Cueva de los Cristales was unveiled. Peñoles Company stopped this illegal trade fairly quickly by locking the cave with two steel doors and the caves and the crystals were reasonably well preserved.

During the development of the Naica Project and later, La Venta Esplorazioni Geografiche explored all possible ways (local, national and international) to maintain accessibility to the caves at the −290 m level after the end of the mining activities. The suggested solutions would have also had a social function: the existence of the village of Naica (with *c.* 4000 inhabitants) depends entirely on the mine and it might have had a future in tourism if the caves remained accessible. Unfortunately, although La Venta Esplorazioni Geografiche's efforts received theoretical acceptance at many levels, no practical solution to maintain access to the caves was found.

The situation deteriorated suddenly in early 2015 when mining works intersected a completely unexpected major thermal vein, which started flooding the lower levels of the mine. Dewatering was increased progressively for 10 months, but even at the maximum possible flow of the hydraulic pumps it was not enough to stop the rise of groundwater. Therefore, at the end of 2015, the Peñoles Company started the procedure to close and abandon the mine. The procedure lasted for a few months and the Naica mine was officially closed in the spring of 2016.

The groundwater has, to date, flooded all the voids to 400 m below the mine entrance; therefore all the caves at −290 are empty and there is still a relatively short time interval in which the preservation of at least part of the giant crystals is possible. A solution could be the extraction of some of the crystals, or even all of them, and the reconstruction of a copy of Cueva de los Cristales using one of the large mine voids above the natural phreatic level, which is located a few metres below the deepest part of Cueva de las Espadas. Probably the best place to realize such an installation would be the huge gallery giving access to the Cueva de las Espadas (Fig. 10), which will be the only cave of Naica, together with the +50 cave, to remain dry. In this way, the replica of the Cueva de los Cristales, hosting its true giant crystals, will be close to the other tourist target of the mine.

Final remarks

The extensive research conducted over the past decade in the framework of the Naica Project has made the thermal caves of Naica probably the most studied cavity systems in the world (Forti 2010; Gázquez *et al.* 2016). So far, more than 80 scientific papers have been published and a few more are still in preparation. The first investigations analysed the genesis and age of the giant crystals and then the mineralogy, geochemistry and microclimatology of the caves. New research lines arose around the gypsum crystals, which have been proposed as palaeoclimatic proxies and have started to be used in fields such as astrobiology and the planetary sciences.

However, in spite of their indubitable scientific and aesthetic value, the future of the Naica caves seems to be once again flooded by thermal water and to become inaccessible to humans. This is disappointing for cavers and scientists, who have lost the possibility of visiting and/or studying this astonishing world. The inhabitants of Naica village will lose much more, however; the town cannot survive without mining activities and tourism, and is likely to become a ghost village.

For the giant crystals, however, this is the best possible solution. During their 15 years of aeration, they not only stopped growing, but also degraded as a result of visits to the cave and, even worse, they started to dissolve and to transform into calcite speleothems as their surface temperature decreased below that of the surrounding atmosphere. Now they will re-grow in the depths of the thermal aquifer and, perhaps, at an unknown future time, someone will again have the chance to see them shining, larger and better-looking than ever. The next generation will have a chance to keep the memory of the Naica caves alive thanks to the photographs, studies and publications of the Naica Project and perhaps to the replica of the Cueva de los Cristales, which will hopefully be realized within the Naica mine.

The author thanks the Peñoles Company for allowing access inside the Naica mine and for their support during fieldwork. Logistics was carried out by the Naica Project of Speleoresearch & Films of Mexico City in co-operation with the La Venta Esplorazioni Geografiche (Italy). The author thanks Georg Kaufmann and another anonymous referee for the useful questions and suggestions, which greatly improved the manuscript.

References

Alva-Valdivia, L.M., Goguitchaichvili, A. & Urrutia-Fucugauchi, J. 2003. Petromagnetic properties in the Naica mining district, Chihuahua, Mexico: searching for source of mineralization. *Earth, Planets, Space*, **55**, 19–31.

Badino, G. 2009. The Cueva de los Cristales micrometeorology. *In*: White, W.B. (ed.) *Proceedings: 15th International Congress of Speleology*, Kerrville, Texas, United States of America, July 19–26 2009. Vol. **3**. National Speleological Society, Huntsville, AL, 1407–1412.

Badino, G., Ferreira, A. *et al.* 2009. The Naica caves survey. *In*: White, W.B. (ed.) *Proceedings: 15th International Congress of Speleology*, Kerrville, Texas, United States of America, July 19–26 2009. Vol. **3**. National Speleological Society, Huntsville, AL, 1764–1769.

Badino, G., Calaforra, J.M., Forti, P., Garofalo, P. & Sanna, L. 2011. The present day genesis and evolution of cave minerals inside the Ojo de la Reina cave (Naica Mine, Mexico). *International Journal of Speleology*, **40**, 125–131.

Bernabei, T., Forti, P. & Villasuso, R. 2007. Sails: a new gypsum speleothem from Naica, Chihuahua, Mexico. *International Journal of Speleology*, **26**, 23–30.

Bernabei, T., Casagrande, G. *et al.* 2009. The Naica project. *In*: White, W.B. (ed.) *Proceedings: 15th International Congress of Speleology*, Kerrville, Texas, United States of America, July 19–26 2009. Vol. **1**. National Speleological Society, Huntsville, AL, 283–288.

Beverly, M. & Forti, P. 2010. L'esplorazione della Grotta Palacios nella miniera di Naica. *Speleologia*, **63**, 46–49.

Boston, P.J., Spilde, M.N., Northup, D.E. & McMillan, C. 2012. Long-term persistence of microorganism in geological materials: Lazarus, Rip van Winkle, and the walking dead. Paper presented at the Life Detection in Extraterrestrial Samples Conference, 13–15 February 2012, San Diego, CA, USA, **6038**.

Calaforra, J.M., Forti, P., Gázquez, F. & Sanna, L. 2013. La gradazione dei cristalli di gesso nella Cueva de las Espadas (Naica, Messico): Un classico esempio del controllo della soprasaturazione sui processi di nucleazione. *In*: *Proceedings of the XXI National Congress of Speleology*, Trieste, Italy, 2011, 325–332.

Canevese, E.P., Tedeschi, R. & Forti, P. 2009. Laser scanner in extreme environments: the caves of Naica. *The American Surveyor*, February, 3–10.

Degoutin, N. 1912. Les grottes a cristaux de gypse de Naica. *Societad Cientifica Antonio Alzate*, **32**, 35–38.

Erwood, R.J., Kesler, S.E. & Cloke, P.L. 1979. Compositionally distinct, saline hydrothermal solutions, Naica Mine, Chihuahua, Mexico. *Economic Geology*, **74**, 95–108.

FORTI, P. 2010. Genesis and evolution of the caves in the Naica mine (Chihuahua, Mexico). *Zeitschrift für Geomorphologie*, **54**, 115–135.

FORTI, P. & LO MASTRO, F. 2008. Il laboratorio sperimentale di −590 nella Miniera di Naica (Chihuahua, Messico). *Mondo Sotterraneo*, **31**, 11–26.

FORTI, P. & SANNA, L. 2010. The Naica project. A multidisciplinary study of the largest gypsum crystal of the world. *Episodes*, **33**, 23–32.

FORTI, P., GALLI, E. & ROSSI, A. 2007. Mineralogy of Cueva de las Velas (Naica, Chihuahua, Mexico). *Acta Carsologica*, **36**, 379–388.

FORTI, P., GALLI, E. & ROSSI, A. 2008. Il sistema Gesso-Calcite-Aragonite: nuovi dati dalle concrezioni del Livello -590 della Miniera di Naica (Messico). *Memorie Istituto Italiano di Speleologia*, **21**, 139–149.

FORTI, P., GALLI, E. & ROSSI, A. 2009. Le grotte di Naica: non solo giganteschi cristalli di gesso. *Rivista Italiana di Mineralogia*, **3/2009**, 180–196.

FOSHAG, W. 1927. The selenite caves of Naica, Mexico. *American Mineralogist*, **12**, 252–256.

FRANCO-RUBIO, M. 1978. Estratigrafía del Albiano-Cenomaniano en la región de Naica, Chihuahua. *Revista del Instituto de Geología (México)*, **2**, 132–149.

GARCÍA-RUIZ, J.M., VILLASUSO, R., AYORA, C., CANALS, A. & OTÁLORA, F. 2007. Formation of natural gypsum megacrystals in Naica, Mexico. *Geology*, **35**, 327–330.

GAROFALO, P.S., GÜNTER, D., FORTI, P., LAURITZEN, S.-E. & CONSTANTIN, S. 2007. The fluids of the giant selenite crystals of Naica (Chiuhahua, Mexico). Paper presented at ECROFI-XIX, 17–20 July 2007, Switzerland [abstract].

GAROFALO, P.S., FRICKER, M., GÜNTHER, D., MERCURI, A.M., LORETI, M., FORTI, P. & CAPACCIONI, B. 2010. A climatic control on the formation of gigantic gypsum crystals within the hypogenic caves of Naica (Mexico). *Earth and Planetary Science Letters*, **289**, 560–569.

GÁZQUEZ, F., CALAFORRA, J.M., SANNA, L. & FORTI, P. 2011. Espeleotemas de yeso: ¿Un nuevo proxy paleoclimático? *Boletín de la Real Sociedad Española de Historia Natural. Sección Geológica*, **105**, 15–24.

GÁZQUEZ, F., CALAFORRA, J.M., FORTI, P., RULL, F. & MARTÍNEZ- PHREATICÍAS, J. 2012. Gypsum-carbonate speleothems from Cueva de las Espadas (Naica mine, Mexico): mineralogy and palaeo-hydrogeological implications. *International Journal of Speleology*, **41**, 211–220.

GÁZQUEZ, F., CALAFORRA, J.M. ET AL. 2013*a*. Isotope and trace element evolution of the Naica aquifer (Chihuahua, Mexico) over the past 60,000 yr revealed by speleothems. *Quaternary Research*, **80**, 510–521.

GÁZQUEZ, F., CALAFORRA, J.M., HODELL, D., SANNA, L. & FORTI, P. 2013*b*. Isotopes of gypsum hydration water in selenite crystals from the caves of the Naica mine (Chihuahua, Mexico). Paper presented at the 16th International Congress of Speleology, 21–28 July, Brno, Czech Republic, Vol. **1**, 388–393.

GÁZQUEZ, F., CALAFORRA, J.M., MARTÍNEZ-FRÍAS, J. & RULL, F. 2013*c*. Gypsum speleothems of mine caves as potential Martian analogues [abstract]. Paper presented at the Second International Symposium on Mine Caves, 27–29 April, Iglesias, Italy, 48–49.

GÁZQUEZ, F., CALAFORRA, J.M., FORTI, P. & BADINO, G. 2016. The caves of Naica: a decade of research. *Boletín Geológico y Minero*, **127**, 147–163.

GIOVINE, G., BADINO, G., DE VIVO, A., LO MASTRO, F., CASAGRANDE, G., DAVILA, A. & HIDALGO, G. 2009. The Naica caves and physiology. *In*: WHITE, W.B. (ed.) *Proceedings: 15th International Congress of Speleology*, Kerrville, Texas, United States of America, July 19–26 2009. Vol. **3**. National Speleological Society, Huntsville, AL, 1980–1984.

HOLDEN, C. 2008. Life in crystal. *Science*, **319**, 391.

KLIMCHOUK, A. 2009. Morphogenesis of hypogenic caves. *Geomorphology*, **106**, 100–117.

LAURITZEN, S.-E., CONSTANTIN, S. & FORTI, P. 2008. Chronology and growth rate of the Naica giant gypsum crystals. Paper presented at the 3rd International Geological Congress, 6–14 August, Oslo, Norway.

LONDON, D. 2003. New 'cave of the crystals' at Naica, Chihuahua, Mexico. *Earth Scientist Magazine of the School of Geology and Geophysics, University of Oklahoma*, 24–27.

MEGAW, P.K.M., RUIZ, J. & TITLEY, S.R. 1988. High-temperature, carbonate-hosted Pb-Zn-Ag (Cu) deposits of northern Mexico. *Economic Geology*, **83**, 1856–1885.

PALMER, A. 2011. Distinction between epigenic and hypogenic caves. *Geomorphology*, **134**, 9–22.

PANIERI, G., FORTI, P., GASPAROTTO, G. & SOLIANI, L. 2008. Studio delle inclusioni solide presenti nei cristalli di gesso della Grotta delle Spade (Naica, Messico). *Memorie dell'Istituto Italiano di Speleologia*, **19**, 335–343.

SANNA, L., SAEZ, F., SIMONSEN, S., COSTANTIN, S., CALAFORRA, J.M., FORTI, P. & LAURITZEN, S.E. 2010. Uranium-series dating of gypsum speleothems: methodology and examples. *International Journal of Speleology*, **39**, 35–46.

SANNA, L., FORTI, P. & LAURITZEN, S.E. 2011. Preliminary U/Th dating and the evolution of gypsum crystals in Naica caves (Mexico). *Acta Carsologica*, **40**, 17–28.

VAN DRIESSCHE, A.E.S., GARCÍA-RUIZ, J.M., TSUKAMOTO, K., PATIÑO-LÓPEZ, L.D. & SATOH, H. 2011. Ultraslow growth rates of giant gypsum crystals. *Proceedings of the National Academy of Sciences of the United States of America*, **108**, 15721.

Index

Page numbers in *italics* refer to Figures. Page numbers in **bold** refer to Tables.